P9-CCI-571

SUPRAMOLECULAR POLYMERS

SUPRAMOLECULAR POLYMERS

EDITED BY

ALBERTO CIFERRI

University of Genoa
Genoa, Italy

Duke University
Durham, North Carolina

MARCEL DEKKER, INC. NEW YORK • BASEL

ISBN: 0-8247-0252-2

This book is printed on acid-free paper.

Headquarters
Marcel Dekker, Inc.
270 Madison Avenue, New York, NY 10016
tel: 212-696-9000; fax: 212-685-4540

Eastern Hemisphere Distribution
Marcel Dekker AG
Hutgasse 4, Postfach 812, CH-4001 Basel, Switzerland
tel: 41-61-261-8482; fax: 41-61-261-8896

World Wide Web
http://www.dekker.com

The publisher offers discounts on this book when ordered in bulk quantities. For more information, write to Special Sales/Professional Marketing at the headquarters address above.

Current printing (last digit):
10 9 8 7 6 5 4 3 2 1

PRINTED IN THE UNITED STATES OF AMERICA

Preface

Contributions from various noncovalent, supramolecular interactions are present in both low molecular weight and polymeric organic materials. Even conventional macromolecules, stabilized by mainchain covalent bonds as originally described by Staudinger, display a variety of supramolecular effects that control their intramolecular conformation and their intermolecular interactions. The selection of systems to be included in a book on supramolecular polymers was therefore a delicate task.

In the early stages of planning the book the tentative title *Supramolecular Polymerizations* was adopted. It was thus intended to restrict the content of the book to the new class of self-assembled polymers that undergo reversible growth by the formation of noncovalent bonds. This class (Part II) is wider than expected: not only mainchain assemblies of hydrogen-bonded repeating units, but also planar organization of S-layer proteins, micellar and related three-dimensional structures of block copolymers may be described as a result of supramolecular polymerization.

The title that was ultimately chosen allowed for the inclusion of polymers conforming to the preceding definition and also the new class of supramolecular polymers that may be stabilized by covalent bonds but nevertheless exhibit *novel* supramolecular features. The latter class (Part III) is exemplified by covalent chains based on 2-concatenane units, dendrimers, monolayers, and some engineered assemblies (Part IV).

Included in the book is a general introduction to supramolecular interactions and assembly processes (Part I) along with two theoretical chapters detailing liquid crystalline phases and micellar-like aggregation, two important driving forces in supramolecular polymerization. Expected developments in supramolecular polymer chemistry leading to functional *materials* and *systems* are highlighted in Part V.

The book focuses on the assembly of synthetic polymers and only relatively simple biopolymers are included. Attempts to control the structure and the growth of supramolecular polymers have been largely empirical. It is hoped that the highlighting in this book of the polymers so far investigated—and of the relevant theoretical mechanisms—will guide the organic chemist in designing chemical units that can be assembled into tailored materials and eventually functional systems.

Several applications as functional materials are already considered in the individual chapters. Hierarchical assembling allows the scaling-up to dimensions exceeding the micrometer range in several instances. Supramolecular polymers include dynamic structures endowed with internal mobility and the potential for storing information. Yet, we are still struggling to learn how to couple these characteristics of supramolecular polymers with their possible functioning as micro-engines allowing, for instance, molecular computing or the motility of some biological systems. A typical case is the long known globular↔fibrous (G↔F) transformation of actin. A basic mechanism for the G→F polymerization has been theoretically described. However, the overall functioning of actin as a system involves a coupling of the association process to a chemical reaction and to additional proteins that control polymerization. Understanding and reproducing coupled mechanisms is a road to future development in supramolecular science and functional systems.

The book was not intended to be a collection of unrelated chapters and efforts were made to achieve at least a partly coordinated outlook. This was made possible by the commitment and patience of the authors. Several discussions followed the preliminary draft, and some of the authors participated in meetings to arrive at a consensus over controversial issues. We were saddened by the premature death of Professor Raymond Stadler shortly after he had confirmed his participation in the book. His symbolic presence and enthusiasm are assured by the contributions of two of his coworkers.

Special thanks are due to Ms. Anita Lekhwani, Acquisitions Editor at Marcel Dekker, Inc., who first saw the importance of the emerging field of supramolecular polymers and insisted that the book should be written. The use of facilities at the Chemistry Departments of both Duke University and the University of Genoa greatly helped the editorial task. The book could not have been written without the encouragement and active cooperation of my wife, Cinzia.

Alberto Ciferri

Contents

Preface *iii*
Contributors *vii*

Part I General Formalism and Theoretical Approaches

1. Mechanism of Supramolecular Polymerizations 1
   Alberto Ciferri
2. Theory of the Supramolecular Liquid Crystal 61
   Reinhard Hentschke and Bernd Fodi
3. Polymeric vs. Monomeric Amphiphiles: Design Parameters 93
   Avraham Halperin

Part II Linear, Planar, and Three-Dimensional Reversible Self-Assemblies

4. Hydrogen-Bonded Supramolecular Polymers 147
   Perry S. Corbin and Steven C. Zimmerman
5. Crystalline Bacterial Cell Surface Layers (S-Layers): A Versatile Self-Assembly System 177
   Uwe B. Sleytr, Margit Sára, and Dietmar Pum
6. Assemblies in Complex Block Copolymer Systems 215
   Volker Abetz

7. Microstructure and Crystallization of Rigid-Coil Comblike Polymers and Block Copolymers 263
Katja Loos and Sebastián Muñoz-Guerra

Part III Assemblies Stabilized by Covalent Bonds

8. Polymers with Intertwined Superstructures and Interlocked Structures 323
Françisco M. Raymo and J. Fraser Stoddart

9. Dendrimeric Supramolecular and Supra*macro*molecular Assemblies 359
Donald A. Tomalia and István Majoros

10. Self-Assembled Monolayers (SAMs) and Synthesis of Planar Micro- and Nanostructures 435
Lin Yan, Wilhelm T. S. Huck, and George M. Whitesides

Part IV Engineered Planar Assemblies

11. Architecture and Applications of Films Based on Surfactants and Polymers 471
Masatsugu Shimomura

12. Supramolecular Polyelectrolyte Assemblies 505
Xavier Arys, Alain M. Jonas, André Laschewsky, and Roger Legras

13. Functional Polymer Brushes 565
Jürgen Rühe and Wolfgang Knoll

Part V Conclusions and Outlook

14. Supramolecular Polymer Chemistry—Scope and Perspectives 615
Jean-Marie Lehn

15. Protein Polymerization and Polymer Dynamics Approach to Functional Systems 643
Fumio Oosawa

Index 663

Contributors

Volker Abetz Makromolekulare Chemie II, Universität Bayreuth, Bayreuth, Germany

Xavier Arys Unité de Physique et de Chimie des Hauts Polymères, Université Catholique de Louvain, Louvain-La-Neuve, Belgium

Alberto Ciferri Department of Chemistry and Industrial Chemistry, University of Genoa, Genoa, Italy

Perry S. Corbin Department of Chemistry, University of Illinois, Urbana, Illinois

Bernd Fodi Max-Planck-Institute for Polymer Research, Mainz, Germany

Avraham Halperin Departement de Recherche Fondamentale sur la Matiere Condensee, UMR 5819 (CEA-CNRS-Université J. Fourier), SI3M, CEA–Grenoble, Grenoble, France

Reinhard Hentschke Department of Physics, Bergische Universität-Gesamthochschule, Wuppertal, Germany

Wilhelm T. S. Huck Department of Chemistry and Chemical Biology, Harvard University, Cambridge, Massachusetts

Alain M. Jonas Unité de Physique et de Chimie des Hauts Polymères, Université Catholique de Louvain, Louvain-La-Neuve, Belgium

Wolfgang Knoll Department of Materials and Surface Science, Max-Planck-Institute for Polymer Research, Mainz, Germany

André Laschewsky Department of Chemistry, Université Catholique de Louvain, Louvain-La-Neuve, Belgium

Roger Legras Unité de Physique et de Chimie des Hauts Polymères, Université Catholique de Louvain, Louvain-La-Neuve, Belgium

Jean-Marie Lehn Université Louis Pasteur, Strasbourg, and Collège de France, Paris, France

Katja Loos Makromolekulare Chemie II, Universität Bayreuth, Bayreuth, Germany

István Majoros Center for Biologic Nanotechnology, University of Michigan, Ann Arbor, Michigan

Sebastián Muñoz-Guerra Departamento d'Enginyeria Química, Universitat Politècnica de Catalunya, Barcelona, Spain

Fumio Oosawa Aichi Institute of Technology, Toyota-Shi, Japan

Dietmar Pum Center for Ultrastructure Research and Ludwig Boltzmann Institute for Molecular Nanotechnology, Universität für Bodenkultur, Wien, Austria

Françisco M. Raymo Department of Chemistry and Biochemistry, University of California, Los Angeles, California

Jürgen Rühe Max-Planck-Institute for Polymer Research, Mainz, Germany

Margit Sára Center for Ultrastructure Research and Ludwig Boltzmann Institute for Molecular Nanotechnology, Universität für Bodenkultur, Wien, Austria

Masatsugu Shimomura Research Institute for Electronic Science, Hokkaido University, Sapporo, Japan

Uwe B. Sleytr Center for Ultrastructure Research and Ludwig Boltzmann Institute for Molecular Nanotechnology, Universität für Bodenkultur, Wien, Austria

J. Fraser Stoddart Department of Chemistry and Biochemistry, University of California, Los Angeles, California

Donald A. Tomalia Center for Biologic Nanotechnology, University of Michigan, Ann Arbor, and Michigan Molecular Institute, Midland, Michigan

George M. Whitesides Department of Chemistry and Chemical Biology, Harvard University, Cambridge, Massachusetts

Lin Yan Discovery Chemistry, Pharmaceutical Research Institute, Bristol-Myers Squibb, Princeton, New Jersey

Steven C. Zimmerman Department of Chemistry, University of Illinois, Urbana, Illinois

1

Mechanism of Supramolecular Polymerizations

Alberto Ciferri
University of Genoa, Genoa, Italy

I. THE WIDE MEANING OF SUPRAMOLECULAR POLYMERIZATION

A. Supramolecular vs. Molecular Polymerization

The concept of covalent polymerization of bifunctional monomeric units was pioneered by Staudinger [1] in the early 1920s. He introduced the term ''macromolecule'' and proposed structural formulas for natural rubber, polystyrene, and polyoxymethylene that are still valid today. The invariance of the colloidal properties of these compounds in different solvents was one of the proofs adduced to support the concept of a linear sequence of covalent bonds exhibiting a large degree of polymerization (DP). The alternative model of a colloid-type aggregate stabilized by weaker secondary interaction was dismissed. Polymer science was then born, growing—distinct from colloid science—to produce the outstanding developments that have affected our lives.

Given the emphasis on the methods of synthesis and on the desirable properties of molecular polymers which continues to the present day, it is all too natural that the concept of a polymer composed by a long sequence of repeating units linked by noncovalent bonds received only limited attention. Thus, in spite of the fact that very long linear assemblies of globular proteins [2] or micelles [3] have been known for quite some time, the recognition of the potential of noncovalently bonded polymers had to wait until the early 1990s when significant developments in *supramolecular chemistry* occurred [4].

According to accepted current definitions, supramolecular chemistry is the chemistry of the intermolecular bond distinct from molecular chemistry, which is the chemistry of the covalent bond. Thus, it is now appropriate to regard a long sequence of units connected by secondary bonds as a supramolecular polymer [4], distinct from a molecular polymer when repeating units are linked by covalent bonds as originally shown by Staudinger.

Correspondingly, one can introduce the term *supramolecular polymerization* [5] to indicate the assembly processes of supramolecular polymers, distinct from the well-known molecular polymerization processes (e.g., condensation, radical, ring opening, . . .) leading to molecular polymers [6]. The scheme in Fig. 1A emphasizes the difference between the type of repeating units and the type of mainchain bonds in supramolecular and molecular polymers of large DP. In the latter case, a chemical reaction between functional groups performed on rather simple low molecular weight monomers leads to a stable linear chain. The extent of growth is controlled by the product of the probabilities of single additions [6]. In the latter case, a *molecular recognition* process [4] leads to a reversibly *self-assembled* chain based on either low or high molecular weight complex repeating units. This chapter is primarily concerned with the analysis of parameters that control supramolecular polymerization through the formation of *supramolecular mainchain bonds in linear self-assembled polymers growing to a large DP*.

It is essential to appreciate from the outset that the concept of supramolecular polymerization entails a much broader significance of the types of repeating units and of their spatial arrangement than is the case with molecular polymers. For instance, the H-bonded supramolecular polymers [7–11] in Fig. 2 (cf. also this volume, Chapter 4) bear similarity to molecular polymers in both the relative

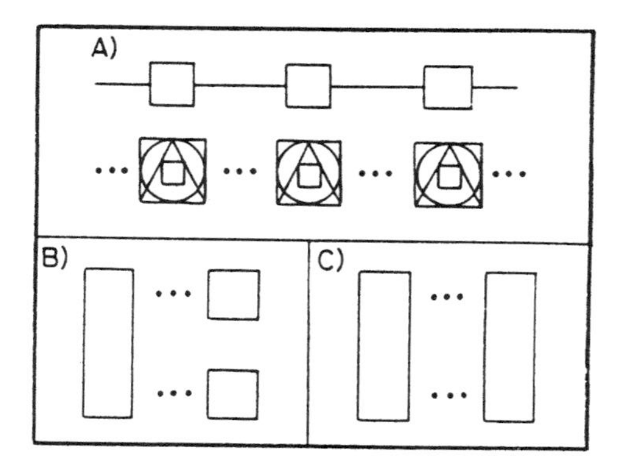

Figure 1 Schemes for: (A) linear molecular (small units, strong covalent bonds) and supramolecular (units of various size and shape, secondary bonds) polymerization; (B) supramolecular binding of small units to a molecular polymer; (C) supramolecular binding of two polymer chains.

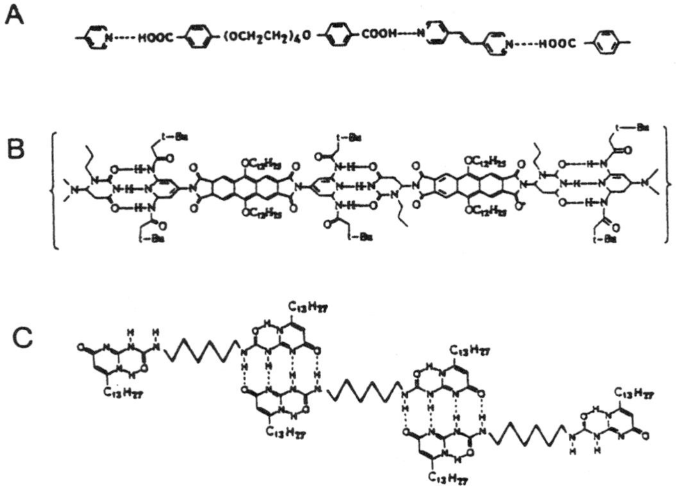

A

B

C

Figure 2 Supramolecular polymers stabilized by one (A, Ref. 7), three (B, Ref. 9), and four (C, Ref. 10) hydrogen bonds for repeating unit. Polymer A exhibits thermotropic liquid crystalline behavior (T_{KN} = 168.5°C, T_{NI} = 180°C). Polymer B forms a nematic lyotropic phase in 1,1,2,2-tetrachloroethane at RT. Polymer C is stable in isotropic solutions of chloroform.

small size of monomers and their chainlike assembly. Polymer A is stable in undiluted thermotropic mesophases. The rigid structure based on antracene segments terminated by uracil or pyridine residues allows the stabilization of the supramolecular polymer B even in lyotropic mesophases in the presence of a solvent. Polymer C attains large DP even in isotropic solutions. The pseudopolyrotaxane assembly in Fig. 3 (cf. also this volume, Chapter 8) has H-bonded repeating units based on a dicarboxylic acid that contains a π-rich hydroquinone ring threaded through a π-poor tetracationic cyclophane cavity. Occurrence of linear pseudopolyrotaxane and their planar assemblies could be revealed by X-ray in the undiluted state [12].

The discotic 2,3,6,7,10,11-hexa-(1,4,7-trioxaocetyltriphenylene) amphiphilic molecule [13] in Fig. 4 (cf. also Section II.E and this volume, Chapter 2) can form single polymeric columnar assemblies stabilized by supramolecular bonds perpendicular to the monomeric surface in nematic solutions containing up to 50% solvent. The tapered 12-ABG-B15C5 compound [14] in Fig. 5, complexed with 0.4 M triflate salt per mole of the crown ether receptor B15C5, self-assembles into an hexagonal columnar (ϕ_h) mesophase revealed by X-ray and optical microscopy in the undiluted system. The column cross-section is based on disks

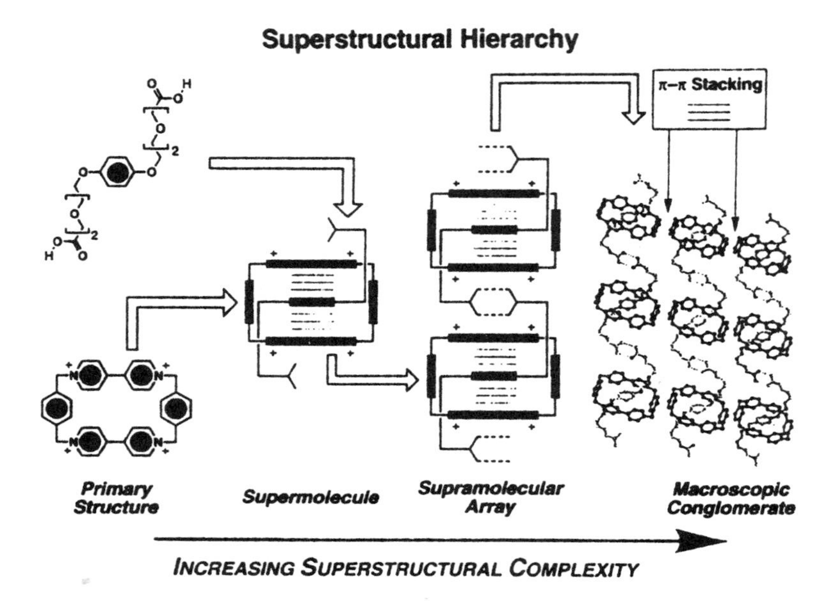

Figure 3 A dicarboxylic acid is threaded through a tetracationic cyclophane macrocycle to form a repeating unit which is connected to other units by a double H-bond. The evolution toward a pseudopolyrotaxane and a planar assembly stabilized by π–π stacking interaction is supported by X-ray data for the undiluted system. (From Ref. 12. Copyright 1997 Am. Chem. Soc.)

of five molecular units, the crown ether *endo*receptors in the center being surrounded by the aromatic moieties, in turn surrounded by the melted alkyl tails of the *exo*receptor. The latter may fold within each column, or interdigitate with adjacent columns. Electrostatic interactions between complexed and uncomplexed crown ether may assist the stacking of these units into channels parallel to the column axis. The disk can thus be regarded as the repeating unit of a tubular ''polymeric'' assembly.

Biological assemblies (Fig. 6A, B, C) such as fibrin [15,16], actin filaments [17], or microtubules [18] (cf. also Section III.C) are based on helical or tubular supramolecular polymers of globular proteins of high molecular weight, or of their oligomeric assemblies. These rigid assemblies attain lengths in the micrometer range. In the case of TMV, the tobacco mosaic virus (Fig. 6D), a molecular polymer (RNA) is hosted within the cavity of a supramolecular cylindrical assembly composed of 2,130 tapered proteins [19]. The above biological assemblies are stable in isotropic solutions even at rather low concentration. The linear assembly of cylindrical micelles [3,20] in the nematic phase (Fig. 7) should also

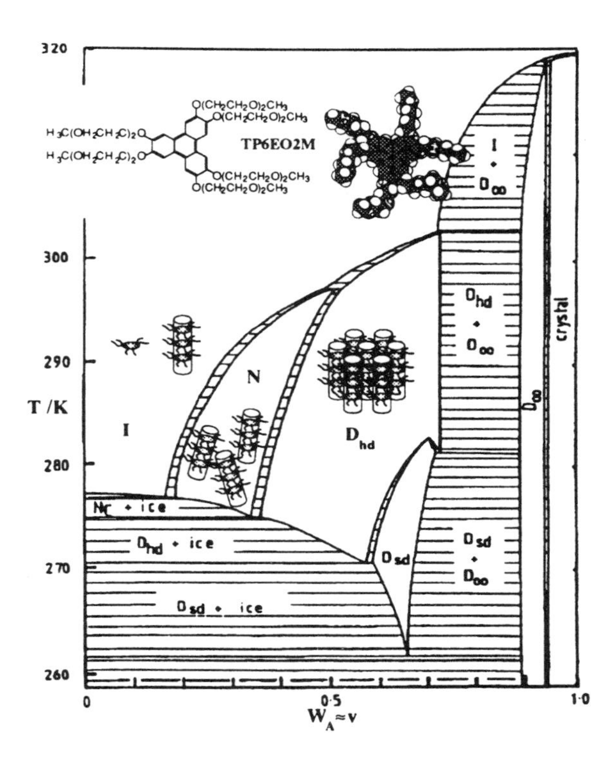

Figure 4 The discotic amphiphile 2,3,6,7,10,11-hexa-(1,4,7-trioxaocethyltriphenylene) is the repeating unit in a columnar stack stabilized by solvophobic interactions in D_20. The temperature versus volume fraction phase diagram shows that single stacks occur in the nematic phase (N) at volume fraction as low as 0.2. Hexagonal columnar and eventually crystalline phases evolve upon increasing concentration. (From Ref. 13.)

be regarded as the result of a supramolecular polymerization driven by hydrophobic forces coupled with the orientational field of the mesophase.

The examples in the preceding figures show that only for some of the supramolecular polymers single assemblies of large DP are stable in diluted, isotropic solutions. In other cases (Fig. 4), single assemblies of large DP are stable only in relatively concentrated lyotropic solutions, when the orientational nematic field contributes to growth (cf. Sections II.E and III.B). In additional cases, the stabilization of single assemblies is restricted to undiluted thermotropic mesophases. These observations immediately suggest the occurrence of a variety of growth mechanisms that depend upon the strength and distribution of attractive interactions and the flexibility of the assembly.

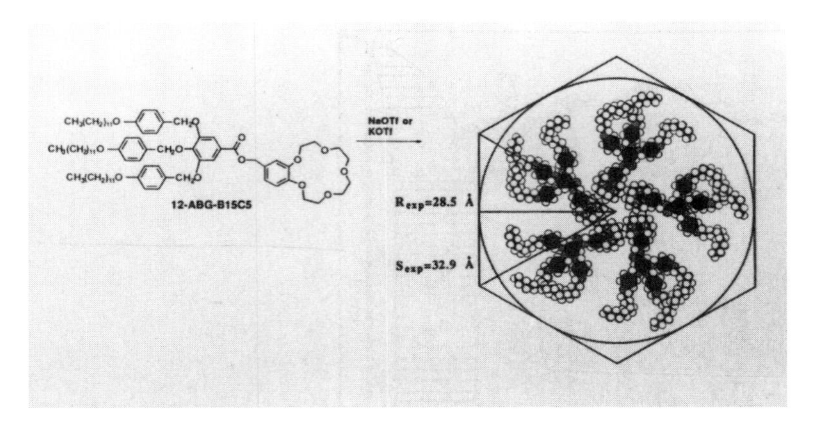

Figure 5 The assembly of the complex between 12-ABG-B15C5 and trifilate salts produces undiluted hexagonal columnar mesophases based on stacks of disks each containing five molecular units (see also Fig. 8). (From Ref. 14.)

The concept of supramolecular polymerization need not be limited to linear or helical growth. Supramolecular interactions can develop into planar or three-dimensional structures capable of indefinite growth. In fact, organic crystals have been regarded as supramolecular entities [21]. A few examples of mesoscopic planar and three-dimensional supramolecular assembly are discussed in this chapter (Section III.D). More are to be found in other chapters. In particular, the solid state morphologies exhibited by block copolymers may also be conceived as a result of supramolecular polymerization with the axis representing the direction of molecular polymerization distinct from that representing the direction of supramolecular order. The concept of supramolecular polymerization can also be extended to organized aggregates such as closed micelles and open bilayers.

To further clarify the aims of this chapter, it should be pointed out that in addition to the mainchain interactions that control the growth of a supramolecular polymer, other types of supramolecular interactions do occur in both molecular and supramolecular polymers. Supramolecular effects in molecular polymers have been the focus of extensive study starting immediately after Staudinger's demonstration of the covalent nature of molecular chains. Conformation (or secondary structure) is controlled by supramolecular interactions among nonbonded atoms within a single molecular chain or supramolecular polymer. Sidechain binding, formation of multiple helices, liquid crystalline, and crystal morphologies are likewise controlled by supramolecular forces between chains or supramolecular assemblies. The current usage of the term supramolecular polymers also includes self-assembled systems in which the supramolecular interaction is not

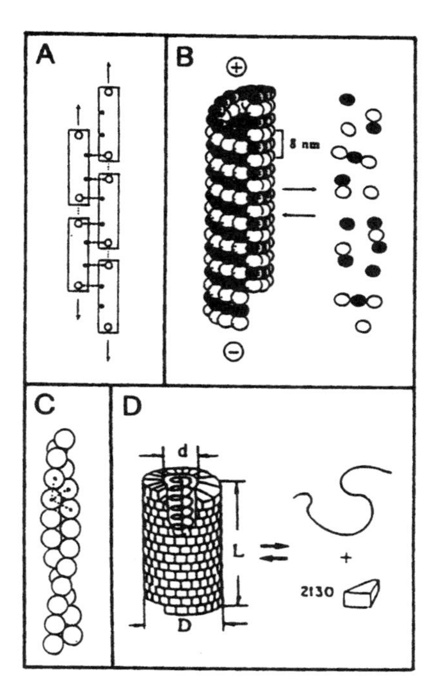

Figure 6 Biological assemblies. (A) Scheme for the assembly of fibrinogen residues during the early stage of conversion to linear fibrin filaments. To the A-type sites (•) uncovered by trombin on the central domain of each residue, two terminal sites (○) of two similar residues bind with strong interaction. A weaker head-to-tail attraction between residues also stabilizes the assembly. (From Ref. 5.) (B) Microtubules are produced by the reversible assembly of 13–16 longitudinal rows (protofilaments) of dimers of α (dark) and β (light shade) tubulin molecules (dimer spacing is 8 nm). A slight stagger produces a helical pattern (3-start helical family). A longitudinal polarity is exhibited. GTP is hydrolyzed to GDP during polymerization. The assembled state is stabilized by increasing temperature (hydrophobic bonds). Length may exceed 50 μm. (From Ref. 18 with permission from Elsevier Science.) (C) Simple representation of F-actin filaments in terms of a two-stranded helix based on the self-assembly of G-actin monomeric units. Each unit interacts with four neighbors. The assembled state is stabilized by increasing ionic strength and is assisted by the ATP $\rightarrow$ ADP hydrolysis. Length may reach 20 μm. (From Ref. 5. Copyright 1999 Taylor & Francis.) (D) TMV is a tubular assembly of 2,130 identical tapered proteins following a helical pattern with 16.3 units per turn and axial ratio = 16.6 (L = 3,000, D = 180 Å). The system can be disassembled and reassembled in vitro by pH changes with or without the RNA hosted in the hollow cavity. Without RNA the assembly is a stack of disks each comprising 17 protein units. (From Ref. 19.)

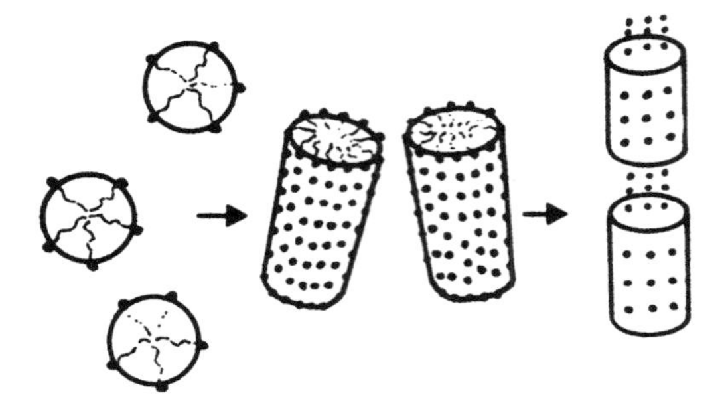

Figure 7 The transformation spherical micelles → cylindrical micelles → linear assemblies occurs upon increasing the surfactant concentration up to formation of a nematic solution.

restricted to mainchain bonds; actually even engineered, nonspontaneous assemblies (cf. Section III.E) are often described (sometimes with questionable usefulness) as supramolecular polymers. Examples of self-assembling involving sidechain and polymer–polymer interactions are schematized in Figs. 1B and 1C, respectively. Examples of covalent polymers exhibiting novel supramolecular effects are discussed in Section I.C. The emphasis given in this chapter to those supramolecular interactions that cause the growth of a linear assembly is dictated by the following considerations.

Supramolecular polymerization is a new event in the area of supramolecular interactions, and it is based on growth mechanisms specifically developed for supramolecular systems. In addition to the cases of the G → F transformation in biological systems and of the linear assembly of micelles already mentioned, the formation of linear sequences of noncovalently bonded repeating units remained unexplored since Staudinger's discovery. Moreover, the wide meaning of supramolecular polymerization invites the use of unifying concepts that cut across the traditional boundaries between colloid, polymer, and solid state science and, in particular, between synthesis and physicochemical properties of polymers. For instance, the occurrence of growth may be simultaneous with the formation of a liquid crystal (cf. Section III.B), at striking variance with the case of molecular polymers. Furthermore, the field of supramolecular polymers is still in its infancy and it is natural, as was the case with molecular polymers in the early twenties, that the current emphasis be on the study of supramolecular polymerization and on ways to control the length of linear assemblies. Developments in areas such as conformations, solution behavior, or mechanical properties are to be expected.

In spite of the vast difference between molecular and supramolecular polymers, some concepts and questions that are relevant in molecular polymerization need to be considered also in its supramolecular counterpart. In particular, a key question is: how large will the degree of linear supramolecular polymerization be? Moreover, how wide will its polydispersity be? There will be a termination step: what will the supramolecular polymerization mechanism be? These points are analyzed in detail in Section III following a consideration of the supramolecular interactions detailed in Section II.

B. Open vs. Closed Assemblies

All the examples cited above refer to supramolecular systems for which the addition of successive repeating units exposes sites at the end of the growing chain to which additional units can bind, a situation (Fig. 1A) that also characterizes the polymerization of molecular polymers. These systems are classified as *open assemblies* and are the specific objective of the present analysis of supramolecular polymerization. Open assemblies are also the planar and three-dimensional assemblies growing by a mechanism similar to a phase transition (cf. Section III.D).

The examples illustrated in Figs. 1B and C represent instead *closed assemblies* for which the binding sites are internally compensated and the complexes have a definite stochiometry. The assembly mechanism of closed oligomers (e.g., DNA, collagen), of complexes involving the sidechains of a molecular polymer, and of host–guest complexes is outside the scope of this chapter. A too strict classification of open or closed assemblies should, however, be avoided. For instance, some closed assemblies can undergo supramolecular polymerization if residual sites allowing growth occur. For instance, although the stochiometry of the triple helix of collagen is strictly defined, it is difficult to observe a stable solution of tropocollagen at neutral pH; side-by-side or head-to-tail aggregation continues until phase separation occurs.

On the other hand, some supramolecular assemblies may have a large number of repeating units but exhibit a fixed stochiometry. These systems are not oligomeric ones and it is convenient to classify them as open systems undergoing supramolecular polymerization once a proper termination step is recognized. Several situations may occur. For instance, a size limitation and a size distribution may be the natural result of a step-by-step nature (cf. Section III.A) of the assembly mechanism. Size limitation may result from termination by a unit unable to further growth. Situations of this type are encountered even in biological polymerizations characterized by a high degree of cooperativity (e.g., termination of actin growth by gelsolin [22]; cf. Section III.C). The growth of a cylindrical micelle is retarded when the length of the assembly attains the value of the persistence length (cf. Section III.B). Giant mesoscopic vesicles based on either simple surfactants [23] or block copolymers [24] (cf. this volume, Chapter 7) are structures that may be conceived of as arising from the closing up of extended bilayer sheets exhibiting randomly fluctuating local curvature [25,26]. Spherical surfactant micelles have typical aggregation number 100, their growth to infinite size being prevented by the peculiar nature of amphiphilic molecules (cf. Sections II.C and III.D).

C. Superimposition of Chemical Bonds to a Supramolecular Structure

Although one of the functional characteristics of supramolecular polymers is their ability to undergo dynamic association–dissociation processes, the possibility of stabilizing the self-assembled polymers by *subsequent* formation of chemical bonds also exists. In this way the organic chemist can stabilize extremely complex functional structures that would have been practically impossible to assemble by ordinary synthesis [27]. Nature has also followed the strategy of self-assembling followed by the superimposition of chemical bonds in the case of some large structures such as, for instance, the keratin fiber [28].

It is the conversion of a supramolecular linear polymer into a covalent one (without the assistance of a template) that should have the greatest interest, as well as important applications. This supramolecularly assisted polymerization could, in fact, allow the preparation of molecular polymers exhibiting significantly different characteristics with respect to the conventional ones. For instance, unusual functionalities, sequences, molecular weights, distributions, and orientation could be achieved. At variance with direct polymerization, no positional entropy loss would accompany the establishment of the mainchain bonds. The degree of association during self-assembly is expected to far exceed the DP obtainable with conventional polymerization (cf. Sections III.B, C, D). These extremely long assemblies could be converted into molecular chains with extremely large persistence lengths. Only a few experiments along these lines have been reported.

The interesting stabilization of a linear supramolecular aggregate by subsequent formation of chemical bonds was achieved by the solid state crosslinking of the ordered cylindrical phase of a block copolymer [29]. Dissolution of the continuous uncrosslinked phase resulted in the isolation of nanofibers (cf. Section III.D). The polymerization of supramolecular assemblies such as lipid bilayers has also been described [30]. Reactive groups (i.e., acryloyl, sorbyl) were incorporated to the apolar tail of the amphiphiles and radical polymerization occurred in the core of the bilayer. The growth of dendrimers through successive generations likely results from a self-assembling process followed by covalent bond formation (this volume, Chapter 9). An example of molecular polymerization assisted by supramolecular effects is the replication of nucleic acids which starts as a recognition of complementary bases along the template by individual nucleotides, followed by their covalent bonding into a complementary molecular chain. In this case, each nucleotide is supramolecularly bonded as a sidechain to the template and therefore the sequence, molecular length, and distribution are controlled by the template and not by the mechanisms of supramolecular polymerization discussed in Section III.

Several situations in which the coexistence of a polymer chain and of a supramolecular assembly occurs have also been considered. Percec and Schlueter [31] have described the system in Fig. 8 based on the polymerization of a 7-oxanorbornene monomer. Two tapered monodendrons are the sidechains of each repeating unit. Monomers, oligomers, and polymers (DP up to 120) exhibited hexagonal columnar (ϕ_h) thermotropic mesophases (fan-shaped textures) in the undiluted state. From X-ray data it was concluded that the columnar cross-section (stratum) is a disk that includes four monodendrons (or two repeating units) with diameter 45.01 Å and thickness 3.74 Å. Disks stack up to form columns of increasing length as the DP increases. Diluted solution properties can be measured for this "polymer jacketed with monodendrons." It was speculated that its persistence length corresponds to a cylinder length of 120 Å (DP = 64). The latter appears too small to confer lyotropic behavior to this system (cf. Section II.E). In earlier work Percec and coworkers [14] had replaced the crown ether B15C5

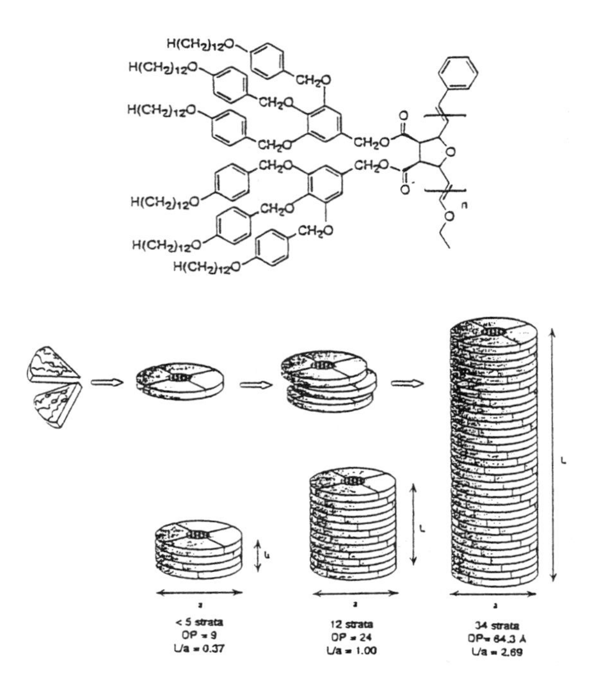

Figure 8 Top: polymer resulting from living ring opening metathesis polymerization of a 7-oxanorbornene monomer substituted with two tapered monodendrons. Bottom: schematization of the assembly of four monodendrons (two monomers) in strata (diameter 45.01 Å, thickness 3.74 Å) and in columns for polymers with DP = 24 and 64. (From Ref. 29. Copyright 1996 Am. Chem. Soc.)

in Fig. 5 with a flexible oligoxyethylenic *endo*-receptor (nEO-PMA) connected as a sidechain to a polymethylacrylate chain. The latter adopted a suitably folded conformation to allow the packing of six 12 ABG-nEO-PMA units into disks arranged along a slow rising helicoidal stack.

The pseudopolyrotaxane schematized in Fig. 3 has a supramolecular main-chain bond based on the dimerization of carboxyl groups. Philp and Stoddart had earlier suggested [27] that a molecular chain with appropriately spaced π-rich rings could thread its way through several macrocycles containing π-acceptors (Fig. 9A) and subsequently be capped with large stoppers to form a permanent rotaxane polymer. The structure may be viewed as a molecular abacus with rotaxane units noncovalently linked to the molecular chain. Newer polyrotaxane polymers are described in this volume in Chapter 8.

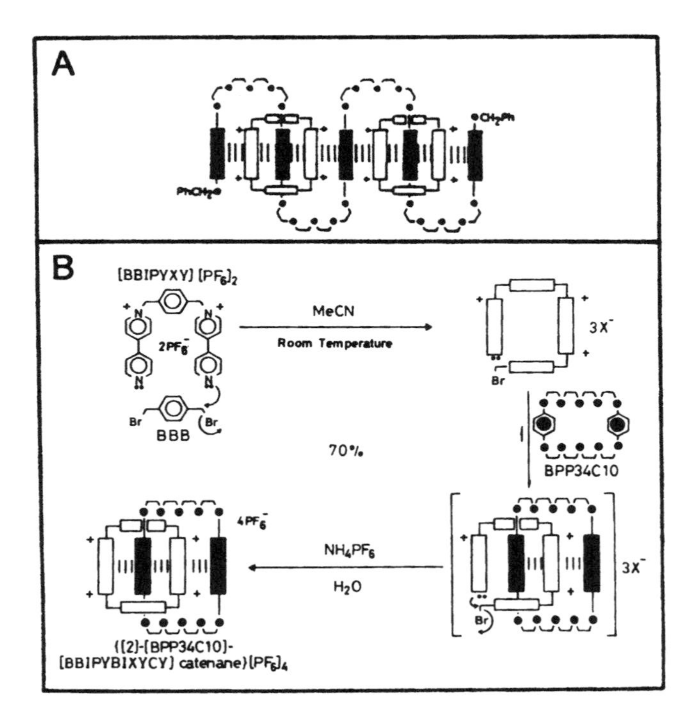

Figure 9 (A) Polyrotaxane-like superstructure based on a polymer chain with π-donors hydroquinone rings threaded through π-acceptor cavities of a tetracationic cyclophane. (B) Formation of a closed 2-concatenane chain from the recognition between the di-cation $[BBIPYXY][PF_6]_2$ and the crown ether BPP34C10, followed by the chemical reaction of 1,4-bis (bromomethyl) benzene (BBB). (From Ref. 27.)

Interlocking of the tetracationic cyclophane with a crown ether through the superimposition of a chemical to a supramolecular bond according to the scheme in Fig. 9B produces a 2-concatenane [27]. The latter is another example of the exciting concept of a ''mechanical bond'' between two closed structures not directly linked by covalent bonds. The structures must remain constrained but are otherwise endowed with the possibility of movement, as in a macroscopic chain. Continuous catenane sequences with DP > 5 appear difficult to prepare. However, newer polymers based on 2-concatenane units connected by segments of a covalent chain are described in this volume in Chapter 8.

Earlier studies of supramolecular order of mesogenic sidechains chemically linked to coiling polymers threw light on the way the conformation of the flexible chain adapts to the order of the sidechains [32]. Neutron scattering revealed a microsegregation of the polymer in interlayers within the smectic layers of the sidechains, with specific interlayer defects. The problem is relevant also to cases in which mesogenic sidechains are noncovalently linked to a flexible chain. This class of systems has been reviewed by Imrie [33] and by Kato and Fréchet [34] (cf. also this volume, Chapter 4). In the TMV assembly schematized in Fig. 6D, the polymer (RNA) and the tapered protein units are also connected via supramolecular interactions.

The self-assembled monolayers (SAM) are prepared by chemically linking ordered rigid molecules to complementary sites on a solid surface. Examples are described in this volume in Chapter 10 (cf. also Section III.D). The topochemical polymerization of diacetylenes should also be regarded as a supramolecularly assisted process requiring very specific arrangements of the monomers within the crystalline state [35].

II. FORCES ASSISTING SUPRAMOLECULAR ORGANIZATION

A. The Supramolecular Bond

The classical noncovalent forces involved in the formation of supramolecular bonds are those based on Coulombic interactions (permanent charges and dipoles), van der Waals interactions (induced dipoles, charge transfer, dispersive attractions), hydrogen, and solvophobic bonds. Shape effects also have an important role in determining the best fit of complementary sites, and in favoring growth of supramolecular assemblies. Additional effects specifically relevant to supramolecular polymerization include microsegregation of chemically incompatible segments even in the absence of a conventional diluent, and liquid crystalline mesophases.

The organic chemist has therefore an ample choice for designing and synthesizing molecules having proper sites, site distribution, and shape allowing the formation of supramolecular assemblies of various design. Extensive and detailed quantitative analysis of the relative role of Coulombic, van der Waals, hydrogen, and solvophobic bonds is available in the literature dealing with supramolecular chemistry and organic host–guest complexes [36–40]. The analysis has reached a level of high sophistication with good agreement in some instances reported between experimental and simulated structures using molecular-mechanics, molecular-dynamics, and Monte Carlo methods. The results obtained for closed host–guest complexes are very important for the design of open polymeric assemblies: only a larger functionality is required for growth. In this section a brief summary of the classical noncovalent interaction is therefore presented, with more emphasis being placed here, and in this volume, Chapters 2 and 3, on effects specifically relevant to supramolecular polymerization.

General conclusions emerging from the literature of host–guest complexes reveal that supramolecular structures are stabilized by a combination of several of the above interactions; the quantitative assessment of their relative contributions is sometimes a difficult task. A single specific interaction may, however, play a prevalent role in several instances. Additivity of the various contributions to the free energy of formation of a given complex is often verified, which is equivalent to saying that the overall formation constant is the product of the constants for single pairwise interactions [36–41]. Thus, the occurrence of several sites for complementary interaction can magnify the effect of single weak interactions. Cooperative effects (when the formation of the first pairwise interactions increases the binding constant at successive sites) are also common. The formation of a supramolecular bond is driven by primarily enthalpic attraction in the case of Coulombic and van der Waals interactions, translational entropy being lost [42] when a supramolecular bond is established between two molecules. However, entropy effects may play a fundamental role when the formation of the assembly is driven by hydrophobic effects, or by the formation of a mesophase.

1. Coulombic Bonds

Interactions between permanent charges may be of the ion–ion, ion dipol, or ion–quadrupol type. These have been used to increase the stability of host–guest complexes, and are also common in biological supramolecular polymers. Attractive interaction (ion pair) occurs between fixed and complementary ionizable groups and is modulated by co- and counter-ions in line with the general features of the Debye–Hückel theory. The ion–ion attraction is the strongest one, with binding free energies that may exceed 10 KJ mol^{-1} in solvents of low polarity. In fact, an increase of association constants of host–guest complexes with a decrease in solvent polarity was observed [43], even though the strict inverse proportionality to the dielectric constant predicted by Debye–Hückel was not followed. Other features of the above theory have been semiquantitatively verified. For instance, complexation of I in Fig. 10A with the oppositely charged $^{+}N\ R_4$ ($R = (CH_2)_n\ Me$) was accompanied [44] by a decrease of the binding free energy with charge separation r. The binding constant of a similar complex was also shown to decrease with ionic strength up to 0.2 M salt in good agreement with Debye–Hückel [45].

Partial charge interactions are exemplified by the case of benzene that displays a partial negative charge in the plane of the ring within the π system, and a partial positive charge on the hydrogen atoms. The parallel stacking of rings is hindered by the repulsive π–π interaction while attractive forces develop for

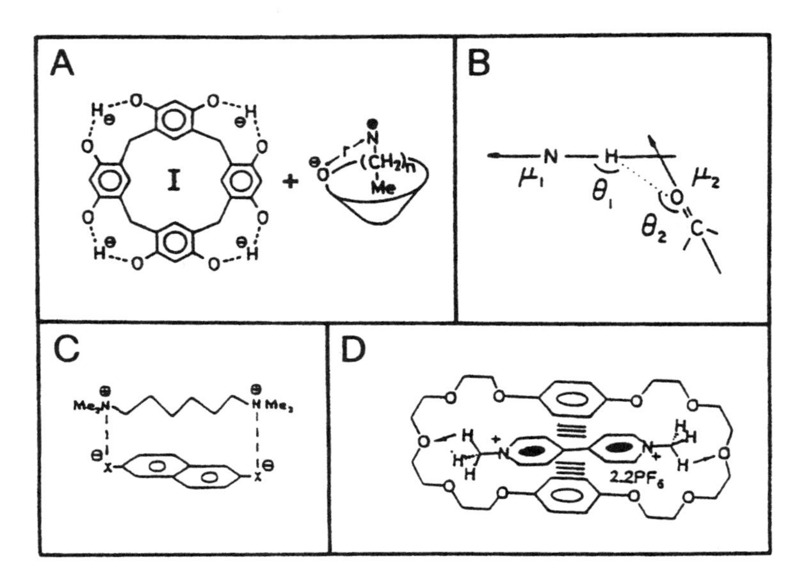

Figure 10 (A) The ion pair $I^{-+}N\ R_4$ [$R = (CH_2)_nMe$]. An increase of n causes an increase of the separation distance and a decrease of ΔF of formation. (From Ref. 44. Copyright 1988 Am. Chem. Soc.) (B) Dipol–dipol interaction parameters for H-bonds. (C) Induced dipoles. ΔF for ion pair formation increases not only with the number of salt bridges but also with the number of aromatic rings in a series of di- and trications and di- and trianions. (From Ref. 36.) (D) π–π stacking interaction plus H-bonds contribute to the stabilization of a complex between paraquat $[PQT]^{2+}$ and a cyclophane-like macrocyclic polyether with hydroquinol rings. (From Ref. 12. Copyright 1997 Am. Chem. Soc.)

an edge-to-face arrangement of successive rings. The occurrence of this conformation is predicted by calculations [46] and has been verified in the crystal structure of benzene [47]. However, in solution the energy difference between the edge-to-face and face-to-face complexation appears to be vanishing small [49]. Complexes with a face-to-face arrangement in which the repulsive π–π interaction is surmounted by proper electrostatic compensation of the substituents are also known [14]. Stabilization of ion pairs may be enhanced by specific salt effects and other non-Coulombic contributions such as induction of dipoles in systems with easily polarizable electron clouds, or by the release of highly ordered solvation shells amounting to a favorable entropy contribution to the binding free energy.

Interaction between permanent dipoles is maximal when a head-to-tail sequence of two dipoles occurs. The net interaction is given by the sum of the Coulombic terms for the four charges, modulated by the directional nature of dipole interaction. For instance, the hydrogen bond energy contains a term related to the interaction of the two dipoles associated with N—H and C=O bonds (Fig. 10B). The directional dependence of the interaction is described [50] by the angles θ_1 and θ_2 that the two dipoles make with the line joining them, and by their ϕ_1 and ϕ_2 azimuthal orientation [cf. Eq. (1)].

2. Hydrogen Bonds

These bonds occupy a major role in the assemblies of both synthetic (Fig. 2) and biological molecules. The H-bond involves a basic donor (C—H, O—H, N—H, F—H) and a basic acceptor (O, N) atom with distance of separation of ~3 Å and a strong directionality that primarily reflects the anisotropy of charge distribution (lone pair) of the acceptor atom. The bond results from an interplay of Coulombic and van der Waals interactions with the former playing a predominant role. Potential functions have been given using either the point charge or the dipole interaction. The expression [50]

$$V = C\left[\left(\frac{r_0}{r}\right)^{12} - \left(\frac{r_0}{r}\right)^{6}\right] - \left(\frac{\mu_1\mu_2}{r^3}\right) g\,(\theta_1\theta_2,\ \phi_1\phi_2) \qquad (1)$$

includes the steric or dispersion interaction terms of the Lennard–Jones potential where r_0 and r are, respectively, the van der Waals and the actual NH · · · O separation distances. The second term represents the Coulombic interaction for the dipoles attached to the NH and O=C bonds, as illustrated in Fig. 1B. The NH · · · O distance is thus a predictable function of the NH · · · O=C angle.

The strength of a H-bond can be determined from simple dimers forming a single bond, or by the analysis of data for complexation of compounds forming multiple H-bonds. The observed linear increase of ΔF for the complexation of amide-type complexes in $CDCl_3$ with the number of H-bonds is a verification of the important ''principle of additive binding increments'' mentioned above [36]. The strength of a single H-bond in the particular solvent was evaluated to be 7.9 KJ $mole^{-1}$. The parallel or antiparallel arrangement of multiple H-bonds in a given complex may increase or reduce the above value due to secondary electrostatic interaction (cf. this volume, Chapter 4 and also Ref. 51). Other data quoted by Schneider [36] suggest a broader range of strength between ~2 and ~20 KJ $mole^{-1}$ per amide link.

3. van der Waals Bonds

Induction of Dipoles. The free energy for ion pair association has been found to be larger than that related to the sole Coulombic interaction when the interacting substances carry easily polarizable electron clouds [36]. The effect is attributed to a supplemental attraction due to the induction of dipoles by the fixed charges. An example is afforded by the association of di- and trications with corresponding di- and trianions containing an increasing number m of benzene rings (one compound is shown in Fig. 10C). The excess free energy of binding (over the Coulombic component) increases with m. This particular interaction decays with the fourth power of the separation distance.

Charge Transfer. The alternation of sites having electron donating and electron withdrawing power may result in attractive interaction due to the transfer of electrons from high-level occupied to low-level unoccupied molecular orbitals (π–π stacking interaction). An example is given in Fig. 10D schematizing a 1:1 complex between the crown ether (host) bisparaphenylene-34-crown-10 having electron-rich hydroquinol rings, and the bipyridinium derivative paraquat $[PQT]^{2+}$ (guest) with electron-poor rings [12]. The complex is a good example of stabilization due to a variety of interactions such as charge transfer, hydrogen bonds (involving a hydrogen atom of $[PQT]^{2+}$ and the polyether oxygen atom of the crown ether), ion-dipole, and dispersive forces. The occurrence of charge transfer bands in the electronic spectra should be verified for a conclusive proof of charge transfer contributions to a given complex in a given solvent.

Dispersive or London Attractions. These occur between all atoms and molecules including nonpolar ones and are due to the difference between the instantaneous conformations of electrons and nuclei allowing the occurrence of transient dipoles that may oscillate in phase. According to the Lennard–Jones or the Buchingham potential functions [cf. Eq. (1)], the attractive energy component decays with the sixth power of the interatomic distance and it is therefore effective only for short site separation, thus requiring a precise fit between partners geometry. At even shorter separation distances, closer to the van der Waals radii, the repulsive (r^{-12}) or steric component becomes effective. The role of dispersive forces in host–guest supramolecular chemistry does not have the same all-important significance that it has on the conformation of molecular polymers. This type of interaction lacks, in fact, the selectivity of other attractive interactions such as the H-bond. Nevertheless, dispersive interactions can be rather strong. The gas-phase calculated interaction between two C—H bonds amounts to ~0,2 K cal/mole, but the cumulative effect of the bonds occurring in two n-hexane molecules amounts to ~6 K cal/mole [36]. The interaction is reduced in polarizable organic solvents but is not affected much by water. Dispersive interactions can therefore be expected to play an essential role in situations (cf. seq.) in which long aliphatic segments undergo a molecular recognition that is not as specific, or pointlike directed, as other types of interactions.

Anisotropic Attractions. Dispersion forces between asymmetrical molecules exhibiting anisotropic polarizability have attractive components that depend upon the mutual molecular orientation. These ''soft'' anisotropic attractions play a main aligning effect in thermotropic (undiluted) liquid crystals, including thermal transitions in lipid bilayers [26]. For the evaluation of these interactions the anisotropy of polarizability must be known (cf. Section II.E). Even a simple C—H bond shows considerable anisotropy: the ratio of polarizabilities in the direction parallel and perpendicular to the bond is $\alpha_{\parallel}/\alpha_{\perp} \sim 1.4$ [26].

B. Solvophobic and Incompatibility Effects

The presentation of the various forces that contribute to host–guest assembly usually includes [36] a discussion of the hydrophobic effect [25] which is based on the poor affinity of water for nonpolar hydrocarbon molecules. Liphophilic segments become capable of molecular recognition in aqueous environments and may stabilize host–guest complexes to which they are connected, or become organized into micellar structures (cf. Section II.C). The ''hydrophobic bond'' has peculiar thermodynamic characteristics: the transfer of hydrocarbons from water to hydrocarbon solvents occurs with a large entropy increase, an often negligible enthalpy contribution, and a decrease of heat capacity [25]. The solubility limit is, in fact, reached upon increasing temperature, a phenomenon that has been described as an ''inverted'' transition [53] distinct from ''normal'' transitions when precipitation occurs on decreasing temperature. The entropy gain that appears to be the driving force for the hydrophobic bond has been attributed to a fluidification of organized water shells surrounding the dispersed liphophobic units. Some controversial features of the above descriptions have been pointed out [36], in particular the disregard of the role of attractive dispersive interaction among liphophobic segments [54]. Earlier work by Debye [55] equated the energy gain in micelle formation with the difference in energy between liquid and gaseous hydrocarbons. A related enthalpic effect was proposed by Sinanoglu [56] in terms of the cohesive energy of the diluent that increases with the release of solvent molecules when a solvophobic bond is formed. An increased stability of a given host–guest complex in solvents with increasing cohesive density was in fact determined [57].

Phase separation and aggregation phenomena occurring upon an increase of temperature are observed for a variety of systems including collagen, poly-L-proline, TMV, and single substances such as NaI $2H_2O$ in water [58]. The phenomenon has also been observed in nonaqueous diluents [59]. In the present context it is relevant to consider not only the interaction of a particular segment of a supramolecular polymer with a poor solvent, but also the interaction of two incompatible segments resulting in their macroscopic segregation (or microsegregation, cf. Section II.C).

The features of both normal and inverted phase separations for these systems are therefore described in terms of an approximate treatment of binary solutions originally developed for mixtures of poorly interacting apolar polymers [60]. A liquid–liquid phase separation is expected to occur at a critical temperature $T = \theta > 0$ at which a balance of the enthalpy (κ) and entropy (ψ) components of dilution is achieved [60]

$$\theta = T\kappa/\psi \tag{2}$$

The condition $\theta > 0$ is fulfilled provided both the heat and entropy of dilution parameters are of the same algebraic sign. Under normal situations the phases separate on cooling and an upper critical consolute temperature (UCST) corresponds to $\psi = \kappa$, with $\psi > 0$ and $\kappa > 0$. However, under the "inverted" situation [53,59] the phases separate on heating and the lower critical consolute temperature (LCST) requires $\psi < 0$ and $\kappa < 0$. In the latter case more order may be said to occur in the solution than in the phase-separated system. Some degree of aggregation must therefore have occurred at least for one of the dissolved components. Moreover, the entropy gain due to the breaking of the aggregation must prevail over the entropy loss resulting from demixing. Liquid–liquid phase separation involving an isotropic and a liquid crystalline phase can also be of the normal or inverted type [61] and conform to the above general principles [62].

The mutual solubility, or compatibility, of two components (a and b, polymer/solvent, or polymer/polymer) may also be described in terms of cohesive energy density using the interaction parameter defined as [60]

$$\chi = z\frac{\Delta\omega}{kT} \tag{3}$$

where $\Delta\omega = \omega_{ab} - \frac{1}{2}(\omega_{aa} + \omega_{bb})$ is the difference in energy between the formation of ab contacts when aa and bb contacts are broken (z is the coordination number, k is the Boltzmann constant, and T the absolute temperature). A critical value of the interaction parameter ($\sim$0.5) again corresponds to either a lower or an upper critical solution temperature Θ since

$$\tfrac{1}{2} - \chi = \psi(1 - \theta/T) \tag{4}$$

The relationship between χ and the cohesive energy density (CED) of two components is [60,61]

$$\chi = V_{seg}\frac{(\delta_a - \delta_b)^2}{RT} \tag{5}$$

where δ^2 = CED is the Hildebrand solubility parameter (calculated with QSPR methods) and V_{seg} the volume of a segment of *a* or *b*. When large differences in the cohesive energy densities of the components occur, χ exceed the critical value and the two components segregate into two macroscopic phases.

Inverted melting transitions in the presence of a diluent have also been described [53]. These are characterized by an ''increase'' of the melting temperature with the amount of diluent, in contrast to the normal case of a depression of the melting temperature by a diluent. Even in this case entropy and enthalpy changes at the transition must be viewed as the sum of the contributions due to two components. The total enthalpy exchange is the sum [53]

$$\Delta H_{tot} = \Delta H^0 + \Delta H_{dil} \tag{6}$$

where $\Delta H^0 > 0$ is the melting enthalpy of a pure component, and ΔH_{dil} is its dilution enthalpy. The condition for normal melting requires $\Delta H_{tot} > 0$, whereas the condition for inverted melting is $\Delta H_{tot} < 0$ [53]. Therefore the inverted transition requires not only $\Delta H_{dil} < 0$ (or $\kappa < 0$), but also $|\Delta H_{dil}| > \Delta H^0$.

Macroscopic phase separations involving two incompatible components can produce at least one phase endowed with significant supramolecular organization [62]. For instance, the solvophobic effect in the system (hydroxypropyl)cellulose in H_2O results in an inverted transition leading to the formation of a liquid crystalline phase at the smallest concentration ($\sim$0.4% v/v, Fig. 11) at which cholesteric order was ever detected [63,64]. More important, however, is the fact that the preceding thermodynamic features are the driving forces for the microstructurization and microsegregation occurring in supramolecular assemblies formed by surfactants and by block copolymers when a complete phase separation of the two incompatible components is prevented by the amphiphilic nature of the molecules (cf. Section II.C).

C. Micellization, Microsegregation, Interdigitation

The poor affinity of nonpolar hydrocarbons with water is manifested in a migration to the air–water interface and in a macroscopic phase separation equivalent

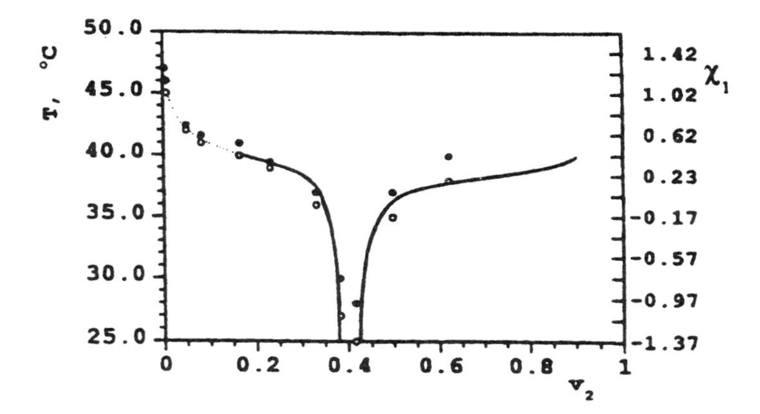

Figure 11 Inverted phase transition for (hydroxypropyl)cellulose in H_2O. An isotropic solution with polymer volume fraction $>\sim 0.05$ produces a biphasic (isotropic + liquid crystalline) mixture at $T > 42°$. The condition $T = \Theta$ ($\chi \sim 0.5$) occurs at $T = 41°C$. (From Ref. 63. Copyright 1980 Am. Chem. Soc.)

to the formation of aggregates of infinite size, as discussed above. However, if a hydrophilic head group is attached to the hydrocarbon molecule the apolar tails can avoid phase separation by forming, above a critical concentration, micellar-like supramolecular structures stabilized by the exposure of the head group to water. Similar structures are expected if two incompatible segments *A* and *B* are chemically connected in an amphiphilic block copolymer and dissolved in a selective solvent for either *A* or *B*. Moreover, if the same *AB* block copolymer is studied in the absence of a solvent, the chemical bond prevents the macroscopic phase separation expected for unconnected *A* and *B*. A supramolecular structure will instead occur in which all *A*-type and *B*-type segments microsegregate in domains separated by a surface that contains the intersegmental bonds.

Self-assembled supramolecular structures formed by surfactants, lipids [25,26], and block copolymers [65,66] in selective solvents have been extensively investigated. Structures involve spherical, cylindrical, and inverted micelles as well as bilayers and vesicles (Fig. 12). In all cases the solvophilic groups point toward the solvent and in the case of vesicles there is also a solvent-filled cavity. The simplest structures in Fig. 12 can be regarded as the repeating building blocks of larger assemblies. The relevance of these structures and their mechanism of formation to biological membranes has been extensively discussed [25,26], and their relevance to supramolecular polymerization has been pointed out [5].

The distinctive feature of block copolymer micelles is the occurrence of a polymer segment as the head group protruding from the core. Figure 13A illus-

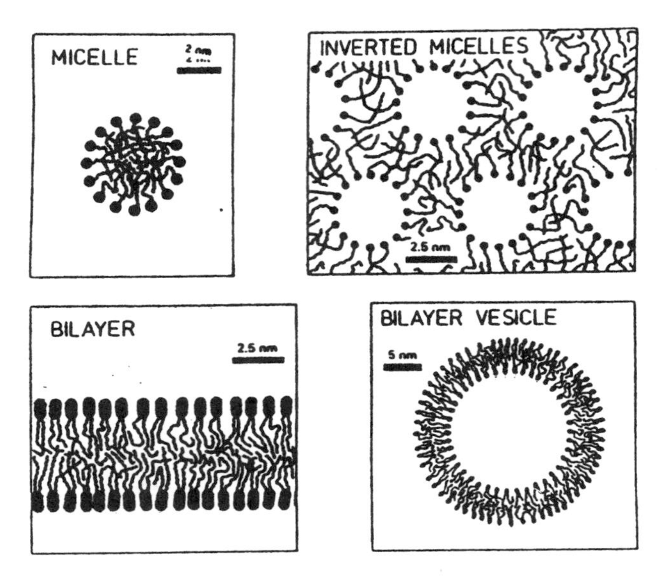

Figure 12 Micelles, bilayers, and vesicles formed by single- or double-chained surfactants in water. (From Ref. 26.)

trates cases in which the relative length of the two flexible blocks determines a large core and a thin corona, or vice versa. Cases in which the selective solvent was a homopolymer of the *A* or *B* type have been described [67]. In the case of micelles formed by a rod/coil copolymer (Fig. 13B), a spherical core formed by the rigid block is not favored. Cylindrical cores or bilayers should be favored instead [66–69]. A lamellar sheet typical of the microsegregation occurring with an *ABA* undiluted triblock copolymer is schematized in Fig. 13C. The formation of bilayers and giant vesicles has also been reported [23,70]. Micellar and microdomain structures are described in more detail in this volume, Chapters 3, 6, and 7.

The formation of spherical micelles occurs with a significant degree of cooperativity at very low amphiphile concentrations but should be regarded as a true phase transition only in the case of an infinite micelle ([25,26], cf. Section III.D). As discussed in more detail in this volume in the chapter by Hentschke, the critical micelle concentration (CMC) marks the limit at which micellar aggregates are formed. Further addition of surfactant does not cause a large increase of free amphiphile molecules but rather an increase of number and average size of micelles (cf. Figs. 4 and 5 in this volume, Chapter 2). Spherical micelles thus exhibit a size distribution that corresponds to variable numbers (s) of constit-

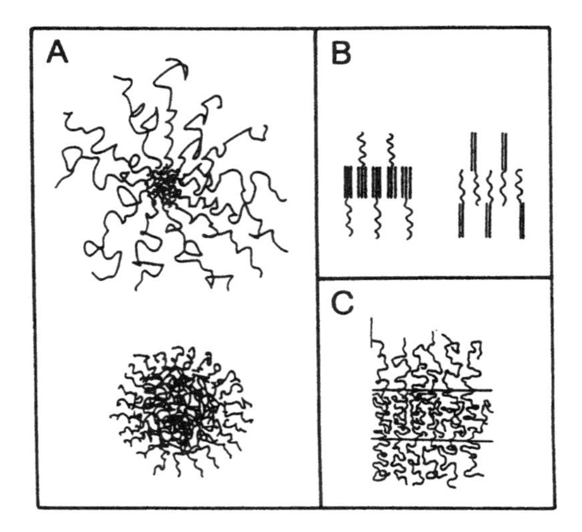

Figure 13 (A) Micelles in solutions of coil/coil diblock copolymers having different length of the solvophilic and solvophobic block. (From Ref. 65.) (B) Bilayers of rod-coil diblock copolymers in solvents affine for either block. (From Ref. 68.) (C) Microdomain structurization for a triblock copolymer in absence of a diluent. (From Ref. 65. Copyright 1991 Springer-Verlag.)

uent amphiphilic molecules (M), and broadens with total surfactant concentration.

Minimum $\langle s \rangle$ values range from 50 to 100 for typical ionic surfactants, average sizes are in the order of nm, and CMC are in the order of mM, decreasing with charge screening and increasing with the size of the apolar tail. Micellar parameters can be predicted from detailed theoretical approaches based on the minimization of the free energy of an isolated micelle or of a multicomponent system for both low MW surfactant and block copolymers [71].

Upon increasing the amphiphile concentration an evolution toward more asymmetric shapes (rodlike or disklike) and decreasing surface/volume ratio is observed. Eventually cylindrical (capped) micelles, bilayers (extended open sheet with rounded edges), and closed vesicles are formed. The prediction of the stability of the various geometrical shapes has been one of the outstanding goals of micellar studies. A simple treatment of the solvophobic core as a structureless continuum provides a justification for all micellar shapes, including vesicles [25]. More complex, however, is the detailed description of how amphiphilic molecules can pack within the micellar structures. For instance, the external surface area of a vesicle is larger than the internal one, requiring a larger number of molecules in the section of the curved bilayer that is pointing outward.

A most successful attempt in using geometrical parameters of the amphiphile to explain micellar shapes (without a detailed knowledge of specific interactions) is due to Israelachvili [26]. He describes the geometrical constraints that affect the interfacial surface area in terms of the area of the solvophilic headgroup (a_0), the volume (v), and extended length (l) of the aliphatic tail (Fig. 14). The parameter v/a_0l controls the critical packing shape and the most stable structure for a given amphiphile in a given solvent environment. For instance, large headgroup areas (e.g., ionic amphiphiles in low salt) favor a conical packing shape ($v/a_0l < 1/3$) and spherical micelles (Fig. 14). On the other hand, cylindrical packing shapes (i.e., double chained lipids with a small head group, $v/a_0l \sim 1$) favor a planar bilayer, whereas a truncated cone ($v/a_0l \sim 1/2 - 1$) favors a vesicle. A generalization of design parameters to polymeric surfactants is presented in this volume, Chapter 3.

Intermicellar forces are generally of a repulsive nature (i.e., charged amphiphiles) and a reduction of such repulsion accompanies the transformation from spherical to cylindrical micelles. Further increase of concentration results in the formation of linear assemblies and liquid crystalline lyotropic mesophases (cf. Section III.B). Not only nematic (N_c and N_d from rodlike or disklike shapes, respectively), hexagonal, and smectic phases, but also biaxial (mixtures of N_c and N_d) and complex cubic phases (bicontinuous networks or plastic crystals)

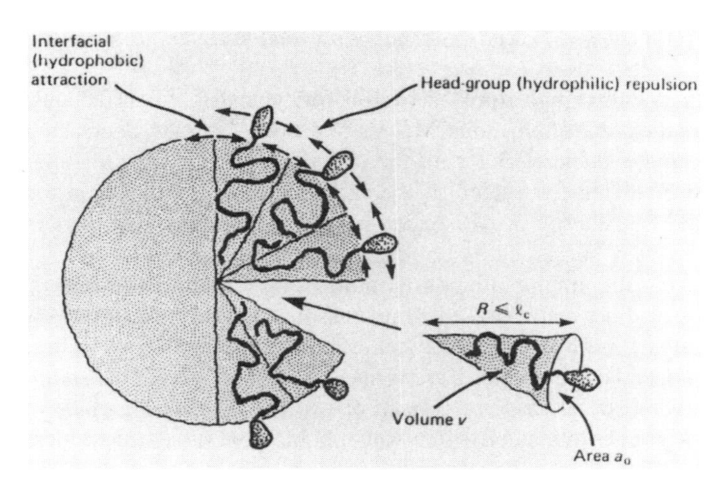

Figure 14 The geometrical shape of the amphiphile, expressed by the ratio v/a_0l, determines the stability of micelles and bilayers. Spherical micelles are stabilized by conical shapes: $v/a_0l < 1/3$ (single chained, large headgroup area). (From Ref. 26.)

were reported [72,73]. For block copolymers with long segments protruding from the core, interlocking may instead occur upon increasing concentrations.

The interdigitation of long aliphatic segments is a typical mode of microsegregation observed for systems having two incompatible components. Interdigitation may promote soluble micellar-like cylindrical assemblies in solutions of a rigid polymer (i.e., DNA; cf. this volume, Chapter 7) with a complementary surfactant. Self-assembled organization of fibrillar elements separated by layers of an interdigitating matrix in comblike or block copolymers was also observed (cf. Section III.D).

D. Shape Effects and Site Distribution

Molecular recognition involving the site interactions described in Section II.A results in the formation of supramolecular bonds which are of the same nature as those occurring with weaker complexes and systems incapable of developing a large structural organization. For the development of more organized functional assemblies, site distribution and geometrical complementarity (shape recognition) also need to be considered. The scheme in Fig. 15A illustrates the assembly of triangular shaped units having mono- or bi-functional binding sites. Although in the former case only dimeric complexes are possible, in the latter case a linear chain or a closed disk can be formed. A closed disk assembly of tapered molecules is found in TMV (cf. Fig. 6D). In Fig. 15B linear assemblies of cylindrical, spherical, and discotic units are illustrated. Interaction sites must occur on the north and south surfaces (or poles) for linear assemblies to grow to an appreciable size. A planar assembly of spheres (Fig. 15C) can grow to a large size even without site interaction [26]. Parallel stacks of such planar assemblies and large three-dimensional assemblies are likewise possible. An ''equatorial'' distribution of binding sites, as indicated in Fig. 15C, does, however, stabilize the planar assembly. An assembly of this type is found in S-layers (cf. this volume, Chapter 5 and Section III.D). The nonsymmetrical site distribution schematized in Figs. 6A and C, respectively, for fibrinogen and for actin, is consistent with the formation of helical assemblies (cf. Section III.C). Fig. 15D details the interaction pattern for the formation of the helical structure in the case of actin [2]. An example of the usefulness of shape concepts was discussed in Section II.C in connection with micellar assemblies (cf. Fig. 14). A practical example showing the importance of simultaneous site and shape recognition is afforded by the technique known as molecular imprinting [74] in which suitable cavities for host molecules of any possible shape are formed, with the aid of a template, within a molecular network.

Although molecular and shape recognition are closely related effects, it is often convenient to distinguish them. With supramolecular polymers it is further

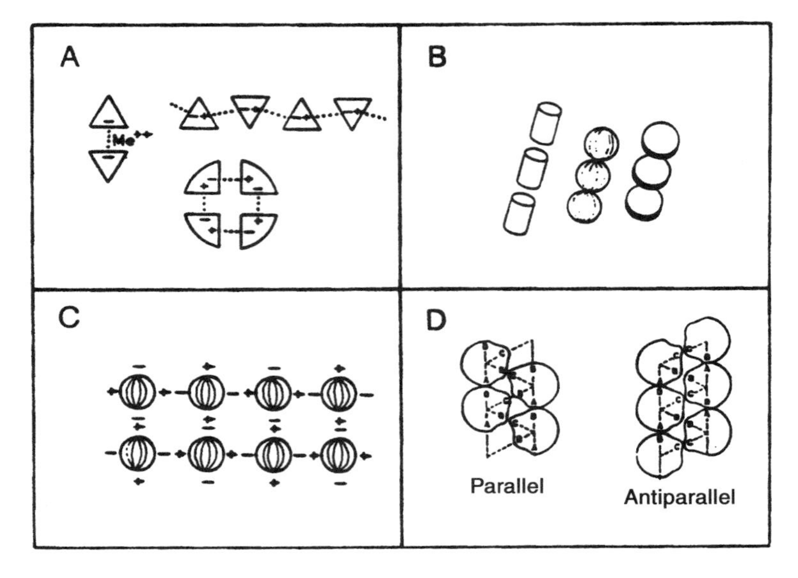

Figure 15 Schematic assemblies of (A) triangular units with monofunctional and bifunctional binding sites; (B) cylindrical, spherical, disklike units forming linear sequences; (C) spherical units with an equatorial distribution of binding sites forming a planar assembly; (D) asymmetrical site distribution generating helical assemblies. (From Ref. 2.)

convenient to distinguish two types of shape effects. The first type (shape I effects) pertains to the shape of the molecules or of the assemblies that form the repeating unit. Shape II effects pertain instead to the assembled polymer. The overall assembly process may start as a molecular and shape I recognition. Eventually this process is translated into well-defined shapes. Thereafter the evolution toward more complex structures may be simply described by shape II recognition using geometrical parameters (cf. Section II.B).

Shape I effects are important at the molecular recognition level. *Endo* recognition occurs when binding sites are oriented into a molecular concavity. A well-known example is the enzyme-substrate catalysis when the specificity toward the hydrolysis of a particular peptide bond is controlled by the binding of the adjacent sidechain inside a pocket of the enzyme [75]. The great selectivity and complex stability observed in closed cavities is attributed to an enhancement of the strength of pairwise attraction (particularly of the dispersive and hydrophobic type) with respect to binding over flat surfaces [36]. Site distribution within a cleft may favor the convergence of groups capable of directional binding, thus enhancing the strength of H-bonds [38]. On the other hand, *exo* recognition occurs

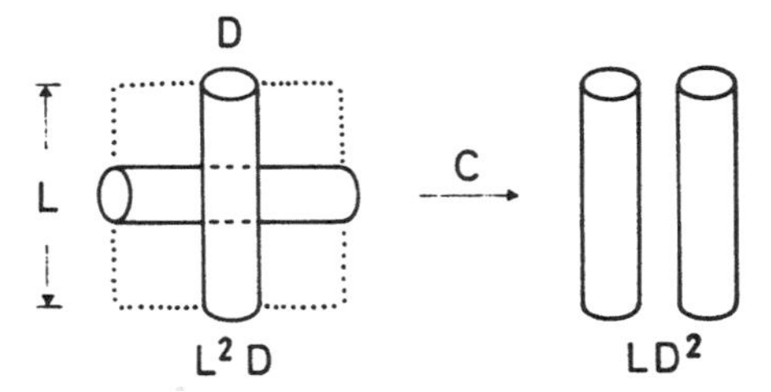

Figure 16 The formation of a parallel assembly of rods is favored by a decrease of excluded volume (shape II effect) even in the absence of binding sites.

when binding sites are directed outward flat surfaces. Similarity of the size of the surfaces and multiple pairwise interactions enhance the binding free energy.

Shape II effects suggest intriguing correlations between macroscopic and molecular design concepts. Shape recognition based on excluded volume effects for rodlike objects is illustrated by the scheme in Fig. 16. Large rods with $L \gg D$ assembly in a parallel rather than a random way because there is a significant reduction of excluded volume (from L^2D to LD^2) upon increasing concentration (C). The model also applies to large disks ($D \gg L$) and is of general validity, from molecules, to assemblies, to macroscopic objects [76]; cf. Section II.E). No interacting sites need to occur on the surface of the rods (or disks); the assembling force may have a purely entropic origin and it controls the organization of macroscopically ordered supramolecular structures in a liquid crystalline field. Furthermore, it may also have a direct effect on increasing the degree of supramolecular polymerization (cf. Section II.E).

The dendrimers (cf. this volume, Chapter 9) may be regarded as a growing sequence of similar chemical steps exhibiting fractal geometries around an initiator core [77]. Dendrimers in the nanometric range may further assemble through a repetition and composition of the basic geometrical design. Macroscopic analogies may be found in the growth of corals, branching of trees, and nesting of spheres into a large sphere, and so on [78].

Intracellular networks reflect a supramolecular assembly design of protein chains that display occasional pentagonal meshes within a hexagonal topology [79]. The pentagons determine the curvature of the network and stabilize the discocythe shape of the erythrocytes. Similar design strategies are found in both molecular systems such as fullerenes [80], and in macroscopic objects such as bamboo vases and football spheres [81].

E. Liquid Crystallinity

The phase transition from the isotropic to the liquid crystalline state brings about a structurization of the nematic or smectic type in solutions of rigid polymers or in melts of low MW and segmented polymers (a segmented polymer is based on low MW mesogens connected by flexible spacers along the mainchain; [82,83]. As a result of the transition the molecular size and shape of the above compounds remain essentially unchanged. At least two types of interactions assist this transition: *soft* anisotropic attraction (chemical recognition) which is the prevailing component for low MW mesogens and segmented polymers (thermotropic melts), and *hard* repulsion (shape II recognition) which is the prevailing component for rigid polymers in lyotropic solutions. Soft attraction results from the orientation-dependent intermolecular energy $\varepsilon(\theta)$ of nonspherical molecules related to their anisotropy of polarization $\Delta\alpha$ [84,85]

$$\varepsilon(\theta) = C\left(\frac{\Delta\alpha}{\overline{\alpha}}\right)^2 \varepsilon_{\text{iso}}\overline{V}S\left(1 - \frac{3}{2}\sin^2\theta\right) \tag{7}$$

where C is a constant, θ is the angle between the molecular and the domain axes, $\overline{\alpha}$ is the mean polarizability, ε_{iso} the isotropic intermolecular energy, $\overline{V}$ the ratio of actual to hard core volume, and S is the order parameter. Hard interactions reflect instead the shape-dependent geometrical anisotropy of the molecule illustrated in Fig. 16. The relevant parameter is the length to diameter (axial) ratio [50]

$$X = \frac{L}{D} \tag{8}$$

Extensive theoretical and experimental investigation [82,83] has shown the limits under which the experimental behavior of low and high MW mesogens is described by the corresponding theoretical approaches. The following important conclusions are relevant to the present discussion.

i). A critical value of the axial ratio (X^i), varying from $\sim$4 to $\sim$8 for different theories of rigid chains [86] determines the limit at which an undiluted mesophase becomes ''absolutely stable'' [84], implying that when $X > X^i$ the mesophase is primarily stabilized by hard interaction and compositional changes (lyotropic systems). In this case a critical solute volume fraction (v^i) can be defined, decreasing with X according to

$$v^i \approx \frac{X^i}{X} \tag{9}$$

For the above system the nematic $\rightarrow$ isotropic transition is not influenced by temperature changes: $T_{NI} \rightarrow \infty$ (unless a large temperature coefficient dX/dT of the rigid conformation does occur). Theory is well developed for large rods ($L/D \gg 1$), but not for large disks ($L/D \ll 1$). On the other hand, when $X < X^i$ soft interaction prevails, the low MW mesogens or the segmented polymers are thermotropic ($T_{NI} > 0$) and may admit only a small amount of isotropic diluent. The small geometrical anisotropy of these systems may include both small rods ($L/D > 1$) or small disks ($L/D < 1$).

ii). The partial rigidity of long chain polymers is characterized by the persistence length P which assumes the role of the limiting rigid segment stabilizing the mesophase. This implies that ν^i will decrease with L until an asymptotic value of the axial ratio is reached. For the model of the freely jointed chain equation (8) is expressed in terms of the Kuhn segment [86]:

$$X = \frac{2P}{D} \tag{10}$$

Persistence lengths for molecular polymers are in the range 10 ÷ 200 nm corresponding to rather large values [83] of the critical volume fractions (in the range 0.02 ÷ 0.2 taking $X^i = 6$ and $D = 5 \div 10$ Å). Within the mesophase, a wormlike chain may be forced to assume a more extended profile than in the isotropic phase due to the restriction imposed on the director by the order parameter [86,87]. Semirigidity in the nematic state is characterized by the "deflection length" λ

$$\lambda = \frac{P}{\alpha} \tag{11}$$

where α is a parameter larger than unity, increasing with concentration and inversely related to the width of the angular distribution of the chain tangent vectors.

The above general considerations, valid for unassociated (molecularly dispersed) low or high MW compounds forming what has been defined [5] as a molecular liquid crystal (MLC), indicate that the free energy of the mesophase may be approximated by the sum of three terms: $F = F_{ster} + F_{\varepsilon} + F_{el}$ representing, respectively, the hard interparticle interactions, the soft interparticle interactions of both anisotropic and isotropic nature (including solvent effects, these have been discussed in Section II.B), and any conformational rearrangement of the semirigid mesogen within the nematic field.

Extension to the case of liquid crystals formed by supramolecular assemblies, the *supramolecular liquid crystals* (*SLC*), entails consideration of the contact or intraparticle interactions that stabilize the assembled units. Figure 17 schematizes the transition from the isotropic to the nematic phase for molecularly dispersed polymers (A), and for closed (B) and open (C) supramolecular assemblies. If, as discussed in Section I.B, no further association → dissociation equilibria accompany the transition of closed assemblies, their liquid crystalline behavior is indistinguishable from that of a MLC. The relevant axial ratio will be determined by the geometry of the assembly with no need to otherwise account for contact interaction. Among the examples are DNA [88] and some synthetic polymers [89,90]. For instance, poly(*p*-benzamide)(PBA) in N,N-dimethylacetamide/LiCl isotropic or nematic solutions may be represented by an assembly of seven PBA molecules with a side-by-side shift of ¼th the molecular length. In this

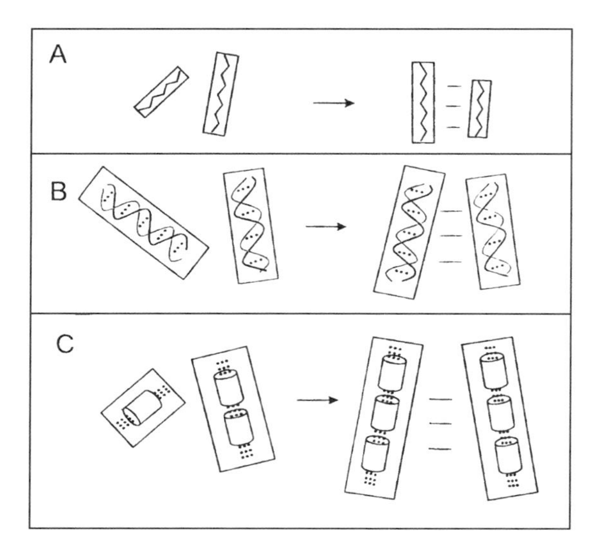

Figure 17 Schematization of the isotropic → nematic transition for molecularly dispersed polymers (A); closed supramolecular polymers (B); open (linear) supramolecular assemblies (C); —hard (soft) interaction among assemblies; . . .contact interactions within each assembly; coupling of the two interactions causes growth simultaneous to orientation for case C. (From Ref. 5. Copyright Taylor & Francis, 1999.)

example the axial ratio of the assembly (~104), is indistinguishable from the axial ratio (~100) of molecularly dispersed PBA.

Quite different is the case of open assemblies such as the supramolecular polymers in Fig. 2, or the micelles in Fig. 7, for which a coupling may occur between the contact interactions that stabilize the assembly and the hard/soft interactions that stabilize the mesophase. Formally, it is necessary to consider a term F_{intra} to account for the stabilization of the assembly through the contact energy so that the expression for the free energy of the supramolecular liquid crystal becomes

$$F' = F'_{\text{ster}} + F'_{\varepsilon} + F'_{\text{el}} + F_{\text{intra}} \tag{12}$$

The result is an enhancement of growth of the assembly [13,91–97] occurring simultaneously with the formation of the ordered mesophase (cf. scheme in Fig. 17C).

A detailed theory for *growth coupled to orientation* was proposed by Herzfeld and Briehl [91] and by Gelbart et al. [92] to describe the assembly of micelles into linear particles. Odijk [20,93] revised the mechanism by recognizing that catastrophic growth in the nematic state is prevented by the flexibility of the linear assembly, resulting in a decoupling of growth in correspondence to the persistence or deflection length of the assembly [94]. Hentschke [13,95] expanded the theory to predict the occurrence of various mesophases for discotic amphiphiles. An important feature of these theories is that a nematic solution of dispersed assemblies occurs only for relatively large values of the contact energy and of chain rigidity. Thus, the ability of a supramolecular polymer to form a nematic lyotropic phase is a good indication of a strong assembly able to grow to a large DP. A detailed account of the theory is given in this volume in the chapter by Hentschke. The mechanism of growth coupled to orientation has been suggested as a main driving force in supramolecular polymerization [96]. Accordingly, its implications are further considered in Section III.B.

III. GROWTH MECHANISMS

A. Multistage Open Association

The scheme of multistage open association (MSOA) of *s* monomeric units (''unimers'') M_1 into a linear sequence M_s is [6,97–100]

$$
\begin{aligned}
&M_1 + M_1 \Leftrightarrow M_2 && C_2 = K\,(C_1)^2 \\
&M_2 + M_1 \Leftrightarrow M_3 && C_3 = K\,C_2\,C_1 = K^2(C_1)^3 \\
&\quad\ldots && \quad\ldots \\
&M_{s-1} + M_1 \Leftrightarrow M_s && C_s = K^{-1}\,(KC_1)^s
\end{aligned}
\tag{13}
$$

where C_s is the concentration of the *s*-mer and identical equilibrium constants K for each step (no cooperation) are assumed. The extent of growth may be related to the probability that a unimer has reacted according to the Carothers equation (6)

$$
DP_n = \frac{1}{(1-p)} \tag{14}
$$

where DP_n is the number-average degree of polymerization and p is the percent conversion of monomer. The extent of growth will be controlled (through p) by either K or the monomer concentration [100]. The probability of a sequence of *s*-mer is given by the product of single step probabilities $[p^{s-1}(1-p)]$ showing that the length distribution widens with p. The width of the distribution, expressed in terms of the number and weight average degree of polymerization, is [6]

$$
\frac{DP_w}{DP_n} = 1 + p \tag{15}
$$

The direct relationship between DP_w and the total unit concentration C_0 ($= \Sigma_s C_s = C_1/(1 - KC_1)^2$) is

$$C_0 \approx \left[\frac{M_0}{4KN_A}\right]\left[\left(\frac{M_w}{M_0}\right)^2 - 1\right] \tag{16}$$

where N_A is Avogadro's number and M_0 the monomer molar mass.

A plot of Eq. (16) is shown in Fig. 18 [98,99]. If each consecutive binding step occurs with a standard free energy change of −7.8 Kcal/mol (comparable to a strong H-bond), growth appears severely limited. At the relatively large unimer concentration of 0.01 g/ml, $M_w/M_0 \approx 4$. The experimental points in Fig. 18 refer to the association of the enzyme glutamate dehydrogenase and appear to conform to the theoretical prediction up to ~1% concentration [99]. Small angle X-ray data confirm that the association of the unimers (each a rounded cylinder with $L = 126$ Å, $D = 84$ Å, $L/D \approx 1.5$) occurs along a linear sequence [100]. The expected size distribution is shown in Fig. 19. Similar low values of DP are evident upon consideration of the effective chain length calculated by Hilger and Stadler [101] for linear supramolecular polymers stabilized by single H-bonds in nonliquid crystalline phases. They predict the occurrence of DP on the order of 15 at room temperature. Even in molecular polycondensation the relatively low K values are known to preclude the attainment of large DP under equilibrium conditions [6]. Griffin and coworkers [7] have reported supramolecular polymers

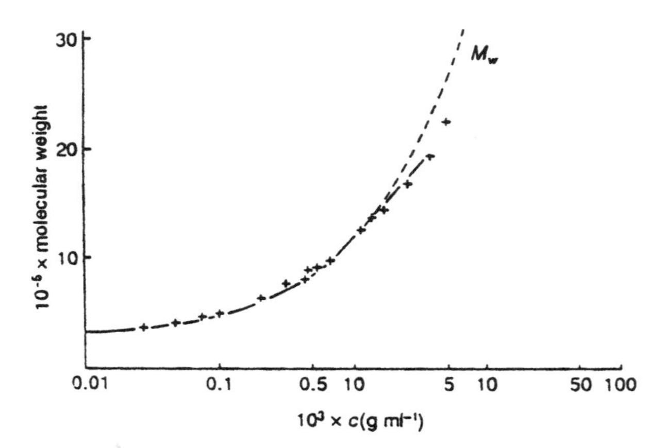

Figure 18 Weight average molar mass M_w as a function of monomer concentration calculated from Eq. (16) for $K = 9.0 \times 10^5$ l/mol and $M_0 = 307{,}000$. Experimental points correspond to the association of glutamate dehydrogenase in phosphate buffer at pH = 7.6, T = 20°C. (From Ref. 98.)

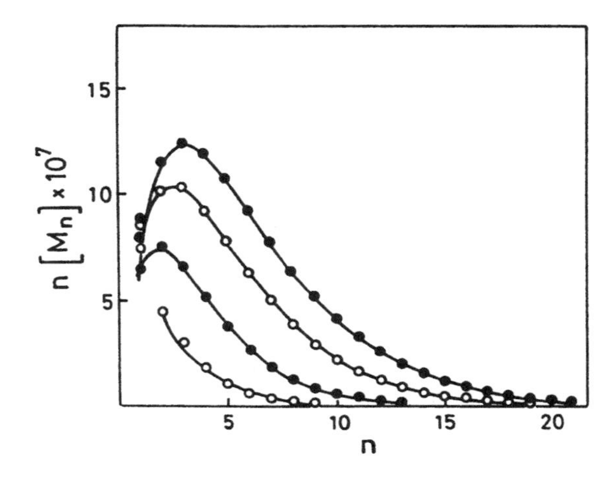

Figure 19 The expected n-mer distribution corresponding to Fig. 18 (K = 9.0×10^5 l/mol) widens with increasing unimer concentration. Curves for 0.51, 1.17, 2.19, and 3.15 mg/ml are shown. (From Ref. 98.)

based on single H-bonds (cf. Fig. 2A) and presented evidence supporting a MSOA mechanism. The supramolecular polymerization of tropomyosin [102] is also satisfactorily represented by Eq. (1).

The most interesting polymers based on four H-bonds per repeating unit reported by Meijer and coworkers ([10]; cf.Fig 2C) appear also to result from a step polymerization process characterized by the lack of reaction products (other than polymer) and by the unusually large dimerization constant of ureidopyrimidone ($K > 10^6\ M^{-1}$ in chloroform). It is easy to show that values of p [Eq. (14)] larger than 0.99 can be expected under equilibium conditions [6]. They estimated a value of DP on the order of 700 in $CHCl_3$ using an empirical equation relating DP, K, and the variation of specific viscosity with the amount of a monofunctional agent (cf. this volume, Chapter 4).

It should be pointed out that excluded volume effects contribute to an enhancement of growth even in the isotropic phase. This effect is described by Hentschke for a spherocylindrical assembly (this volume, Chapter 2; cf. Fig. 5). However, in order to explain still larger DP values it is necessary to consider alternative growth mechanisms that include cooperative binding of the unimers. The cooperative effect described in Section III.B is based on the formation of liquid crystalline phases and is of an "interassembly" nature since it requires the simultaneous growth of neighboring assemblies. The cooperative effect described in Section III.C is based instead on the growth of a single helical or tubular particle in isotropic solutions and it is of an "intra-assembly" nature.

B. Growth Coupled with Liquid Crystalline Orientation

Often the formation of supramolecular dimers, or low DP oligomers between similar or dissimilar components, is accompanied by the formation of a liquid crystalline phase. For instance, hydroxypyridine dimers [103] or H-bonded complexes between adenine and thymine [104] are capable of forming liquid crystals. Nonmesogenic pyridine and carboxylic acid derivatives have also been shown to develop liquid crystallinity upon complexation [105]. Complexation at the sidechain of a nonmesogenic, flexible polymer [106,107] may also result in the formation of mesophases (Fig. 20A). The linear segmented assemblies between a dipyridyl and a diacid reported by Griffin and coworkers ([7]; Fig. 2A) also exhibit a thermotropic nematic phase. In terms of the discussion in Section II.E, liquid crystallinity in segmented chains is a reflection of the soft interaction occurring for the low MW mesogens incorporated in the mainchain. In fact, both the polymer in Fig. 2A and the compound in Fig. 20B exhibit comparable T_{NI} temperatures. For this class of systems occurrence of polymerization may be evidenced by larger transition temperatures with respect to the corresponding low MW mesogen [108]. It is, however, difficult to quantitatively assess the value of DP starting from the increased T_{NI} [109].

In principle, bonding in thermotropic systems such as the compound in Fig. 20B and the polymer in Fig. 2A should be enhanced by the formation of the nematic phase, just as in the case of the lyotropic systems discussed below. Bladon and Griffin [7] have indeed theoretically shown that growth is expected for a wormlike chain forming a nematic phase stabilized by soft interactions of the Maier–Saupe type. The modest effect expected would not be easily revealed by

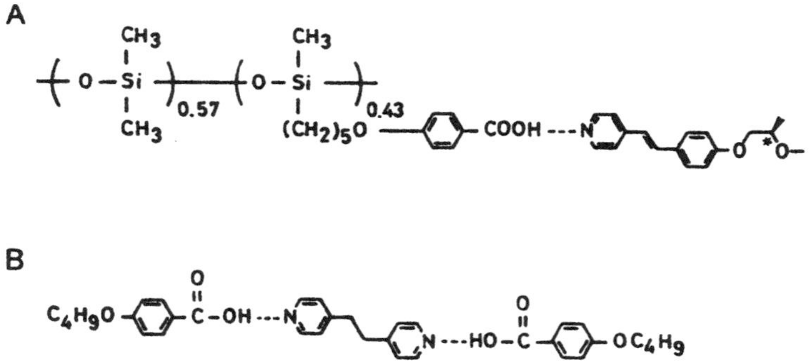

A

B

Figure 20 (A) A polysiloxane copolymer having benzoic acid sidechains complexed with a mesogenic chiral stilbazole derivative [105]. (B) A dipyridil complexed with two monofunctional acids showing $T_{SN} = 166.5°$, $T_{NI} = 178.5°C$. Compare with the open assembly in Fig. 2A (Ref. 7a).

changes in the IR band associated with the H-bond. In fact, in all the above examples growth appears limited to the formation of supermolecules (this volume, Chapter 14) or of short open sequences and the salient hard interactions typical of rodlike or wormlike assemblies, leading to the stabilization of lyotropic mesophases, are absent. This conclusion does not limit the potential significance of either sidechain or segmented supramolecular polymers [107] but serves to highlight the advantage that lyotropic systems have in the assessment of a large DP. The occurrence of the lyotropic mesophase is direct evidence of elongated particles interacting by excluded volume effects according to Eq. (9).

The formation of a lyotropic liquid crystalline phase for supramolecular polymers (this volume, Chapter 2; cf. also Section II.E and Fig. 17C) will dramatically reinforce a modest growth expected from Eq. (16). When the concentration of unimers is such that a linear aggregate—if formed—would have an axial ratio corresponding to its critical volume fraction ν^i, growth becomes coupled to the development of the long-range order of the SLC [5,20,91–95]. A general feature of excluded volume treatments is that the critical volume fraction is inversely related to the axial ratio of either a molecular or a supramolecular polymer. Therefore, the assembly suddenly growing when concentration reaches ν^i is necessarily the one with the largest possible geometrical anisotropy as expressed by its persistence length P [20,93–95].

The superimposition between growth according the MSOA theory and growth coupled to orientation is illustrated in Fig. 21 schematizing the expected

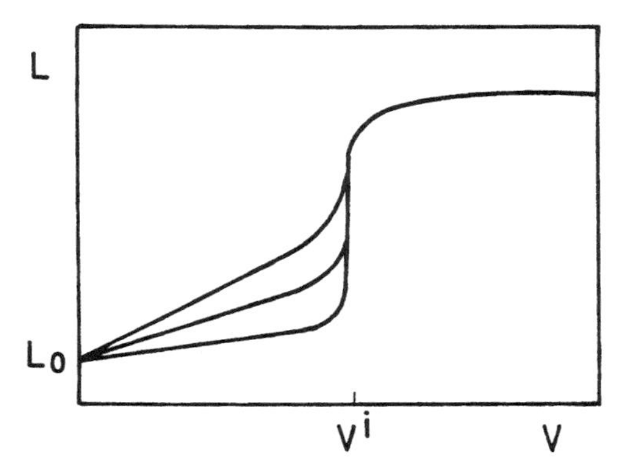

Figure 21 Schematic variation of the length L of a supramolecular polymer with the volume fraction of unimers having length L_o. At the critical volume fraction ν^i sudden growth is simultaneous with the formation of the nematic phase. At $\nu < \nu^i$-growth occurs according to the MSOA theory [Eq. 16], three curves for increasing values of the contact energy being represented.

variation of the length of the growing assembly with the concentration of unimers [3]. For $\nu < \nu^i$ three curves for different values of the binding constant in Eq. (16) represent growth due to MSOA ($L < P$). As a result of interassembly cooperative growth, the length attains at $\nu = \nu^i$ a value reflecting the semirigidity of the assembly. When $\nu > \nu^i$ decoupling of growth does not allow the occurrence of values of L much larger than P, although some further lengthening upon increasing concentration above ν^i is predicted [20,94,95]. For polydisperse molecular polymers, fractions with $L > P$ are known to enter the mesophase at an asymptotic value of the critical conventration [101]. A similar situation for the supramolecular polymer has not been described. Moreover, the parameters that decouple growth, persistence, or deflection length do not exhibit a size distribution. Therefore, a tendency toward a reduced polydispersity could be read into the theory of growth coupled to orientation [5]. However, calculations reported by Hentschke in this volume (Chapter 2, cf. Figs. 6 and 8) do not show substantial differences between the distributions for rodlike aggregates at the transition boundaries. The persistence length appears to slow down growth, but does not terminate it.

Unambiguous evidence for the mechanism of growth coupled to orientation so far reported is based on the discotic amphiphile 2,3,6,7,10,11-hexa-(1,4,7-trioxaocetyltriphenylene) in water (Fig. 4 and Ref. 13). The details of the phase diagram in Fig. 4 appear in line with theory and the nematic phase is stabilized at rather low volume fractions (~0.2). Hentschke (this volume, Chapter 2) attributes the formation of the hexagonal columnar phase to its better packing efficiency as compared to the nematic phase. Theory predicts that the critical volume fraction will be in the accessible range provided the flexibility of the assembly is not too large (P too small) in which case a direct isotropic → columnar transition occurs without the intermediate nematic phase [94,95].

Another evidence for growth coupled to orientation comes from the work of Hentschke and Herzfeld [111]. Osmotic pressure versus concentration data for normal (nonaggregating) and sickle cell hemoglobin showed that the onset of linear growth for the latter was directly coupled to the onset of alignment (data are illustrated in this volume, Chapter 2; cf. Fig. 9). In the case of micelles, Odijk [93] has critically reviewed studies aiming at an experimental verification of the theory. The broad features of the phase diagram are explained. However, the anisometry of micellar aggregates appears smaller than predicted [112], and the critical concentration is also smaller than expected [113,114].

A guideline for the design of single supramolecular assemblies of large DP should therefore be the synthesis of assemblies characterized by large rigidity and persistence length. These features are expected for assemblies based on intrinsically rigid units that are coupled by strong multifunctional contacts. The polymer in Fig. 2B should exhibit these features, even though its persistence length has not been reported. Values of P, determined only for a few supramolecular assemblies, reveal a much larger rigidity than reported for molecular polymers. Table 1 compares data for elongated assemblies having different shapes (linear, helical, tubular) with data for poly(*p*-benzamide), a typical rigid polymer [115,116]. The scatter of determinations performed with various techniques is indicated in the table. For micelles, light scattering data for the nonionic surfactant dimethyloleylamine oxide are reported [117]. The study evidenced the formation of monodisperse wormlike micelles having aggregation number and persistence length affected by the concentration of added NaCl. Alternative determinations for other types of micelles provided values of P ranging from 0.02 to 10 μm. An a priori evaluation based on elasticity arguments provided a value of 1 μm [20]. The P values for the cytoskeleton assemblies were assessed from their flexural rigidity (Fig. 22) measured from thermally driven fluctuations in shape [118], or from fluctuations of the end-to-end distance [122]. The persistence length of microtubules (reaching 5.2 mm!) is larger than the length of the samples used for its determination (Fig. 22) and of microtubules found in cells. Its larger value with respect to actin is a reflection of its larger cross-section [118]. The tubular shape, achieved by lateral association of 13 to 16 protofilaments composed of linear sequences of α- and β-tubulin molecules, appears to be a most efficient way to maximize the bending stiffness of a structure.

Although growth coupled to orientation is not necessarily restricted to linear sequences, it is doubtful that this mechanism controls the growth of the helical and tubular polymers in Table 1. The mechanism of helical and tubular growth described in the following section is likely to promote cooperative growth of actin filaments even before the oriented mesophase appears (cf. Section III.C). In the case of microtubules, a synchronous occurrence of growth and liquid crystallinity was observed by Hitt et al. [123]. Figure 23 illustrates dynamic association–dissociation cycles of α- and β-tubulin into microtubules growing to a size visible under the polarizing microscope. The polymerization → depolymerization cycles are modulated by the reversible GTP → GDP reaction that uncovers sites on β-tubulin (see Chapter 15 for details). These experiments do not, however, confirm a direct coupling between growth and alignment since they were carried out at a tubulin concentration (15 mg/ml) likely larger than either the critical concentration at which the helix nucleates (tubulin polymerizes already at 80 mM Pipes; cf. Ref. 118), or the mesophase appears.

C. Helical or Tubular Growth

Cooperative growth occurring within a single assembly can occur without the simultaneous formation of a mesophase. Site distribution is instead the most relevant parameter. Consider for instance the multiple H-bonded systems illustrated in Fig. 2. For polymer C the formation of the first H-bond tends to favor the formation of the parallel ones. With respect to the single H-bond of Fig. 2A, the

Table 1 Persistence Length for Supramolecular and Molecular Polymers

System	Shape	P µm	D µm	P/D	DP[a]	References
Poly(p-benzamide)/DMAc/3%LiCl	Linear	7.5×10^{-2}	5.10^{-4}	150	110	115,116
Dimethyoleylamine Oxide/H_2O + 10^{-2}M NaCl	Linear	(0,2 →)1.7	6.10^{-3}	280	20.000	117
Actin/22 Phalloidin stabilized + buffer	Helical	(6 →)17.7	5.10^{-3}	3.500	6.500	119, 120, 118
Microtubules/Taxol stabilized + buffer	Tubular	(79 →)5.200	3.10^{-2}	170.000	17.10^6	121, 118

[a] Degree of polymerization referred to persistence length.

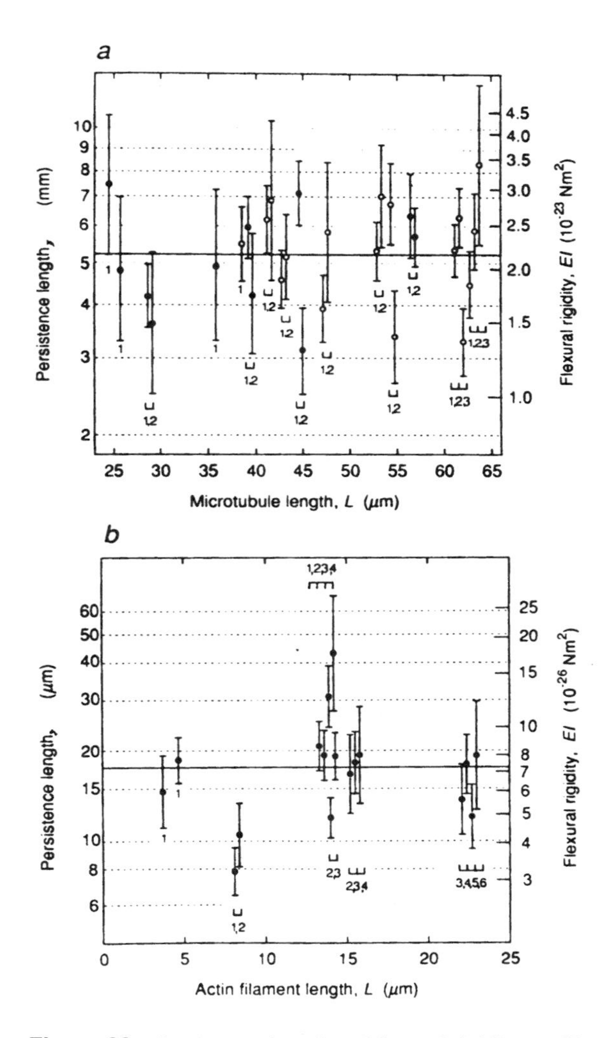

Figure 22 Persistence length and flexural rigidity vs. filament length for taxol-stabilized microtubules (a) and phalloidin–stabilized actin (b) in buffer solution. Labeled and nonlabeled filaments included show L/P <1 (microtubules) and L/P ≥ 1 (actin). Shape fluctuations were determined by microscopy and analyzed through Fourier decomposition. (From Ref. 118, by copyright permission of the Rockefeller University Press.)

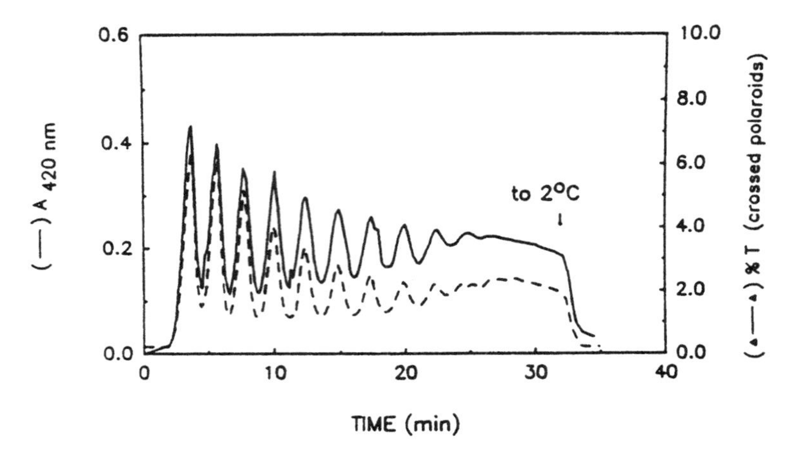

Figure 23 Dynamic assembly and disassembling of microtubules. Time variation of turbidity (—) and birefringence (---) at 420 nm for solutions of tubulin (15 mg/ml) in pH 6.9 buffer + 12 m/M Mg SO_4 + 2 mM GPT. Measurements at 37° followed by quenching at 2°C. (From Ref. 123.)

contact-free energy for the repeating unit of polymer C is larger due to the occurrence of four H-bonds possibly augmented by cooperative effects. A cooperation of this type is localized to the junctions between donors and acceptors but does not directly promote binding of successive units. To be sure, the larger formation constant will promote enhanced growth of polymer C with respect to polymer A but this enhanced growth could still be predicted in terms of the MSAO theory.

A new situation can be expected, however, when the cooperative effect is such that binding of one unit promotes binding of successive units along a helical pattern. This type of cooperation was introduced by Oosawa [2] to interpret the occurrence of extremely large linear assemblies during the G → F (globular → fibrous) transformation in proteins. Oosawa's model was inspired by the Zimm and Bragg [124] model for the coil → helix transformation in linear polypeptides. In the α-helix, each amino acid residue is bound to two residues by mainchain covalent bonds and to two additional residues by H-bonds. In analogy, Oosawa imagines a helical assembly in which each unit can make two kinds of supramolecular bonds with four neighboring units: two along a linear sequence and two along a helical pattern. In this case the shortest oligomer having a helical sequence is composed of four units (Fig. 24 and also Fig. 6C). Under conditions disfavoring the formation of one type of bonds only linear aggregates (or a transformation from helical to linear assemblies) are expected.

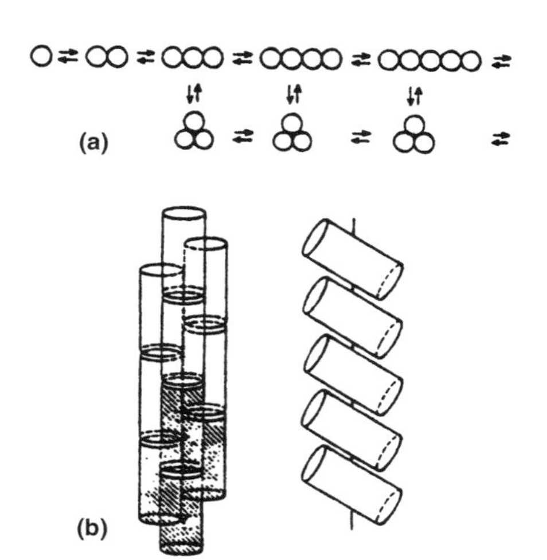

Figure 24 Scheme for the formation of linear and helical assemblies for units having four interacting sites: two along the linear sequence and two along the transversal direction; (a) top view; (b) side view. (From Ref. 2.)

The distinctive features of the Oosawa mechanism, in particular the large DP of the supramolecular polymer, emerge upon detailed consideration of the model. The formation and the growth of the basic helical nucleus of four units (Fig. 24) is characterized by the fact that the addition of the fourth (and successive) unit(s) involves a larger number of bonds per unit, and a larger binding constant (K_h) relative to the constant K of the linear sequence. Thus, in analogy with Eq. (13), the concentration of incipient helical trimers may be written as

$$C_{3h} = \gamma C_3 = \gamma K^{-1}(KC_1)^3 \tag{17}$$

where the factor γ (<1) accounts for the excess free energy needed to distort a linear trimer into the incipient helical sequence. The concentration of helical s-mers having $s \geq 4$ becomes

$$C_{sh} = K_h C_{(s-1)} C_1 = \sigma K_h^{-1}(K_h\ C_1)^s \tag{18}$$

where the front factor $\sigma = \gamma(K/K_h)^2$ is a key parameter of the theory. Usually σ will be $\ll 1$ (e.g., for $K = 0.1$, $K_h = 100$ and $\gamma \sim 0.1$, $\sigma \sim 10^{-8}$) accounting for the low probability of initiating the helical sequence and for the cooperativity of the growth process [2]. From Eqs. (13) and (17) the total unit concentration C_0 may be approximated as (cf. Ref. 2)

$$C_0 \approx C_1 + \frac{\sigma C_1}{(1 - K_h C_1)^2} \tag{19}$$

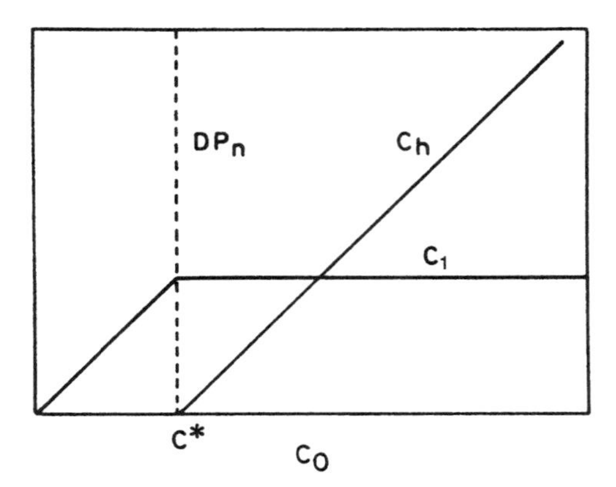

Figure 25 Features of helical supramolecular polymerization. Ordinate: unimer + linear oligomer concentration C_1; helical polymer concentration C_h; average degree of polymerization DP_n. Abscissa: total unit concentration (unimers + polymers). (From Ref. 2.)

A schematic representation of the expected trend is given in Fig. 25. At low C_0, almost all units occur as dispersed unimers and short linear polymers ($C_0 = C_1$) since the second terms on the right of Eq. (19) can be neglected on account of the small value of σ. Upon further increase of C_0, a critical concentration C^* is reached when

$$C^* = K_h^{-1} \tag{20}$$

For $C_0 > C^*$ the second term in Eq. (19) increases with C_0 and all excess units form helical supramolecular polymers coexisting with monomers and short linear sequences having constant concentration C^*.

The average degree of polymerization of the helical polymer

$$DP_n = \frac{\Sigma\, DP_s C_{sh}}{\Sigma\, C_{sh}} = \frac{1}{(1 - K_h C_1)} = \left(\frac{C_h}{C^*}\right)^{1/2} \sigma^{-1/2} \tag{21}$$

becomes very large near the critical concentration. For instance if $C_h/C^* \sim 0.1$ and $\sigma = 10^{-8}$, $DP_n = 10^3$. Thus, the helical supramolecular polymerization exhibits a high degree of cooperativity and involves essentially two extremes: dissociated units (G) and very long assemblies (F). A broad equilibrium distribution can be expected, however, from Eq. (18) with an exponential decrease of C_s with DP, and a DP_w/DP_n ratio ≈ 2. Sharp, Poisson-type distributions can nevertheless be observed under nonequilibrium conditions in nucleation-controlled polymerization, or in vivo. In the former case, the uniform length is controlled by the ratio nuclei/monomers and by a slow reverse (depolymerization) reaction that prevents a length redistribution [125]. In the latter case special controlling mechanisms may be involved [2]. The critical concentration is expected to depend upon temperature and solvent type through the value of K_h in Eq. (8). The G → F transformation is thus predicted to occur in isotropic solution before the liquid crystalline phase is formed.

Experimental support for the Oosawa model may be derived from studies with biological systems in vitro. G-actin is a globular protein made up of four subdomains having overall dimensions $5.5 \times 5.5 \times 3.5$ nm and a distinct polarity [126]. The reversible polymerization to F-actin occurs by increasing ionic strength or temperature, and may or may not be accompanied by the dephosphorylation of ATP bound to actin subunits. The helical actin filament is preferentially described in terms of two right-handed helices of relatively long pitch exhibiting contacts between adjacent G-units that are stronger along the helical strands than in the lateral direction. An early model for the nucleation of dimers, trimers, and the ensuing polymerization is given in Fig. 26 [127]. The determination of the length of F-actin filaments by electron microscopy allows an evaluation of the DP since each G-actin residue corresponds to a length of 27 Å (1 μm filament contains ~370 G-proteins) [128]. Polymerization has been shown to occur at G-actin concentration below 0.04 mg/ml producing filament lengths in excess of 11 μm [129]. Length control and a prescribed DP can also be achieved by poly-

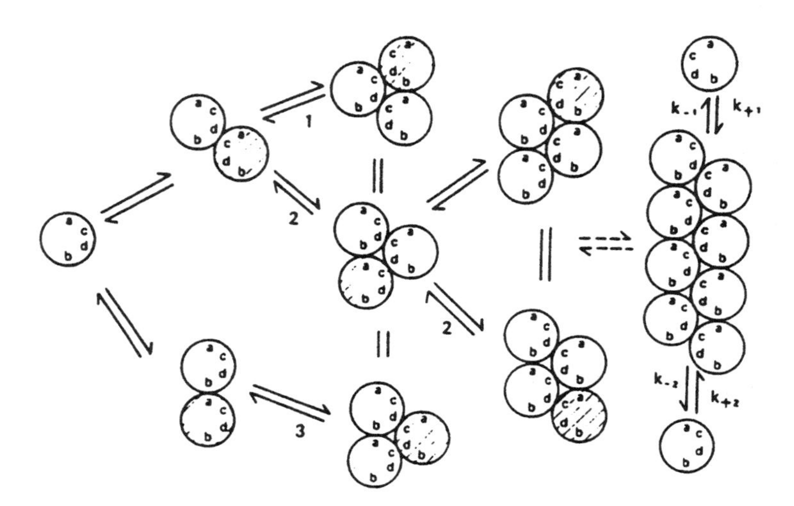

Figure 26 Scheme for the polymerization of actin-ADP. Two dimers (*cd* and *ab* bonds) originate identical trimers that may elongate by addition of unimers at either end through the simultaneous formation of *ab* and *cd* bonds. A steady-state situation is eventually reached with each end independently at equilibrium with unimers. (From Ref. 127.)

merizing G-actin in the presence of gelsolin, a protein that acts as a nucleus for filament growth and binds 1:1 to the fast growing end. In this case the actin/gelsolin mole ratio coincides with the DP, allowing an evaluation of filament length. Figure 27 illustrates the length distribution of F-actin filaments polymerized at different actin/gelsolin ratios [129]. The ratio L_w/L_n is ~1.7 and the theoretical exponential distribution was verified [129]. DP_n (=L_n/27 Å) is usually close to the actin/gelsolin ratio.

The two ends of the actin filament (the ''barbed'' and the ''pointed'' ends) have different kinetic constants for association and dissociation [130]. Additional complications are manifested when ATP hydrolysis occurs. The critical growth

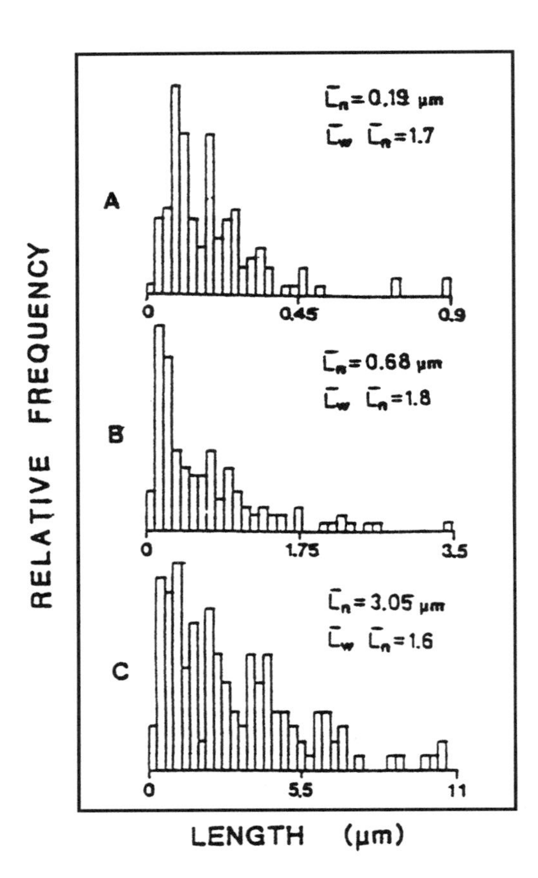

Figure 27 Length distribution determined from electron micrographs for filaments polymerized from solutions having actin/gelsolin mole ratio 61:1 (A); 256:1 (B); 2,048:1 (C). (From Ref. 129.)

concentration at the two ends is modified and eventually a flow of actin subunits from one end of the growing filament to the other is observed (treadmilling effect) [131]. In vivo assembly includes regulatory mechanisms by accessory proteins that direct or inhibit formation of spontaneous assemblies and orchestrate the formation of higher-order structures (see Chap. 15).

The polymerization of actin does produce the kind of DP and distribution theoretically expected in terms of the Oosawa model. Since the growing filament will eventually achieve a length at which a lyotropic phase must appear it is important to explore a possible correlation between the Oosawa mechanism and the mechanisms of growth coupled to orientation. Pending a more detailed investigation of the problem, an analysis of recent data for the formation of mesophases by actin filaments suggests no correlation between the two growth mechanisms. In fact, the lowest critical concentration for appearance of the mesophase reported by Furukawa et al. [132] was ~2 mg/ml for a filament length of ~5 μm (DP ~ 1780). However, as indicated above, Janmey et al. [129] were able to grow filaments with a comparable length at a concentration of ~0.04 mg/ml, suggesting that the critical concentration C^* [Eq. (20)] is substantially smaller than the critical concentration (C^i) at which liquid crystallinity develops.

The fibrinogen → fibrin transformation is another example of a supramolecular polymerization based on a site distribution consistent with Oosawa's model. During the early stage of linear polymerization (cf. schematization in Fig. 6A), two binding sites (A-type; [133]), are uncovered by trombin on the same side of the central region of fibrinogen. These sites bind to the two terminal sites of two similar units. Recently it has been demonstrated [16] that the terminal sites are also capable of direct head-to-tail attraction. Linear cooperative growth ensues. Later, a second stage of trombin activation uncovers two additional sites (B-sites; [133]) on the other side of the central domain of fibrinogen. At this point the assembly process can continue along the lateral direction and a network is formed [134].

Silk fibers formed by the secretion of the major ampullate glands of the spider *Nephila clavipes* also involve the linear supramolecular aggregation of a globular protein (fibroin). The formation of rodlike aggregates is consistent with the observation of a precursor liquid crystalline phase before fiber solidification [135,136]. Details of this supramolecular polymerization are still unclear.

There is no definite evidence of synthetic supramolecular polymers able to produce the extremely large DP achieved in the helical polymerization of biological systems. The polymer in Fig. 2B has been reported to display a triple helical structure in the solid state [9]. However, occurrence of large DP in isotropic solution was not reported. The H-bonded assembly involving pimelic acid and bis-2-aminopyridine has been suggested to exhibit helical structure in both solid state and diluted solutions ([137]; cf. also this volume, Chapter 4). Guidelines for future attempts to synthesize helical supramolecular polymers stable even in isotropic solutions can be inferred from Oosawa's theory. An appropriate site distribution will favor helical growth, but it is also necessary to devise structures not allowing extensive lateral growth. Moreover, the organic chemist should aim for situations characterized by extremely small values of the parameter σ, requiring low values of γ [cf. eq. (18)] or high values of K_h. The former reflects [2] the excess free energy for the deformation of a linear nucleus to start the curvature of the helical polymer: $\gamma = \exp(-\Delta F^*/kT)$. Situations under which σ is not so small should tend to favor the MSOA mechanism.

D. Planar and Three-Dimensional Growth

Three-dimensional and planar assemblies are expected when shape and site distribution allow head-to-tail and/or lateral growth (cf. Fig. 15C and this volume, Chapter 5). These assemblies have a tendency to grow to large sizes revealing the occurrence of cooperative effects and ruling out a simple growth mechanism such as MSOA. The general thermodynamic conditions governing the growth of multidimensional assemblies were summarized by Israelachvili [26]. The standard chemical potential per unit in an aggregate of N units (μ_n^0) decreases with increasing N toward the "bulk" free energy of a unit in an infinite aggregate (μ_∞^0) according to

$$\mu_n^0 = \mu_\infty^0 + \frac{\alpha\, kT}{N^p} \tag{22}$$

where α is a positive constant reflecting the strength of the contact energy, and p is a number depending upon the dimensionality of the aggregate: $p = 1$ for rods; $p = 1/2$ for disks; $p = 1/3$ for spheres (note that different but related dimensionality indices are given by Hentschke in this volume, Chapter 2). Accounting for the formation of a critical aggregate nucleus at a critical aggregation concentration (CAC), Eq. (10) can be rearranged to show that whenever $p < 1$ (disks or spheres) a macroscopic aggregate ($N \rightarrow \infty$) appears at the CAC and increases to infinite size upon further increase of concentration, while the concentration of dispersed units remains constant at CAC. An abrupt phase transition is thus expected and both two- and three-dimensional supramolecular polymerizations have the character of a condensation (i.e., crystallization) phenomenon. The concept of a size distribution loses its significance.

The above situation should be clearly distinguished from the case of rods when $p = 1$ and the mean aggregation number can be shown [26], at variance with the case of disks and spheres, to be concentration-dependent producing broad size distributions (as observed in the case of actin, Fig. 27). In fact, as recognized early by Oosawa and Asakura [2], helical or tubular growth has features intermediate between the continuous growth described by MSOA and a true phase transition.

Spherical micelles also share some of features of the $p = 1$ systems due to the characteristics of amphiphilic molecules (i.e., the nonpolar tails control micellar size) causing μ_n^0 to reach a minimum at a finite value of N. Note, in fact, the similarity between the diagram in Fig. 25, illustrating helical polymerization, and the diagram illustrating micellar formation above CMC in isotropic solution (Fig. 4 in this volume, Chapter 2 or Fig. 16.5 of Ref. 26).

A beautiful example of planar assemblies is described in this volume, Chapter 5 (cf. their Fig. 2). Monomolecular films of globular proteins known as S-layers were shown to grow to a very large size even over noninteracting substrata. The authors interpret the formation of S-layers as a two-dimensional crystallization [138]. Although no thermodynamic analysis of the system has yet been made, Sleytr and coworkers' interpretation appears to be quite consistent with the general thermodynamic considerations discussed above. As schematized in Fig. 15C, an equatorial distribution of weak binding sites can stabilize a planar assembly of quasispherical units. On the other hand, a three-dimensional assembly of spheres would be stabilized if a north–south distribution of complementary sites also occurred. A film anchored to a surface is instead expected if, in addition to the equatorial distribution, sites for complementary interaction with the substrate occur only at one of the poles. A film of this type has an intrinsic self-assembling ability due to the equatorial distribution and derives additional stabilization through interaction with the substrate. S-layers show these features (this volume, Chapter 5) because they are bound in vivo to cells but may exist even over noninteracting substrata. The forces holding the S-layer together (prevalently H-bonds) are stronger than the electrostatic bonds anchoring the layer to the cell surface and the overall assembly process is completely reversible.

Planar synthetic systems include the self-assembled monolayers (SAMs, this volume, Chapter 10) that result from physisorption or chemisorption processes of small molecules such as $CH_3(CH_2)_nX$ over a planar substrate [30]. Acid-functionalized n-alkanes [139] or alkanethiolates [140,141] can form bonds of increasing strength between the head group X and alumina or gold surfaces, respectively. Most extensively studied is the case of decanethiol over the 111 face of gold single crystals between which a covalent bond actually occurs. With increasing coverage, the assemblage of the monolayer evolves from an initial phase with molecular axes of decanethiol parallel to the gold surface (Fig. 28) to a final phase characterized by bound axes nearly perpendicular to the surface. The lateral intermolecular interaction between grafted alkyl segments is the supramolecular force in this two-dimensional assembly stabilized by covalent bonds.

Alignment of liquid crystalline layers over a surface may also be described [142]. Spontaneous growth of highly organized films of amphiphiles over various surfaces is described in this volume in Chapter 11.

The phase transition predicted by Israelachvili is usually described as a

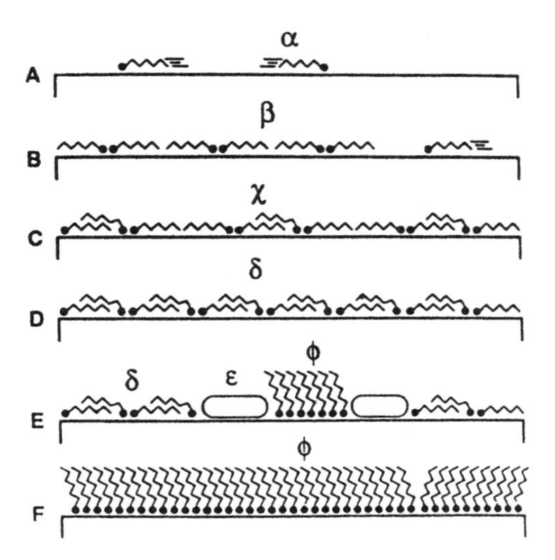

Figure 28 Sequence of monolayer phases with increasing coverage of decanethiol on Au(111). (From Ref. 141. Copyright 1999 Am. Chem. Soc.)

crystallization, but does the transition involve a liquid crystalline phase, perhaps as a precursor step of a two-dimensional crystallization? This question underlines the possible occurrence of a two-dimensional variation of the growth coupled to orientation mechanism described in Section II.B. Large planar structures could, in fact, be expected to result from supramolecular polymerization driven by entropy effects (hard interaction; cf. Section II.C) and by only a small contact energy. In this case a critical concentration should exist at which an isotropic solution of dispersed units cooperatively orients into a nematic phase formed by large planar assemblies. Nematic phases formed by planar, covalent macromolecules are indeed known (i.e., nonpolar, fused-ring aromatic hydrocarbons; cf. Ref. 143) and a relationship ought to exist between the critical volume fraction and the relevant axial ratio ($L/D \ll 1$). However, the theory for large molecular discs is not well developed due to the difficulty of neglecting the higher virial coefficient in Onsager's approximation [86]. Likewise, difficulties would be encountered in the assessment of a persistent length in the case of large disks, or equivalent effects such as surface tension and membrane-bending modulus [144]. In any event, the experimental features of S-layers do not suggest that a two-dimensional growth coupled to an orientation mechanism is operative. In fact, single layers are stabilized (even over indifferent surfaces) whereas a simultaneous growth of several nematic layers would be expected if the latter mechanism were operative.

An outstanding example of three-dimensional assembly was described by

Muñoz-Guerra and coworkers ([145], this volume, Chapter 7). They synthesized poly(α-n-alkyl-β-aspartate)s, a family of comb-like polyamides characterized by a rigid (helical) backbone and sidechains of flexible aliphatic segments having up to 22 methylene units. The assembly produced a solid-state composite structure, characterized by a regular distribution of rigid and flexible components, having similarities to the structure of keratin fibers (Fig. 29) [146]. They observed the occurrence of one crystalline and two liquid crystalline phases upon increasing temperature (cf. their Fig. 13). In the crystalline phase of the sample with 18 methylene units the sidechains are crystallized in a separate hexagonal lattice with an interlayer spacing of 3.1 nm. In the high temperature phase the sidechains are completely molten (spacing = 2.3 nm) but the features of a LC (nematic) phase attributed to the molecular helices diluted by the flexible components are retained. The intermediate phase may be described as a cholesteric supramolecular liquid crystal in which the rigid helices are correlated by interdigitation [96] with partly molten sidechains, as schematized by the model on the right of Fig. 29.

In the latter model interdigitation represents the mode of growth along the lateral direction. Growth also occurs along the longitudinal direction through head-to-tail recognition of the rods, primarily based on the hydrophobic recognition of exposed alkyl segments on the north and south surfaces. The basic micellar-like unit (the hexagon of 6 + 1 rods) cannot be isolated even in dilute solution (above the CAC) due to the all-or-none character of the phase transition. One might expect that if longitudinal growth could be prevented (i.e., by end-capping

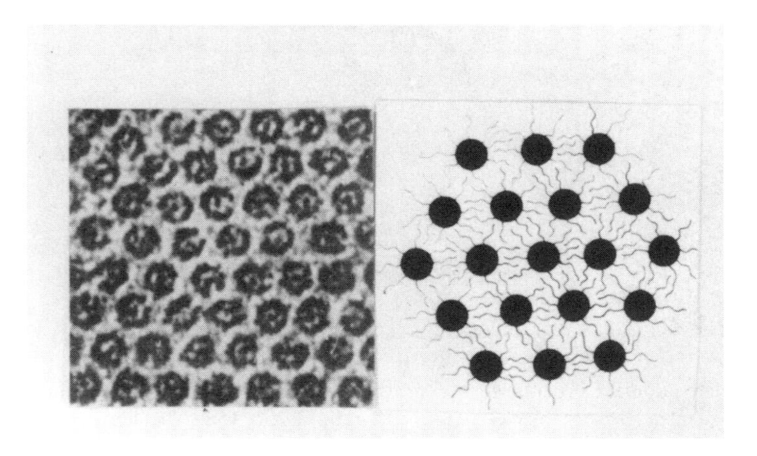

Figure 29 Electron micrograph of a cross-section of Australian fine merino wool (G.E. Roger's micrographs adapted from Ref. 146.) The scheme on the right represents a section of a cylindrical assembly based on rigid rods and interdigitating sidechains (From Ref. 28.)

the north and south surfaces), a planar assembly having a thickness corresponding to the length of the rod might be produced. A system in which longitudinal vs. lateral growth could be controlled is based on rigid DNA to which dodecylpyridinium cations are electrostatically bound [147]. In this case linear micellar-like structures could be stabilized by controlling the composition of the two components. Described in this volume in Chapter 7 are the solution and solid-state morphology of systems based on helical, charged polypeptides having long alkyl sidechains electrostatically bound.

Another approach by which linear aggregates could be isolated out of an ordered three-dimensional structure was presented by Liu et al. [29]. Diblock copolymers $(A)_n - (B)_m$ form ordered structures (this volume, Chapter 6) that for suitable values of the n/m ratio are composed of cylindrical domains of one block dispersed in a continuous phase of the other block [148,149]. Liu et al. prepared diblock copolymers using as a block for the cylindrical phase poly(2-cinnamoylethylmethacrylate) that could be crosslinked by irradiation. Dissolution of the continuous uncrosslinked phase of the other block in THF resulted in the isolation of cylindrical aggregates (diblock nanofibers) up to 60 nm thick and several μm long (Fig. 30). The nanofibers had a core-shell structure formed, respectively, by the piled crosslinked and uncrosslinked blocks.

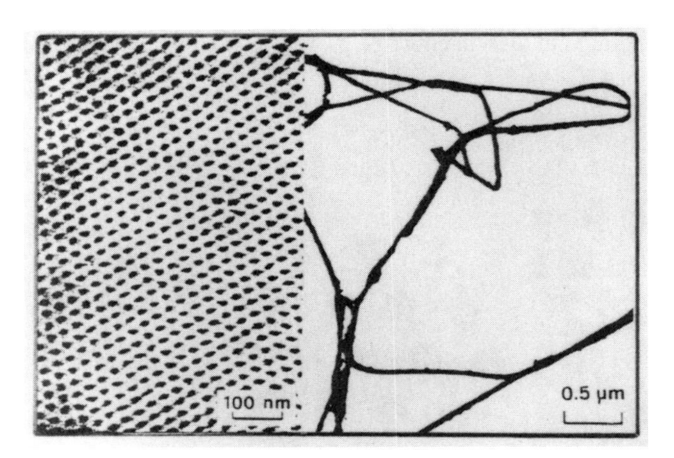

Figure 30 Ordered domain structure for a copolymer based on poly (styrene) and poly (2-cinnamoylethyl methacrylate) blocks with a DP ratio 1,250:158. The hexagonally packed dark ovals represent PCEMA cylinders oriented perpendicularly to the micrograph. The TEM micrograph on the right reveals nanofibers of the cylindrical crosslinked component following dissolutions of the continuous matrix. (From Ref. 29. Copyright 1996 Am. Chem. Soc.)

It is worth noticing that in all the structures discussed above an ordered distribution of a flexible component within an ordered phase is achieved. It is therefore possible to directly self-assemble comb or block copolymers bypassing the well-known incompatibility between a flexible component and an ordered distribution of rigid elements [150]. The model on the right of Fig. 29 is also representative of the organization of microfilaments in the keratin fiber [151,152]. Each microfilament (length ~1 μm, diameter ~8 nm) is composed of eight protofibrils which are left-hand cables of two strands, each composed of two right-handed α-helices [152]. The less-ordered matrix is based on a conformationally disordered protein rich in cystine residues and –S-S crosslinkages (these occur to a smaller degree even within each microfilament, and likely between filaments and matrix [151]. A mechanism of the assembly of keratin fibers in vivo might include the sequence: (i) extrusion of the low-sulfur protein into extracellular fluids and protofibril assembly; (ii) site recognition and formation of –S-S bonds stabilizing the microfibrils; (iii) mesophase formation by microfibrils accompanied by lateral and longitudinal growth; and (iv) establishment of –S-S crosslinkages in the S-rich matrix [28].

Finally mention should be made of two- and three-dimensional self-assembled supramolecular networks that do not exhibit a large structural organization comparable to that shown by the above systems. Purely H-bonded networks [7,153,154] are described in this volume in Chapter 4. Particularly significant for potential application as materials with a strongly temperature-dependent rheology are the systems described by Meijer and coworkers [10]. These are based on tetrafunctionalized copolymers of propylene oxide and ethylene oxide reacted with methyl isocytosine producing the strong four H-bond scheme in Fig. 2C. The mechanical spectra of this network attain a plateau modulus of 5×10^5 Pa while showing a frequency-dependent transition from viscous to viscoelastic behavior. The value of the modulus is six times larger than that of a corresponding chemically crosslinked network showing viscoelastic behavior over the whole frequency (temperature) range.

There is an extensive literature on thermoplastic supramolecular networks (or associating polymers) based on copolymers. The mesogel features of *ABA* block copolymers are analyzed by Halperin in this volume in Chapter 3. Supramolecular assemblies resulting from the interaction of polyelectrolyte networks with surfactants have been reported [155–159]. For flexible polyelectrolytes gel collapse was described in terms of a reduced Donnan effect resulting from micellization within the meshes of the network [157–159]. For rigid polyelectrolytes with large charge density, ordered cylindrical assemblies (cf. Fig. 29) were postulated. Alternative models were, however, suggested for networks formed by rigid chains with a low charge density, such as collagen [156] or chitin with low degree of acetylation [155].

E. Engineered Growth

Assembling by controlled addition of components, by superimposition of appropriate chemical reactions, or by using external fields produces engineered or fabricated assemblies. Elements of supramolecular interaction such as site and shape recognition are usually detectable in engineered assemblies, but their growth may be based on mechanisms entirely different from the ones described in the preceding sections. Some engineered systems (not considered here) can be mechanically assembled even from incompatible components. Such is the case of high-performance composites and even of some natural systems such as connective tissue or the vitreous body of the eye [28]. Use of compatibilizers, fast quenching techniques, or covalent crosslinking may prevent or retard phase separation in the latter cases.

The complexation of complementary polyelectrolytes can be exploited for the formation of skins of various shapes. For instance, Larez et al. [155] reported the formation of spherical assemblies (diameters up to 500 μm) by letting drops of a chitosan solution free fall into a solution of oxidized scleroglucan (Fig. 31). The skin of the assembly, separating the external polyanion from the internal polycation solution, acted as a semipermeable membrane. The possibility of encapsulating a micelle within the primary spherical skin was suggested as a model for the organization of cells containing a nucleus where a variety of reactions could be performed with a release of products in the external solution.

In the Languimir–Blodgett assemblies described by Shimomura (this volume, Chapter 11) the recognition between the substrate and the first layer, and between subsequently transferred layers, represents the supramolecular component. A chemical reaction can in some cases occur between the monolayer and the substrate. In other cases a phase change from a two-dimensional liquid crystalline phase on the water surface to a crystalline phase over the substrate may occur [160,161]. The engineered growth is instead controlled by the applied deposition pressure, the type of deposition (Y-, Z-, and X-type), and by the number of transferred layers [160]. Variations in the orientation of consecutive layers may also be obtained [162].

In the assemblies described by Jonas, Laschewsky, and coworkers (this volume, Chapter 12), the self-assembling component is the recognition of complementary charges of oppositely charged polyelectrolytes, colloids, or bolatype amphiphiles that takes place under kinetic control. The stepwise, layer-by-layer adsorption of oppositely charged elements, assisted by uncompensated sites, enables the engineered growth of these assemblies on suitable substrates. Due to the kinetically controlled assembly, interfacial complexation gives rise to stratified supramolecular structures that strongly differ from the ones obtained by direct polyelectrolyte complexation as discussed by Larez et al. [155].

An interesting example of growth promoted by an electrolytic process was

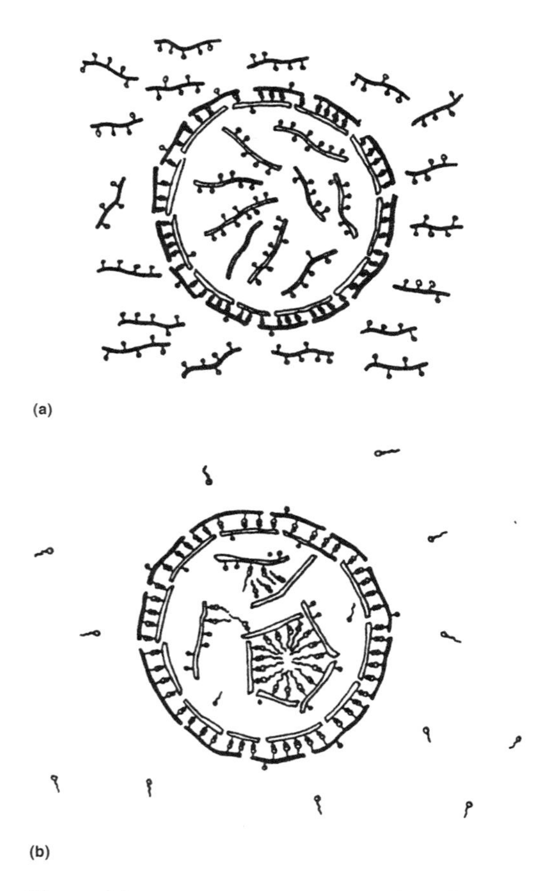

Figure 31 (a) Schematic representation of a spherical assembly formed by cationic chitosan (light mark) and anionic, oxidized scleroglucan (dark mark). (b) Same assembly in the presence of surfactant molecules forming a micelle in the interior of the skin. (From Ref. 155. Courtesy of Elsevier Science.)

reported by Marquez and coworkers [163,164]. As illustrated in Fig. 32, when a solution of 1,2 dimethoxy benzene **I** (veratrole) in CH_3CN/Bu_4NBF_4 is subjected to anodic oxidation (1.6 V), the discotic compound **II** (1,2,5,6,9,10-hexamethoxytriphenylene) is formed. The positive charge of the cation radical **II** is compensated by the tetrafluoroborate counterion favoring the next and subsequent additions of disklike units to the growing end, resulting in the formation of polyveratrole **III**. Fibers emanating from the electrode surface can be detected and isolated with no loss of properties (including paramagnetism) after several years.

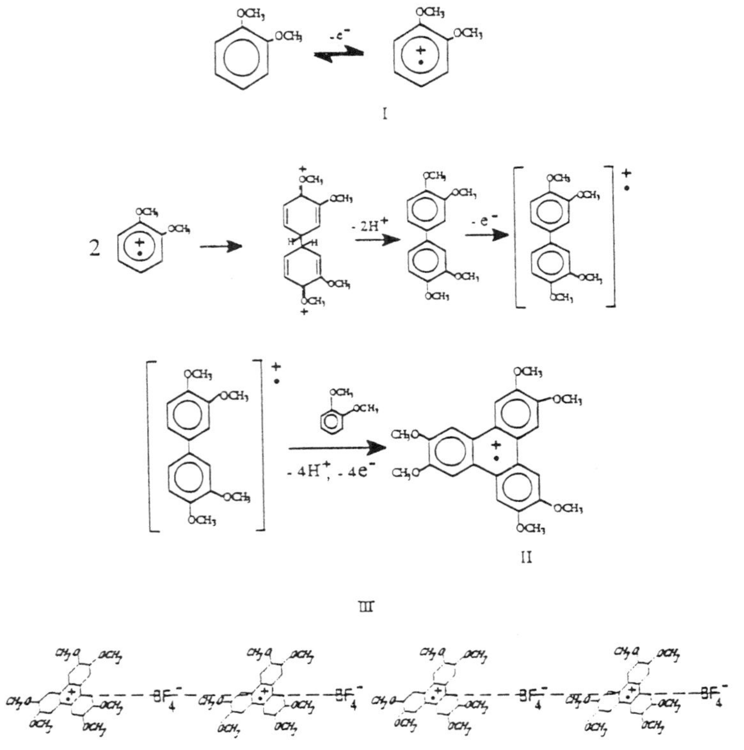

Figure 32 A solution of 1,2 dimethoxybenzene **I** in Bu_4NBF_4 subjected to electrolysis under an applied voltage of 1.6 V produces compound **II** which assembles into stacks of **III** growing out of the electrode until visible fibrils are observed. (From Ref. 164. Reproduced by permission of The Electrochemical Society, Inc.)

Application of pressure to the growing fibers caused their transformation into an anisotropic liquid phase. The unpaired electrons along the structure are responsible for the conductivity of the material (band gap of 0.39 eV).

In the above example the supramolecular component that stabilizes polyveratrole is again the recognition of the complementary charges of the stacked disks. However, charge distribution and thus growth may be regarded as being engineered by the application of the electric field to the solution. In fact, compound **II** would not grow outside the electrolytic cell. Spontaneous self-assembly was instead observed for a molecule related to **II**, the discotic amphiphile 2,3,6,7,10,11-hexa-(1,4,7-trioxaocethyltriphenylene) (cf. Fig. 4 and Section

III.B) and the effect was primarily attributed to the hydrophobic recognition of the long alkyl chains.

REFERENCES

1. H Staudinger. Ber 53:1073, 1920; 57:10203, 1924;59:3019, 1926.
2. F Oosawa, S Asakura. Thermodynamics of the Polymerization of Protein. London: Academic Press, 1975, p 25.
3. WM Gelbart, A Ben-Shaul, D Roux, eds. Micelles, Membranes, Microemulsions and Monolayers. New York: Springer-Verlag, 1994.
4. J-M Lehn. Angew Chem Int Ed Engl 29:1304, 1990.
5. A Ciferri. Liquid Crystals 26:489, 1999.
6. G Odian. Principles of Polymerization. New York: McGraw-Hill, 1991.
7. a) P Bladon, AC Griffin. Macromolecules 26:6004, 1993; b) C Alexander, CP Jariwals, CM Lee, AC Griffin. Macromol Symp 77:283, 1994.
8. C Fouquey, J-M Lehn, AM Levelut. Adv Mat 2:254, 1990.
9. M Kotera, J-M Lehn, J-P Vigneron. J Chem Soc, Chem Commun 197, 1994.
10. RP Sijbesma, FH Beijer, BJB Folmer, JHKK Hirschberg, RFM Lange, JKL Lowe, EW Meijer. Science 278:1601, 1997.
11. CM Paleos, D Tsiourvas. Angew Chem Int Ed Engl 34:1696, 1995.
12. MCT Fyfe, JF Stoddart. Acc Chem Res 30:393, 1997.
13. R Hentschke, R Edwards, PJB Boden, RJ Bushby. Macromol Symp 81:361, 1994.
14. V Percec, J Heck, G Johansen, G Tomazos, M Kawagumi. J Macromol Sci—Pure Appl Chem A11:1031, 1994.
15. JW Weisel. Biophys J 50:1079, 1986.
16. S Bernocco, M Rocco, A Profumo, C Cuniberti, F Ferri. Biophys J in press.
17. LA Amos, WB Amos. Molecules of the Cytoskeleton. New York: MacMillan, 1991, p 43.
18. E Mandelkov, EM Mandelkov. Curr Opinion Struct Biol 4:71, 1994.
19. A Klug. Angew Chem Int Ed Engl 22:565, 1983.
20. T Odijk. J Phys 48:125, 1987.
21. GR Desiraju, Ed. The Crystal as a Supramolecular Entity. New York: Wiley, 1996.
22. HL Yin, KS Zaner, TP Stossel. J Biol Chem 255:9494, 1980.
23. R Wick, P Walde, PL Luisi. J Am Chem Soc 117:1435, 1995.
24. SA Jenecke, XL Chen. Science 279:1903, 1998.
25. C Tanford. Formation of Micelles and Biological Membranes. New York: Wiley, 1980.
26. JN Israelachvili. Intermolecular and Surface Forces. London: Academic Press, 1992.
27. D Philp, JF Stoddart. Synlett 7:445, 1991.
28. A Ciferri. Prog Polym Sci 20:1081, 1995.
29. G Liu, L Qiao, A Guo. Macromolecules 29:5508, 1996.
30. F O' Brien. Trends Polym Sci 2:183, 1994.
31. V Percec, D Schlueter. Macromolecules 30:5783, 1997.

32. VV Tsukruk, JH Wendorff. Trends Polym Sci 3:82, 1995.
33. CT Imrie. Trends Polym Sci 3:22, 1995.
34. T Kato, JM Fréchet. Macromol Symp 98:311, 1995.
35. cf. e.g., H-J Cantow, ed. Polyacethylenes. Adv Polymer Sci 63, 1984.
36. H-J Schneider. Angew Chem Int Ed Engl 30:1417, 1991.
37. P Hobza, R Zahradnik. Intermolecular Complexes. Amsterdam: Elsevier, 1988.
38. Chem Rev 88:813–988, 1988.
39. WJ Jorgensen. Acc Chem Res 22:184, 1989.
40. H Bruning, D Feil. J Comput Chem 12:1, 1991.
41. RD Hancock, AE Martel. Chem Rev 89:1875, 1989.
42. MI Page. The Chemistry of Enzyme Action. MI Page ed. Amsterdam: Elsevier, 1984.
43. A Marcus. Ion Solvation. New York: Wiley, 1986.
44. H-J Schneider, D Güttes, V Schneider. J Am Chem Soc 110:6449, 1988.
45. H-J Schneider, R Kramers, S Simova, V Schneider. J Am Chem Soc 110:6642, 1988.
46. SL Price, AJ Stone. J Chem Phys 86:2859, 1987.
47. SK Burley, GA Petsko. Adv Protein Chem 39:125, 1988.
48. WJ Jorgensen, DL Severance. J Am Chem Soc 112:4268, 1990.
49. HL Anderson, CA Hunter, MN Meach, JKM Sanders. J Am Chem Soc 112:5780, 1990.
50. P De Santis. Nature 206:456, 1965.
51. J Sartorius, H-J Schneider. Chem Eur J 2:1446, 1996.
52. D Philp, JF Stoddart. Synlett 7:445, 1991.
53. TA Orofino, A Ciferri, JJ Hermans. Biopolymers 5:773, 1967.
54. MH Abraham. J Am Chem Soc 104:2085, 1982.
55. P Debye. Ann NY Acad Sci 51:575, 1949.
56. O Sinanoglu. Intermolecular Forces. B. Pullman, ed. Dordrecht: Reidel, 1981.
57. DB Smithrud, F Dietrich. J Am Chem Soc 93:7239, 1990.
58. A Ciferri, TA Orofino. J Phys Chem 70:3277, 1966.
59. C Balbi, E Bianchi, A Ciferri, A Tealdi, WR Krigbaum. J Polym Sci Polym Phys 18:2037, 1980.
60. PJ Flory. Principles of Polymer Chemistry. New York: Cornell Univ Press, 1953.
61. FH Case, JD Honeycutt. Trends Polym Sci 2:259, 1994.
62. PJ Flory. Adv Polym Sci 59:1, 1984.
63. CV Larez, V Crescenzi, A Ciferri. Macromolecules 28:5280, 1995.
64. RS Werbowyj, DG Gray. Macromolecules 13:69, 1980.
65. A Halperin, M Tirrel, TP Lodge. Adv Polym Sci 100:31, 1991.
66. A Halperin. Macromolecules 23:2724, 1990.
67. D Whitmore, J Noolandi. Macromolecules 18:657, 1985.
68. A Gabellini, M Novi, A Ciferri, C Dell'Erba. Acta Polymerica 50:127, 1999.
69. LH Radzilowski, SI Stupp. Macromolecules 27:7747, 1994.
70. SA Jenekhe. XL Chen. Science 279:1903, 1998.
71. R Nagarajan, K Ganesh. J Chem Phys 90:5843, 1989.
72. JN Israelachvili, DJ Mitchell, BW Niham. J Chem Soc Faraday Trans 2. 72:1525, 1976.

73. A Stroobants, HNW Lekkerkerker. J Phys Chem 88:3699, 1984.
74. KJ Shea. Trends Polym Sci 2:166, 1994.
75. J Rebeck. Angew Chem Int Ed Engl 29:245, 1990.
76. SF Edwards. Molecular Fluids. R Balian, G Weil, eds. New York: Gordon and Breach, 1976.
77. DA Tomalia, AM Naylor, WA Goddart III. Angew Chem Int Ed Engl 29:138, 1990.
78. BB Mandelbrot. The Fractal Geometry of Nature. New York: Freeman, 1983.
79. E Sackman. Macromol Chem Phys 195:7, 1994.
80. J Baggott. Perfect Symmetry: The Accidental Discovery of a New Form of Carbon. London: Oxford University Press, 1994.
81. E Osawa, M Yoshida, M Fujita. MRS Bull 11:23, 1994.
82. A Ciferri, WR Krigbaum, RB Meyer, eds. Polymer Liquid Crystals. New York: Academic Press, 1982.
83. A Ciferri, ed. Liquid Crystallinity in Polymers. New York: VCH, 1991.
84. PJ Flory, G Ronca. Mol Cryst Liq Cryst 54:311, 1979.
85. W Maier, A Saupe. Z Naturforsch 15:287, 1960.
86. AR Khokhlov. Liquid Crystallinity in Polymers. A. Ciferri, ed. New York: VCH, 1991.
87. GJ Vroege, T Odijk. Macromolecules 21:2848, 1988.
88. K Merchant, RL Rill. Macromolecules 27:2365, 1994.
89. W Wegner, personal communication, 1998.
90. P Cavalleri, A Ciferri, C Dell'Erba, M Novi, B Purevsuren. Macromolecules 30: 3513, 1997.
91. J Herzfeld, RW Briehl. Macromolecules 14:1209, 1981.
92. WM Gelbart, WE McMullen, A Ben-Shaul. J Phys 46:1137, 1985.
93. T Odijk. Curr Opin Coll Interface Sci 1:337, 1996.
94. P van der Schoot. J Phys II 5:243, 1995.
95. R Hentschke. Liq Cryst 10:691, 1991.
96. A Ciferri. Trends Polym Sci 5:142, 1997.
97. K Markau, J Schneider, A Sund. Eur J Biochem 24,393, 1972.
98. H Sund, K Markau. Intern J Polymeric Mat 4:251, 1976.
99. W Burchard, Trends Polym Sci 1:192, 1993.
100. F Oosawa, M Kasai. J Mol Biol 4:10, 1962.
101. C Hilger, R Stadler. Makromol Chem 192:805, 1991.
102. T Ooi, K Mihashi, H Kobayashi. Arch Biochem Biophys 98:1, 1962.
103. S Hoffmann, W Witkowski, G Borrman, H Shubert, WZ Weissflog Chem 18:403, 1978.
104. CM Paleos, J Michas. Liq Cryst 11:773, 1992.
105. T Kato, H Adachi, A Fujishima, JMJ Fréchet. Chem Lett 265, 1992.
106. U Kumar, JMJ Fréchet, T Kato, S Ujiie, K Timura. Angew Chem Int Ed Engl 31: 1531, 1992.
107. CT Imrie, Trends Polym Sci 3:22, 1995.
108. A Sirigu. Liquid Crystallinity in Polymers. A Ciferri, ed. New York: VCH, 1991.
109. G Ronca, A Ten Bosch. Liquid Crystallinity in Polymers. A Ciferri, ed. New York: VCH, 1991.

110. E Marsano, G Conio, A Ciferri. Mol Cryst Liq Cryst 154:69, 1988.
111. R Hentschke, T Herzfeld. Phys Rev A 43:7019, 1991.
112. R Itri, LQ Amaral. Phys Rev E 47:255, 1993.
113. P van der Shoot, ME Cates. Europhys Lett 25:515, 1994.
114. T Shikata, DS Pearson. Langmuir 10:4027, 1994.
115. Q Ying, B Chu. Macromolecules 20:871, 1987.
116. GL Brelsford, WR Krigbaum. Liquid Crystallinity in Polymers. A Ciferri, ed. New York: VCH, 1991.
117. T Imae, S Ikeda. Coll Polym Sci 262:497, 1984.
118. F Gittes, B Mickey, J Nettleton, J Howard. J Cell Biol 120:923, 1993.
119. S Burlacu, PA Janmey, J Borejdo. Ann J Physiol 262:C569, 1992.
120. T Yanagida, T Nakase, K Nishiyama, F Oosawa. Nature 307:58, 1984.
121. J Mizushima-Sangano, T Maeda, T Miki-Noumura. Biochim Biophys Acta 755: 257, 1983.
122. LD Landau, EM Lifshitz. Statistical Physics. Tarrytown, NY: Pergamon Press, 1980.
123. AL Hitt, AR Cross, CR Williams Jr. J Biol Chem 265:1639, 1990.
124. B Zimm, JK Bragg. J Chem Phys 31:526, 1959.
125. S Asakura, G Eguchi, T Iino. J Mol Biol 10:42, 1964.
126. MO Steinmetz, D Stoffler, A Hoenger, A Bremer, U Aebi. J Struct Biol 119:295, 1997.
127. ED Korn. Physiological Rev 62:672, 1982.
128. J Hanson, J Lowy. J Mol Biol 6:46, 1963.
129. PA Janmey, J Peetermans, KS Zanert, TP Stossel, T Tanaka. J Biol Chem 261: 8357, 1986.
130. F Oosawa. Biophys Chem 47:1010, 1993.
131. A Wegner. J Mol Biol 108:139, 1976.
132. R Furukawa, R Kundra, M Fechheimer. Biochemistry 32:12346, 1993.
133. JA Shafer, DL Higgins. CRC Critical Reviews in Clinical Laboratory Science 26, 1988.
134. JW Weisel, C Nagaswami. Biophys J 63:111, 1992.
135. C Viney. Supramol Sci 4:75, 1997.
136. BL Thiel, KB Guess, C Viney. Biopolymers 41:111, 1997.
137. SJ Geib, C Vincent, E Fan, AD Hamilton. Ang Chem Int Ed Engl 32:119, 1993.
138. UB Sleytr. Nature 257:400, 1975.
139. Y-T Tao, GD Hietpas, DL Allara. J Am Chem Soc 118:6724, 1996.
140. N Camillone III, TYB Leung, G Scoles. Surface Science 373:333, 1996.
141. GE Poirier. Langmuir 15:1167, 1999.
142. WJ Miller, VK Gupta, NL Abbott, H Johnson, MW Taso, J Rabolt. Liq Cryst 23: 175, 1997.
143. LS Singer. Ultra-High Modulus Polymers. A Ciferri, IM Ward, eds. London: Applied Science, 1979.
144. D Sornette, N Ostrowsky. Micelles, Membranes, Microemulsions and Monolayers. WM Gelbart, A Ben-Shaul, D Roux, eds. New York: Springer-Verlag, 1994.
145. F Lopez-Carrasquero, S Monserrat, A Martinez de Harduya, S Muñoz-Guerra. Macromolecules 28:5535, 1995.

146. S Seifter, PM Gallop. The Proteins. H Neurath, ed. New York: Academic Press, 1966.
147. A Ciferri. Macromol Chem Phys 195:457, 1994.
148. Y Matsushita, M Nomura, J Watanabe, Y Mogi, I Noda, M Imai. Macromolecules 28:6007, 1995.
149. AN Semenov. Sov Phys-JETP 61:733, 1985.
150. RR Matheson, PJ Flory. Macromolecules 14:954, 1981.
151. MK Hartzer, Y-YS Pang, RM Robson. J Mol Biol 365:376, 1985.
152. K Arai, F Hirata, S Nishimura, M Hirano, S Naito. J Appl Polym Sci 47:1973, 1993.
153. C Hilger, M Drager, R Stadler. Macromolecules 25:2498, 1992.
154. CB St Pourcain, AC Griffin. Macromolecules 28:4116, 1995.
155. C Larez, V Crescenzi, M Dentini, A Ciferri. Supramol Sci 2:141, 1995.
156. M Henriquez, E Lissi, E Abuin, A Ciferri. Macromolecules 27:6834, 1994.
157. H Okuzahi, Y Osada. Macromolecules 28:380, 4554, 1995.
158. A Khokhlov, EY Kramarenko, EE Makaeva, SG Starodubtzev. Makromol Chem Theory Simul 1:105, 1992.
159. A Khokhlov, EY Kramarenko, EE Makaeva, SG Starodubtzev. Macromolecules 25:4479, 1992.
160. MC Petty, WA Barlow. Langmuir–Blodgett Films. G Roberts, ed. New York: Plenum, 1990.
161. RHG Brinkhuis, AJ Schouten. Macromolecular Assemblies in Polymeric Systems. P Stroeve, AC Balazs, eds. Symposium Series 493. Washington, DC: ACS, 1992.
162. AT Royappa, MF Rubner. Macromolecular Assemblies in Polymeric Systems. P Stroeve, AC Balazs, eds. Symposium Series 493. Washington, DC: ACS, 1992.
163. OP Marquez, B Fontal, J Marquez, R Ortiz, R Castillo, M Choy, C Larez. J Electrochem Soc 144:707, 1995.
164. C Boraz, E Weinhold, W Cabrera, OP Martinez, J Martinez, RO Lenza. J Electrochem Soc 144:3871, 1997.

2
Theory of the Supramolecular Liquid Crystal

Reinhard Hentschke
Bergische Universität-Gesamthochschule, Wuppertal, Germany

Bernd Fodi
Max-Planck-Institute for Polymer Research, Mainz, Germany

I. INTRODUCTION

Generating specific functionality of supramolecular assemblies based on tailor-made molecular building blocks has been quite successful in the past [1,2] and certainly will continue to be of increasing importance in the future. Thus far the underlying design concepts are largely based on intuition and on simple theoretical concepts. More detailed theories of supramolecular assembly face the principal challenge of predicting structural and dynamic properties of very complex macroscopic or mesoscopic systems based on the microscopic interactions of their constituent molecular units. Most likely predictions of this type will not come from analytic theory but rather will become possible in the future using computer modeling. Here we pursue a less ambitious goal. We review theoretical concepts describing the structural phase behavior of reversibly assembling systems forming lyotropic liquid-crystalline phases of polymeric aggregates (cf. [3–6]). In the same context we also take a brief look over the fence and discuss the current possibilities of computer modeling, for both detailed molecular force fields as well as coarse-grained interactions.

Section II discusses the analytic approach. In Section II.A we begin with a brief outline of the well-developed theory for ordinary lyotropic polymer liquid crystals in terms of statistical mechanical models. Starting with Onsager's seminal work we discuss various extensions, mostly based on excluded volume descriptions of molecular interactions, which enable the calculation of at least partial phase diagrams for complex lyotropic mesogens. In Section II.B we discuss some concepts of reversible assembly including factors governing aggregate growth and form at low solute concentrations. Emphasizing polymeric aggregates, the combination of self-assembly and liquid-crystalline ordering is the subject of Section II.C. In the next part, Section III, we address computer modeling as an alternative theoretical approach to the understanding of supramolecular assembly and ordering. We take a look at the current possibilities and limitations of molecular simulations. The focus is hereby on force field approaches in Section III.A, which allow a detailed description of molecular interaction, and on coarse-grained models in Section III.B, which enable us to extend the limited time and spatial scales accessible by the former. Here we do not address methodological details but rather confine ourselves to a number of illustrative examples.

II. ANALYTIC APPROACHES

The following is a discussion of statistical mechanical models for liquid-crystalline phase behavior driven by concentration changes, i.e., for lyotropic solutions. The intermolecular or interaggregate interactions are assumed to be well represented by excluded volume. Clearly, there are cases where this is not a valid assumption, and some of these are discussed. In addition, we are concerned exclusively with structural ordering as a function of concentration; i.e., dynamic phenomena are omitted. Because our focus is on liquid-crystalline behavior, we also omit the discussion of continuous structural phases, which usually do occur in self-assembling systems at high concentrations.

A. Principles of Molecular Lyotropic Liquid Crystals

1. Continuum Approach [7,8]

Consider a solution containing N axial symmetric particles* in a volume V, which in the following are distinguished according to their type ν and their orientation $\Omega = \Omega(\theta, \phi)$. The configuration part of their free energy is $F_{\text{conf}} = -k_B T \ln Q_{\text{conf}}$, where

$$Q_{\text{conf}} = \frac{1}{\prod_{\nu,k} N_{\nu,k}!} \int \prod_{l=1}^{N} \left(\frac{d\Omega_l}{4\pi} d^3 r_l \right) \exp\left[-\frac{1}{k_B T} \sum_{i<j}^{N} w(r_{ij}, \Omega_i, \Omega_j) \right] \quad (1)$$

* Assuming axial symmetry is not crucial for what follows, but it certainly simplifies matters.

$N_{\nu,k}$ is the number of ν-type particles with orientation $\Omega(\theta_k, \phi_k)$. The particles interact via the potential of mean force $w(r_{ij}, \Omega_i, \Omega_j)$, where r_{ij} is the center of mass separation between particles i and j. Using Sterling's approximation we may write $\ln[\prod_{\nu,k} N_{\nu,k}!] \approx N \sum_\nu \sigma(f_\nu) + N(\ln N - 1)$, where $\sigma(f_\nu) = \int_\Omega d\Omega/(4\pi) f_\nu(\Omega) \ln f_\nu(\Omega)$ and $f_\nu(\Omega_k) \approx N_{\nu,k}/N$. The integral, on the other hand, may be expanded in a virial series (e.g., [9]) which yields

$$\frac{F_{\text{conf}}}{Nk_BT} = \ln\rho - 1 + \sum_\nu \sigma(f_\nu) + \rho \sum_{\nu,\nu'} \int_{\Omega,\Omega'} \frac{d\Omega}{4\pi}\frac{d\Omega'}{4\pi} f_\nu(\Omega) f_{\nu'}(\Omega') B_{\nu,\nu'}(\Omega, \Omega') + \cdots, \tag{2}$$

where $\rho = N/V$ is the particle number density, and B is the second virial coefficient for fixed particle orientation. The corresponding part of the chemical potential for ν-particles, $\mu_{\text{conf},\nu} = \partial F_{\text{conf}}/\partial N_\nu|_{T,V,N_{\nu'}}$, is

$$\frac{\mu_{\text{conf},\nu}}{k_BT} = \int_\Omega \frac{d\Omega}{4\pi} f'_\nu(\Omega) \left\{ \ln\left[\frac{N_\nu f'_\nu(\Omega)}{N}\right] + 2\rho \sum_{\nu'} \int_{\Omega'} \frac{d\Omega'}{4\pi} \frac{N_{\nu'} f'_{\nu'}(\Omega')}{N} B_{\nu,\nu'}(\Omega, \Omega') + \cdots \right\} \tag{3}$$

with $f_\nu(\Omega) = (N_\nu/N) f'_\nu(\Omega)$.

Notice that $\{\cdots\} = \mu_{\text{conf},\nu}(\Omega)/k_BT = N^{-1}\partial F_{\text{conf}}/\partial f_\nu(\Omega)|_{T,V,f_{\nu'}(\Omega')}$. In particular $\mu_{\text{conf},\nu} = \mu_{\text{conf},\nu}(\Omega)$, because all ν-particles have the same chemical potential regardless of their orientation in equilibrium. This immediately implies $\{\cdots\} =$ const; i.e.,

$$f_\nu(\Omega) = \exp\left[\text{const} - 2\rho \sum_{\nu'} \int_{\Omega'} \frac{d\Omega'}{4\pi} f_{\nu'}(\Omega') B_{\nu,\nu'}(\Omega, \Omega') + \cdots \right] \tag{4}$$

The nonlinear integral equation (4) can be solved numerically for the equilibrium distribution $f_\nu(\Omega)$, where const is obtained via the normalization condition $\int_\Omega d\Omega/(4\pi) f'_\nu(\Omega) = 1$ (e.g., [10]).

In his seminal paper Onsager [11] showed, based on Eq. (2), that monodisperse, rigid, rodlike particles, which interact via hard core repulsion, spontaneously align at high enough concentration ρ or volume fraction $v = b\rho$, where b is the particle volume. More precisely, for $v < 3.29\ D/L$, where L and D are the length and the diameter of the rods, is the solution isotropic, whereas for $v >$

$4.223D/L$ it is anisotropic (nematic) (e.g., [12]).* In between the solution phase separates. Onsager used a variational approach approximating the orientation distribution via the trial function $f'(\Omega) = (\alpha/\sinh\alpha)\cosh[\alpha\cos\theta]$; i.e., $f'(\Omega) = 1$ for $\alpha = 0$ (isotropic solution) and $f'(\Omega) \approx \alpha\exp[-\alpha\theta^2/2]$ for large α and $0 \leq \theta \leq \pi/2$ (anisotropic solution). Subsequently Eq. (2) is minimized with respect to α. In conjunction with the conditions that the pressures and the chemical potentials at opposite phase boundaries are equal, this yields three equations. From these the values of α at the nematic phase boundary and the phase boundaries themselves can be obtained numerically. The first-order isotropic-to-nematic transition arises due to the competition between the orientation and packing contributions in the above configuration free energy as illustrated in Fig. 1 in terms of the orientation parameter α. In his original paper Onsager also discusses the effect of electric double layer interactions, i.e., the interaction of polyelectrolyte rods in electrolyte solutions (having supramolecular aggregates like the tobacco mosaic virus in mind). The double layer mainly acts as a soft repulsive shell whose thickness is controlled by the Debye length. The corresponding increase of the effective diameter tends to shift the transition to lower particle concentrations. (Cf. Fig. 10 in [8]; in addition there is a twist-effect, due to the softness of the repulsion, which tends to oppose alignment to a lesser extent [7,8].)

2. Extensions of the Continuum Approach

Extensions of Onsager's work mainly have focused on improving the description of the (excluded volume) interaction beyond the second virial term, the investigation of various particle geometries (like ellipsoids or plates) including polydispersity (e.g., [13]), and the inclusion of molecular flexibility for strongly elongated particles.

Carrying the exact virial expansion in Eq. (2) beyond the second order is prohibitively complex except for very simple particle shapes like hard spheres (for a discussion see [7,8]). However, there are closed approximate expressions describing F_{conf} at higher concentrations derived for instance via scaled particle theory SPT [14,15] or the comparatively ad hoc decoupling approximation (DA) [16]. Overall SPT probably yields the best compromise between simplicity and accuracy for the interaction of hard convex bodies. Excluded volume approaches like SPT and DA are likewise useful ingredients for the study of transitions in translationally ordered lyotropic liquid crystals exhibiting smectic, columnar, or crystalline ordering. Translational ordering frequently is modeled using density

* Note that the second virial approximation thus always holds in the vicinity of the transition provided the particles are sufficiently elongated.

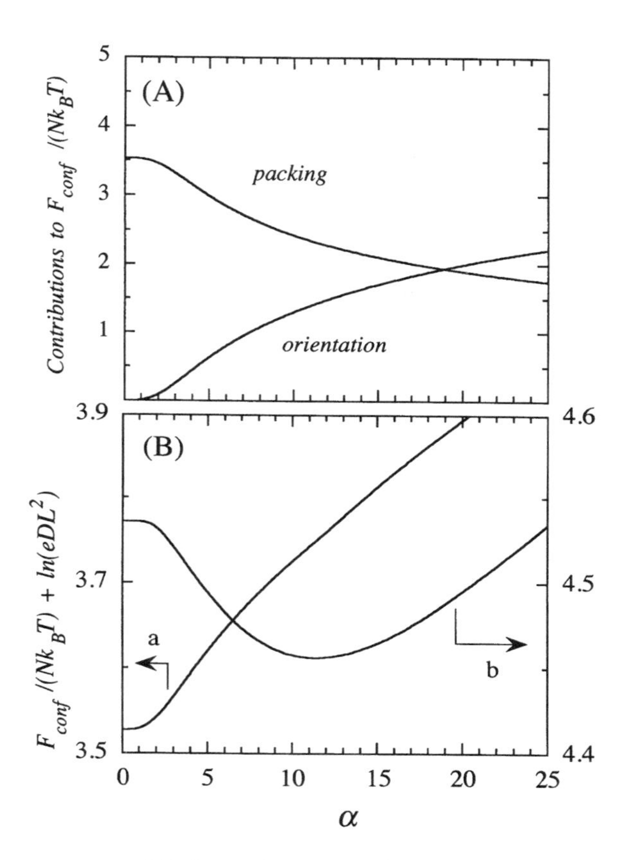

Figure 1 (A) Competing contributions to F_{conf} computed with the trial function mentioned in the text. Orientation: $\sigma(f')$; packing: $\ln \nu - 1 + \cdots$ second virial term . . . . Here $\nu \approx 3.93D/L$. (B) Sum of the above two contributions for $\nu \approx 3.14D/L$ (a; minimum of F_{conf} corresponds to isotropic orientation) and $\nu \approx 3.93D/L$ (b; minimum of F_{conf} corresponds to anisotropic orientation).

functional theory (DFT) considering thermodynamic potentials like the Helmholtz free energy* functional of the inhomogeneous one-particle density $\rho(\boldsymbol{r})$; e.g.,

$$F_{\text{conf}}[\rho(\boldsymbol{r})] = k_B T \int_V d^3 r \rho(\boldsymbol{r})(\ln \rho(\boldsymbol{r}) - 1) + F_{\text{ex}}[\rho(\boldsymbol{r})] \tag{5}$$

* Here and in the remaining section we consider one type of particle only.

Descriptions of the excess part F_{ex} include expansion around a uniform liquid with density ρ_R; i.e.,

$$F_{ex}[\rho(\boldsymbol{r})] = F_{ex}[\rho_R] - k_B T \int_V d^3 r_1 c^{(1)}(\rho_R) \Delta\rho(\boldsymbol{r}_1) - \frac{1}{2} k_B T \int_V d^3 r_1 \int_V d^3 r_2 c^{(2)}(|\boldsymbol{r}_1 - \boldsymbol{r}_2|; \rho_R) \Delta\rho(\boldsymbol{r}_1) \Delta\rho(\boldsymbol{r}_2) \tag{6}$$

where $\Delta\rho(\boldsymbol{r}) = \rho(\boldsymbol{r}) - \rho_R$, and $c^{(n)}$ denotes the n-particle direct correlation function, or the weighted density approximation

$$F_{ex}[\rho(\boldsymbol{r})] = \int_V d^3 r \rho(\boldsymbol{r}) \Delta\Psi[\bar{\rho}(\boldsymbol{r})] \tag{7}$$

Here the local free energy $\Delta\Psi$ is a functional of the local average $\bar{\rho}(\boldsymbol{r}) = \int W(\boldsymbol{r}, \boldsymbol{r}')\rho(\boldsymbol{r}')$, where $W(\boldsymbol{r}, \boldsymbol{r}')$ is a weighting function. A detailed discussion of DFT applied to lyotropic liquid crystals can be found in [8]. Another interesting approach in this context is the method of the symmetry breaking potential [17,18]. Here the partition function $Q(\kappa)$ depends, besides on the particle–particle interaction, on an additional external potential κ possessing the symmetry of the higher-ordered phase. Subsequently κ in $Q(\kappa)$ is replaced by the order parameter $\eta = \partial \ln Q(\kappa)/\partial\kappa$. The stable phase follows via minimization of the free energy with respect to η. A third approach to translational ordering approximates the excess configuration free energy via

$$F_{ex}^{(d)} = F_{fluid}^{(d)} + F_{cryst}^{(D_{sp}-d)} \tag{8}$$

where D_{sp} is the space dimension [19]. $F_{fluid}^{(d)}$ and $F_{cryst}^{(D_{sp}-d)}$ are the excess free energy contributions of a d-dimensional fluid and the corresponding $(D_{sp} - d)$-dimensional crystal, which are coupled via the D_{sp}-dimensional density. A smectic phase of thin cylindrical rods, for example, is described as a two-dimensional fluid ($d = 2$) of discs parallel plus a one-dimensional crystal of lines perpendicular to the smectic layers. Expressions for $F_{fluid}^{(d)}$ can be obtained for instance via SPT, whereas cell theories can be used to model $F_{cryst}^{(D_{sp}-d)}$ (e.g., [20]). Figure 2 illustrates an application of this approach to the phase diagram of parallel rigid spherocylinders, which turned out to be surprisingly rich. Note that the increasing crystallinity with increasing density is favored by a gain in free volume per particle (better packing efficiency) at a given density which is offsetting the loss of translational freedom. This system received considerable attention following a computer simulation by Frenkel et al. [21] (see also [15]). Their Monte Carlo work was a direct extension of the very early computer simulations by Alder and Wainwright [23], who discovered a freezing transition in concentrated hard spheres. It is important to note that the above approaches to translational ordering have in common that they do not a priori predict the symmetry of the stable

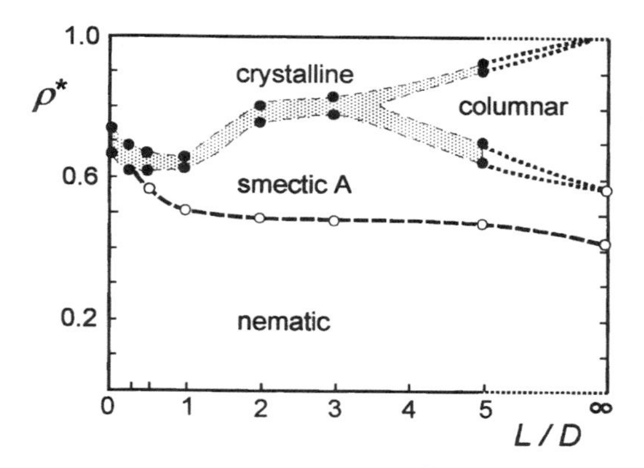

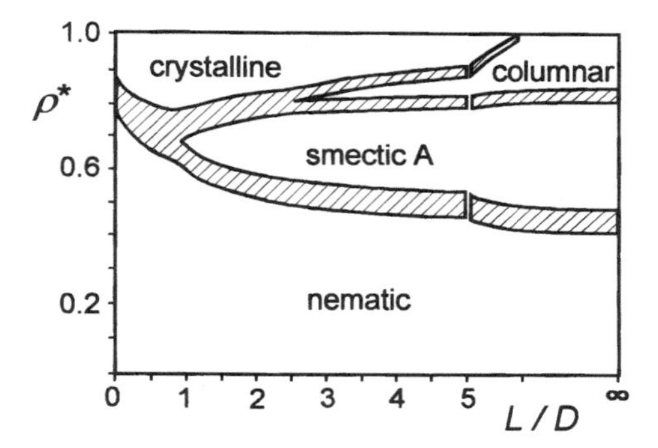

Figure 2 Phase diagram of a fluid of parallel hard spherocylinders. Note that L is the length of the cylinders; i.e., $L = 0$ corresponds to spheres. Here ρ^* is the number density divided by the number density at close packing. Top: Monte Carlo simulation. (From Ref. 21.) Shaded areas indicate phase coexistence. The dashed line indicates a continuous transition. Bottom: theoretical phase diagram of Ref. 19 resulting from the comparison of the free energies for $d = 3$ (nematic), $d = 2$ (smectic), $d = 1$ (columnar), and $d = 0$ (crystal). Shaded areas again indicate phase coexistence. A subsequent revision of the Monte Carlo phase diagram was published in Ref. 22 replacing the columnar phase with a phase of a different symmetry.

phase. Instead, various symmetries are ''offered'' via a suitable parameterization of the density or the free energy and the methods ''select'' the most stable of the ''offered'' symmetries (cf. the caption of Fig. 2).

Ordering of molecules with pronounced shape anisotropy is affected by their flexibility. Khokhlov and Semenov [12] (and references therein; see also [24] for a more recent discussion) derived an expression for the orientation contribution to the free energy $\sigma(f)$ for continuous bend elastic rodlike molecules or rodlike aggregates (cf. below). This expression, however, is rather complicated and concrete calculations usually employ the expansions.

$$\sigma(f) = \begin{cases} \int_{\Omega} \frac{d\Omega}{4\pi} f(\Omega) \ln f(\Omega) + \frac{L}{P}\frac{1}{12} \int_{\Omega} \frac{d\Omega}{4\pi} \frac{\partial f(\Omega)}{\partial\theta} \frac{\partial \ln f(\Omega)}{\partial\theta} + \mathcal{O}((L/P)^2) & L/P \ll 1 \\ \frac{L}{P}\frac{1}{8} \int_{\Omega} \frac{d\Omega}{4\pi} \frac{\partial f(\Omega)}{\partial\theta} \frac{\partial \ln f(\Omega)}{\partial\theta} - 2 \ln\left[\int_{\Omega} \frac{d\Omega}{4\pi} (f(\Omega))^{1/2} \right] + \mathcal{O}(P/L) & L/P \gg 1 \end{cases} \tag{9}$$

or interpolating expressions based on these expansions. P is the persistence length defined via

$$\langle \boldsymbol{u}(0) \cdot \boldsymbol{u}(\tau) \rangle = \exp\left[-\frac{\tau}{P} \right] \tag{10}$$

where the $\boldsymbol{u}$ denote unit vectors tangential to the contour of the molecule or aggregate separated by a distance τ along the contour of length L.* Thus, the above expansions are the leading contributions to $\sigma(f)$ for stiff ''rods'' ($P \gg L$) and for ''wormlike rods'' ($P \ll L$). Within the second virial approximation the interaction term [cf. Eq. (2)] to good approximation is the same as for rigid rods. This is because the second viral term in this case consists of ''independent,'' local, rodlike two-particle interactions; i.e., the quantity $N\rho B$ remains unchanged if N long rods are replaced by $n = N \cdot m$ shorter rods of length $l = L/m$ ($l \gg D$), because B scales with the square of the length of the interacting rods (neglecting end-effects). An application to a concrete system, i.e., a solution of Poly (γ-benzyl-L-glutamate (PBLG) in dimethylformamide (DMF), is illustrated in Fig. 3.†

We conclude this section with a brief look at the hierarchy of forces and their influence on the stability of phases. With increasing concentration orientational ordering of large molecules or long molecular aggregates is affected first by excluded volume, i.e., the transition density scales like $\rho \sim (DL^2)^{-1}$. Within the validity of the Debye–Hückel approximation the main effect of an electrostatic double layer, if twist is neglected, is an increased particle diameter due to a soft repulsive shell whose thickness is essentially the Debye length λ_D, which varies according to salt concentration, c_{salt} (i.e., $\lambda_D \approx 0.304/c_{\text{salt}}^{1/2} nm$ for a 1:1

* Another definition of the persistence length, $a' = \lim_{n\to\infty} \langle \boldsymbol{u}_1 \cdot \sum_{i=1}^{n} \boldsymbol{l}_i \rangle$, is that of a jointed chain of in this case rigid segments $\boldsymbol{l}_i$, of equal length l, where $\boldsymbol{u}_1 = \boldsymbol{l}_1/l$. It is easy to show that $P \approx a'$ if the angle between successive segments is close to π. A quantity often referred to in the context of chain stiffness, which is mentioned here for the sake of completeness, is the Kuhn segment or effective bond length K, defined via $K = \langle (\sum_{i=1}^{n} \boldsymbol{l}_i)^2 \rangle / (nl)$. In the case of freely jointed (!) bonds, $\boldsymbol{l}_i$, it can be shown [25] that the right side equals $2a'$, and thus $K = 2a'$ in this limit.

† Note that the helical structure of PBLG results in colesteric rather than nematic ordering, but this does not lead to significant deviations regarding the osmotic pressure.

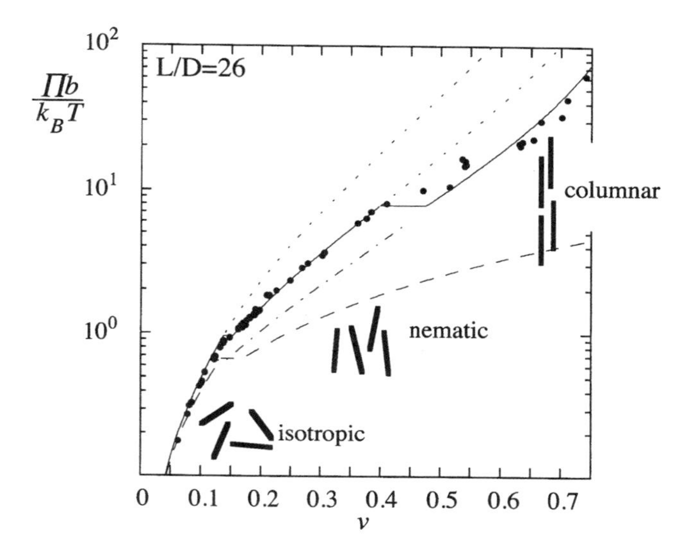

Figure 3 Experimental and theoretical osmotic pressure, Π, of PBLG in DMF versus solute volume fraction. b is the molecular volume. The datapoints are taken from Refs. 27 and 28 (data set B), and encompass measurements between 15°C and 45°C. Solid line: theory based on the Khokhlov–Semenov approach to flexibility in combination with the decoupling approximation and the dimensional separation approach based on Eq. (8) [29]. Upper dotted line: extension of isotropic branch. Lower dotted line: extension of nematic branch. Dashed-dotted line: result for completely rigid molecules. Dashed line: Onsager's second virial coefficient-approximation for rigid rods.

electrolyte or $\lambda_D \approx 0.176/c_{\text{salt}}^{1/2} nm$ for 2:1 and 1:2 electrolytes). The increased diameter tends to lower the concentration necessary for alignment.* Flexibility, on the other hand, tends to postpone the orientation ordering. In leading order for not too flexible rods the transition density behaves as $\rho \sim (DL^2)^{-1} + a_o(DLP)^{-1}$, where $a_o > 0$. For wormlike rods the leading behavior of the transition density is $\rho \sim (DLP)^{-1}$. Notice also that P will tend to decrease with increasing temperature due to conformational softening, and the lyotropic system will acquire thermotropic features [26]. In general, flexibility will favor the least confining phase (e.g., [29]). Confinement of rodlike particles becomes important on the length scale of the deflection length $\lambda = P/\alpha$, where α is the above alignment parameter [7]. For lengths shorter than λ the particle contour undulations are

* Consider the special case of a thin polyelectrolyte rod for which $D \approx 2\lambda_D$. Because $\lambda_D \propto 1/c_{\text{salt}}^{1/2}$ one finds that the transition density scales like $\rho \sim c_{\text{salt}}^{1/2}/L^2$ (e.g., Fig. 10 in Ref. 8).

essentially free, whereas otherwise interaction with neighboring particles increases the (orientation) free energy; i.e., note that $\sigma \propto L/\lambda$ as one can see if the leading term for $L/P \gg 1$ is evaluated with $f(\Omega) \approx \alpha \exp[-\alpha\theta^2/2]$. For hard rods confined to a hard "tube" with mean diameter Δ formed by neighboring molecules, e.g., in a columnar or crystalline phase, the contribution to the free energy due to orientational confinement becomes $\sigma \propto (P^{1/3}\Delta^{2/3})^{-1}$ [30]. This is analogous to the interaction between flexible membranes separated by a mean distance Δ_z, where the free energy contribution due to confinement is $\propto (k_B T)^2/(\varepsilon\Delta_z^2)$, and ε is the bending module. Notice that flexibility always produces long-range repulsion! For more details see for instance Ref. 31. An interesting self-consistent field study of undulatory repulsion in combination with Coulomb interaction for hexagonally packed DNA can be found in Ref. 32. Here the undulatory confinement within a tube of soft Coulomb-walls leads to a doubling of the exponential decay in accord with the experiment. The undulation-enhanced electrostatic forces have also been studied in Ref. 33. Shorter-ranged interactions, like van der Waals forces, hydrogen bonding, and solvation forces, mainly affect ordering through the bonding within aggregates (cf. below) which themselves may interact via the interactions just discussed.

3. The Lattice Approach and Its Extensions

Subsequent to Onsager's work Flory [34,35] introduced a lattice model to study the interplay between concentration and orientation distribution for rodlike particles. Flory employed a cubic lattice, where a rod corresponds to a column of lattice cells. The restriction to three orientations can be overcome to some extent by a stairlike placement of rod segments, where the number of cubes per segment controls the tilt of the rod with respect to the lattice axes. Overall, the results of this model are qualitatively similar to the Onsager model. A comparative discussion of the continuum versus the lattice approach by Grosberg and Khokhlov [36] concludes, however, that the continuum model is the most promising one from the viewpoint of application to liquid-crystalline ordering. Nevertheless, the lattice statistics used to calculate the configurational free energy tends to be somewhat simpler than the continuum expressions, and extensions of Flory's lattice model have been considered for instance by DiMarzio [37] and Herzfeld [38], who studied orientational ordering in a polydisperse population of rigid rectangular particles. We emphasize the latter model because of its extensive use in the context of reversibly assembling systems (cf. below).

B. Reversibly Assembling Molecular Systems—Some Basic Concepts [39–42]

Consider a solution containing different types t of monomer molecules M_t ($t = A, B, \ldots$). The monomer chemical potential is given by

$$\mu_t = \bar{\mu}_t + k_B T \ln[x_t \gamma_t] \tag{11}$$

where x_t denotes the mole fraction t-monomer, and γ_t denotes the respective activity coefficient, which according to Eq. (3) is given by

$$\begin{aligned}\ln \gamma_t &= \int_\Omega \frac{d\Omega}{4\pi} f_t'(\Omega_t) \\ &\quad \times \left\{ \ln f_t'(\Omega_t) + 2\rho_{\text{total}} \sum_\nu \int_\Omega \frac{d\Omega}{4\pi} x_\nu f_\nu'(\Omega) B_{t,\nu} \Omega_t, \Omega) + \cdots \right\} \\ &= \ln f_t'(\Omega_t) + 2\rho_{\text{total}} \sum_\nu \int_\Omega \frac{d\Omega}{4\pi} x_\nu f_\nu'(\Omega) B_{t,\nu} \Omega_t, \Omega) + \cdots \end{aligned} \tag{12}$$

Here x_ν are the mole fraction ν-particles. Note also that $\rho N_\nu / N = (N/V)(N_\nu/N)$ $(N + N_{\text{solvent}}/(N + N_{\text{solvent}}) = \rho_{\text{total}} x_\nu$, where N_{solvent} is the number of solvent molecules in the system. The index ν stands for other monomers as well as for aggregates $A_\nu = A_{s_A, s_B, \ldots}$, reversibly assembled by the monomers, where s_t is the number of t-monomers contained in an s_A, s_B, . . .-aggregate. (Whether it is useful to include the solvent as one explicit component depends on the system of interest; e.g., solvent is built into the aggregates.) The quantity $\bar{\mu}_t$ ''lumps together'' the remaining dependencies of the chemical potential.

The starting point for a quantitative description of the distribution of material in the system may be the assumption of independent reactions, $s_A M_A + s_B M_B + \cdots \rightleftharpoons A_{s_A, s_B, \ldots}$, for every type of aggregate (closed association model [40]) or equivalently the step-wise association, $M_t + A_{s_A, s_B, \ldots, s_t - 1, \ldots} \rightleftharpoons A_{s_A, s_B, \ldots, s_t, \ldots}$ [42]. The independent reaction model is equivalent to $s_A \mu_A + s_B \mu_B + \cdots = \mu_{s_A, s_B, \ldots}$ where

$$\mu_{s_A, s_B, \ldots} = \bar{\mu}_{s_A, s_B, \ldots} + k_B T \ln \left[\frac{1}{s_\zeta} x_{\zeta; s_A, s_B, \ldots} \gamma_{s_A, s_B, \ldots} \right] \tag{13}$$

is the chemical potential of an s_A, s_B, . . .-aggregate in solution, and $x_{\zeta; s_A, s_B, \ldots} = s_\zeta x_{s_A, s_B, \ldots}$ is the mole fraction ζ-monomer contained in the aggregate. Note that the activity coefficient is again of the above form. We thus obtain

$$x_{\zeta; s_A, s_B, \ldots} \gamma_{s_A, \ldots} = s_\zeta \prod_t (x_t \gamma_t)^{s_t} \exp[s_\zeta \Phi_{\zeta; s_A, s_B, \ldots}] \tag{14}$$

with

$$\Phi_{\zeta; s_A, s_B, \ldots} = \frac{1}{k_B T} \left(\sum_t \left(\frac{s_t}{s_\zeta} \bar{\mu}_t \right) - \frac{1}{s_\zeta} \bar{\mu}_{s_A, \ldots} \right) \tag{15}$$

which completely characterizes the material distribution and the structure of the system at equilibrium. Note that the $x_{t;s_A,s_B,\ldots,s_t}$ are coupled both via the monomer activity coefficients and the mass conservation condition

$$x_{\text{solute}} = \sum_t x_t + \sum_t \sum_{s_A,s_B,\ldots} x_{t;s_A,s_B,\ldots} \tag{16}$$

Provided that expressions for the $\bar{\mu}$ and the activity coefficients do exist, which generally is not the case due to the complicated dependence of these quantities on the thermodynamic variables and the systems composition, one may attempt the (iterative numerical) solution of the coupled set of Eqs. (14), (15, and (16).

It is instructive to consider solutions containing one type of monomer only. At low concentrations the monomer and aggregate activity coefficients are (close to) unity, and Eqs. (14) to (16) simplify to

$$x_s = s\beta^s \qquad (\beta = xe^{\Phi_s}) \tag{17}$$

$$\Phi_s = \frac{1}{k_B T}\left(\bar{\mu} - \frac{1}{s}\bar{\mu}_s\right) \tag{18}$$

$$x_{\text{solute}} = x + \sum_{s=m}^{\infty} x_s \tag{19}$$

Here x is the mole fraction of free monomers, x_s is the monomer mole fraction in s-aggregates, and m is a minimum aggregation number, which depends on the monomer "architecture." The latter governs the s-dependence of Φ_s. It is at this point that models for $\bar{\mu}_s$ must be constructed, which describe the growth and the aggregate form at low concentrations. A simple example is the one-parameter expression, $\bar{\mu}_s = s\bar{\mu}_{\text{bulk}} + k_B T\delta s^{(d-1)/d}$.* The first term assigns an s-independent free enthalpy $\bar{\mu}_{\text{bulk}}$ to every monomer in the "bulk" of an aggregate. The second term is a surface term expressing the surface to volume ratio in terms of the aggregation number, where $\delta > 0$† is a constant. Here $d = 1$ corresponds to linear aggregates (e.g., supramolocular polymers), where the end-monomers possess a different free enthalpy compared to the "bulk" monomers. Similarly, $d = 2$ corresponds to disclike aggregates, where monomers along the rim feel a different environment. Finally, $d = 3$ corresponds to spherelike aggregates. Inserting this expression for $\bar{\mu}_s$ into Eq. (18) yields $\beta = x\exp[\Phi_* - \delta s^{-1/d}] \approx x\exp[\Phi_*]$,‡ where $\Phi_* = (\bar{\mu} - \bar{\mu}_{\text{bulk}})/k_B T$. In combination with Eqs. (17) and (19) we obtain $x_{\text{solute}} \approx x + \beta^m(m - (m-1)\beta)/(1-\beta)^2$ if $\beta < 1$. Thus, if m is large (in micellar systems m ranges from 50 to 100) we have $x_{\text{solute}} \approx x$, because $\beta^m \ll 1$. However,

* Note that here the meaning of d differs from the meaning of d in Eq. (8). Note also that some references (e.g., [39]) discuss $\bar{\mu}_s/s$ instead of $\bar{\mu}_s$, which accounts for the different exponent of s.

† Disfavoring the surface compared to the bulk.

‡ Not necessarily a good approximation, but it simplifies the discussion.

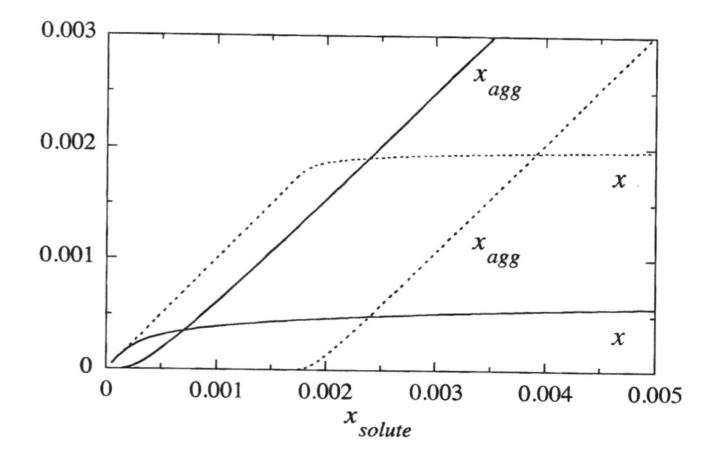

Figure 4 Mole fraction free monomer x and aggregated monomer mole fraction $x_{\text{agg}} = \sum_{s=m}^{\infty} x_s$ versus solute mole fraction x_{solute}. Solid lines: $m = 5$ and $\Phi_* = 6$; dashed lines: $m = 50$ and $\Phi_* = 6$. Notice that the sudden break in the increase of x, which corresponds to x_{CAC}, easily can be identified for the large m-value, whereas for $m = 5$ this is much less the case.

for β very close to unity there must be an abrupt change, because otherwise the rhs of the equation for x_{solute} diverges. Increasing the monomer concentration beyond this point leads to spontaneous formation of aggregates while x remains essentially constant (cf. Fig. 4). The mole fraction of free monomers at which this happens, $x_{\text{CAC}} \approx \exp[-\Phi_*]$, is called the critical aggregate (or micelle) concentration. Simple as it is, this model for reversible assembly captures experimental observations in surfactant systems like the dependence of x_{CAC} on hydrocarbon chain lengths ($\ln x_{\text{CAC}} \approx -\Phi_* \propto -n_{\text{CH}_2}$, where n_{CH_2} is the number of methylene units) or on ionic strength (e.g., Fig. 4.12 in [40]).

What can be said about the sizes and the shapes of the aggregates above the CAC? For $d > 1$, still referring to the above model, one can show that as soon as aggregates form they will grow without bound akin to a phase transition (e.g., [39,42]). Generally, however, this is not the case, because the model does not include any details of the monomer architecture that greatly affect the aggregate shape. Simple surfactant for instance may form spherical micelles, but they are not described by the $d = 3$-model, because their diameter is essentially fixed by the tail-length.* Only by forming vesicles, which is not compatible with the

* The shape of surfactant aggregates largely depends on the critical packing parameter, i.e., the quantity $b/(Ra_h)$, where b is the volume of the surfactant molecule, R is its tail length, and a_h is the head-group area. Certain values of $b/(Ra_h)$ produce spherical aggregates ($<1/3$), cylinders (between 1/3 and 1/2), bilayers (≈ 1), continuous cubic structures (≥ 1), etc.; see, e.g., Fig. 3.27 in [41].

above simple $d = 3$-model, can they expand into larger spherical aggregates. In the literature complex models for $\bar{\mu}_s$ are discussed, which are tailored according to the specific system [39–42]. However, it is beyond the scope of this chapter to discuss details. Instead we return to the above model for $d = 1$, i.e., reversibly assembled linear polymers. An s-polymer is stabilized by its $(s - 1)$ monomer–monomer contacts each contributing a free enthalpy Φ_o (in units of $k_B T$); i.e., Φ_s is given by $\Phi_s = \Phi_o(s - 1)/s$ [here: $\Phi_o = \delta = (\mu - \bar{\mu}_{\text{bulk}})/k_B T$ and $m = 2$]. The particle size distribution in terms of the aggregation number is $x_s = sx^s \exp[\Phi_o(s - 1)]$. A straightforward calculation yields the maximum of the distribution at $s_M = (\ln[(\langle s\rangle + 1)/(\langle s\rangle - 1)])^{-1}$, the average aggregation number, $\langle s\rangle = \sum_{s=1}^{\infty} sx_s/\sum_{s=1}^{\infty} x_s = (1 + 4x_{\text{solute}} \exp[\Phi_o])^{1/2}$ (for a recent experimental study see [43]; see also Fig. 1 in Section III of Chapter 1, this volume), and the polydispersity, $\sigma = (\langle s^2\rangle - \langle s\rangle^2)^{1/2} = (2x_{\text{solute}} \exp[\Phi_o])^{1/2}$.* Note that for large $\langle s\rangle$ we have $s_M \sim (x_{\text{solute}} \exp[\Phi_o])^{1/2}$ and $\sigma \sim \sqrt{2}s_M$. It is worth emphasizing the strong dependence of $\langle s\rangle$ on the contact-free enthalpy Φ_o. Notice that in rodlike aggregates with large diameter (e.g., many of the biological rodlike aggregates) growth is significantly enhanced due to the increase of the effective monomer coordination number.

Figure 5 compares the ideal behavior, i.e., $\langle s\rangle \sim x_{\text{solute}}^{1/2}$, to the result when excluded volume interactions are included. We can understand the shape of the excluded volume curve again in terms of our simple model for linear aggregates. Using $\ln \gamma_s \approx (a_1 s + a_2)x_{\text{solute}}$ from [44] (second virial approximation neglecting polydispersity†) and $x_s = s(x\gamma_1 \exp[\Phi_s])^s\gamma_s^{-1}$ [cf. Eq. (14)] yields $\langle s\rangle \sim (x_{\text{solute}} \exp[\Phi_o + a_2 x_{\text{solute}}])^{1/2}$ and $\sigma \sim \langle s\rangle$ for large $\langle s\rangle$‡; i.e, we obtain the same excluded volume-induced growth behavior as shown in Fig. 5. Notice that excluded volume affects the size through a_2, which means that the aggregate ends drive the growth. Another noteworthy aspect is the continuous growth of the average aggregation number with increasing solute concentration as predicted by the present model. In contrast to this a number of experimental studies observe a pronounced maximum of the average aggregation number as a function of solute concentration at constant temperature [45]. The height of the maximum as well as $\langle s\rangle$ at fixed x_{solute} both depend on T. Both decrease upon heating. The decrease of $\langle s\rangle$ at fixed x_{solute} could be attributed to a decrease of Φ_o with increasing T (e.g., [46]). The

* In the polymer field polydispersity often is characterized via the polydispersity coefficient $U = \langle s\rangle_w/\langle s\rangle_n - 1$, where $\langle s\rangle_w$ is the aggregate weight average, and $\langle s\rangle_n$ is the aggregate number average; i.e., $U = 0$ in the monodisperse limit. Here $\langle s\rangle_w = \langle s\rangle$ and $\langle s\rangle_n = \sum_{i=1}^{\infty} x_s/\sum_{i=1}^{\infty} x_s/s$ so that $U = (\langle s\rangle - 1)/(\langle s\rangle + 1)$.

† Note that in this case $B_{\nu\nu'} = L^2 D\,|\sin(\Omega, \Omega')| + 2\pi D^2 L + 4\pi D^3/3$ for spherocylindrical ν-aggregates of length $L + D$ and diameter D. Assuming that almost all monomers are contained in large polymeric aggregates with $L \propto s$ and $D \sim \mathcal{O}(1)$ we obtain γ_s as stated using Eq. (12) with $x_\nu = x_s/s \approx x_{\text{solute}}/s$ and $\int d\Omega/(4\pi) d\Omega'/(4\pi)\,|\sin(\Omega, \Omega')| = \pi/4$ (isotropic phase).

‡ Here we use $\sum_{i=1}^{\infty} s^{\lambda+1}\beta^s \approx \int_0^{\infty} s^{\lambda+1} \exp[s \ln \beta]ds = \Gamma[\lambda + 2](\ln \beta)^{\lambda+2}$ with $\beta \approx 1 - z$ and $z \ll 1$ assuming on average large aggregation numbers.

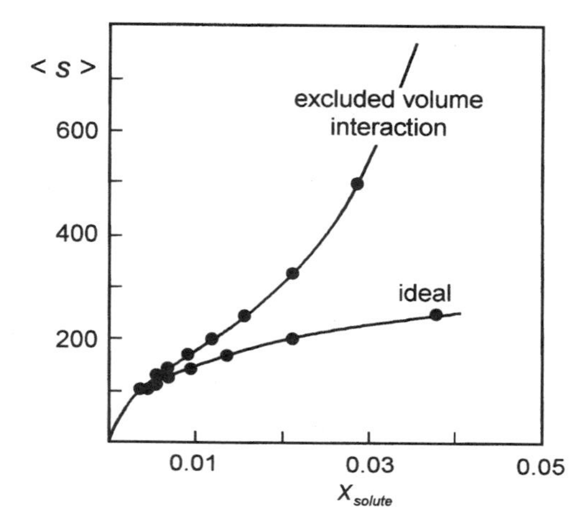

Figure 5 Mean aggregation number $\langle s \rangle$ versus solute concentration x_{solute} for polydisperse rodlike aggregates. (Adapted from Ref. 44.) Ideal means that no intermicellar interactions are included. The excluded volume interaction is calculated based on the assumption that the aggregates are spherocylinders using the so-called ''y-expansion.''

occurrence of the maximum, i.e., the decrease of $\langle s \rangle$ with increasing x_{solute} may be explained by the presence of attractive interactions between the aggregates, e.g., by the addition of a mean field term $\ln \gamma_s^{\text{attract}} = -(1/T)g(x_{\text{solute}})$ with $g > 0$ to $\ln \gamma_s$. An in-depth discussion, however, requires a detailed model of the functional dependence $g(x_{\text{solute}})$, i.e., an exact knowledge of the interactions. Again we stress that our present discussion is based on assuming ''reasonable'' approximations to Φ_s, because it is prohibitively complicated to calculate $\bar{\mu} - \bar{\mu}_s/s$ exactly for any real molecular system. Another simple model of a reversibly assembling linear aggregate is that of monomers at positions $\boldsymbol{r}_i$ subject to the constraints $\langle \boldsymbol{l}_1^2 \rangle = \langle \boldsymbol{l}_{s-1}^2 \rangle = l^2$, $\langle \sum_{i=2}^{s-2} \boldsymbol{l}_i^2 \rangle = (s-3)l^2$, and $\langle \sum_{i=1}^{s-2} \boldsymbol{l}_i \cdot \boldsymbol{l}_{i+1} \rangle = (s-2)l^2 t$, where $\boldsymbol{l}_i = \boldsymbol{r}_i - \boldsymbol{r}_{i-1}$, and l is a constant [47,48]. The quantity t controls the stiffness; i.e., $t = 0$ corresponds to wormlike flexibility, whereas $t = 1$ creates a stiff chain. Unfortunately, the nonlocal form of the constraints reduces the transparency of the t-dependence. Nevertheless, the resulting aggregate size distribution is $x_s \sim sx^s \exp[(E_{sc}/k_B T)(s-1)](1 - t^{2)3(s-2)/2})$ for $s \geq 2$, where E_{sc} is a bond scission energy analogous to the above quantity Φ_o. Thus, this model introduces an additional intra-aggregate interaction that results in the t-dependent factor even in the isotropic phase.* For $t = 0$, i.e., wormlike aggregates, this factor

* Note that Eq. (9) does not come into play here because it describes the loss of orientational entropy of a persistent flexible chain due to a confining (anisotropic) environment, which is missing in the present case.

is unity and the above ideal linear aggregate model is obtained. For $t > 0$, i.e., with increasing stiffness, a strong suppression of the larger aggregates results; i.e., $\langle s \rangle \sim (1 - t^2)^{3/2}(x_{\text{solute}} \exp[E_{sc}/k_B T])^{1/2}$ and $\sigma \sim \langle s \rangle$ for large $\langle s \rangle$. It is worth noting that the nonlocal form of the above constraints is motivated by the desire to obtain a simple analytic partition function—at the expense of unphysical behavior in the limit $t \to 0$. For instance, for $s = 3$ it is easy to see that $t = 1$ implies $\mathbf{l}_1 = \mathbf{l}_2$, which means complete internal rigidity of the trimer. The question is therefore how much of the pronounced coupling between aggregate flexibility and aggregate size predicted by this model can be observed in real experimental systems.

Another model of the aggregate interior takes an opposite viewpoint. It assumes shape-persistent aggregates whose (single-aggregate) partition function is approximated by the product $Q_s = q_t(s)q_r(s)q_v(s)$; i.e., the aggregate is a ''solid'' possessing translational (t), rotational (r), and vibrational (v) degrees of freedom. Each term contributes an s-dependence, and thus couples aggregate size and shape. Consider, for example, the rotational partition function $q_r(s) \propto (I_A I_B I_C)^{1/2}$, where the I are the three main moments of inertia of the aggregate (e.g., [49]). For cylinders we obtain $q_r(s) \propto s^{7/2}$ (i.e., $I_A \sim I_B \propto ss^2$ and $I_C \propto s$) whereas for spheres we have $q_r(s) \propto s^{5/2}$. Thus $\bar{\mu}_s$ now contains a contribution $\propto - \ln q_r(s)$. In general terms $\sim -\lambda \ln s$ in $\bar{\mu}_s$ yield a mean aggregate size $\langle s \rangle \sim (x_{\text{solute}} \exp[\Phi_o])^{1/(2+\lambda)}$ (in the ideal solution) and $\sigma \approx \langle s \rangle / \sqrt{2 + \lambda}$ reducing the 1/2-power in the above $\sqrt{x_{\text{solute}}}$-growth behavior.* We do not want to expand this point, and rather refer the reader to the more extensive discussion in Ref. 42.

C. Self-Assembly and Liquid-Crystalline Ordering

The same mechanisms that lead to liquid-crystalline phases in inert lyotropic systems are also responsible for liquid crystallinity in reversibly assembling systems. Here the new and interesting aspect is the coupling between the size and shape distribution of the aggregates and the formation of ordered phases. Early extensive studies were carried out by Gelbart and coworkers, using continuum models along the lines of the Onsager approach [50–52] (for a more complete listing see [42]), and by Herzfeld and coworkers, using lattice models [38,53,54] (for a more complete listing see [55,56]). A critical discussion of the present state theories can be found in [57].

1. Continuum Models of the Onsager Type

The development of continuum models for the structural phase behavior of reversible assembly largely parallels the Onsager approach to inert lyotropic sys-

---

* Again evaluating the summations via the above integral approximation.

tems and its extensions as discussed above. Inserting the activity coefficient according to Eq. (12) into Eq. (14) yields

$$x_s(\Omega) = \exp\left[(\Phi_s + \ln[x\gamma_1])s + \ln s - 2\rho_{\text{total}} \sum_{s'} \int_{\Omega'} \frac{d\Omega'}{4\pi} \frac{x_{s'}(\Omega')}{s'} B_{s,s'}(\Omega, \Omega') + \cdots\right] \tag{20}$$

where $x_s(\Omega) = x_s f_s'(\Omega)$, and $x\gamma_1$ is the activity of the free monomers. This extension of Eq. (17), which is analogous to Eq. (4) for inert lyotropic systems, can be solved numerically. Figure 6 shows a plot of x_s comparing the "typical" size distributions on the isotropic and the nematic side of the coexistence region for a system forming spherocylindrical micelles [50]. The calculation is based on Eq. (20) assuming a detailed model of the micelle interior along the lines of the last paragraph of Section II.B. Note that the general behavior, i.e., the shift of s_M to higher s-values and the corresponding broadening of the distribution, is in accord with our above simple model calculations (neglecting the effects of anisotropy, of course).

Calculations based on the full size distribution, $x_s(\Omega)$, are tedious. A number of authors therefore replace $x_s(\Omega)$ by $x_L(\Omega)$, where L is a mean aggregate length. Using this monodispersity assumption in conjunction with Onsager's trial function approach Odijk [58] has investigated the effect of micellar flexibility on the isotropic-to-nematic phase transition in solutions of linear aggregates. Φ_L is modeled in terms of the simple $d = 1$-model discussed above. In addition

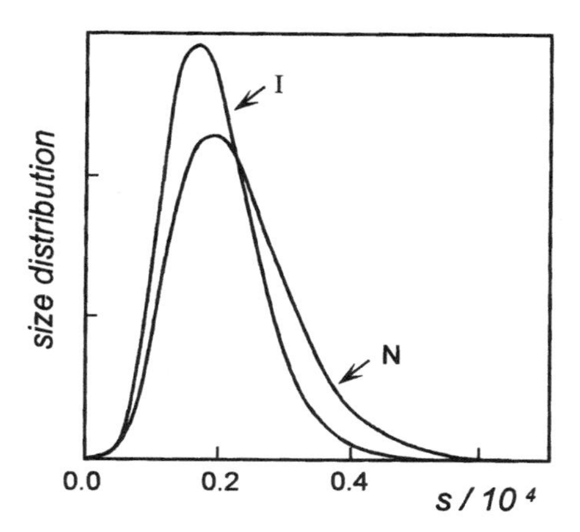

Figure 6 Aggregation number distribution for rodlike aggregates at isotropic-nematic coexistence. (From Ref. 50.)

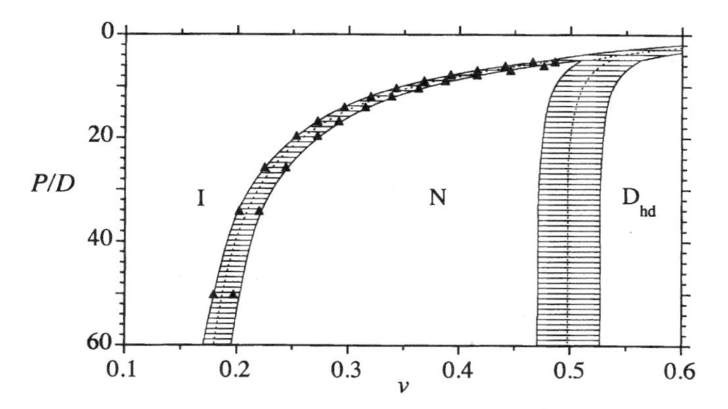

Figure 7 Theoretical phase diagram of a system of monodisperse self-assembling linear aggregates. (From Ref. 61.) I: isotropic phase; N: nematic phase; D_{hd}: hexagonal columnar phase. Here P is the persistence length, D is the aggregate diameter, and v is the solute volume fraction. Shaded areas indicate phase coexistence. The dotted lines mark the crossover of the free energies calculated according to Eq. (8). The triangles are experimental coexistence volume fractions, which here were fitted by adjusting certain model parameters.

Odijk replaces the ln s-term in Eq. (20) by 6 ln s, where the additional 5 ln s are due to translational (3/2) and rotational (7/2) degrees of freedom of the rodlike L-polymers (cf. the above discussion of the intramicellar partition function Q_s). Finally, the orientational confinement of the persistent flexible polymers is included via Eq. (9). The main result of this investigation is the strong inhibition of growth of flexible rodlike micelles in the anisotropic phase. This calculation was extended by Hentschke [59] taking the possibility of hexagonal ordering* at high concentrations via Eq. (8) into account. The main result of this calculation is the possible existence of an isotropic-nematic-hexagonal columnar triple point in the P-x_{solute}-plane; i.e., the reduction of the persistence length P suppresses the nematic phase, which intervenes between the isotropic and the columnar phase, as shown in Fig. 7. Note that compared to the (metastable) nematic phase under identical conditions the aggregate ends are more tightly packed in the columnar phase, whereas the average side-by-side separation is increased. The resulting reduction of undulatory confinement favors the columnar phase. On a qualitative level this prediction is in accord with experimental observations (cf. [6]). A triple point of this type is found for instance in the phase diagram of a stacking tripheny-

* A discussion of micellar growth in hexagonal phases can be found in Ref. 60.

lene derivative, which was studied in detail by Boden and coworkers [61] (cf. Fig. 4 in Chapter 1, this book). One problem of course is that temperature does not appear explicitly in the model. It enters indirectly via the (unknown) T-dependence of P and Φ_o.* The model of Ref. 59 was refined by van der School and Cates [62,63] taking polydispersity into account. Qualitatively, however, the results remain the same.

The functional dependence of aggregate size on alignment and flexibility, within the above simple polymeric aggregate model, can be understood roughly as follows. Rewriting Eq. (20) yields

$$x_s = \exp[-\ln f'_s(\Omega) + [\cdots]] = \int_\Omega \frac{d\Omega}{4\pi} f'_s(\Omega)\exp[-\ln f'_s(\Omega) + [\cdots]] \quad (21)$$

where $[\cdots]$ denotes the term in corresponding brackets in Eq. (20). Using the cumulant expansion, $\langle e^A\rangle = e^{\langle A\rangle}(1 + \sigma_A^2/(2\langle A\rangle) + \cdots)$, we may write

$$x_s \approx \exp\left[-\sigma_s(f'_s) + \int_\Omega \frac{d\Omega}{4\pi} f'_s(\Omega)[\cdots]\right] \quad (22)$$

To include flexibility we replace $\sigma_s(f'_s)$ by Eq. (9). In addition it is useful to replace f'_s by the above trial function and to assume large α (strong alignment). The resulting leading behavior in α is†

$$\sigma_s = \begin{cases} \ln\alpha - 1 + \dfrac{L_s}{6P}\alpha + \cdots & L_s/P \ll 1 \\ \dfrac{L_s}{4P}\alpha + \ln\left[\dfrac{\alpha}{4}\right] + \cdots & L_s/P \gg 1 \end{cases} \quad (23)$$

Correspondingly, $2\rho_{\text{total}} \sum_{s'} (x_{s'}/s') \int_{\Omega,\Omega'} (d\Omega/4\pi)(d\Omega'/4\pi) f'_s(\Omega) f'_{s'}(\Omega') B_{s,s'}(\Omega, \Omega') \approx (b_1 s\alpha^{-1/2} + b_2)x_{\text{solute}}$, using the same approximations as in the above isotropic case,‡ where b_1 and b_2 are positive constants. Analogous to the above isotropic case we thus obtain $\langle s\rangle \sim (x_{\text{solute}} \exp[\Phi_o + b_2 x_{\text{solute}} + \ln\alpha - 1])^{1/2}$ in the limit of slightly flexible aggregates and $\langle s\rangle \sim (x_{\text{solute}} \exp[\Phi_o + b_2 x_{\text{solute}} + \ln[\alpha/4]])^{1/2}$ in the limit of wormlike aggregates. Because α increases upon in-

---

* For the above triphenylene-derivative the T-dependence of P and Φ_o was studied in a computer simulation [46], which showed a strong decrease of both quantities with increasing temperature ($280K \leq T \leq 300K$). For an experimental study of $P(T)$ see [6].

† Assuming that all s have the same orientation distribution.

‡ Here $\int_{\Omega,\Omega'} d\Omega/(4\pi)d\Omega'/(4\pi)\,|\sin(\Omega, \Omega')| \approx (\pi/\alpha)^{1/2}$ for large α (e.g., [7]).

creasing alignment, we find that alignment progressively increases the aggregate size, where the effect of flexibility enters indirectly through α. Notice that the minimization of the free energy [cf. Eq. (2)] with respect to α, i.e., $d(\sigma_s + \text{const}\ Lx_{\text{solute}}\alpha^{-1/2})/d\alpha = 0$, yields $\alpha \sim (Lx_{\text{solute}})^2$ for stiff aggregates, where L is the mean aggregate size, and $\alpha \sim (Px_{\text{solute}})^{2/3}$ for wormlike aggregates.

We remark that a similar isotropic-nematic-columnar triple point as mentioned above is also obtained in the Φ-v-plane of the phase diagram for hard, rigid, polydisperse, self-assembling rods [66]. The same reference also studies the phase behavior of disclike aggregates. Both phase diagrams are shown in Fig. 8. As indicated, Φ is expected to decrease with increasing temperature. Notice the qualitative agreement of the size distributions at the isotropic-to-nematic transition in the rodlike system with the result of Fig. 6.

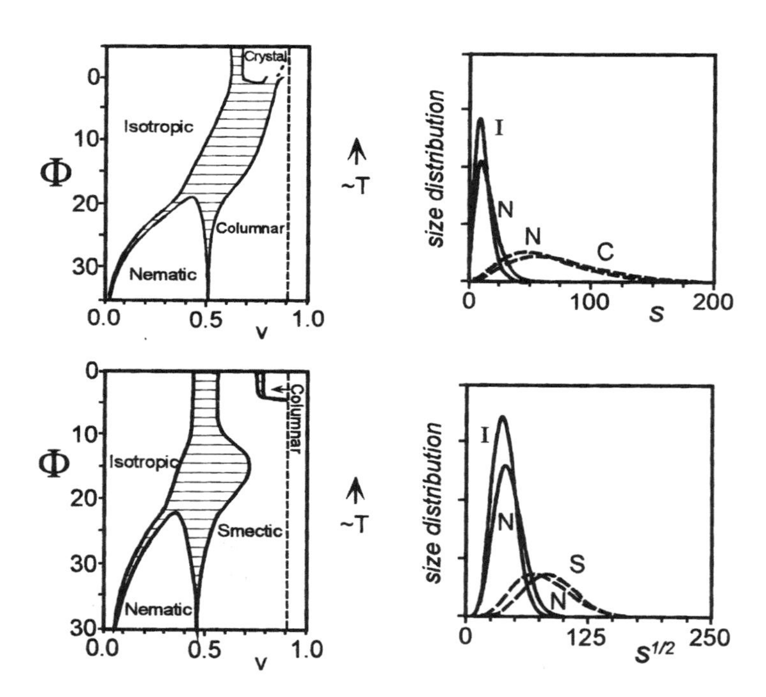

Figure 8 Left: Φ-v-phase diagrams of self-assembling rodlike (top) and disclike (bottom) systems. (Adapted from Ref. 66.) Shaded areas indicate phase coexistence. The vertical dashed lines mark the respective close packing limits. Right: corresponding orientationally averaged aggregation number distributions for $\Phi = 25$ (top) and $\Phi = 30$ (bottom) at the indicated transition boundaries (I: isotropic; N: nematic; C: columnar; S: smectic).

Here we have considered effective one-component systems only. Lyotropic ternary mixtures containing self-assembling proteins forming rodlike aggregates, nonaggregating protein, and solvent have been considered by Herzfeld and coworkers [64] (and references therein). The technical approach is the same as we have discussed above. The additional computational effort, however, is considerable. Another example of a binary mixture containing polymer and surfactant that form complexes exhibiting lyotropic behavior is discussed in Ref. 65 in terms of scaling arguments.

2. Lattice Models

Lattice calculations of liquid-crystallinity in self-assembling systems are certainly less realistic than continuum calculations. This is due to the limited orientational freedom, the crude representation of molecular shape and thus molecular interaction, as well as the inferior description of excluded volume at high concentrations. In turn these simplifications allow easier handling of highly polydisperse aggregate populations if necessary. Herzfeld et al. were the first who extensively applied lattice models to the equilibrium phase behavior of systems forming simple convex aggregates [38,53,54]. The aggregates, which consist of cubic monomers, are rectangular parallelepipeds with independently self-adjustable edges. This allows the competition between different particle shapes; i.e., rod- and platelike aggregates may exist simultaneously. The shape of the aggregates at fixed solute concentration is controlled by monomer–monomer contact-free energies analogous to Φ_o, which are different in the bulk, along the edges, and at the corners of the aggregates. A compilation of the results obtained with this model can be found in Ref. 67. Figure 9 shows an example of a lattice calculation applied to the alignment transition observed in sickle cell hemoglobin [68] (This topic was recently revisited in Ref. 69. The globular protein monomers (typical dimension $\approx$60 Å) reversibly self-assemble into protein polymers,* i.e., long, rigid rods (typical diameter 180 Å to 240 Å and lengths in the μm range), which spontaneously align at high concentrations. The alignment of these extremely stiff protein polymers† here is indicated by a temperature-dependent decrease of the osmotic pressure. In this example association and alignment are tightly coupled. In order to obtain agreement between the model and the experiment it was necessary to allow the inclusion of water into the aggregates as well as their concentration-dependent dehydration.

---

* Many general aspects of protein self-assembly are discussed in great detail in Ref. 70.

† Assuming continuous bend elastic rods one can show that $P \sim \varepsilon D^4/(k_B T)$, where ε is the rod's bending modulus and D is its diameter (for a discussion see [71]). Thus, the stiffness increases rapidly with increasing rod diameter.

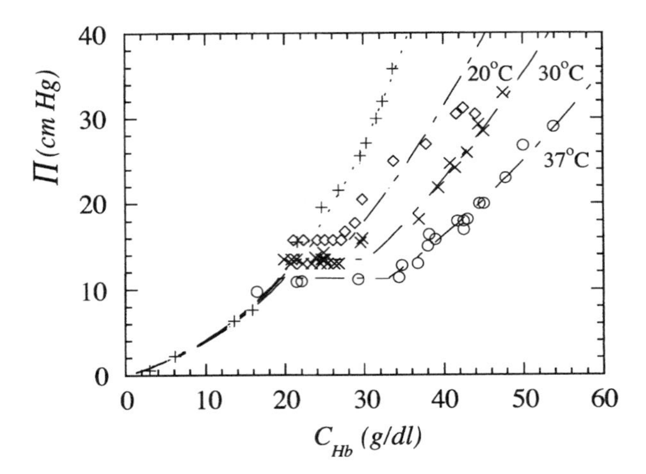

Figure 9 Osmotic pressure, Π, versus concentration, C_{Hb}, for deoxygenated hemoglobin. (From Ref. 68.) Symbols: experimental data for normal hemoglobin (plusses) and sickle hemoglobin [37°C (circles), 30°C (crosses), and 20°C (diamonds)]. Dotted line: theoretical result for normal hemoglobin (T = 37°C). Dashed-dotted lines: theoretical fits for highly concentrated deoxygenated sickle cell hemoglobin.

III. COMPUTER SIMULATIONS

Application of computer modeling techniques to the study of supramolecular phenomena is appealing, because in contrast to the above analytic approaches it is possible, at least in principle, to study systems of arbitrary complexity. Despite the advances in processor technology and the sophistication of current algorithms, however, the accessibility of molecular phenomena is still limited to fairly small system sizes and short times. Molecular modeling ranges from quantum chemical ab initio calculations to phenomenological force field models. Corresponding linear dimensions and time scales for atomistic models vary from 10 Å and 10^{-12} s to 100 Å and 10^{-8} s on current workstation computers. In the case of force field models with additive nonbonded interactions, the simplest of the atomistic models, the computational effort increases in proportion to the number of interaction sites M, for large M and short-range interaction potentials. For long-range interaction potentials the effort increases in proportion to $M \ln M$. Consider, for example, the system composed of the triphenylene derivative in aqueous solution of Fig. 4 in Chapter 1, which we mention also in the context of Fig. 7. If we want to study 10^4 monomers (e.g., 100 100-mers) in aqueous solution at volume fractions around 0.2 to 0.3, which is close to the isotropic-to-nematic transition at T = 290 K, we need to consider roughly 10 million interaction sites on the

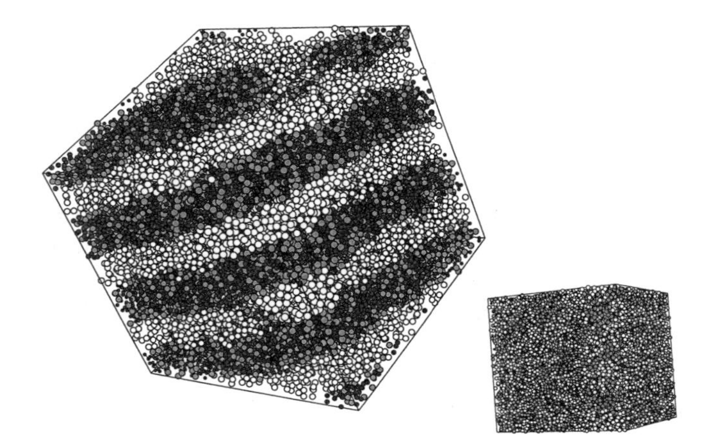

Figure 10 Molecular dynamics simulation of a surfactant system using a coarse-grained model. A surfactant molecule consists of four head (light grey) and four tail sites (white); a water molecule is represented by one site (dark grey). The inset on the lower right shows the random distribution of the molecules at the beginning of the simulation.

level of the simplest atomistic model (cf. Ref. 46). For comparison, Fig. 10 shows a lamellar morphology found in a simulation of a simple coarse-grained surfactant containing $M = 3 \cdot 10^4$ sites corresponding to 3350 surfactant molecules (8 sites each) and 3200 "water" molecules (1 site each). The simulation, which was started from the random mixture shown on the lower right, took approximately 400 CPU hours on a DEC ALPHA 433 MHz-workstation to complete.

Even though more abstract or "coarse-grained" models allow significantly larger systems and larger time scales, mapping a particular molecular architecture onto a coarse-grained model, performing the simulation, and relating the results to the original atomistic model is an as yet unresolved challenge. In the following we present a brief illustration of both approaches in terms of examples, which thus far have mainly focused on micellar assemblies. Recent review articles of Larson [72] and Karaborni and Smit [73] give some insight into the history and progress of the modeling of self-assembling systems. Readers interested in the details of the molecular modeling methodology are referred to Refs. 74–76.

A. Molecular Models

On the atomically realistic level, the molecular dynamics method, i.e., the numerical solution of Newton's equations of motion, often is the method of choice. Nevertheless, tracking the individual motion of the atoms requires time steps on the order of 1 fs limiting the overall time window to a few nanoseconds on a workstation. Therefore, the self-assembly process itself and equilibrium fluctuations within the aggregate distribution usually cannot be simulated using molecular dynamics. Instead most atomistic simulation studies start from educated guesses of the equilibrium structure of either a single small aggregate or a cutout section of a continuous phase morphology like a section of a lamella. The authors then focus on the structure and dynamics of the aggregate–solvent interface and on fast processes within the molecular assembly. Clearly the quality of such work strongly depends on the initial guess. A typical example of this approach is Ref. 77, where the authors study the sodium-decanoate/decanol/water system. Their membrane model uses an atomistic description for the amphiphile, except that CH_2-groups are combined into effective or united atoms. The employed force field encompasses valence contributions (due to bond and valence angle deformations as well as torsional rotations) and nonbonded interactions (Coulomb and Lennard–Jones). The system in this example contains 52 decanoate ions and 76 decanol molecules forming a bilayer in contact with 526 water molecules. The simulation time was 80 ps (after about 100 ps of preliminary equilibration). Obviously, the narrow time window does not allow the self-assembly of the amphiphiles, and the bilayer was initially chosen as the starting configuration. In this example the authors analyze the distribution of the different components in the system as well as certain dynamic properties such as the diffusion constants of the components. A similar 275 ps molecular dynamics study in Ref. 78 focuses on a single 30-monomer n-decyltrimethylammonium micelle in water (2166 molecules). Here the main objective of the authors is to analyze the shape and structure of the micelle in addition to the micelle–water interface.

A more recent example is a simulation by Griffiths and Heyes [79], who observe the self-assembly of inverse micelles with small aggregation numbers of six or eight. Based on an atomistic description of oil-soluble sulfonate and phenate micelles (both types of surfactants are important commercial detergents) containing a calcium carbonate core, they examine the influence of the surfactant type on the assembled microstructure. Most of the quoted simulations are carried out in vacuum, but the effect of a hydrophilic surfactant is also considered. From their simulations the authors conclude that the shape and the properties of micelles are strongly determined by the geometry of the building blocks. They find micellar structures that are both relatively insensitive to temperature changes and quite rigid showing negligible fluctuations with time.

Another recent molecular dynamics simulation [80] investigates the formation of liquid crystal phases using atomistic potentials. Here the authors model the mesogen, 4,4′-di-n-pentyl-bibicyclo [2,2,2]octane, using the united atom approach (CH_3, CH_2, and CH-groups are condensed into effective atoms). To the best of our knowledge, this is the first time that the spontaneous growth of a (nematic) mesophase starting from an isotropic state using a fully atomistic model

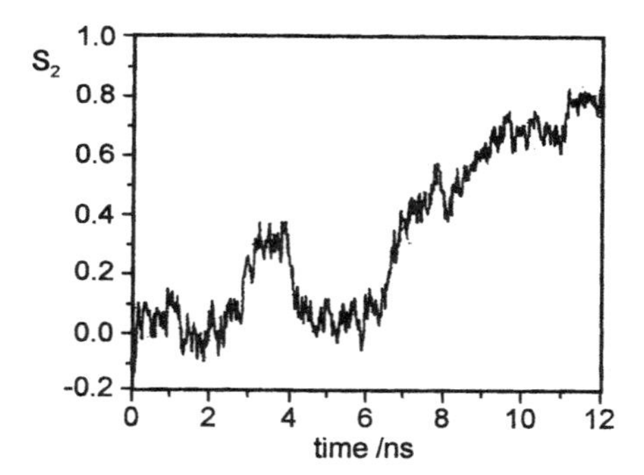

Figure 11 Spontaneous growth of the orientational order parameter S_2 for 64 4,4′-di-n-pentyl-bibicyclo[2,2,2]octane molecules at 300 K. (From Ref. 80.)

is reported. Figure 11 shows the time evolution of the orientation order parameter $S_2 = \langle P_2(\cos(\phi))\rangle$, where ϕ is the angle between a molecular axis and the average molecular orientation within the system defining the director. Notice that even though the increase of the order parameter is obvious its equilibrium value under the given conditions could not be established within the simulation time of 12 ns. McBride et al., who model systems consisting of up to 125 mesogen molecules for simulation times of up to 12 ns, analyze the molecular structure in dependence of the observed phase and the dynamics of the molecular motion.

B. Coarse-Grained Models

A nice example of coarse-grained modeling is documented in a series of papers by Larson (see [81] and references therein), who carries out a systematic study of the phase behavior of single-tail surfactants, H_iT_j, with $4 \leq i, j \leq 16$ (note the analogy to the system in Fig. 10). H denotes the head-group and T the tail of an amphiphile. In his studies Larson uses a Monte Carlo lattice model with each head or tail site occupying one site on a cubic lattice. Likewise the solvent (water, oil) is modeled as a single site. At high concentrations he finds a sequence of liquid-crystalline phases as illustrated in Fig. 12, like the hexagonal cylindrical, the bicontinuous gyroid, and the lamellar phase. The general dependence of the calculated phase diagrams on amphiphile volume fraction, temperature, and amphiphile architecture (head-to-tail ratio) are found to be in good accord with corresponding experimental surfactant systems. Of course, effects tied to the details of the molecular architecture are not captured by the lattice model.

An illustrative example of a molecular dynamics simulation using a coarse-grained model is the study of the self-assembly of a ‘‘gemini’’ surfactant by

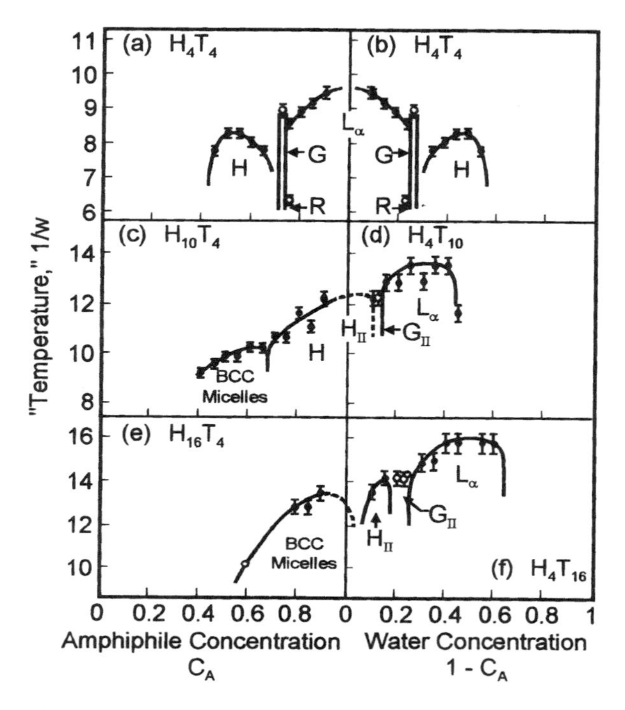

Figure 12 Phase diagrams of (a) H_4T_4, (c) $H_{10}T_4$, (e) $H_{16}T_4$, and the complementary surfactants (b) H_4T_4, (d) H_4T_{10}, and (f) H_4T_{16} in "water." H and L_α are hexagonal and lamellar, G and R are cubic gyroid and rhombohedral-like mesh intermediate phases. BCC-Micelles is a body-centered cubic packing of spherical micelles. The index II indicates inverse symmetry. (From Ref. 81.)

Karaborni et al. [82]. A gemini surfactant has two hydrophobic chains and two hydrophilic head-groups linked by a spacer. This study examines the influence of the surfactant concentration and the spacer length on the aggregation behavior. The authors find that in the concentration regime where single-chain surfactants form spherical micelles, gemini surfactants with a small spacer form linear threadlike micelles. With increasing concentration, their shape changes to treelike aggregates. For larger spacers, the formation of spheroids and branched treelike structures is observed. These simulations are performed under NVT-conditions, i.e., constant number of monomers, constant volume, and constant temperature, with $M = 3.2 \cdot 10^4$ sites in total. We should again emphasize that it is the use of an idealized model that permits the self-assembly of such threadlike or tree morphologies. Another example addressing self-assembly of simple single-tail

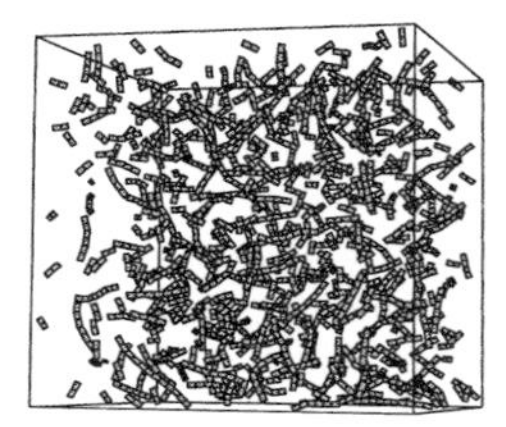
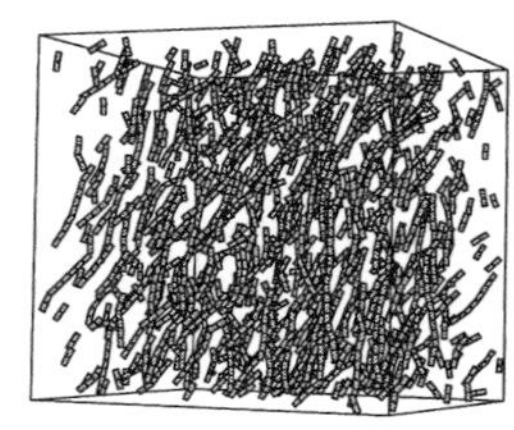
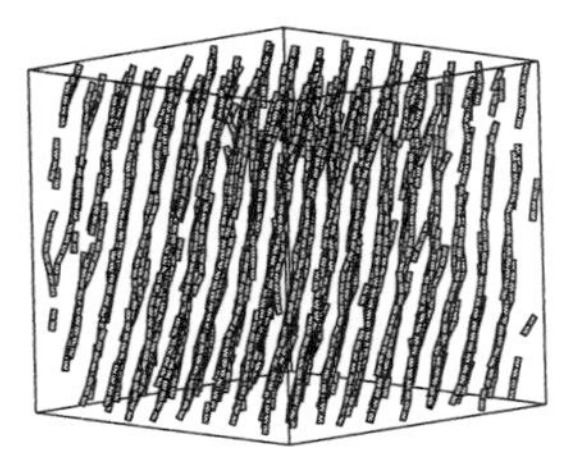

Figure 13 Snapshots of a ''living polymer'' system at different densities. Left: $\rho^* = 0.34$ (isotropic ordering). Middle: $\rho^* = 0.47$ (nematic ordering). Right: $\rho^* = 0.67$ (hexagonal ordering). The potential parameters are $\sigma = 1$, $\varepsilon = 1$, $\mu = 3.3$, and $\nu = 2$.

surfactants is discussed in Ref. 83. Here the authors investigate the effects of surfactant chain length and concentration on the micelle formation using simple model interactions and the molecular dynamics approach.

The final example, which we discuss in some detail, is a molecular dynamics study of the self-assembly and phase behavior of reversibly aggregating linear model polymers.* Here we consider the coupling between the polymer length distribution and orientational ordering as a function of monomer concentration including the influence of aggregate flexibility (cf. above). In this model, a detailed description is given elsewhere [87]; monomers interact via a Lennard–Jones-like potential with anisotropic attraction: i.e.,

$$u_{ij} = 4\varepsilon_{ij}\left[\left(\frac{\sigma_{ij}}{r_{ij}}\right)^{12} - \mu\left(\frac{(\boldsymbol{n}_i \cdot \boldsymbol{r}_{ij})(\boldsymbol{n}_j \cdot \boldsymbol{r}_{ij})}{r_{ij}^2}\right)^{\nu}\left(\frac{\sigma_{ij}}{r_{ij}}\right)^{6}\right] \tag{24}$$

where ε_{ij} and σ_{ij} are the usual Lennard–Jones parameters. $\boldsymbol{n}_i$ is a unit vector assigning an orientation to monomer i. Note that the factor $(\cdots)^{\nu}$ is unity if both $\boldsymbol{n}_i$ and $\boldsymbol{n}_j$ are parallel to $\boldsymbol{r}_{ij}$, the vector connecting the monomers (or sites) i and j. If either $\boldsymbol{n}_i$ or $\boldsymbol{n}_j$ or both are perpendicular to $\boldsymbol{r}_{ij}$ then $(\cdots)^{\nu}$ is zero. If either $\boldsymbol{n}_i$ or $\boldsymbol{n}_j$ is antiparallel to $\boldsymbol{r}_{ij}$ then $(\cdots)^{\nu}$ depends on ν; i.e., $(\cdots)^{\nu} = -1$ for ν odd and $(\cdots)^{\nu} = 1$ for ν even. Notice that the magnitude of ν controls the angular width of the attraction, whereas μ controls the strength of the attraction. Both parameters thus control the stiffness, i.e., the persistence length, of the polymer aggregates. Figure 13 shows simulation snapshots of the system at three different densities (here * indicates Lennard–Jones units). Below a certain monomer density the system is isotropic, whereas at higher densities the aggregates are aligned,

* Recently, Rouault and coworkers have carried out a series of Monte Carlo simulations of living polymers using abstract models. The authors give special attention to the influence of flexibility, constraining geometries, and concentration on the polymer size distribution [84–86].

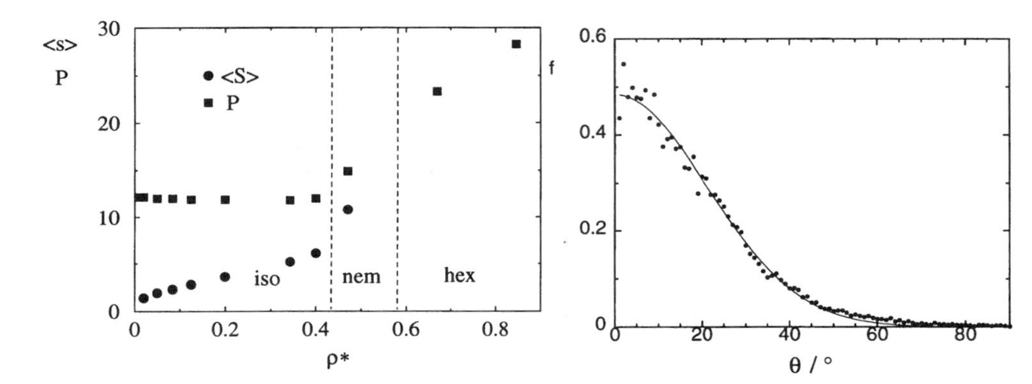

Figure 14 Left: average aggregation number $\langle s \rangle$ and persistence length P versus monomer number density ρ^*. The density regions of the different phases are indicated by dotted lines (see Fig. 13 for the potential parameters). Right: average angular distribution of the $\mathbf{n}_i$ fitted via const $\cdot$ $\exp(-\alpha\theta^2/2)$. Here $\rho^* = 0.467$ and $\alpha = 7.5$.

and the average aggregation number increases. Further increase of the density drives the system into a hexagonal columnar phase.

The mean aggregation number $\langle s \rangle$ together with the persistence length P as a function of number density is shown in Fig. 14. Here, the persistence length is calculated via $P = -1/(\ln \langle \boldsymbol{n}_i \cdot \boldsymbol{n}_{i+1} \rangle)$. Note that in this definition the persistence length depends on density; i.e., P is increased by the interaction with neighboring aggregates.* Also shown in Fig. 14 is the nematic orientation distribution obtained at $\rho^* = 0.47$. Increasing the density from $\rho^* = 0.47$ to $\rho^* = 0.67$ results in a strong growth of the aggregates, with average aggregation numbers exceeding the simulation box length. Therefore the analysis of the polymer length distribution is not possible for these densities. Our results for the dependence of the mean aggregation number on the density agree with the behavior obtained by the theoretical model for linear aggregation that is discussed in the context of Fig. 5.

IV. CONCLUSION

Our intention was to illustrate the current level of understanding of self-assembly and lyotropic ordering based on statistical mechanical models. Clearly, analytical

* Strictly speaking P measures the stiffness of the unperturbed aggregates. Apparently this is the case for $p^* < 0.4$.

models can describe simple systems with likewise simple aggregate architectures, and thus may guide our understanding of the more complex systems; i.e., the simple models do provide important insight regarding the relation between molecular interaction and aggregate phase behavior as it couples with size and shape of the aggregate population. However, there remains a long way to go to quantitatively predict the properties and functions of complex multicomponent supramolecular assemblies in biological systems or in technical applications. Using a number of examples we have illustrated the current state of computer modeling as it applies to this field. Here we are rapidly approaching size and time scales, which will allow us to understand the mesoscopic and macroscopic behavior of supramolecular assemblies based on the molecular level even for complicated systems.

ACKNOWLEDGMENTS

We are grateful to Professor A. Ciferri and Dr. Y. Rouault for a number of useful discussions and comments.

REFERENCES

1. H Ringsdorf, B Schlarb, J Venzmer. Angew Chem Int Ed Engl 27:113–158, 1988.
2. J-M Lehn. Supramolecular Chemistry. New York: VCH, 1995.
3. A Ciferri. Macromol Chem Phys 195:457–461, 1994.
4. A Ciferri. TRIP 5:142–145, 1997.
5. A Ciferri. Supramolecular Liquid Crystallinity as a Mechanism of Supramolecular Polymerization. Liquid Crystals, 26:489–494, 1999.
6. JR Mishic, RJ Nash, MR Fisch. Langmuir 6:915–919, 1990.
7. T Odijk. Macromolecules 19:2313–2329, 1986.
8. GJ Vroege, HNW Lekkerkerker. Rep Prog Phys 55:1241–1309, 1992.
9. MA Cotter, DC Walker. Phys Rev A 18:2669–2675, 1978.
10. J Herzfeld, AE Berger, JW Wingate. Macromolecules 17:1718–1723, 1984.
11. L Onsager. Ann NY Acad Sci 51:988–1014, 1949.
12. AN Semenov, AR Khokhlov. Sov Phys Usp 31:627–659, 1989.
13. TJ Sluckin. Liquid Crystals 6:111–131, 1989.
14. MA Cotter. In: GR Luckhurst, GW Gray, eds. The Molecular Physics of Liquid Crystals. London: Academic Press, 1979, Chapter 7.
15. D Frenkel. In: JP Hansen, D Levesque, J Zinn-Justin, eds. Les Houches 1989—Session LI (Part II): Liquids, freezing and glass transition. Amsterdam: Elsevier Science, 1991, pp 689–762.
16. SD Lee. J Chem Phys 87:4972–4974, 1987.

17. H Nakano, M Hattori. Prog Theor Phys 49:1752–1754, 1973.
18. M Hosino, H Nakano, H Kimura. J Phys Soc Japan 51:741–748, 1982.
19. MP Taylor, R Hentschke, J Herzfeld. Phys Rev Lett 62:800–803, 1989.
20. JO Hirschfelder, CF Curtiss, RB Bird. Molecular Theory of Gases and Liquids, New York: Wiley, 1964.
21. A Stroobant, HNW Lekkerkerker, D Frenkel. Phys Rev A 36:2929–2945, 1987.
22. JA Veerman, D Frenkel. Phys Rev A 43:4334–4343, 1991.
23. BJ Alder, TE Wainwright. J Chem Phys 27:1208–1209, 1957.
24. B Mulder. Macromol Symp 81:329–331, 1994.
25. H Yamakawa. Modern Theory of Polymer Solutions. New York: Harper & Row, 1971.
26. A Ciferri. In: A Ciferri, ed. Liquid Crystallinity in Polymers. New York: VCH, 1991, pp 209–260.
27. K Kubo, K Ogino. Mol Cryst Liq Cryst 53:207–228, 1979.
28. K Kubo. Mol Cryst Liq Cryst 74:71–87, 1981.
29. R Hentschke, J Herzfeld. Phys Rev A 44:1148–1155, 1991.
30. W Helfrich, W Harbich. Chem Scr 25:32–36, 1985.
31. D Sornette, N Ostrowsky. In: WM Gelbart, A Ben-Shaul, D Roux, eds. Micelles, Membranes, Microemulsions, and Monolayers. New York: Springer-Verlag, 1994.
32. R Podgornik, VA Parsegian. Macromolecules 23:2265–2269, 1990.
33. T Odijk. Biophys Chem 46:69–75, 1993.
34. PJ Flory. Proc R Soc London Ser A 234:73–103, 1956.
35. PJ Flory, G Ronca. Mol Cryst Liq Cryst 54:289–330, 1979.
36. AYu Grosberg, AR Khokhlov. Soc Sci Rev A Phys 8:147–258, 1987.
37. EA DiMarzio. J Chem Phys 35:658–669, 1961.
38. J Herzfeld. J Chem Phys 76:4185–4190, 1982.
39. JN Israelachvili. Intermolecular and Surface Forces. New York: Academic Press, 1992.
40. DF Evans, H Wennerström. The Colloidal Domain. New York: VCH, 1994.
41. B Jönsson, B Lindman, K Holmberg, B Kronberg. Surfactants and Polymers in Aqueous Solution. New York:Wiley, 1998.
42. A Ben-Shaul, WM Gelbart. In: WM Gelbart, A Ben-Shaul, D Roux, eds. Micelles, Membranes, Microemulsions, and Monolayers. New York: Springer-Verlag, 1994.
43. JM Biltz, MR Fisch. Langmuir 11:3595–3597, 1995.
44. WM Gelbart, A Ben-Shaul, WE McMullen, A Masters. J Phys Chem 88:861–866, 1984.
45. N Boden. In: WM Gelbart, A Ben-Shaul, D Roux, eds. Micelles, Membranes, Microemulsions, and Monolayers. New York: Springer-Verlag, 1994.
46. T Bast, R Hentschke. J Phys Chem 100:12162–12171, 1996.
47. RG Winkler, P Reineker, L Harnau. J Chem Phys 101:8119–8129, 1994.
48. W Carl, Y Rouault. Macromol Theory Simul 7:497–500, 1998.
49. TL Hill. An Introduction to Statistical Thermodynamics. New York: Dover, 1960.
50. WM McMullen, WM Gelbart, A Ben-Shaul. J Chem Phys 82:5616–5623, 1985.
51. WM McMullen, WM Gelbart, A Ben-Shaul. J Phys Chem 88:6649–6654, 1984.
52. WM Gelbart, WM McMullen, A Ben-Shaul. J Physique 46:1137–1144, 1985.
53. J Herzfeld, RW Briehl. Macromolecules 14:1209–1214, 1981.

54. J Herzfeld, MP Taylor. J Chem Phys 88:2780–2787, 1988.
55. MP Taylor, J Herzfeld. J Phys Condens Matter 5:2651–2678, 1993.
56. J Herzfeld. Acc Chem Res 29:31–37, 1996.
57. T Odijk. Current Opinions Colloid Interface Science 1:337–340, 1996.
58. T Odijk. J Physique 48:125–129, 1987.
59. R Hentschke. Liq Crystals 10:691–702, 1991.
60. P Mariani, LQ Amaral Phys Rev E 50:1678–1681, 1994.
61. R Hentschke, PJB Edwards, N Boden, R Bushby. Macromol Symp 81:361–367, 1994.
62. P van der Schoot, ME Cates. Europhys Lett 25:515–520, 1994.
63. P van der Schoot. J Chem Phys 104:1130–1139, 1996.
64. DT Kulp, J Herzfeld. Biophys Chem 57:93–102, 1995.
65. GH Fredrickson. Macromolecules 26:2825–2831, 1993.
66. MP Taylor, J Herzfeld. Phys Rev A 43:1892–1905, 1991.
67. MP Taylor. Statistical Mechanical Models of Liquid Crystalline Ordering in Reversibly Assembling Lyotropic systems. PhD thesis, Brandeis University, 1991.
68. R Hentschke, J Herzfeld. Phys Rev A 43:7019–7030, 1991.
69. J Han, J Herzfeld. Biopolymers 45:299–306, 1998.
70. F Oosawa, S Asakura. Thermodynamics of the Polymerization of Protein. New York: Academic Press, 1975.
71. J Helfrich, U Apel, R Hentschke. Macromolecules 27:472–482, 1994.
72. RG Larson. Current Opinion Colloid Interface Science 2:361–364, 1997.
73. S Karaborni, B Smit. Current Opinion Colloid Interface Science 1:411–415, 1996.
74. KB Lipkowitz, DB Boyd, eds. Reviews in Computational Chemistry. Weinheim: VCH, 1990.
75. P von Ragué Schleyer, ed. Encyclopedia of Computational Chemistry. New York: Wiley, 1988.
76. AR Leach. Molecular Modeling. Harlow: Addison-Wesley, 1996.
77. E Egberts, H Berendsen. J Chem Phys 89:3718-3731, 1988.
78. J Böcker, J Brickmann, P Bopp. J Phys Chem 98:712–717, 1994.
79. JA Griffiths, DM Heyes. Langmuir 12:2418–2424, 1996.
80. C McBride, MR Wilson, JAK Howard. Molecular Phys. 93:955–964, 1998.
81. RG Larson. J Phys II France 6:1441–1463, 1996.
82. S Karaborni, K Esselink, PAJ Hilbers, B Smit, J Karthäuser, NM van Os, R Zana. Science 266:254–256, 1994.
83. BJ Palmer, J Liu. Langmuir 12:746–753, 1996.
84. Y Rouault. Eur Phys J B 6:75–81, 1998.
85. Y Rouault, A Milchev. Macromol Theory Simul 6:1177–1190, 1997.
86. Y Rouault. Phys Rev E 58:6155–6157, 1998.
87. B Fodi, R Hentschke. 1999, (Submitted).

3

Polymeric vs. Monomeric Amphiphiles: Design Parameters

Avraham Halperin
UMR 5819 (CEA-CNRS-Université J. Fourier), SI3M, CEA–Grenoble, Grenoble, France

I. INTRODUCTION

Examples of polymeric amphiphiles include diblock copolymers and proteins. These two examples represent the two poles of this family, the simplest and the most complex. Intermediate cases include, among others, block copolymers of various architectures, graft or comb copolymers, and polysoaps. The aim of this chapter is to present a didactic, though incomplete, theoretical viewpoint of polymeric amphiphiles. In particular, we address two questions: what are the features that distinguish polymeric surfactants from monomeric, low molecular weight, amphiphiles and how is the behavior of polymeric amphiphiles related to their molecular architecture? As we discuss, some broad features are common to monomeric and polymeric surfactants. Both self-assemble into supramolecular structures such as micelles, lamellae, etc. In turn, these supramolecular aggregates can form ordered mesophases. However, the polymeric surfactants do exhibit qualitatively distinctive features with respect to their monomeric counterparts. These include the formation of physical gels and of distinctively polymeric mesophases. Certain polymeric amphiphiles can also undergo intrachain self-assembly that affects their configurations and elasticity. In any case, the underlying physics that controls the size and the shape of the self-assembled structures are significantly modified.

Polymeric surfactants constitute a currently emerging field in spite of their long history and wide range of practical applications. The synthesis of various polymeric surfactants was pioneered more than 50 years ago. Among their applications, two are especially important. Polymeric surfactants serve as colloidal stabilizers and viscosity modifiers in a great variety of industrial complex fluids such as lubricants, paints, and drilling muds. They are also used as structural materials, primarily as thermotropic elastomers and as compatilizers for polymer alloys. Some topics within this field, for example, diblock copolymers, constitute well-established areas and are studied intensively from both fundamental and practical points of view. Other members of this family, such as graft copolymers and polysoaps, have received much less attention. Furthermore, the different systems are studied individually, with little attention to their common features. This situation may be attributed, in part, to the multidisciplinary nature of this field. If one focuses only on synthetic polymeric surfactants, it is necessary to consider contributions originating from supramolecular chemistry [1], polymer science [2], and surface science [3]. Within polymer science, this field involves at least three specialities, i.e., block copolymers, associating polymers, and ionomers. Imperfect communications among the various communities has hindered the recognition of the common features and the development of the field as a whole.

This review makes no claims for completeness in the coverage of the subject or of the relevant literature. Such undertaking is ruled out by the combination of space limitations and the wide scope of the field. Further information and a wealth of references can be found in a number of reviews and treatises concerning block copolymers [2,4,5], polymeric surfactants [6,7], ionomers [8,9], polysoaps [1,10], and amphiphilic systems in general [11–14].

The body of this article consists of a systematic exploration of certain design parameters and their signatures and focuses on the structure of single micelles and their interactions. Some general considerations concerning the design parameters of monomeric and polymeric amphiphiles and the features that distinguish the two families are presented in Section II. In particular, it examines the relative role of translational entropy, configurational entropy, and interaction energies. Diblock copolymers may be considered as the polymeric analogues of shortchain, monomeric surfactants. Models of the self-assembly of these two types of amphiphiles are presented in Section III. The self-assembly of monomeric surfactants is described and compared to that of three different types of *AB* diblock copolymers. Triblock copolymers no longer have a counterpart among the ''classical amphiphiles.'' *ABA* triblock copolymers can form physical gels due to exchange interactions. *ABC* triblock copolymers form novel mesophases. The self-assembly of triblock copolymers is discussed in Section IV. Multiblock $(AB)_n$ block copolymers and polysoaps are capable of undergoing purely intrachain self-assembly. Their self-assembly and its effects on the configurations and elasticity of the chains are described in Section V. The current understanding of hydrophilic polymers that are water soluble because of hydrogen bonds is incomplete. The implications concerning waterborne polymeric aggregates are briefly discussed in Section VI. Nomenclature and taxonomy are not the main topic of this review. It is, however, useful to consider the position occupied by polymeric surfactants within the ''polymer kingdom.'' The partial overlap with surface active polymers and associating polymers is discussed in Section VII. Brushes of grafted, terminally anchored, chains are a recurring structural motif in aggregates of polymeric surfactants. A brief summary of brush theory is provided in the Appendix.

Before proceeding it is of use to comment on notation and terminology. The following discussion is concerned with the interplay of free energies and constraints that control the self-assembly of polymeric surfactants. For simplicity, numerical prefactors are ignored. The symbol $\approx$ is used to mean ''equal within a numerical prefactor'' and the symbol $\sim$ is used to signify ''proportional to.'' The term ''configuration''* is utilized to denote the spatial arrangement of the monomers within the chain [15].

II. GENERAL CONSIDERATIONS: MOLECULAR DESIGNS, ENTROPY, AND THE "COMBINATORIAL EFFECT"

Amphiphilic molecules [11–14] consist of mutually incompatible components. Since these components are chemically joined, complete segregation is impossible. It is replaced by various forms of microphase separation. These involve formation of segregated domains such that at least one of their dimensions is comparable to the molecular size. The domains are formed by spontaneous, thermodynamically driven aggregation of the amphiphiles. The process is thus often referred to as self-assembly. The resulting structures, micelles, lamellae, etc. can also form ordered mesophases. The microphase separation can take place in a solvent that selectively solubilizes one component or in a ''melt'' of neat amphiphiles. These characteristics are common to both polymeric and monomeric, low molecular weight amphiphiles. For the purposes of our discussion monomeric amphiphiles are defined, somewhat arbitrarily, as those consisting of $\sim 10^2$ atoms. Polymeric amphiphiles, on the other hand, can incorporate 10^6–10^8 atoms. The consequences of this difference are the topic of this article.

As a first step it is useful to recall the molecular design parameters of monomeric amphiphiles [1]. In this case one component is typically hydrophilic and the other hydropohobic. The hydrophilic group can be ionic or neutral. The hydrophobic component is typically a hydrocarbon or fluorocarbon chain. The hydrocarbon chain may contain double bonds. ''Classical'' amphiphiles are comprised of a single ionic head-group attached to one or two hydrophobic tails. The number of possible designs is much larger (Fig. 1). Bipolar, or ''bolaform,'' amphiphiles incorporate two ionic head-groups joined by a hydrophobic chain. The head-groups can be identical or different. Multipolar surfactants consist of

* Note from the Editor: the term ''conformation'' is used in the balance of the book.

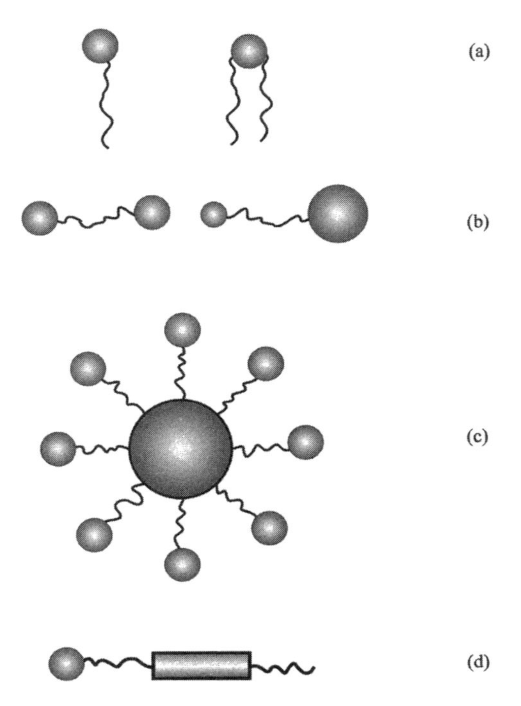

Figure 1 Possible molecular architectures of monomeric amphiphiles: classical amphiphiles (a); bolaform, bipolar amphiphiles (b); multipolar (c); and amphotropic (d) amphiphiles. The circles represent the ionic head-groups and the hydrophobic chains are depicted by the wavy lines. (Adapted from Ref. 1.)

a number of polar head-groups joined to a single ''core'' by hydrophobic chains. Amphotropic amphiphiles can incorporate a mesotropic moiety in the hydrophobic tail. This incomplete catalogue already allows us to identify a number of molecular design parameters: the number and the length of the hydrophobic chains; the number, the size, and the charge of the polar head-groups; the chemical nature of the hydrophilic head-groups and the hydrophobic tails; the molecular ''topology,'' etc.

What happens when the number of atoms in the amphiphiles increases from $\sim 10^2$ to 10^6–10^8? As we shall see, the design parameters do not change qualitatively. Two important differences do, however, develop. First, the number of possible molecular architectures grows enormously. This ''combinatorial effect'' is accompanied by a change in the underlying physics due to the change in the relative importance of the translational and configurational entropies.

The molecular design of polymeric amphiphiles is based on two extreme strategies (Fig. 2). One utilizes incompatible but nonamphiphilic monomers; i.e., monomers that on their own do not exhibit self-assembly or form mesophases. In this case the amphiphilicity is obtained by grouping the monomers into blocks [4]. Diblock copolymers are the best known example of this class. In the second strategy amphiphilic monomers are polymerized, thus producing ''polysoaps'' [1,10]. When both amphiphilic and nonamphiphilic monomers are copolymerized, the polysoaps incorporate spacer chains that join the amphiphilic monomers. Many intermediate strategies are of course possible.

The architectural diversity of polymeric amphiphiles is enormous. To illustrate this point, the ''combinatorial effect,'' consider the design parameters of block copolymers (Fig. 3). It is possible to change the overall polymerization degree as well as the relative size of the blocks. One may also change the number of blocks and their topology. The polymer can be linear or branched. In turn, the branching can take many forms: various star block copolymers, comb copolymers, etc. In addition, one may vary the chemical nature of the blocks. Thus, linear *ABA* triblock copolymers differ from *BAB* triblocks and from *ABC* triblocks. Finally, the chemistry can modify other attributes of the blocks such as rigidity, crystallinity, charge, or polarity.

Polymerization is accompanied by a change in the relative importance of the interaction energies, the translational entropy, and the configurational entropy. Consider a solution of N free monomers prior to polymerization (Fig. 4). For

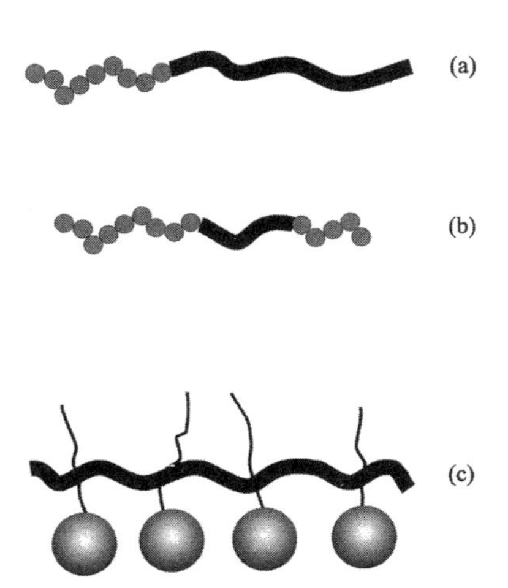

Figure 2 Two extreme strategies for polymeric amphiphiles: grouping nonamphiphilic monomers into blocks (a, b) and incorporating amphiphilic monomers into the chain (c).

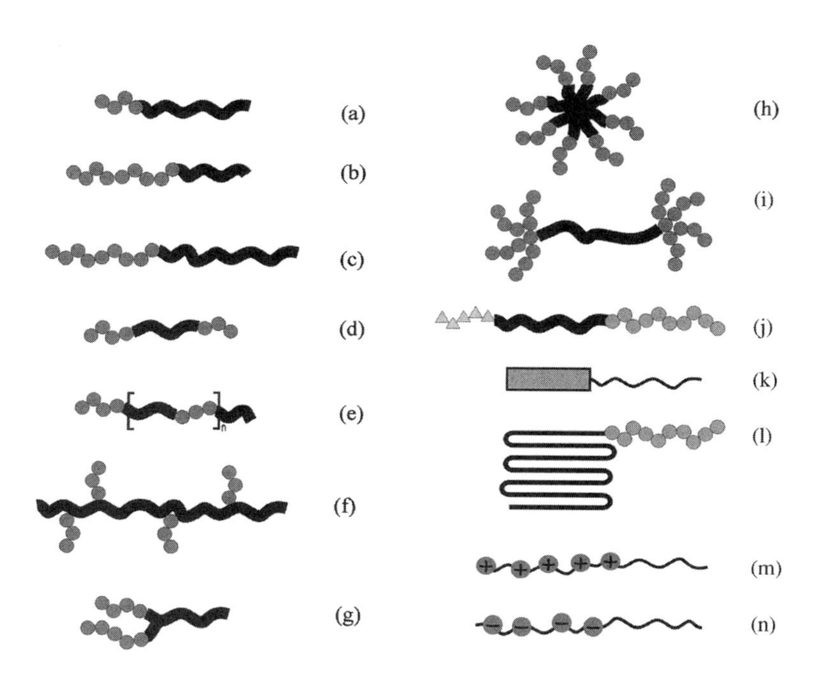

Figure 3 The architectural diversity, ''combinatorial effect,'' of polymeric surfactants as illustrated for block copolymers. The design parameters include the relative and absolute size of the blocks (a–c), their number (d, e), ''topology'' (f–i), the chemical nature of the blocks (j), their rigidity or crystallinity (k, l) and their charge (m, n).

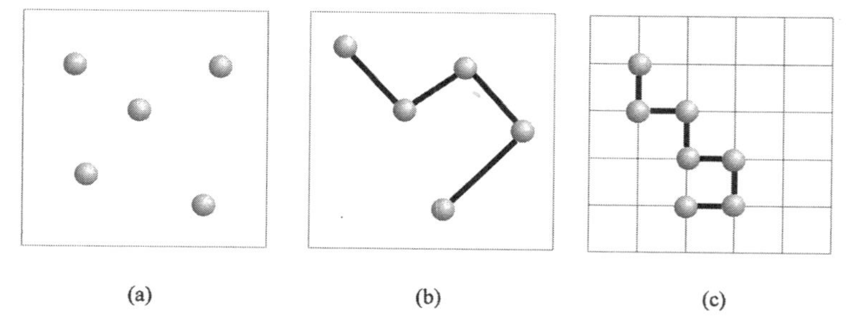

Figure 4 The effects of polymerization: prior to polymerization the monomers can move independently (a). The translational entropy of the polymerized monomers is associated with the center of mass of the chain (b). At the same time, a flexible chain possesses an entropy due to the multitude of different possible configurations (c).

simplicity we consider the monomers as rigid structures possessing no configurational entropy. Each monomer can move independently and is thus endowed with $\sim kT$ of translational free energy. In addition, each monomer is characterized by a set of interaction energies ε_i that specify the strength of its interactions with other monomers, surfaces, etc. What happens when the N monomers are polymerized into a single flexible chain? The interaction energies of the monomers ε_i are essentially unmodified [16]. In marked contrast, the translational entropy per monomer is greatly diminished. After the polymerization the individual monomers can no longer move independently. Only the chain as a whole can undergo translational motion. Accordingly, the translational free energy of the chain as a whole is $\sim kT$ where k is the Boltzmann factor and T is the temperature. As a result, the translational free energy per monomer is $\sim kT/N$. On the other hand, the polymerized chain does possess configurational entropy. To estimate it we inscribe the chain on a lattice such that each site has z neighbors. The number of possible configurations of the unperturbed chain, not allowing for self-avoidance, is $\Re \approx z^N$. Self-avoidance decreases the number of allowed configurations without affecting the leading behavior [17]. The total configurational entropy is thus $k \ln \Re \approx kN \ln z$ while the configurational free energy per monomer is constant, $kT \ln z$. Altogether, the interaction energies per monomer ε_i and the configurational entropy per monomer are roughly independent of N. In marked distinction, the translational entropy per monomer scales as $1/N$ and thus approaches zero as $N \to \infty$. This picture is oversimplified: (i) in many situations, such as adsorption, only a fraction of the monomers interact; (ii) the interaction energy of terminal monomers may differ from that of monomers within the chain; and (iii) this discussion does not allow for possible chain deformation and the associated free energy penalty. These caveats do not, however, change the main conclusion: the polymerization decreases the relative importance of the translational entropy vis-à-vis the configurational entropy and the interaction energies. This gives rise to the propensity of polymers to adsorb and undergo phase separation. As we discuss, this feature is also responsible for many distinctive aspects of polymeric amphiphiles.

III. DIBLOCK COPOLYMERS VS. MONOMERIC AMPHIPHILES

Diblock copolymers may be viewed as the macromolecular counterparts of single-tailed ‘‘classical’’ amphiphiles. Both the molecular architecture and the phenomenology of the two families are similar. The common features in the phenomenology include the self-assembly into micelles, lamellae, etc. and the general aspects of the mesophase they form. Some of the differences between the two systems are transparent, reflecting the immediate consequences of the size disparity. The aggregates formed by the diblocks are much bigger than those formed by the monomeric amphiphiles. A related difference concerns the time scales: the polymeric systems are characterized by much longer relaxation times. Finally, the greater size of the diblocks gives rise to higher bending rigidity of their lamellas which tends to decrease the importance of shape fluctuations in these systems.

The differences listed above are important but, in a sense, obvious. A more delicate point concerns the selectivity of the two families. We use this term to characterize the dissimilarity between selective solvents for the two incompatible components of the amphiphile, i.e., the copolymeric blocks or, in the monomeric case, the head-groups and the tails. The selectivity of monomeric surfactants is rather low. As a result, they can distinguish between highly polar media, such as water, and nonpolar organic solvents. Polymeric surfactants are much more selective and may distinguish between polymeric melts that differ only in isotopic composition. This is a direct result of our earlier considerations concerning the consequences of polymerization. Since the translational free energy diminishes as kT/N, the price of spatial nonuniformities is low. The two important forms of spatial nonuniformities are adsorption and phase separation. When N is large enough the energy gain involved will dominate since the ε_i are independent of N. Thus all polymers are prone to adsorption even if the surface is only weakly attractive. Similarly, weak incompatibility is sufficient to cause phase separation providing N is large enough. An extreme example of this type is the phase separation of hydrogenated (H) and deuterated (D) polymers, which are otherwise identical [19]. The corresponding HD diblock copolymers will act as amphiphiles that will straddle the resulting HD interface. Thus, by increasing N it is possible to enhance enormously the selectivity of polymeric surfactants to an extent that is unimaginable among monomeric amphiphiles.

Apart from their inherent importance, *AB* diblock copolymers afford a number of important advantages as model amphiphiles. First, it is possible to vary the polymerization degree of the two blocks N_A and N_B over a wide range and to study the ensuing trends. Second, the theoretical interpretation of the results is relatively simple. This is because the length of the chains enables their description in terms of their asymptotic behavior. The resulting scaling-type description circumvents the detailed molecular structure of the chains.

In the following we discuss, in some detail, the self-assembly of monomeric surfactants and of three types of diblock copolymers: flexible and neutral diblocks, flexible diblocks incorporating one neutral block and one polyelectrolyte block, and rod–coil diblocks in which one of the blocks is rigid. This discussion brings out the features that distinguish the self-assembly of block copolymers from that of monomeric surfactants in particular, the key role of the configurational free energy of the chains. In addition, monomeric amphiphiles and diblock copolymers are structural motifs that appear in polymeric surfactants of greater complexity such as polysoaps and multiblock copolymers. Consequently, an understanding of the self-assembly of these simpler surfactants sets the stage for the modeling of the more complicated polymeric amphiphiles.

A. Micelles of Monomeric Amphiphiles

The most familiar micelles among monomeric amphiphiles are those formed in water by ionic single-tail surfactants [11–14]. These serve as a ''reference'' system for comparison purposes. Such micelles can be spherical, ellipsoid, or cylindrical. They are comprised of a dense core formed by the hydrophobic tails with the ionic head-groups straddling the interface (Fig. 5). The amphiphiles are fully dissociated when their concentration is low enough. Micelles begin to form once the concentration reaches the critical micelle concentration, denoted the CMC. Further increase in the number of amphiphiles increases the number of micelles with little change in the concentration of free surfactants. A minimal description of this system must specify its main characteristics: (i) the CMC; (ii) the aggregation number (i.e., the number of amphiphiles p within a micelle); and (iii) the shape of the micelle.

The micellization is associated with a loss of translational entropy. This entropy favors the dissociated state and smaller aggregation numbers. The transfer free energy of the hydrophobic tails from water into the hydrophobic cores, $-\delta kT$, is the driving force for the micellization. The micellization process in-

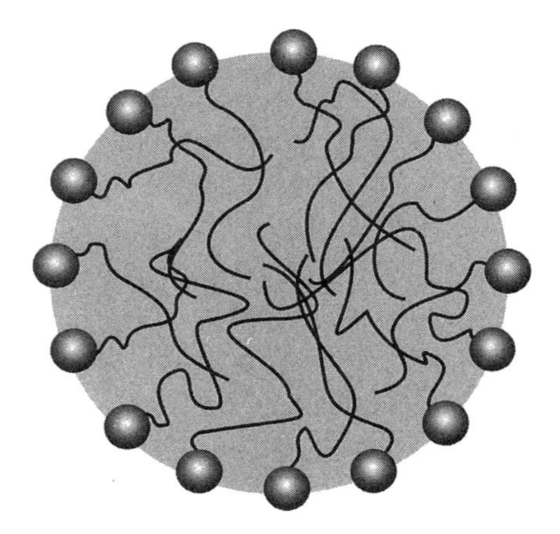

Figure 5 A schematic cross-section of a spherical micelle formed by monomeric amphiphiles with the ionic head-groups straddling the interface of the hydrophobic core. (Adapted from Ref. 11.)

volves a complex equilibrium of micelles/clusters of different size. A full description of this process specifies the population of all species involved. It is determined by $kT\epsilon_p$, the free energy per amphiphile incorporated into a micelle comprised of p surfactants. The crucial property of ϵ_p is a minimum at the average aggregation number $p_{eq} > 1$. The distribution function, in equilibrium, is determined by the equality of the chemical potentials of the amphiphiles in the various clusters together with the constraint on the total number of amphiphiles. The chemical potential of an amphiphile in a micelle comprising p surfactants is

$$\frac{\mu}{kT} = \epsilon_p + \frac{1}{p} \ln \frac{X_p}{p} \tag{1}$$

Here X_p is the mole fraction of surfactants in micelles whose aggregation number is p. Since the total number of amphiphiles is constant,

$$\sum_p X_p = X_T \tag{2}$$

where X_T is the mole fraction of surfactants in the solution, irrespective of the aggregation state. To identify the CMC within this scheme one argues that the number of micellized amphiphiles at the CMC is comparable to the number of free amphiphiles. A simpler analysis is sufficient when only the CMC is of interest. If we choose the free amphiphile as a reference state, the chemical potential of the free dissociated surfactants in dilute solution reflects only their translational entropy, $\mu/kT \approx \ln X$. It is helpful to consider the grand-canonical potential of a micelle coexisting with free amphiphiles, $\Omega = p(\epsilon_p - \ln X)$. The equilibrium size of the micelle p_{eq} and the corresponding equilibrium ϵ are determined by the condition $\partial\epsilon_p/\partial p = 0$. The onset of micellization occurs when $\partial\Omega/\partial p = \Omega = 0$ thus leading to $\Omega = \epsilon - \ln X_{CMC} = 0$. This balances the loss of translational entropy due to the micellization of p_{eq} surfactants, $p_{eq}kT \ln X$, with the most favorable micellization bonus, $p_{eq}kT\epsilon$. Below the CMC when $\Omega > 0$, the loss of translational entropy upon micellization overcompensates $\epsilon_{p_{eq}}$ and the dissociated state is thus preferred. Micelles are favored when $\Omega < 0$.

An explicit form for ϵ_p is necessary in order to proceed further. A simple model yielding such an expression was proposed by Israelachvili et al. [11,20]. Within it, ϵ_p reflects three contributions. (i) A surface free energy associated with the interface between the hydrophobic core and the surrounding water: this term favors aggregation so as to minimize the surface free energy per aggregated amphiphile γkTa. Here a is the surface area per head-group and γkT is the surface tension of the hydrophobic core. (ii) This term is opposed by the repulsive interactions among the head-groups. One may argue, heuristically, that this contribution is proportional to the number of binary contacts among the head-groups. The number of binary contacts per head-group is proportional to the surface density of head-groups, $1/a$. The associated term is thus KkT/a where K, for a given T and ionic strength, is a constant. We elaborate on the nature of this term later. (iii) The transfer free energy of the hydrophobic tails from water into the core, $-\delta kT$: in the case of linear monomeric surfactants with saturated hydrocarbon tails, δ is roughly proportional to the number of CH_2 units in the tail, $\delta \sim n_{\text{tail}}$. Altogether

$$\frac{\epsilon_p}{kT} \approx \gamma a + \frac{K}{a} - \delta \tag{3}$$

At equilibrium, when $\partial \epsilon_p / \partial a = 0$, the first two terms are comparable thus setting an optimal area per head-group, $a_o \approx (K/\gamma)^{1/2}$. The corresponding free energy per amphiphile is

$$\frac{\epsilon}{kT} \approx \gamma a_o - \delta \tag{4}$$

This free energy is, however, insensitive to the geometry of the micelles. It cannot thus specify their shape. Within this model the micellar shape is determined by packing constraints and the translational entropy. Two packing constraints are involved: (i) a constant "melt" density within the core. Thus, for a hydrophobic tail of volume v, the surface area per head-group in a spherical micelle obeys $pa \approx (pv)^{2/3}$ or $p \approx v^2/a^3$. (ii) Since the head-groups straddle the interface, the constant density constraint does not allow the radius of the core to exceed the maximal extended length of the hydrophobic tail l. The possible micellar shapes for a given a_o, in view of the above considerations, are determined by the packing parameter, $v/a_o l$. The packing considerations indicate which geometries are impossible for a given a_o but cannot, on their own, determine the equilibrium shape. This is set by the translational entropy that favors the smallest possible aggregate that satisfies the packing constraints. Altogether, spherical micelles are found when $v/a_o l < 1/3$, cylindrical micelles for $1/3 < v/a_o l < 1/2$, and vesicles or bilayers when $1/2 < v/a_o l < 1$.

The analysis presented above accounts for many observations. It is attractive because of its simplicity. Clearly, many issues are not addressed within this approach. A more rigorous account of the head-group repulsion requires solution of the Boltzmann–Poisson equation allowing for the distribution of the counterions and their entropy. Within this approach the K/a term corresponds to high ionic strength situations, when the electrostatic free energy is due to a narrow electric double layer. The "dressed micelle" model presents a more complete treatment of this aspect [12]. In addition, a complete analysis should account for the configurational entropy of the tails [21]. Finally, a full understanding of γ and δ requires a better understanding of the structure of water and of the hydrophobic effect [22]. The hydration of the head-groups and the counterions is not explicitly reflected in the model described above.

For comparison purposes, it is useful to stress some important features of micelles formed by monomeric amphiphiles within the model described above. (i) Translational entropy plays a crucial role in this picture. In particular, the translational entropy of the free amphiphiles gives rise to the CMC. In addition, the shape of the micelles is controlled, in part, by their translational entropy. (ii) In this model the micellar size is determined by two ''energies'' i.e., the surface term and the head-group repulsion. There is no explicit account of entropic contributions. The configurational entropy of the tails and the conformational entropy of the water as well as of the counterions are not accounted for explicitly. (iii) Finally, ε_p is insensitive to the micellar geometry.

B. Micelles of Neutral, Flexible, Diblock Copolymers

The configurational entropy of the chains plays a crucial role in polymeric systems. This is well illustrated in the case of micelles formed by flexible and neutral *AB* diblock copolymers in a selective solvent [23,24]. We have in mind organic nonpolar solvents. The case of waterborne micelles formed by diblocks with polyelectrolyte blocks is considered in the next section. The description of micelles involving neutral hydrophilic blocks is discussed briefly in Section VI. For specificity, we consider the case of a selective solvent for the *A* block that is a precipitant for the *B* block. For simplicity, the two blocks are also assumed to be strongly incompatible. The polymerization degrees of the two blocks are denoted, respectively, by N_A and N_B. Spherical copolymeric micelles that are formed in this situation consist of two regions. An inner dense core comprised of the solvophobic *B* blocks is surrounded by a swollen corona of *A* blocks (Fig. 6). This micellar structure is similar to that of nonionic monomeric surfactants. As in the case of low molecular weight amphiphiles, the driving force for the micellization is the transfer free energy of the *B* blocks from the solvent into the core. The micellization is again opposed by the associated loss of translational entropy which gives rise to a CMC. The distinctive features of such diblock micelles are largely due to the corona. This has no counterpart in micelles formed by conventional ionic surfactants. When thermodynamic equilibrium is attained, the micellar size is essentially dominated by the coronal contribution. The free energy penalty associated with crowding of the coronal chains replaces the head-group repulsion in determining the micellar structure. However, as opposed to the head-group repulsion, the coronal penalty is geometry dependent. As a result, it determines both the shape and the size of the micelles.

In the following we consider spherical micelles with a sharp core–corona interface associated with a surface tension γkT that is independent of N_B and N_A. Such is the case when the solvent is highly selective and the blocks strongly

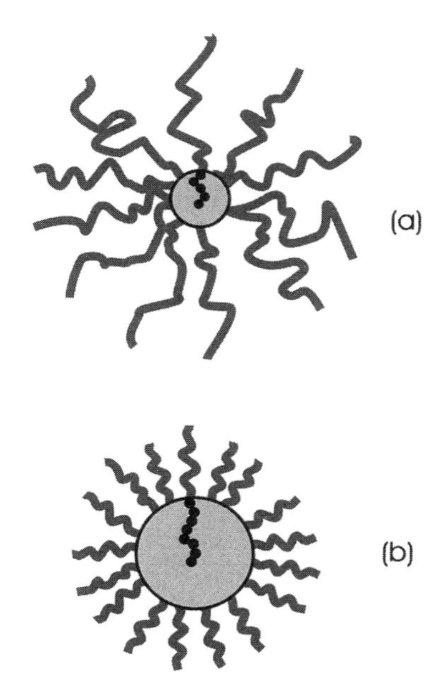

Figure 6 Micelles formed by flexible diblock copolymer in a selective solvent comprised of a dense core surrounded by a corona of swollen coronal blocks. The coronal contribution is dominant for starlike micelles (a) while the core dominates in the case of crewcut micelles (b). (Adapted from Ref. 23.)

incompatible. The corresponding ϵ_p reflects four contributions. The total surface free energy of a core of radius R_{core} is $\gamma kTR_{\text{core}}^2$. The surface free energy per diblock $F_{\text{surface}}/kT \approx \gamma R_{\text{core}}^2/p$ favors micellar growth so as to minimize the surface free energy per diblock. This term is the counterpart of the γa term in the model of Israelachvili et al. The volume of a dense melt-like core is $R_{\text{core}}^3 \approx pN_B b^3$, where b is the monomer size. Accordingly,

$$\frac{F_{\text{surface}}}{kT} \approx \gamma b^2 p^{-1/3} N_B^{2/3} \tag{5}$$

This term is complemented by two penalty terms that oppose micellar growth. One penalty is due to the stretching of the core blocks. As p increases, $R_{\text{core}} \approx p^{1/3} N_B^{1/3} b$ exceeds the unperturbed size of the core block $N_B^{1/2} b$. In order to maintain a constant melt density within the core it is necessary to stretch some of the B blocks. This effect is well approximated by assuming that all the B blocks are uniformly stretched thus giving rise to an elastic penalty $kTR^2_{\text{core}}/N_B b^2$ per chain or

$$\frac{F_{\text{core}}}{kT} \approx p^{2/3} N_B^{1/3} \tag{6}$$

The counterpart of this contribution, the configurational entropy of the tails, is neglected within the model of Israelachvili et al. Large p also cause crowding of the coronal chains when the area per chain $a \approx R^2_{\text{core}}/p \approx p^{-1/3} N_B^{2/3} b^2$ decreases to below the cross-section of the unperturbed coronal block $N_A^{6/5} b^2$. The crowded coronal chains form a ''polymer brush''; i.e., the chains stretch out in order to lower the monomer concentration in the corona and thus the number of repulsive monomer–monomer contacts (Appendix). The resulting penalty for spherical micelles, as given by the Daoud–Cotton model, is

$$\frac{F_{\text{corona}}}{kT} \approx p^{1/2} \ln \frac{R_{\text{core}} + L}{R_{\text{core}}} \tag{7}$$

where L is the width of the corona. In this system, the coronal term is the counterpart of the head-group repulsion in micelles of ionic monomeric amphiphiles. The last term, as in the case of monomeric micelles, is the transfer free energy of the B block from the solvent into the core δkT. However, in the polymeric case this term scales as $\delta \approx \gamma N_B^{2/3} b^2$ rather than as $\gamma N_B b$. The difference reflects the configuration of the B block in the isolated AB diblock. The B block collapses onto a dense spherical globule of radius $N_B^{1/3} b$ thus imparting a ''tadpole''-like configuration to the diblock. The surface free energy of the collapsed B block gives rise to δ. Altogether

$$\frac{\epsilon_p}{kT} \approx \gamma b^2 p^{-1/3} N_B^{2/3} + p^{2/3} N_B^{1/3} + p^{1/2} \ln \frac{R_{\text{core}} + L}{R_{\text{core}}} - \gamma N_B^{2/3} b^2 \tag{8}$$

The equilibrium size of the micelle is determined by $\partial \epsilon_p / \partial p = 0$. The equilibrium condition can be easily solved in two limits (Fig. 6). When $N_A \gg N_B$ the size of the core is negligible compared to the span of the corona. In this limit, of starlike micelles, we may approximate the logarithmic term in F_{corona} as constant thus leading to $F_{\text{corona}}/kT \approx p^{1/2}$. Furthermore, $F_{\text{corona}} \gg F_{\text{core}}$ and the equilibrium state of the micelle thus reflects the competition between F_{corona} and F_{surface}. In the opposite limit, $N_A \ll N_B$, the corona is, in effect, a flat brush. Its thickness L is negligible compared to R_{core}. In this case, of ''crewcut'' micelles, $F_{\text{corona}}/kT \approx N_A (b^2/a)^{5/6} \approx p^{5/18} N_A N_B^{-5/9}$ is negligible compared to F_{core} and the equilibrium state reflects a competition between F_{core} and F_{surface}. In equilibrium the two competing terms are comparable. The equilibrium state of starlike micelles is thus given by the condition $F_{\text{corona}} \approx F_{\text{surface}}$ or

$$p_{\text{eq}} \approx (\gamma b^2)^{6/5} N_B^{4/5} \tag{9}$$

leading to $R_{\text{core}} \approx N_B^{3/5} b$ while the overall span of the micelle is

$$R_{\text{micelle}} \approx L \approx N_B^{4/25} N_A^{3/5} b \tag{10}$$

In the opposite limit, of crewcut micelles, the equilibrium state is specified by $F_{\text{core}} \approx F_{\text{surface}}$ leading to

$$p_{\text{eq}} \approx (\gamma b^2) N_B \tag{11}$$

and, as result, to $R_{\text{micelle}} \approx R_{\text{core}} \approx N_B^{2/3} b$.

It is important to stress that the discussion presented above does not consider the relative stability of spherical micelles as compared to other aggregates of different geometries such as cylinders and lamellae. It focuses on the behavior of ϵ_p of an isolated spherical micelle in order to bring out the factors that determine the equilibrium structure of *AB* aggregates in general. Systematic analysis of this system suggests that crewcut micelles are not stable with respect to other types of aggregates [25].

C. Copolymeric Micelles with Charged Coronas

The copolymeric micelles considered in the previous section were neutral and thus similar to micelles formed by nonionic monomeric amphiphiles. The monomeric micelles considered in Section III.A incorporated charged head-groups. In that case, electrostatic interactions played a crucial role in determining the micellar behavior. In addition, the head-groups were rigid objects, possessing no configurational degrees of freedom. Finally, the size and charge of the head-groups were rather small. Waterborne copolymeric micelles incorporating polyelectrolyte coronal blocks [26] exhibit an intermediate structure. Here, one may compare the polyelectrolyte block to the ionic head-group. However, in this case the head-group is not rigid. The polyelectrolyte blocks are deformable and their configurational free energy plays an important role. Furthermore, both the span and the charge of the coronal blocks can be large. As in the case of neutral copolymeric micelles, the coronal penalty is the analogue of the head-group repulsion in that it opposes micellar growth. However, in this case F_{corona} must allow for two additional contributions: repulsive electrostatic interactions between the dissociated ionic groups within the corona, and the mixing entropy of the free mobile counterions and the associated osmotic pressure. Two distinctive polymeric scenarios are expected. One involves strong stretching of the coronal blocks due to the electrostatic interactions. In the second, the counterions are trapped, condensed within the corona, and the coronal swelling is due to their osmotic pressure [27].

N_A and N_B, the polymerization degrees of the two blocks, were the two primary design parameters of neutral micelles formed by flexible diblocks. A third parameter appears in the present case. This is the number of ionizable groups in a coronal block n. The micellar characteristics depend on the fraction of ionizable groups in the coronal block n/N_A. In the case of weak polyelectrolyte A blocks, $n/N_A \ll 1$, counterion condensation does not play a role; i.e., the counterions are uniformly distributed throughout the solution. In this case we limit the discussion to starlike micelles where $N_A \gg N_B$ and thus $L \gg R_{\text{core}}$. The coronal chains are stretched by excluded volume interactions and by electrostatic repulsion between the ionic monomers. The electrostatic repulsions are dominant when $L/b \gg p^{1/5}N_A^{3/5}$. We further limit the discussion of weak A polyelectrolytes to this regime, when monomer–monomer interactions are negligible. The electrostatic energy of the corona as a whole may be approximated as $(pne)^2/\varepsilon R$ where ε is the dielectric constant of the solvent and e is the elementary charge. In terms of the Bjerrum length $l_B \approx e^2/\varepsilon kT$, the electrostatic energy per coronal block may be expressed as $kTpn^2 l_B/R$. This term favors stretching of the coronal blocks. The extension is opposed by the elasticity of the chains. For simplicity this is described by the Gaussian forms $kTL^2/N_A b$. For starlike micelles $L \approx R$ and the coronal free energy for a given p is

$$\frac{F_{\text{corona}}}{kT} \approx \frac{R^2}{N_A b^2} + \frac{pn^2 l_B}{R} \tag{12}$$

The corresponding R is set by the equilibrium condition $\partial F_{\text{corona}}/\partial R = 0$. This yields $R \approx p^{1/3}n^{2/3}N_A^{1/3}(b^2 l_B)^{1/3}$ and leads to $F_{\text{corona}}/kT \approx p^{2/3}n^{4/3}N_A^{-1/3}(l_B/b)^{2/3}$. In a starlike micelle the core penalty is negligible and the free energy per copolymer is $\epsilon_p \approx F_{\text{corona}} + F_{\text{surface}}$ or

$$\frac{\epsilon_p}{kT} \approx \gamma b^2 p^{-1/3} N_B^{2/3} + p^{2/3} n^{4/3} N_A^{-1/3} \left(\frac{l_B}{b}\right)^{2/3} \tag{13}$$

where we ignored δ because it does not affect the equilibrium structure. The equilibrium state of the micelle is specified, as before, by $\partial\epsilon_p/\partial p = 0$. This yields the equilibrium characteristics

$$p_{\text{eq}} \approx \gamma b^2 N_B^{2/3} N_A^{1/3} n^{-4/3} \left(\frac{b}{l_B}\right)^{2/3} \tag{14}$$

and

$$\frac{R}{b} \approx \left(\frac{l_B}{b}\right)^{1/9} (\gamma b^2)^{1/3} N_B^{2/9} N_A^{4/9} n^{2/9} \tag{15}$$

This result does not however apply to strong polyelectrolyte A blocks for which counterion condensation occurs; i.e., the counterions are trapped within the corona. In this case the electrostatic interactions within the corona are screened and the stretching of the coronal chains is due to the osmotic pressure of the condensed counterions.

The essential features of the counterion condensation for spherical micelles can be obtained from a simple argument proposed, in a different context, by Alexander et al. [28]. The net charge of the micelle, $pn^* < pn$, is the charge that is not screened by the trapped counterions. Roughly speaking, it may be estimated by equating the electrostatic reassociation energy to the thermal energy kT. The chemical potential of the pn^* fully dissociated ions is $\mu_{\text{free}} \approx kT \ln \phi_{\text{ion}}$, where $\phi_{\text{ion}} \approx pn^* v\phi/b^3$, ϕ is the volume fraction of the micelles, and v is the volume of the individual counterion. The electrostatic potential of a condensed counterion at the surface of a sphere of radius R_S and a charge $pn^* e$ is $\mu_{\text{el}} \approx -pn^* e^2/\varepsilon R_S \approx -kTpn^* l_B/R_S$. Equating μ_{free} and μ_{el} leads to

$$pn^* \approx -\frac{(\ln \phi_{\text{ion}})R_S}{l_B} \approx \frac{R_S}{l_B} \tag{16}$$

where the $\ln \phi_{\text{ion}}$ is approximated as a constant. The estimates of n^* in the two micellar limits differ because of the corresponding differences in R_S. For a starlike micelle $R_S \approx R \gg R_{\text{core}}$ while for crewcut micelles $R_S \approx R \approx R_{\text{core}}$.

F_{corona} of micelles incorporating strong polyelectrolyte A blocks should allow for the mixing entropy of the $N_A - n^*$ condensed counterions within the corona. It is thus necessary to supplement Eq. (12) by an extra term $kT(N_A - n^*)\ln \phi_{c\text{-ion}}$, where $\phi_{c\text{-ion}} \approx p(N_A - n^*)/R^3$ is the concentration of trapped counterions within the corona.

$$\frac{F_{\text{corona}}}{kT} \approx \frac{R^2}{N_A b^2} + \frac{pn^2 l_B}{R} + (N_A - n^*)\ln \frac{p(N_A - n^*)}{R^3} \tag{17}$$

The corresponding equilibrium structure of the corona, for a given p, is specified by $\partial F_{\text{corona}}/\partial R = 0$, or

$$\frac{R}{N_A b^2} - \frac{pn^{*2} l_B}{R^2} - \frac{N_A - n^*}{R} = 0 \tag{18}$$

Since $n^* \approx R/pl_B$, the second, electrostatic, term is negligible in comparison to the third, mixing entropy, term. As a result the micellar radius exhibits two interesting features: (i) R is independent of p and (ii) it reflects extremely strong stretching of the coronal chains

$$\frac{R}{b} \approx N_A \tag{19}$$

The number of free counterions is

$$n^* \approx \frac{N_A b}{p l_B} \tag{20}$$

indicating that for large p the majority of the counterions are condensed; i.e., $n^* \ll N_A$.

This brings us to the structure of the micelle in the strongly charged limit. As usual, the micellar structure in equilibrium reflects the interplay of F_{surface} and F_{corona}. In the present case, with $R/b \approx N_A$ and $n^*/N_A \ll 1$, the elastic term in F_{corona} is a constant while the electrostatic term is negligible. The dominant contribution is due to the mixing entropy of the condensed counterions $F_{\text{corona}}/kT \approx kTN_A \ln pN_A$ and thus

$$\frac{\epsilon_p}{kT} \approx \gamma b^2 p^{-1/3} N_B^{2/3} + N_A \ln pN_A \tag{21}$$

As a result the equilibrium condition $\partial\epsilon_p/\partial p = 0$ yields

$$p_{\text{eq}} \approx (\gamma b^2)^3 N_B^2 N_A^{-3} \tag{22}$$

Imposing the requirement that $p_{\text{eq}} \gg 1$ leads to $N_B \gg \gamma b^2 N_A^{2/3}$. In other words, this scenario involves crewcut micelles with short, highly stretched, coronal blocks surrounding a relatively large core.

D. On Rod–Coil Block Copolymers, Tilt, and the Overspill Effect

In the examples previously considered, the insoluble block was flexible. When the insoluble block is rodlike the underlying physics is modified in two respects. First, rigid rodlike blocks have no configurational entropy. While flexible coils assume a multitude of different configurations rodlike blocks possess a uniquely defined geometry. As a result, the core penalty term F_{core} no longer plays a role. Second, the packing of the core blocks is strongly modified because the insoluble rods favor parallel orientation. Consequently, the core is cylindrical rather than spherical. Some of the consequences of these features are apparent in lamellae and platelets rather than micelles. Accordingly, while the beginning of this section is devoted to micelles [29], the later part focuses on two effects, tilt [29,30] and overspill [31], that can only occur in more extended aggregates.

For simplicity we ignore the molecular structure of the rodlike block and model it as a cylinder of length $L_{\text{rod}} \sim N_B$ and diameter d such that $L_{\text{rod}} \gg d$. The minimal surface area per rod is attained when the micellar core assumes a cylindrical shape. The B blocks are close packed with their axes aligned and their tips towing a single basal plane. The cylindrical core thus formed carries two

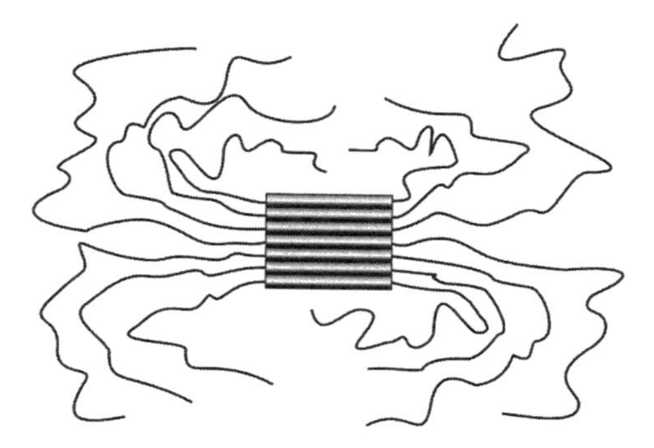

Figure 7 Micelles of rod–coil diblock copolymers can be stable in the starlike limit when the corona is large in comparison to the core. In this case the overall shape is roughly spherical but the core is cylindrical. (Adapted from Ref. 29.)

"coronal mops" instead of a single spherical corona (Fig. 7). The core geometry also affects the scaling behavior of F_{surface} and, consequently, the equilibrium characteristics of the micelles. The height of the cylindrical core is L_{rod}, irrespective of the aggregation number. Only the radius of the cylinder $p^{1/2}d$ changes with p. The basal surface area is pd^2 and the lateral surface area is $p^{1/2}dL_{\text{rod}}$. For simplicity we assume that the basal and lateral surfaces of the cylinder are associated with the same surface tension γkT. The surface free energy, omitting constant terms, is thus $F_{\text{surface}}/kT \approx \gamma d^2 p^{-1/2}(L_{\text{rod}}/d)$. Note that in the case of a spherical core formed by flexible B blocks, $F_{\text{surface}} \sim R^2_{\text{core}}/p \approx p^{-1/3}N_B^{2/3}$. As discussed previously there is no core penalty in this case and thus $\epsilon_p = F_{\text{surface}} + F_{\text{corona}}$. Two limits are, again, easy to analyze. One is the limit of starlike micelles $N_A \gg N_B$, when the coronal span dominates the size of the micelle $R \gg R_{\text{core}}$. The details of the monomer distribution around the core only modify a small inner region of the corona. Its overall structure is starlike and thus $F_{\text{corona}}/kT \approx p^{1/2}$. In this case $\epsilon_p/kT \approx p^{1/2} + \gamma d^2 p^{-1/2}(L_{\text{rod}}/d)$. At equilibrium $\partial\epsilon_p/\partial p = 0$, the two terms are comparable leading to

$$p_{\text{eq}} \approx N_B \tag{23}$$

The radius of the cylindrical core is

$$R_{\text{core}} \approx N_B^{1/2} d \tag{24}$$

and the overall micellar size is

$$R \approx L \approx N_B^{1/5} N_A^{3/5} b \tag{25}$$

In the opposite limit $N_A \ll N_B$, of crewcut micelles with $R \approx R_{\text{core}}$, the corona forms a flat brush. In this case the coronal penalty, neglecting edge effects, is independent of p, $F_{\text{corona}}/kT \approx N_A(b/d)^{5/3}$. As a result $\partial\epsilon_p/\partial p = 0$ yields $p_{\text{eq}} = \infty$, indicating a tendency to form lamellae. Note that in the case of flexible diblocks ϵ_p of crewcut micelles exhibits a minimum at a finite p because of the core's elastic penalty.

This brings us to the behavior of extended lamellar sheets. As was noted before, the coronal chains in such systems form a flat brush. This, together with the packing mode of the core, enables a new mechanism for lowering the free energy of the system. It involves the tilting of the rods' axes with respect to the normal to the lamella (Fig. 8). The tilt increases the surface area per coronal chain thus lowering F_{corona} at the price of a higher surface penalty. When there is no tilt the angle between the axis of the rod and the normal is $\theta = 0$. In this case only the bases of the rods are exposed to the solvent and the surface free energy per rod is $F_{\text{surface}}/kT \approx \gamma d^2$ while $F_{\text{corona}}/kT \approx N_A N_A(b/d)^{5/3}$. When the rods tilt, their exposed surface increases from d^2 to $d^2(1 + |\tan\theta|)$ with a corresponding increase in F_{surface}. At the same time, the grafting density decreases. While the shortest distance between grafting sites remains d, the longest distance becomes $d/\cos\theta$ and the area per chain becomes $a \approx d^2/\cos\theta$. In turn, this is associated with a lower coronal penalty $F_{\text{corona}}/kT \approx N_A(b/d)^{5/3}\cos^{5/6}\theta$. Altogether, the free energy per chain is thus

$$\frac{\epsilon}{kT} \approx \gamma d^2 + N_A\left(\frac{b}{d}\right)^{5/3}(\Delta\,|\tan\theta| + \cos^{5/6}\theta) \tag{26}$$

where $\Delta \approx \gamma d^2 N_A^{-1}(b/d)^{-5/3}$ is the ratio $F_{\text{surface}}(\theta = 0)/F_{\text{corona}}(\theta = 0)$. When $\Delta >$ 1 the surface term is dominant and no tilt is expected. Tilt becomes favorable

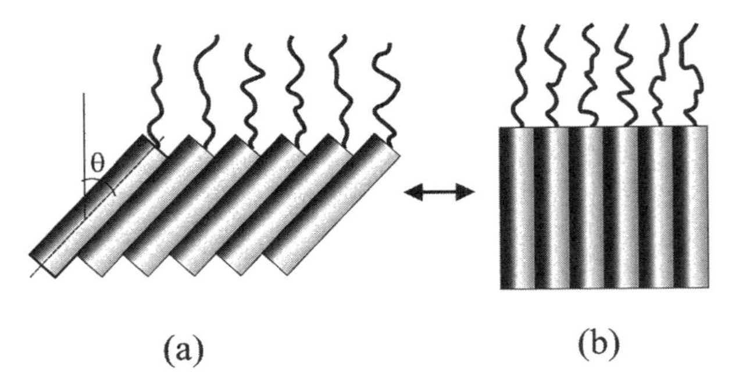

Figure 8 Lamellae of rod–coil diblock copolymers can lower their free energy by tilt. It is favored when the brush penalty is dominant (a). The nontilted state is preferred when the surface free energy is dominant (b). (Adapted from Ref. 29.)

when the coronal penalty gains importance, when Δ decreases, as a way of lowering the crowding of the coronal chains. The onset of the tilt within this picture involves a first-order phase transition. This is because for weak tilts $\theta \ll 1$, the surface penalty grows as $\gamma kTd^2(1 + |\theta|)$ while the coronal penalty decreases less steeply as $kTN_A(b/d)^{5/3}(1 - \theta^2)$. A more detailed analysis suggests that the transition occurs at $\Delta = 0.26$ and $\theta = 53.30°$.

It is possible to vary Δ in two ways. One involves a systematic change of N_A. In this case the phase diagram is explored by preparing samples of appropriately modified polymers. It is also possible to attain the transition by changing the surface tension γkT. This may be attained, for example, by change of solvent or of temperature. The discussion as presented is valid in the strong segregation limit. The mechanism proposed above assumes that the lamella is immersed in a good selective solvent for the *A* blocks. However, it should also be valid, with some modification, for the melt case. In the analysis as presented above the tilt is uniform throughout the layer. It is, however, entropically beneficial to have domains characterized by different orientation of the tilt. The origin of zigzag structures observed in melts of rod–coil copolymers [32] was attributed to this mechanism.

The above discussion of lamellae and platelets formed by rod–coil copolymers is applicable as long as the line tension associated with the periphery of the platelet, τkT is positive. A positive line tension favors the growth of dense circular platelets and its contribution to ϵ diminishes as $p^{-1/2}$. In such a case, rod–coil platelets may attain ''local equilibrium structure,'' tilt, etc. while the lateral span is finite. A negative τ changes this picture since it favors a high ratio of boundary length to area; i.e., for example, rectangles are preferred over circles. Is $\tau < 0$ possible? Two contributions are involved. One, $\tau_{rod} \approx \gamma L_{rod} > 0$, arises because of the positive surface tension of the rods. The second term, τ_{brush}, due to the brush, is negative. The negative τ_{brush} reflects the weaker stretching of the peripheral coronal chains. These chains may ''overspill'' over the edge of the cylindrical core thus avoiding crowding and lowering their free energy (Fig. 9). Introducing a negative τ_{brush} to allow for the overspill effect is only meaningful

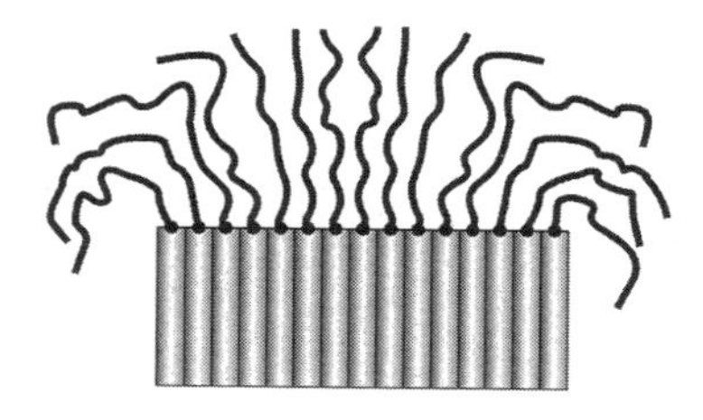

Figure 9 The effective line tension of platelets formed by rod–coil copolymers includes a negative term due to the overspill of the peripheral chains. (Adapted from Ref. 33.)

when the core is large enough so that the central part of the corona can be viewed as an unperturbed flat brush. A detailed analysis of this effect may be found in the original paper of Raphael and de Gennes [31] and in a review [33]. Briefly, it is possible to interpret τ_{brush} as the osmotic work done by a brush of unit width as it overspills. The characteristic width of the region affected by the overspill is comparable to the height of the unperturbed brush $L/b \approx (b^2/a)^{1/3}N_A$. The associated osmotic work is thus πL^2, where π/kT is the osmotic pressure in the center of the brush. Since within the blob model $\pi \approx kT/a^{3/2}$, this argument yields

$$\frac{-\tau_{\text{brush}} b}{kT} \approx N_A^2 \left(\frac{b}{a}\right)^{13/6} \tag{27}$$

The net line tension $\tau = \tau_{\text{rod}} + \tau_{\text{brush}}$ may thus be tuned down by increasing N_A and if N_A is large enough, a $\tau < 0$ is possible.

Finally, note that the overspill effect is not specific to rod–coil aggregates. It is expected to play a role in the phase separation of membranes incorporating both monomeric lipids and polymerized lipids or lipids attached to a hydrophilic chain.

E. Thermodynamics vs. Kinetics

The preceding sections focused on equilibrium structures, i.e., structures corresponding to a minimum in the appropriate free energy. Such thermodynamic equilibrium implies that the aggregate structure is independent of time and of the preparation method. Attainment of thermodynamic equilibrium requires freedom of exchange of diblocks between the self-assembled structures, micelles, lamellae, etc. The required exchange process cannot take place when the solvophobic cores are in a glassy state. The bulk glass transition temperature T_g gives a rough indicator for the onset of this effect. When $T < T_g$ the core may be in a glassy state and, as a result, equilibration is impossible. Equilibration may be effectively frozen even when $T > T_g$. This is because the exchange of diblocks between different aggregates is an activation process. Such an exchange involves, as an intermediate step, the expulsion of the solvophobic block into the solvent [34,35]. This gives rise to an activation barrier $\Delta F_{\text{barrier}}$ and the characteristic time τ_{exchange} scales as $\tau_{\text{exchange}} \sim \exp(-\Delta F_{\text{barrier}}/kT)$. $\Delta F_{\text{barrier}}$ is due to the surface free energy of the expelled block. When the core blocks are flexible $\Delta F_{\text{barrier}}/kT \approx \gamma N_B^{2/3} b^2$ because the solvophobic block collapses upon expulsion. For rodlike core blocks $\Delta F_{\text{barrier}}/kT \sim \gamma N_B bd$ since there is no configurational change upon expulsion. It both cases, when N_B is large enough the expulsion is effectively frozen thus preventing exchange and equilibration.

F. An Interim Summary

Some general features of the self-assembly of diblock copolymers and of ionic amphiphiles are identical. The aggregation is driven by the transfer free energy of the solvophobic component and opposed by the translational entropy. In both systems, the core–solvent surface tension favors larger aggregation numbers p. The two systems differ, however, with respect to the terms that oppose an increase in p. In the copolymeric case the configurational free energy is the leading penalty while in the monomeric case the head-group repulsion dominates. In turn, the configurational contribution may lead to distinctive features such as tilt and to the overspill effect.

The three copolymeric cases considered do not exhaust the possible copolymeric scenarios. For example, when the solvophobic blocks are crystallizable, fold crystallization affords a mechanism for reducing the coronal crowding (Fig. 10) [36]. In such systems fold crystallization can be thermodynamically preferred. In this situation, as well as in the case of rod–coil self-assembly, it may be necessary to allow for differences between the surface tensions associated with the different facets of the anisotropic core, i.e., folds vs. stems in the fold crystalline case or basal vs. lateral surfaces in the rod–coil case. The number of possible scenarios is even larger if one considers systems incorporating ionomeric [8,9] or liquid-crystalline components [37].

The preceding discussion focused on the simplest situations and on the leading contributions. For example, in the case of flexible diblock copolymers, it was assumed that the size of the monomers in the two blocks is identical, $b_A = b_B = b$. Similarly, the persistence length l_p of the flexible chains was assumed to be equal to the monomer size $l_p = b$. In the general case it is necessary to

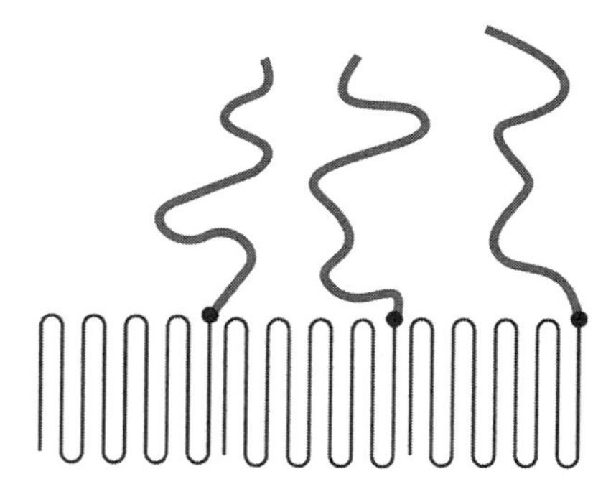

Figure 10 Fold crystallization in lamella of block copolymers incorporating crystallizable blocks provides a route for lowering the coronal penalty. The orientation of the folds depends on the surface tension of the stems and the folds. (Adapted from Ref. 36.)

allow for monomers of different size $b_A \neq b_B$ [39], and for $l_p > b$. The persistence length can affect both the packing constraints and the configurational entropy. A chain with $l_p \gg b$ behaves as a rod when $l_p > Nb$ but as a random coil when $Nb \gg l_p$ [40]. However, in this last case the chain elasticity can be significantly modified. The effect is easy to understand for ideal chains. The span of a long persistent chain is $R_o \approx (Nb/l_p)^{1/2} l_p \approx (Nbl_p)^{1/2}$ instead of $N^{1/2}b$. Thus when $l_p \gg b$ the chain span is significantly larger and the elastic penalty $F_{el}/kt \approx (R/R_o)^2$ considerably weaker. In addition, our analysis of the flexible *AB* micelles focused on the case of a perfectly selective solvent with a meltlike core. In realistic situations the solvent is not fully excluded from the core which can then be described as a close-packed globule of collapse blobs [17,33]. Finally, the presence of free *A* homopolymers in the solution can screen out the correlations within the micellar corona, i.e., decrease the dimensions of the blobs. This, in turn, modifies F_{corona} and L thus giving rise to changes in p_{eq} and R [38].

IV. TRIBLOCK COPOLYMERS: MESOGELS AND NOVEL MESOPHASES

How is the situation described above modified when the number of blocks increases from two to three? Two types of linear triblock copolymers are possible, *ABA* and *ABC*. In both cases, there is little effect on the single-chain behavior. However, the many-chain behavior exhibits qualitatively different features. Two effects are most noticeable. *ABA* triblock copolymers are capable of forming physical networks while *ABC* triblock copolymers form distinctive mesophases. The two scenarios have no counterpart in the behavior of diblock copolymers or of monomeric shortchain amphiphiles. Since the bulk phase behavior of *ABC* block copolymers is discussed at length in Chapter 10 we mostly focus on the behavior of *ABA* triblock copolymers.

A. *ABA* Triblock Copolymers: Exchange, Bridging, and Mesogels

1. Qualitative Aspects

The mesophases and aggregates formed by *AB* diblocks and *ABA* triblocks are similar. So are the single-chain properties of the two species. However, the interactions between aggregates formed by *AB* diblocks and by *ABA* differ qualitatively. The coronal interactions in the first case are purely repulsive while attractive interactions are possible in the second. As a result *ABA* triblocks can form physical gels and networks, while *AB* diblocks cannot [41,42]. This leads to major differences in the mechanical and rheological properties of the two systems.

The selectivity of the solvent, whether it is a highly selective precipitant for the A block or for the B block, does not affect the single-chain behavior of AB diblock copolymers or their aggregation behavior. The single-chain adopts a tadpole-like configuration comprising a collapsed globule of the solvophobic block decorated by a tail formed by the soluble block. The aggregates formed by the diblocks consist of an inner, dense, solvophobic core with an outer corona of swollen solvophilic blocks. The interactions between coronas of AB micelles are purely repulsive because overlap of two coronas increases the number of unfavorable monomer–monomer contacts.

In marked distinction, the behavior of ABA triblocks in the two solvents is qualitatively different. ABA triblock copolymers in a highly selective precipitant for the B block behave, essentially, as AB diblocks. The single-chain adopts a "two-tail tadpole" configuration in which the globule formed by the B block is decorated by two A tails. This similarity extends also to the aggregation behavior. The only difference is that the number of coronal chains in an aggregate comprising p triblocks is $2p$ rather than p. The interactions between two coronas are, again, purely repulsive. New features emerge in a highly selective precipitant for the A blocks. The effect on the configurations adopted by a single-chain is not dramatic. In this case the triblock may form a single A globule decorated with a loop of B or a "dumbbell" of two globules joined by a B chain. The coronas of isolated aggregates, in the high selectivity limit, consist of loops of swollen B blocks (Fig. 11). The formation of the loops affects the dimensions of the isolated chains and of the corona. It also gives rise to a logarithmic free energy penalty. Neither of these effects affects the scaling behavior of the system. However, the resulting interactions between different aggregates are qualitatively different. The net interaction now reflects two contributions: (i) the "osmotic" repulsion due to the overlap between the coronas and the associated increase in the number of monomer–monomer contacts, and (ii) an attractive interaction due to the exchange of solvophobic blocks between two different cores. When the aggregates are in grazing contract, the ABA may adopt two states. In one state the two A blocks inhabit a single core while in the other state each A block is situated in a different core. This exchange of A blocks is associated with a gain of entropy of roughly $k \ln 2$ per aggregated triblock [43]. Grazing contact specifies the optimal distance for this process. Closer approach gives rise to an osmotic penalty. Larger separations give rise to an elastic penalty due to the stretching of the B blocks. The attraction disappears when the distance between the surfaces of the cores exceeds the length of the fully extended B block $N_B b$. Within this rough argument, the free energy of two aggregates at grazing contact is lower by $\sim pkT$ at a price of kT due to loss of translational entropy. Altogether, the exchange attraction favors phase separation involving a coexistence of a dilute phase and of a dense phase consisting of closely packed aggregates. A more refined analysis of this effect is presented in the following subsection.

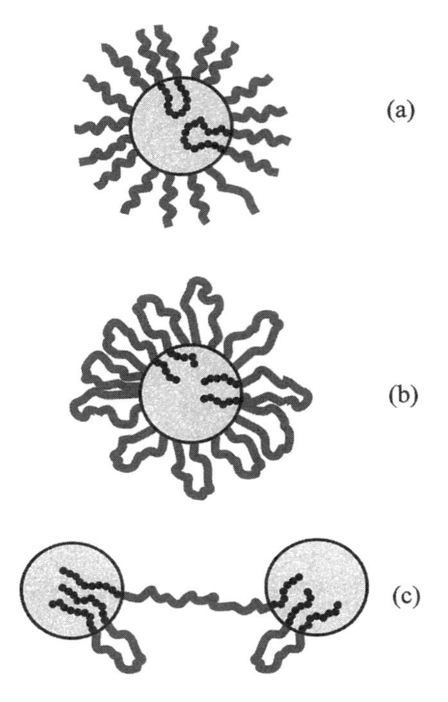

Figure 11 The structure of micelles of *ABA* triblock copolymers changes qualitatively with the selectivity of the solvent. A corona of singly anchored *A* chains is formed in a precipitant for the *B* blocks (a). A corona of *B* loops is expected when the solvent is a precipitant for the *A* blocks (b). In this case, exchange interactions and bridging are possible. When two micelles are at grazing contact, the coronal block can adopt two possible states: a loop (b) or a bridge (c).

At this point it is important to note that the exchange interaction depends on the residence time $\tau_{exchange}$ of the *A* blocks within the core. When $\tau_{exchange}$ is long in comparison to the duration of the experiment the exchange is, in effect, frozen and the resulting entropic attraction is not operative. Such is the case when the cores are in a glassy state or when N_A is very high. In this situation the behavior of the system depends on its history, i.e. method of preparation. One may consider two extreme situations [44]. (i) The exchange interactions are frozen when the coronas of the aggregates are comprised solely of *B* loops. In this case the interactions between the aggregates are purely repulsive. (ii) Bridging *B* chains were established before the exchange was frozen; i.e., the aggregates are joined by *ABA* triblock copolymers whose solvophobic *A* blocks reside in different cores. The aggregates are then connected by long-lived bridges giving rise to a strong physical gel. The strength of the physical bond between the aggregates in this case is related to N_A and to the strength of the chemical bonds in the chain. This is in marked contrast to the case of exchange attraction where these factors do affect the strength of the interaction. When the initial state is an ordered mesophase of *ABA* triblock copolymers it is possible to produce ''mesogels'' that retain the symmetry of the parent phase. The swelling behavior and mechanical properties of such mesogels are discussed later.

2. The Exchange Attraction

If $\tau_{exchange}$ is short, two aggregates of *ABA* triblock copolymers may gain entropy when the distance between them allows the *A* blocks to reside in two different cores. The resulting gain in entropy is large in comparison to the loss of translational entropy incurred when the two aggregates maintain close proximity. The resulting ''exchange attraction'' between the aggregates favors phase separation involving coexistence of a dense phase of closely packed aggregates, and a dilute phase [42,45]. The quantitative characteristics of the exchange attraction depend on the form of the aggregates. The geometry affects the fraction of chains that can undergo exchange via certain routes. (i) It determines the size of the contact area between two aggregates. For example, the ratio of contact area to total surface area is larger for two aligned lamellae than for two spherical micelles. (ii) The spatial distribution of midpoints in a corona of loops changes with the geometry. In starlike coronas the midpoints are expelled from the interior of the brush while in planar coronas they are distributed throughout the layer. The exchange attraction between starlike micelles is discussed in this section. Since the analysis of the lamellar case is more complicated it is discussed only briefly.

A simple analysis of the exchange interaction between two starlike micelles is possible when the micellar corona is described by the Daoud–Cotton model [53]. The validity of this assumption is discussed later. The first step is to estimate the contact area between two micelles. The depth of interpenetration of two coronas is set by the size of the outermost coronal blob $\xi_{out} = \xi(L) \approx L/p^{1/2}$. The associated repulsive free energy is of order of kT and the corresponding contact area is $A \approx L\xi_{out}$. Stronger interpenetration is unlikely because the resulting osmotic penalty [54] is much higher,

$$\frac{\Delta F_r}{kT} \approx p^{3/2} \ln \frac{L}{D} \tag{28}$$

where D is the distance between the two micellar centers. The second step is to estimate the number of chains that can undergo exchange in this situation. It is assumed that this number is proportional to the number of midpoints located within the overlap zone. Within the Daoud–Cotton model the midpoints are localized at the outer edge of the corona. Each ξ_{out} blob contains a single midpoint.

The number of chains that participate in the exchange interaction is thus $A/\xi_{out}^2 \approx p^{1/2}$. This number is much smaller than p because the spherical geometry imposes a relatively small contact area. Of the total surface area of the corona $R^2 \approx L^2$, only a small fraction $\xi_{out}/L \approx p^{-1/2}$ is involved. The exchange free energy per micelle is estimated by assigning an attractive energy of kT to each participating chain

$$\frac{\Delta F_{exchange}}{kT} \approx -p^{1/2} \tag{29}$$

The intermicellar potential U reflects the superposition of an osmotic repulsion and the exchange attraction. To obtain the second virial coefficient of micelle–micelle interactions, B, we approximate U as a square-well potential of width ξ_{out} and depth $|\Delta F_{exchange}|$ having an infinite wall at $L - \xi_{out}$. Accordingly, $B = \int_o^\infty [1 - \exp(-U/kT)] r^2 dr$ is

$$B \approx L^3 - L^2 \xi_{out} \exp(|\Delta F_{exchange}/kT|) \tag{30}$$

The first term represents the excluded volume of an impenetrable micelle. The exchange attraction gives rise to the second, negative, term. This term dominates whenever $p^{1/2} > \ln p^{1/2}$, as is the case when $p > 1$. Accordingly, solutions of micelles formed by *ABA* triblock copolymers in a selective precipitant for the *A* blocks are characterized by $B < 0$, irrespective of the aggregation number. In other words, such micellar solutions experience poor solvent conditions even though the medium is a good solvent for the *B* blocks. As a result, such solutions undergo phase separation leading to coexistence of dense and dilute phases. The dense phase, comprised of close-packed micelles, is a weak physical gel that will dissolve in an excess of solvent. Note further that the present situation differs from the familiar behavior of homopolymers in a poor solvent. In the present case there is no analogue to the collapse of a single chain. This feature will reemerge in polysoaps and in multiblock copolymers. Furthermore, the exchange interaction is an activation process. When the activation barriers are high enough the exchange is frozen, allowing for long-lived metastable states. In marked distinction, the attraction between monomers in a poor solvent does not involve an activation barrier.

In the case of spherical starlike micelles it is possible to estimate the exchange attraction on the basis of a ''blobological'' picture, the Daoud–Cotton model. Two ingredients are involved. The concentration profile predicted by this model $\phi \sim r^{-4/3}$ has been confirmed by computer simulations, self-consistent field (SCF) calculations, and scattering experiments [82]. A more fundamental issue concerns the *assumed* distribution of endpoints, i.e., their localization at the outer boundary of the corona. This assumption is consistent with the results of numerical studies that reveal an inner ''dead zone'' of width comparable to L from which endpoints are excluded. The situation is qualitatively different for lamellar systems endowed with planar brush coronas. In this case, there is no dead zone in the isolated brush and the endpoints are distributed throughout the layer. As a result, the Alexander model, the planar counterpart of the Daoud–Cotton model, does not provide a useful starting point for the analysis of the exchange interaction. Instead, it is necessary to base such discussion on SCF theory. In the case of a lamellar melt this leads to an equilibrium bridging fraction of [46–49]

$$q \approx (a/b^2)^{2/3} N^{-1/3} \tag{31}$$

This result agrees with the experimental measurements of Watanabe [50]. It is, however, important to stress that the equilibrium state may not always be attained due to kinetic reasons.

3. Mesogels and Bridging

The preceding discussion focused on the case of short $\tau_{exchange}$. In the following we consider the opposite case, when the exchange is effectively frozen. The behavior of such a system depends on the method of preparation. For concreteness we discuss gels obtained from mesophases of symmetric *ABA* triblock copolymers with identical *A* blocks. The starting point is a homogeneous melt of *ABA* triblock copolymers. As the temperature is lowered, the melt undergoes microphase separation. The nature of the mesophase thus formed, lamellar, cylindrical, or micellar, depends on the polymerization degrees of the blocks N_A and N_B. The phase diagram is essentially that of *AB* diblocks with polymerization degrees N_A and $N_B/2$. The cylindrical and lamellar domains can be aligned by application of shear. The shear aligned sample is then quenched to below the glass transition temperature T_g of the *B* blocks. The swelling behavior of this system differs qualitatively from the one expected of the corresponding *AB* diblock analogue. In this last case, one expects a dispersion of lamellae, cylinders, or micelles, depending on the initial mesophase. Each of the free aggregates is endowed with a corona of singly anchored *B* blocks. In the *ABA* case a long-lived gel is formed. The coronal *B* chains in this system can assume two states. When the two *A* blocks reside in the same core domain the *B* block forms a loop. If the two *A* blocks reside on different domains the *B* block forms a bridge. The bridging blocks give rise to a physical gel in which the crosslinks can be micelles, cylinders, or lamellae. Since $T < T_g$ the *A* blocks cannot exchange, $\tau_{exchange}$ is very long, and the bridging is effectively permanent. The gel will not dissolve in the presence of excess solvent, nor will it undergo a morphological change; i.e., the geometry of the domains as well as their orientation are retained. For future reference, note that the area per coronal chain, as set in the melt state, is also fixed. We refer to the gel thus formed as ''mesogel'' because it retains the morphology of the parent mesophase. The glassy state of the *A* blocks plays a crucial part in this scenario. It renders the crosslinks effectively permanent. When $T > T_g$ the procedure outlined above would yield a system of the type described in the previous section. In this case the morphology of the gel will change with the polymer concentration. Furthermore, such a gel will dissolve when the polymer concentration becomes sufficiently low.

The crosslinks in mesogels are the A domains. According to their morphology one may distinguish between lamellar, cylindrical, and micellar gels. In all three types of mesogels the crosslinks are of high functionality; i.e., many bridging chains are anchored to each crosslink. This is in marked contrast to the familiar, chemically crosslinked, gels where the functionality of the crosslinks is small, typically three to four. In addition, the spatial distribution of crosslinks in lamellar and cylindrical gels is anisotropic. The crosslinks are constrained, respectively, to planes and lines. Furthermore, the application of shear allows the production of macroscopic samples of well-aligned cylinders and lamellae. Such ''single crystal'' mesogels exhibit especially distinctive features. One qualitative distinctive feature is anisotropic swelling. Single crystal lamellar mesogels swell uniaxially, along the normal to the aligned A domains. On the other hand, cylindrical gels exhibit biaxial swelling while micellar gels swell isotropically. Other distinctive aspects concern the deformation behavior of the mesogels: anisotropic compressibility as well as asymmetry between compression and extension. The following discussion focuses on the simplest and most distinctive case, of lamellar, single crystal mesogels formed by neutral and flexible ABA triblocks. Other scenarios, involving polyelectrolytes or liquid-crystalline blocks, are possible and were reviewed in Ref. 51.

The quantitative aspects of the swelling equilibrium of mesogels are also distinctive. The equilibrium swelling of simple gels is often rationalized in terms of the c^* theorem [17]. Denoting the number of monomers in the chain segments between trifunctional crosslinks by N, this theorem states that the monomer concentration in a gel in equilibrium with a reservoir of good solvent c_e is

$$c_e \approx c^* \sim N^{-4/5} \tag{32}$$

Here c^* denotes the overlap threshold of a semidilute solution of free chains of polymerization degree N and radius $R_F \approx N^{3/5}b$; i.e., $c^* \approx N/R_F^3$. The swollen gel is thus viewed as a close-packed array of swollen coils of size N at grazing contact. This picture also specifies the compression and shear moduli of the gel K and μ, since both scale as the osmotic pressure $\sim kT/\xi^3$; that is, $K \sim \mu \sim c_e^{9/4}$. One may also formulate a c^* theorem for mesogels. However, the scaling behavior and the underlying picture are both modified. A lamellar mesogel in equilibrium with a reservoir may be viewed as a stack of swollen lamellae at grazing contact. In the symmetric case $N_A \approx N_B \approx N$ the area per coronal chain a as imposed by the preparation method is $a/b^2 \approx (\gamma b^2)^{-1/3}N^{1/3}$. The corresponding equilibrium thickness of the corona is thus $L/b \approx (\gamma b^2)^{1/3} N^{2/3}$ and the concentration of B monomers is

$$c_e \approx (b^2/a)^{2/3} \sim N^{-2/9} \tag{33}$$

As a result the compression modulus of lamellar gels scales as $K \sim N^{-1/2}$ instead of $K \sim N^{-9/5}$ as expected in simple gels. Similar c^* theorems apply to cylindrical and micellar mesogels. In these, the mesogel is viewed as a close-packed array of cylinders or micelles at grazing contact between their swollen coronas. The anisotropic swelling of well-aligned mesogels was experimentally observed by Folkes et al. [52] although the results were interpreted in a different language.

The deformation behavior of lamellar mesogels is qualitatively different from that of simple isotropic gels. Extension or compression of simple gels is accompanied by a change of cross-section. When the deformation rate is fast compared to the deswelling rate, the volume of the sample remains constant. In any case, the elastic properties of the gel are isotropic. The elastic behavior of a single crystal mesogel is clearly anisotropic. Extension or compression along the axes of the oriented lamellae are opposed, primarily by the rigidity of the glassy A domains. Strong strain gives rise to irreversible deformation due to domain breakage. The counterpart of the elastic behavior of gels is encountered when the deformation, compression, extension, or shear affects only the swollen B layer. To bring out the distinctive aspects of this scenario it is helpful to consider the deformation of ideal mesogels having perfectly rigid A domains. Within this picture, a single crystal lamellar mesogel can only be compressed along the lamellar normal. No change in the lateral cross-section is allowed. Similarly, compression of cylindrical mesogel may only involve a change of cross-section along the normal to the cylindrical domains. Consequently, these deformations are accompanied by a change of volume. Another characteristic aspect of mesogels concerns the state of the individual chains in the undeformed swollen gel: the B chains in mesogels are strongly stretched. This last feature, together with the geometrical constraints, gives rise to another distinctive feature: a qualitative difference between the compression and the extension. Compression of lamellar mesogels is opposed by the osmotic pressure of the B chains. The fraction of bridging chains does not affect the restoring force. Consequently the restoring pressure for strong compression is $P \approx \pi \approx kT/\xi^3$, where $\xi \sim \phi^{-4/3}$ and $\phi \approx Nb^3/aL$ leading to

$$P \sim L^{-9/4} \tag{34}$$

On the other hand, extension is opposed by the elasticity of the bridging B chains. B loops are only weakly involved. As a result, the extension behavior depends on the bridging fraction. Since the coronal chains are already stretched in equilibrium the extension is described by the Pincus force law. For strong extensions, the osmotic contribution is negligible, and the restoring pressure is due to the Pincus force of the bridging chains and

$$P \sim qL^{3/2} \tag{35}$$

where q is the fraction of bridging chains. Finally, the concept of affine deformation is often invoked in the description of simple gels. The relative deformation of an individual chain is assumed to be identical to that of the sample whose volume as a whole is maintained constant. This situation, although not necessarily realized, is possible for simple gels. For strained lamellar and cylindrical gels the chain deformation is nonaffine.

B. A Little on *ABC* Triblock Copolymers

The number of blocks, or amphiphilic motifs, is an important architectural design parameter of copolymeric surfactants. It is possible to change the number of blocks while maintaining the number of incompatible chemical moieties involved. Pursuing this route one proceeds from *AB* diblocks to linear *ABA* or *BAB* triblocks and then to various linear $(AB)_n$ multiblocks. The addition of a third block gives rise to qualitative change in the solution phase behavior and in the mechanical properties of the mesophases. However, there is no change in the morphology of the phases and the aggregates. Alternatively, it is possible to change the chemical identity of the third block and to obtain an *ABC* instead of *ABA* triblock copolymer. This gives rise to qualitatively novel features, the most outstanding being the phase diagram of the melt. Distinctive mesophases, having no counterpart in the phase diagram of *AB* diblock copolymers or of low molecular weight amphiphiles, emerge. This topic is addressed at length in Chapter 10. Accordingly the discussion of these systems is limited to a brief commentary on the phase behavior, the distinctive features of *ABC* amphiphiles, and the design strategies afforded by these copolymers.

The underlying factors determining the melt phase behavior of *ABC* triblock copolymers and of *AB* diblock copolymers are essentially identical. In the strong segregation limit, when all the blocks are highly incompatible, the relative stability of the mesophases is determined by the interplay of three main factors: the interfacial free energies of the boundaries between the domains, the elastic penalty due to the stretching of the blocks whose junctions straddle the domain boundaries, and the constraint of constant melt density. Yet, the repertoire resulting from the presence of a third, chemically different, *C* block is much richer. Certain morphologies incorporate the familiar elements of lamellae, cylinders, and spheres. In the case of *ABC* copolymers different elements may coexist; for example, cylinders of *B* can be embedded at the boundary between lamella of *A* and *C*. Furthermore there appear hitherto unfamiliar motifs, such as cylinders threading rings. These were not anticipated prior to their discovery by Stadler et al. [see Chap. 6]. In addition to these features, mesophases of *ABC* triblock copolymers are noteworthy because of the suppression of loops. When the three blocks are incompatible the terminal blocks cannot reside in the same domain and all *B* blocks form bridges.

The suppression of loops is an example of the effect of the incompatibility of the terminal blocks. The incompatibility of the *A* and *C* blocks may also provide a mechanism for the control of other material properties. The considerations involved are illustrated by a strategy proposed by Petschek and Wiefling [55] for the design of ferroelectric liquid crystals [29]. In its simplest form, this involves *ABC* triblock copolymers incorporating a rodlike *B* block endowed with a dipole moment aligned with its axis. In addition, the synthesis should impose a consistent alignment of the *AB* and *BC* junctions with respect to the electrical dipole of the *B* block. In other words, the electrical dipole is consistently aligned with the "chemical dipole." Such polymers self-assemble into aggregates, micelles, etc. having a "liquid crystalline" core of rods with their geometrical axes aligned. A local alignment of the electrical dipoles is expected when incompatibility between the *A* and *C* blocks is sufficiently strong so as to favor *AC* segregation (Fig. 12). Unfortunately, the implementation of this scheme to align the electrical dipoles of macroscopic domains faces difficulties best illustrated for

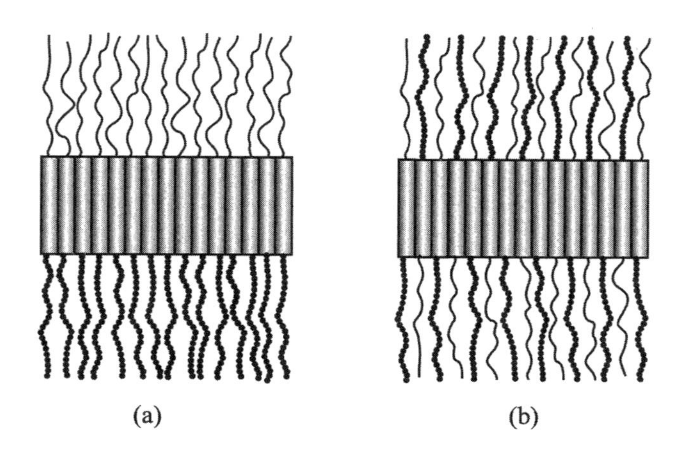

Figure 12 The Petschek–Wiefling design for ferroelectric liquid crystals. The incompatibility between the *A* and *C* blocks can result in microphase separation of the coronal blocks thus aligning the electric dipole residing on the middle, rodlike, *B* block. The aligned state is depicted in (a) and the unaligned situation, with mixed coronas, in (b). (Adapted from Ref. 29.)

the case of a melt lamellar phase. In this situation the *ABC* lamellae are expected to form *ABCBA* bilayers of zero net dipole. Furthermore, the competition between the long-range electrostatic interactions and the short-range chemical interactions opposes the global alignment of the dipoles.

The term ''amphiphile'' implies an affinity to two different media. Familiar amphiphilic molecules incorporate two incompatible components that give rise to this behavior. Similarly, in *AB* and *ABA* block copolymers there are two incompatible blocks of different solubility. However, *ABC* triblock copolymers incorporate three chemically different blocks. When the three blocks are mutually incompatible and of different solubilities the *ABC* surfactants can exhibit affinity to three different media rather than two. The consequences of this higher ''functionality'' have not been explored in detail. For example, little attention has been given to their behavior at interfaces. Linear *ABC* triblock copolymers or the corresponding star copolymers may be able to form two-dimensional mesophases [56]. This can occur at the interface between two fluids *I* and *II* such that the *B* block is selectively solubilized in *I* while *A* and *C* are only soluble in *II*. In this situation, the *A* and *C* blocks are constrained to the surface and bound to each other. A two-dimensional amphiphile is obtained when the *A* and *C* blocks are incompatible. A dense monolayer of this type should undergo microphase separation leading to the formation of circular and striped mesophases. Note that cylindrical and lamellar mesophases are indistinguishable in this case. A mixed monolayer comprised of *BC*, *BA*, and *ABC* block copolymers will mimic the behavior of amphiphiles in the presence of two two-dimensional and incompatible fluids. When the *ABC* copolymers are a minority component, they should straddle the boundary line between the two-dimensional *A* and *C* phases.

V. MULTIBLOCK COPOLYMERS AND POLYSOAPS

As we have discussed, the phase behavior of triblock copolymers differs qualitatively from that of diblock copolymers and of short-chain surfactants. However, there is no corresponding gap between the singlechain behavior of diblock and triblock copolymers. $(AB)_n$ multiblock copolymers, like *ABA* triblock copolymers, can form physical gels. However, multiblock copolymers exhibit, in addition, a qualitatively distinct single-chain behavior: a hierarchial intrachain self-assembly that modifies their configurations and elasticity. Interestingly, when the self-assembly involves only components of a single-chain there is no associated loss of translational entropy and thus no CMC. The complete repertoire of this category is yet to be explored. Our discussion focuses mainly on polysoaps [44]. However, much of the discussion applies also to the two related families of *AB* graft copolymers and linear $(AB)_n$ multiblock copolymers [57]. Note that intrachain self-assembly is not unique to linear chains. It can occur, in a variety of different architectures such as in ''super H'' multiblock copolymers [58] where two A stars are joined by a linear B chain (Figure 3i).

Polysoaps are flexible hydrophilic polymers that incorporate, at intervals, covalently bound amphiphilic monomers. As a model system, polysoaps afford two useful advantages: First, models of micelles of free unpolymerized amphiphiles provide a natural starting point for the analysis of this system. Second, it is possible to draw on the extensive experimental data accumulated on the interaction energies and aggregation numbers of monomeric surfactants. In addition, the synthesis of long polysoaps with numerous amphiphilic comonomers and well-controlled architecture is possible. In marked contrast, the synthesis of $(AB)_n$ with large n poses problems yet to be overcome. Finally, the modeling of polysoaps and the corresponding graft copolymers differ only in the details of the micellar structures. The following discussion concerns polysoaps in which the m amphiphilic monomers are joined by monodispersed, flexible, hydrophilic spacer chains comprising $n \gg 1$ monomers of size b. It is mostly limited to the case of polysoaps that form spherical intrachain micelles. The overall polymerization degree is $N \approx mn \gg 1$. We focus primarily on the case of $m \gg 1$ so that a single polysoap can form numerous intrachain micelles. The amphiphilic monomers are characterized by the volume v and the length l of their hydrophobic tails. For simplicity only amphiphilic monomers that do not adsorb onto the backbone are considered. To this end we also limit the discussion to solutions of high ionic strength where long-range electrostatic interactions are screened out. The case of short spacers and low ionic strength was considered by Turner and Joanny [59].

A. Intrachain Micelles and Their Interactions

The structure of the intrachain micelles is reminiscent of that of micelles formed by monomeric, nonpolymerized amphiphiles. The hydrophobic tails form an inner dense core and the ionic head-groups are localized at the core–water interface. The intrachain micelles are, however, surrounded by a swollen corona of loops formed by the flexible hydrophilic spacers joining the amphiphiles (Fig. 13). The corona of spherical micelles is similar to the corona of star polymers. It is thus convenient to describe the intrachain micelles by combining two models [60]: the phenomenological model of Israelachvili et al. for ''simple'' micelles [11,20], and the Daoud–Cotton model for the corona of star polymers [53]. In particular, it is necessary to supplement the free energy per surfactant, as given by the model of Israelachvili et al., by a fourth term reflecting the repulsion between the coronal loops. For spherical micelles this term is $kTp^{1/2} \ln R_{\text{corona}}$, where $R_{\text{corona}} \approx n^{3/5} p^{1/5} b$ is the span of the corona in a good solvent. This term, like the head-group repulsion, favors smaller micelles. The first three terms, on their own, describe a free micelle formed by unpolymerized amphiphiles. It is thus expedient to use the equilibrated free micelles as a reference state. It specifies a size scale a_o or $p_o \approx$

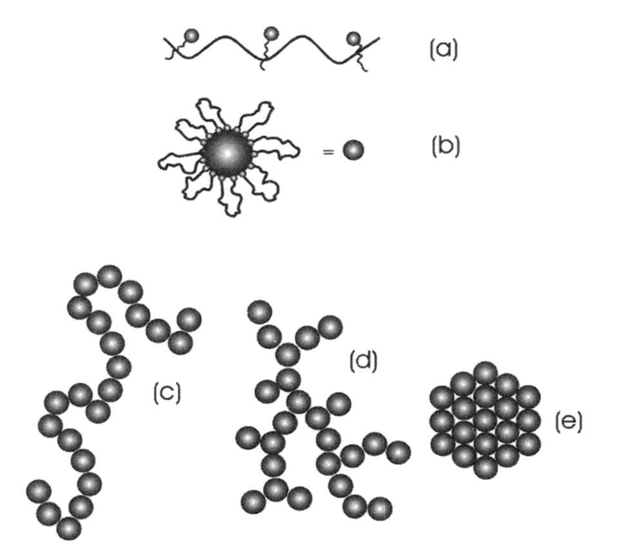

Figure 13 The structural hierarchy of polysoaps involves three levels. The primary structure, the monomer sequence, is set by the chemistry (a). The intrachain micelles introduce a secondary structure (b). The configurations of the micellar string, linear (c), branched (d), or globular (e), define the tertiary structure. (Adapted from Refs. 45 and 66.)

v^2/a_o^3 and an energy scale γa_o. In turn, this suggests a dimensionless variable $u = p/p_o = (a_o/a)^3$. Utilizing the packing condition $p \approx v^2/a^3$ it is possible to express the free energy per amphiphile in a spherical intrachain micelle as

$$\epsilon_p/kT \approx -\delta + \gamma a_o(u^{-1/3} + u^{1/3} + \kappa u^{1/2}) \tag{36}$$

The $u^{-1/3}$ term reflects the surface free energy, $u^{1/3}$ allows for the head-group repulsion, while the last term $\kappa u^{1/2}$ is due to the coronal contribution. Here $\kappa \approx kTp_o^{1/2} \ln n/\gamma a_o$ is a dimensionless parameter, the ratio of the coronal and the head-group penalties when $a = a_o$. κ measures the relative importance of the coronal term compared to γa_o. Because the corona is large compared to the core, the intrachain micelles are significantly larger than the corresponding free micelles. However, the equilibrium aggregation number of intrachain micelles p_{eq} is smaller because of the coronal penalty. The importance of this last contribution is determined by κ. When $\kappa \ll 1$ the coronal penalty is negligible and $u_{eq} \approx 1$. In this case the equilibrium area per head-group in the intrachain micelle a_{eq} is essentially that of a free micelle; that is, $a_{eq} \approx a_0$. The aggregation numbers are also comparable: $p_{eq} \approx p_o$. In the opposite limit, when $\kappa \gg 1$, the equilibrium aggregation number is much smaller. The coronal term $\kappa u^{1/2}$ is comparable to the surface term $u^{-1/3}$ thus leading to $u_{eq} \approx \kappa^{-6/5}$.

The analysis presented above focused on spherical intrachain micelles. Cylindrical intrachain micelles may occur when the amphiphilic comonomers form cylindrical micelles in their free state. However, this geometry is thermodynamically stable only for short spacer chains. Spherical intrachain micelles are favored when n increases because F_{corona} of starlike coronas is lower.

The structure of the corona of spherical intrachain micelles can be compared to that of star polymers or that of starlike micelles formed by *AB* diblock copolymers. Within this view, the role of the loops is overlooked. In effect, one considers a corona where the loops were cut in half. The logarithmic correction thus ignored has little effect on the equilibrium structure of the micelle. This aspect is, however, important when considering the interactions between micelles. In this context, the intrachain micelles bear closer similarity to micelles of *ABA* triblock copolymers in a selective solvent for the *B* blocks. In both systems, exchange interactions and bridging play an important role. As was discussed in Section IV.A, the exchange free energy between two spherical micelles is attractive [45] and scales as $\Delta F_{ex}/kT \approx -p^{1/2}$. In turn, ΔF_{ex} gives rise to a negative second virial coefficient for micelle–micelle interactions $B < 0$, whenever $p > 1$. Thus polysoaps experience poor solvent conditions even though the backbone is hydrophilic. This has two consequences. One is the possibility of phase separation involving the formation of physical gel, as in the case of *ABA* triblock copolymers. The second, to be discussed shortly, concerns the configuration of isolated polysoaps. In particular, the equilibrium configuration of long polysoaps is collapsed; that is, $R \sim N^{1/3}$. Before we discuss this issue, it is helpful to note an important difference between the situation described above and that of "simple" flexible homopolymers in poor solvents. The attraction between monomers in a poor solvent is instantaneous. The exchange attraction between micelles is not. The exchange is an activation process involving the expulsion of a hydrophobic tail out of the micellar core and into the aqueous medium. The associated characteristic time scales as $\tau \approx \exp(-\delta)$ and is thus tunable by varying the length of the tail. The exchange interaction is only relevant when the observation time is much longer than τ.

B. Large Scale Tertiary Structure

There is a clear hierarchy in the intrachain self-assembly of polysoaps (Fig. 13). A short polysoap forms a single intrachain micelle. Long polysoaps can form numerous intrachain micelles. In such a case it is useful to distinguish between three levels in the hierarchy of self-assembly [45,60]. As in protein science, it is convenient to refer to the chemical sequence as primary structure. The next level, the secondary structure, consists of the intrachain micelles. The configuration of the string of micelles defines the tertiary, large scale, structure. Three extreme scenarios may be envisioned for the tertiary structure: a linear string, a branched one, and a spherical globule of close-packed micelles. Intermediate situations are also possible. The globular collapsed state is thermodynamically favored when the exchange attraction is operative. It is nevertheless useful to consider the different configurations because they may occur experimentally as metastable states. This is because the transition between the different configurations involves an activation process, that is, repartitioning of amphiphiles among the different micelles.

A linear string of micelles is expected to behave as a self-avoiding chain where micelles play the role of monomers. The size of the linear string is

$$R_{\text{string}} \approx \left(\frac{m}{p}\right)^{3/5} R_{\text{micelle}} \tag{37}$$

Typically the span of the corona is much larger than the radius of the core and the size of the micelle is $R_{\text{micelle}} \approx R_{\text{corona}}$. The linear configuration is actually least stable. When bridging interactions do not play a role, the most favorable configuration is branched. Each micelle can be a branching site thus leading to a gain in configurational entropy. In this case

$$R_{\text{branched}} \approx \left(\frac{m}{p}\right)^{1/2} R_{\text{micelle}} \tag{38}$$

When bridging interactions operate, the stable configuration is globular. This state is analogous to the collapsed configuration of a flexible homopolymer. Because the second virial coefficient for micelle–micelle interactions is strongly negative the polysoap forms a spherical globule of close-packed micelles with a radius

$$R_{\text{globule}} \approx \left(\frac{m}{p}\right)^{1/3} R_{\text{micelle}} \tag{39}$$

In this context the intrachain micelles are analogous to collapse blobs. It is possible to assign an effective surface tension $kT\gamma_g$ to the boundary of the globule. It is traceable to the inability of the outermost micelles to undergo exchange interactions at the exterior boundary of the globule. The resulting excess free energy is roughly $p^{1/2}kT$ per micelle leading to

$$\frac{\gamma_g}{kT} \approx \frac{p^{1/2}}{R^2_{\text{micelle}}} \tag{40}$$

It is important to note three features that distinguish the globular form of polysoaps from the collapsed state of flexible homopolymers. First, while the polysoap chain is globally collapsed $R_{\text{globule}} \sim m^{1/3}$, the coronas of the constituting micelles are swollen; i.e., on length scales smaller than R_{micelle} the density is not uniform. Second, as noted previously, the exchange attraction is an activation process and metastable nonglobular configurations are possible. Finally, $\gamma_g \sim p^{1/10}/n^{6/5}$ depends not only on T but also on n. It thus possible to tune γ_g by changing the length of the spacer chains.

C. Extension of Globular Polysoaps

The distinctive features of the elasticity of polysoaps arise because of the coupling of the strain to the "internal" degrees of freedom associated with the secondary and tertiary structure. Both the tertiary and secondary structures can re-equilibrate thus affecting the corresponding force law according to the Le Chatelier principle [61–63]. In the the case of globular polysoaps it is possible to distinguish between two regimes. For weak deformations only the tertiary, large scale, structure is affected. In particular, the globule unfolds into a string of micelles (Fig. 14). Stronger deformations couple with the secondary structure, the intrachain mi-

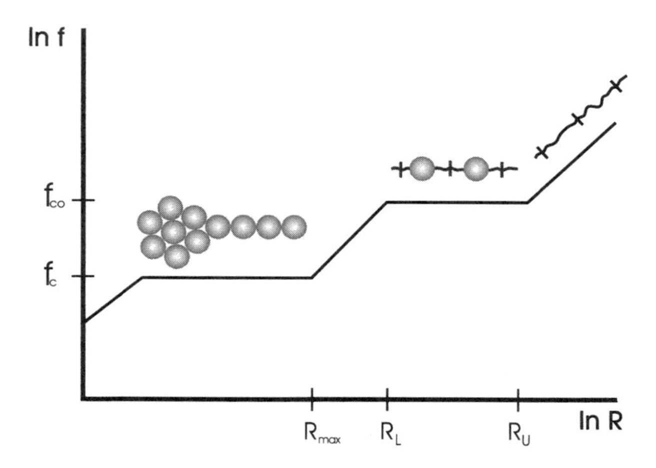

Figure 14 The main features of the force law associated with the extension of a globular polysoap. Two coexistence regimes are involved. For weak deformations, a globule of closely packed micelles coexists with a stretched string of micelles. Stronger deformations involve coexistence of dissociated amphiphiles and weakly perturbed intrachain micelles. Eventually, all the amphiphiles are dissociated. The intrachain micelles are depicted as spheres while the dissociated amphiphiles are represented as rods. (Adapted from Ref. 63.)

celles, by favoring micellar dissociation. When the number of intrachain micelles m/p is large and $R_{\text{globule}} \gg R_{\text{micelle}}$, the two processes involve well-separated length and force scales.

The extension of the globular configuration is similar to the stretching of a collapsed flexible chain [64,65]. The spherical form of the globule is initially deformed into an ellipsoid while maintaining constant volume corresponding to close packing of the intrachain micelles. Within this linear response regime the free energy penalty incurred is due to the increase of the surface free energy

$$F/kT \approx \gamma_g \Delta A \approx \gamma_g (R - R_{\text{globule}})^2 \tag{41}$$

where $\Delta A \approx (R - R_{\text{globule}})^2$ is the surface area increment associated with the deformation. The corresponding restoring force $f = -\partial F/\partial R$ is proportional to the strain $(R - R_{\text{globule}})$,

$$f/kT \approx -\gamma_g (R - R_{\text{globule}}) \tag{42}$$

This type of process cannot proceed indefinitely. If pursued, the distorted globule will assume a cylindrical shape and, eventually, form a string of micelles. This scenario gives rise to a van der Waals loop in the fR diagram. This is indicative of instability with respect to a coexistence of a weakly elongated globule and a stretched string of micelles. This effect is reminiscent of the Rayleigh–Plateau instability involving the breakup of a fluid jet into a succession of droplets. Within the coexistence regime the chain is comprised of a string of m'/p micelles and a roughly spherical globule of $(m - m')/p$ micelles. The free energy of this configuration is approximately

$$\frac{F}{kT} \approx \gamma_g R^2_{\text{micelle}} \left[\left(\frac{m - m'}{p} \right)^{2/3} + \frac{m'}{p} \right] \tag{43}$$

where the two terms reflect, respectively, the surface free energy of the globule and of the stretched micellar string. The end-to-end distance is

$$R \approx R_{\text{micelle}} \left[\left(\frac{m - m'}{p} \right)^{1/3} + \frac{m'}{p} \right] \tag{44}$$

Here the first term is the radius of the globule and the second is the span of the stretched string of micelles. When m' is sufficiently large we may approximate dR as $R_{\text{micelle}} dm'/p$ and the corresponding force law $f = -\partial F/\partial R$ is

$$f/kT \approx -\gamma_g R^2_{\text{micelle}} (R^{-1}_{\text{micelle}} - r^{-1}_{\text{globule}}) \tag{45}$$

where $r_{\text{globule}} \approx R_{\text{micelle}}((m - m')/p)^{1/3}$ denotes the radius of the partially depleted globule. It is important to note that f decreases as r_{globule} approaches R_{micelle}. As a result different scenarios are expected for the f = const′ and the R = const′ ensembles. In the first case the globule unravels completely once a critical force $f_c/kT \approx -\gamma_g R_{\text{micelle}} \approx p^{1/2}/R_{\text{micelle}}$ is applied. No globule–coil coexistence is expected. Such a coexistence is, however, expected when the end-to-end distance is imposed. In this case the tension in the string is f_c. The onset of the coexistence occurs when $R \approx R_{\text{globule}} + R_{\text{micelle}}$ as seen by equating f_c/kT to Eq. (42). The upper boundary of this regime corresponds to a fully extended string of micelles $R_{\max} \approx (m/p)R_{\text{micelle}}$.

Some insight regarding f_c may be gained by noting that the corresponding tensile energy per micelle f_cR is comparable to the exchange free energy of a micelle at the surface of the globule $p^{1/2}kT$. It is useful to note that f_c may also be expressed as $f_c/kT \approx 1/\xi_o$ where ξ_o is the size of the outermost coronal blob $\xi_o \approx R_{\text{micelle}}/p^{1/2}$. Accordingly, f_c is comparable to the tension at the periphery of the unperturbed corona due to the crowding-induced stretching of the loops. Hence, subjecting the string of micelles to a tension f_c does not perturb the micellar structure.

Further extension of the unraveled string of micelles is accommodated first by stretching the bridges between the intrachain micelles. Eventually it enforces the dissociation of some of the micelles. This coexistence regime ends when all the micelles are dissociated and the chain deformation proceeds as in a ''simple'' flexible chain (Fig. 14). The free energy of the chain in the stretched bridges regime is

$$\frac{F_B}{kT} \approx m\epsilon_o + \left(\frac{R}{R_B}\right)^{5/2} \tag{46}$$

The first term allows for the lowered free energy of amphiphiles incorporated into unperturbed micelles while the second accounts for the Pincus elastic penalty [17] due to the strong stretching of the micellar string. This regime involves strong stretching of m/p bridges comprising each of n monomers. Consequently, the ''elastic constant'' is specified by $R_B \approx (mn/p)^{3/5}b$. The corresponding force law is

$$\frac{f}{kT} \approx \left(\frac{R}{R_B}\right)^{3/2} R_B^{-1} \tag{47}$$

This regime lasts while the elastic penalty $(R/R_B)^{5/2}$ is small compared to $m\epsilon_o$, i.e., when $\tau_b \approx (m\gamma a_o)^{-1}(R/R_B)^{5/2} \ll 1$. The micellar structure within this regime is only weakly perturbed. Upon further extension of the chain the elastic penalty couples strongly with the micellar structure. In this situation it is no longer possible to view the chain as a uniform string of intrachain micelles. Rather, the extension is associated with a coexistence of intrachain micelles and dissociated amphiphiles. The detailed analysis of this scenario is somewhat tedious. The essential features of this regime may be recovered, however, using a simple argument. Two free energies are involved: the free energy [Eq. (46)] of a string of micelles in the stretched bridges regime, F_B; and the free energy of a fully dissociated, strongly stretched chain $F_{dis}/kT \approx (R/R_F)^{5/2}$, where $R_F \approx (mn)^{3/5} b \approx p^{3/5} R_B$ is the Flory radius of the flexible swollen backbone. The bottom of the free energy curve F_B is located at $R_B < R_F$ and is $m\,|\epsilon_o|$ below the minimum of F_{dis}. For small R, F_B is lower than F_{dis}. However, since $R_B < R_F$ the "spring constant" of the fully dissociated chain is weaker. Consequently, the two curves cross at $R_{co} \approx R_F[m\,|\epsilon_o|/(p^{3/2} - 1)]^{2/5}$. For $R > R_{co}$ the fully dissociated chain is of lower free energy. The crossover of the free energy curves is a rough indicator for a first-order phase transition involving a coexistence between micellized and dissociated amphiphiles. This simple view suggests that the coexistence regime is associated with a plateau in the force law with $f_{co} \sim R^o$. Strictly speaking, this simple view is wrong. Since the interactions in this system are short-ranged, the mixing entropy of the one-dimensional mixture of micelles and dissociated amphiphiles disallows a first-order phase transition [40]. As a result the tension in the plateau regime is not independent of R. It exhibits instead a weak logarithmic dependence. The crossover regimes at the boundaries of the coexistence regime are also smoothed out. Nevertheless, the force diagram obtained by ignoring the mixing entropy, the $S_{mix} = 0$ approximation, yields the correct tension and length scales. These can be obtained by the following argument. The onset of micellar dissociation is expected to occur when the elastic energy of a stretched bridge is comparable to the micellization free energy of an amphiphile incorporated into an unperturbed micelle. Denoting the span of a stretched bridge by r_{co} we have

$$f_{co} r_{co} \approx \epsilon_o \approx kT\,(\gamma a_o - \delta) \tag{48}$$

Since the bridge is strongly stretched, the Pincus force law applies [17]. Consequently,

$$r_{co} \approx nb\left(\frac{f_{co} b}{kT}\right)^{2/3} \tag{49}$$

The combination of Eqs. (48) and (49) yields

$$\frac{f_{co} b}{kT} \approx \left(\frac{|\epsilon_o|}{n}\right)^{3/5} \tag{50}$$

Within the $S_{mix} = 0$ approximation the coexistence regime, when $f_{co} \sim R^o$, occurs in a sharply defined range $R_L < R < R_U$. The boundaries are smeared out when allowing for mixing entropy. Nevertheless, R_L and R_U provide a good approximation for the boundaries of this regime. To obtain R_L we utilize Eq. (47) in the form $f_{co}/kT \approx (R_L/R_B)^{3/2}R_B^{-1}$ leading to

$$R_L \approx R_B\left(\frac{f_{co}R_B}{kT}\right)^{2/3} \tag{51}$$

Similarly we obtain R_U from the force law of the fully dissociated chain $f_{co}/kT \approx (R_U/R_F)^{3/2}R_F^{-1}$ leading to

$$R_U \approx R_F\left(\frac{f_{co}R_F}{kT}\right)^{2/3} \approx pR_L \tag{52}$$

For $R > R_U$ all amphiphiles are fully dissociated, there is no trace of the intrachain self-assembly, and the corresponding force law is

$$\frac{f}{kT} \approx \left(\frac{R}{R_F}\right)^{3/2} R_F^{-1} \tag{53}$$

Our initial considerations, concerning the globule and its unraveling, followed the discussion of the deformation behavior of a collapsed homopolymer. The collapsed globule is implicitly assumed to be a closely packed linear string of micelles. Such a view is meaningless in light of the exchange interactions favoring the globular state. It is thus impossible to assign an amphiphile to a specific micelle and the ''topology'' of the micellar string is not defined. However, the validity of our analysis does not depend on the topology of the micellar string within the globule. It only requires that the extension be slow enough to allow for the repartitioning of the amphiphiles into the ''tadpole'' configuration obtained in the coexistence regime. This is certainly the case for the equilibrium force laws considered.

D. A Little on the Interaction of Polysoaps and Free Amphiphiles

The configurations of polysoaps change in the presence of free amphiphiles. In this case, mixed micelles (Fig. 15), comprising both free and polymerized amphiphiles, can form [45,60]. Pure intrachain micelles coexist with free surfactants while the concentration of the free surfactants is low enough. When their concentration exceeds the critical aggregation concentration CAC, mixed micelles appear. The fraction of polymerized amphiphiles in these micelles α decreases as

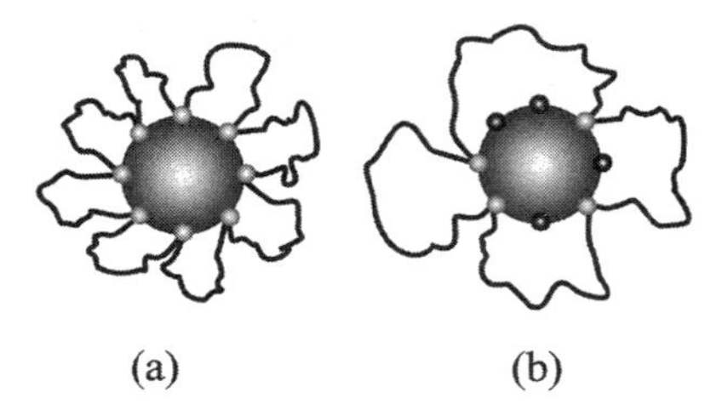

Figure 15 Below the CAC, the intrachain micelles consist solely of polymerized amphiphiles that are covalently bound to the chain (a). Mixed micelles, comprised of both free and polymerized amphiphiles appear above the CAC (b). The head-groups of the polymerized and free amphiphiles are depicted, respectively, as light and dark spheres. (Adapted from Ref. 66.)

the concentration of free surfactants increases. Eventually, a saturation concentration is reached when every mixed micelle contains, roughly, a single polymerized amphiphile. The formation of mixed micelles strongly affects the configurations and the dimensions of the chain. Three effects are involved: (i) the number of coronal loops decrease with α thus causing shrinking of the micellar corona; (ii) the exchange interactions decrease with α and eventually vanish at the saturation; and (iii) the total number of micelles increases as α decreases and, as a result, the chain span increases. At saturation the chain is completely unfolded. All traces of the intrachain self-assembly are lost and the chain exhibits the swollen configurations of a simple homopolymer. This effect clearly modifies the deformation behavior of the polysoaps [66]. The intrachain self-assembly dominates the elastic behavior below the CAC. In the saturation range there is no trace of the intrachain self-assembly and the deformation behavior follows the elastic behavior of a simple homopolymer.

VI. ON WATER SOLUBLE POLYMERS

The theoretical description of polymeric surfactants, as portrayed above, is based on the Flory theory. In turn, this is a modification of regular solution theory. Each type of monomer is assigned a *single* Flory χ interaction parameter that characterizes the solvent quality. The enthalpic term is assumed to be of the form $\chi\phi(1 - \phi)$. A second characteristic, the dielectric constant ε, appeared in the discussion of polyelectrolytes. It is important to note the limitations of this picture. It provides a useful description of many polymer–solvent and polymer–

polymer systems. In particular, it is effective in the case of nonpolar organic solvents. It has, however, clear shortcomings in the case of waterborne systems. In these, the polymeric surfactants must incorporate hydrophilic blocks. In turn, hydrophilic polymers belong to one of two categories: polyelectrolytes and neutral hydrophilic polymers that are water soluble because of the formation of hydrogen bonds. While the behavior of polyelectrolytes remains to be fully elucidated, the Flory type approach does not appear to pose problems. Such difficulties do appear in the case of neutral, hydrogen bonding, hydrophilic polymers. The most striking example is poly(ethylene oxide) (PEO). The following observations are of particular importance. (i) Aqueous solutions of PEO exhibit both upper and lower critical solution temperatures. By contrast, the Flory theory predicts only an upper critical point [67]. (ii) Analysis of calorimetric data of PEO solutions in terms of the Flory free energy yields a ϕ-dependent χ parameter [68]. Within the Flory type theories the χ parameter is independent of ϕ. (iii) There are indications that PEO segments may assume helical configurations in water [69]. A single χ parameter cannot fully describe the interactions with water of both helical and random chain segments. While most theoretical models of PEO focused on the first point, it appears that all incorporate some form of a ‘‘two state’’ assumption. In other words, the PEO monomers are allowed to exist in two different states. In the models of Tanaka and Matsuyama [70] and of Bekiranov et al. [71] the distinction is between a bare hydrophobic backbone and segments that bind water. Karlström postulates two different monomeric, interconverting monomeric states of different hydrophilicity [72]. Within the *n*-cluster model, proposed by de Gennes, the distinction is between monomers that are part of stable clusters of *n* monomers and those that are not [73]. These refinements of the Flory theory can lead to significant modification of the coronal structure. The effect is most apparent in the case of a flat brush when the grafted chains have no translational entropy, i.e., in the limit of $N \to \infty$. It occurs when the free energy of the free polymer solution in the limit of $N \to \infty$ allows for phase separation involving two phases of *finite* concentration; that is, $0 < \phi_- < \phi_+ < 1$ [74]. The chemical potential μ of the monomers in the brush varies parabolically with the distance from the grafting surface. Consequently, the coexistence condition $\mu(\phi_-) = \mu(\phi_+)$ can lead to discontinuity in the concentration profile of the brush [75]. This corresponds, in effect, to vertical phase separation of an inner dense phase which is hydrophobic and an outer dilute phase which is hydrophilic. No such effect occurs within the familiar Flory theory which predicts, for this case, a phase separation involving a neat solvent $\phi_- = 0$. The consequences of this effect on the structure of micelles and other aggregates have not yet been fully analyzed. However, it provides a possible explanation for the deviations of micelles with PEO coronas from the scaling behavior described in Section III.B [76].

VII. POLYMERIC SURFACTANTS VS. ASSOCIATING POLYMERS AND SURFACE ACTIVE POLYMERS

Polymeric surfactants are, by definition, surface active. On the other hand, not all surface active polymers are amphiphilic. In effect, every polymer is surface active in the sense that it shows a strong tendency to adsorb. This, as we discussed, is a consequence of the polymerization: the translational entropy per monomer decreases as $1/N$ while the interaction energy between a monomer and the surface is roughly independent of N. Polymeric surfactants are distinguished by their ability to self-assemble because of microphase separation.

The term ''associating polymers'' is often used to denote polymers capable of forming reversible physical networks [77–79]. Some polymeric surfactants, such as *ABA* triblock copolymers and polysoaps, fall into this category. However, the overlap between these two families is clearly only partial since *AB* diblock copolymers are polymeric surfactants that are incapable of forming physical gels. The boundary between these two families is determined, in part, by the number of amphiphilic motifs, ''stickers,'' within the chain. This is the number of elements within the chain that are capable of self-assembly on their own. One can imagine cutting an *ABA* triblock copolymer into two *AB* diblock copolymers and the number of stickers is thus two. Physical gels can only form when the number of amphiphilic stickers is two or higher.

In addition, there are associating polymers that are not polymeric surfactants in the sense that their self-assembly does not involve microphase separation of incompatible components of the same chain. Association behavior can result from the presence of ''bonding stickers,'' for example, structural elements capable of forming hydrogen bonds. Bonding stickers differ from amphiphilic stickers in two important respects. One distinction relates to the functionality of the stickers, i.e., the number of stickers that bind to a given sticker. The functionality of bonding stickers is typically small, two to four, and sharply defined. It is determined by the chemical design of the sticker, for example, the number of functionalities capable of forming hydrogen bonds and their relative orientation. The functionality of amphiphilic stickers is much higher. It is roughly comparable to the aggregation number of the self-assembled units, micelles, etc. In this case, the functionality is determined, in principle, by the thermodynamic equilibrium conditions that set the aggregation number. By contrast to the bonding stickers, the functionality of the amphiphilic stickers is not sharply defined because of fluctuations around the equilibrium state. Note that the condition for association behavior in the two cases may be different. Formation of physical gels by polymers bearing bifunctional stickers can only occur if the minimal number of stickers is three rather than two. The two association modes give rise to qualitative differences in the behavior of the system. Amphiphilic stickers can lead to formation of the mesophases observed for the unpolymerized amphiphiles. In distinction, the self-assembly of associating polymers bearing bonding stickers is not expected to give rise to such ordered states. In addition, amphiphilic stickers can interact with free amphiphiles thus giving rise to a rich repertoire that is absent in the case of bonding stickers.

The applicability of thermodynamic theory of self-assembly, as described above, is limited to systems that reach an equilibrium state. The attainability of equilibrium depends largely on the strength of the stickers as measured by their association energy as compared to the thermal energy kT. When the stickers are too strong thermodynamic equilibrium may become unattainable. In this case the aggregation is in effect irreversible and thus subject to kinetic control. Note, however, that equilibrium structures may obtain if the sample history allows for equilibration at an intermediate stage.

Altogether, the qualitative aspects of the behavior of polymeric surfactants are determined by the interplay of the three factors described above: the number of stickers, their functionality, and their strength.

VIII. CONCLUDING REMARKS

Polymeric surfactants display a rich repertoire of molecular architectures and physical properties. The models discussed above explored a very limited selection of the synthetic members of this family. Most of the discussion focused on chains comprising two types of monomers. Of the chains incorporating three different monomers, only the simplest possible case, linear *ABC* triblocks, was considered. The overall behavior of the simplest polymeric surfactants, *AB* diblock copolymers, resembles that of the familiar monomeric amphiphiles. The two systems differ, however, in important respects: the characteristic length and time scales, their selectivity, and the role of configurational entropy in determining the equilibrium aggregation state. Triblock copolymers exhibit qualitatively distinctive features. Aggregates of *ABA* triblock copolymers are subject to exchange interactions that lead to attraction and the formation of physical gels. The presence of a chemically different third block, in *ABC* triblock copolymers, leads to a much richer phase behavior involving a number of novel phases. When the number of amphiphilic motifs is larger, in polysoaps and multiblock copolymers, the polymer can undergo intrachain self-assembly that drastically modifies the single-chain behavior.

APPENDIX: A REMINDER ON POLYMER BRUSHES

Polymer brushes are a recurrent structural motif in self-assembled aggregates of polymeric surfactants. The theory of polymer brushes has been reviewed extensively [33,56,80–82]. A brief reminder is provided in view of their central role in modeling aggregates formed by macromolecular amphiphiles. It concerns brushes formed from chains that are terminally attached, grafted, to the surface of a sphere, cylinder, or plane. In the context of our discussion, the grafting results from the microphase separation. The junction of two blocks is constrained to the interface between the two phases. As a result, each block is, in effect, grafted to this interface. It is important to note that other grafting modes are possible, for example, chemical grafting involving covalent bonds. The grafting due to microphase seperation is often thermodynamically controlled. On the other hand, chemical grafting is inherently irreversible. This distinction does not, however, affect the consideration presented in the following. At low grafting densities, in the ''mushroom'' regime, the chains do not overlap. In this situation they retain, essentially, the configurations and dimensions of free chains. When the grafting density increases so as to impose chain overlap, the equilibrium configurations of the chains are stretched along the normal to the surface. The stretching increases the layer thickness, thus decreasing the monomer concentration and the number of repulsive monomer–monomer interactions. A complete description of the brushes, allowing for the endpoint distribution, requires a self-consistent field theory. The main characteristics, thickness and free energy, may be obtained via a simpler ''blobological'' argument. A unified presentation proceeds as follows. The grafted layer is envisioned as a stratified array of blobs. All blobs in a given strata are of equal size ξ related to the local concentration ϕ as $\xi \approx \phi^{-3/4}b$. Any single chain contributes a single blob to each of the sublayers. The area of a given sublayer S depends on the distance from the grafting surface r and on the geometry. In every case, $S(r) \approx p\xi^2$, where p is the number of chains in the brush. In a flat layer S is independent of r. In this case $S \approx p\xi^2$ or $\xi \approx a^{1/2}$ where a is the surface area per chain (Fig. 16) [83]. For cylindrical layers of length H the area of a given stratum depends on r as $S \approx rH$. This leads to $\xi \approx (rH/p)^{1/2}$ [84]. Finally, for spherical brushes, $S \approx r^2$ and $\xi \approx r/p^{1/2}$ (Fig. 17) [53].

Knowing $\xi(r)$, it is straightforward to obtain the brush thickness L and the free energy per chain F_{corona}. L is determined by monomer conservation. Within this picture each grafted chain is a fully stretched string of blobs and the number of monomers per chain N is

$$N \approx \int_{R_{\text{in}}}^{R_{\text{in}}+L} \left(\frac{\xi}{b}\right)^{5/3} \xi^{-1}\, dr \tag{54}$$

Here R_{in} is the radius (in the case of a flat layer, altitude) of the grafted surface, $(\xi/b)^{5/3}$ is the number of monomers per blob, and $(\xi/b)^{5/3}\xi^{-1}$ the corresponding linear density of monomers. F_{corona} is obtained by assigning kT to each blob or

$$\frac{F_{\text{corona}}}{kT} \approx \int_{R_{\text{in}}}^{R_{\text{in}}+L} \xi^{-1}\, dr \tag{55}$$

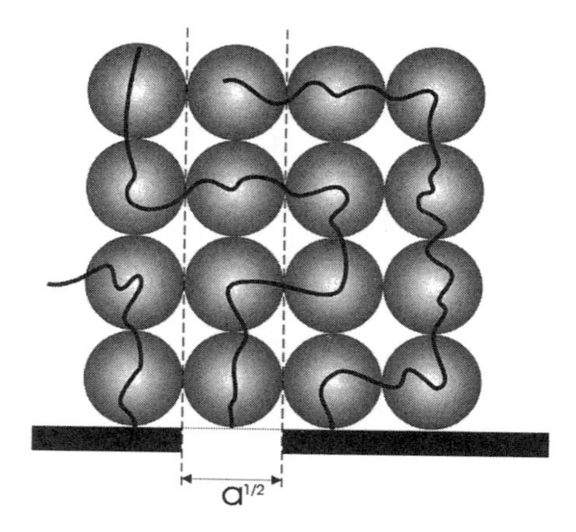

Figure 16 A schematic cross-section of a planar brush within the Alexander model. All the blobs, represented by circles, are of equal size. (Adapted from Ref. 29.)

where ξ^{-1} is the number of blobs per unit length along the normal to the surface. For a flat brush, where $\xi \approx a^{1/2}$, this expression leads to

$$\frac{L}{b} \approx N\left(\frac{b^2}{a}\right)^{1/3} \tag{56}$$

and

$$\frac{F_{\text{corona}}}{kT} \approx N\left(\frac{b^2}{a}\right)^{5/6} \tag{57}$$

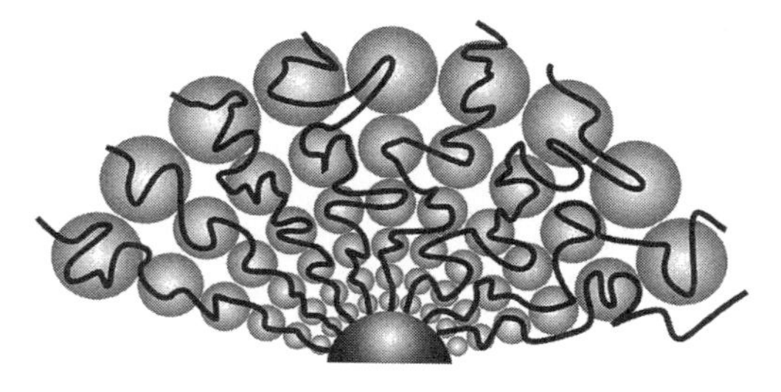

Figure 17 A schematic cross-section of a spherical brush according to the Daoud–Cotton model. The blob size scales the distance from the center. (Adapted from Ref. 29.)

In the limit of $L \gg R_{\text{in}}$ the thickness of a cylindrical brush of length H is

$$\frac{L}{b} \approx \left(\frac{pb}{H}\right)^{1/4} N^{3/4} \tag{58}$$

while the corresponding F_{corona} is

$$\frac{F_{\text{corona}}}{kT} \approx \left(\frac{pb}{H}\right)^{5/8} N^{3/8} \tag{59}$$

In the case of spherical starlike brushes

$$\frac{L}{b} \approx p^{1/5} N^{3/5} \tag{60}$$

and

$$\frac{F_{\text{corona}}}{kT} \approx p^{1/2} \ln \frac{R_{\text{in}} + L}{R_{\text{in}}} \tag{61}$$

The blobological approach, as described above, is an elaboration of the Alexander model of flat brushes [83,56]. Within this model the concentration within the brush is constant; that is, the concentration profile is a step function. Furthermore, the chains are viewed as uniformly stretched strings of blobs with the chains' ends straddling the sharp brush boundary. The free energy per chain reflects the monomer–monomer interactions and a stretching penalty. At equilibrium this leads to the picture of a flat brush as a stratified array of blobs of constant size. This model overconstrains the brush in that it does not allow for spatial distribution of the ends and for nonuniform stretching. Both factors are incorporated into the SCF theory. Some of the important conclusions of the SCF theory can be recovered via a simple approximation due to Pincus [27]. The Pincus approximation occupies the middle ground between the Alexander model and the full SCF theory of the brush [80]. In this picture the chain ends are distributed throughout the brush and the mean free energy per unit area is $\gamma = \int_0^L F_{\text{brush}}(z)\,dz$ where z is the distance from the grafting surface and the free energy density is

$$\frac{F_{\text{brush}}}{kT} \approx b^{-3}\left[\phi^2 + \frac{z^2}{Na^2}\Psi(z) - \lambda\phi(z)\right] \tag{62}$$

The first term is the interaction free energy associated with repulsive monomer–monomer contacts as given by the Flory theory. For a brush, this free energy is valid for any N since the chains lose their translational entropy upon grafting. The handling of the elastic free energy penalty is the core of the Pincus approximation. A chain having its endpoint at altitude z is assumed to be a uniformly extended Gaussian spring with a free energy penalty $F_{el}/kT \approx z^2/Nb^2$. In marked distinction, in the full SCF theory the chain stretching varies along the chain. The endpoints are assumed to be distributed throughout the layer with a density $\Psi(z)$. The key assumption is that the local fraction of chain ends $\Psi(z)$ is proportional to the local concentration and to the fraction of chain ends within the chain itself $1/N$. In other words

$$\Psi(z) \approx \frac{\phi(z)}{N} \tag{63}$$

Note that in this approximation the endpoint distribution is *assumed* rather than derived as is the case in the SCF theory [80]. Furthermore, $\Psi(z)$ is wrong for small altitudes. Nevertheless, this approximation yields the correct concentration profile because $F_{el} \sim z^2$ and the large z contribution, where this assumption is reasonable, dominates. Finally, λ is a Lagrange parameter fixing the number of monomers per chain N. The equilibrium state of the brush is specified by $\delta\gamma/\delta\phi = \partial F_{brush}/\partial\phi = 0$ leading to $\mu[\phi(z)]/kT = \lambda - Bz^2$, where $B \approx 1/N^2b^2$. λ is determined by the constraint $Nb^3 = \sigma \int_0^{L_o} \phi dz$ together with the requirement that $\phi(L_o) = 0$ where L_o is the height of the unperturbed brush. Accordingly

$$\frac{\mu}{kT} = B(L_o^2 - z^2) \tag{64}$$

That is, the chemical potential is parabolic. When the brush is semidilute and no coexistence is involved, $\mu \sim \phi$ and the concentration profile is parabolic as well.

ACKNOWLEDGMENT

The author benefited from the critical comments of V. Abetz, A. Ciferri, and M. Schick.

REFERENCES

1. H Ringsdorf, B Schlarb, J Venzmer. Angew Chem Int Ed Engl 27:113, 1988.
2. Z Tuzar, P Kartochvil. Colloids and Surfaces 15:1, 1991.
3. AW Adamson. Physical Chemistry of Surfaces. New York: Wiley, 1982.
4. IW Hamley. Block Copolymers. Oxford: Oxford University Press, 1999.
5. F Bates, GH Fredrickson. Ann Rev Phys Chem 41:525, 1990.
6. I Piirma. Polymeric Surfactants, New York: Marcel Dekker, 1992.

7. ED Goddard, KP Ananthapadmanabhan, eds. Interactions of Surfactants with Polymers and Proteins. Boca Raton, FL: CRC Press, 1993.
8. A Eisenberg, J-S Kim. Introduction to Ionomers. New York: Wiley, 1998.
9. S Schlick, ed. Ionomers: Characterization, Theory, and Applications. Boca Raton, FL: CRC Press, 1966.
10. A Laschewsky. Adv Polym Sci 124:1, 1995.
11. JN Israelachvili. Intramolecular and Surfaces Forces. 2nd ed. London: Academic Press, 1991.
12. DF Evans, H Wennerström. The Colloidal Domain. New York: VCH, 1994.
13. S Safran, Statistical Thermodynamics of Surfaces and Interfaces. New York: Addison-Wesley, 1994.
14. G Gompper, M Schick. Self-Assembling Amphiphilic Systems. In C Domb, JL Lebowitz, eds. Phase Transitions and Critical Phenomena, vol 16. London: Academic Press, 1994.
15. PJ Flory. Statistical Mechanics of Chain Molecules. New York: Wiley, 1969.
16. PJ Flory. Principles of Polymer Chemistry. Ithaca, NY: Cornell University Press, 1953.
17. PG de Gennes. Scaling Concepts in Polymer Physics. Ithaca, NY: Cornell University Press, 1979.
18. PG de Gennes, J Prost. The Physics of Liquid Crystals. Oxford: Oxford University Press, 1993.
19. FS Bates, GD Wignall, WC Koehler. Phys Rev Lett 55:2425, 1985.
20. JN Israelachvili, DJ Mitchel, BW Ninham. J Chem Soc Faraday Trans 2 72:1525, 1976.
21. W Gelbart, A Ben-Shaul, D Roux. eds. Micelles, Microemulsions, Membranes and Monolayers. New York: Springer Verlag, 1993.
22. C Tanford. The Hydrophobic Effect. 2nd ed. New York: Wiley, 1980.
23. A Halperin. Macromolecules 22:1943, 1987.
24. C Marques, JF Joanny, L Leibler. Macromolecules 21:1051, 1988.
25. D Izzo, CM Marques. Macromolecules 26:7189, 1993.
26. JF Marko, Y Rabin. Macromolecules 25:1503, 1992.
27. P Pincus. Macromolecules 24:2912, 1991.
28. S Alexander, PM Chaikin, P Grant, GJ Morales, P Pincus, D Hone. J Chem Phys 80:5776, 1984.
29. A Halperin. Macromolecules 23:2724, 1990.
30. A Halperin. Europhys Lett 10:549, 1989.
31. E Raphael, PG de Gennes. Physica A177:294, 1991.
32. JT Chen, EL Thomas, CK Ober, G-P Mao. Science 273:343, 1996.
33. A Halperin. In: Y Rabin, R Bruinsma, eds. Soft Order in Physical Systems. NATO ASI Series B 323. New York: Plenum Press, 1994.
34. A Halperin. Europhys Lett 8:351, 1989.
35. A Halperin, S Alexander. Macromolecules 22:2403, 1989.
36. T Vilgis, A Halperin. Macromolecules 24:2090, 1991.
37. DRM Williams, A Halperin. Phys Rev Lett 71:1557, 1993.
38. A Halperin. Macromolecules 22:3806, 1989.
39. FS Bates, GH Fredrickson. Macromolecules 27:1065, 1994.

40. LL Landau, EM Lifshitz. Statistical Physics. 3rd ed. Oxford: Pergamon Press, 1980.
41. A Halperin, EB Zhulina. Europhys Lett 16:337, 1991.
42. AN Semenov, JF Joanny, AR Khokhlov. Macromolecules 26:1066, 1995.
43. TA Witten. J Phys France 49:1056, 1988.
44. OV Borisov, A Halperin. Curr Opin Colloid Interface Sci 3:415, 1998.
45. OV Borisov, A Halperin. Macromolecules 29:2612, 1996.
46. EB Zhulina, A Halperin. Macromolecules 25:5730, 1992.
47. ST Milner, TA Witten. Macromolecules 25:5495, 1992.
48. B Li, E Ruckenstein. Macromol Theory Simul 7:333, 1998.
49. M Matsen. J Chem Phys 102:3884, 1995.
50. H Watanabe. Macromolecules 28:5006, 1995.
51. A Halperin. J Adhesion 58:1, 1996.
52. A Keller, JA Odell. In: MJ Folkes, ed. Processing, Structure and Properties of Block Copolymers. New York: Elsevier, 1985.
53. M Daoud, JP Cotton. J Phys Fr 43:531, 1982.
54. TA Witten, PA Pincus. Macromolecules 19:2509, 1986.
55. RG Petschek, KM Wiefling. Phys Rev Lett 59:343, 1987.
56. A Halperin, M Tirrell, TP Lodge. Adv Polym Sci 100:31, 1992.
57. A Halperin. Macromolecules 24:1418, 1991.
58. H Iatrou, L Willner, N Hadjichristidis, D Richter, A Halperin. Macromolecules 29: 581, 1996.
59. MS Turner, JF Joanny. J Phys Chem 97:4825, 1993.
60. OV Borisov, A Halperin. Langmuir 28:2911, 1995.
61. OV Borisov, A Halperin. Europhys Lett 34:657, 1996.
62. OV Borisov, A Halperin. Macromolecules 30:4432, 1997.
63. OV Borisov, A Halperin. Euro Phys J B 9:251, 1999.
64. A Halperin, EB Zhulina. Europhys Lett 15:417, 1991.
65. A Halperin, EB Zhulina. Macromolecules 24:5393, 1991.
66. OV Borisov, A Halperin. Phys Rev E 30:812, 1998.
67. P Molyneux. Water Soluble Polymers: Properties and Behavior. Boca Raton, FL: CRC Press 1983.
68. P Molyneux. In: F Franks, ed. Water, vol. 4. New York: Plenum, 1975.
69. R Kjellander, E Florin. J Chem Soc Faraday Trans I 77:2053, 1981.
70. A Matsuyama, F Tanaka. Phys Rev Lett 65:34, 1990.
71. S Bekiranov, R Bruinsma, P Pincus. Phys Rev E 55:577, 1993.
72. G Karlström. J Phys Chem 89:4962, 1985.
73. PG de Gennes. CR Acad Sci Paris II 313:1117, 1991.
74. M Wagner, F Brochard-Wyaert, H Hervet, PG de Gennes. Colloid Polym Sci 271: 621, 1993.
75. A Halperin. Eur Phys J B 3:359, 1998.
76. A Jada, G Hurtrez, B Siffert, G Reiss. Macromol Chem Phys 197:3697, 1996.
77. JE Glass, ed. Polymers in Aqueous Media: Performance Through Association. Washington, DC: ACS Press, 1989.
78. JE Glass, ed. Hydrophilic Polymers: Performance with Environmental Acceptance. Washington, DC: ACS Press, 1996.
79. M Rubinstein, AV Dobrynin. Trends Polym Sci 5:181, 1997.

80. ST Milner. Science 251:905, 1991.
81. I Szleifer, MA Carignano. In: I Prigorgine, SA Rice, eds. Advances in Chemical Physics, vol XCIV. John Wiley, New York, 1996.
82. GS Grest, LJ Fetters, JS Huang, D Richter. In: I Prigorgine, SA Rice, eds. Advances in Chemical Physics, vol XCIV. John Wiley, New York, 1996.
83. S Alexander. J Phys (France) 14:1637, 1977.
84. ALR Bug, ME Cates, SA Safran, TA Witten. J Chem Phys 87:1824, 1987.

4

Hydrogen-Bonded Supramolecular Polymers

Perry S. Corbin and Steven C. Zimmerman
University of Illinois, Urbana, Illinois

I. INTRODUCTION

Of the innumerable chemical advances in the twentieth century, one could argue that the development of polymers has had the most dramatic impact upon everyday living. Spurred by an overwhelming accumulation of fundamental knowledge in chemistry and physics, polymer science has blossomed into a field that has shaped, in part, both industry and academia. As part of the evolution of this field, numerous strategies have been developed for synthesizing polymers. These polymers, viewed simplistically, consist of covalently linked repeat units derived from single molecule precursors (monomers). However, recently there have been reports of ''supramolecular polymers'' [1] constructed from noncovalently associated monomers. Certainly, these studies challenge Carothers' original notion that structural repeat units ''are not capable of independent existence'' [2]. Taken more broadly, supramolecular polymers encompass specific noncovalent interactions between repeat units and sidechains of classical covalent polymers, as well as interactions of these polymers with small molecules.

One of the most formidable challenges in developing supramolecular polymers is the design of ''sticky'' monomers, or polymer substituents, that associate strongly in a predictable manner. Because of their strength and directionality, hydrogen bonds are an ideal ''glue'' for assembling such polymers. Although a variety of elegant hydrogen-bonded supramolecules have been reported [3], ranging from solution aggregates to engineered crystals, the study of hydrogen-bonded supramolecular polymers is still in its infancy. Progress in this area is reported herein. For related reviews of supramolecular polymers see Ref. 4.

One should be aware that studies of the type described are multidisciplinary, representing an opportune merging of organic and materials chemistries [4a]. Thus, a report of this nature could be written from varying perspectives. We have taken the relatively broad view of supramolecular polymers alluded to above (see Fig. 1). However, this account primarily focuses on hydrogen-bonded polymers that have been rationally designed to form in liquid-crystalline and isotropic solution phases and is preceded by a general description of hydrogen-bonding subunits for supramolecular polymer construction. Although hydrogen-bonded tapes, ribbons, helices, and sheets are relatively common structural motifs found in crystals, and such solid-state aggregates may be classified as hydrogen-bonded polymers, an indepth description of these structures is not given herein.

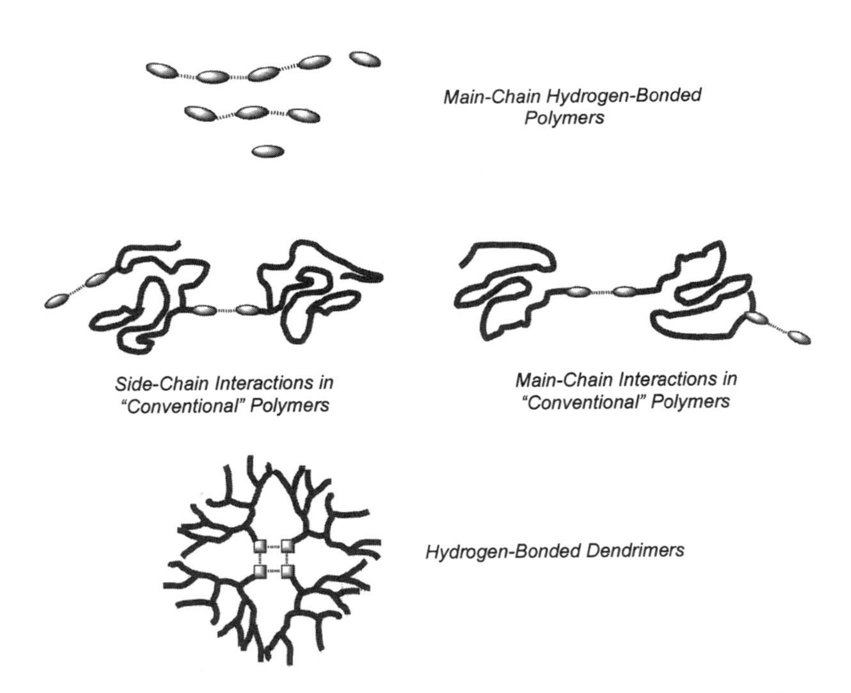

Figure 1 A schematic representation of different classes of hydrogen-bonded supramolecular polymers.

Readers are referred to pertinent reviews for a discussion of this topic (see, e.g., Ref. 5).

II. HYDROGEN-BONDING SUBUNITS FOR SUPRAMOLECULAR POLYMER CONSTRUCTION

A. General Considerations

One cannot initiate an account of hydrogen bonding without first recognizing the significance of this noncovalent, primarily electrostatic, force in biological systems [6]. From a molecular standpoint, biology offers several excellent examples of supramolecular hydrogen-bonded polymers, including double- and triple-helical DNA, as well as protein β-sheets (see, e.g., Ref. 7). As architects of artificial assemblies, supramolecular chemists will always be challenged to contemplate and create systems that are as intricate and complex, yet as simple and efficient in function, as those found in nature.

When considering the assembly of nonnatural, supramolecular, hydrogen-bonded polymers, two questions immediately arise: why use hydrogen-bonds and what types of building blocks should be used? As mentioned above, hydrogen bonds are relatively strong and directional, but hydrogen bonding is also reversible. Therefore, from a practical standpoint, hydrogen bonds are ideal for supramolecular polymer construction because they facilitate the assembly of polymers with substantial lengths and with well-defined, thermodynamically controlled, primary structures. Moreover, because of the reversible nature of hydrogen bonding, polymers can be produced with lengths and, thus, properties that are highly dependent upon concentration, temperature, pH, and additives. As a result, supramolecular hydrogen-bonded polymers may represent an important new class of functional materials.

Addressing the second question, a variety of building blocks can be envisioned to ''synthesize'' hydrogen-bonded polymers. Singly hydrogen-bonded complexes have been remarkably effective in affording liquid-crystalline polymers (vide infra). Typically, the subunits of these complexes have the advantage of being easy to synthesize. Moreover, individual hydrogen bonds, although weak, may sum in multivalent contacts to provide high stability. Nonetheless, robust, multiple hydrogen-bonded complexes are also attractive targets. It is especially critical to use such building blocks when designing solution aggregates because unfavorable entropic forces must be overcome in dilute solution. Furthermore, multiply hydrogen-bonded, liquid-crystalline polymers may have more stable mesophases than their singly hydrogen-bonded counterparts. Polymers may also be created from multiple hydrogen-bonding subunits that are stable at relatively high temperatures and under significant strain, but yet have reversible, thermally dependent bulk properties. Solution studies of some recently reported hydrogen-bonding building blocks are discussed below.

B. Models for Predicting the Stability of Multiple Hydrogen-Bonded Complexes

Before describing specific building blocks, it is important to make note of several issues that influence complex stability. Although it is difficult to predict the strength (i.e., association constant, K_{assoc}) of a multiply hydrogen-bonded complex a priori, theoretical and experimental investigations have aided in the dissection of components that affect binding strength in solution. Primary hydrogen-bond strength, i.e., the energy (enthalpy) of a single hydrogen bond between a hydrogen-bond acceptor (A) and donor (D), is determined by a number of factors, including the acidity and basicity of A and D, geometry, and substituent effects ([8,9] and references therein). Thus, complexes with the same number of hydrogen bonds have been found to have very different stabilities. In many cases, the differences depend upon the arrangement of the donors and acceptors within the complex. In short, complex strength does not depend solely upon the number of hydrogen bonds in the complex.

This observation, along with computational studies, led Jorgensen and Pranata to propose that secondary electrostatic interactions are critical in determining complex stability [10,11]. Their hypothesis is illustrated in Fig. 2, with hydrogen-bond donors represented as positive charges and hydrogen-bond acceptors as negative charges. The model specifically suggests that in addition to attractive primary electrostatic interactions there are important secondary interactions (attractive or repulsive) between neighboring hydrogen-bonding sites. This model explained the observation that a guanine:cytosine (AAD • DDA) base pair is over one hundred times stronger in terms of its K_{assoc} than a uracil: 2,6-diaminopyridine (ADA • DAD) complex and predicted the general stability series AAA • DDD > AAD • DDA > ADA • DAD for triply hydrogen-bonded complexes with equivalent primary interactions.

Experimental studies on a series of triply hydrogen-bonded complexes by Zimmerman and Murray are also consistent with the secondary hydrogen-bonding hypothesis [8,12]. For example, AAA • DDD complex **1** • **2**, has a K_{assoc} in chloroform that is more than 10^4 times larger than the K_{assoc} for **3** • **4**, an ADA • DAD complex (Fig. 3). The primary hydrogen bonds in **1** • **2** and **3** • **4** are expected to be similar in nature. Therefore, the difference in stability most likely arises because of the varied A/D arrangement and, thus, differences in secondary hydrogen-bonding interactions.

Subsequently, Sartorius and Schneider have proposed that the free energy of association (ΔG_{assoc}) for a hydrogen-bonded complex in chloroform can be

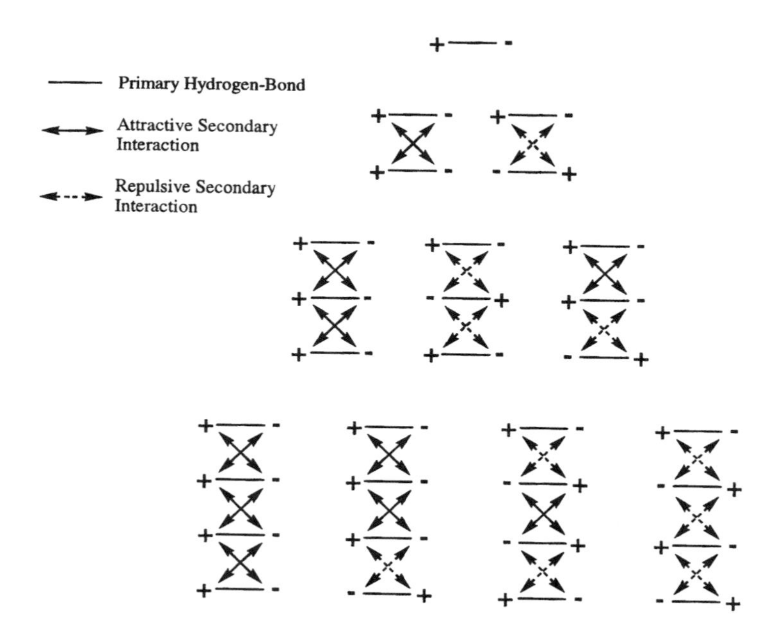

Figure 2 A schematic representation of Jorgensen's secondary hydrogen-bonding hypothesis for complexes containing one to four primary hydrogen-bonds. (Adapted from Refs. 10, 11.)

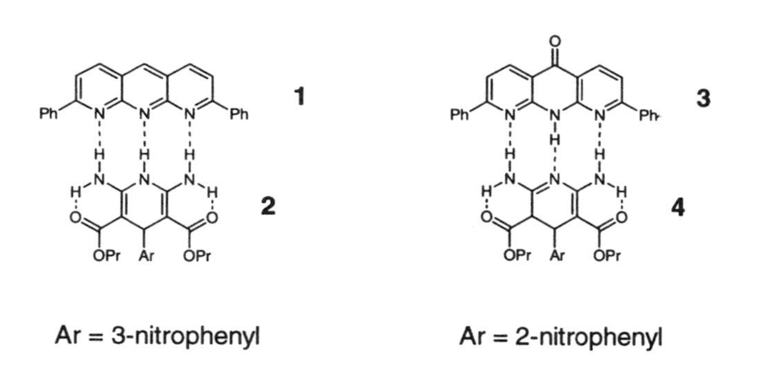

Figure 3 An AAA · DDD (**1 • 2**) and ADA • DAD (**3 • 4**) complex studied by Zimmerman and Murray. (From Refs. 8, 12.)

described by two increments, one representing primary interactions and another representing secondary interactions [13]. Based upon this linear free energy relationship, the energetic contribution of a single primary hydrogen bond is estimated to be 7.9 kJ/mol (1.9 kcal/mol), and the contribution from a secondary interaction, either attractive or repulsive, is estimated to be approximately 2.9 kJ/mol (0.7 kcal/mol). In an earlier analysis, the energy of a hydrogen bond was estimated to be 5 $\pm$ 1 kJ/mol using only a single increment [14].

The models above appear to be useful tools for predicting relative stabilities of rigid heterocyclic complexes with equivalent primary hydrogen-bonding sites. However, the situation becomes even more complicated when considering conformationally flexible systems where entropic factors are even more significant.* Moreover, based upon studies of flexible tetrapeptide analogues, Gardner and Gellman have demonstrated that secondary interactions do not necessarily determine the optimal arrangement of hydrogen-bonding sites in flexible systems [17]. Thus, one cannot expect all hydrogen-bonded complexes to have predictable complexation energies. However, the models described are clearly worth considering when designing hydrogen-bonded aggregates. These studies, along with those described below, have also led to a database of stability constants and synthetic procedures for the supramolecular polymer chemist.

C. Examples of Multiple Hydrogen-Bonding Building Blocks

Recently, there have been several reports of building blocks that are capable of forming exceptionally robust complexes. Examples include **1 • 2** and the complexes shown in Fig. 4. A description of single hydrogen-bonded complexes is reserved for the discussion on hydrogen-bonded liquid-crystalline polymers. Being cued by nature, the majority of subunits, including those in Fig. 4, are heterocycles. Two structurally related self-complementary hydrogen-bonded dimers of interest are the ureidopyrimidine dimer **6 • 6** reported by Meijer, Sijbesma, and coworkers [19,20] and the dimers formed from urylpyridopyrimidine **7** and its tautomers, which was studied by Corbin and Zimmerman [21]. Dimer **6 • 6** associated strongly in chloroform with a dimerization constant (K_{dim}) > 10^6 M^{-1} and was also formed in the solid state as evidenced in the crystal structure of **6**. Likewise, Meijer and coworkers have demonstrated that ditopic monomers containing **6** are useful for constructing hydrogen-bonded polymers with intriguing properties (vide infra) [19].

---

* For a study of the effect of "freely rotating bonds" upon host–guest complexation see, e.g., Ref. 15: for a study of the effect of these bonds upon hydrogen-bond complexation see, e.g., Ref. 16.

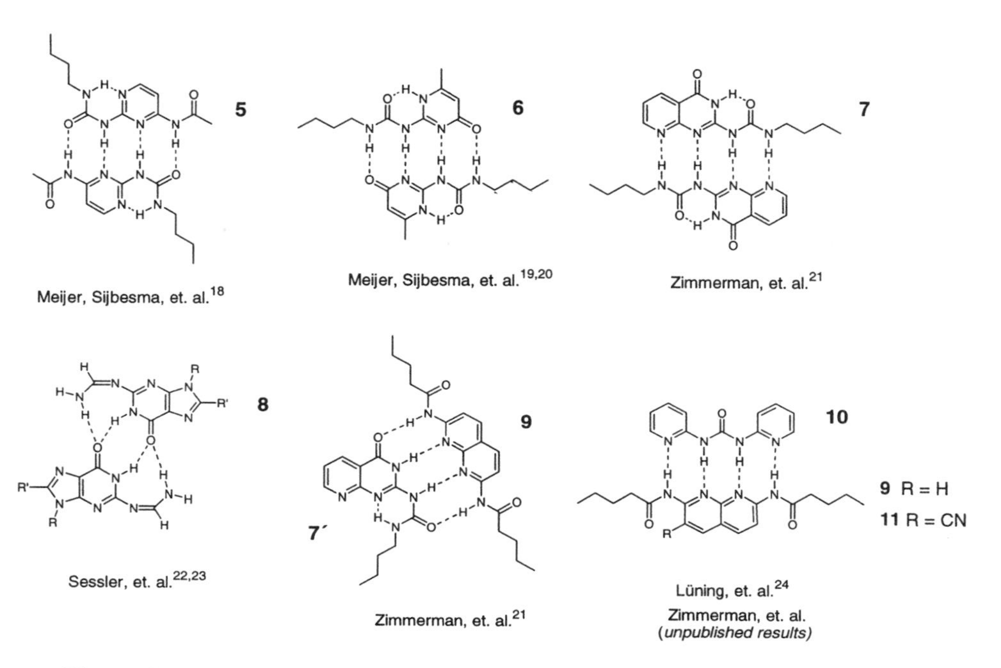

Figure 4 Recently reported examples of multiple hydrogen-bonded complexes.

Corbin and Zimmerman have pointed out that prototropy (i.e., tautomerism involving proton shifts) can be detrimental to hydrogen-bonded complex formation [21]. Therefore, heterocycle **7** was designed to contain a self-complementary hydrogen-bonding array irrespective of its protomeric form. The dimer formed from the N-3(H) protomer, **7 • 7**, is shown in Fig. 4. An N-1(H) homodimer and N-1(H)/N-3(H) heterodimer were also formed. Interestingly, all three of these dimers maintained a similar spatial arrangement of an alkyl substituent, and association was strong, with $K_{dim} \geq 10^7\ M^{-1}$ estimated for the dimers in chloroform. Another example of a self-complementary hydrogen-bonding subunit is the modified guanosine **8** reported by Sessler and Wang [22,23]. Upon linking two units of **8** with a rigid spacer, a robust ‘‘tetrameric’’ array formed in a number of solvents.

In addition to the homodimeric units described above, which form by self-complementary hydrogen bonding, heterodimeric complexes may also be effective for supramolecular polymer construction. Three such complexes are **1 • 2**, **7′ • 9**, and **10 • 11(9)**. Complex **7′ • 9** $K_{assoc} = 3500$ (M^{-1}, 5% DMSO-d_6/$CDCl_3$) is formed by a conformational switch in the aforementioned urylpyridopyrimidine **7**.

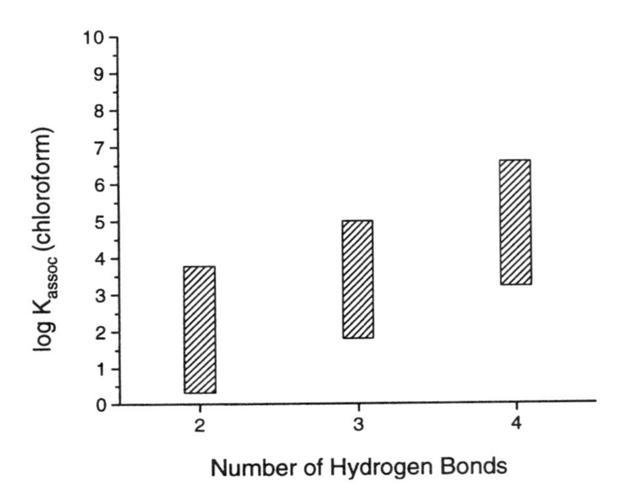

Figure 5 A graphical representation of the variation of K_{assoc} (chloroform) with the number of hydrogen bonds for some recently reported complexes.

The heterocycles described above are just a few examples of the types of building blocks that have the potential to be useful in the construction of supramolecular polymers. Most of the studies described in this section are solution studies. Thus, when using such units to design supramolecular aggregates in different states, for example, liquid-crystalline and melt phases, one must consider whether solution studies provide adequate precedent for the formation of hydrogen-bonded complexes in these phases. Indeed, if similar hydrogen-bonded complexes are formed both in the solid state and in solution, it seems reasonable to conclude that comparable complexes may be formed, for example, in liquid-crystalline mesophases. However, one should be aware of this "phase-problem." See Fig. 5.

III. HYDROGEN-BONDED, SUPRAMOLECULAR LIQUID-CRYSTALLINE POLYMERS

A. General Considerations

The general premise of using hydrogen bonds to construct low-molecular weight [25] mesogenic supramolecules (e.g., dimers and trimers) has been demonstrated in a number of accounts and was initially evoked to explain the mesomorphic

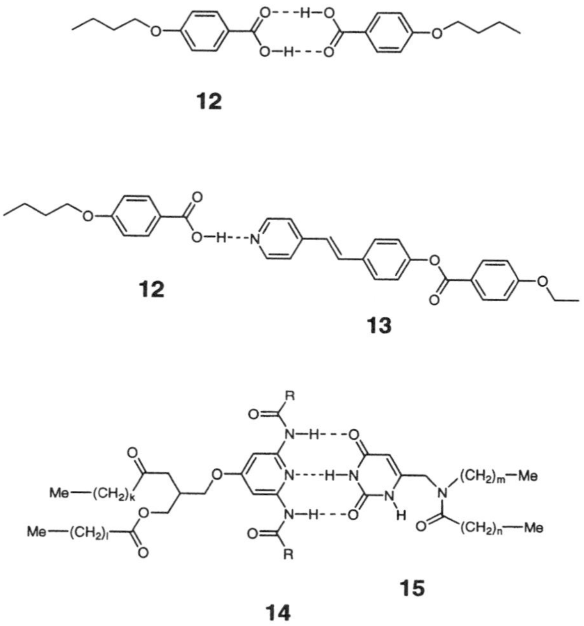

12

12

13

14

15

Figure 6 Typical examples of low-molecular weight, hydrogen-bonded, liquid-crystalline complexes.

behavior of carboxylic acids,* e.g., 4-alkoxybenzoic acids [27]. In this case, as well as other examples described below, intermolecular hydrogen bonding between two subunits is believed to afford a rigid core which, along with a flexible sidechain(s), promotes liquid crystallinity.

Analogously, Fréchet and Lehn have demonstrated the ability to construct heterodimeric mesogens from independent, complementary hydrogen-bonding units (Fig. 6). An example of Fréchet's and Kato's work is the mesogenic unit arising from the singly hydrogen-bonded complex of 4-butoxybenzoic acid (**12**) and stilbazole **13** [28].† A 1:1 mixture of these compounds exhibited a nematic phase over a significantly broader temperature range than did **12** and **13** alone and had a smectic phase that was not observed for the individual molecular com-

* An excellent review of hydrogen-bonded liquid crystals (including polymers) has been recently reported; see Ref. 26 and references therein.

† For an early review of Fréchet's work on hydrogen-bonded, supramolecular liquid crystals see Ref. 29 and references therein.

ponents. Hydrogen-bonded complex formation was verified by infrared spectroscopy (IR) [30]. Furthermore, Lehn's triply hydrogen-bonded complex **14 • 15** exhibited a ''metastable'' mesophase depending upon the length of the sidechains [31].* This particular liquid-crystalline complex is attractive because it is formed from nonmesomorphic subunits.

The examples above clearly demonstrated the viability of using hydrogen bonds to construct novel mesogens and raised the question as to whether hydrogen-bonding units could be used to construct mesomorphic hydrogen-bonded polymers. Indeed, the majority of work in the area of supramolecular hydrogen-bonded polymers, to date, has been devoted to the study of thermotropic liquid crystals. These polymers can be generically classified as either mainchain or sidechain polymers [4d]. A third class consisting of networked, hyperbranched, and dendritic polymers may also be considered [4a]. Examples of these structures are dispersed within the discussions of mainchain and sidechain, liquid-crystalline systems.

B. Mainchain Polymers

One could envision assembling supramolecular polymers from ditopic units containing hydrogen-bonding complements A and B similar to those described in the preceding sections. The ditopic subunits could, conceivably, be constructed with two units of either A or B (i.e., A—A and B—B) or with a single unit containing both A and B (i.e., A—B). Hydrogen bonding between A and B would then serve as the glue, or perhaps more appropriately the molecular ''velcro'' (suggesting a greater ease in reversibility) that holds the polymer together (Fig. 7). Ditopic units containing a self-complementary hydrogen-bonding component C could be used in a similar manner (Fig. 7). Polymers similar to those in Fig. 7 are what we and others refer to as mainchain polymers [4d,33].

As an example, Griffin and coworkers have described liquid-crystalline, mainchain supramolecular polymers constructed from hydrogen-bonded dipyridyl and diacid units (Fig. 8) [25,33]. Complex 8A, for instance, showed a nematic, as well as a monotropic, smectic phase. Maximum transition temperatures were obtained with a 1:1 mixture of the two components, which the authors suggest is an indication of linear chain formation and reflects a step-growth polymerization process [25]. Further evidence (IR, small-angle X-ray scattering, etc.) supporting the formation of liquid-crystalline hydrogen-bonded polymers from dipyridyl and diacid units has been recently reported [34]. However, it is difficult to estimate the average degree of polymerization (DP) for the 1 : 1 liquid-crystalline complexes.

* For studies of a related complex in the solid state and solution see, e.g., Ref. 32 and references therein.

A
A B
B
A B
A B
C
C
C

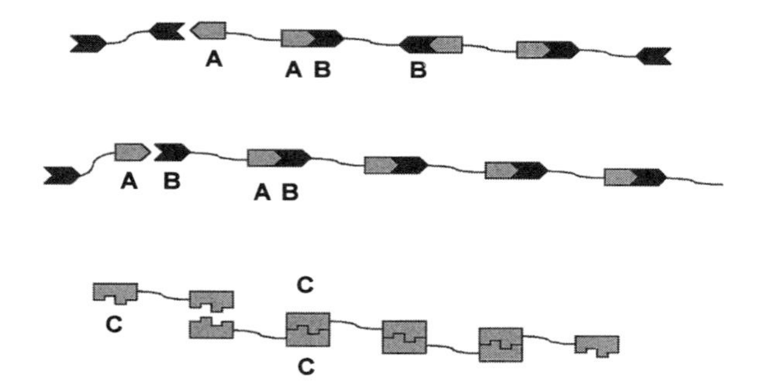

Figure 7 A schematic representation of mainchain hydrogen-bonded polymers constructed from ditopic building blocks.

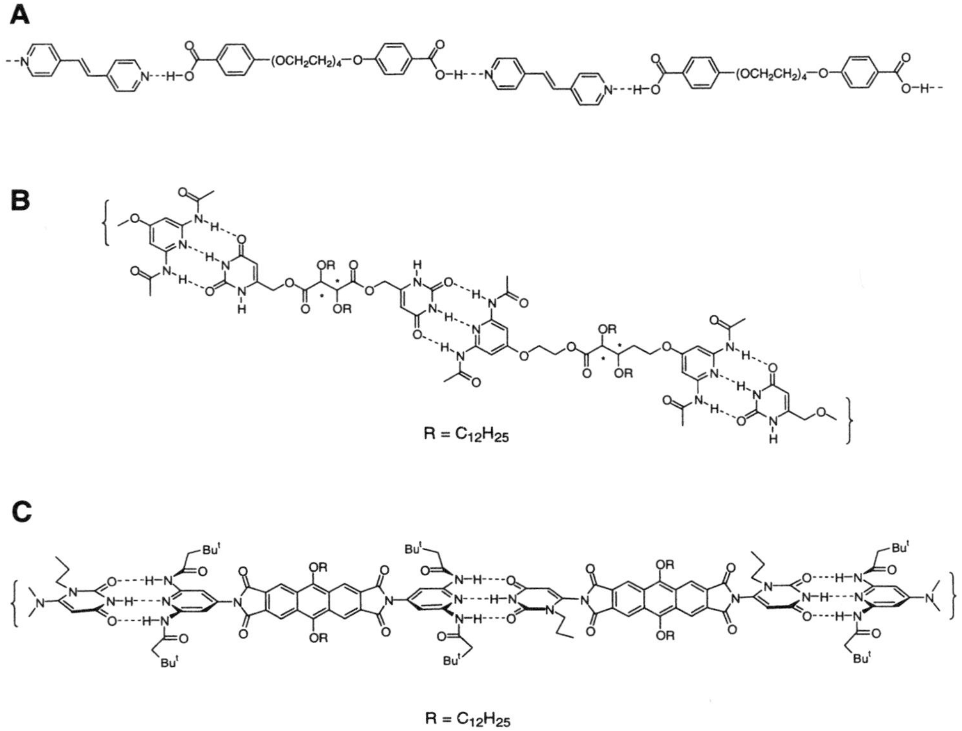

A
(OCH2CH2)4
B
R = C12H25
C
R = C12H25

Figure 8 Representative examples of mainchain hydrogen-bonded, liquid-crystalline polymers.

As an extension of the dimeric liquid-crystalline complex **14 • 15** described above, Lehn et al. have characterized polymeric systems assembled from ditopic units containing these two components [1,35]. Multiple hydrogen-bonding interactions between the diamidopyridine and uracil building blocks lead to polymeric structures (Fig. 8B) that have thermotropic (hexagonal columnar) mesophases over broad temperature ranges from below room temperature to greater than 220°C. The isotropization temperature for the complex is significantly higher than the isotropic to metastable mesophase transition temperature of 73°C observed for **14 • 15** and appears to support supramolecular polymer formation. This is analogous to classical liquid-crystalline polymers, which display increasing isotropization temperatures with increasing molecular weight [36]. Moreover, X-ray diffraction studies led to the proposal of a triple-helical superstructure for the polymer in which both ditopic units contain the linker derived from L (+) tartaric acid [35]. Likewise, Lehn has reported the self-assembly of rigid rods from a related ditopic pyridine and uracil unit with rigid aromatic spacers (Fig. 8C) [37]. A lyotropic mesophase is formed upon cooling a hot solution of the two components (1 : 1 mixture, 50 mM in 1,1,2,2,-tetrachloroethane) and can be attributed to hydrogen-bonded polymer formation.

At this point, it is worth mentioning that the possibility exists to assemble mainchain hydrogen-bonded polymers from covalent polymers with sticky ends. For example, Lenz et al. have proposed that the liquid-crystalline behavior of polymeric glycols terminated with diacids might be explained by dimerization of the carboxylic-acid termini [38]. Although not liquid-crystalline, Lillya and coworkers have also demonstrated that dimerization of the carboxyl termini of modified poly(tetramethyloxides) has interesting effects upon the polymer's bulk properties [39].

The polymers represented in Fig. 7 are based upon ditopic hydrogen-bonding subunits. However, multisite (>2 sites) building blocks might also be used. A schematic representation of an aggregate resulting from interaction of hydrogen-bond complementary tritopic and ditopic components is shown in Fig. 9. If the linker of one or both molecular components is flexible then crosslinking should arise and networked polymers would be produced. On the other hand, if rigid

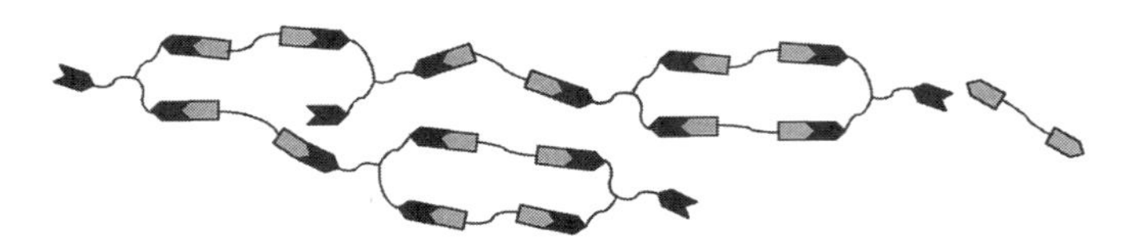

Figure 9 A schematic representation of crosslinked polymers formed from a ditopic and tritopic hydrogen-bonding unit. (Adapted from Ref. 40.)

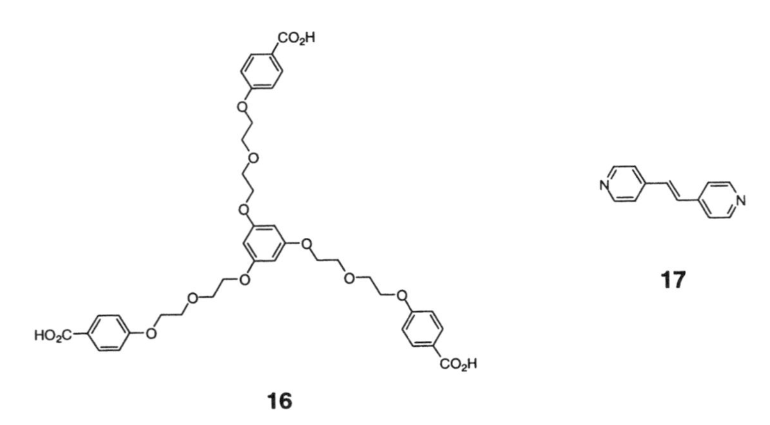

Figure 10 Subunits used by Fréchet and coworkers to construct liquid-crystalline network polymers. (From Ref. 40.)

linkers are used then crosslinking should be inhibited and a hyperbranched structure should form. Along these lines, Fréchet has reported the formation of liquid-crystalline network polymers from triacids and bipyridine and bipyridine-derivatives (Fig. 10). For example, mixtures of **16** and **17** were found to have a smectic A phase [40]. The fluidity of the proposed liquid-crystalline networks has been ascribed to the dynamic nature of hydrogen bonding, i.e., the breaking and re-forming of hydrogen bonds [40].

In related work, Griffin and coworkers have reported hydrogen-bonded polymers constructed from **18**, **19**, and **20** (Fig. 11) [41]. In these studies, a mixture of **18** and **20** was found to have temperature-dependent rheological properties consistent with the formation of reversible hydrogen-bonded networks. Furthermore, complex **19** • **20** exhibited a mesophase, which has been attributed to a ladder-type structure in which **19** is in a rodlike conformation [41]. Interestingly, the mesogenicity of the complex is lost upon extensive heating above the isotropization temperature. This is believed to arise because of a conformational switch of **19** from the rodlike form to a pseudotetrahedral conformation—thus leading to the formation of extended hydrogen-bonded networks.

C. Sidechain Polymers

The second generic class of hydrogen-bonded, liquid-crystalline polymers is sidechain polymers. The general design of these structures entails incorporation of hydrogen-bonding sites into the sidechains of covalent polymers and is based upon the desire to enhance the "mesomorphicity" of small molecule mesogens

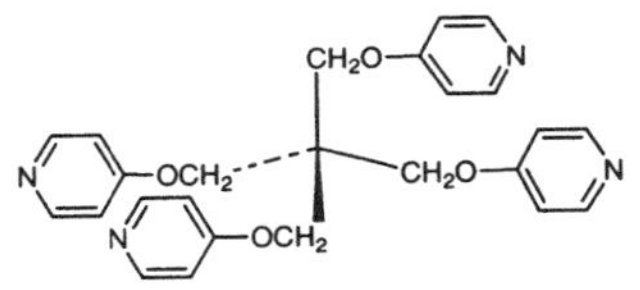

18

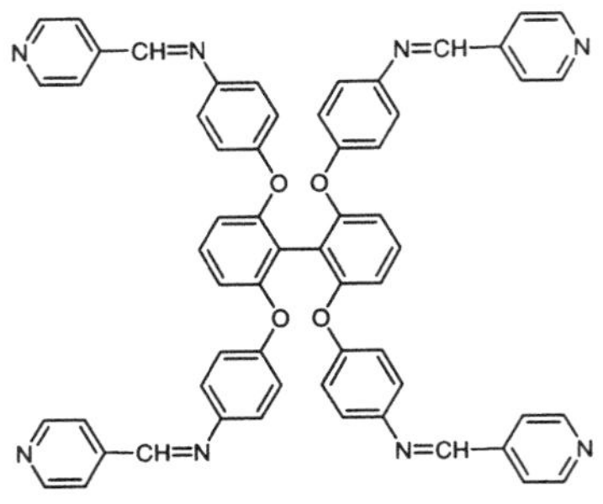

19

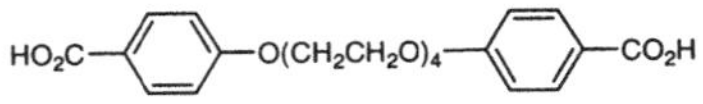

20

Figure 11 Subunits used by Griffin and coworkers to construct supramolecular networks. (From Ref. 41.)

and polymers by hydrogen-bonding interactions between the two [42]. A schematic representation of this process is shown in Fig. 12. The concept of liquid-crystalline sidechain polymers, and thus most of the work in this area, is a product of Fréchet, Kato, and coworkers. Subsequently, an excellent summary of these types of aggregates has been reported elsewhere [29]; and, therefore, an example suffices to demonstrate this concept.

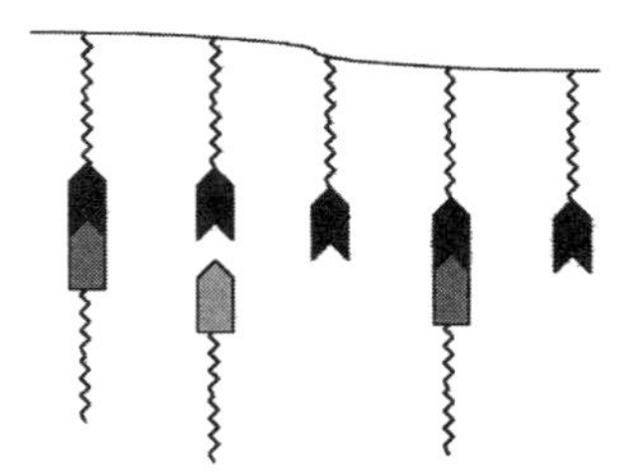

Figure 12 A schematic representation of sidechain/small molecule hydrogen-bonding interactions in liquid-crystalline polymers.

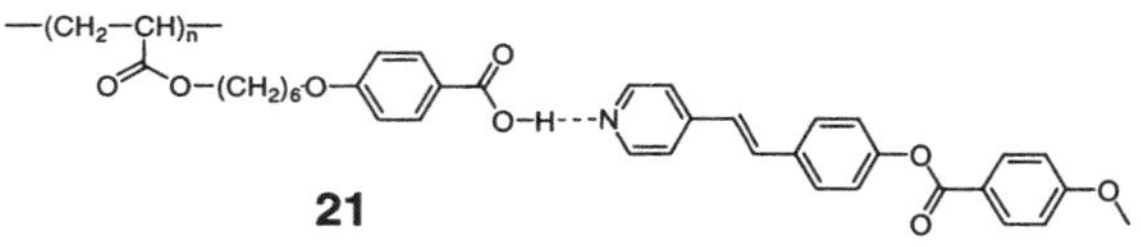

21

22

Figure 13 A representative example of a sidechain liquid-crystalline polymer. (From Refs. 29, 42.)

The first hydrogen-bonded, liquid-crystalline sidechain polymer reported by Fréchet used a polyacrylate covalent polymer with pendant, hexyloxybenzoic acid sidechains (**21**), and stilbazole **22** (Fig. 13) [42]. As a result, a 1 : 1 mixture (with respect to the sidechain) formed a nematic phase over a broad temperature range from 140–252°C, exceeding that of the individual mesomorphic components. This strategy has been extended to polysiloxane polymers, as well as to systems with varying small molecule mesogens [29]. In addition, when using a ditopic hydrogen-bonding unit, network polymers, represented schematically in Fig. 14, have been proposed to form [43].

D. Liquid Crystals from Cyclic Hydrogen-Bonded Aggregates

In addition to the liquid-crystalline polymers described in previous sections, liquid crystals may also form by the self-organization of discrete, cyclic, hydrogen-

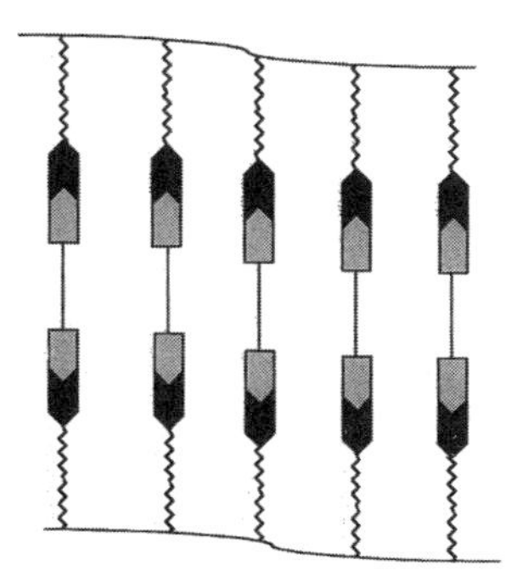

Figure 14 A schematic representation of crosslinking in liquid-crystalline sidechain polymers. (Adapted from Ref. 43.)

bonded aggregates. Along these lines, there have been several excellent examples of cyclic hydrogen-bonded aggregates in solution [3b, 44–46] and in mesophases. Although these species are typically oligomeric, containing 3 to 10 hydrogen-bonding subunits, they can be considered as supramolecular hydrogen-bonded polymers. Two examples of hydrogen-bonded disks that form mesophases are described below.

The first example, reported by Lehn, Zimmerman, and coworkers is based upon a phthalhydrazide subunit [47]. In these studies, lactim–lactam protomer **23**, containing Fréchet-type dendrons, trimerized in solution, as evidenced by ^{1}H NMR and size-exclusion chromatography (SEC). Furthermore, columnar, discotic mesophases were observed for alkyl-substituted versions of **23**. Depending upon the length of the peripheral sidechains on the heterocycle, hexagonal, and/or rectangular columnar mesophases were observed. The formation of these thermotropic discotic liquid crystals can be attributed to the formation of hydrogen-bonded trimers and subsequent stacking of these disks.

Similarly, Gotarelli and coworkers have described lyotropic mesophases formed from stacked guanosine tetramers (**24**) in water (Fig. 15) [48]. This strategy has been extended to a series of oligomers containing two to five guanosines [48–51]. Lyotropic phases have also been reported to form from an oligomer d(GpGPApGpG) in which the central guanosine is replaced by an adenosine [52].

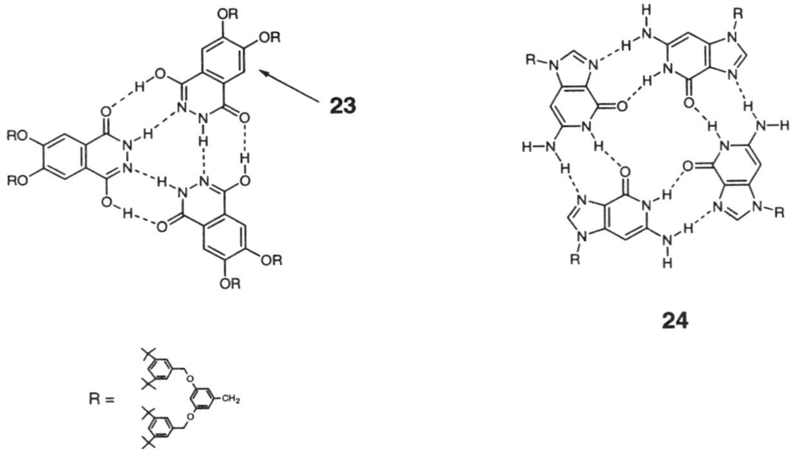

23

24

R = $CH_2(CH_2)_nCH_3$, n = 6,10,14

Figure 15 Examples of liquid-crystalline, disk-shaped aggregates.

$$DP = \frac{1}{1 - p}$$

DP = degree of polymerization (number average)

p = percent conversion or percent complexation

Figure 16 Equation describing the number average degree of polymerization for a step-growth polymerization process. (See Ref. 33 and references therein.)

IV. HYDROGEN-BONDED SUPRAMOLECULAR POLYMERS IN ISOTROPIC SOLUTIONS

A. Mainchain Polymers

The same general classification scheme used to describe hydrogen-bonded liquid-crystalline polymers may also be used to describe hydrogen-bonded polymers in isotropic solution. However, this area is less well developed, and few examples have been reported. The first examples described are mainchain polymers constructed from ditopic monomers (Fig. 7).

As pointed out by Griffin and coworkers, the production of mainchain hydrogen-bonded polymers can be described as a step-growth process,* with the number-average DP defined as in Fig. 16. If the stepwise polymerization of ditopic units is noncooperative, i.e., the association constant for each step is equal, and the association constant for the corresponding monotopic subunits is known, then the percent conversion (complexation) and average DP can be estimated. To obtain a high DP in ''dilute'' solution, the association constant of the monomers should be large.

Along these lines, the hydrogen-bonded polymers of Meijer and coworkers are distinct in that they are the first examples of supramolecular polymers with significant lengths in a dilute isotropic solution [19]. The specific polymers studied were assembled from ditopic unit **25** (Fig. 17), which contains the quadruply hydrogen-bonding subunit **6** (vide supra). As mentioned above, ureidopyrimidine **6** associates very strongly in chloroform. Likewise, a chloroform solution of **25** had a highly concentration-dependent viscosity, which was attributed to changes in polymer length. Viscosities were also lowered upon addition of **6**, as the result of **6** acting as a ''capping'' agent to decrease the average DP. All of these observations were consistent with the formation of hydrogen-bonded polymers in chloroform. The average DP for **25** was especially notable—estimated to be 700 at a

* See Ref. 33 and references therein.

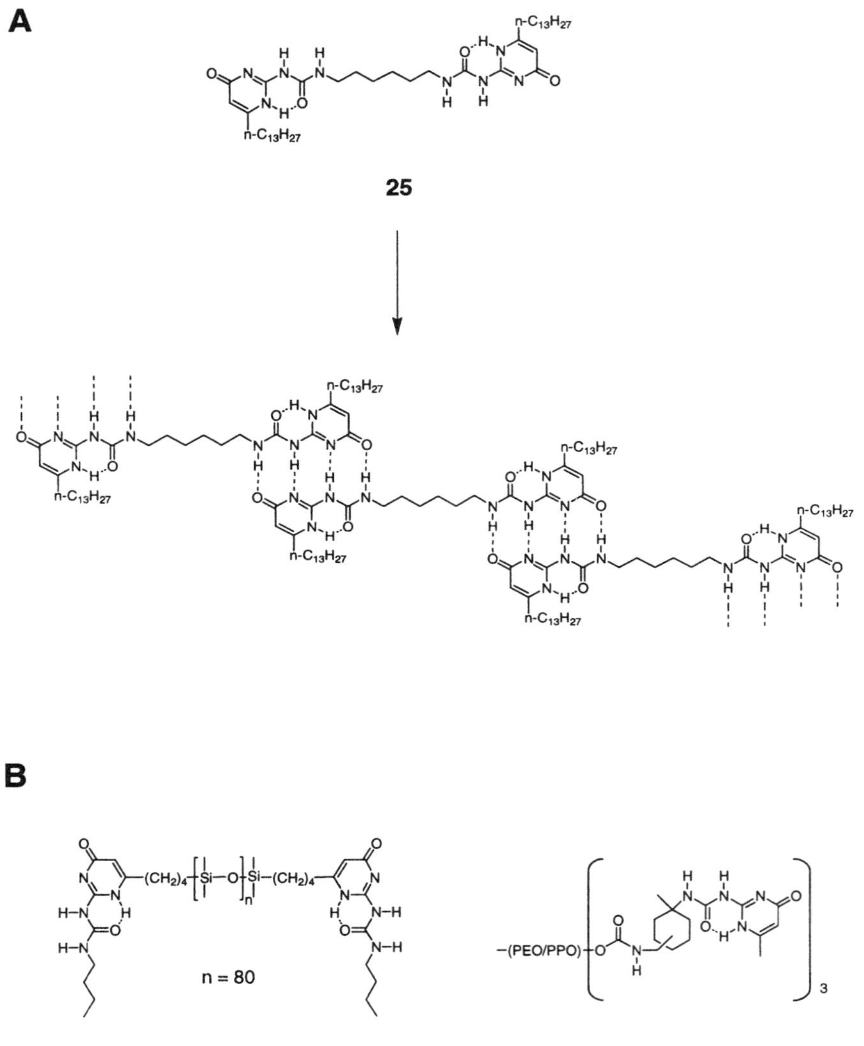

A

25

B

26

n = 80

27

Figure 17 (A) Supramolecular polymer formation from ditopic ureidopyrimidine **25**. (B) A ditopic siloxane polymer **26** and a trifunctionalized copolymer **27** (PEO = polyethylene oxide, PPO = polypropylene) containing ureidopyrimidine **6**.

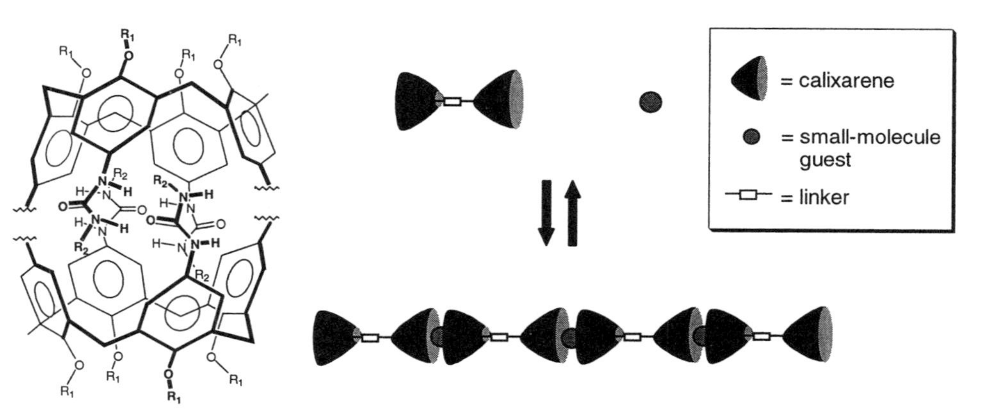

Figure 18 Calixarene dimer (left) and homopolymer (right). (Adapted from Refs. 53, 54.)

concentration of 40 mM—and is consistent with the large dimerization constant observed for **6**.

In addition to solution studies, a related siloxane polymer, **26** (Fig. 17B), had bulk properties that were reminiscent of thermoplastics and differed from an unmodified siloxane of similar molecular weight, as a result of reversible association of the hydrogen-bonding end-groups. Likewise, networked polymers have been assembled from a trifunctionalized copolymer **27**. This polymer had interesting mechanical properties (e.g., a higher plateau modulus in dynamic mechanical analysis than a control polymer) as a result of the formation of reversible hydrogen-bonded crosslinks [19]. Certainly, this work points to the power and potential of using multiply hydrogen-bonding units to construct supramolecular polymers with useful material properties.

In addition to the mainchain supramolecular polymers of Meijer, Rebek and coworkers have described the assembly of ditopic calixarene units into polymeric capsules, ''polycaps,'' in chloroform, as a result of homo- and heterodimerization of urea and sulfonyl-urea groups on the upper rim of the calixarenes [53,54]. Moreover, each calixarene dimer along the polymer chain encapsulates solvent or other small molecule guests (e.g., 4-fluorodibenzene). Polymer formation was supported by ^{1}H NMR, including capping studies, and by SEC. A schematic representation of a calixarene dimer and homopolymer are shown in Fig. 18.*

The ability to vary the linker and, thus, the orientation of hydrogen-bonding

* For a report of the formation of a calixarene dimer through multiple hydrogen-bonding interactions as in **6** see Ref. 55.

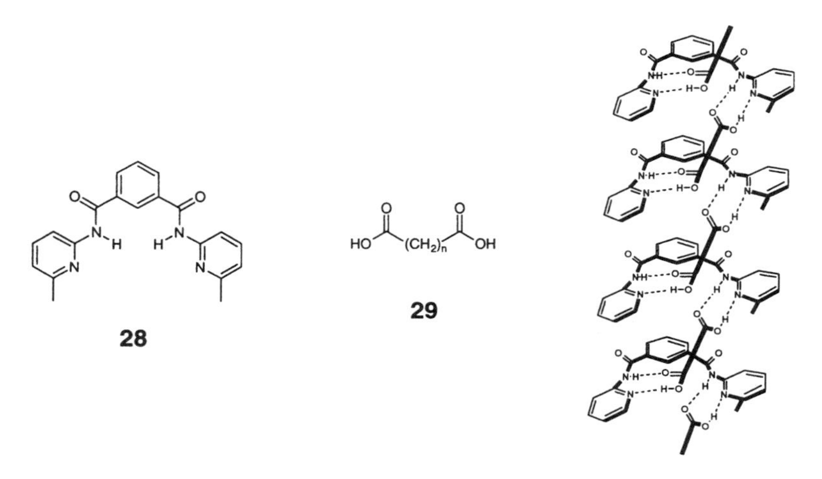

Figure 19 Schematic representation of a helical aggregate formed from bis-amidopyridine **28** and pimelic acid (**29**). (From Ref. 56.)

groups in ditopic building blocks similar to those described above is very powerful. For instance, appropriate linkers may lead to polymers with well-defined, potentially predictable, secondary structures. The aggregate formed from bis-2-amidopyridine unit **28** and pimelic acid (**29**) is such an example [56]. With regard to this aggregate, Hamilton and coworkers have reported a helical superstructure in the solid state that is formed by hydrogen bonding between pimelic acid and **28** in a ''syn–syn'' orientation (Fig. 19). ^{1}H NMR chemical shifts (5% THF-d_8/CD_2Cl_2), as well as variable temperature ^{1}H NMR and NOE investigations, are consistent with the formation of a helical aggregate in solution. However, these studies are not conclusive, and the DP has not been estimated. The cyclic-peptide nanotubes of Ghadiri, which form in the solid state and in lipid bilayers, also have intriguing secondary structures and can be regarded as supramolecular polymers (see Refs. 57 and 58 and references therein).

B. Hydrogen-Bonded Dendrimers

The synthesis and study of dendrimers has received significant attention recently [59]. As part of the growth of this field, there has been sharpened interest in the self-assembly of dendrimers and their use in molecular recognition [60,61]. Along these lines, hydrogen bonding has proven to be very effective for constructing self-assembled dendrimers. The aggregates arising from the assembly

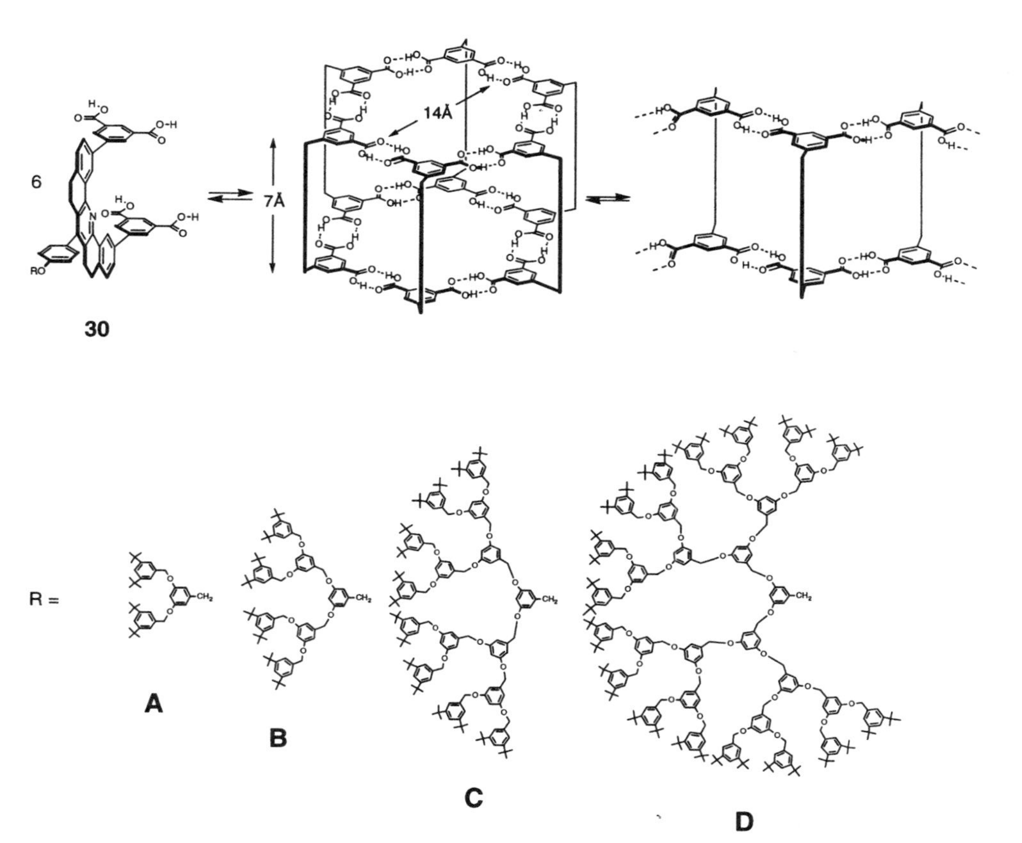

Figure 20 Self-assembled dendrimers studied by Zimmerman and coworkers. (From Refs. 62, 63.)

of bis-isophthalic acid units **30A-D** are such examples [62,63] (Fig. 20). In addition, robust hydrogen-bonded self-assembled dendrimers have been constructed from heterocyclic subunits [64,65]. The bis-isophthalic-acid unit **30** is particularly intriguing because subunits **30B-D**, containing second-through fourth-generation Fréchet-type dendrons [59], formed cyclic hexamers in nonpolar solvents as a result of carboxylic-acid dimerization. On the other hand, the bis-isophthalic acid unit **30A** containing a less bulky, first-generation dendron is believed to form polymeric aggregates. Based upon these observations and modeling, it has been suggested that peripheral dendron interactions, specifically those that would arise in the linear aggregates of the second-through fourth-generation isophthalic acid subunits, aid in the formation of hexameric species. Discrete aggregates such as those described rival the size (MW of the fourth-generation hexamer = 34.6 kD), shape, and, potentially, function of small proteins [62]. For example, one could envision catalysis in the core of a dendrimer similar to the action of an enzyme, as well as reversible guest complexation reminiscent of holoprotein complex formation.

V. OTHER EXAMPLES OF SUPRAMOLECULAR HYDROGEN-BONDED POLYMERS

A. Polymer Blends and Networks

Hydrogen bonding has been used to fruition in the preparation of polymer blends* and is especially effective in improving the miscibility of polymer components within a blend. Numerous accounts of hydrogen-bonded polymer blends have been reported—over 120 reports over the past four years—and a complete description is beyond the scope of this review. However, a few examples are mentioned to illustrate the concept.

Meftahi and Fréchet have recently described polymer blends prepared from poly(4-vinylpyridine) (**31**) and poly(4-hydroxystyrene) (**32**) [67]. In these studies poly(4-vinylpyridine) was immiscible with various weight percentages of polystyrene, a polymer lacking hydrogen-bond acceptor groups, as evidenced by the observation of two discrete glass-transition temperatures. To the contrary, a 1:1 mixture of poly(4-vinylpyridine) and poly(4-hydroxystyrene) precipitated from methanol, a solvent in which the individual components are soluble, is miscible and displays a single glass-transition temperature higher than that of the individual polymers [67]. Hydrogen bonding between the two polymers, as represented schematically in Fig. 21, is believed to give rise to blend compatibility. In addition to this example, there are many other blends prepared from phenolic-type polymers and polymers containing a variety of hydrogen-bond acceptors.† Meijer and coworkers have also briefly described blends prepared from melaimide and cyanuric acid polymers that rely on multiple hydrogen-bonding interactions between repeat units [71].

In addition to improving the miscibility of polymers in blends, hydrogen-bonding interactions between the side-groups of covalent polymers may also lead to network polymers with interesting properties. For example, polybutadienes modified with 4-(3,5-dioxo-1,4,4-triazlidine-4-yl) benzoic acid (**33**) form such structures. The mechanical and thermal properties of these polymers differ from unmodified polybutadienes as the result of hydrogen-bonded crosslinks (Fig. 22)

* For a general description of polymer blends see, e.g., Ref. 66.
† See, e.g., Refs. 68–70.

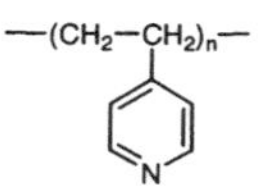

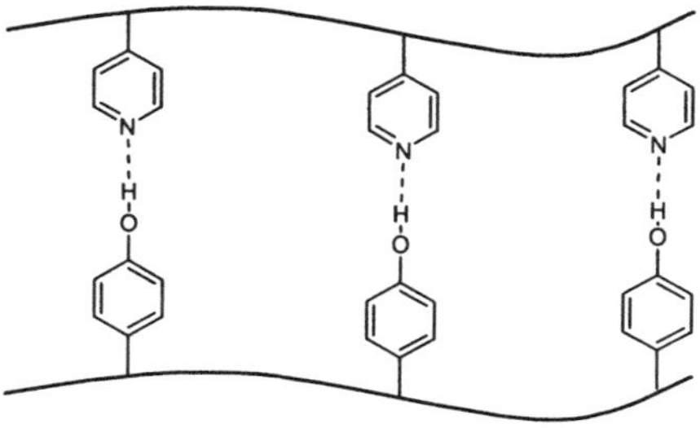

Figure 21 A schematic representation of hydrogen-bonding interactions between polymers **31** and **32** in a polymer blend. (From Ref. 67.)

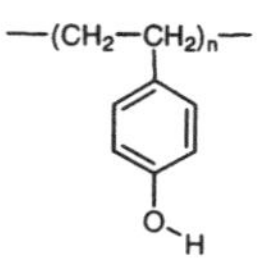

and further organization of the ‘‘association polymers,’’ through dipole–dipole interactions, into ordered domains (see Ref. 72 and references therein). Cross-linked polymers have also been reported to form from polybutadienes derivatized with a related phenyl-substituted urazole; see for example, Ref. 73 and references therein.

B. From Dispersions to Monolayers

In addition to the solution aggregates described in the previous section, stable dispersions of supramolecular hydrogen-bonded polymers have also been reported. One example is the aggregate arising from diimide **34** and melamine **35**

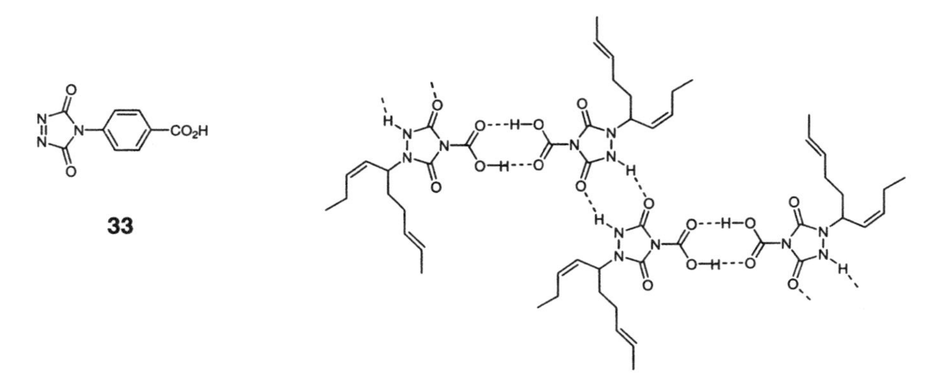

Figure 22 When polybutadiene is modified with **33** by an ene reaction, network polymers arise that rely on hydrogen-bond crosslinks (structure to the right) and dipole–dipole interactions to form ordered domains. (From Ref. 72.)

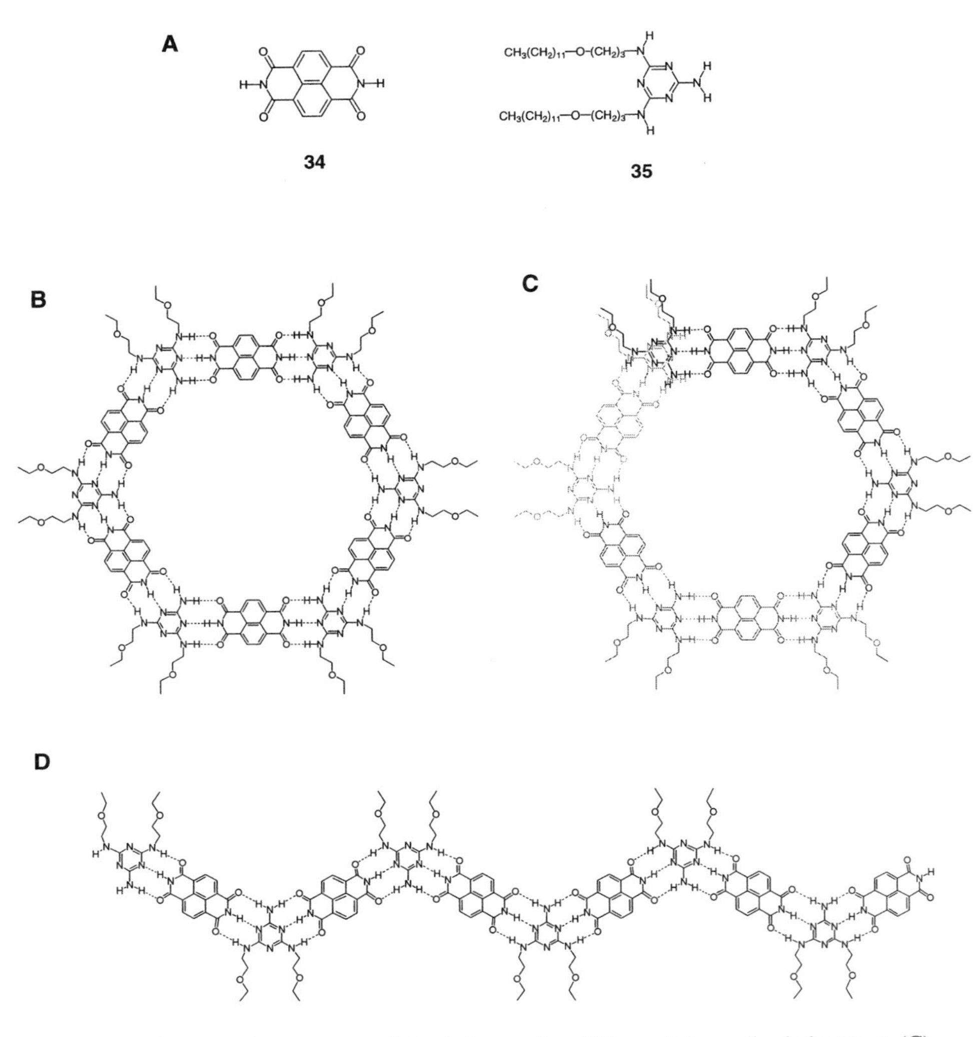

Figure 23 (A) Components of the 1:1 complex; (B) possible cyclic dodecamer; (C) possible helical aggregate; (D) possible linear tape. Peripheral alkyl chains are truncated in B–D for clarity.

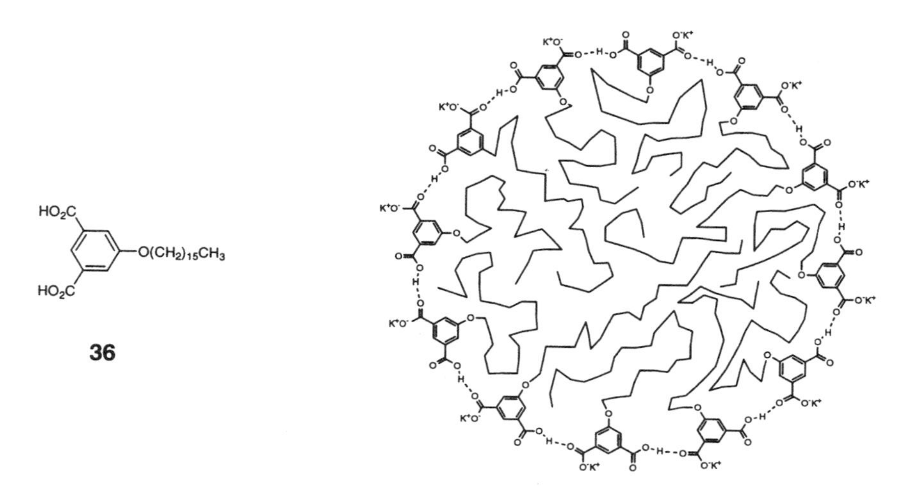

Figure 24 Schematic representation of a proposed cyclic aggregate that self-organizes into a fibrous material. (From Refs. 77, 78.)

[74].* Kimizuka and coworkers have reported that a dispersion of a 1 : 1 complex of **34** and **35** forms in cyclohexane. Interestingly, bundles of long stands were observed in both transmission- and scanning-electron micrographs. The exact nature of the supramolecular aggregate that leads to these strands has not been determined. However, the authors have suggested three possible supramolecular structures (Fig. 23): stacks of dodecameric disks, helical tapes, or extended linear tapes. In a somewhat related report, Menger and coworkers have reported the formation of fibers from the monopotassium salt of **36** when ''precipitating'' the salt from water [77,78]. In this case, hydrogen bonding and hydrophobic interactions are proposed to lead to disks (Fig. 24), which form stacks that, subsequently, produce the fibrous material.

Kimizuka and coworkers have extended their own work to the ''assembly'' of bilayer membranes. For example, multiple hydrogen-bonding interactions between **35** and cyanuric acid derivative **37** containing a polar head-group leads to hydrogen-bonded tapes that organize into bilayers [79,80].† The proposed supramolecular structure of the bilayer, represented schematically in Fig. 25, was verified by IR and X-ray diffraction studies of films cast from an aqueous dispersion of the bilayer, as well as by electron microscopy of the dispersions.

* For a slightly varied example see Ref. 75, and for an example of a gel constructed from related building block's see, Ref. 76.

† For related studies see Refs. 81 and 82.

35

37

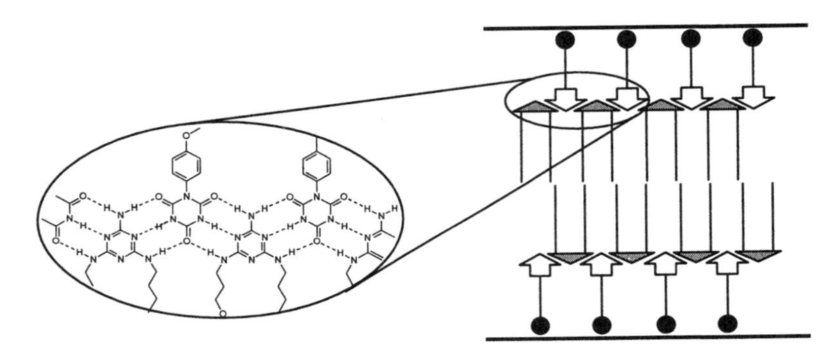

Figure 25 Schematic representation of the formation of a bilayer membrane. (Adapted from Refs. 79, 80.)

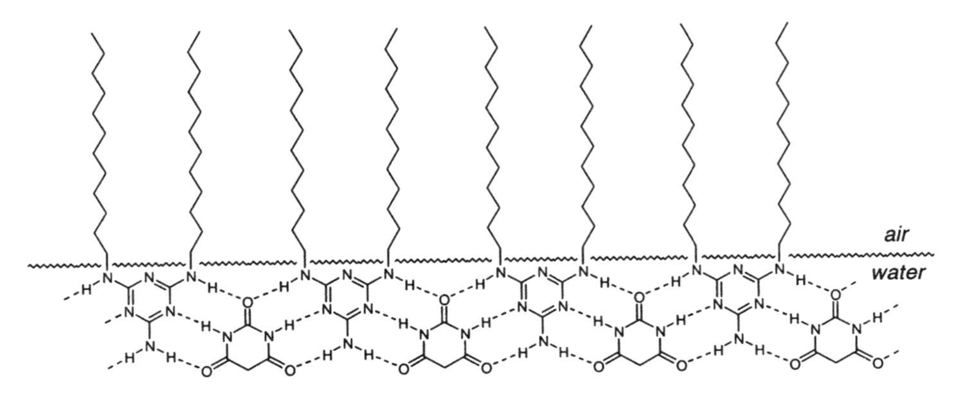

Figure 26 Schematic representation of monolayer formation from a substituted melamine and barbituric acid. (From Ref. 83.)

In related studies, Kunitake and coworkers have described the preparation of monolayers consisting of hydrogen-bonded tapes formed from a subsituted melamine and barbituric acid (Fig. 26) [83]. Hydrogen bonding in the resulting Langmuir–Blodgett film was verified by infrared and X-ray photoelectron spectroscopy. This strategy has recently been extended to the preparation of monolayers from tritopic melamine units, with hopes of producing network structures within the monolayer [84].

VI. CONCLUSIONS

The study of hydrogen-bonded polymers is, clearly, a burgeoning area of research. A set of building blocks is continuing to be compiled which will facilitate expansion of this field and allow the construction of even more new and interesting hydrogen-bonded polymers. One can assume that further development of these polymers will lead to new assemblies and materials with applications in a number of different areas. Studies of this nature will also continue to answer important fundamental questions concerning how and why molecules interact. Whether the development of these ''supramolecular polymers'' will have as great an impact as the development of conventional polymers is yet to be seen. However, the field of supramolecular hydrogen-bonded polymers unquestionably has a promising future.

ACKNOWLEDGMENTS

Financial support from the National Institutes of Health (NIH GM39782) is gratefully acknowledged.

REFERENCES

1. C Fouquey, J-M Lehn, A-M Levelut. Adv Mater 2:254–257, 1990.
2. WH Carothers. J Am Chem Soc 51:2548–2559, 1929.
3. (a) DS Lawrence, T Jiang, M Levett. Chem Rev 95:2229–2260, 1995. (b) GM Whitesides, EE Simanek, JP Mathias, CT Sato, DN Chin, M Mammen, DM Gordon. Acc Chem Res 28:37–44, 1995. (c) MM Conn, J Rebek Jr. Chem Rev 97:1647–1668, 1997.
4. (a) J-M Lehn. Makromol Chem, Macromol Symp 69:1–17, 1993. (b) J-M Lehn. Pure Appl Chem 66:1961–1966, 1994. (c) V Percec, J Heck, G Johannson, D Tomazos, M Kawasumi. J Mater Sci—Pure Appl Chem A 31:1031–1070, 1994. (d) N Zimmerman, JS Moore, SC Zimmerman. Chem Industry 15:604–610, 1998.

5. JC Macdonald, GM Whitesides. Chem Rev 94:2383–2420, 1994.
6. GA Jeffrey. An Introduction to Hydrogen Bonding. New York: Oxford University Press, 1997, pp 184–212.
7. A Aggeli, M Bell, N Boden, JN Keen, PF Knowles, TCB McLeish, M Pitkeathly, SE Radford. Nature 386:259–262, 1997.
8. SC Zimmerman, TJ Murray. Phil Trans R Soc Lond A 345:49–56, 1993.
9. CS Wilcox, E Kim, D Romano, LH Kuo, AL Burt, DP Curran. Tetrahedron 51: 621–634, 1995.
10. WL Jorgensen, J Pranata. J Am Chem Soc 112:2008–2010, 1990.
11. J Pranata, SG Wierschke, WL Jorgensen. J Am Chem Soc 113:2810–2819, 1991.
12. TJ Murray, SC Zimmerman. J Am Chem Soc 114:4010–4011, 1992.
13. J Sartorius, H-J Schneider. Chem Eur J 2:1446–1452, 1996.
14. H-J Schneider, RK Juneja, S Simiova. Chem Ber 122:1211–1213, 1989.
15. SC Zimmerman, M Mrksich, M Baloga. J Am Chem Soc 111:8528–8530, 1989.
16. F Eblinger, H-J Schneider. Angew Chem Int Ed Engl 37:826–829, 1998.
17. RB Gardner, SH Gellman. J Am Chem Soc 117:10411–10412, 1995.
18. FH Beijer, H Kooijman, AL Spek, RP Sijbesma, EW Meijer. Angew Chem Int Ed Engl 37:75–78, 1998.
19. (a) RP Sijbesma, FH Beijer, BJB Folmer, JHKK Hirschberg, RFM Lange, JKL Lowe, EW Meijer. Science 278:1601–1604, 1997. (b) BJB Folmer, E Cavini, RP Sijbesma, EW Meijer. Chem Commun 17:1847–1848, 1998.
20. FH Beijer, RP Sijbesma, H Kooijman, AL Spek, EW Meijer. J Am Chem Soc 120: 6761–6769, 1998.
21. PS Corbin, SC Zimmerman. J Am Chem Soc 120:9710–9711, 1998.
22. JL Sessler, R Wang. Angew Chem Int Ed Engl 37:1726–1729, 1998.
23. JL Sessler, R Wang. J Org Chem 63:4079–4091, 1998.
24. U Lüning, C Kühl. Tetrahedron Lett 39:5735–5738, 1998.
25. P Bladon, AC Griffin. Macromolecules 26:6604–6610, 1993.
26. CM Paleos, D Tsiourvas. Angew Chem Int Ed Engl 34:1696–1711, 1995.
27. E Bradfield, B Jones. J Chem Soc: 2660–2661, 1929.
28. T Kato, JMJ Fréchet. J Am Chem Soc 111:8533–8534, 1989.
29. T Kato, JMJ Fréchet. Macromol Symp 98:311–326, 1995.
30. T Kato, T Uryu, F Kaneuchi, C Jin, JMJ Fréchet. Liq Cryst 14:1311–1317, 1993.
31. M-J Brienne, J-M Lehn. J Chem Soc Chem Commun 24:1868–1870, 1989.
32. AD Hamilton, D Van Engen. J Am. Chem. Soc 109:5035–5036, 1987.
33. C Alexander, CP Jariwala, AC Griffin. Macromol Symp 77:283–294, 1994.
34. C He, AM Donald, AC Griffin, T Waigh, AH Windle. J Polym Sci Part B: Polym Phys 36:1617–1624, 1998.
35. T Gulik-Krzywicki, C Fouquey, J-M Lehn. Proc Natl Acad Sci USA 90:163–167, 1993.
36. A Sirigu. In A Ciferri, ed. Liquid Crystallinity in Polymers. New York: VCH, 1991, pp 261–313.
37. M Kotera, J-M Lehn, J-P Vigneron. J Chem Soc Chem Commun 2:197–199, 1994.
38. H. Hoshino, JI Jin, RW Lenz. J Applied Poly Sci 29:547–554, 1984.
39. CP Lillya, RJ Baker, S Hütte, HH Winter, Y-G Lin, LC Dickinson, JCW Chien. Macromolecules 25:2076–2080, 1992.

40. H Kihara, T Kato, T Uryu, JMJ Fréchet. Chem Mater 8:961–968, 1996.
41. CB St. Pourcain, AC Griffin. Macromolecules 28:4116–4121, 1995.
42. (a) T Kato, JMJ Fréchet. Macromolecules 22:3818–3819, 1989. (b) T Kato, JMJ Fréchet. Macromolecules 23:360, 1990.
43. T Kato, H Kihara, U Kumar, T Uryu, JMJ Fréchet. Angew Chem Int Ed Engl 33: 1644–1645, 1994.
44. A Marsh, M Silvestri, J-M Lehn. Chem Commun 13:1527–1528, 1996.
45. JP Mathias, EE Simanek, GM Whitesides. J Am Chem Soc 116:4326–4340, 1994.
46. SC Zimmerman, BF Duerr. J Org Chem 57:2215–2217, 1992.
47. M Suarez, J-M Lehn, SC Zimmerman, A Skoulios, B Heinrich. J Am Chem Soc 120:9526–9532, 1998.
48. G Gottarelli, GP Spada, A Garbesi. In: Comprehensive Supramolecular Chemistry, vol 9. New York: Pergamon Press, 1996, 483–506.
49. P Mariani, C Mazabard, A Garbesi, GP Spada. J Am Chem Soc 111:6369–6373, 1989.
50. S Bonazzi, M Capobianco, MM de Morais, A Garbesi, G Gottarelli, P Mariani, MGP Bossi, GP Spada, L Tondelli. J Am Chem Soc 113:5809–5816, 1991.
51. G Gottarelli, G Proni, GP Spada, S Bonazzi, A Garbesi, F Ciuchi, P Mariani. Biopolymers 42:561–574, 1997.
52. G Gottarelli, G Proni, GP Spada. Liq Cryst 22:563–566, 1997.
53. RK Castellano, DM Rudkevich, J Rebek Jr. Proc Natl Acad Sci USA 94:7132–7137, 1997.
54. RK Castellano, J Rebek, Jr. J Am Chem Soc 120:3657–3663, 1998.
55. JJ González, P Prodos, J de Mendoza. Angew Chem Int Ed Engl 38:525–528, 1999.
56. SJ Geib, C Vicent, E Fan, AD Hamilton. Angew Chem Int Ed Engl 32:119–121, 1993.
57. MR Ghadiri, JR Granja, RA Milligan, DE McRee, N Khazanovich. Nature 366: 324–327, 1993.
58. HS Kim, JD Hartgerink, MR Ghadiri. J Am Chem Soc 120:4417–4424, 1998.
59. GR Newkome, CN Moorefield, F Vögtle. Dendritic Molecules—Concepts, Syntheses, Perspectives. New York: VCH, 1996.
60. F Zeng, SC Zimmerman. Chem Rev 97:1681–1712, 1997.
61. SC Zimmerman. Current Opinion Colloid Interface Science 2:89–99, 1997.
62. SC Zimmerman, F Zeng, DEC Reichert, SV Kolotuchin. Science 271:1095–1098, 1996.
63. P Thiyagarajan, F Zeng, CY Ku, SC Zimmerman. J Mater Chem 7:1221–1226, 1997.
64. SV Kolotuchin, SC Zimmerman. J Am Chem Soc 120:9092–9093, 1998.
65. Y Wang, F Zeng, SC Zimmerman. Tetrahedron Lett 38:5459–5462, 1997.
66. JA Manson, LH Sperling. Polymer Blends and Composites. New York: Plenum, 1976.
67. MV Meftahi, JMJ Fréchet. Polymer 29:477–482, 1988.
68. MR Landry, DJ Massa, DM Teegarden, CJT Landry, RH Colby, PM Henrichs. Macromolecules 26:6299–6307, 1993.
69. CJT Landry, DJ Massa, DM Teergarden, MR Landry, PM Henrichs, RH Colby, TE Long. J Appl Polym Sci 59:991–1011, 1994.

70. DJ Massa, KA Shriner, SR Turner, BI Voit. Macromolecules 28:3214–3220, 1995.
71. RFM Lange, EW Meijer. Macromolecules 28:782–783, 1995.
72. C Hilger, R Stadler. Macromolecules 25:6670–6680, 1992.
73. R Stadler, L de Luca Freitas. Colloid Polym Sci 264:773–778, 1986.
74. N Kimizuka, T Kawasaki, K Hirata, T Kunitake. J Am Chem Soc 117:6360–6361, 1995.
75. N Kimuzuka, S Fujikawa, H Kuwahara, T Kunitake, A Marsh, J-M Lehn. J Chem Soc Chem Commun 20:2103–2104, 1995.
76. K Hanabusa, T Miki, Y Taguchi, T Koyama, H Shirai. J Chem Soc Chem Commun 18:1382–1384, 1993.
77. FM Menger, SJ Lee. J Am Chem Soc 116:5987–5988, 1994.
78. FM Menger, SJ Lee, X Tao. Adv Mater 7:669–671, 1995.
79. N Kimizuka, T Kawasaki, T Kunitake. J Am Chem Soc 115:4387–4388, 1993.
80. N Kimizuka, T Kawasaki, K Hirata, T Kunitake. J Am Chem Soc 120:4094–4104, 1998.
81. N Kimizuka, T Kawasaki, T Kunitake. Chem Lett 1:33–36, 1994.
82. N Kimizuka, T Kawasaki, T Kunitake. Chem Lett 8:1399–1402, 1994.
83. H Koyano, K Yoshihara, K Ariga, T Kunitake, Y Oishi, O Kawano, M Kuramori, K Suehiro. Chem Commun 15:1769–1770, 1996.
84. H Koyano, P Bissel, K Yoshihara, K Ariga, T Kunitake. Langmuir 13:5426–5432, 1997.

5

Crystalline Bacterial Cell Surface Layers (S-Layers): A Versatile Self-Assembly System

Uwe B. Sleytr, Margit Sára, and Dietmar Pum
Center for Ultrastructure Research and Ludwig Boltzmann Institute for Molecular Nanotechnology, Universität für Bodenkultur, Wien, Austria

I. INTRODUCTION

Self-assembly of molecules into monomolecular arrays is a new and rapidly growing transdisciplinary scientific and engineering field. Particularly the immobilization of biomolecules in an ordered fashion on solid substrates and their controlled confinement in definite areas of nanometer dimensions are key requirements for many applications in molecular nanotechnology and nanobiotechnology including the development of bioanalytical sensors, molecular electronics, biocompatible surfaces, and signal processing between cells and integrated circuits.

In the last decade several techniques and strategies emerged for creating two-dimensional arrays of proteins on surfaces and for patterning such structures down to the submicrometer scale [1–3].

In this review we describe the basic principles and application potential of crystalline bacterial cell surface layers (S-layers), a self-assembly system optimized through time, selection, and evolution as a unique nanostructure [3–6]. S-layers are fascinating model systems for studying the dynamic process of assembly of a biological supramolecular structure [7]. The broad application potential of S-layers is based on the specific intrinsic features of the monomolecular arrays composed of identical protein or glycoprotein subunits. Since S-layers are periodic structures they exhibit identical physicochemical properties on each molecular unit down to the subnanometer scale and possess pores identical in size and morphology. Most important, functional groups are aligned on the surface and within the pore areas of S-layer lattices in well-defined position and orientation. Many applications of S-layers depend on the capability of isolated subunits to recrystallize into monomolecular lattices in suspension or on suitable surfaces or interfaces [3,7,8].

Finally, S-layers represent a unique structural basis for generating more complex biological supramolecular assemblies, involving the essential ''building blocks'' for life, such as proteins, lipids, and carbohydrates [3,6].

II. LOCATION AND ULTRASTRUCTURE OF S-LAYERS

Among the most commonly observed surface structures on prokaryotic organisms are monomolecular crystalline arrays of proteinaceous subunits termed S-layers [9,10]. S-layers have now been identified in hundreds of different species of every taxonomical group of walled bacteria and are an almost universal feature of archaeal cell envelopes (for compilation see Refs. 8, 10, 11).

The location and ultrastructure of S-layers were investigated by different electron microscopical procedures including freeze-etching, ultrathin-sectioning, freeze-drying in combination with metal-shadowing, and negative-staining (for review see Refs. 9, 12–14).

Although there exists considerable variation in the molecular architecture and supramolecular complexity of prokaryotic envelopes, it is reasonable to classify cell wall profiles containing S-layers into three main groups (Fig. 1). In Gram-negative archaea the S-layer represents the only cell wall component and can be so closely associated with the plasma membrane that it is actually integrated into the lipid layer [15]. On both Gram-positive bacteria and archaea S-layers assemble on the surface of the rigid wall matrix (e.g., peptidoglycan or pseudomurein). In the more complex Gram-negative bacterial cell envelopes S-layers are linked to the surface of the outer membrane.

Unless abundant glycocalyces are present, freeze-etching is the most suitable technique for identifying S-layers on cell surfaces (Fig. 2). High-resolution studies on the mass distribution of the lattices are generally performed on negatively stained S-layer fragments or unstained, thin frozen foils [16–18]. More recently, S-layer lattices also have been studied by scanning probe microscopy. Both scanning tunneling microscopy and atomic force microscopy have been applied [3,19,20]. The topographical images obtained strongly resemble the three-dimensional reconstructions of S-layer lattices derived from the tilt series

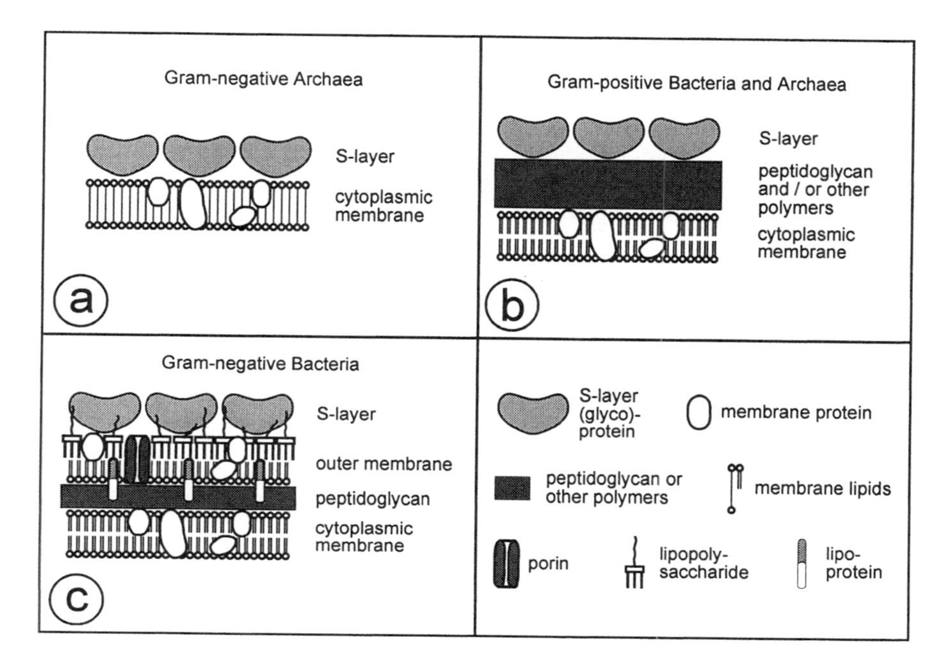

Figure 1 Schematic illustration of the supramolecular architecture of the three major classes of prokaryotic cell envelopes containing crystalline bacterial cell surface layers (S-layers). (a) Cell envelope structure of Gram-negative archaea with S-layers as the only cell wall component external to the cytoplasmic membrane. (b) Cell envelope as observed in Gram-positive archaea and bacteria. In bacteria the rigid wall component is primarily composed of peptidoglycan. In archaea other wall polymers (e.g., pseudomurein or methanochondroitin) are found. (c) Cell envelope profile of Gram-negative bacteria, composed of a thin peptidoglycan layer and an outer membrane. If present the S-layer is closely associated with the lipopolysaccharide of the outer membrane. (Modified from Ref. 8.)

of electron microscopical images. In particular, atomic force microscopy in liquid on unfixed S-layer proteins recrystallized on a solid support revealed structural resolution down to the 1 nm range.

S-layer lattices show oblique (p1, p2), square (p4), or hexagonal (p3, p6) symmetry [8] (Fig. 3). Hexagonal symmetry of S-layers is predominant among the archaea [15]. Depending on the lattice type, the morphological units constituting the regular lattices consist of one, two, three, four, and six monomers. The morphological units may have center-to-center spacings from approximately 3

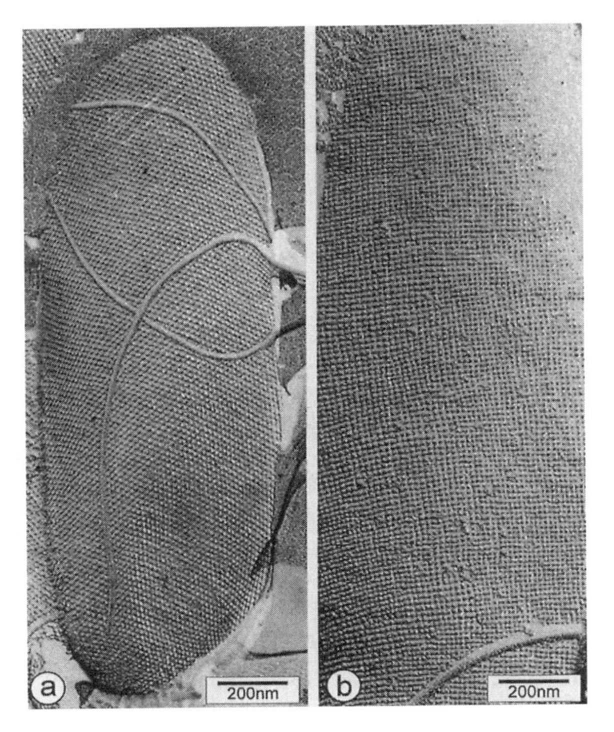

Figure 2 Electron micrographs of freeze-etched preparations of whole cells from (a) *Thermoanaerobacter thermohydrosulfuricus* L111-69 (hexagonal S-layer lattice) and (b) *Desulfotomaculum nigrificans* NCIB 8706 (square S-layer lattice).

to 30 nm. The monomolecular lattices are generally 5 to 25 nm thick. A feature seen with many S-layers is a rather smooth outer and a more corrugated inner surface. In S-layers of archaea frequently pillar-like domains on the inner corrugated surface are observed which are associated or even integrated in the cytoplasma membrane [15].

Due to their crystalline nature S-layers exhibit uniform pore morphologies. Individual lattices can display more than one pore size. Pore sizes between 2 and 8 nm have been estimated [4,21]. The porosity of the protein meshwork may range approximately from 30 to 70%.

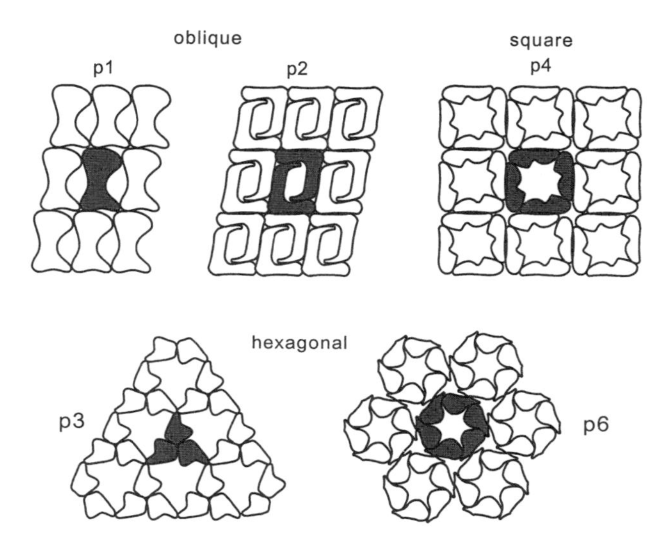

Figure 3 Schematic drawing of different S-layer lattice types. The regular arrays exhibit either oblique (p1, p2), square (p4), or hexagonal (p3, p6) lattice symmetry. The morphological units are composed of one, two, three, four, or six identical subunits. (Modified from Ref. 8.)

III. ISOLATION, CHEMICAL CHARACTERIZATION, AND MOLECULAR BIOLOGY

The subunits of most S-layers interact with each other and with the supporting cell envelope layers through noncovalent forces. In Gram-positive bacteria, a complete disintegration of the protein or glycoprotein lattices into the constituent subunits can be achieved by treatment of intact cells or cell wall fragments with high concentrations of hydrogen-bond breaking agents (e.g., urea or guanidinium hydrochloride) [7,11]. S-layers from Gram-negative bacteria frequently disrupt upon application of metal-chelating agents (e.g., EDTA, EGTA), cation substitution (e.g., Na^+ to replace Ca^{2+}), detergents [8,16,22,23], or by changing the pH value (e.g., pH < 4.0). From conditions required for extraction and disintegration of S-layer lattices it was derived that bonds holding the subunits together are stronger than those between the crystalline array and the supporting cell envelope layers. There are some S-layers that are very resistant to extraction and disintegration suggesting that the subunits are held together by covalent bonds [24–26]. During removal of the disrupting agents, e.g., by dialysis, isolated S-layer subunits frequently reassemble into lattices identical to those observed on intact cells (self-assembly in suspension) [7].

Amino acid analysis of S-layer proteins of organisms from all phylogenetic branches revealed a rather similar overall composition [4,11,16]. Sequencing of genes encoding the S-layer proteins supported data from isoelectric focusing that with a few exceptions (e.g., *Lactobacillus* and *Methanothermus*), S-layers are composed of an acidic protein or glycoprotein species with an isoelectric point between pH 3 and 6. Accordingly, S-layer proteins have a high amount of glutamic and aspartic acid which together resemble about 15 mol%. The lysine content of S-layer proteins is in the range of 10 mol%. Thus, approximately one quarter of the amino acids is charged indicating that ionic bonds play an important role in intersubunit bonding and/or in attaching the S-layer subunits to the underlying cell envelope layer. S-layer proteins have no or only a low content of sulfur-containing amino acids and a high proportion of 40 to 60 mol% of hydrophobic amino acids. Interestingly, hydrophilic and hydrophobic amino acids do not form extended clusters but, instead, the hydrophobic and hydrophilic segments alternate with a more hydrophilic region at the very N-terminal end [27].

Information regarding the secondary structure of S-layer proteins is either derived from the amino acid sequence or from circular dichroism measurements. In most S-layer proteins, 40% of the amino acids are organized as β-sheet and 10 to 20% occur as α-helix. Aperiodic foldings and β-turn content may vary between 5 and 45%. Posttranslational modifications of S-layer proteins include removal of the signal peptide [27,28], phosphorylation [29], and glycosilation [30,31]. With the S-layer protein from the extremely halophilic archeon *Halobacterium halobium* it could be demonstrated for the first time that prokaryotes are capable of producing glycoproteins [32]. Structural analysis of the carbohydrate chains, cloning and sequencing of the S-layer gene, and biosynthesis of this S-layer glycoprotein have been described in detail [33,34]. Chemical analysis and data from nuclear magnetic resonance studies revealed that the carbohydrate chains from bacterial S-layer glycoproteins consist of up to 50 repeating units. Among the monosaccharide constituents, sugars have been detected that are typical of O-antigens of lipopolysaccharides such as quinovosamine, D-rhamnose, N-acetylfucosamine, and heptoses [30,31]. Different novel carbohydrate–amino acid linkage types such as asparagine–rhamnose or tyrosine–galactose have been identified [30,35].

Although it is quite evident that common structural principles must exist in S-layer proteins (e.g., the ability to form intersubunit bonds and to self-assemble, the formation of hydrophilic pores with low unspecific adsorption, the interaction with the underlying cell envelope layer), sequencing of S-layer genes from organisms of all phylogenetic branches led to the conclusion that sequence identities are rare [27,28]. Further sequencing of S-layer genes from strains belonging to the same species such as *Campylobacter fetus* [36], *Lactobacillus acidophilus*

[28], or *Bacillus stearothermophilus* [37–39] revealed that evolutionary relationship plays an important role for the sequence identity of functionally homologous domains. Actually, for some species the N-terminal part of the S-layer protein was found to represent the conserved structural element responsible for anchoring the S-layer subunits to the underlying cell envelope layer. For example, the N-terminal part of the S-layer proteins from the Gram-negative bacteria *Caulobacter crescentus, Campylobacter fetus*, and *Aeromonas salmonicida* recognizes specific lipopolysaccharides in the outer membrane as binding site [36,40,41], while the C-terminal part contains most of the surface-located amino acids. The S-layer proteins SbsA and SbsC from two *B. stearothermophilus* wild-type strains are bound via their N-terminal region to an identical high molecular weight secondary cell wall polymer that is covalently linked to the peptidoglycan backbone [42]. The N-terminal part of these S-layer proteins shows an identity of 85% and more than 70% of the N-terminal 240 amino acids are organized as short α-helices. The sequence identity for the larger part of these S-layer proteins including those domains being involved in the self-assembly and which are located on the outer S-layer surface is below 25%. According to the low sequence identity observed for the C-terminal part, SbsA and SbsC assemble into S-layer lattices with either hexagonal or oblique symmetry and only the outer surface of the SbsC functions as binding site for a cell-associated exoamylase [43]. A peptidoglycan and a secondary cell wall polymer binding region were identified on the S-layer protein SbsB [44] produced by an oxygen-induced strain variant from *B. stearothermophilus* [45,46]. In the case of *Corynebacterium glutamicum* the hydrophobic C-terminal part was found to anchor the S-layer protein to the rigid cell wall layer possessing a high content of hydrophobic mycolic acids [47]. The S-layer proteins from archaea lacking a rigid cell wall layer integrate with their hydrophobic C-terminal part into the cytoplasmic membrane.

By sequence comparison S-layer homologous (SLH)-motifs have been identified at the N-terminal part of many S-layer proteins and at the C-terminal end of cell-associated coenzymes and other exoproteins [48]. Typically, S-layer proteins possess three repeats of SLH-motifs, each consisting of 50 to 60 amino acids. Due to their wide distribution among Gram-positive bacteria, SLH-motifs have been suggested to anchor cell-associated exoproteins permanently or transiently to the cell surface. In contrast to the original assumption that peptidoglycan functions as binding site, it is now evident that secondary cell wall polymers serve as anchoring structures for S-layer proteins [44,49,50,50a].

Since at a generation time of 20 min about 500 S-layer subunits have to be produced per second to keep the bacterial cell surface completely covered with an S-layer lattice, S-layer protein expression must be very efficient and regulatory circuits are necessary to ensure that its synthesis is coordinated with cell growth. Actually, most S-layer protein mRNAs have a long leader sequence and an exceptionally high stability [51,52]. In addition, two or even more promoters have been identified for S-layer gene expression which are turned on at different growth stages [53–55]. With exception of S-layer proteins from *Campylobacter* and *Caulobacter*, all others are produced with a signal peptide suggesting the classical route of secretion [27,28].

Important for understanding S-layer gene regulation was the observation that single bacterial strains can express different (silent) genes. In pathogens such as *C. fetus* S-layer variation can be considered an antigenic variation that is induced in response to the lytic activity of the immune system and leads to modified cell surface properties. Eight to nine S-layer gene cassettes have been identified in *C. fetus* wild-type strains [36] which are tightly clustered on the genome. Several studies confirmed that only a single promoter exists and that antigenic variation is due to recombination of partial coding sequences. Thereby, the N-terminal part of the S-layer protein remains conserved while the C-terminal part is exchanged. In case of *B. stearothermophilus* PV72, S-layer variation could be induced by changing the growth conditions, e.g., oxygen supply, during continuous cultivation [46].

IV. DYNAMIC PROCESS OF ASSEMBLY OF S-LAYERS DURING CELL GROWTH

A. Incorporation of New Subunits into Closed Surface Crystals

Numerous in vitro and in vivo studies have been performed to elucidate the dynamic process of assembly of coherent S-layers during cell growth.

Electron micrographs of freeze-etched rod-shaped bacteria generally reveal a characteristic orientation of the lattice with respect to the longitudinal axis of the cylindrical part of the cell (Fig. 2). This defined alignment of lattices supports the notion that S-layers are ''dynamic closed surface crystals'' with the intrinsic tendency to assume continuously a structure in a state of low free energy during cell growth [7,56].

Information about the development of coherent S-layer lattices on growing cell surfaces also came from reconstitution experiments with isolated S-layers on cell surfaces from which they had been removed (homologous reattachment) or on those of other organisms (heterologous reattachment) [57,58]. These experiments clearly demonstrated that the formation of the regular patterns entirely resides in the subunits themselves and is not affected by the matrix of the supporting cell envelope layer. Lattices reconstituted on cell envelopes that had maintained their cylindrical shape frequently revealed the orientation with respect to the longitudinal axis of the cell as observed by freeze-etching of intact cells. These in vitro experiments confirmed that the curvature of the cylindrical part of the cell induces an orientation of the lattice with least strain between the constituent subunits. On the other hand, the spherical curvature on cell poles and septation sites or on the whole surface of coccoidal cells allows a random orientation of the lattices [56,59].

Knowing the mechanisms that govern S-layer assembly and recrystallization, it became evident that the only requirement for maintaining the highly ordered monomolecular arrays with no gaps on a growing cell surface is a continuous synthesis of a surplus of subunits and their translocation to sites of lattice growth. Although only limited data are available, S-layers revealed a highly anisotropic charge distribution [21]. For example, in Bacillaceae the inner surface is negatively charged, whereas the outer face is charge neutral. Such dipol characteristics of protomeric units may contribute to the proper orientation during local insertion in the course of lattice growth [60,61].

In most organisms the rate of synthesis of S-layer protein appears to be strictly controlled since only small amounts are detectable in the growth medium. On the other hand, studies on a variety of Bacillaceae have demonstrated that a pool of S-layer subunits, at least sufficient for generating one complete S-layer on the cell surface, may be present in the peptidoglycan containing wall matrix (Fig. 1b) [62].

Labeling experiments with fluorescent antibodies and colloidal gold/antibody marker methods showed that different patterns of S-layer lattice extensions exist for Gram-positive and Gram-negative bacteria. In Gram-positive bacteria (Fig. 1b) lattice growth occurs primarily by insertion of multiple bands of S-layers on the cylindrical part of the cell and at new cell poles. In Gram-negative bacteria (Fig. 1c), however, S-layer lattices grow by insertion of new subunits at diffuse sites over the main cell body [7].

Very few data are available about the accurate incorporation sites of constituent subunits on closed surface crystals. As discussed by Harris and coworkers [63–65] dislocations can serve as sites for incorporation of new subunits in crystalline arrays which grow by ''intussusception'' (Fig. 4). Furthermore, as a geometrical necessity, closed surface crystals must contain local wedge disclinations (Fig. 4) which themselves can act as sources of edge dislocations [63]. Consequently, the rate of growth of a closed surface crystal by the mechanism of non-conservative climb of dislocations will depend on the number of dislocations present and the rate of incorporation of new subunits at these sites. High-resolution electron micrographs of freeze-etched preparations revealed the presence of both dislocations and disclinations (Fig. 5) on S-layers of intact cells. Nevertheless, the final proof that dislocations are sites of intussusceptive growth of S-layer lattices will require labeling procedures at a resolution where individual S-layer subunits can be detected.

Whereas in Gram-positive bacteria and archaea (Fig. 1b) the dynamic process of assembly and recrystallization of S-layers occurs on the surface of a rigid cell shape determining a supramolecular envelope structure, Gram-negative arch-

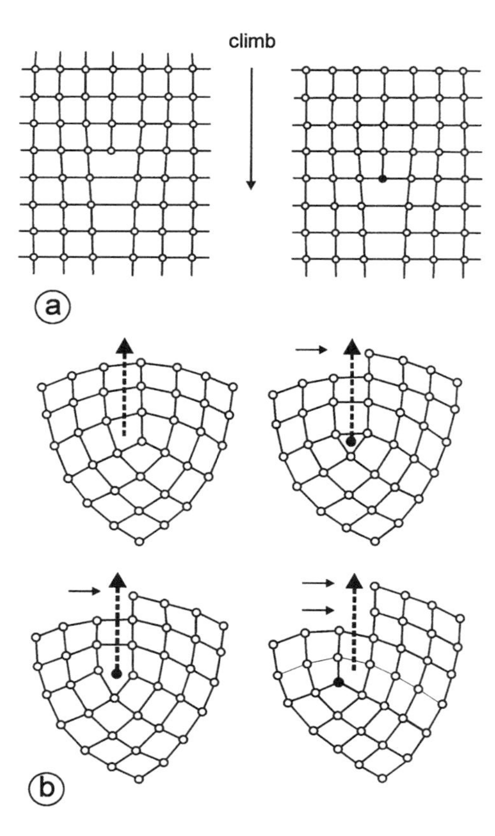

Figure 4 Schematic drawing of dislocations and disclinations as observed on S-layers. Intussusceptive growth of a two-dimensional crystal with square lattice symmetry by non-conservative climb of a dislocation (a): By adding a new subunit (solid dot) the dislocation climbs one lattice spacing to a new position. Movement of a local wedge disclination in a square lattice (b): A wedge disclination may be constructed by cutting into the crystal and rotating one face of the cut into the other (positive wedge disclination), or alternatively by inserting a wedge into the cut instead of removing it (negative wedge disclination). When moving the disclination is shifted diagonally across a distorted square, and in the course of this process generates two edge dislocations (arrows). (Modified from Ref. 7.)

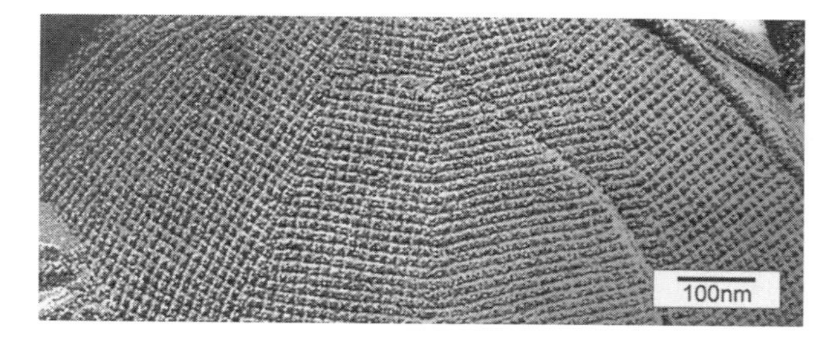

Figure 5 Electron micrograph of a freeze-etched preparation showing several edge dislocations in the S-layer with square lattice symmetry at the cell pole of *Aneurinibacillus thermoaerophilus* DSM 10155. The edge dislocations become visible as line imperfections in the regular array. (Courtesy P. Messner.)

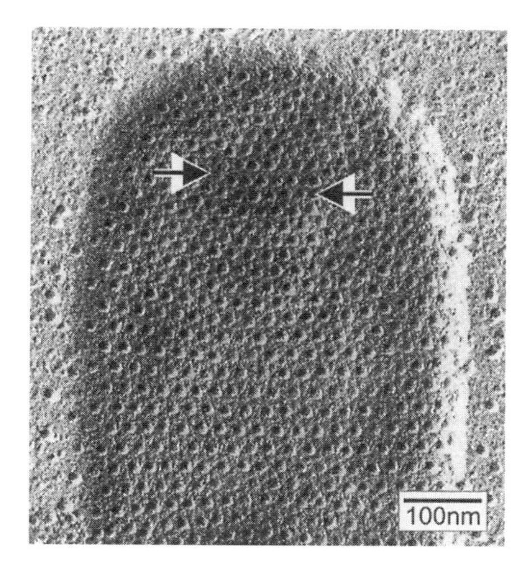

Figure 6 Freeze-dried and shadowed preparation of envelopes of *Thermoproteus tenax* labeled with polycationic ferritin (PCF). The marker molecule binds to the hexagonal S-layer in a regular fashion. Because one PCF molecule is bound per hexameric unit cell of the S-layer lattice, lattice faults in the cell envelope preparations become clearly visible. Based on theoretical considerations six wedge disclinations have to be present at each cell pole. Two of the six wedge disclinations are marked. (Modified from Ref. 66.)

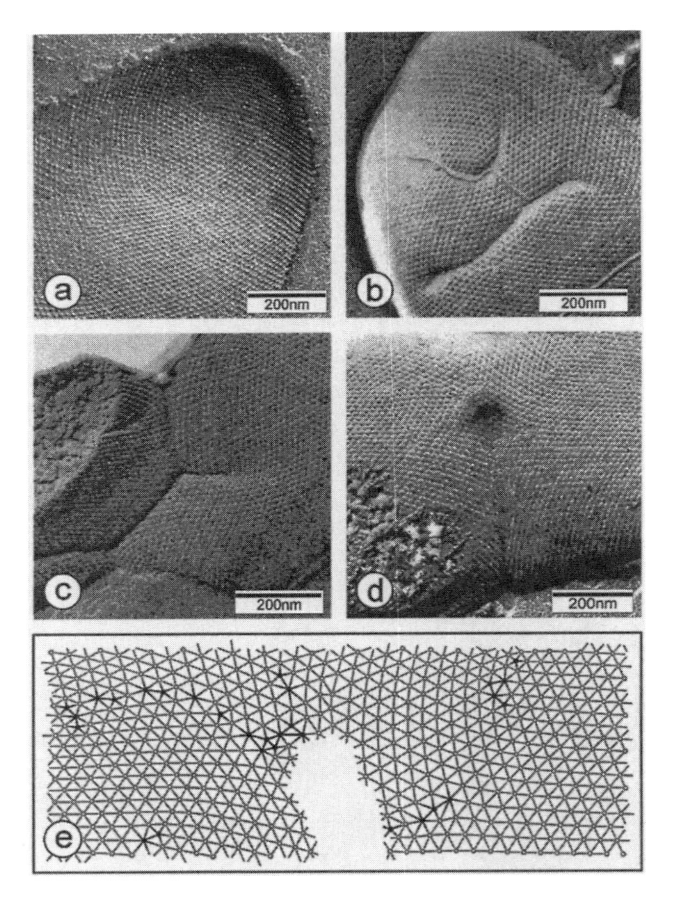

Figure 7 Freeze-etched preparation of *Methanocorpusculum sinense* (a) to (d). The hexagonally ordered S-layer shows several lattice faults (a). Wedge disclinations and edge dislocations are seen as point imperfections in the crystalline array. Consecutive stages in the invagination of the cell wall and cell septation are shown in (b) to (d). Initially shallow invaginations are formed (b) which become longer and deeper as new S-layer material is incorporated (c) and (d). The division of deeper invaginations shows that they can also fuse or branch (c). A far-advanced stage in the cell fission process is shown in (d). Neighborhood graph of the central region of panel (d) is shown in (e). The alignment of lattice faults (pentagons and heptagons; marked in bold with solid dots) in line with the septation direction indicates the route of the progressing cell septation. (Modified from Ref. 68.)

aea (Fig. 1a) possess S-layers as an exclusive wall component. Analysis of cell morphology and lattice fault distribution in Gram-negative archaea provided strong evidence that S-layers can define cell shape and are involved in cell fission. *Thermoproteus tenax*, an extremely thermophilic archeon, has a cylindrical shape with constant diameter but is variable in length and hemispherical cell poles (Fig. 6) [66]. Whereas no dislocations could be observed on the hexagonal array covering the cylindrical part, six wedge disclinations could be visualized on each hemispherical cap. This number resembles the minimum of lattice faults required for covering the rounded surface as known from viruses with icosahedral symmetry [67]. Thus it appears feasible that elongation of the cylindrical part of the cell only requires insertion of S-layer subunits at these distinct lattice faults [63]. More detailed studies on the involvement of an S-layer in cell morphology and division was reported for *Methanocorpusculum sinense* [68]. Cells of this organism reveal, like in many other Gram-negative archeons, a highly lobed cell structure with a hexagonally ordered S-layer. In freeze-etched preparations of intact cells numerous positive and negative 60° wedge disclinations could be detected (Fig. 7) which form pentagons and heptagons in the hexagonal array. Since complementary pairs of pentagons and heptagons are the termination points of edge dislocations they can be expected to function both as sites for incorporation of new morphological units into the lattice (Fig. 8) in the formation process of the lobed structure, and as inition points for the cell division process. The latter was shown to be determined by the ratio between the increase of protoplast volume and the increase in actual S-layer surface area during cell growth [68].

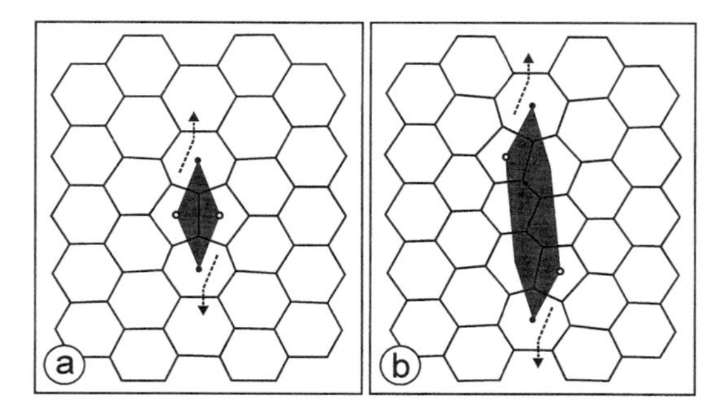

Figure 8 Schematic drawing of the incorporation of a single morphological unit (shaded) in a perfect hexagonal lattice creates a double pair of five- and seven-fold wedge disclinations (a). The two pairs move away from each other by gliding or climbing. One possibility is shown in (b) where the incorporation of new morphological units (shaded) along the arrows pushes the two pairs apart and results in an invagination that becomes longer and deeper. (Modified from Ref. 68.)

V. SELF-ASSEMBLY OF ISOLATED S-LAYER SUBUNITS

A. Self-Assembly in Suspension

Isolated S-layer subunits from a great variety of bacteria show the ability to reassemble into lattices identical to those observed on intact cells. Depending on the intrinsic properties of the S-layer protein and the reassembly conditions (e.g., pH value, ionic strength, ion composition) isolated S-layer subunits may recrystallize into flat sheets, open-ended cylinders, or closed vesicles (Figs. 9, 10) [7,59,69]. Several studies confirmed that S-layers are self-assembly systems in which all the information for crystallization resides within the individual monomers [57] (see this volume, Chapter 1). Detailed studies on the kinetics of the in vitro self-assembly in suspension, the shape of self-assembly products, the charge distribution, and the topographical properties of the outer and inner surface were carried out with S-layers from Bacillaceae.

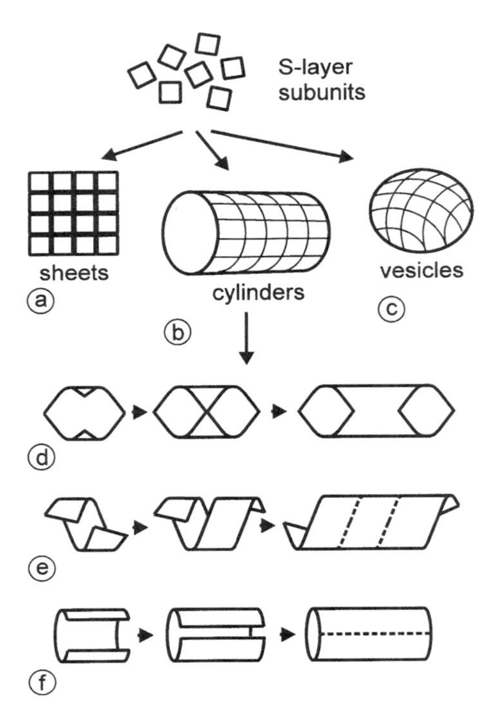

Figure 9 Diagram illustrating different self-assembly routes of S-layer subunits leading to the formation of flat sheets (a), cylinders (b) and (d) to (f), and spheres (c).

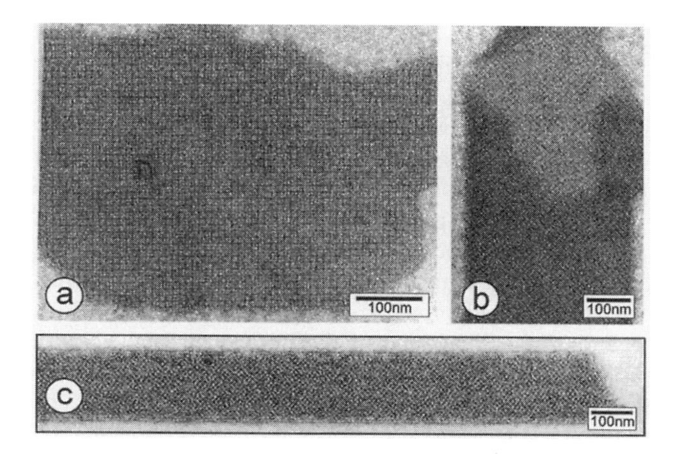

Figure 10 Electron micrographs of negatively stained preparations of S-layer self-assembly products representing double-layers: (a) flat sheet; (b) (c) open-ended cylinders; (a) (b) square S-layer lattice; (c) oblique S-layer lattice.

Studies on the kinetics of the in vitro self-assembly of the S-layer subunits from *B. stearothermophilus* NRS 1536/3c included light scattering and crosslinking experiments and it could be demonstrated that a rapid initial phase and a slow consecutive process of higher than second order exists [70]. The rapid initial phase led to the formation of oligomeric precursors composed of 12 to 16 subunits ($M_r > 10^6$) which fused and recrystallized during the second stage. The sheetlike self-assembly products attained a size of 1 μm and clearly exhibited the square lattice symmetry observed on S-layers on intact cells [70].

Labeling of S-layer self-assembly products with polycationic ferritin revealed that S-layers from most Bacillaceae are highly anisoptropic structures [71–73]. The inner surface carries a net negative charge due to an excess of carboxylic acid groups, while on the outer surface an equimolar amount of amino and carboxylic acid groups is arranged. After blocking the amino groups on the outer S-layer surface from *B. sphaericus* CCM 2120 with glutaraldehyde, the negative charge density was determined to be 1.6/nm^2 [74,75]. By crosslinking experiments it could be further demonstrated that amino and carboxylic acid groups from adjacent S-layer subunits are involved in direct electrostatic interactions. Those chemical modification reactions that changed the native charge distribution in the S-layer lattice led to a complete disintegration or at least to the loss of its structural integrity [71–73]. Most S-layer proteins from *B. stearothermophilus* wild-type strains assemble into mono- or double-layer sheets. This assembly process is independent of the presence of mono- or bivalent cations and could even be observed after addition of metal chelating agents. In double-layer self-assembly products the individual monolayers are bound to each other with the outer charge neutral surface [71]. However, an exception was found for the S-layer glycoprotein from *B. stearothermophilus* NRS 2004/3a which assembles into an oblique lattice (Fig. 11). In case of this glycosilated S-layer protein, the formation of double-layer self-assembly products depended on the presence of bivalent cations during the dialysis procedure [76]. Under these conditions, double-layer sheets or cylinders were formed in which the individual S-layers faced each other with the net negatively charged inner surfaces (Fig. 11). Five different possibilities regarding the orientation of the individual S-layers with respect to each other were identified [76]. In the absence of bivalent cations the S-layer glycoprotein assembled into two types of monolayer cylinders in which the charge neutral outer surface was exposed to the ambient environment (Fig. 11). In the case of S-layer subunits isolated from *B. sphaericus* species, the self-assembly process strongly depended on the presence of bivalent cations. In the absence of bivalent cations, most of the S-layer protein stayed in the water-soluble state and could only be used for recrystallization on solid supports or lipid films [77].

First studies on the self-assembly properties of proteolytic cleavage fragments were carried out with the S-layer protein from *B. sphaericus* P-1 [78]. After removing an 18,000 molecular weight cleavage fragment from this S-layer protein by limited proteolysis, cylindrical self-assembly products were formed instead of flat sheets. By treating the S-layer protein SbsC from *B. stearothermophilus* ATCC 12980 with endoproteinase Glu-C, the N-terminal 227 amino acids were cleaved off and the remaining 100,000 molecular weight cleavage fragment reassembled into flat sheets with a size of 1 to 3 μm clearly exhibiting the oblique lattice structure [42]. By contrast, the whole S-layer protein reassembled into small cylinders with an average diameter of 100 nm. The production of recombinant S-layer proteins revealed that the signal peptide did not disturb the self-assembly process [38]. The recombinant S-layer proteins SbsA, SbsB, and SbsC reassembled even in the cytoplasm of Gram-positive or Gram-negative host cells [38,39,45,79]. These results indicated that the S-layer protein had assumed the correct tertiary structure without the necessity of appropriate chaperones.

The S-layer protein SbsB reassembles into flat mono- and double-layer sheets with a size of 1–3 μm [80]. The addition of the purified high molecular weight secondary cell wall polymer which functions as an anchoring structure for the S-layer protein in the bacterial cell wall [81] inhibited the in vitro self-assembly of the isolated S-layer protein and kept it in the water-soluble state. Interestingly, the soluble monomeric and/or oligomeric S-layer protein recrystallized into closed monolayers on poly-L-lysine coated EM-grids to which the S-

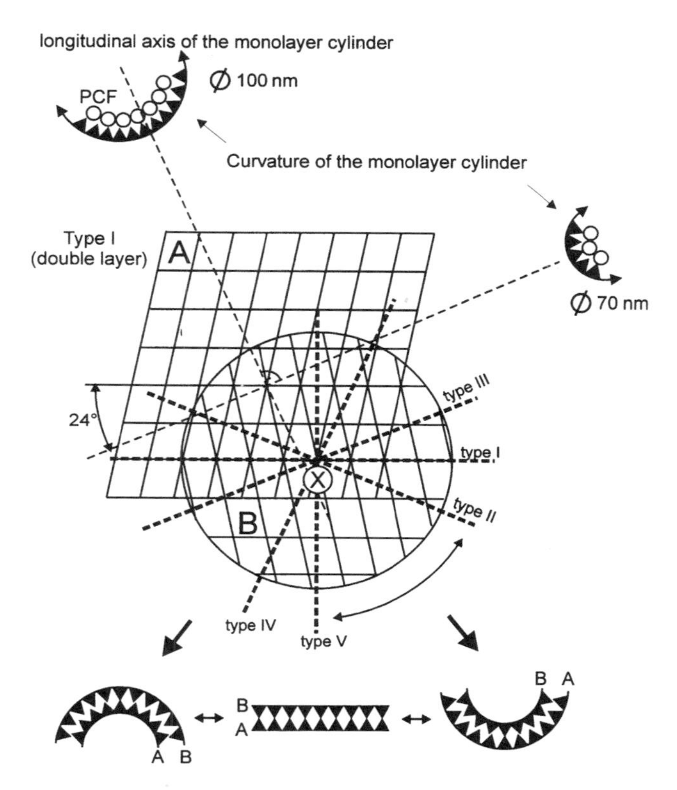

Figure 11 Schematic representation of the formation of mono- and double-layer assembly products as described with S-layer subunits isolated from *Bacillus stearothermophilus* NRS 2004/3a. This S-layer shows oblique lattice symmetry with center-to-center spacings of the morphological units of 9.4 nm and 11.6 nm and a base angle of 78°. The oblique lattice symmetry allows us to unambiguously determine the orientation of the constituent monolayer sheets in double-layer self-assembly products. On the oblique monolayer sheet A the axes of the two types of small (70 and 100 nm diameter) monolayer cylinders are formed as indicated. One of the axes includes an angle of 24° to the short base vector of the oblique S-layer lattice. The second axis is perpendicular to the first. Both monolayer cylinders have an identical direction of curvature. Due to differences in the charge distribution on both S-layer surfaces polycationic ferritin (PCF) is only bound to the inner surface of both types of monolayer cylinders. Five types of double-layer self-assembly products with back-to-back orientation of the inner surface of the constituent monolayers have been found. The superimposition of sheets A and B in the double-layer assembly products of type I is demonstrated and the angular displacement of sheet B with respect to A around point X for the assembly products of type II to V is indicated. (Modified from Ref. 76.)

layer subunits bind with their outer charge neutral surface. Several data indicate that the polymer chains prevent the formation of monomolecular crystalline arrays by acting as spacers between the individual S-layer subunits [80].

B. Self-Assembly on Lipid Films and at the Air–Water Interface

Reassembly of isolated S-layer proteins at the air–water interface and on lipid films have proven to be a very useful way to generate coherent S-layer lattices in large scale (Fig. 12).

The anisotropy in the physicochemical surface properties of the S-layer lattices allows control of the orientation of the monolayer against different surfaces and interfaces. Since, in S-layers used for recrystallization studies the outer surface is more hydrophobic than the inner one, the protein lattices are generally oriented with their outer face against the air–water interface [82,83]. In the case of S-layer protein recrystallized on Langmuir lipid films the orientation was

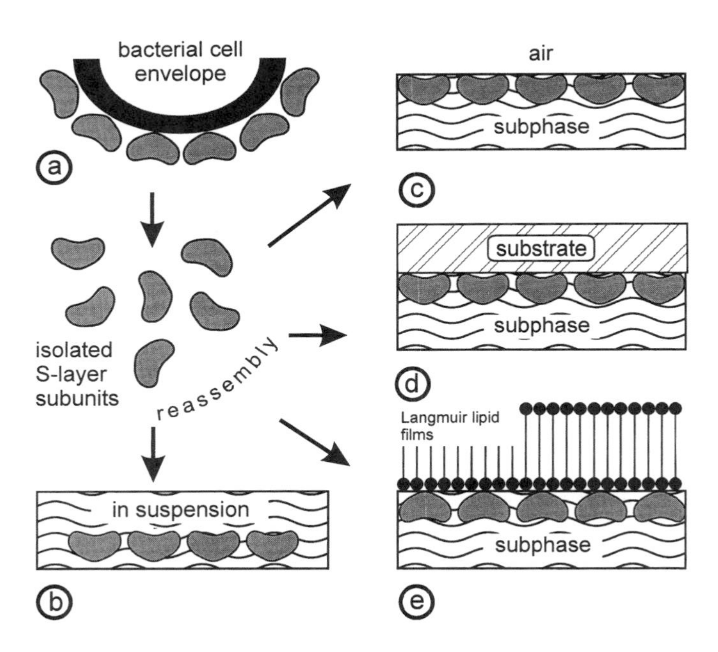

Figure 12 Schematic illustration of recrystallization of isolated S-layer subunits (a) into crystalline arrays. Formation of self-assembly products in suspension (b). The self-assembly process can occur in defined orientation, at the air–water interface (c), on solid supports (d), and on Langmuir lipid films (e). (Modified from Ref. 2.)

shown to depend on the nature of the lipid head-group, the phase state of the surface monolayer, as well as the ionic content, and the pH of the subphase. S-layer lattices formed at different interfaces were studied by electron microscopy and X-ray reflectivity measurements [82–85]. Coherent S-layer monolayers could be obtained on lipid films composed of lipids with zwitterionic head-groups in the presence of calcium ions if the lipid chains possessed a high degree of order, i.e., if the lipid films were in the liquid condensed phase. Under these conditions, the S-layer subunits attached to the lipid with their net negatively charged inner surface. In contrast, the S-layer protein recrystallized poorly under most lipids with negatively charged head-groups and under lipids with unsaturated chains. At monolayers of cationic lipids, reconstituted S-layers were observed in which the protein was attached to the lipid with its outer surface [84,85].

The orientation of the S-layer lattice with respect to the plane of the interface is routinely determined by high-resolution electron microscopical studies of negatively stained S-layer lattices that have been lifted from the air–water interface onto a carbon-coated electron microscope grid by horizontal deposition (Langmuir–Schaefer technique) [82,83]. The determination of the orientation of S-layers with oblique lattice symmetry is particularly easy since the handedness of the base vector pairs can be compared directly to that observed on the bacterial cell. In the case of S-layers with higher lattice symmetry (square or hexagonal), image processing is required to obtain unambiguous results.

It was previously demonstrated that S-layers have a stabilizing effect on the associated lipid layer [83,84,86]. The proportion of lipid molecules in the monolayer that is bound to repetitive domains in the S-layer lattice modulates the lateral diffusion of the free lipid molecules and consequently the fluidity of the whole membrane. Subsequent lipid layers can be deposited on such ''semifluid membranes'' by standard Langmuir–Blodgett techniques or by fusion of lipid vesicles [86]. The stabilizing effect of S-layers on lipid films has already been demonstrated by covering apertures several micrometers in size on holey carbon grids (Fig. 13). These composite structures strongly resemble those archaeal envelope structures that are exclusively composed of an S-layer and a closely associated plasma membrane (Fig. 1a). Since many of these organisms dwell under extreme environmental conditions (e.g., high temperatures, low pH values, high salt concentrations) their S-layers must have a strong stabilizing effect on lipid membranes. Apparently the main reason for this is the reduction of horizontal vibrations which are considered to be the main cause for disintegration of unsupported lipid membranes.

Crystal growth of S-layer protein lattices at different surfaces and interfaces was studied by high-resolution electron microscopy and scanning force microscopy (Fig. 14) [20,83]. Generally crystal growth is initiated simultaneously at many randomly distributed nucleation points and proceeds in plane until the crystalline domains meet leading to a closed, coherent mosaic of crystalline areas

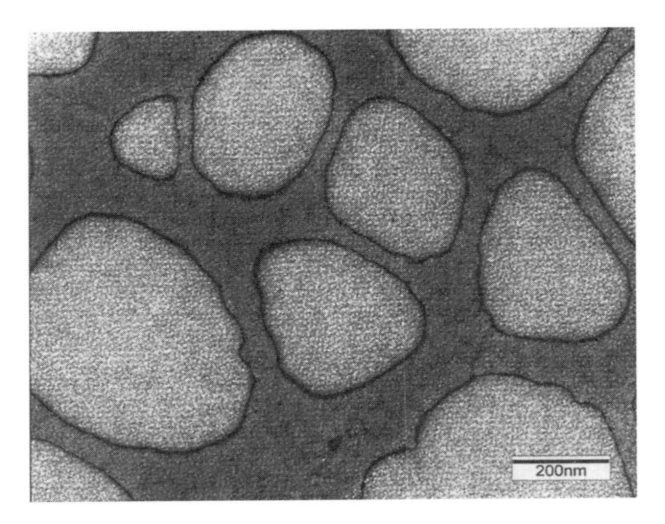

Figure 13 Electron micrograph of an S-layer supported lipid film spanning freestanding the holes in a holey carbon film (on an electron microscope grid). Apertures up to 15 μm in diameter may be covered by composite S-layer/lipid films.

with mean diameters of one to several tens of micrometers. This crystal growth process was commonly observed at liquid–air interfaces, lipid films, and at solid supports [86].

The recrystallization of S-layer proteins at phosphatidylethanolamine monolayers on aqueous subphases was also studied on a mesoscopic scale by dual label fluorescence microscopy and Fourier-Transform-Infra-Red spectroscopy (FTIR) [87]. It was shown that the phase state of the lipid exerts a marked influence on the protein crystallization. When the surface monolayer was in the phase separated state between the fluid and crystalline phases, the S-layer protein was preferentially absorbed at the boundary line between the two coexisting phases and crystallization proceeded underneath the crystalline phase. Crystal growth was much slower under the fluid lipid and the entire interface was overgrown only after prolonged protein incubation. In turn, as indicated by characteristic frequency shifts of the methylene stretch vibrations on the lipids, protein crystallization affected the order of the alkane chains and drove the fluid lipid into a state of higher order. It was also found that the protein did not interpenetrate the lipid monolayer as confirmed by X-ray reflectivity studies [85].

The importance of the ionic strength of the subphase was demonstrated in the recrystallization of the S-layer protein of *B. sphaericus* CCM 2177, which assembles into a lattice with square symmetry [88]. Depending on the calcium

Figure 14 High-resolution scanning force microscopical images of S-layers with oblique (p1) (a) and square (p4) lattice symmetry (c) on silicon surfaces. The corresponding computer image reconstructions obtained by cross-correlation averaging are shown in (b) and (d), respectively. Crystal growth is initiated at randomly distributed nucleation points from which crystalline domains grow (e) until the front edges meet and a closed monolayer is formed (f). (Modified from Ref. 89.)

concentration, a broad range of crystal morphologies varying from tenuous fractal-like structures to large monocrystalline patches was found. Although all these structures looked like fractals obtained by diffusion limited aggregation, they were not aggregates of randomly oriented protein subunits. Image processing revealed that all morphological units followed the orientation of the crystal lattice.

C. Self-Assembly on Solid Supports

Reassembly of isolated S-layer proteins into larger crystalline arrays can be also induced on solid supports [86,88,89]. In particular, the recrystallization of S-layer proteins on technologically relevant substrates such as silicon wafers (Fig. 14), carbon-, platinum-, gold-, or silver-electrodes, and on synthetic polymers already revealed a broad application potential for the crystalline arrays in micro- and nanotechnology [86]. The formation of coherent crystalline arrays depended strongly on the S-layer species, the environmental conditions of the subphase (e.g., temperature, pH value, ion composition, and ionic strength), and, in particular, on the surface properties of the substrate. In general, highly hydrophobic surfaces are better suited for the formation of large-scale closed S-protein monolayers than less hydrophobic or hydrophilic supports [89]. Silanization procedures with different compounds such as octadecyltrichlorosilane (OTS) or hexamethyldisilazane (HMDS) can be used to produce hydrophobic surfaces on silicon or glass [89,90]. Due to practical reasons, most investigations have been carried out on silicon wafers with a native or plasma induced oxide layer, or photoresist coated silicon wafers [89,90]. The recrystallization process follows the same rules as those already described for the formation of coherent large-scale monolayers at liquid–surface interfaces. Scanning force microscopy is the only tool that allows the imaging of S-layer protein monolayers on solid supports at molecular resolution (Fig. 14) [20,90]. In particular, scanning force microscopy in contact mode with loading forces in the range of 100 to 200 pN leads to an image resolution in the subnanometer range (0.5 to 1.0 nm). All these high resolution investigations are only possible in a fluid cell.

S-layers that have been recrystallized on solid surfaces are the basis for a variety of applications in nanomanufactoring, and in the development of miniaturized biosensors [86]. For this purpose supports have to be selected that fulfill specific requirements such as stiffness or flatness. Furthermore, the availability of functional groups becomes important if a specific orientation of the recrystallized S-layer lattice or crosslinking of the protein array is required. Silicon and gallium arsenide have proven to be the most potent materials for all those applications where S-layer technology will be combined with micromachining, nanoelectronics, or nanooptics [3].

D. Solid Supported Lipid Membranes

Phospholipid bilayers or tetraether lipid films incorporating functional molecules (e.g., ion channels, carriers, pore-forming proteins, proton pumps) represent key elements in the development of biomimetic membranes. Unfortunately, plain lipid membranes are highly susceptible to damage during manual handling procedures and prolonged storage, and are thus usually not considered for practical devices. Therefore, much effort has been concentrated on the development of stabilized or supported lipid membranes in order to increase their long-term stability [91]. It was recently proposed that soft polymer cushions be placed between the substrate and the functionalized lipid membrane in order to maintain the thermodynamic and structural properties of the system [91]. As an alternative to this approach, S-layers can be used as supports and stabilizing structures for lipid-bilayer and tetraetherlipid-monolayer films (Figs. 15, 16) [86]. As previously demonstrated, in comparison with unsupported mono- or bimolecular lipid layers, lipid films associated with S-layers are much more stable structures [83].

S-layer stabilized solid supported lipid membranes have been fabricated as follows (Fig. 15) [20]. After compressing a phospholipid monolayer on a Langmuir trough into the liquid-crystalline phase, a thiolipid-coated solid support (glass cover slip or, alternatively, a silicon wafer) is placed horizontally onto the monolayer and left in this position until the S-layer protein, which is subsequently injected into the subphase, has assembled into a closed crystalline monolayer. The whole assembly is then removed from the liquid–air interface (Langmuir–Schaefer technique) and rinsed with deionized water. For demonstrating that the recrystallization process is complete the support can subsequently be transferred to a scanning force microscope (Fig. 15). The crystallization process of the S-layer proteins follows the same kinetics as described for the reassembly on solid supports. It has also recently been shown that S-layer supported lipid membranes reveal a reduced tendency to rupture especially in the presence of ionophores or pore-forming proteins [92,93].

The electrophysical features of S-layer supported lipid membranes were studied both by voltage clamp and black lipid membrane techniques. In the first set of experiments S-layer protein isolated from *Bacillus coagulans* E38-66 was recrystallized on monolayers of glyceroldialkylnonitol tetraether lipid (GDNT). Voltage clamp examinations were applied for determining the barrier function of the lipid film before and after recrystallization of the S-layer protein and the effect of incorporation of the potassium selective ion channel valinomycin [92]. Upon recrystallization of the S-layer protein a decrease in conductance of the GDNT-monolayer was observed. Furthermore, it was found that the valinomycin-mediated increase in conductance was less pronounced for the S-layer supported than for the plain GDNT-monolayer confirming differences in the accessibility and/or in the fluidity of the lipid membranes. In contrast to plain GDNT-monolayers, S-layer supported GDNT-monolayers with high valinomycin mediated conductance persisted over much longer periods of time indicating enhanced stability of the composite structure. The effects of a supporting S-layer from *Bacillus coagulans* E38-66 on lipid bilayers was also investigated by comparative voltage clamp studies on plain and S-layer supported 1,2-diphytanoyl-*sn*-glycero-3-phos-

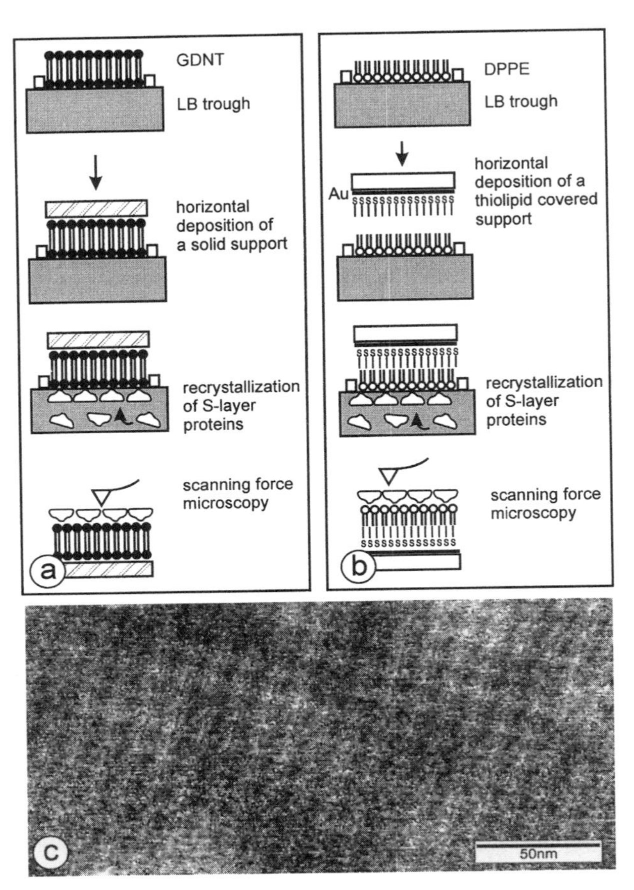

Figure 15 Schematic illustration of the preparation steps required for generating an S-layer stabilized tetraether lipid film (a) and lipid bilayer (b) on a solid support. Top-to-bottom: after compressing a glycerol dialkylnonitol tetraether lipid (GDNT) monolayer (a), or a 1,2-dipalmitoyl-*sn*-glycero-3-phosphatidylethanolamine (DPPE) monolayer (b) to a surface pressure of 25–30 mN/m, a solid support (silicon wafer: plain or for attachment of thiolipids coated with gold) is placed horizontally onto the monolayer and left in this position until the S-layer protein which was injected afterwards in the subphase has recrystallized as a closed monolayer. The whole assembly is subsequently removed, rinsed in deionized water, and transferred into a scanning force microscope (c). Scanning force micrograph of the surface topography of the outer S-layer face of an S-layer with square lattice symmetry. (Modified from Ref. 20.)

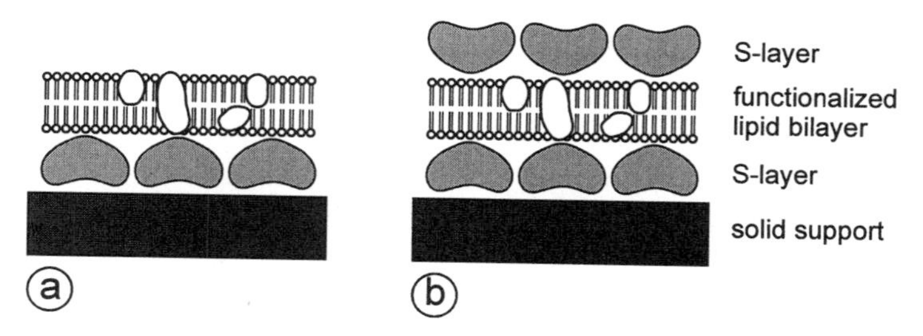

Figure 16 Schematic drawing illustrating S-layer stabilized solid supported lipid membranes. The S-layer functions as a water-carrying layer into which membrane integrated molecules may protrude (a). An S-layer assembled on top of the bilayer which may serve as a protective and stabilizing cover (b).

phatidyl-choline (DPhPC) bilayers [93]. Upon S-layer recrystallization no significant changes but a slight decrease in conductance could be observed. Thus, both the GDNT monolayer and DPhPC bilayer studies indicated that the recrystallized S-layer protein did not penetrate or rupture the lipid film.

The effect of a supporting S-layer on the incorporation and self-assembly of a pore-forming protein was also studied with DPhPC bilayers formed over a thin teflon aperture and the staphylococcal exotoxin α-hemolysin [93]. When added to the lipid-exposed side the assembly of the heptameric α-hemolysin pore was slow compared to unsupported membranes. According to the semifluid membrane model [83] this phenomenon is explained by the altered fluidity of the lipid bilayer. On the other hand, no pore assembly could be detected upon adding α-hemolysin monomers to the S-layer-faced side of the composite membrane. This is due to the intrinsic molecular sieving properties of the S-layer lattice preventing passage of α-hemolysin monomers through the S-layer pores to the lipid bilayer. In comparison to plain lipid bilayers, the S-layer supported lipid membrane had a decreased tendency to rupture in the presence of α-hemolysin.

Based on these experiments, studies are in progress to exploit the functionality of transmembrane proteins in S-layer stabilized, solid-supported lipid membranes in which the S-layer is directly attached to the solid support and the lipid layer associated with the S-layer (Fig. 16). Most recently lateral diffusion of fluorescence lipid probes in S-layer supported lipid membranes on solid supports have been investigated with fluorescence recovery after photobleaching FRAP [94]. It was demonstrated that the S-layer cover induced an enhanced mobility of the probe in the lipid layer which also supported the semifluid membrane concept [83,86]. Furthermore it was noticed that the S-layer lattice cover could prevent the formation of cracks and other inhomogeneties in the bilayer.

E. Self-Assembly on Liposomes

Artificial lipid vesicles, termed liposomes, are colloid particles in which phospholipid bilayers or tetraether monolayers encapsulate an aqueous medium. Because of their physicochemical properties, liposomes are widely used as model systems for biological membranes and as delivery systems for biologically active molecules. In general, water-soluble molecules are encapsulated within the aqueous compartment whereas water-insoluble substances may be intercalated into the liposomal membrane [95].

Isolated monomeric and/or oligomeric S-layer protein from *B. coagulans* E38–66 [96–98], *B. sphaericus* (Fig. 17), and the SbsB from *B. stearothermophilus* PV72/p2 [99] could be recrystallized into the respective lattice type on posi-

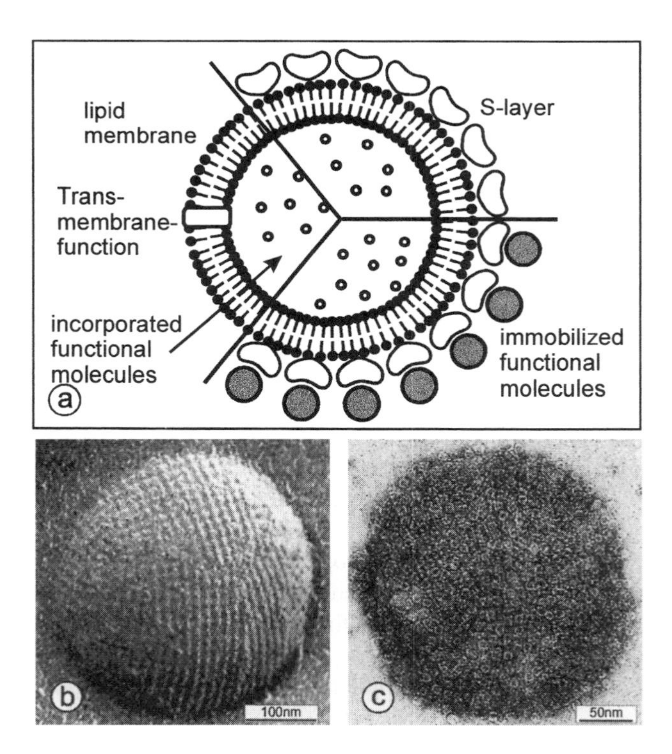

Figure 17 Schematic drawing of a liposome coated with an S-layer lattice used as a matrix for immobilizing functional macromolecules (a). Freeze-etched preparation of a liposome completely covered with a square S-layer lattice (courtesy S. Küpcü) (b). Negative-staining of a liposome coated with an oblique S-layer lattice which was subsequently exploited for the covalent attachment of ferritin (c).

tively charged liposomes composed of dipalmitoylphosphatidylcholine (DPPC), cholesterol, and hexadecylamine. In the case of the SbsB, Zeta-Potential measurements strongly indicated that the S-layer subunits had bound with their outer charge neutral surface to the positively charged liposomes [99]. This was identical to the orientation of the S-layer lattice obtained by recrystallization of the SbsB on poly-L-lysine coated EM grids. On the contrary, the SbsB had bound with its inner net negatively charged surface to Langmuir films composed of DPPC [81]. Crosslinking of the S-layer lattice on liposomes revealed that most of the hexadecylamine incorporated into the bilayer could react with the S-layer subunits which supported the hypothesis of the semifluid membrane model that at least part of the membrane lipids are fixed to discrete positions of the S-layer subunits [3,83]. More recently, microcalorimetric [97] and sound velocity studies [98] on S-layer coated liposomes strongly supported the proposed semifluid membrane model in demonstrating that S-layers increase intermolecular order in lipid membranes.

The presence of S-layer lattices significantly enhanced the stability of the liposomes against mechanical stress such as shear forces or ultrasonication and against thermal challenges [99]. The S-layer lattices were further exploited as a template for chemical modifications and as a matrix for the immobilization of macromolecules, such as ferritin (Fig. 17) [96]. After biotinylation of histidine or tyrosine residues which did not disturb the structural integrity of the oblique S-layer lattice formed by the SbsB, streptavidin could be immobilized as a dense monomolecular layer that was capable of binding biotinylated antibodies in high packing denity [100].

VI. PATTERNING OF S-LAYERS RECRYSTALLIZED ON SOLID SUPPORTS

Most applications of S-layers, particularly in nanotechnology, involve procedures generating S-protein monolayers on solid supports and subsequent patterning of these layers [86]. For example, if biologically functional molecules have to be bound to S-layers recrystallized on electrodes, a procedure must be available that allows in a first step controlled removal of the S-layer in certain areas. Patterning of S-layers by exposure to deep ultraviolet radiation (DUV) has proven to be a powerful technique for imposing structures in recrystallized S-layer lattices on solid supports (Fig. 18) [89,90]. In this approach S-layer protein recrystallized on a silicon wafer is brought in direct contact with a microlithographic mask and exposed to the deep ultraviolet radiation of a pulsed Argon fluoride (ArF) excimer laser (wavelength of emitted light 193 nm). The S-layer is removed specifically from the silicon surface in the exposed regions but retains its structure and functionality in the unexposed areas. The crucial step in this procedure is drying of the protein layer. While excess water causing interference fringes in the course

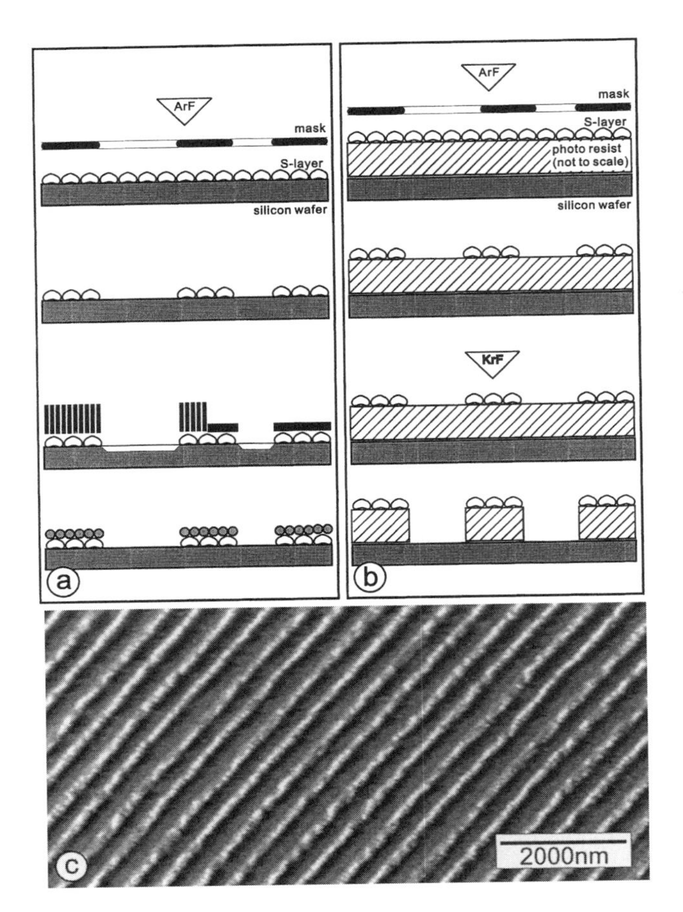

Figure 18 Schematic drawing of patterning S-layers by exposure to deep ultraviolet radiation. (a) A pattern is transferred onto the S-layer by exposure to ArF excimer laser radiation (wavelength 193 nm) through a microlithographic mask. The S-layer is specifically removed from the silicon surface in the exposed regions but retains its crystalline and functional integrity in the unexposed areas. Unexposed S-layer areas can be used either to bind enhancing ligands or to enable electroless metallization. In both cases a layer is formed that allows a patterning process by reactive ion etching. Alternatively, unexposed S-layers may also be used for selectively binding biologically active molecules that would be necessary for the fabrication of miniaturized biosensors or biocompatible surfaces. (b) In the two-layer resist approach, the S-layer that was formed on top of a spin-coated polymeric resist (on a silicon wafer) is first patterned by ArF radiation and subsequently serves as a mask for a blank exposure of the resist by irradiation with KrF-radiation. Due to the thinness of the S-layer very steep sidewalls are obtained in the developed resist. (c) Scanning-force microscopical image of a patterned S-layer on a silicon wafer. The ultimate resolution is determined by the wavelength of the excimer laser radiation. (Modified from Ref. 90.)

of DUV exposure has to be eliminated, enough water has to be retained for the structural integrity of the S-layer protein. Drying the S-layer in a stream of dry nitrogen at room temperature has proven to be a feasible method. The patterning process was performed in several shots of 100 to 200 mJ/cm^2 each (pulse frequency 1 Hz, pulse duration ~ 8 nsec). Subsequently the remaining unexposed S-layer areas could be used either to bind enhancing ligands or to enable electroless metallization in order to form a layer that allows a final patterning process of the silicon by reactive ion etching. Since S-layers are only 5 to 10 nm thick and consequently much thinner than conventional resists a considerable improvement in edge resolution in the fabrication of submicron structures can be expected. As an alternative to the application in the microelectronic sciences, the unexposed S-layer areas may also be used for selectively binding intact cells (e.g., neurons), lipid layers, or biologically active molecules as required for the development of biosensors [101,102].

Finally, it is interesting to note that under exposure to krypton fluoride (KrF) excimer laser radiation supplied in several shots of ~350 mJ/cm^2 the S-layer is not ablated but carbonized in the exposed areas [88,103]. This result has already been used for high-resolution patterning of polymeric resists (Fig. 18). S-layers that had been formed on top of a spin-coated polymeric resist (on a silicon wafer) were first patterned by ArF radiation and subsequently served as a mask for a blank exposure of the resist by irradiation with KrF-pulses. This two-step process was possible because S-layers are less sensitive to KrF radiation than polymeric resists. The thinness of the S-layer mask causes very steep side walls in the developed polymeric resist.

Preliminary experiments with electron beam writing or ion beam projection lithography have demonstrated that the S-layer may also be patterned by these techniques in the sub 100 nm range (unpublished results). Most recently, S-layers have also been exploited as matrices for electrochemical deposition and electron beam deposition for generating calibration standards in the 10 nm range for scanning probe microscopy [104].

VII. S-LAYERS AS MATRICES FOR THE IMMOBILIZATION OF FUNCTIONAL MOLECULES

The high density and defined position and orientation of surface located functional groups on S-layer lattices was exploited for the immobilization of different (macro)molecules [6,8]. For introducing covalent bonds between the S-layer subunits and for enhancing the stability properties, S-layer lattices were crosslinked with homobifunctional amino group specific crosslinkers of different bridge length. For immobilization of foreign (macro)molecules, vicinal hydroxyl groups from the carbohydrate chains of S-layer glycoproteins were either cleaved with periodate or they were activated with cyanogen bromide [105]. Carboxylic acid groups from the S-layer protein were activated with carbodiimide which, regarding the modification rate or the binding density, turned out to be the best method [105]. Most enzymes such as invertase, glucose oxidase, β-glucosidase, or naringinase formed a monolayer of densely arranged molecules on the outer surface of the S-layer lattice (Fig. 19) [105,106]. For example, the binding capacity of the square S-layer lattice from *B. sphaericus* for IgG was determined to be 375 ng/cm^2. As derived from the saturation capacity of a planar surface and the molecular dimension of IgG, this corresponds to a monolayer of randomly oriented antibody molecules [107,108]. S-layer microparticles with immobilized Protein A were used as escort particles for affinity crossflow filtration for isolating IgG from serum or hybridoma culture supernatants [109,110].

By depositing S-layers on microporous supports and crosslinking the S-layer lattice with glutaraldehyde, mechanically stable composite structures became available which after immobilization of various monoclonal antibodies were used as a reaction zone for S-layer based dipsticks [107]. So far, different

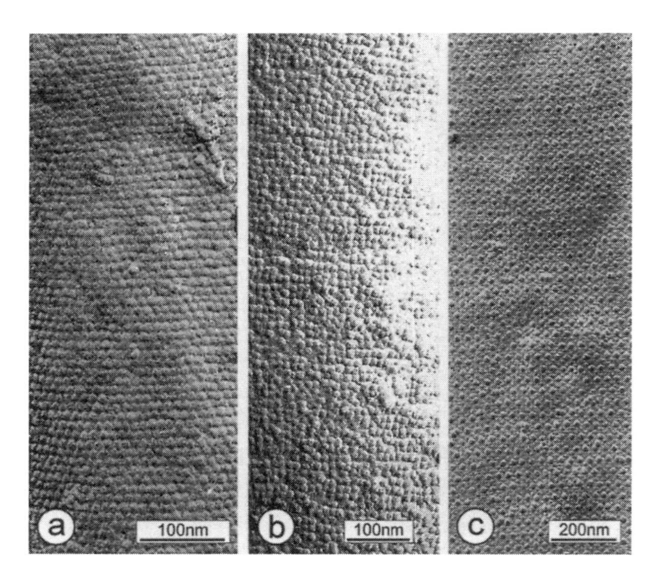

Figure 19 Freeze-etched preparations showing ferritin molecules covalently bound (a) (b) or immobilized by electrostatic interactions (c) to S-layer lattices with different lattice symmetry. The immobilized ferritin with a molecular size of 12 nm reflects the periodicity of the underlying S-layer lattice type over a wide range [(a), (c) hexagonal; (b) square]. The center-to-center spacings of the morphological units in the S-layer lattices are 14 nm (a) (b) and 30 nm (c).

types of S-layer based dipsticks have been developed that work according to the principle of solid-phase immunoassays [111]. Currently they are used for the determination of the concentration of tissue-type plasminogen activator (t-PA) in whole blood or plasma, IgE as a marker for type I allergies in serum, and interleukines in serum or blood to differentiate between septic and traumatic shock. In general, the advantages of S-layers as an immobilization matrix and especially as a reaction zone for solid-phase immunoassays in comparison to amorphous polymers can be summarized as follows.

1. S-layers have well-defined surface properties and immobilization of functional macromolecules or the catching antibody can only occur on the outermost surface of the crystal lattice preventing diffusion controlled reactions;
2. since the functional macromolecules are covalently linked to the S-layer protein, no leakage can occur;
3. S-layers generally show a low unspecific adsorption and S-layer based dipsticks can therefore be incubated in whole blood, plasma, or serum without the necessity of blocking steps.

Future developments of S-layer-based diagnostic test systems are focused on the use of recombinant S-layer fusion proteins with integrated functional domains [6,112].

VIII. S-LAYERS AS TEMPLATES IN THE FORMATION OF REGULARLY ARRANGED NANOPARTICLES

Currently there is great interest in fabricating nanostructures for the development of a new generation of electronic and optic devices. In particular, the formation of metal clusters for nanoelectronic digital circuits requires a well-defined size and arrangement of the particles. As an alternative to approaches in which colloidal crystallization was used to make close-packed nanoparticle arrays [113], the use of S-layers as organic templates allows the synthesis of a wide range of inorganic nanocrystal superlattices. Recently, it was demonstrated that S-layer proteins recrystallized on solid supports or S-layer self-assembly products which were deposited on such substrates may be used to induce the formation of CdS particles [114] or gold nanoparticles (Fig. 20) [115]. CdS inorganic superlattices with either oblique or square lattice symmetries of approximately 10 nm repeat distance were fabricated by exposing self-assembled S-layer lattices to cadmium ion solutions followed by slow reaction with hydrogen sulfide. Precipitation of the inorganic phase was confined to the pores of the S-layers with the result that CdS superlattices with prescribed symmetries were prepared.

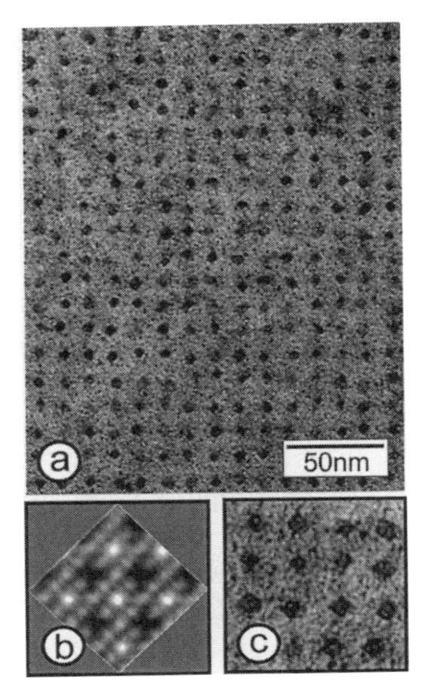

Figure 20 Electron micrograph of a gold superlattice with square lattice symmetry and a lattice constant of 12.8 nm consisting of monodisperse gold nanoparticles with mean diameters of 4 to 5 nm. The point pattern resembles the lattice parameters of the underlying S-layer. (b) Scanning force microscopic image of the native S-layer. (c) Magnified subregion of (a).

In a similar procedure a square superlattice of uniform 4 to 5 nm-sized gold particles with 12.8 nm repeat distance was fabricated by exposing a square S-layer lattice in which thiol groups had been introduced before, to a tetrachloroauric (III) acid solution. Transmission electron microscopical studies showed that the gold nanoparticles were formed in the pore region during electron irradiation of an initially grainy gold coating covering the whole S-layer lattice. The shape of the gold particles resembled the morphology of the pore region of the square S-layer lattice. By electron diffraction and energy dispersive X-ray analysis the crystallites were identified as gold (Au(0)). Electron diffraction patterns revealed that the gold nanoparticles were crystalline but in the long-range order not crystallographically aligned.

It should be stressed that with S-layers as molecular templates the formation of superlattices with a wide range of interparticle spacings as well as with oblique, square, or hexagonal lattice symmetry becomes possible. This is particularly important for the development of nanometric electronic or optical devices since isolated S-layer subunits have shown the inherent capability to recrystallize on a great variety of solid supports including structured semiconductors [89,90].

IX. CONCLUSIONS

Two-dimensional crystalline arrays of protein or glycoprotein subunits are now recognized as one of the most common cell surface structures in Archaea and Bacteria. The intrinsic assembly and recrystallization properties of S-layer subunits enable the maintenance of a closed lattice during cell growth and division. From a structural and morphogenetic point of view S-layers represent the simplest biological (protein) membranes developed during evolution. Therefore, it was suggested that S-layer-like dynamic membranes could have fulfilled barrier and supporting functions as required by self-reproducing systems (progenotes) during the early periods of biological evolution [58,59].

The wealth of information obtained on the general principles of crystalline bacterial cell surface layers, particularly on their structure, assembly, surface, and molecular sieving properties has revealed a broad application potential. Above all, the repetitive physicochemical properties down to the subnanometer scale make S-layer lattices unique self-assembly structures for functionalization of surfaces and interfaces down to the ultimate resolution limit. S-layers that have been recrystallized on solid substrates can be used as immobilization matrices for a great variety of functional molecules or as templates for the fabrication of ordered and precisely located nanometer-scale particles as required for the production of biosensors, diagnostics, molecular electronics, and nonlinear optics [2,3,6].

Biomimetic approaches copying the supramolecular cell envelope structure in those archaea that have S-layers as the only, and occasionally quite rigid, wall component should lead to new technologies exploiting functional lipid membranes at meso- and macroscopic scale. In particular the technology of S-layer supported lipid membranes in the long term may be exploited for generating defined junctions between living cells (e.g., neurons) and integrated circuits for signal processing. Although numerous applications for S-layers have already been demonstrated, many other areas may yet emerge. In the near future genetic modifications of S-layer proteins and specific chemical modifications will significantly influence the development of basic and applied S-layer research [2].

REFERENCES

1. AS Blawas, WM Reichert. Biomaterials 19:595–609, 1998.
2. UB Sleytr, P Messner, D Pum, M Sára. Angew Chemie Int Ed 38:1034–1054, 1999.

3. D Pum, UB Sleytr. Trends Biotechnol 17:8–12, 1999.
4. M Sára, UB Sleytr. Prog Biophys Mol Biol 65:83–111, 1996.
5. UB Sleytr. FEMS Microbiol Rev 20:5–12, 1997
6. UB Sleytr, M Sára. Trends Biotechnol 15:20–26, 1997.
7. UB Sleytr, P Messner. In: H Plattner, ed. Electron Microscopy of Subcellular Dynamics. Boca Raton, FL: CRC, 1989, pp 13–31.
8. UB Sleytr, P Messner, D Pum, M Sára. Crystalline Bacterial Cell Surface Proteins. Austin, TX: Landes/Academic, 1996.
9. UB Sleytr. Int Rev Cytol 53:1–26, 1978.
10. UB Sleytr, P Messner, D Pum, M Sára. Crystalline Bacterial Cell Surface Layers. Berlin: Springer-Verlag, 1988.
11. P Messner, UB Sleytr. Adv Microb Physiol 33:213–275, 1992.
12. UB Sleytr, AM Glauert. In: JR Harris, ed. Electron Microscopy of Proteins vol. 3. London: Academic, 1982, pp 41–76.
13. S Hovmöller, A Sjögren, DN Wang. Prog Biophys Mol Biol 51:131–163, 1988.
14. UB Sleytr, P Messner, D Pum. In: F Mayer, ed. Methods in Microbiology vol. 20. London: Academic, 1988, pp 29–60.
15. W Baumeister, G Lembke. J Bioenerg Biomembr 24:567–575, 1992.
16. TJ Beveridge. Curr Opin Struct Biol 4:204–212, 1994.
17. W Baumeister, I Wildhaber, BM Phipps. Can J Microbiol 35:215–227, 1989.
18. S Hovmöller. In: TJ Beveridge, SF Koval, ed. Advances in Bacterial Paracrystalline Surface Layers. New York: Plenum, 1993, pp 13–21.
19. DJ Müller, W Baumeister, A Engel. J Bacteriol 178:3025–3030, 1996.
20. B Wetzer, D Pum, UB Sleytr. J Struct Biol 119:123–128, 1997.
21. M Sára, UB Sleytr. J Bacteriol 169:2804–2809, 1987.
22. DP Bayley, SF Koval. Can J Microbiol 40:237–242, 1993.
23. SF Koval. Can J Microbiol 34:407–414, 1988.
24. TJ Beveridge, M Stewart, RJ Doyle, GD Sprott. J Bacteriol 162:728–736, 1985.
25. H König. Can J Microbiol 34:395–406, 1988.
26. S Schultze-Lam, TJ Beveridge. Appl Environ Microbiol 60:447–453, 1994.
27. B Kuen, W Lubitz. In: UB Sleytr, P Messner, D Pum, M Sára, eds. Crystalline Bacterial Cell Surface Proteins. Austin TX: Landes, 1996, pp 77–111.
28. H Boot, P Pouwels. Mol Microbiol 21:1117–1123, 1996.
29. SR Thomas, TJ Trust. J Mol Biol 245:568–581, 1995.
30. P Messner. Glycoconj J 14:3–11, 1997.
31. P Messner, C Schäffer. In: RJ Doyle, ed. Glycomicrobiology. New York: Plenum 1999 (in press).
32. MF Mescher, JL Strominger. J Biol Chem 252:2005–2014, 1976.
33. J Lechner, M Sumper. J Biol Chem 262:9724–9729, 1987.
34. M Sumper. In: TJ Beveridge, SF Koval, eds. Advances in Paracrystalline Bacterial Surface Layers. New York: Plenum, 1993, pp 109–118.
35. C Schäffer, T Wugeditsch, C Neuninger, P Messner. Microbial Drug Resistance 2:17–23, 1996.
36. J Dworkin, MJ Blaser. Mol Microbiol 26:433–440, 1997.
37. B Kuen, UB Sleytr, W Lubitz. Gene 145:115–120, 1994.

38. B Kuen, M Sára, W Lubitz. Mol Microbiol 19:495–503, 1995.
39. M Jarosch, EM Egelseer, D Mattanovich, UB Sleytr, M Sára. Microbiol (in press).
40. P Doig, L Emödy, TJ Trust. J Biol Chem 267:43–51, 1992.
41. SG Walker, DN Karunaratne, N Ravenscroft, J Smit. J Bacteriol 176:6312–6323, 1994.
42. EM Egelseer, K Leitner, M Jarosch, C Hotzy, S Zayni, UB Sleytr, M Sára. J Bacteriol 180:1488–1495, 1998.
43. EM Egelseer, I Schocher, UB Sleytr, M Sára. J Bacteriol 178:5602–5609, 1996.
44. M Sára, EM Egelseer, C Dekitsch, UB Sleytr. J Bacteriol 180:6780–6783, 1998.
45. B Kuen, A Koch, E Asenbauer, M Sára, W. Lubitz. J Bacteriol 179:1664–1970, 1997.
46. M Sára, B Kuen, HF Mayer, F Mandl, KC Schuster, UB Sleytr. J Bacteriol 178: 2108–2117, 1996.
47. M Chami, N Bayan, JL Peyret, T Guli-Krzywicki, G Leblon, E Shechter. Mol Microbiol 23:483–492, 1997.
48. A Lupas, H Engelhardt, J Peters, U Santarius, S Volker, W Baumeister. J Bacteriol 176:1224–1233, 1994.
49. M Chauvaux, M Matuschek, P Béguin. J Bacteriol 181:2455–2458, 1999.
50. S Mesnage, E Tosi-Couture, A Fouet. Mol Microbiol 31:927–936, 1999.
50a. N Ilk, P Kosma, M Puchberger, EM Egelseer, UB Sleytr, M Sára. J Bacteriol (in press).
51. S Chu, S Chavaignac, J Feutrier, B Phipps, M Kostrzynska, WW Kay. J Biol Chem 266:15258–15265, 1991.
52. JA Fisher, J Smit, N Agabian. J Bacteriol 170:4706–4713, 1988.
53. T Adachi, H Yamagata, N Tsukagoshi, S Udaka. J Bacteriol 171, 1010–1016, 1989.
54. S Ebisu, A Tsuboi, H Takagi, Y Naruse, H Yamagata, N Tsukagoshi, S Udaka. J Bacteriol 172:614–620, 1990.
55. M Kahala, K Savijoki, A Palva. J Bacteriol 179:284–286, 1992.
56. UB Sleytr, AM Glauert. J Ultrastruct Res 50:103–116, 1975.
57. UB Sleytr. Nature 257:400–402, 1975.
58. UB Sleytr. In: O Kiermayer, ed. Cell Biology Monographs vol. 8. Wien and New York: Springer-Verlag, 1981, pp 1–26.
59. UB Sleytr, R Plohberger. In: W Baumeister, W Vogell, eds. Microscopy at Molecular Dimensions. Berlin: Springer-Verlag, 1980, pp 36–47.
60. UB Sleytr, P Messner. Annu Rev Microbiol 37:311–339, 1983.
61. UB Sleytr, P Messner. J Bacteriol 170:2891–2897, 1988.
62. A Breitwieser, UB Sleytr, K Gruber. J Bacteriol 174:8008–8015, 1992.
63. WF Harris, LE Scriven. Nature 228:827–828, 1970.
64. FRN Nabarro, WF Harris. Nature 232:423–425, 1971.
65. WF Harris. Sci Am 237:130–145, 1977.
66. P Messner, D Pum, M Sára, KO Stetter, UB Sleytr. J Bacteriol 166:1046–1054, 1986.
67. DLD Caspar, A Klug. Cold Spring Harbor Symp Quant Biol 27:1–24, 1962.
68. D Pum, P Messner, UB Sleytr. J Bacteriol 173:6865–6873, 1991.
69. UB Sleytr, P Messner, D Pum. Methods Microbiol 20:29–60, 1988.

70. R Jaenicke, R Welsch, M Sára, UB Sleytr. Biol Chem Hoppe–Seyler 366:663–670, 1985.
71. M Sára, UB Sleytr. J Bacteriol 169:2804–2809, 1987.
72. M Sára, I Kalsner, UB Sleytr. Arch Microbiol 149:527–233, 1988.
73. M Sára, D Pum, UB Sleytr. J Bacteriol 174:3487–3493, 1992.
74. S Weigert, M Sára. J Membrane Sci 106:147–159, 1995.
75. S Weigert, M Sára. J Membrane Sci 121:185–196, 1996.
76. P Messner, D Pum, UB Sleytr. J Ultrastruct Mol Struct Res 97:73–88, 1986.
77. UB Sleytr, D Pum, M Sára. Adv Biophys 34:71–79, 1997.
78. AT Hastie, CC Brinton Jr. J Bacteriol 138:999–1009, 1979.
79. S Howorka, M Sára, W Lubitz, B Kuen. FEMS Microbiol Lett 172:187–196, 1999.
80. M Sára, C Dekitsch, HF Mayer, E Egelseer, UB Sleytr. J Bacteriol 180:4146–4153, 1998.
81. W Ries, C Hotzy, I Schocher, UB Sleytr. J Bacteriol 179:3892–3898, 1997.
82. D Pum, M Weinhandl, C Hödl, UB Sleytr. J Bacteriol 175:2762–2766, 1993.
83. D Pum, UB Sleytr. Thin Solid Films 244:882–886, 1994.
84. B Wetzer, A Pfandler, E Györvary, D Pum, M Lösche, UB Sleytr. Langmuir 14: 6899–6906, 1998.
85. M Weygand, B Wetzer, D Pum, UB Sleytr, K Kjaer, PB Howes, M Lösche. Biophys J 76:458–468, 1999.
86. D Pum, UB Sleytr. In: UB Sleytr, P Messner, D Pum, M Sára, eds. Crystalline Bacterial Cell Surface Proteins. Austin, TX: Landes/Academic, 1996, pp 175–209.
87. A Diederich, C Hödl, D Pum, UB Sleytr, M Lösche. Colloids and Surfaces B: Biointerfaces 6:335–346, 1996.
88. D Pum, UB Sleytr. Colloids and Surfaces A: Physicochem Eng Aspects 102:99–104, 1995.
89. D Pum, UB Sleytr. Supramolec Sci 2:193–197, 1995.
90. D Pum, G Stangl, C Sponer, W Fallmann, UB Sleytr. Colloids and Surfaces B: Biointerfaces 8:157–162, 1996.
91. E Sackmann. Science 271:43–48, 1996.
92. B Schuster, D Pum, UB Sleytr. Biochim Biophys Acta 1369:51–60, 1998.
93. B Schuster, D Pum, O Braha, H Bayley, UB Sleytr. Biochim Biophys Acta 1370: 280–288, 1998.
94. E Györvary, B Wetzer, UB Sleytr, A Sinner, A Offenhäuser, W Knoll. Langmuir 15:1337–1347, 1999.
95. DD Lasic. Trends Biotechnol 16:307–321, 1998.
96. S Küpcü, M Sára, UB Sleytr. Biochim Biophys Acta 1235: 263–269, 1995.
97. S Küpcü, K Lohner, C Mader, UB Sleytr. Mol Membrane Biol 15:151–175, 1998.
98. T Hianik, S Küpcü, UB Sleytr, P Rybar, R Krivanek, U Kaatze. Colloid and Surfaces A 1999 (in press).
99. C Mader, S Küpcü, M Sára, UB Sleytr. Biochim Biophys Acta 1418:106–116, 1999.
100. C Mader, S Küpcü, UB Sleytr, M Sára. Biochim Biophys Acta (in press) 1999.
101. P Fromherz, A Offenhäuser, T Vetter, J Weis. Science 252:1290–1293, 1991.
102. P Fromherz, H Schaden. Eur J Neurosci 6:1500–1504, 1994.

103. D Pum, G Stangl, C Sponer, K Riedling, P Hudek, W Fallmann, UB Sleytr. Microelectron Eng 35:297–300, 1997.
104. A Neubauer, W Kautek, S Dieluweit, D Pum, M Sahre, C Traher, UB Sleytr. PTB Berichte F-30:188–190, 1997.
105. S Küpcü, C Mader, M Sára. Biotechnol Appl Biochem 21:275–286, 1995.
106. M Sára, UB Sleytr. Appl Microbiol Biotechnol 30, 184–189, 1989.
107. A Breitwieser, S Küpcü, S Howorka, S Weigert, C Langer, K Hoffmann-1 Sommergruber, O Scheiner, UB Sleytr, M Sára. Bio Techniques 21:918–925, 1996.
108. S Küpcü, M Sára, UB Sleytr. J Immunol Methods 196:73–84, 1996.
109. C Weiner, M Sára, UB Sleytr. Biotechnol Bioeng 43:321–330, 1994.
110. C Weiner, M Sára, G Dasgupta, UB Sleytr. Biotechnol Bioeng 44:55–65, 1994.
111. A Breitwieser, C Mader, I Schocher, UB Sleytr, M Sára. Allergy 53:786–793, 1998.
112. M Truppe, S Howorka, G Schroll, S Lechleitner, B Kuen, S Resch, W Lubitz. FEMS Microbiol Rev 20:88–91, 1997.
113. K Nagayama. Nanobiology 1:25–37, 1992.
114. W Shenton, D Pum, UB Sleytr, S Mann. Nature, 389:585–587, 1997.
115. S Dieluweit, D Pum, UB Sleytr. Supramolec Sci 5:15–19, 1998.

6

Assemblies in Complex Block Copolymer Systems

Volker Abetz
Universität Bayreuth, Bayreuth, Germany

I. INTRODUCTION

Block copolymers are macromolecules composed of two or more polymer blocks of chemically different monomers that are linked together by chemical bonds. The resulting chain topologies can be linear, branched, or cyclic. Table 1 shows a few examples for block copolymers containing two or three different types of monomers. Systematic studies of these materials became possible through developments in polymerization techniques, which made possible the synthesis of well-defined block copolymers with a very small polydispersity.

During the last decades block copolymers have attracted increasing interest both from scientific and commercial points of view [1–4]. This is due to their unique morphological behavior, namely, the formation of crystallike order on a length scale in the range between a few nanometers up to several hundred nanometers. Figure 1 shows some of the typical morphologies found in amorphous diblock copolymers, where the different blocks self-assemble into different microphases (microdomains). The dispersed microdomains (spheres, cylinders) form the ''core'' and the surrounding matrix forms a ''shell'' or ''corona.'' Note that the crystallike order occurs on a supramolecular length scale in these systems. On a molecular level the blocks may be disordered (liquidlike, amorphous), but they can also be liquid-crystalline or crystalline.

In memory of Prof. Dr. Reimund Stadler.

Table 1 Examples of Binary and Ternary Block Copolymers with Different Block Distributions[a]

		Independent system variables
Binary block copolymers		
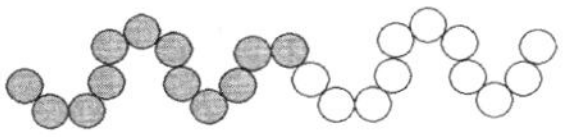	AB diblock copolymer	ϕ_A, χ_{AB}
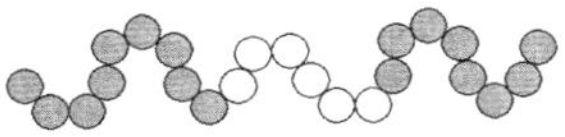	ABA triblock copolymer	ϕ_{A1}, ϕ_{A2}, χ_{AB}
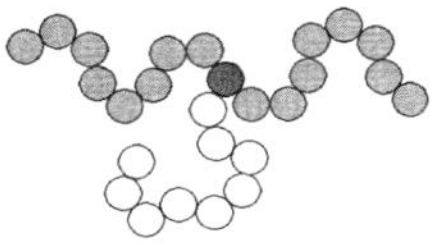	A_2B miktoarm star copolymer	ϕ_{A1}, ϕ_{A2}, χ_{AB}
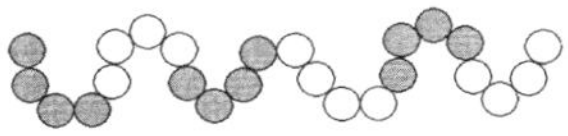	AB multiblock copolymer	ϕ_A, χ_{AB}
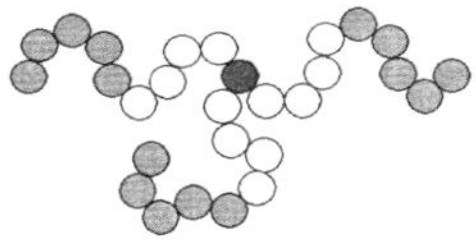	AB starblock copolymer	ϕ_A, χ_{AB}
Ternary block copolymers		
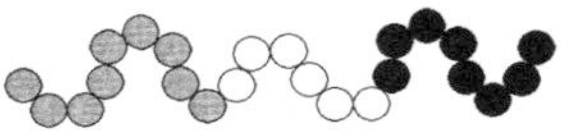	ABC triblock copolymer	ϕ_A, ϕ_B, χ_{AB}, χ_{BC}, χ_{AC}
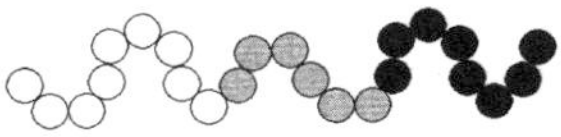	BAC triblock copolymer	ϕ_A, ϕ_B, χ_{AB}, χ_{BC}, χ_{AC}
	ACB triblock copolymer	ϕ_A, ϕ_B, χ_{AB}, χ_{BC}, χ_{AC}
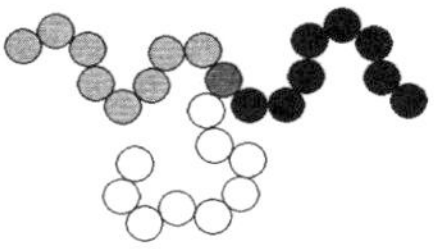	ABC miktoarm star copolymer	ϕ_A, ϕ_B, χ_{AB}, χ_{BC}, χ_{AC}

[a] ϕ: volume fraction, χ: segmental interaction parameter.

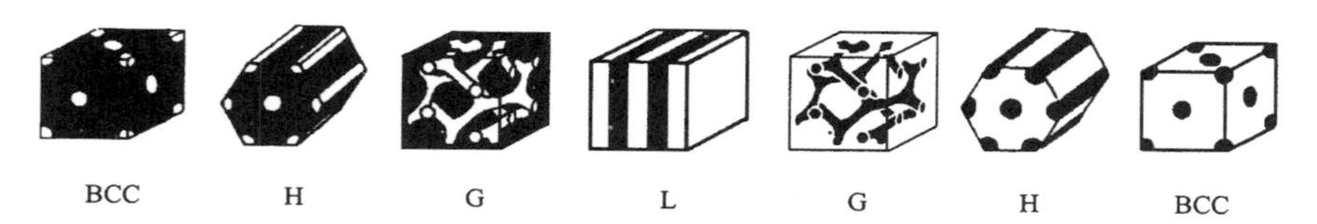

Figure 1 Schemes for different diblock copolymer morphologies. From left to right the volume fraction of one component increases. The morphologies are body-centered cubic spheres (BCC), hexagonally packed cylinders (H), gyroid (G), and lamellae (L).

In addition to the investigations on block copolymers in the bulk state, a lot of work on block copolymers in solution has been published. Diblock and triblock copolymers are well known for their association into micelles when dissolved in a selective solvent. These works deal mainly with the phase behavior of amphiphilic block copolymers (micellization) [5–11] and the rheological behavior [12–14]. Rather closely related to the phase and rheological behavior of block copolymers are the corresponding properties of low molecular weight surfactants in solution [15–17]. Amphiphilic molecules play a fundamental role in biology and also find widespread technological applications because of their unique ability to self-organize at interfaces, which leads to modification of interfacial properties and enhances the compatibility between two phases. Amphiphiles are also the subject of other chapters in this volume.

Figure 2 shows various morphologies observed in micellar solutions of amphiphilic diblock copolymers in a water–oil system. The micelles self-assemble into similar morphologies as they are observed in the bulk state of amorphous diblock copolymers. These morphologies arise because the different blocks are chemically linked with one another and thus cannot undergo a macroscopic phase separation. In order to minimize the free energy such a system has to find a compromise between the repulsive enthalpic interactions between chemically different segments, and the reduction of conformational entropy that arises by the suppression of direct contacts between chemically different segments. The enthalpic interactions between different segments are often described by the Flory–Huggins–Staverman interaction parameter χ [18–20] (a positive value of χ indicates repulsion, while a negative value indicates attraction between the segments; see also Chapter 1). The conformational entropy of the block copolymer chain has a maximum when $\chi = 0$, i.e., when contacts between chemically different segments and chemical similar segments have the same enthalpy. Since usually chemically different segments show repulsive interactions, the conformational entropy will be reduced, because the different blocks will self-assemble into different domains, the so-called microphases. The junction points between the blocks are located on a common interface or within an interfacial region of finite

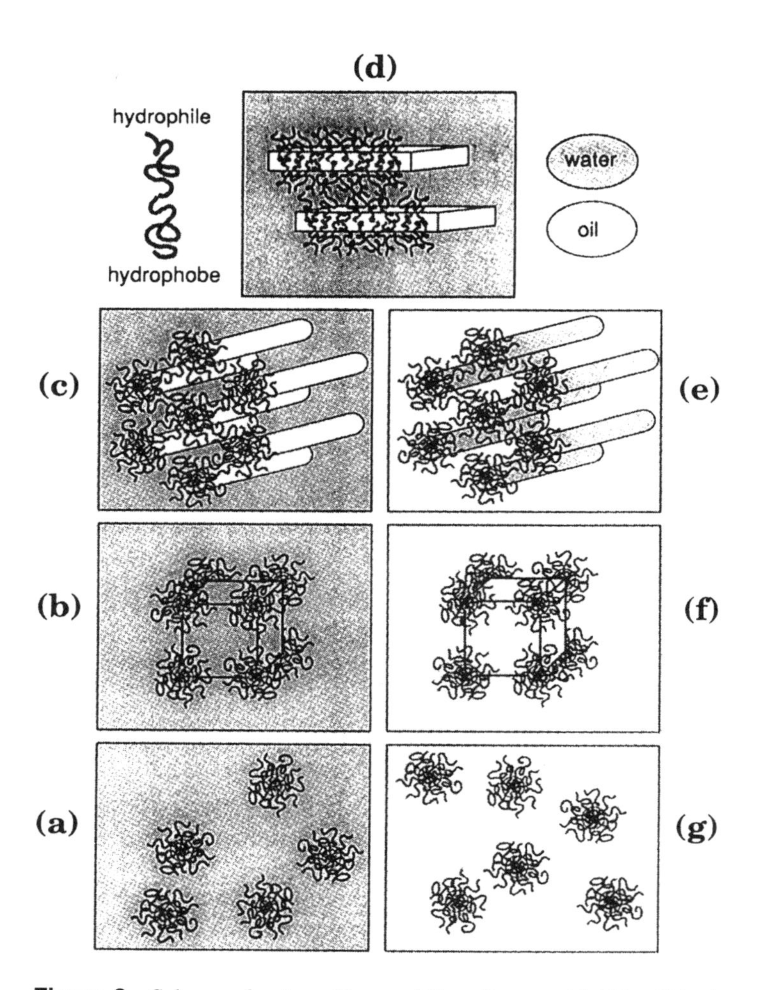

Figure 2 Schemes for the self-assemblies of an amphiphilic diblock copolymer in the presence of the solvents water and oil selective for the two blocks: (a) micellar solution, (b) micellar cubic lyotropic liquid crystal (LLC), (c) hexagonal LLC, (d) lamellar LLC, (e) reverse hexagonal LLC, (f) reverse cubic LLC, (g) reverse micellar solution. (From Ref. 8, Copyright 1997 American Chemical Society.)

thickness. The larger the thickness of the interface, the less reduced is the conformational entropy. Systems with sharp interfaces (where the thickness of the interface is much smaller than the domain size or long period of the morphology) are called strongly segregated systems. The concentration profile perpendicular to the microphase domain boundary can be described by a hyperbolic tangent (or less precisely by a step function). The two different microdomains contain only one type of segments; i.e., the microdomains are composed of only one component. If the interfacial width is comparable to the length scale of the long period and the concentration profile is only a weak fluctuation from the mean value, the system is called a weakly segregated system and the concentration profile perpendicular to the interface can be described by a sine function. The two extrema are thus called the strong segregation limit (SSL) and the weak segregation limit (WSL), respectively [1]. Schemes of the concentration profiles perpendicular to an interface between two blocks of a diblock copolymer in the bulk state are shown in Fig. 3. Different microphase separated morphologies occur depending on the relative compositions of the different components and the total molecular weight as expressed by the degree of polymerization. In addition, the aggregation state of the blocks (amorphous, liquid-crystalline, or crystalline) largely influences the morphology, too.

Thus microphase separated block copolymers can be considered as a class of supramolecular polymers that form large regular structures via a self-assembling process without being chemically linked to one another. In this chapter most exciting morphologies of such supramolecular assemblies are shown.

Their thermodynamic properties make block copolymers interesting for applications such as thermoplastic elastomers [2], surfactants [21], compatibilizers

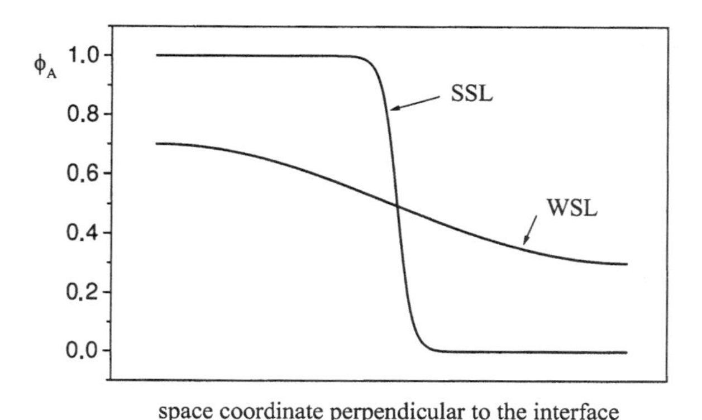

Figure 3 Scheme of the concentration profile perpendicular to a microdomain boundary; SSL: strong segregation limit; WSL: weak segregation limit.

in multiphase polymer blends [22], and others such as topologically controlled hosts for catalysts (e.g., transition metal complexes) [23], colloidal metals [24–27], nonlinear optical moieties [28,29], or light-emitting devices [30], etc. Liu et al. synthesized nanofibers by crosslinking the cylindrical domains of a diblock copolymer in the bulk and subsequent dissolution of the system leading to a kind of ''hairy rods'' [31]. Ikkala et al. synthesized diblock copolymers showing microphase separation on two different length scales. By using hydrogen-bonding complexes they could attach low molecular weight amphiphiles to a polar block of a microphase separated diblock copolymer, leading to a second microphase separation between the hydrogen complexes and the hydrophobic parts of the amphiphiles within the domains of the polar block [32,33]. Block copolymers have also attracted interest recently as templates to create mesoscale structures in inorganic materials, which never would self-organize in such a way by themselves. Templin et al. have synthesized ceramic materials with spherical, cylindri-

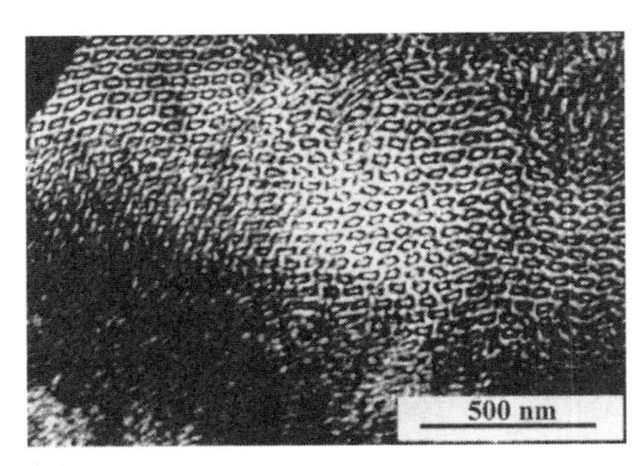

(a)

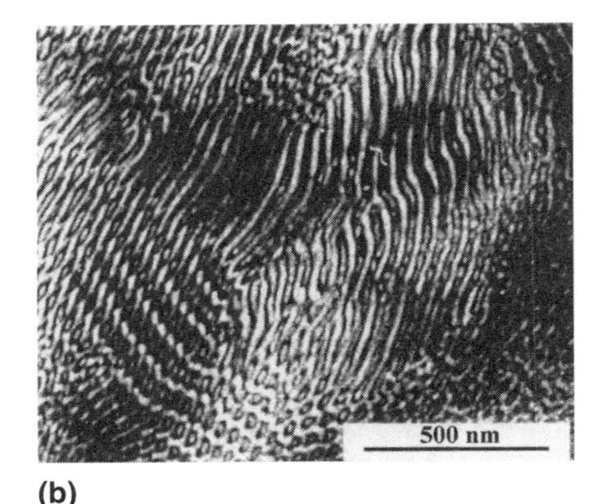

(b)

cal, and disclike structures [34]. Krämer et al. used block copolymers as templates to create nanoporous silica [35].

There are different possibilities for classifying block copolymers in the bulk state, which depend on the choice of a particular property such as the number of components, chain topology, or the aggregation state of the blocks. Since the field is too large to be covered completely in this chapter, we concentrate on

(c)

(d)

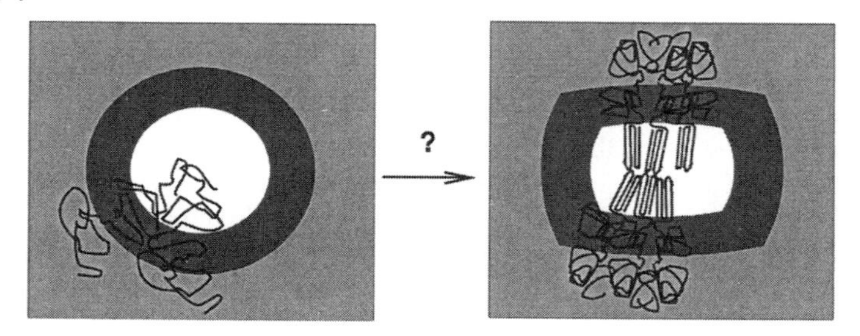

Figure 4 Transmission electron micrographs of polystyrene-block-polybutadiene-block-polycaprolactone (S-B-Cl), stained with OsO_4. B appears dark, Cl appears light, and S appears grey. (a) Top view onto the cylindrical domains; (b) side view onto the cylindrical domains; (c) scheme of the morphology; (d) scheme of the distortion of round cylinders into cylinders with edges by the crystallization of the Cl-core. (From Ref. 43, Copyright 1999 American Chemical Society.)

amorphous (coil/coil) block copolymers in the bulk state. In the following we just refer to some works in other areas of block copolymers, which are not further discussed in this chapter.

For linear semicrystalline block copolymers the reader is referred to experimental works carried out on binary [36–39] and ternary [40–43] block copolymers where the influence of morphology on crystallization (and vice versa) has been investigated. The crystallization process can disturb an already formed microphase separated structure, inhibit the formation of self-organized microphases on larger length scales, or induce morphological transitions. Nojima et al. investigated a diblock copolymer of polybutadiene and polycaprolactone, which microphase separated above the crystallization temperature of polycaprolactone [39]. Lowering the temperature leads to a change from spherical polycaprolactone domains into lamellar ones upon crystallization of the latter. Crosslinking of the polybutadiene matrix above the crystallization temperature suppresses a morphological transition from spheres to lamellae and only a deformation of the spheres upon crystallization of polycaprolactone could be observed. Balsamo et al. studied a linear triblock copolymer of polystyrene, polybutadiene, and polycaprolactone (S-B-Cl) with a polystyrene matrix and a short polycaprolactone block (16 wt%) [43]. Here the crystalline polycaprolactone block self-organizes into cylindrical domains, surrounded by a polybutadiene shell. Due to the crystallinity of the polycaprolactone block, the cylinders display a noncircular cross-section. Figure 4 shows this morphology. The mechanical properties of this material are very interesting: most likely due to the crystalline nature of the core-block, the whole material behaves ductile and can be extended to large strains (~900%) in a tensile-testing experiment at room temperature, although the polystyrene matrix is glassy. Crystallization in star copolymers was recently investigated by Floudas and coworkers [44]. They studied systems with two crystallizable blocks and could show that depending on the thermal treatment the crystallization of one block can be largely suppressed in favor of the other one.

Theoretical treatments of the problem of crystallization in microdomains have been presented by DiMarzio et al. [45] and Whitmore and Noolandi [46].

The first work on block copolymers with a block capable of forming liquid-crystalline phases was published by Gronski's group [47]; later work of the same group dealt with the behavior of liquid-crystalline block copolymers under shear [48]. Möller's group investigated block copolymers containing mesomorphous polysiloxane blocks in the bulk [49] and the groups of Ober and Thomas investigated the influence of the block composition on the phase transition between liquid-crystalline and isotropic states in a rigid/coil block copolymer [50]. A theoretical description of the phase diagram of diblock copolymers consisting of a liquid-crystalline and an amorphous block was given by Williams and Halperin [51]. The same authors also investigated the mechanical properties of ABA triblock copolymers with glassy A and a B block capable of forming liquid-crystalline phases [52,53]. More rod–coil block copolymers are described in detail in Chapter 7.

Very interesting, but also not further treated in this chapter, are the properties of block copolymers prepared as thin films. Here the contacting media (i.e., substrate, air) play a major role in the self-assembling process [54–56]. The self-assembly induced by surfaces has been reviewed by Krausch [57].

This chapter is organized in the following way. In the next section a short overview of synthetic work in the field of block copolymers is given, followed by two sections about linear and star copolymers. In the last section blends of block copolymers are discussed.

II. SYNTHESIS OF BLOCK COPOLYMERS

There are different ways to synthesize block copolymers. It depends on the kind of monomers which chemical route can be chosen. The chemical route also determines the polydispersity of the product. Principally, a block copolymer can be formed by subsequently adding the various types of monomers to the chain, as happens in some chain reactions. Or it can be formed by linking different polymers with one another, as happens in step-growth reactions. This can be achieved by using polymers carrying functional groups at their chain ends or somewhere along the chain. In the first case linear block copolymers will be obtained and in the latter case grafted or starlike block copolymers will occur. Often this type of block copolymer is the result of a polycondensation reaction. Polycondensation is important for commercial multiblock copolymers, like polyurethanes, polyureas, and polyesters with hard and soft segments, but leads to polydisperse materials. Some recent work in the field of polycondensation was related to linear and hyperbranched block copolymers [58–61]. Blocking is usually performed by reacting a telechelic prepolymer carrying hydroxyl or amino end-groups with other telechelic prepolymers carrying isocyanate, acid, or ester end-groups. The overall molecular weight thereby depends on the stoichiometry between the two different functional groups as well as upon the extent of reaction. Removing the side product (water or alcohol in the cases of polyesters and polyamides) moves the chemical equilibrium towards the formation of the block copolymers.

In chain reactions the different types of monomers can be added subsequently to an active chain end. The most important techniques here are sequential living polymerization techniques, such as anionic or cationic polymerization. Also radical polymerization can lead to the formation of block copolymers. In deviation from the living polymerization techniques, here all comonomers need to be present simultaneously in the reaction medium, since in this type of polymerization chain initiation, growth, and termination of growing chains happen simultaneously. To obtain block copolymers via radical polymerization, the copolymerization parameters (reactivity ratios) have to be much larger than unity (i.e., an active chain end favors the addition of a similar monomer as compared to another type of monomer). Certain metallocenes can be used in coordination polymerization of olefins leading to stereo block copolymers, like polypropylene where crystalline and amorphous blocks alter with each other due to the change of tacticity along the chain [62]. In comparison to living polymerization techniques, radical and coordination polymerization lead to rather polydisperse materials in terms of the number of the blocks and their degree of polymerization.

Living cationic polymerization has been used for the synthesis of block copolymers, which are often based on polyisobutylene [63–65], polysiloxanes [66], or polytetrahydrofurane [67,68]. Cationic polymerization also was used to synthesize block copolymers containing organic and phosphazene blocks [69]. In addition, much progress has been achieved in the field of controlled radical polymerization [70,71]. However, living anionic polymerization is still the most important way to synthesize well-defined block copolymers and most basic investigations on the properties of linear and star copolymers are based on systems prepared by living anionic polymerization [72–87]. In recent years combinations of various polymerization techniques have been used in order to synthesize block copolymers of different monomers that could not be polymerized by the same technique. Besides combining cationic and radical polymerization [88], the combination of anionic and cationic polymerization [89,90] and the combination of anionic polymerization with enzymatic polymerization also have been used [91].

Since in all the techniques based on living polymerization the different monomers are added sequentially, deactivation of some living chains cannot be suppressed completely when the next monomer is added (due to impurities). These deactivated polymers (homopolymers in the case of diblock copolymers or homo and diblock copolymers in the case of triblock copolymers) can sometimes be separated from the desired block copolymer by fractionation or preparative size exclusion chromatography.

III. LINEAR BLOCK COPOLYMERS

A. Binary Block Copolymers

In the field of amorphous, linear, binary block copolymers a lot of work has been done and extensively reviewed [1,2,92–97]. We address here recent developments in the understanding of the phase behavior of linear binary block copolymers including some theoretical work that was not adequately reviewed before.

Among the binary linear block copolymers the diblock copolymers have been studied in great detail. They can be considered as model systems for more complicated block copolymers, such as block copolymers with more than two components, or block copolymers with other block distributions.

The connectivity of the different incompatible blocks restricts the assembly mode only to certain morphologies, the main variable being the relative composition. Well known are the results on the morphology of polystyrene-block-polyisoprene (S-I) diblock copolymers by means of transmission electron microscopy (TEM) as a function of composition [1]. Three different morphologies were found: spheres on a cubic lattice, hexagonally arranged cylinders, and lamellae (see Fig. 1). In 1972 Aggarwal [98] found a new morphology which in 1986 was identified as a "cocontinuous morphology" [99], where the minority component forms two interpenetrating networks. This morphology is observed within a relatively small composition range between the lamellar and cylindrical regions of the phase diagram. The cocontinuous networks, which were observable both in linear and star block copolymers, were believed to have the symmetry of a diamond and thus the morphology was named "ordered bicontinuous double diamond" (OBDD) structure [99]. However, a few years later investigations by small angle X-ray scattering (SAXS) on star block copolymers showed that rather than two interpenetrating tetrapod-diamond lattices two interpenetrating tripod lattices with a mirror-symmetry characterize this structure, which was then named gyroid structure [100] (see Fig. 1). A lot of work has been carried out on this type of cocontinuous morphology [101–104]. It could be shown theoretically that there exists a whole class of gyroid morphologies rather than only one type [105,106]. However, these morphologies differ only in detail from one another and so far no experimental results on different gyroid morphologies have been reported for binary block copolymers.

Other morphologies such as perforated lamellae [107–109], which have been discovered more recently, are considered to be metastable [110].

1. Temperature Dependent Phase Behavior

Different techniques have been used for the investigation of the phase behavior of diblock copolymers. Small angle X-ray scattering (SAXS) and small angle neutron scattering (SANS) have been widely used for the investigation of the order–disorder transition temperature (ODT) [111–115], and the characterization of the ordered morphology. In addition, SANS was used for the investigation of the singlechain behavior in the disordered [116] and ordered melt [116–118]. In the latter it was found that the block chains are stretched along the normal of a lamella, while they are not stretched in the disordered state. Neutron reflection has been used for studies of the interfacial widths of block copolymers with lamellar morphology [119]. NMR spin-diffusion experiments were used also for the investigation of the interfacial width and are not restricted to lamellar morphologies [120]. Dynamic mechanical analysis has been proven to be very sensitive for the ODT [121–124]. Balsara et al. used time-resolved depolarized light scattering to follow the ordering kinetics of a cylindrical diblock copolymer after a temperature quench from the disordered state and found a nucleation and growth mechanism, which they could explain by a Ginzburg–Landau model [125].

After the first systematic morphological studies on amorphous block copolymers, different approaches to interpret the morphologies were carried out. Most of these are based on the assumption that the block copolymers are incompressible. Giving up that assumption would lead to a large complication of the theoretical treatments that were developed in two opposite limits. The weak segregation limit (WSL) was explored after the understanding by Leibler [126], de Gennes [127], and Erukhimovich [128] that microphase separation of an incompressible AB block copolymer is caused by an instability of its homogeneous state with respect to certain spatial composition fluctuations of a finite period L and wavenumber $q = 2\pi/L$. It is this value of q that corresponds to the maximum of the scattering intensity I(q) even for the disordered block copolymer.

The problems solvable within the WSL approach are the prediction of the scattering behavior and stability of the disordered state (spinodals) as well as the phase diagrams near the order–disorder transition temperature. In particular, the scattering behavior and spinodals of diblock copolymers with due regard for their polymer structure were first described using the Random Phase Approximation (RPA) by Leibler [126]. Erukhimovich first calculated the spinodals for the asymmetric triblock and trigraft ABA and polyblock and polygraft $(AB)_nA$ copolymers [129]. Consistently with the latter work [129], Benoît and Hadziioannou [130] showed that in the limit $k \rightarrow \infty$ polyblock and polygraft copolymers $(AB)_k$ have a larger disordered phase as compared to linear diblock copolymers AB. They also described the effects of polydispersity on the lengths of the blocks. Moreover, unlike the cases where only flexible blocks were considered, Benoît and Hadziioannou also found spinodals for some block copolymers with flexible A and semiflexible B blocks.

The phase diagram for diblock copolymers was first derived in the Mean Field Approximation by Leibler [126]. He found that for these materials the same succession of phase transitions occurs with increase of temperature (i.e., lamellae, cylinders, spheres, and disordered state) as calculated phenomenologically (without taking into account the material structure) by Alexander and McTaque [131]. However, while according to the Leibler theory a continuous transition between the disordered and the lamellar state could be expected only for a symmetrically composed diblock copolymer at a critical value of $\chi N = 10.495$, experiments showed this transition to be discrete and also to occur between asymmetrically composed diblock copolymers within a certain composition range. χ is the segmental interaction parameter introduced before and N is the overall degree of polymerization. In the same seminal work [126] Leibler explained qualitatively this and other deviations between the basic theory and experimental data in terms of fluctuations (coupling between the concentration waves with different values of q) neglected in the Mean Field Approximation. The corresponding quantitative changes in the phase diagram for diblock copolymers were studied first by Fredrickson and Helfand [132] on the base of the procedure properly taking the fluctuation corrections into account developed earlier by Brazovskii [133]. Fredrickson and Helfand found that the fluctuation corrections include a molecular weight-dependent term (which vanishes in the limit of infinitely large chains) resulting in an increase of the critical value of χN upon decreasing the degree of polymerization in the regime of shorter chains. The phase diagrams of the ABA triblock copolymers were considered first in the WSL by Mayes and Olvera de la Cruz [134] (in the symmetric case) and for an arbitrary ratio of the side blocks A by Dobrynin and Erukhimovich [135,136]. They found a critical value for $\chi N = 18$ when coupling a symmetric diblock copolymer. The same authors also studied the fluctuation corrections to the phase diagrams of the aforementioned systems [137,138]. The relationship between the critical values of linear diblock and triblock copolymers was confirmed experimentally [121].

Within the so-called strong segregation limit, after basic works by Meier [139] and Helfand and coworkers [140–142], the phase behavior was also described by Semenov who gave an analytical solution for the elastic part of the free energy originated by the reduced conformational entropy of the different blocks [143]. Since strong segregation theories overestimate the degree of stretching of the blocks, their prediction of the scaling bahavior of the degree of polymerization N with the long period of the morphology L cannot be a correct one. Khokhlov and coworkers introduced the super-strong segregation limit (SSSL), in which the scaling predictions of the SSL-theories ($L \propto N^{2/3}$ as compared to $L \propto N^{1/2}$ in the (WSL) become correct due to a very large degree of incompatibility ($\chi N \gg 100$) [144].

Matsen and Bates studied the phase behavior in the intermediate segregation regime [102] and also gave a self-consistent description of the whole phase behavior from the disordered state through the weakly segregated to the strongly segregated state [110] (Fig. 5). When one of the two blocks is liquid crystalline, at larger values of χN spherical and cylindrical morphologies are predicted to become more stable with the liquid-crystalline block forming the core domains [51].

Although there are uncertainties in the calculations in a certain range of incompatibility as expressed by χN, it follows from these calculations that the gyroid morphology is only stable in a finite range of χN. Further increase of incompatibility leads to a lamellar morphology. The influence of dissimilarities

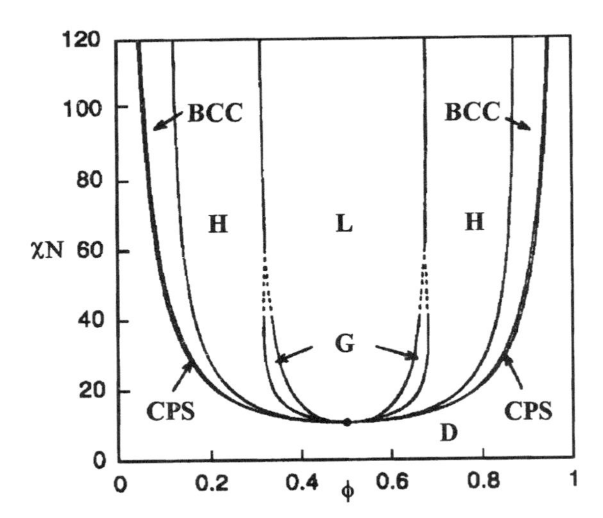

Figure 5 Phase diagram of diblock copolymers with equal segmental lengths and segmental volumes of both block components. χ: Flory–Huggins–Staverman interaction parameter, N: degree of polymerization, ϕ: volume fraction, D: disordered phase, CPS: close packed spheres, BCC: body-centered cubic spheres, H: hexagonally packed cylinders, G: gyroid, L: lamellae. (From Ref. 110, Copyright 1996 American Chemical Society.)

between the segmental lengths of the different blocks also has been studied and it was found that it only affects the symmetry of the phase diagram, with no qualitative change of it [145].

2. Behavior of Block Copolymers Under External Fields

In the following we give a short presentation of the influence of pressure and shear on block copolymers. This is a short overview with only few important results being presented. For further reading we refer to the mentioned literature.

(a). Pressure Dependent Phase Behavior. In the preceding section the incompressibility condition was used for the theoretical description of the temperature-dependent phase behavior. In this section, however, some experimental results are presented, showing that block copolymers are to some extent compressible; i.e., their phase behavior can be manipulated by pressure. So far there have not been many studies on this problem. Some authors reported on pressure-induced miscibility of diblock copolymers, both for systems with an upper critical [146–

148] and lower critical disordering temperature [149]. For some systems with an UCDT it was shown that a minimum of the ODT occurs upon increasing pressures [150,151]. Stühn explained this effect by the creation of free volume in the mixing process, forming the disordered state at lower pressures. At higher pressures the larger compressibility of the ordered state reduces the mixing tendency and thus leads to the microphase separated state.

(b). Behavior Under Large Shear. In this part some recent and important results for structuring of block copolymers subjected to external shear fields are reported. One basic goal in polymer science is to relate macroscopic properties to the molecular structure or to the morphology. In the case of microphase separated block copolymers it is necessary to use macroscopically aligned systems, i.e., single crystals without defects. Morphological defects might otherwise govern the macroscopic properties and prevent insight into the inherent anisotropic properties of the morphology. Thomas's group developed a method called "roll casting," which enables preparation of films with a size of several square centimeters by a slow solvent evaporation while the solution is subjected to a shear field between two rotating cylinders [152]. In a study comparing different methods for macroscopic alignment of block copolymers, roll casting proved to be most effective [153]. However, this method must be optimized for each block copolymer system by using the right solvent, rotation speeds of the cylinders, gap width between the cylinders, etc.

Block copolymers subjected to large shear fields in the bulk state have been of interest since the first studies by Keller and coworkers [154]. They oriented a cylindrical S-B-S triblock copolymer of polystyrene (S) and polybutadiene (B) by extrusion. The cylindrical domains of the polystyrene blocks were found to orient parallel to the direction of shear flow. Then several years passed before the influence of shear on the alignment and phase behavior of block copolymers became the subject of further scientific investigations. The experimental and theoretical works on diblock copolymers subjected to shear have been reviewed recently [96,97].

Using large amplitude oscillatory shear (LAOS), Koppi and coworkers investigated the orientation behavior of lamellar poly(ethylene-alt-propylene)-block-poly(ethyl ethylene) EP-EE diblock copolymers and found at lower frequencies a parallel alignment of the lamellae to the shear field, while at larger frequencies the lamellae and their normals orient perpendicular to the shear field [155]. Winey et al., Riise et al., and Chen et al. made similar experiments on lamellar S-I diblock copolymers and found just the opposite behavior [156–158]. Wiesner investigated the influence of LAOS as a function of shear amplitude and frequency on the orientation behavior of lamellar S-I diblock copolymers of various molecular weights and confirmed that both the above-described situations may occur [159]. At low strain amplitudes parallel alignment seems to be favored all over the frequency regime. However, at larger strains an intermediate frequency regime exists for the perpendicular orientation. Kornfield's group investigated the time-resolved strain birefringence of S-I diblock copolymers in an oscillatory shear field at various strain amplitudes and frequencies. They studied the morphology after various times of the orientation process by SAXS and TEM [160] and found at a low frequency (1 rad/s) a perpendicular alignment of the lamellae and a monotonic increase of the birefringence (which is mainly form birefringence). At a larger frequency (4 rad/s) the form birefringence went through a maximum and at a very large frequency (100 rad/s) it went through a minimum. Also under these conditions transient perpendicular (4 rad/s) and transversal (100 rad/s) orientations were found on the pathway to a parallel alignment of the lamellae to the shear flow direction. Figure 6 shows the different orientations with respect to the shear field.

It was pointed out in their work that morphological defects play an important role during the orientation process [160]. While S-I diblock copolymers showed different orientations under LAOS for various frequencies, a lamellar SIS-triblock copolymer showed only perpendicular orientation [157]. The perpendicular orientation was found to be only a transient state for EP-EE-EP triblock copolymers [161]. Moreover, the ODT itself is influenced by shear. In some triblock copolymers, like cylindrical S-I-S [162] or lamellar EP-EE-EP [161], shear stabilizes the disordered state; in other systems like cylindrical S-B-S it stabilizes the ordered phase [163]. Jackson and coworkers also found indi-

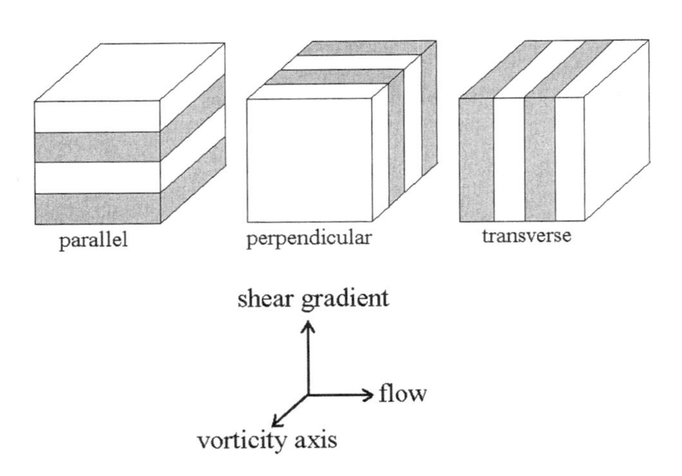

Figure 6 Scheme of the different orientations of a lamellar diblock copolymer in a shear field.

cations of a martensitic-like transition at very high shear rates for a cylindrical S-B-S triblock copolymer [163,164].

B. Ternary Block Copolymers

In comparison to binary block copolymers relatively little work on ternary block copolymers has so far been published. There are more independent variables in ternary block copolymers as compared to binary block copolymers. While in the latter only one independent composition variable and one interaction parameter exist, in ternary systems there are two independent composition variables and three interaction parameters. This leads to a richer phase diagram. In addition, the block sequence also can be changed, which introduces another tool to influence the morphology [165]. As mentioned before in the case of diblock copolymers, systematic studies of triblock copolymers became possible with the development of sequential polymerization techniques with living anionic polymerization being still the most important one.

Riess et al. gave a first description of the phase behavior of linear ternary triblocks [87]. While in microphase separated diblock copolymers only one structural feature can exist such as lamella, cylinder, or sphere, ternary block copolymers can simultaneously exhibit different features in the microphase separated state (e.g., spheres within a lamella, etc.).

It was established at an early stage of the research on ternary block copolymers that the morphology of the sample can be influenced by the solvent, when films are cast from solution [166–168]. While for binary block copolymers a nonselective solvent could be used for film casting, in the case of ternary (or higher) block copolymers a selectivity of the solvent will always be involved. This results in a different degree of swelling of the different blocks and thus their effective volume fractions are modified at the point where the morphology forms during evaporation of the solvent. Thus great care must be taken when discussing thermodynamic stability of the observed morphologies. Possibilities for assessing the thermodynamic stability include annealing of samples or using solvents with different selectivities and comparing the obtained morphologies. While the first strategy might work for block copolymers with relatively low molecular weights and low degree of entanglements, it usually fails for samples with large molecular weights.

Mogi et al. [169] and Gido et al. [170] studied triblock copolymers based on polystyrene (S), polyisoprene (I), and polyvinylpyridine (VP) with different block sequences. The difference in block sequence resulted in a different morphology for a similar overall composition of the systems. While polyisoprene-block-polystyrene-block-polyvinylpyridine I-S-VP with similar amounts of all three components forms lamellar stacks (Fig. 7a) [169], polystyrene-block-poly-

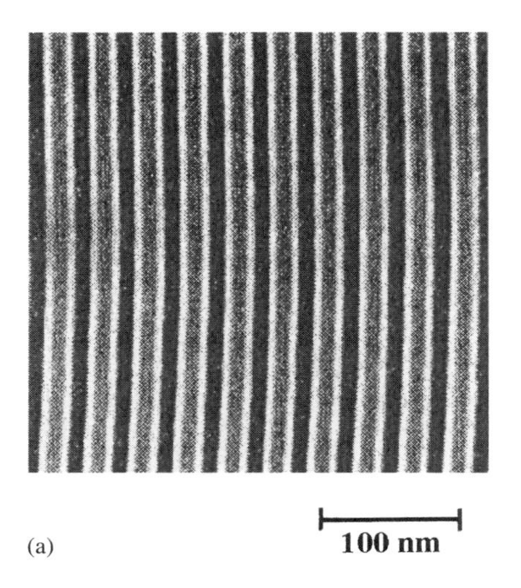

(a)

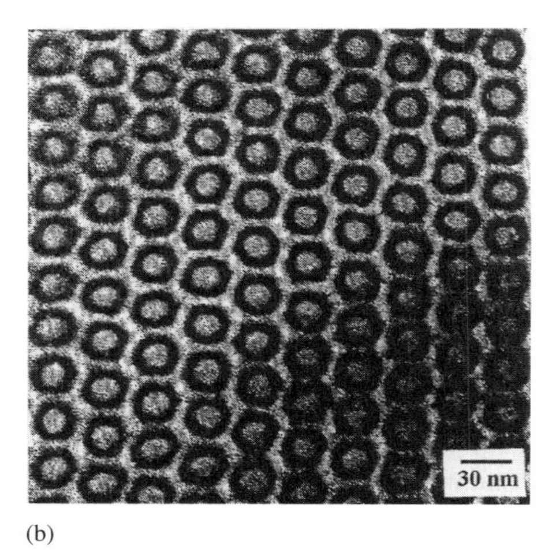

(b)

Figure 7 Transmission electron micrographs of (a) polyisoprene-block-polystyrene-block-poly(2-vinylpyridine) I-S-VP. (From Ref. 169, Copyright 1994 American Chemical Society.); (b) polystyrene-block-polyisoprene-block-poly(2-vinylpyridine) S-I-VP. (From Ref. 170, Copyright 1993 American Chemical Society.) Polyisoprene appears black due to staining with OsO_4.

isoprene-block-polyvinylpyridine S-I-VP forms hexagonally packed core shell cylinders (Fig. 7b) [170].

This behavior can be understood as a consequence of the competition between the different interfacial tensions of adjacent blocks: while the interfacial tensions between S and I on one side and S and VP on the other side are of approximately similar magnitude, the interfacial tension between I and VP is much larger than that between I and S. As a consequence the system favors a smaller interface between I and VP as compared to I and S, which leads to a morphology with different interfacial areas on both ends of the middle block in the case of S-I-VP. In comparison, the system I-S-VP will form lamellae due to the fact that the interfacial areas on both ends of the middle block are of approximately the same size.

Kane and Spontak developed a self-consistent field theory for lamellar ABC triblock copolymers [171] based on Semenov's approach for diblock copolymers [143]. They also described the scaling behavior of the periodicity L with the degree of polymerization N, which was found to be similar to diblock copolymers ($L \propto N^{2/3}$). The periodicity of an ABC triblock copolymer was found to be slightly larger than the periodicity of an AC-diblock copolymer with the same overall degree of polymerization. The theoretical results confirm systematic SAXS and SANS studies by Mogi et al. on lamellar I-S-VP block copolymers [172].

Zheng and Wang [173] published a theoretical description of different morphologies in ABC triblock copolymers based on a strong segregation approach following earlier work by Ohta and Kawasaki [174] and Nakazawa and Ohta [175]. Lyatskaya and Birshtein [176] described different cylindrical and lamellar morphologies within the SSL, including the possibility of different segmental volumes and persistence lengths of the different blocks. They could show that these segmental properties greatly influence the borderline of stability between different morphologies, a result which also had been obtained by Matsen and Schick for diblock copolymers [145]. Phan and Fredrickson [177] studied symmetric ABC triblock copolymers and confirmed former results of Nakazawa and Ohta [175], according to which the square lattice arrangement of A and C cylinders in a B-matrix should be more stable than a hexagonal arrangement. Also the CsCl-type of packing for A and C spheres in a B matrix was confirmed to be more stable than other types of spherical morphologies. They found that even for ABC-triblock copolymers the gyroid should be more stable than the ordered tricontinuous double diamond (OTDD) morphology; however; both cocontinuous morphologies are unstable with respect to cylindrical or lamellar morphologies in the SSL.

Matsen showed the gyroid morphology to be stable for symmetric ABC triblock copolymers with a B matrix in the intermediate segregation regime [178]. Its stability range is extended towards the cylindrical region, because the tetragonal packing of A and C cylinders leads to a stronger chain frustration (and thus larger energy). Based on his theory, Matsen simulated different projections of TEM images of the gyroid and found strong evidence for the morphology found before by Mogi et al. [169,179] (and identified as an OTDD) to be gyroid. Dormidontova and Khokhlov [180] investigated spherical micelles of the asymmetric ABC triblock copolymers in the SSL and SSSL. According to their prediction the inner structure of the micelle core (formed by A- and B-blocks) can contain either a single A-aggregate at the center of a micelle or several A-aggregates depending on the interaction parameters and the A-block length. The shape of the A-aggregates was found to be spherical in the SSL and disclike in the SSSL. It was shown that the presence of strongly associating A-blocks intensifies the segregation tendency between B and C blocks [180].

Stadler et al. studied triblock copolymers based on polystyrene S polybutadiene B, and poly(methyl methacrylate) M. This system was investigated systematically and a number of new morphologies were discovered [165,181–186]. For symmetric systems with the block sequence S-B-M, i.e., where the end blocks have similar size, lamellar morphologies were found. Varying the volume fraction of the middle block from 0.03 up to approximately 0.3 it forms spheres, cylinders, or a lamella between the lamellae of the outer blocks [181] (Fig. 8).

Upon further increase of the volume fraction of the middle block it forms the matrix embedding cylindrical [186] or spherical domains of the S and M end blocks. Depending on the molecular weight, the end blocks can be either located in different domains or form mixed microdomains. This should have consequences on the symmetry of the morphology and, more important, on the mechanical properties. While in the case of mixed microdomains the middle block can form either loop (both end blocks of a particular chain are located within the same microdomain), it can only form links (bridges) between two different microdomains when the end blocks stay incompatible. In ABA triblock copolymers the question, what fraction of the B blocks forms bridges, has been investigated theoretically by Zhulina and Halperin [187] and motivated experimental investigations by Watanabe [188]. Using dielectric spectroscopy he determined the frac-

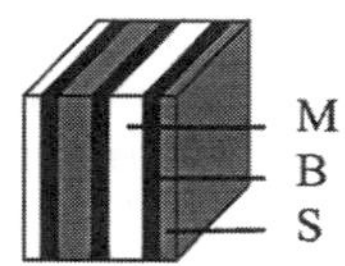

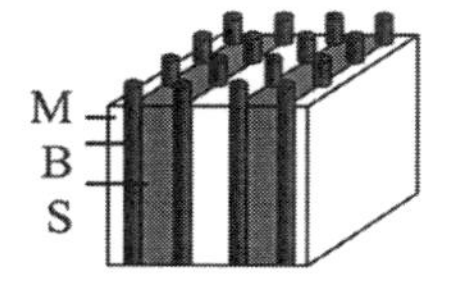

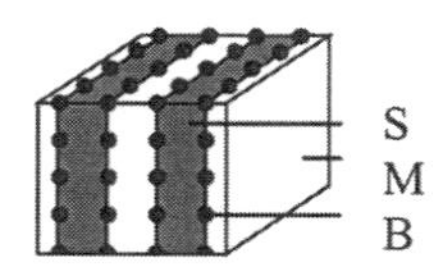

Figure 8 Schemes for different lamellar morphologies of ABC triblock copolymers. Upon decreasing the volume fraction of the middle block it changes from lamellae via cylinders to spheres.

tion of loop chains for S-I-S triblock copolymers to be on the order of 60%. Later on the theory was improved by Matsen [189]. A comparison between S-B-S and S-B-M triblock copolymers of same molecular weight with lamellar or cylindrical morphology shows in fact a higher modulus of the S-B-M systems [190]. As mentioned before, the casting solvent might influence the morphology of ABC triblock copolymers [168]. In the case of an S-B-M triblock copolymer with a B matrix a new hexagonal morphology was found, which had not been predicted before by theory [186]. While a square-like ordering of the S and M cylinders may be expected for an S-B-M triblock copolymer with similar amounts of S and M, in this case a hexagonally packed array of S cylinders was found, where each S cylinder was surrounded by six M cylinders (Fig. 9). This special morphology was obtained from a solution with mixed solvents (30% benzene and 70% cyclohexane).

When changing the relative composition in S-B-M triblock copolymers in such a way that the volume fraction of one end block is above approximately 0.6, lamellar morphologies are no longer obtained. In such cases one end block forms the matrix and the other end block forms a core sphere or core cylinder. Depending on the relative volume fraction of the middle block with respect to the core-forming end block, different morphologies are found [184], which are schematically shown in Fig. 10. The most spectacular morphology in Fig. 10 is the helical morphology [182], which is the only noncentrosymmetric morphology in ABC triblock copolymers observed so far. However, due to helix reversals, the sample contains about the same number of right- and left-handed helices and thus is not chiral on a large scale. An electron micrograph of that morphology is shown in Fig. 11. The problem of noncentrosymmetry is addressed again in the section about block copolymer blends.

Besides changes of the volume fractions discussed so far, chemical modification of one or two blocks is another way to influence morphology. For example, by hydrogenation of the B block in symmetric S-B-M block copolymers, where B forms spheres [181,191] or cylinders at the lamellar interface of S and M, the corresponding polystyrene-block-poly(ethylene-co-butylene)-block-poly(methyl methacrylate) S-EB-M forms a hexagonal morphology, where S cylinders are surrounded by EB rings in an M matrix [181,192] (Fig. 12). This morphological transition is induced by a change of the interfacial tensions between the middle block and the end blocks. While the interfacial tension between S and B is close to the one between B and M, the situation changes strongly for S and EB and EB and M. This leads to a displacement of the spheres or cylinders at the lamellar interface in the S-B-M block copolymers and induces curvature into the interface between the outer blocks. This scenario is schematically shown for an ABC triblock copolymer in Fig. 13.

A similar observation was made for an S-B-M triblock copolymer with

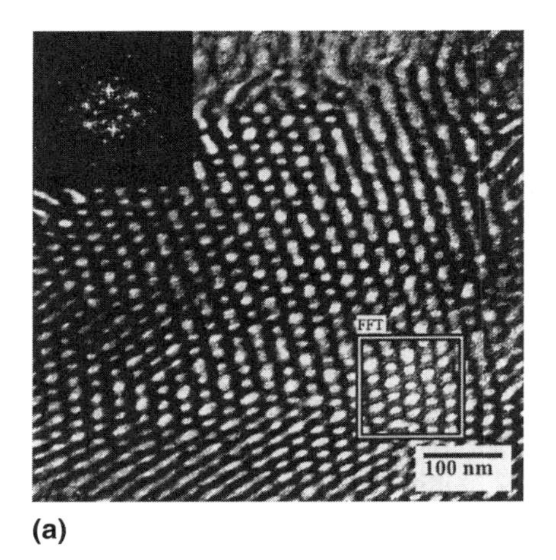

(a)

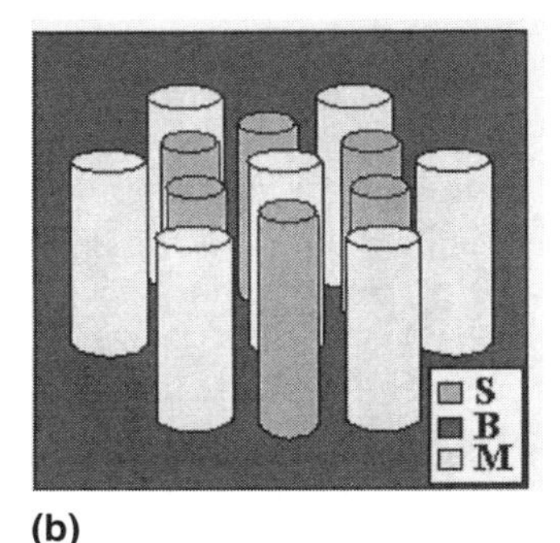

(b)

Figure 9 Polystyrene-block-polybutadiene-block-poly(methyl methacrylate) S-B-M with a B matrix embedding hexagonally packed cylindrical domains of S and M. (a) Transmission electron micrograph stained with OsO_4; (b) Scheme of the morphology. (From Ref. 186, Copyright 1998 American Chemical Society.)

spheres at the lamellar interface, where the B block was partly modified with different transition metal complexes [23]. In that case a morphological transition to a cocontinuous morphology or a hexagonal morphology was observed. Comparison of these findings with the morphological behavior of diblock copolymers leads to the conclusion that the stability region of different morphological classes (such as the cylindrical morphology) may be extended to block copolymers with almost symmetric compositions by adjusting the relative interactions of a short middle block with respect to the end blocks.

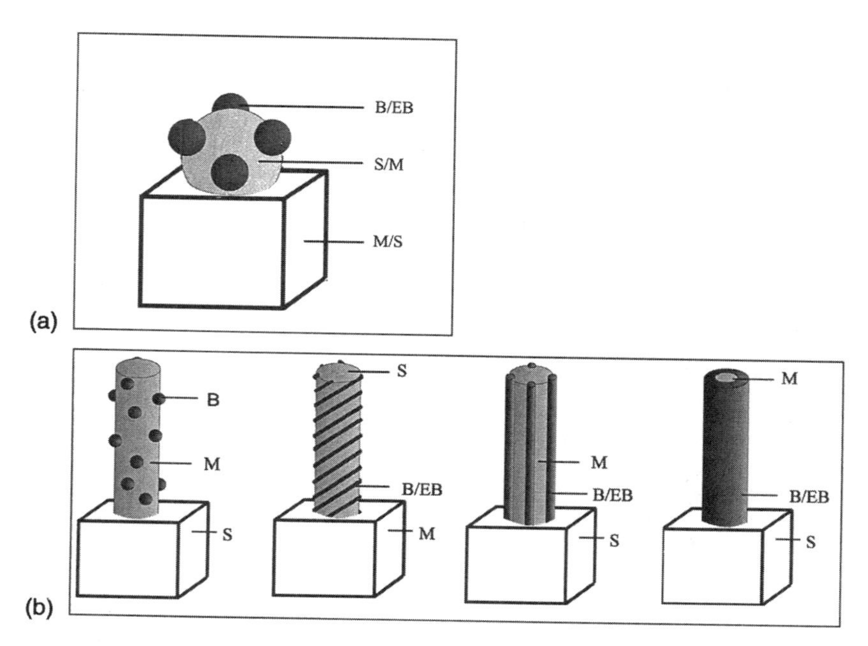

Figure 10 Schemes for (a) the spheres-on-sphere morphology. (From Ref. 185, Copyright 1998 Springer-Verlag GmbH & Co. KG.) (b) From left to right: spheres-on-cylinder, helices-around-cylinder, cylinders-at-cylinder, core-shell-cylinder morphology. (From Ref. 184, Copyright 1997 Wiley-VCH Verlag GmbH, Weinheim.)

A very interesting example of morphological change induced by hydrogenation of the B block in an S-B-M triblock copolymer was obtained for symmetric systems with approximately 27% B. While the S-B-M triblock copolymer shows a lamellar morphology, the analogous S-EB-M self-organizes into the so-called ''knitting pattern'' morphology [193,194] (Fig. 14). This is a nice example of a block copolymer morphology with a highly nonconstant mean curvature of the interfaces between the domains. This morphology is supposed to be located between the lamellar morphology where all blocks are localized in lamellae, and the morphology where the middle block forms a cylinder at the lamellar interface between the outer blocks. A support for this assumption is the dependence of morphology upon the casting solvent: while the knitting pattern morphology is obtained when casting the polymer film from chloroform, a lamellar morphology is obtained from toluene solution. In other samples with a larger volume fraction of the B block where lamellae were obtained, the choice of the solvent had no

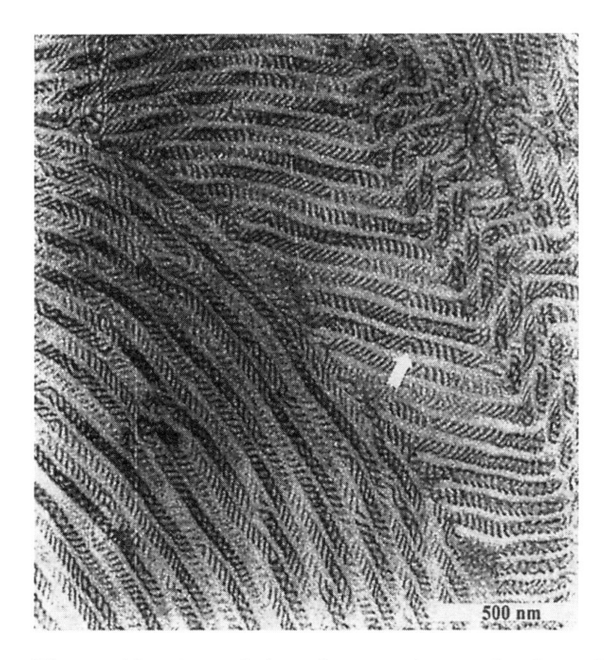

Figure 11 Transmission electron micrograph of a polystyrene-block-polybutadiene-block-poly(methyl methacrylate) S-B-M showing the helical morphology. Cylindrical domains of S are surrounded by black-stained helices of B. M forms the matrix. The arrow points to a helix reversal. (From Ref. 182, Copyright 1995 American Chemical Society.)

influence on the morphology and thus a larger separation towards morphological transition within the phase diagram can be assumed.

Changing the sequence of the three blocks from S-B-M to B-S-M leads to changes in the morphology, too. This is due to a situation similar to that seen for I-S-VP and S-I-VP mentioned before: while in S-B-M with approximately equal amounts of all blocks a lamellar morphology is obtained, the change of the sequence of the blocks into B-S-M leads to a significant dissimilarity between the interfacial tensions between subsequent blocks. Thus in this case a curved morphology is obtained. While in the case of I-S-VP a core-shell cylinder has been reported, in the case of B-S-M we have cocontinuous morphologies [195]. In the same composition range a cocontinuous morphology was observed for polystyrene-block-poly(2-vinylpyridine)-block-poly(tert. butyl methacrylate) [196], where the relative interfacial tensions should be comparable to the ones in B-S-M. When the volume fractions of both B and M in a B-S-M triblock copolymer are approximately 0.2, S forms the matrix. Both B and M tend to

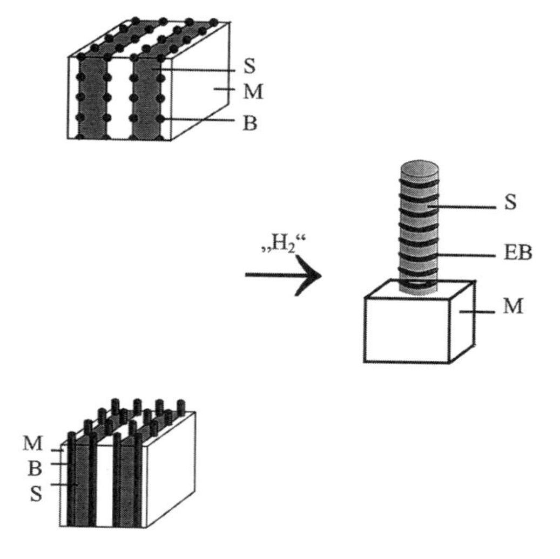

Figure 12 Scheme for the change of the morphological behavior of a polystyrene-block-polybutadiene-block-poly(methyl methacrylate) S-B-M induced by hydrogenation of B poly(ethylene-co-butylene) EB: from a lamellar morphology with B spheres or B cylinders between S and M lamellae in an S-B-M triblock copolymer to a hexagonal morphology, where EB rings surround S cylinders in an M matrix after hydrogenation.

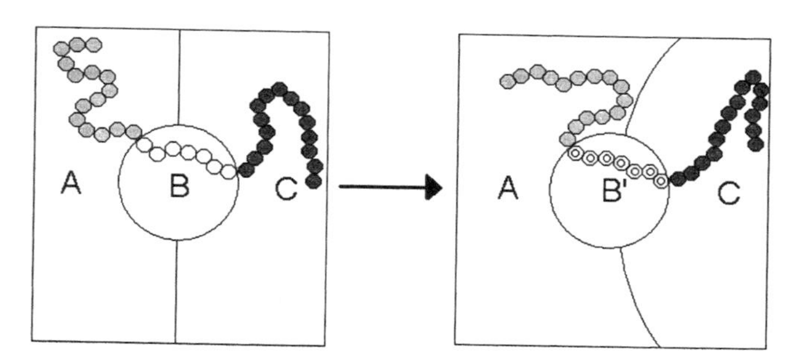

Figure 13 Scheme for the change of curvature of the intermaterial dividing surfaces by changing the relative interactions between the middle and the outer blocks in an ABC triblock copolymer via chemical modification of B to B′ ($\chi_{AB} = \chi_{BC}$, $\chi_{AB'} < \chi_{B'C}$).

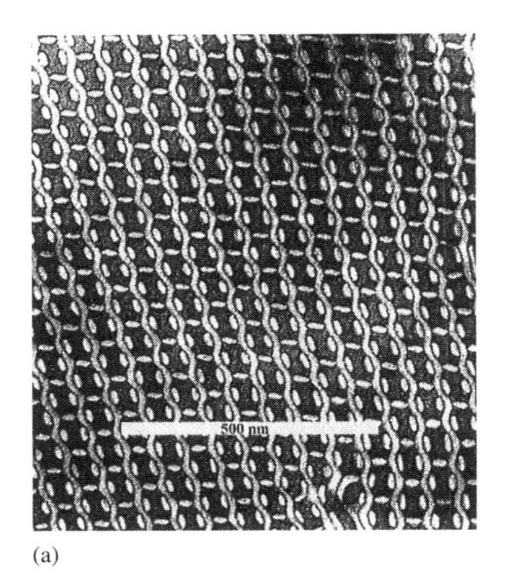

(a)

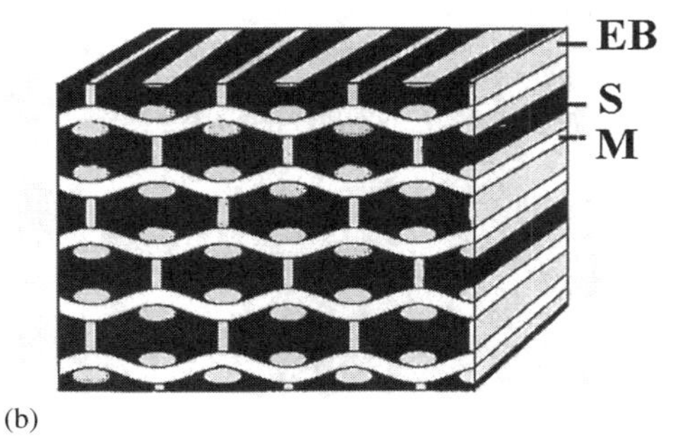

(b)

Figure 14 Polystyrene-block-poly(ethylene-co-butylene)-block-poly(methyl methacrylate) S-EB-M showing the ''knitting pattern'' morphology: (a) transmission electron micrograph stained with RuO_4 (S appear dark, M and EB appear light); (b) scheme. (From Ref. 194, Copyright 1996 Wiley-VCH Verlag GmbH, Weinheim.)

form cylinders. However, due to the dissimilarity of the interfacial tensions between the end blocks and the S matrix, the interfacial areas between B/S and S/M are not the same. This leads to the formation of cylinders with different diameters and different long periods for both lattices of B and M cylinders. Since such kind of simultaneous packing into different hexagonal or tetragonal lattices with different long periods is impossible, an irregular microphase separated ''banana''-shaped morphology is obtained [183] (Fig. 15).

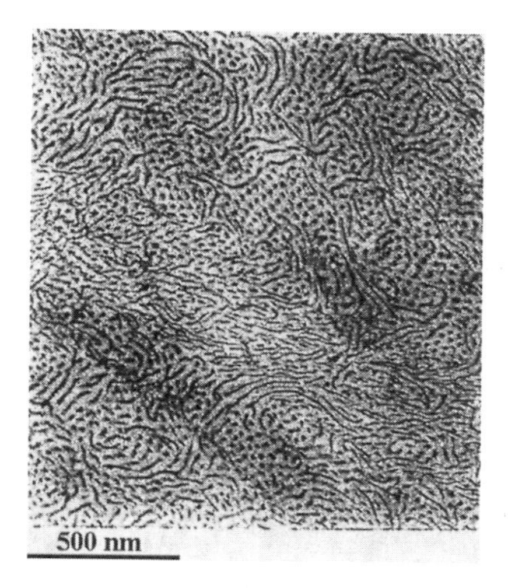

Figure 15 Transmission electron micrograph of polybutadiene-block-polystyrene-block-poly(methyl methacrylate) B-S-M forming the ''banana'' morphology. Due to staining with OsO_4 only the curved B cylinders are visible. (From Ref. 183, Copyright 1996 American Chemical Society.)

That situation can be regarded as a mesoscale glass, generated by the competing tendencies within this block copolymer to form two cylindrical morphologies with different periodicity. An interesting question relating AC diblock and ABC triblock copolymers is the influence of the B block on the microphase separation between A and C. Annighöfer and Gronski [197,198], as well as Hashimoto et al. [199], reported on the morphological properties of ABC triblock copolymers where B consisted of a random or tapered block of A and C. Kane and Spontak found in their theoretical work that a random A/C middle block can enhance the mixing of the outer blocks due to an increase of the conformational entropy of the middle block [171]. A similar result was obtained for symmetric ABC triblock copolymers, where B forms spheres, cylinders, or a lamella between the lamellae of the A and C blocks [200]. Erukhimovich and coworkers studied the influence of a very short strongly incompatible C block on the order–disorder transition of an ABC and ACB block copolymer within the weak segregation limit [201]. It was found that in both cases, for certain compositions and certain relative incompatibilities among C and the other two blocks, a stabilization of the disordered phase can occur as compared to the pure AB diblock copolymer.

While only the influence of a very short strongly interacting third block on the microphase separation of a diblock copolymer was theoretically investigated, Neumann and coworkers studied the influence of curvature between an incompatible C block on the phase behavior of the adjacent AB diblock copolymer. These studies were performed on four different poly(1,4-isoprene)-block-poly(1,2-butadiene)-block-polystyrene (I-B-S) [202] and their hydrogenated analogues, poly (ethylene-alt-propylene)-block-poly(ethyl ethylene)-block-polystyrene (EP-EE-S) [84,124]. In these block copolymers the volume fractions of the two elastomeric components I and B or EP and EE were similar and only the relative amount of polystyrene was changed with respect to the other two blocks. It is well known that blends of I and B are highly compatible [203] and thus the I-B diblock copolymer also will be most likely disordered. In the I-B-S triblock copolymers a phase behavior similar to that of a diblock copolymer was found, where polystyrene forms one microphase and the two elastomeric components together form the other microphase. In the EP-EE-S triblock copolymers the behavior was different. While the EP-EE diblock precursors were found to show an ODT around room temperature (as shown by dynamic mechanical properties), the corresponding EP-EE-S triblock copolymers showed a more complicated picture. While I-B-S with 16% of S showed a spherical morphology, the corresponding EP-EE-S showed a cylindrical morphology, which is orthorhombic, as evidenced by SAXS. I-B-S with 26% of S showed a hexagonally packed cylindrical morphology, and the corresponding EP-EE-S showed a cocontinuous morphology. These morphological transitions via hydrogenation indicate an increase of the incompatibility between the two elastomeric blocks and also a shift of the ODT of EP-EE towards higher temperatures when being grafted on an incompatible surface (polystyrene).

IV. STAR COPOLYMERS

Polymers with a starlike topology have attracted interest for many years. The rheological behavior in the melt and in solution of starpolymers differs from the behavior of linear polymers [204]. Polystyrene starpolymers with selectively deuterated core or corona chains were investigated by SANS and it was found that the chains are more stretched within the core (or close to the branching point), while the outer parts of the chains follow the singlechain behavior of linear polymers [205]. This result confirmed theoretical predictions by Daoud and Cotton [206] and Birshtein et al. [207]. A similar behavior was found for the chain conformations in starlike block copolymer ionomer micelles, which were also studied by SANS [208].

There are different types of starblock copolymers. In one kind of these polymers, several diblock copolymers are connected at one of their chain ends. In other systems different homopolymers are connected at a chain end. The different situations are schematically shown in Table 1.

One of the basic questions is how the topological restriction of the different blocks by a common junction point influences the morphology. While in microphase separated binary miktoarm star copolymers (simple graft copolymers, Table 1) the common junction points between different blocks are located on a common interface, in completely microphase separated ternary miktoarm star copolymers (miktoarm star terpolymers) an interfacial line might be expected rather than an interfacial surface.

As in the case of linear block copolymers, we concentrate here on the phase behavior and morphologies of amorphous miktoarm star copolymers. Other interesting problems such as the influence of the chain topology on the local chain dynamics [209] or on the crystallization of individual blocks [44] are beyond the scope of this chapter.

A. Binary Star Copolymers

The first binary star copolymers were synthesized by Fetters' group more than two decades ago [210]. Hadjichristidis' group has done a lot of work in the field of star copolymers with various architectures [77,79,211].

A few studies on miktoarm star copolymers A_nB_m have been carried out in order to compare their morphological behavior with that of linear diblock copolymers having similar composition. Whereas similar morphology was found for $n = m$ [212,213], differences were found for $n \neq m$. The finding was attributed to the bending energy arising from the reduction of chain stretching of the component with more arms [214]. Hadjichristidis and coworkers synthesized compositionally symmetric star copolymers of various types and found that a tricontinuous morphology could be induced by the chain architecture [215]. This is in agreement with the results on compositionally similar AB, AB_2, or AB_3 block copolymers, where the increasing asymmetry of the numbers of chemically different blocks also leads to an increasing curvature of the interface between the different domains [76], thus shifting the system in the direction from lamellar towards spherical or even disordered morphology.

Matsushita and coworkers found a similar morphological behavior for 4- and 12-armed miktoarm star copolymers of polystyrene and polyisoprene as compared to linear diblock copolymers of the same composition [216]. For a polystyrene-polyisobutylene-polyisoprene triblock copolymer S-IB-S and a 3-miktoarm star copolymer with IB core and S corona of the same composition, the same morphology was also found [217]. A cocontinuous morphology in a star copolymer, where each arm consists of a diblock copolymer, is attributed to be stable in a wider composition range as compared to a linear diblock copolymer [218].

Erukhimovich [129] and Olvera de la Cruz and Sanchez [219] developed a theory for the order–disorder transition of AB_2-miktoarm star copolymers, which predicts that branched block copolymers have a larger disordered phase as compared to linear block copolymers of the same overall degree of polymerization and composition. The same result was obtained for linear $(AB)_n$ multiblock copolymers and A_nB_n miktoarm star copolymers [219]. The phase diagrams of monodisperse miktoarm star copolymers were developed in the WSL by Dobrynin and Erukhimovich both in the Mean Field Approximation [135,136] and with due regard to fluctuation corrections [138]. It was also found that for star copolymers with diblock copolymer arms the general topological behavior of the phase diagram is similar to that for linear diblock copolymers and reveals the same succession of the transitions disordered state–body-centered cubic phase—hexagonally packed cylindrical phase—lamellar phase [138]. For the miktoarm star copolymers with different homopolymer arms this behavior prevails only if the arm numbers m and n are not too large ($m,n < 5$). In the opposite case the hexagonal and body-centered cubic morphologies were found to be less stable than the lamellar one for compositions close to symmetrical. This was explained by a rather specific behavior of the 4-point-correlations involving the monomers of many miktoarm star copolymers. Phase diagrams of star copolymers consisting of four diblock copolymers were also studied [220] for which a restricted composition range of a stable gyroid phase was predicted via a WSL theoretical approach taking into account the higher harmonics contribution. The fluctuation corrections were shown to affect mainly the body-centered cubic and hexagonal phases.

B. Ternary Star Copolymers

The first ternary star copolymers containing three different arms were presented by Fujimoto and coworkers [81]. They used a special technique to combine three different blocks (tert-butyl methacrylate, styrene, and dimethylsiloxane) which were all synthesized by living anionic polymerization. PDMS was synthesized by initiation with a functionalized 1,1,-diphenyl ethylene (DPE). In a second step living polystyrene anions were reacted with the DPE linked to the PDMS, before the tert-butyl methacrylate was added to give the 3-miktoarm star terpolymer. Several years later the morphological properties of these star copolymers were published as well. TEM and SAXS support a threefold symmetry where the junction points between the three incompatible blocks are confined on one line [221].

Iatrou and Hadjichristidis synthesized a 3-miktoarm star copolymer containing polystyrene, polyisoprene, and polybutadiene by terminating these living polyanions in a sequential mode by a trichlorosilane. In these systems the microphase separation seems mainly to occur between polystyrene and the two elastomers [78]. Hückstädt et al. investigated a 3-miktoarm star terpolymer of styrene, butadiene, and methyl methacrylate, where three glass transition temperatures were found indicating three different microphases [82]. Due to the asymmetric composition a hexagonal morphology was found, where the polybutadiene block forms the cylinders in a PMMA matrix (unpublished). Most likely the PS block forms a shell around the PB cylinders, due to a reduction of interfacial energy.

Hadjichristidis' group later worked on 3-miktoarm star terpolymers of polyisoprene, polystyrene, and poly(methyl methacrylate), where they also investigated an asymmetric system. They found a cylindrical morphology with a nonconstant mean curvature; i.e., the cylinders were deformed to rhombohedral structures [222,223]. In symmetric systems they found threefold morphologies [224], similar to the results of Hashimoto's group [221] (Fig. 16a,b).

It is interesting to note that some of their electron micrographs exhibit features similar to those of a linear polystyrene-block-poly((4-vinylbenzyl)dimethylamine)-block-polyisoprene triblock copolymer with almost equal amounts of the three components, when cast from benzene [168] (Fig. 16c,d). Also a quaternary star copolymer consisting of polystyrene, polyisoprene, polybutadiene, and poly(4-methylstyrene) was reported, but no morphological characterization was given [79].

V. BLENDS OF BLOCK COPOLYMERS

Most blends of different polymers are incompatible, i.e., do not form a homogeneous phase. This is due to the usually repulsive enthalpic interactions between the different species, which can easily balance the entropy of mixing, and which is small compared to low molecular weight materials. Thus macrophase separation occurs with a minimization of the interfacial area between the components if the system is in thermodynamic equilibrium (see Chapter 1). Reduction of the interfacial tension between the components may reduce the extent of macrophase separation and this can be achieved by the addition of compatibilizers such as block copolymers. Adding an AB diblock copolymer to a blend of A and B will not necessarily lead to an increase of the internal surface between A and B, because the diblock copolymer can also self-aggregate into micelles in the A and B phases. In fact, in these cases there are no enthalpic attracting forces between the block copolymer and the other blend components, thus there is hardly any driving force for the block copolymer to localize itself at the interface between A and B. A way out of this problem is to synthesize in situ during processing graft block copolymers at the interface between A and B.

Stadler et al. used S-M diblock copolymers to compatibilize blends of poly(styrene-stat-acrylonitrile) (SAN) and poly(2,6-dimethyl phenylene ether) (PPE). In this case there are enthalpic attractive forces between S and PPE, and between M and SAN (the corresponding χ-parameters are favorable to mixing). It could be shown by TEM-investigations that in fact the domain sizes in these blends are much smaller compared to the SAN/PPE-blend without S-M diblock copolymer [225]. To improve the mechanical properties of the blend, the S-M diblock copolymer was replaced by S-EB-M triblock copolymers, where the poly(ethyl-

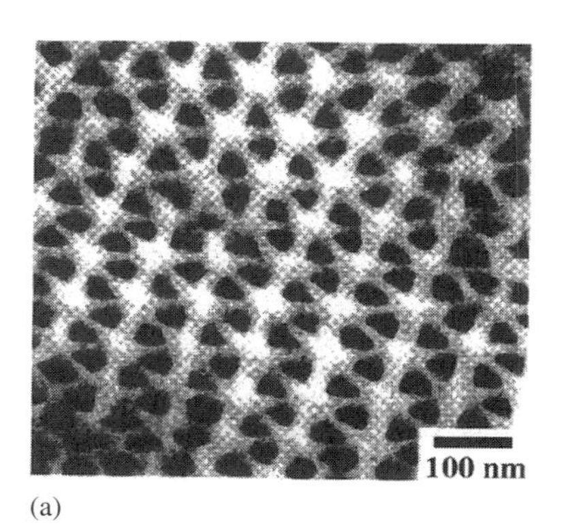

(a)

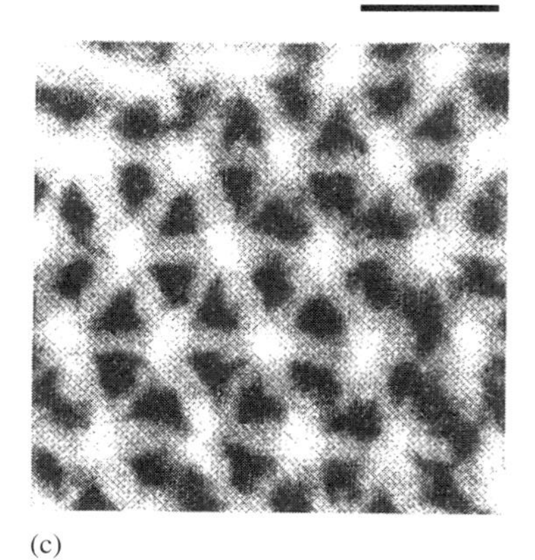

(c)

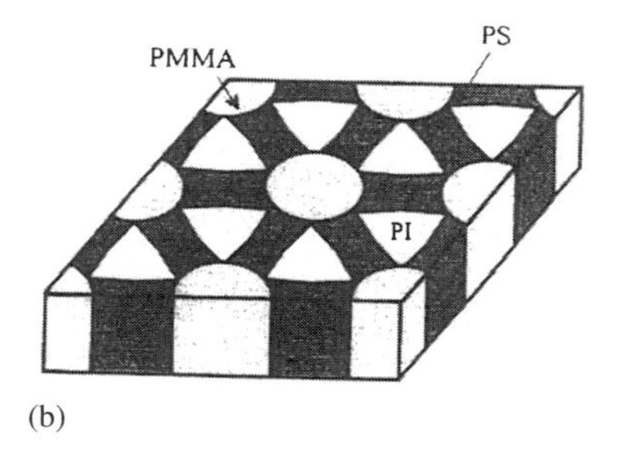

(b)

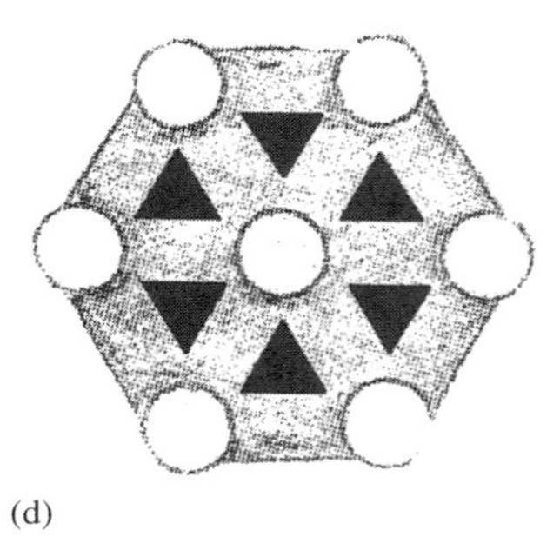

(d)

Figure 16 (a) Transmission electron micrograph of a polyisoprene-polystyrene-poly (methyl methacrylate) star copolymer stained with OsO_4; (b) scheme of the threefold symmetry. (From Ref. 224, Copyright 1998 American Chemical Society.) (c) Transmission electron micrograph of a polystyrene-block-poly((4-vinylbenzyl)dimethylamine)-block-polyisoprene; (d) scheme of the threefold symmetry. (From Ref. 168, Copyright 1983 American Chemical Society.)

ene-co-butylene) EB middle block avoids fracture at the domain boundaries between PPE and SAN [22].

Balsara et al. studied by SANS the thermodynamic behavior of a blend of AB-diblock with A and B homopolymers in the homogeneous disordered melt [226]. They found the Random Phase Approximation for multicomponent systems [227,228] to work for their systems of polyolefines, where only van der Waals interactions are present.

In this section we concentrate on the question of how to control the morphology of microphase separated block copolymers by adding a homopolymer or another block copolymer. Some of the work on blends of diblock copolymers with a homopolymer being chemically identical to one of the blocks was motivated by the investigation of the stability range of the ordered bicontinuous phase between the cylindrical and lamellar phases. Winey et al. [229] investigated S-I and S-B diblock copolymers that were blended with corresponding homopolymers. The stability window of the cocontinuous morphology (in their paper still erroneously named OBDD) was found to be comparable to the stability window of the diblock copolymers at a corresponding overall composition. Macrophase separation was shown to occur when the homopolymer was of larger size than the corresponding block in the copolymer. Macrophase separation was also observed when a large amount of the homopolymer corresponding to the shorter block was added. Koberstein et al. blended a short poly(ethylene-co-butylene) with different degrees of deuteration into an S-EB-S triblock copolymer and studied the location of the homopolymer within the midblock domain with SANS and SAXS. They found an enrichment of the homopolymer in the center of the midblock domain, where the triblock most likely has more loop chains, and a uniform distribution within the midblock domain, and where the midblock mostly forms tie-chains between neighboring PS-domains [230]. Hashimoto et al. investigated blends of a lamellar starblock copolymer $(S\text{-}B)_4$, with a linear polystyrene having a smaller molecular weight as compared to the S-blocks in the starblock copolymer. They found two new morphologies with mesh and strut topology [231] which have hyperbolic interfaces and are similar to perforated lamellae and cocontinuous networks, respectively. Hashimoto et al. also investigated blends of S-I with PPE [232]. Using an S-I diblock copolymer forming I-spheres, the addition of PPE (which homogeneously mixes with the S corona) does not change the morphology and no macrophase separation occurs from solution in toluene. Blends of a lamellar S-I diblock copolymer with PPE, however, besides microphase separation of the I also show a long wavelength fluctuation of the morphology, indicating a macroscopic demixing into regions with variable ratio of the two polymers. Xie and coworkers studied S-B-S triblock copolymers with S cylinders, which they blended with poly(vinyl methylether) (PVME). The PVME mixed only with the S domains and a change from cylindrical to a cocontinuous morphology was found [233].

Ways to control morphology of block copolymers have also been investigated via blending of different block copolymers. Hadziioannou and Skoulios were the first to investigate various blends of linear S-I, S-I-S, and I-S-I block copolymers. They found mixing at a molecular level and observed a dependence of the morphology on the overall chemical composition, rather than on chain architecture [234]. Hashimoto et al. investigated blends of two S-I diblock copolymers with different molecular weights, but mainly symmetric compositions [235]. When the molecular weights differed by more than a factor of 10, macrophase separation into partly mixed phases of long and short chains was observed. The domains with larger chains contained a larger amount of small chains, and vice versa. It was found that the presence of one asymmetric composed block copolymer influences the interface curvature, thus leading to morphologies different from the lamellar one for the mixed system. Mayes et al. studied thin films of blends containing symmetric long and short S-M diblock copolymers by neutron reflectometry. They found the short chains at the lamellar interface, while the segments of the larger chains also filled the center regions of the lamellar domains [119].

In macroscopically demixed lamellar phases, the phase boundaries between the lamellae with larger and smaller long periods were investigated and the possibility of a macrophase separation initiated by microphase separation was discussed [236]. Spontak et al. studied a blend of a symmetric S-I diblock copolymer and a $(S\text{-}I)_4$ multiblock copolymer with the same overall molecular weight by electron tomography and found macroscopic phase separation into two different lamellar phases [237]. Vilesov et al. [238] mixed two antisymmetric cylindrical S-B diblock copolymers with each other and obtained a lamellar phase. Schulz et al. did a similar experiment by blending a cylindrical and a lamellar S-VP diblock copolymer to obtain a gyroid morphology [239]. Spontak et al. used S-I diblock copolymers of various composition but similar molecular weight to tune the morphology by changing the relative amounts of the two block copolymers [240]. Sakurai et al. studied a similar system and in addition they found an order–order transition between the lamellar and the gyroid phase in a block copolymer blend upon increasing temperature [241,242].

The mixing behavior of two diblock copolymers with different relative composition has been treated theoretically. Sakurai and Nomura analyzed the phase behavior of diblock mixtures using the Random Phase Approximation. They found the disordered phase to be suppressed in mixed diblock blends and a tendency to macrophase separation for such blends when the composition of the involved blocks becomes very asymmetric. They conclude that there are limitations in the comparison between blends of diblock copolymers with a diblock copolymer having the same overall composition, since the χ-parameter determined for the blend may differ from the one determined for a single block copolymer [243]. Shi and Noolandi [244] developed phase diagrams for different ratios of two diblock copolymers and found regions where single mixed phases should be stable, besides regions where two different phases should coexist. A similar result was obtained by Matsen and Bates [245], who found a considerably large two-phase region. Birshtein et al. [238,246] and Borovinskii and Khokhlov [247] also studied blends of diblock copolymers in the strong segregation limit.

Blending of an ABC triblock copolymer of S-I-VP (showing the honeycomb-like core-shell morphology with a nonconstant mean curvature of the interface between I and S earlier discussed by Gido et al. [170]) with linear polystyrene leads to a morphology with a constant mean curvature, i.e., a core-shell cylindrical morphology [248]. If the degree of polymerization of the linear polystyrene is too large, however, macrophase separation occurs because the free polystyrene chains cannot swell the S-corona of the triblock copolymer.

In the last part of this section results on blends of ABC with BC or ABC with AC block copolymers are presented. We consider here only block copolymers that form lamellae by themselves and are based on the individual blocks of components all having approximately the same length. As mentioned before, the way of preparing the blend may have a great influence on the final morphology (choice of casting solvent, casting speed, etc.).

When blending ABC with BC block copolymers, different situations can occur, which are schematically shown in Fig. 17. Besides a macrophase separation between the two block copolymers, also blends with the sequence ABC CB BC CBA can occur (a centrosymmetric structure of double layers of both diblock and triblock copolymers). Another possibility is a kind of random sequence between BC and ABC block copolymers, which will occur when the C blocks of both diblock and triblock copolymer do not show any preferential mixing with either C block. In this case an aperiodic superstructure will be obtained. Another possible superstructure is the incorporation of the BC diblock with the same molecular orientation into the ABC structure, which will lead to a real effective increase of the volume fractions of both C and B with respect to A. Thus, a lamellar superstructure may only be expected for small volume fractions of diblock copolymer. For larger amounts of diblock chains a lamellar superstructure will be disfavored with respect to a superstructure with curved intermaterial dividing surfaces, such as cocontinuous, cylindrical, or spherical morphologies. Experiments on blends of lamellar S-B-M with lamellar B-M [249], and also of lamellar S-B-T with lamellar B-T, when all blocks were of about the same length proved the existence of the last case in Fig. 17.

Depending on the blend ratio, either a core-shell cylindrical morphology (Fig. 18a,c) or a core-shell cocontinuous morphology (Fig. 18b,d) was obtained for both blend systems. The driving force for the formation of this kind of blend is of entropic origin. Due to the same chemical nature of the mixing blocks there is no favored enthalpic interaction in this kind of blend. In recent theoretical work it could be shown that the swelling of the AB domains of an ABC triblock with an AB diblock copolymer can reduce the stretching of the blocks, in addition to the gain of mixing entropy for the system [250].

In blends of ABC with AB block copolymers a mixing at the molecular level of both blocks of the diblock copolymer with the corresponding blocks of the triblock copolymer leads to a centrosymmetric superstructure. However, in

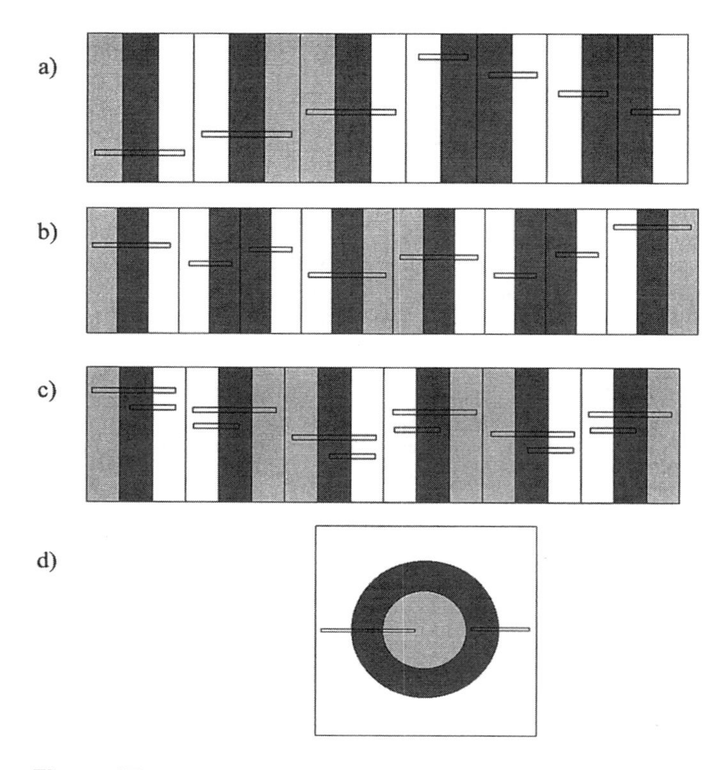

Figure 17 Schemes for blends of lamellar ABC and BC block copolymers: (a) macrophase separation; (b) centrosymmetric double layers of diblock and triblock copolymer; (c) centrosymmetric mixed layers of diblock and triblock copolymer; (d) induction of curvature into the intermaterial dividing surfaces due to the increased effective volume of B and C.

the case of a blend of ABC and AC block copolymers a noncentrosymmetric structure is obtained when all diblock and triblock copolymer molecules prefer to form common domains with the corresponding block of the other species. Before discussing this situation in some detail, other general possibilities for blend formations need to be considered. Besides the trivial case of macrophase separation and the random sequence of ABC and AC block copolymers due to equal energetic situation between the different A blocks and C blocks, two different centrosymmetric double-layered structures are also possible. Either the A blocks prefer the formation of common microdomains with A blocks of the other species, while the corresponding C blocks tend to phase separate from each other, or vice versa. The different possibilities are shown in Fig. 19.

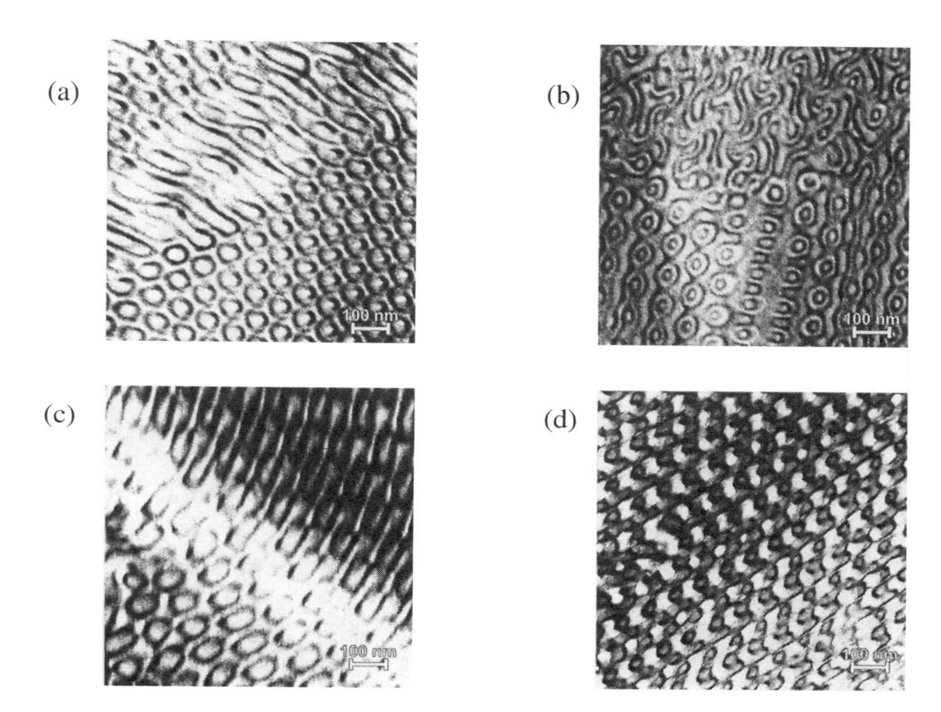

Figure 18 Transmission electron micrographs (stained with OsO_4). Blends of polystyrene-block-polybutadiene-block-poly(methyl methacrylate) S-B-M with polybutadiene-block-poly(methyl methacrylate) B-M: (a) 40 mol% S-B-M: 60 mol% B-M; (b) 70 mol% S-B-M: 30 mol% B-M. Blends of polystyrene-block-polybutadiene-block-poly(tert.butyl methacrylate) S-B-T with polybutadiene-block-poly(tert.butyl methacrylate) B-T: (c) 40 mol% S-B-T:60 mol% B-T; (d) 70 mol% S-B-T:30 mol% B-T.

The formation of mixed domains of different A blocks or C blocks is due to an entropic gain caused by the reduction of chain stretching for one species. This leads to a depression of the overall free energy as compared to the macrophase separated state (where only similar blocks from the same species form common microdomains). In fact, in ABC triblock copolymers there are AB and BC interfaces, while in AC diblock copolymers there is an AC interface. The interfacial tensions are different and thus the interfacial areas per chain are different as well, resulting in a different degree of chain stretching. It is known for the bulk state that the free chain ends interdiffuse with each other, causing a certain interpenetration between blocks from opposite layers (these blocks have one free chain end and are connected with the other end to an interface; they can be thus considered as polymer brushes). Note that this situation differs from co-

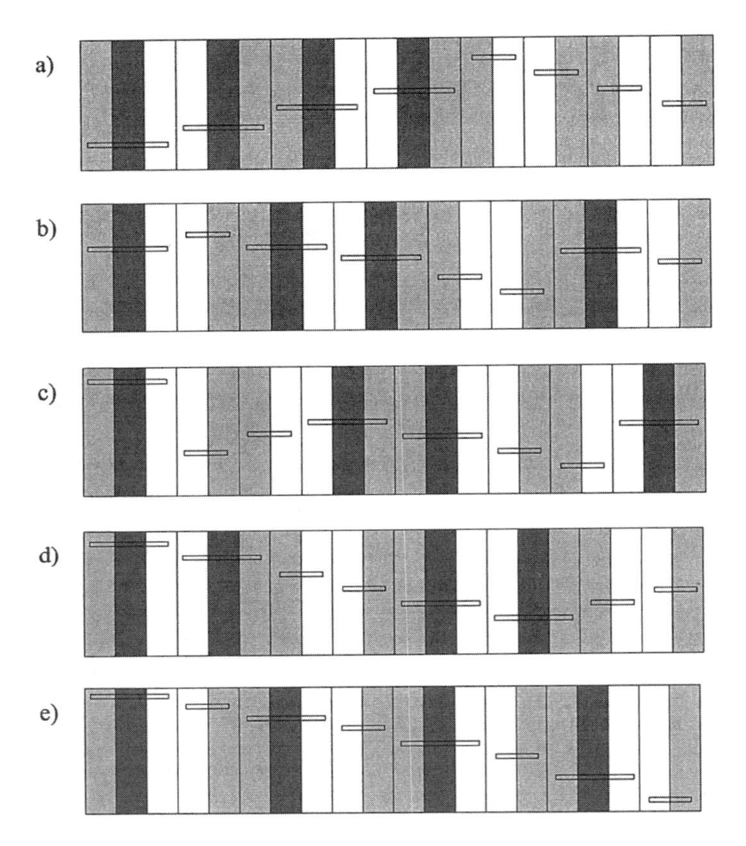

Figure 19 Schemes for blends of ABC and AC block copolymers: (a) macrophase separation; (b) random sequence of layers of diblock and triblock copolymers; (c) and (d) centrosymmetric double layers of diblock and triblock copolymers; (e) noncentrosymmetric array of alternating layers of diblock and triblock copolymers.

rona chains of diblock copolymer micelles in good solvents which do not interdiffuse with similar chains of other micelles (see Chapter 3). The interpenetration zone will be symmetric when the end blocks coming from opposite sides belong to similar block copolymers, and asymmetric when end blocks from AC and ABC block copolymers form common microdomains. Leibler et al. have shown theoretically that a free energy gain results from the formation of this noncentrosymmetric layering [251]. Figure 20a shows a transmission electron micrograph of a blend of S-B-T and S-T block copolymers. The different sequences of colors on the left and right part of that picture indicate opposite polarities. A special

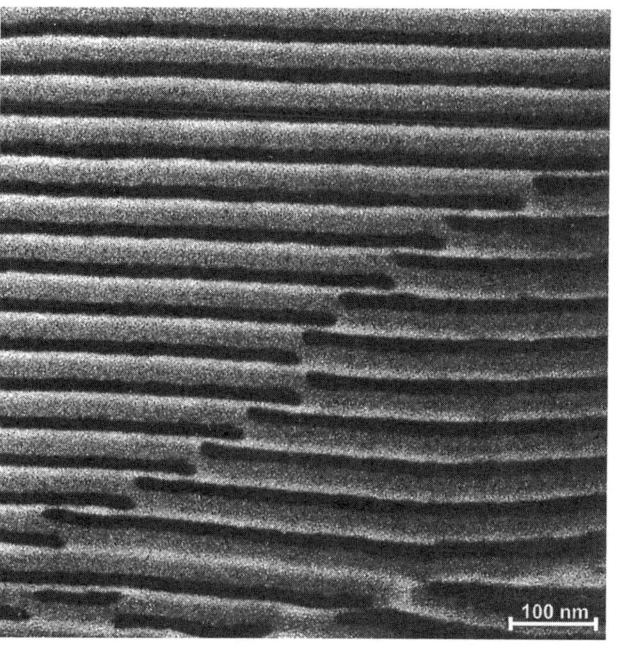

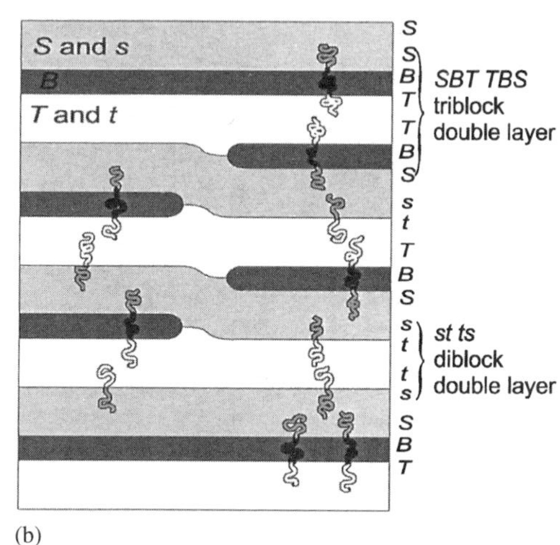

Figure 20 Periodic noncentrosymmetric lamellar superstructure of a blend of polystyrene - block - polybutadiene - block - poly(tert.butyl methacrylate) S-B-T with polystyrene-block-poly(tert.butyl methacrylate) S-T with the composition 50 mol% S-B-T: 50 mol% S-T. (a) Transmission electron micrograph (stained with OsO_4); (b) scheme of the characteristic defect proving the periodic noncentrosymmetry. (From Ref. 252, Copyright 1999 Macmillan Magazines Ltd.)

proof for the existence of the noncentrosymmetric superlattice is the defect in the middle of the TEM micrograph, also shown in Fig. 20b. The system avoids additional interfaces between S and T domains by breaking up the B lamellae. This superlattice shows for the first time the spontaneous formation of a noncentrosymmetric lamellar morphology with a periodicity length scale in the range of 60 nm [252]. The formation of this blend demonstrates in a nice way that superstructures with new symmetries can be obtained by self-assembling of block copolymers in a blend.

Other types of block copolymer blends also have been studied recently. For example the formation of a knitting-pattern morphology via blending of two different S-B-M triblock copolymers [253] or the formation of a core-shell double gyroid morphology via blending of a B-S-M and an S-B-T triblock copolymer show that blending of block copolymers is a promising concept for designing their morphological properties [254].

VI. CONCLUDING REMARKS

This chapter presented an overview of the fascinating variety of microphase morphologies found in binary and ternary coiling block copolymers. The increase in the number of components was shown to lead to an enormous increase of outstanding morphologies consistent with the increase of independent parameters of the system: composition and segmental interaction. Chain topology was also shown to affect morphology. Blends of these materials further enrich the population of self-assembled microphases with regular periodic arrays of the component copolymers.

All the above morphologies are the natural result of molecular recognition of two or more incompatible blocks connected by single chemical bonds. Self-assembly of block copolymers may thus be regarded as a supramolecular polymerization of similar or different macromolecules occurring in a precursor solution, or directly in the bulk.

ACKNOWLEDGMENTS

I wish to thank very much my former advisor and friend Reimund Stadler for many years of fruitful collaboration. He suddenly died in June, 1998. It was he who brought my attention to the fascinating field of block copolymers and I am very grateful for the many stimulating discussions we had on this subject. I am indebted to Igor Erukhimovich, who contributed much information related to the theoretical aspect. I am also grateful to Gudrun Schmidt and the group of Reimund Stadler, in particular, Thorsten Goldacker, Friederike von Gyldenfeldt,

Hanno Hückstädt, Katja Loos, Holger Schmalz, Harald Schmitz, and Stefan Stangler for their help in various phases of the preparation of this chapter. Finally I wish to thank Alberto Ciferri and Avi Halperin for their encouragement to write this chapter, which was originally supposed to be written by Reimund Stadler.

REFERENCES

1. FS Bates, GH Fredrickson. Annu Rev Phys Chem 41:525, 1990.
2. G Riess. In: NR Ledge, G Holden, HE Schroeder, eds. Thermoplastic Elastomers. A Comprehensive Review. Munich: Hanser, 1987.
3. FS Bates, GH Fredrickson. Physics Today 52:32, 1999.
4. MW Matsen, M Schick. Curr Opin Colloid Interface Sci 1:329, 1996.
5. K Mortensen, Y Talmon, B Gao, J Kops. Macromolecules 30:6764, 1997.
6. G Yu, A Eisenberg. Macromolecules 31:5546, 1998.
7. P Alexandridis, B Lindman, eds. Amphiphilic Block Copolymers: Self-Assembly and Applications. The Netherlands: Elsevier Science BV, 1997.
8. P Alexandridis, U Olsson, B Lindman. Langmuir 13:23, 1997.
9. O Glatter, G Scherf, K Schiller, W Brown. Macromolecules 27:6046, 1994.
10. G Wanka, H Hoffmann, W Ulbricht. Macromolecules 27:4145, 1994.
11. Z Tuzar, P Kratochvil. In: E Matijevic ed. Surface and Colloid Science vol. 15, New York: Plenum P, 1993.
12. G Schmidt, W Richtering, P Lindner, P Alexandridis. Macromolecules 31:2293, 1998.
13. H Soenen, H Berghmans, HH Winter, N Overbergh. Polymer 38:5653, 1997.
14. A Gast. Langmuir 12:4060, 1996.
15. H Hoffmann, S Hofmann, U Kästner. Advances in Chemistry Series 248:219, 1996.
16. G Schmidt, S Müller, P Lindner, C Schmidt, W Richtering. J Phys Chem B 102: 507, 1998.
17. MW Matsen, FS Bates. Macromolecules 29:7641, 1996.
18. JP Flory. Principles of Polymer Chemistry. Ithaca: Cornell University Press, 1953.
19. ML Huggins. J Chem Phys 9:440, 1941.
20. AJ Staverman. Rec Trav Chim 60:640, 1941.
21. Hernández-Barajas, DJ Hunkeler. Polymer 38:449, 1997.
22. C Auschra, R Stadler. Macromolecules 26:6364, 1993.
23. L Bronstein, M Seregina, P Valetsky, U Breiner, V Abetz, R Stadler. Polym Bull 39:361, 1997.
24. YNC Chan, RR Schrock, RE. Cohen. Chem Mater 4:24, 1992.
25. M Antonietti, E Wenz, L Bronstein, M Seregina. Adv Mater 7:1000, 1995.
26. JP Spatz, A Roescher, M Möller. Adv Mater 8:337, 1996.
27. JP Spatz, S Sheiko, M Möller. Macromolecules 29:3220, 1996.
28. GSW Craig, RE Cohen, RR Schrock, C Dhenaut, I LeDoux, J Zyss. Macromolecules 27:1875, 1994.
29. GSW Craig, RE Cohen, RR Schrock, RJ Silby, G Puccetti, I LeDoux, J Zyss. J Am Chem Soc 115:860, 1993.

30. Y Heischkel, H-W Schmidt. Macromol Chem Phys 199:869, 1998.
31. G Liu, L Quiao, A Guo. Macromolecules 29:5508, 1996.
32. J Ruokolainen, R Mäkinen, M Torkkeli, T Mäkelä, R Serimaa, G ten Brinke, O Ikkala. Science 280:557, 1998.
33. J Ruokolainen, M Saariaho, O Ikkala, G ten Brinke, EL Thomas, M Torkkeli, R Serimaa. Macromolecules 32:1152, 1999.
34. M Templin, A Franck, A Du Chesne, H Leist, Y Zhang, R Ulrich, V Schädler, U Wiesner. Science 278:1795, 1997.
35. E Krämer, S Förster, C Göltner, M Antonietti. Langmuir 14:2027, 1998.
36. R Unger, D Beyer, E Donth Polymer 32:3305, 1991.
37. P Rangarajan, RA Register, LJ Fetters. Macromolecules 26:4640, 1993.
38. S Nojima, K Kato, S Yamamoto, T Ashida. Macromolecules 25:2237, 1992.
39. S Nojima, K Hashizume, A Rohadie, S Sasaki. Polymer 38:2711, 1997.
40. V Balsamo, AJ Müller, R Stadler. Macromolecules 31:7756, 1998.
41. V Balsamo, AJ Müller, F von Gyldenfeldt, R Stadler. Macromol Chem Phys 199: 1063, 1998.
42. V Balsamo, R Stadler. Macromol Symp 117:153, 1997.
43. V Balsamo, F von Gyldenfeldt, R Stadler. Macromolecules 32:1226, 1999.
44. G Floudas, G Reiter, O Lambert, P Dumas. Macromolecules 31:7279, 1998.
45. EA DiMarzio, CM Guttman, JD Hoffman. Macromolecules 13:1194, 1980.
46. MD Whitmore, J Noolandi. Macromolecules 21:1482, 1988.
47. J Adams, W Gronski. Makromol Chem Rapid Commun 10:553, 1989.
48. J Sänger, W Gronski, H Leist, U Wiesner. Macromolecules 30:7621, 1997.
49. A Molenberg, M Möller, T Pieper. Macromol Chem Phys 199:299, 1998.
50. G Mao, J Wang, SR Clingman, CK Ober, JT Chen, EL Thomas. Macromolecules 30:2556, 1997.
51. DRM Williams, A Halperin. Phys Rev Lett 71:1557, 1993.
52. A Halperin, DRM Williams. Macromolecules 26:6652, 1993.
53. A Halperin, DRM Williams. Phys Rev E Rapid Commun 49:R986, 1994.
54. G Kämpf, M Hoffmann, H Krömer. Ber Bunsenges Phys Chem 74:851, 1970.
55. LH Radzilowski, BL Carvalho, EL Thomas. J Polym Sci Polym Phys Ed 34:3081, 1996.
56. H Elbs, K Fukunaga, R Stadler, G Sauer, R Magerle, G Krausch. Macromolecules 32:1204, 1999.
57. G Krausch. Mater Sci Eng R14:1, 1995.
58. HR Kricheldorf, T Stukenbrok. J Polym Sci Polym Chem 36:31, 1998.
59. U Schulze, HW Schmidt. Polym Bull 40:159, 1998.
60. MV Pandya, M Subramaniam, MR Desai. European Polym. J 33:789, 1997.
61. Y Bourgeois, Y Charlier, R Legras. Polymer 37:5503, 1996.
62. JCW Chien, Y Iwamoto, MD Rausch, W Wedler, HH Winter. Macromolecules 30: 3447, 1997.
63. S Jacob, JP Kennedy. Polym Bull 41:167, 1998.
64. R Faust, D Li. Macromolecules 28:4893, 1995.
65. YC Bae, R Faust. Macromolecules 31:2480, 1998.
66. N Mougin, P Rempp, Y Gnanou. Makromol Chem 194:2553, 1993.
67. G-H Hsiue, Y-L Liu, Y-S Chiu. J Polym Sci Polym Chem 31:3371, 1993.
68. P Van Caeter, EJ Goethals, R Velichkova. Polym Bull 39:589, 1997.

69. JM Nelson, AP Primrose, TJ Hartle, HR Allcock. Macromolecules 31:947, 1998.
70. DA Shipp, J-L Wang, K Matyjaszewski. Macromolecules 31:8005, 1998.
71. A Mühlebach, SG Gaynor, K Matyjaszewski. Macromolecules 31:6046, 1998.
72. M Szwarc. J Polym Sci Polym Chem 36:IX, 1998.
73. JM Yu, Y Yu, R Jérôme. Polymer 38:3091, 1997.
74. G Hild, JP Lamps. Polymer 36:4841, 1995.
75. RP Quirk, T Yoo, B Lee. JMS Pure Appl Chem A31:911, 1994.
76. N Hadjichristidis, Y Tselikas, H Iatrou, V Efstratiadis, A Avgeropoylos. JMS Pure Appl Chem A33:1447, 1996.
77. N Hadjichristidis. J Polym. Sci Part A Polym Chem 37:857, 1999.
78. N Hadjichristidis, H Iatrou. Macromolecules 26:5812, 1993.
79. H Iatrou, N Hadjichristidis. Macromolecules 26:2479, 1993.
80. S Sioula, Y Tselikas, N Hadjichristidis. Macromol Symp 117:167, 1997.
81. T Fujimoto, H Zhang, T Kazama, Y Isono, H Hasegawa, T Hashimoto. Polymer 33:2208, 1992.
82. H Hückstädt, V Abetz, R Stadler. Macromol Rapid Commun 17:599, 1996.
83. C Auschra, R Stadler. Polym Bull 30:257, 1993.
84. C Neumann, V Abetz, R Stadler. Polym Bull 36:43, 1996.
85. O Lambert, S Reutenauer, G Hurtrez, G Riess, P Dumas. Polym Bull 40:143, 1998.
86. O Lambert, P Dumas, G Hurtrez, G Riess. Macromol Rapid Commun 18:343, 1997.
87. G Riess, M Schlienger, S Marti. J Macromol Sci Polym Phys Ed 17:355, 1980.
88. E Yoshida, A Sugita. J Polym Sci Polym Chem 36:2059, 1998.
89. E Ruckenstein, H Zhang. Macromolecules 31:2977, 1998.
90. J Feldthusen, B Iván, AHE Müller. Macromolecules 31:578, 1998.
91. K Loos, R Stadler. Macromolecules 30:7641, 1997.
92. GE Molau. In: GE Molau ed. Colloidal and Morphological Behavior of Block Copolymers. New York: Plenum P, 1971.
93. DJ Meier, ed. Block Copolymers: Science and Technology. Tokyo: Gordon & Breach, 1983.
94. G Riess, G Hurtrez, P Bahadur. Block Copolymers. In: Encyclopedia of Polymer Science and Engineering vol. 2. New York: Wiley, 1985.
95. K Binder. Adv Polym Sci 112:182, 1994.
96. GH Fredrickson, FS Bates. Annu Rev Mater Sci 26:501, 1996.
97. RH Colby. Curr Opin Colloid Polym Sci 1:454, 1996.
98. SL Aggarwal. Polymer 17:938, 1972.
99. EL Thomas, DB Alward, DJ Kinning, DC Martin, DL Handlin, LJ Fetters. Macromolecules 19:2197, 1986.
100. DA Hajduk, PE Harper, SM Gruner, CC Honeker, EL Thomas, LJ Fetters. Macromolecules 28:2570, 1995.
101. H Hasegawa, T Hashimoto, ST Hyde. Polymer 37:3825, 1996.
102. MW Matsen, FS Bates. J Chem Phys 106:2436, 1997.
103. IY Erukhimovich. JETP Lett 63:460, 1996.
104. ST Milner, PT Olmsted. J Phys II 7:249, 1997.
105. W Gozdz, R Holyst. Macromol Theory Simul 5:321, 1996.
106. K Grosse-Brauckmann. J Colloid Interface Sci 187:418, 1997.
107. IW Hamley, KA Koppi, JH Rosedale, FS Bates, K Almadal, K Mortensen. Macromolecules 26:5959, 1993.

108. S Förster, AK Khandpur, J Zhao, FS Bates, IW Hamley, AJ Ryan, W Bras. Macromolecules 27:6922, 1994.
109. AK Khandpur, S Förster, FS Bates, IW Hamley, AJ Ryan, W Bras, K Almdal, K Mortensen. Macromolecules 28:8796, 1995.
110. MW Matsen, FS Bates. Macromolecules 29:1091, 1996.
111. B Stühn, R Mutter, T Albrecht Europhys Lett 18:427, 1992.
112. T Wolff, C Burger, W Ruland. Macromolecules 26:1707, 1993.
113. K Almdal, FS Bates, K Mortensen. J Chem Phys 96:9122, 1992.
114. B Holzer, A Lehmann, B Stühn, M Kowalski. Polymer 32:1935, 1991.
115. VT Bartels, V Abetz, K Mortensen, M Stamm. Europhys Lett 27:371, 1994.
116. VT Bartels, M Stamm, V Abetz, K Mortensen. Europhys Lett 31:81, 1995.
117. G Hadziioannou, C Picot, A Skoulios, M-L Ionesu, A Mathis, R Duplessix, Y Gallot, J-P Lingelser. Macromolecules 15:263, 1982.
118. H Hasegawa, T Hashimoto, H Kawai, TP Lodge, EJ Amis, CJ Glinka, CC Han. Macromolecules 18:67, 1985.
119. AM Mayes, TP Russell, VR Deline, SK Satija, CF Majkrzak. Macromolecules 27: 7447, 1994.
120. WZ Cai, K Schmidt-Rohr, N Egger, B Gerharz, H-W Spiess. Polymer 34:267, 1993.
121. MD Gehlsen, K Almdal, FS Bates. Macromolecules 25:939, 1992.
122. JH Rosedale, FS Bates, K Almdal, K Mortensen, GD Wignall. Macromolecules 28:1429, 1995.
123. CD Han, DM Baek, JK Kim. Macromolecules 23:561, 1990.
124. C Neumann, DR Loveday, V Abetz, R Stadler. Macromolecules 31:2493, 1998.
125. NP Balsara, BA Garetz, MY Chang, HJ Dai, MC Newstein, JL Goveas, R Krishnamoorti, S Rai. Macromolecules 31:5309, 1998.
126. L Leibler. Macromolecules 13:1602, 1980.
127. PG de Gennes. Faraday Disc Chem Soc 68:96, 1979.
128. IY Erukhimovich. Polymer Science USSR 24:2223, 1982.
129. IY Erukhimovich. Polymer Science USSR 24:2232, 1982.
130. H Benoît, G Hadziioannou. Macromolecules 21:1449, 1988.
131. S Alexander, J Mc Taque. Phys Rev Lett 41:702, 1978.
132. GH Fredrickson, E Helfand. J Chem Phys 87:697, 1987.
133. SA Brazovskii. Sov Phys JETP 41:85, 1975.
134. AM Mayes, M Olvera de la Cruz. J Chem Phys 91:7228, 1989.
135. AV Dobrynin, IY Erukhimovich. Vysokomol Soedin 32B:663, 1990.
136. AV Dobrynin, IY Erukhimovich. Macromolecules 26:276, 1993.
137. AM Mayes, M Olvera de la Cruz. J Chem Phys 95:4670, 1991.
138. AV Dobrynin, IY Erukhimovich. Vysokomol Soyed 33A:1100, 1991.
139. DJ Meier. J Polym Sci C26:81, 1969.
140. E Helfand, AM Sapse. J Chem Phys 62:1327, 1975.
141. E Helfand, ZR Wasserman. Macromolecules 9:879, 1976.
142. E Helfand, ZR Wasserman. Macromolecules 11:960, 1978.
143. AN Semenov. Sov Phys JETP 61:733, 1985.
144. IA Nyrkova, AR Khokhlov, M Doi. Macromolecules 26:3601, 1993.
145. MW Matsen, M Schick. Macromolecules 27:4014, 1994.
146. VT Bartels, M Stamm, K Mortensen. Polym Bull 36:103, 1996.

147. H Frielinghaus, D Schwahn, T Springer. Macromolecules 29:3263, 1996.
148. H Ladynski, W De Odorico, M Stamm. J Non-Crystall Solids 235/237:491, 1998.
149. M Pollard, TP Russell, AV Ruzette, AM Mayes, Y Gallot. Macromolecules 31: 6493, 1998.
150. B Steinhoff, M Rüllmann, M Wenzel, M Junker, I Alig, R Oser, B Stühn, G Meier, O Diat, P Bösecke, HB Stanley. Macromolecules 31:36, 1998.
151. D Schwahn, H Frielinghaus, K Mortensen, K Almdal. Phys Rev Lett 77:3153, 1996.
152. RJ Albalak, EL Thomas. J Polym Sci Polym Phys 31:37, 1993.
153. CC Honeker, EL Thomas. Chem Mater 8:1702, 1996.
154. A Keller, E Pedemonte, FM Willmouth. Nature 225:538, 1970.
155. KA Koppi, M Tirrell, FS Bates, K Almdal, RH Colby. J Phys II 2:1941, 1992.
156. KI Winey, SS Patel, RG Larson, H Watanabe. Macromolecules 26:2542, 1993.
157. BL Riise, GH Fredrickson, RG Larson, DS Pearson. Macromolecules 28:7653, 1995.
158. Z-R Chen, JA Kornfield, SD Smith, JT Grothaus, MM Satkowski. Science 277: 1248, 1997.
159. U Wiesner. Macromol Chem Phys 198:3319, 1997.
160. Z-R Chen, AM Issaian, JA Kornfield, SD Smith, JT Grothaus, MM Satkowski. Macromolecules 30:7096, 1997.
161. T Tepe, DA Hajduk, MA Hillmyer, PA Weimann, M Tirrell, FS Bates, K Almdal, K Mortensen. J Rheol 41:1147, 1997.
162. HH Winter, DB Scott, W Gronski, S Okamoto, T Hashimoto. Macromolecules 26: 7236, 1993.
163. AI Nakatani, FA Morrison, JF Douglas, JW Mays, CL Jackson, M Muthukumar, CC Han. J Chem Phys 104:1589, 1996.
164. CL Jackson, KA Barnes, FA Morrison, JW Mays, AI Nakatani, CC Han. Macromolecules 28:713, 1995.
165. V Abetz, R Stadler. Macromol Symp 113:19, 1997.
166. H Funabashi, Y Miyamoto, Y Isono, T Fujimoto, Y Matsushita, M Nagasawa. Macromolecules 16:1, 1983.
167. Y Isono, H Tanisugi, K Endo, T Fujimoto, H Hasegawa, T Hashimoto, H Kawai. Macromolecules 16:5, 1983.
168. Y Matsushita, K Yamada, T Hattori, T Fujimoto, Y Sawada, M Nasagawa, C Matsui. Macromolecules 16:10, 1983.
169. Y Mogi, M Nomura, H Kotsuji, K Ohnishi, Y Matsushita, I Noda. Macromolecules 27:6755, 1994.
170. SP Gido, DW Schwark, EL Thomas, MC Goncalves. Macromolecules 26:2636, 1993.
171. L Kane, RJ Spontak. Macromolecules 27:663, 1994.
172. Y Mogi, K Mori, H Kotsuji, Y Matsushita, I Noda, CC Han. Macromolecules 26: 5169, 1993.
173. W Zheng, Z-G Wang. Macromolecules 28:7215, 1995.
174. T Ohta, K Kawasaki. Macromolecules 19:2621, 1989; 23:2413, 1990.
175. H Nakazawa, T Ohta. Macromolecules 26:5503, 1993.
176. YV Lyatskaya, TM Birshtein. Polymer 36:975, 1995.
177. S Phan, GH Fredrickson. Macromolecules 31:59, 1998.

178. MW Matsen. J Chem Phys 108:785, 1998.
179. Y Mogi, K Mori, Y Matsushita, I Noda. Macromolecules 25:5412, 1992.
180. EE Dormidontova, AR Khokhlov. Macromolecules 30:1980, 1997.
181. R Stadler, C Auschra, J Beckmann, U Krappe, I Voigt-Martin, L Leibler. Macromolecules 28:3080, 1995.
182. U Krappe, R Stadler, I-G Voigt-Martin. Macromolecules 28:4458, 1995.
183. K Jung, V Abetz, R Stadler. Macromolecules 29:1076, 1996.
184. U Breiner, U Krappe, V Abetz, R Stadler. Macromol Chem Phys 198:1051, 1997.
185. U Breiner, U Krappe, T Jakob, V Abetz, R Stadler. Polym Bull 40:219, 1998.
186. S Brinkmann, R Stadler, EL Thomas. Macromolecules 31:6566, 1998.
187. EB Zhulina, A Halperin. Macromolecules 25:2730, 1992.
188. H Watanabe. Macromolecules 28:5006, 1995.
189. MW Matsen. J Chem Phys 102:3884, 1995.
190. S Brinkmann, V Abetz, R Stadler, EL Thomas. Kautschuk Gummi Kunststoffe 52: 806, 1999.
191. C Auschra, J Beckmann, R Stadler. Macromol Rapid Commun 15:67, 1994.
192. C Auschra, R Stadler. Macromolecules 26:2171, 1993.
193. U Breiner, U Krappe, EL Thomas, R Stadler. Macromolecules 31:135, 1998.
194. U Breiner, U Krappe, R Stadler. Macromol Rapid Commun 17:567, 1996.
195. K Jung. Doctoral thesis, Universität Mainz, 1996.
196. E Giebeler. Doctoral thesis, Universität Mainz, 1996.
197. F Annighöfer, W Gronski. Colloid Polym Sci 261:15, 1983.
198. F Annighöfer, W Gronski. Makromol Chem 185:2213, 1984.
199. T Hashimoto, Y Tsukahara, K Tachi, H Kawai. Macromolecules 16:648, 1983.
200. V Abetz, R Stadler, L Leibler. Polym Bull 37:135, 1996.
201. I Erukhimovich, V Abetz, R Stadler. Macromolecules 30:7435, 1997.
202. C Neumann, V Abetz, R Stadler. Colloid Polym Sci 276:19, 1998.
203. CM Roland, CA Trask. Macromolecules 22:256, 1989.
204. M Doi, SF Edwards. The Theory of Polymer Dynamics. Oxford: Clarendon, 1989.
205. CW Lantman, WJ MacKnight, JF Tassin, L Monnerie, LJ Fetters. Macromolecules 23:836, 1990.
206. M Daoud, JP Cotton. J Phys (Les Ulis Fr) 43:539, 1982.
207. TM Birshtein, EB Zhulina, OV Borisov. Polymer 27:1078, 1986.
208. M Moffitt, Y Yu, D Nguyen, V Graziano, DK Schneider, A Eisenberg. Macromolecules 31:2190, 1998.
209. G Floudas, S Paraskeva, N Hadjichristidis, G Fytas, B Chu, AN Semenov. J Chem Phys 107:5502, 1997.
210. L-K Bi, LJ Fetters. Macromolecules 9:732, 1976.
211. Y Tselikas, H Iatrou, N Hadjichristidis, KS Liang, K Mohanty, DJ Lohse. J Chem Phys 105:2456, 1996.
212. CM Turner, NB Sheller, MD Foster, B Lee, S Corona-Galvan, RP Quirk, B Annis, J-S Lin. Macromolecules 31:4372, 1998.
213. FL Beyer, SP Gido, Y Poulos, A Avgeropoylos, N Hadjichristidis. Macromolecules 30:2373, 1997.
214. S Milner. Macromolecules 27:2333, 1994.

215. Y Tselikas, N Hadjichristidis, RL Lescanec, CC Honeker, M Wohlgemuth, EL Thomas. Macromolecules 29:3390, 1996.
216. Y Matsushita, T Takasu, K Yagi. Polymer 35:2862, 1994.
217. RF Story, BJ Chisholm, Y Lee. Polymer 34:4330, 1993.
218. MW Matsen, M Schick. Macromolecules 27:6761, 1994.
219. M Olvera de la Cruz, IC Sanchez. Macromolecules 19:2501, 1986.
220. G Floudas, S Pispas, N Hadjichristidis, T Pakula, I Erukhimovich. Macromolecules, 29:4142, 1996.
221. S Okamoto, H Hasegawa, T Hashimoto, T Fujimoto, H Zhang, T Kazama, A Takano, Y Isono. Polymer 38:5275, 1997.
222. S Sioula, Y Tselikas, N Hadjichristidis. Macromolecules 30:1518, 1997.
223. S Sioula, N Hadjichristidis, EL Thomas. Macromolecules 31:5272, 1998.
224. S Sioula, N Hadjichristidis, EL Thomas. Macromolecules 31:8429, 1998.
225. C Auschra, R Stadler, I Voigt-Martin. Polymer 34:2094, 1993.
226. NP Balsara, SV Jonnalagadda, CC Lin, CC Han, R Krishnamoorti. J Chem Phys 99:10011, 1993.
227. AZ Akcasu, M Tombakoglu. Macromolecules 23:607, 1990.
228. H Benoît, M Benmouna, WL Wu. Macromolecules 23:1511, 1990.
229. KI Winey, EL Thomas, LJ Fetters. Macromolecules 25:422, 1992.
230. S-H Lee, JT Koberstein, X Quan, I Gancarz, GD Wignall, FC Wilson. Macromolecules 27:3199, 1994.
231. T Hashimoto, S Koizumi, H Hasegawa, T Izumitani, ST Hyde. Macromolecules 25:1433, 1992.
232. T Hashimoto, K Kimishima, H Hasegawa. Macromolecules 24:5704, 1991.
233. R Xie, B Yang, B Jiang. Macromolecules 26:7097, 1993.
234. G Hadziioannou, A Skoulios. Macromolecules 15:267, 1982.
235. T Hashimoto, K Yamasaki, S Koizumi, H Hasegawa. Macromolecules 26:2895, 1993.
236. T Hashimoto, S Koizumi, H Hasegawa. Macromolecules 27:1562, 1994.
237. RJ Spontak, CJ Fung, MB Braunfeld, JW Sedat, DA Agard, A Ashraf, SD Smith. Macromolecules 29:2850, 1996.
238. AD Vilesov, G Floudas, T Pakula, EY Melenevskaya, TM Birshtein, YV Lyatskaya. Macromol Chem Phys 195:2132, 1994.
239. MF Schulz, FS Bates, K Almdal, K Mortensen. Phys Rev Lett 73:86, 1994.
240. RJ Spontak, JC Fung, MB Braunfeld, JW Sedat, DA Agard, L Kane, SD Smith, MM Satkowski, A Ashraf, DA Hajduk, SM Gruner. Macromolecules 29:4494, 1996.
241. S Sakurai, H Irie, H Umeda, S Nomura, HH Lee, JK Kim. Macromolecules 31:336, 1998.
242. S Sakurai, H Umeda, C Furukawa, H Irie, S Nomura, HH Lee, JK Kim. J Chem Phys 108:4333, 1998.
243. S Sakurai, S Nomura. Polymer 38:4103, 1997.
244. A-C Shi, J Noolandi. Macromolecules 28:3103, 1995.
245. MW Matsen, FS Bates. Macromolecules 28:7298, 1995.
246. TM Birshtein, YV Lyatskaya, EB Zhulina. Polymer 33:2750, 1992.
247. AL Borovinskii, AR Khokhlov. Macromolecules 31:1180, 1998.

248. RL Lescanec, FJ Fetters, EL Thomas. Macromolecules 31:1680, 1998.
249. T Goldacker, V Abetz. Macromolecules 32:5165, 1999.
250. TM Birshtein, EB Zhulina, AA Polotsky, V Abetz, R Stadler. Macromol Theory Simul 8:151, 1999.
251. L Leibler, C Gay, IY Erukhimovich. Europhys Lett 46:549, 1999.
252. T Goldacker, V Abetz, R Stadler, IY Erukhimovich, L Leibler. Nature 398:137, 1999.
253. T Goldacker, V Abetz. Macromol Rapid Commun 20:415, 1999.
254. V Abetz, T Goldacker. Macromol Rapid Commun 21:16, 2000.

7

Microstructure and Crystallization of Rigid-Coil Comblike Polymers and Block Copolymers

Katja Loos
Universität Bayreuth, Bayreuth, Germany

Sebastián Muñoz-Guerra
Universitat Politècnica de Catalunya, Barcelona, Spain

I. INTRODUCTION

Copolymer systems based on blocks that behave in a coillike fashion (including di- and triblock copolymers) have been widely studied (Chapter 6). Coil–coil multiblock systems built of incompatible coil segments have been found to exist in a wide range of microphase separated supramolecular structures such as spheres, cylinders, double diamond (DD), double gyroid (DG), and lamella. Their phase behavior mostly results from the packing constraints imposed by the connectivity of each block and by the mutual repulsion of the dissimilar blocks. Phase separation and therefore the resulting stable morphology in diblock systems is greatly influenced by the total degree of polymerization ($N = N_A + N_B$), the Flory–Huggins χ parameter, and the composition expressed by volume fractions f_A, f_B, ...

This chapter focuses on block copolymers and comblike polymers in which at least one component is based on a conformationally rigid segment. A measurement of the stiffness of a polymer is afforded by the so-called persistence length which gives an estimate of the length scale over which the tangent vectors along

---

In Memoriam Professor Dr. Reimund Stadler.

the contour of the chain's backbone are correlated. Typical values for persistence lengths in synthetic and biological systems can be several orders of magnitude larger than for flexible, coillike polymers. Rodlike polymers have been found to exhibit lyotropic liquid-crystalline ordered phases such as nematic and/or layered smectic structures with the molecules arranged with their long axes nearly parallel to each other. Supramolecular assemblies of rodlike molecules are also capable of forming liquid-crystalline phases. The main factor governing the geometry of supramolecular structures in the liquid-crystalline phase is the anisotropic aggregation of the molecules (Chapter 1).

By combining these different classes of polymers a novel class of self-assembling materials can be produced since the molecules share certain general characteristics typical of diblock molecules and thermotropic calamitic molecules. The difference in chain rigidity of rodlike and coillike blocks is expected to greatly affect the details of molecular packing in the condensed phases and thus the nature of thermodynamically stable morphologies in these materials. The thermodynamic stable morphology should be originated as the result of the interdependence of microsegregation and liquid crystallinity. From this point of view it is very fascinating to compare the microstructures originating in solution and in the bulk for such materials.

Practical applications in which the block copolymers are characterized by some degree of structural asymmetry have been suggested. For instance a flexible block may be chosen as it donates a flexural compliance, whereas the more rigid portion offers tensile strength. Besides mechanical properties the orientational order and electrical conductance of certain rigid blocks could be exploited in optical and electrical devices. Furthermore one could utilize the special properties of a constituent rod segment such as taking advantage of the chiral information in helical rods or even using certain hybrid systems (copolymers with polypeptides or polysaccharides, etc.) as enhanced biocompatible materials in medical technology.

II. COMBLIKE POLYMERS: HELICAL MAINCHAIN WITH FLEXIBLE SIDECHAINS

A. Polypeptides with Covalently Attached Long Alkyl Sidechains

References to polypeptides bearing long alkyl sidechains are almost confined to the esters of poly(α,L-glutamic acid) (PγAG-n) and poly(α or β,L-aspartic acid) (PβAA-n or PαAA-n) whose chemical structures are depicted in Fig. 1. In these compounds the alkyl sidechain is connected to the mainchain by a carboxylate group that is directly anchored to the backbone in the case of poly(β-peptide)s and through a methylene or ethylene spacer in poly (α-aspartate)s and poly(α-

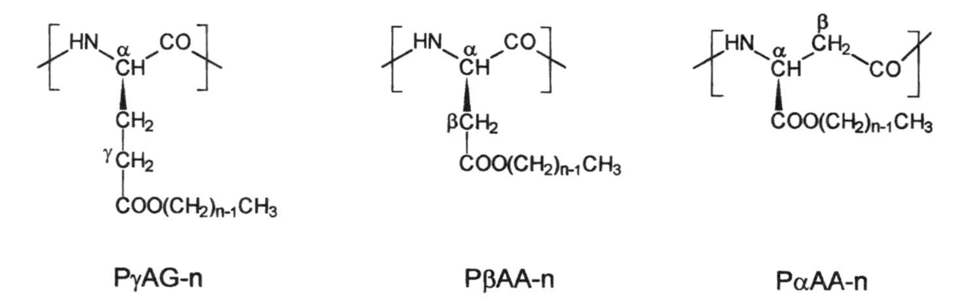

Figure 1 Chemical structures of comblike poly(γ-alkyl-α,L-glutamate)s (PγAG-n), poly(β-alkyl-α,L-aspartate)s (PβAA-n), and poly(α-alkyl-β,L-aspartate)s (PαAA-n). n indicates the number of carbons in the polymethylene sidechain and the carbonyl position relative to the nitrogen atom is indicated by Greek letters.

glutamate)s, respectively. Most of work on comblike polypeptides has been done with polyglutamates with some sporadic incursions in the polyaspartate area. *N*-acyl substituted poly(L-lysine)s constitute the only system investigated aside from polyaspartates and polyglutamates. Although solution studies revealed that comblike poly(*N*-acyl-L-lysine)s are in the α-helical conformation when dissolved in hydrocarbons [1] these polypeptides show a strong tendency to adopt the β-folded conformation in the solid state [2,3].

Until very recently, the α-helical conformation with 3.6 residues per turn (Fig. 2a) was considered to be exclusive of poly(α-peptide)s. In these last two decades it has been demonstrated, however, that poly(β-peptide)s, specifically those derived from aspartic acid, are also well suited for adopting folded secondary structures stabilized by intramolecular hydrogen bonds [4,5]. In these systems, hydrogen bonds are set in a similar way as they are in the genuine α-helix although the existence of one additional methylene in the backbone of the repeating unit gives rise to a 13/4 helix with 3.25 residues per turn (Fig. 2b). The stability of these pseudo α-helices has been shown to be comparable to that of the α-helix in spite of the fact that several polymorphs have been identified in the solid state for different sidechain groups. Nevertheless features relevant to the assembling of supramolecular structures such as polarity, chain stiffness, and molecular shape are substantially identical for the all these types of helices.

A great number of flexible polymers bearing linear long alkyl sidechains have been examined and much evidence has been collected to conclude that the commonplace structure of these systems in the solid state is a biphasic arrangement of alternating layers of mainchain and sidechains [6]. A detailed X-ray diffraction study on poly(α-olefin)s carried out by Turner-Jones [7] demonstrated that the polymethylene sidechain is crystallized only if it contains seven or more

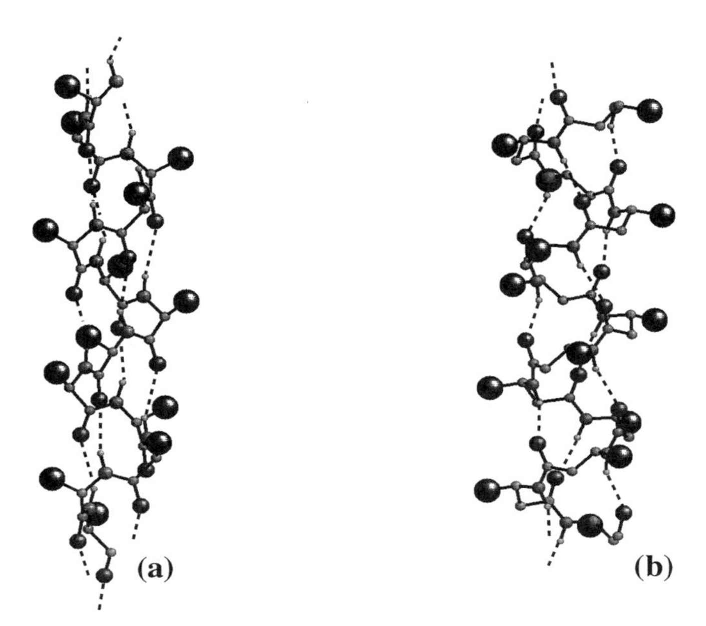

Figure 2 (a) The 18/5 right-handed α-helix typical of poly(α,L-peptide)s. (b) The 13/4 right-handed pseudo-α-helix observed in poly(α-alkyl-β,L-aspartate)s. In both helices the hydrogen bond scheme is set between every third amide group but counted at opposite directions with respect to the directionality of the mainchain. The larger spheres represent the sidechains and hydrogen bonds are indicated as dotted lines.

carbon atoms and that the crystal structure adopted may be either orthorhombic or monoclinic. Today it is widely accepted that the critical chain length for the onset crystallization depends on the chemical constitution of the polymer and turns out to be around 8–10 carbon atoms. In many comblike polymers such as polyacrylates, polyethers, or poly(*N*-alkyl acrylamide)s [8], the sidechain tends to crystallize in a hexagonal lattice with chains lying parallel to each other and the zigzag chain planes oriented at random about the chain axis (Fig. 3a). Such structure is similar to that found in *n*-paraffins at temperatures near above the melting point described as a roughly hexagonal packing of cylinders with an average distance of 4.6–4.8 Å [9]. The number of methylenes included in the sidechain crystallites (which may be estimated by calorimetry using heat fusion data available for *n*-alkanes [10]) is variable and may be only a small fraction of the whole sidechain. This is thought to be due to the conformational distortions created when sidechains approximate each other and to the restrictions to mobility imposed by the functional bridging groups.

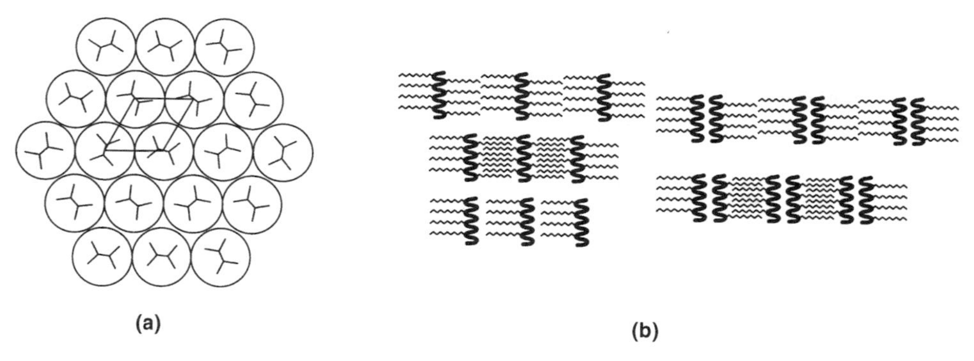

Figure 3 Packing of comblike polymers bearing polymethylene side chains: (a) scheme of the hexagonal packing of polymethylene chains as viewed down the chain axis; (b) one-layer (left) and two-layer (right) models for the comblike polymer lamellar structure.

Two types of arrangements corresponding to one- and two-layer models with sidechains interdigitated to a more or less extent can be observed in comblike polymers (Fig. 3b). Sidechains may be lying either normal or tilted with respect to the layers although the former is the model most frequently observed. The arrangement adopted is mainly determined by the constitutional nature of the polymer and its stereochemistry, although crystallization conditions may play a decisive role too.

It is well known that the polymethylene chain in the amorphous state tends to retain a considerable degree of ordering. In the axial direction, molten chains are described as an average *trans* sequence of 3–4 backbone bonds whereas the radial packing consists of flexible chains arranged in a random manner. In comblike polymers the regular spacing imposed by the mainchain sequence contributes additionally to retain some ordering in the molten state making the fusion process particularly complex. As a result, polymethylene sidechains of comblike polymers at temperatures near above melting tend to be arranged in a quasihexagonal structure with chains in a more or less extended conformation. Such a quasi-ordered structure is thought to act as a very efficient nucleating agent for rapid crystallization usually taking place in these systems upon cooling.

1. Poly(γ-alkyl-α,L-glutamate)s (PγAG-n)

(a). Isotropic and Liquid-Crystalline Solutions. In spite of the fact that the higher homologues of the PγAG-n series are widely soluble in organic solvents, the behavior of these polymers in solution has not received much attention. This is partially explained because the ability of these systems to form crystal liquid phases in the bulk has directed much of the interest to the study of the solid state.

Smith and Woody [11] carried out a viscosimetric and DC-ORD study on poly (γ-*n*-dodecyl-α,L-glutamate) (PγAG-12) dissolved in hydrocarbons and in strong breaking H-bond solvents like dichloroacetic or trifluoroacetic acids. As with polypeptides made of naturally occurring amino acids, the conformation of PγAG-n in solution was shown to be highly dependent upon both solvent and temperature. More recently Poché [12] examined the solution properties of polydisperse poly(γ-stearyl-α,L-glutamate) (PγAG-18) in tetrahydrofuran. Light-scattering and capillary viscosimetry measurements were used to establish a Mark–Houwink equation with an exponent $a \approx 1.3$ applicable in the 35K–250K molecular weight domain. Such value is indicative of a stiff chain, although less stiff than poly(γ-benzyl-α,L-glutamate) and other rodlike polymers for which a has been estimated to be much closer to the theoretical value of 1.8. The apparent diameter of mutual exclusion calculated with the Onsager–Zimm–Schulz equation [13] was 37 ± 6 Å substantially higher than the diameter of 23 Å computed by treating the polymer as a solid cylinder. This implies that the sidechain sheath wrapping the helical backbone is perfectly solvent permeable. It was concluded furthermore that interrod distances less than 30 Å are prevented by the dynamics of the system which includes both sidechain motion and molecular spinning about the rod axis. Cholesteric phases are described for toluene solutions of PγAG-18 with M_w in the 60–200K range at concentrations above 16% [14]. No significant differences from those described for classical polypeptides such as poly(γ-benzyl-α,L-glutamate) are apparent. Homeotropic structures displaying a colorful background are reported to be occasionally observed. The cholesteric arrangement was found to collapse upon heating into a nematic structure [15].

Iizuka et al. [16] have examined the solution properties of a series of comblike poly(γ-*p*-alkylbenzyl-α,L-glutamate)s with alkyl chains containing from 18 up to 22 carbon atoms. In these systems a *p*-benzylene unit has been inserted between the carboxylate group and the polymethylene chain. When these polymers are dissolved in tetrahydrofuran they show the high ellipticity characteristic of α-helix conformation with a helical content similar to that present in poly(γ-benzyl-α,L-glutamate)s of comparable molecular weights. Their concentrated solutions display under crossed polarizers equally spaced retardation lines characteristic of cholesteric mesophases. A new and interesting observation reported by these authors is that line-width and therefore the pitch of the supramolecular helix is a function of the length of the alkyl sidechain although no correlation is apparent. Moreover the concentration dependence of the cholesteric pitch diminishes as the alkyl sidechain becomes longer and vanishes when the number of carbons is larger than 16 (Fig. 4). Solvent molecule exclusion from side interchain space is claimed to be the reason for these striking experimental results.

(b). Crystallized Structures. The first account of the structural behavior in the solid state of comblike polyglutamates was reported by Thierry and Skoulios in a short letter published in 1978 [17]. They identified three ordered phases for

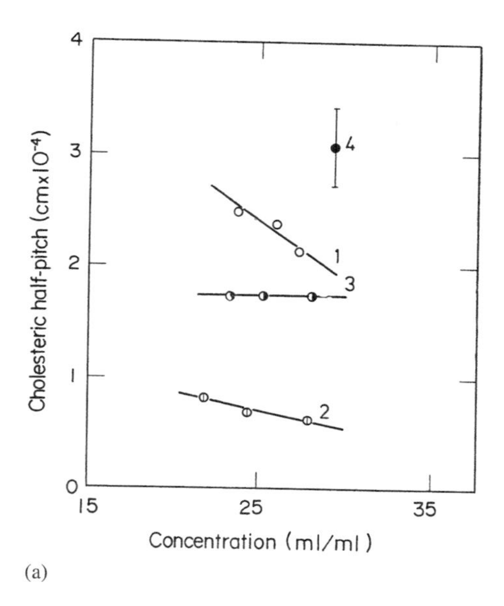

(a)

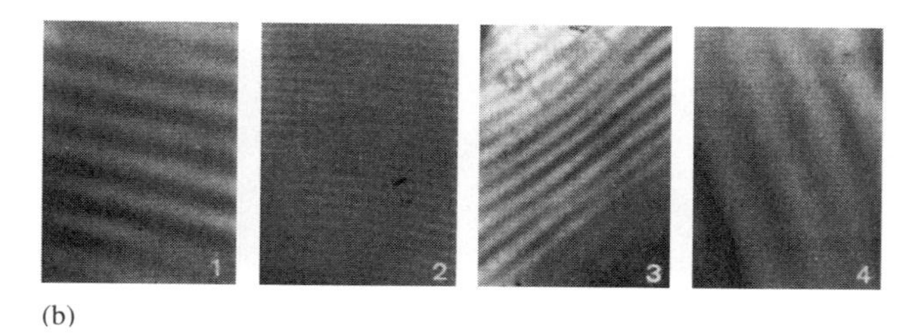
(b)

Figure 4 (a) Changes in the cholesteric pitch of poly(γ-*p*-alkylbenzyl-α,L-glutamate)s concentrated solutions: (a) dependence of the half-pitch on polymer concentration; (b) polarizing optical micrographs (1:octyl, 2:dodecyl, 3:hexadecyl, 4:docosyl). (Adapted from Ref. 16.)

PγAG-12 in the bulk as a function of temperature. The phase occurring at temperatures below 25°C was described as a layered structure with the polypeptide chain in the β-sheet conformation and the alkyl sidechain crystallized in a separate phase with a periodicity of 24 Å. The high temperature phase appeared in the proximity of 125°C and was thought to be a nematic phase with the α-helix polypeptide packed in a hexagonal array. Nothing was said about the structure adopted by the polymer at intermediate temperatures.

Watanabe et al. [18] carried out a systematic research on the structure in the solid state of PγAG-n embracing pentyl to octadecyl sidechains. The study included X-ray diffraction, differential scanning calorimetry, and dynamic-mechanical methods. They concluded that the mainchain assumes the α-helix conformation for all members within the whole range of temperatures examined. Members with sidechains containing 10 or more carbon atoms were found to behave according to the pattern described by Thierry and Skoulios [17] for the dodecyl derivative. In such cases, sidechains are long enough to crystallize separately upon cooling and the crystallized phase induces a layered arrangement with the α-helices aligned in sheets and the paraffin crystallites placed in between (Fig. 5).

Information provided by the X-ray diffraction recorded in the low angle region revealed that sidechains are oriented near normal to the sheet planes and that they have to interdigitate in more or less extension according to their lengths. Unfortunately, no information concerning the packing of the α-helices along the sheets could be obtained due to shear disorder affecting the stacking of the sheets. The melting temperature of the paraffin crystallites increased from −24°C to +62°C when the number of carbon atoms in the alkyl sidechain increased from 10 to 18. Wide-angle X-ray diffraction data indicated that crystallization of the

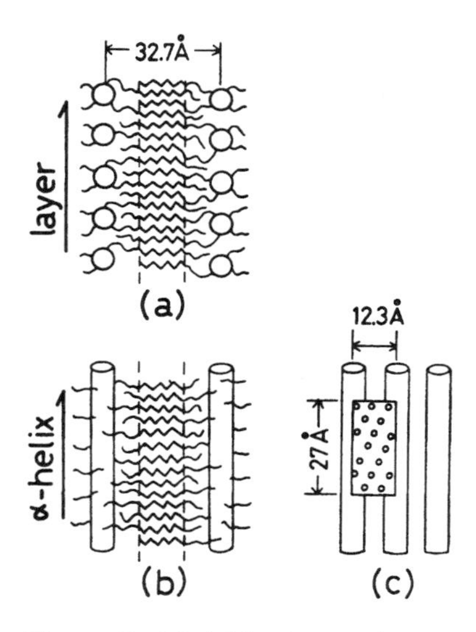

Figure 5 Model for the packing of PγAG-18 at low temperatures: (a) view parallel to the mainchain axes; (b) view parallel to the sidechain axis and perpendicular to the mainchain axis; (c) view perpendicular to both the layer and the chain axis of the α-helices. (After Ref. 18.)

sidechains proceeded as in low molecular weight *n*-alkanes so that a triclinic unit cell with the polymethylene chain in a fully extended conformation was adopted. The crystallinity of the system was evaluated by DSC which showed that the enthalpy of fusion increased linearly with the length of the sidechain. The number of crystallized methylene units was calculated to vary from a minimum of 0.4 up to a maximum of 7.7 for the decyl and the octadecyl derivatives, respectively. These results are also supported by dynamic mechanical and dielectric measurements [18,19]. The fact that the γ-relaxation commonly attributed to small amplitude motions of alkyl chains was observed irrespective of temperature, whereas the presence of the β-relaxation arising from wholechain movements appeared restricted to temperatures at which the sidechains in the molten state are in agreement with the occurrence of partial crystallization of the sidechain.

(c). Liquid Crystallinity in the Bulk. In the pioneer paper published by Thierry and Skoulios [17] it was advanced that polypeptides with sufficient long polymethylene sidechains could display liquid-crystalline phases upon melting of the sidechain. They reported, in fact, a high-temperature nematic phase that reversed upon cooling into a lower-temperature phase whose structure was not elucidated. Since then a good amount of research has been done on comblike polyglutamates mostly focused on the development of thin films with thermochromic properties. The liquid crystal structure anticipated for these systems should be cholesteric according to the chiral nature of the main chain α-helix. However, other mesomorphs have been described under certain conditions and for certain sidechain compositions.

Watanabe et al. [20] dedicated great efforts to understanding the high-temperature behavior of PγAG-n. However, many aspects still remain undisclosed, in particular those concerning the new phase that is formed when the lamellar structure is heated just above the melting point of crystallites formed by the alkyl sidechains. A hexagonal lattice made of a two-strand coiled-coil conformation was put forward for such a phase in PγAG-12 on the basis of both density measurements and low angle X-ray diffraction [21]. Since no similar evidence has been found so far for any other member of the series [18,22], such a situation seems to be rather exceptional and needs corroboration. In this regard one should mention the DSC studies carried out by Daly et al. [15] which show that melting of the lamellar state is strongly influenced by the existence of microheterogeneity in the polymer composition and that the structure of the phase formed after melting is largely conditioned by the thermal history of the sample.

PγAG-18 is likely the most thoroughly investigated comblike polypeptide. The most outstanding feature displayed by PγAG-18, shared also by other PγAG-n with n $\geq$ 10, is the formation of liquid-crystalline phases upon sidechain melting. However, all structural work published so far on the behavior of this polymer is phenomenological and restricted to phase identification and thermal transition characterization. The properties displayed by such phases are reported to be largely dependent on polymer size [19]. In samples with molecular weights above 100K, a cholesteric phase is formed immediately after melting at about 50°C which is then converted into a columnar liquid-crystal phase upon heating at higher temperatures [23]. The latter phase exhibits an extremely high viscosity and can be distinguished by a characteristic fan-shaped texture. X-ray diffraction indicated that such a structure is made of a hexagonal two-dimensional lattice of α-helices. The columnar phase is actually viewed as a kind of micellar structure derived from the differences in polarity and shape existing between the side-chains and the mainchain. This structure had been theoretically predicted to appear as a result of excluded volume effects associated with the lateral packing of hard rods [24,25]. The columnar phase is closely related to the smectic B phase from which it differs in the translational disorder existing along the direction of the helix axis (Fig. 6a). The transition from cholesteric to columnar occurs in the vicinity of 200°C for samples with molecular weights above 60K. For smaller sizes the transition temperature increases to attain values beyond the decomposi-

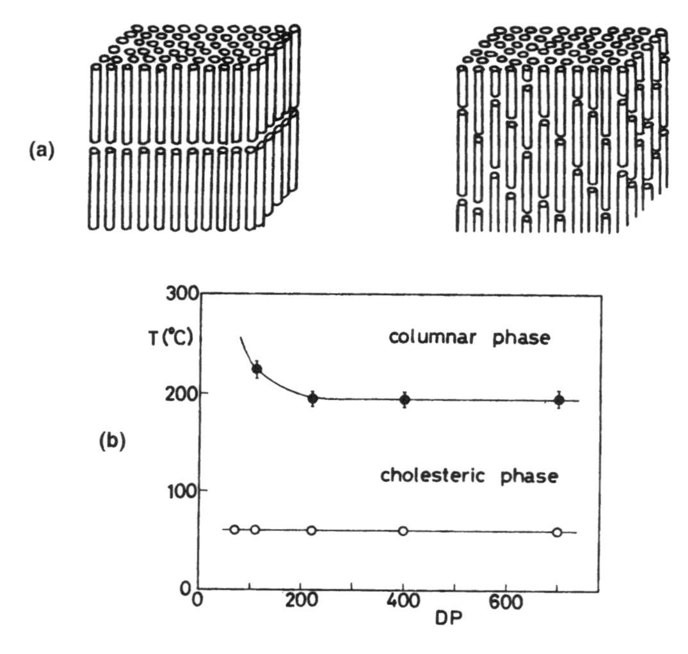

Figure 6 (a) Models for the smectic-B and hexagonal columnar phases of PγAG-18; (b) variation in the cholesteric-to-columnar transition temperature with the polymerization degree. (After Ref. 23.)

tion temperature when molecular weights fall below 20K (Fig. 6b). In fact the liquid-crystalline behavior of low molecular weight PγAG-18 appears to be much more complex [26]. Up to three types of textures were observed under the polarizing optical microscope for a 20K sample of this polymer as a function of temperature (Fig. 7). The low-temperature lamellar crystal phase is found to melt at 60°C into a biphasic wormlike texture composed of alternating dark and bright stripes. Above 110°C the polymer exhibits a pseudofocal conic texture which is attributed to a smectic A structure, and at 180°C a typical cholesteric phase is revealed by the appearance of the characteristic fingerprint pattern.

Stable liquid-crystallinity phases in the bulk may be also attained by combining short and medium size alkyl groups in the same polypeptide. As in the case of homopolypeptides, the flexible sidechains in the molten state play the role of the solvent in lyotropic compounds. Early observations on these systems were made on poly(γ-methyl-α,D-glutamate-*co*-γ-hexyl-α,D-glutamate)s [27] which are able to develop mesophases for comonomer compositions ranging from 30 to 70%. Similar results obtained in the study of other copolymers led to the conclusion that differences in length of the comonomer sidechains of at least five methylene units are required to induce such behavior. Other comblike random copolymers showing a similar interesting behavior are poly(γ-benzyl-α, D-glutamate-*co*-γ-dodecyl-α,D-glutamate)s [28,29]. An additional interest afforded by

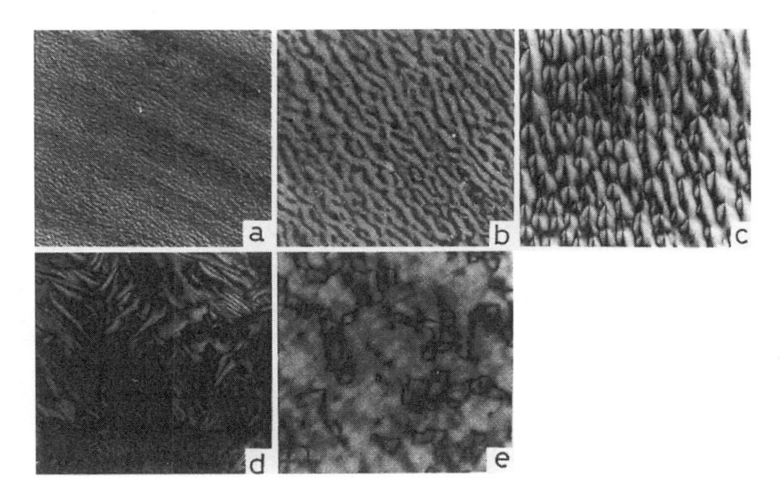

Figure 7 Liquid-crystal textures observed for PγAG-18 under the polarizing microscope as a function of temperature: (a) and (b) wormlike textures found in the 60–110°C range; (c) the pseudofocal conic texture observed at 110°C; (d) the transition phase occurring at 130°C; and (e) the cholesteric fingerprint texture observed at 180°C. (After Ref. 26.)

these systems is that introduction of irregularities under control in the sidechain wrap makes difficult or even suppresses sidechain crystallization upon solidification while the lamellar structure is retained [20].

2. Poly(α-alkyl-β,L-aspartate)s, (PαAA-n)

In the last few years sustained research has been done on the structure of poly(β-peptide)s derived from L-aspartic acid. This research benefits from the expertise accumulated in the investigation of comblike poly(α-peptide)s and has brought into evidence the close resemblance between the two families of polypeptides in both structure and properties. The research covers more than 20 poly(β,L-aspartic) esters differing either in the length or in the shape of the sidechain. All these compounds were prepared by ring-opening polymerization of optically active β-lactams and have molecular weights above 100K. The monomers were synthesized by a general methodology that makes use of cyclization and ester exchange reactions and starts from naturally occurring L-aspartic acid [30]. The most outstanding feature is that an α-helixlike conformation is adopted by the whole series both in solution and in the solid state [31,32]. As with PγAG-n the degree of ordering achieved in the packing of the helices and the model of assembly adopted by the poly(β-peptide) is determined by the length of the alkyl sidechain. This behavior is clearly characterized by X-ray diffraction of both isotropic and uniaxially oriented samples (Fig. 8). Consequently PαAA-n may be classified in three structural groups corresponding to alkyl sidechains of short ($n \leq 5$), medium ($n = 6$ and 8), and large ($n \geq 12$) sizes. The structural and conformational aspects of these three groups in connection with the supramolecular properties exhibited by each of them are discussed in the next paragraphs.

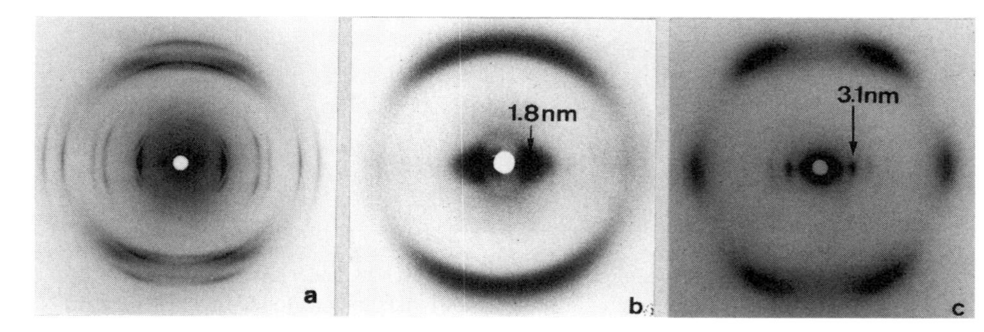

Figure 8 X-ray diffraction patterns of PαAA-n fibers: (a) short sidechain (PαAA-3); (b) medium sidechain (PαAA-6); (c) long sidechain (PαAA-18).

(a). Short Sidechain Systems. Alkyl esters of poly(β,L-aspartic acid) ranging from methyl to pentyl in the solid state tend to be arranged in three-dimensionally ordered arrays with chains in the α-helixlike conformation. Two crystal forms are known to prevail over a few others of minor importance that show only slight differences with respect to the former [33]. The main crystal form is a pseudohexagonal packing of 13/4 helices arranged in antiparallel fashion while the second major form is a tetragonal array of parallel 4/1 helices (Fig. 9). In both cases the structure consists of a tied assembly of packed cylinders with lattice positions

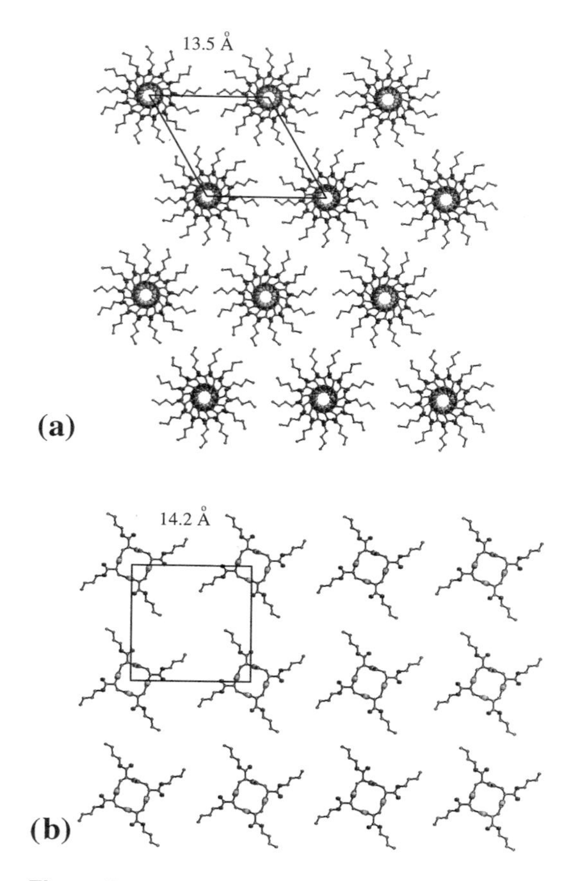

Figure 9 Projection along the chain axis of the crystal lattices of PαAA-4: (a) hexagonal form; (b) tetragonal form. Note that in (a) the helices are up and down whereas in (b) all the chains point in the same direction. (For clarity the lattice scale is exaggerated with respect to the motif.)

and radial orientations precisely determined by the interactions taking place between alkoxicarbonyl side groups.

In analogy with PγAG-n the conformation of PαAA-n in nonpolar solvents is helical and transition to the random coil state is promoted by addition of strong acids. Crystallization of the isobutyl derivative from chloroform dilute solutions yields hexagonal or square-shaped single crystals 200–400 Å thick with helices arranged normal to the surface of the crystals [34]. On the other hand, the slow evaporation from chloroform produces complex microcrystalline aggregates with morphological features suggestive of a supramolecular helical structure (Fig. 10a). Consistently, a continuous mesophase displaying the fingerprint pattern characteristic of cholesteric liquid crystals is generated upon storage of concentrated dichloromethane solutions (Fig. 10b) [35].

(b). Medium Size Sidechain Systems. PαAA-6 failed to crystallize in a three-dimensionally ordered structure. X-ray diffraction from solid films of this polymer indicated a rough hexagonal packing of 13/4 helices with an average interchain distance of 17 Å lacking azimuthal order [36]. Computer simulations based on molecular mechanics calculations revealed that helices tend to tilt with respect to the molecular axis when they are forced to approach each other in a hexagonal lattice with a parameter consistent with experimental data (Fig. 11a). Also the sidechain becomes distorted from the *all-trans* conformation with torsion angles adopting values as low as 120°. Therefore both calculations and experimental results support the conclusion that *n*-hexyl sidechains are too long to

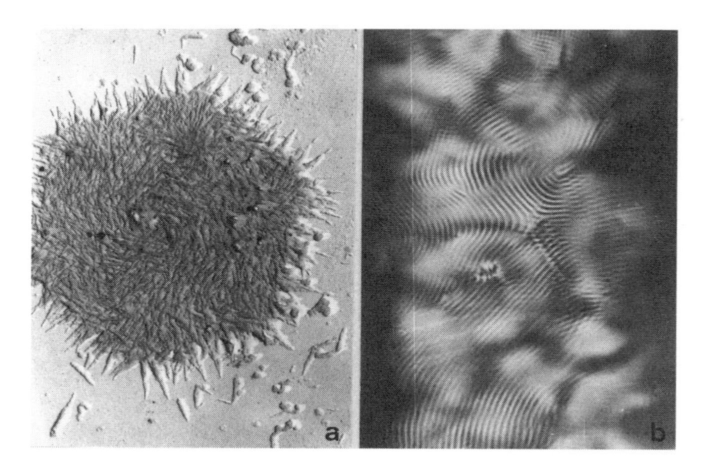

Figure 10 (a) Complex entities of poly(α-isobutyl-β,L-aspartate)s grown upon slow evaporation of a dilute solution in chloroform. (b) Fingerprint pattern displayed by a 23% (v/v) dichloromethane solution. (From Ref. 35.)

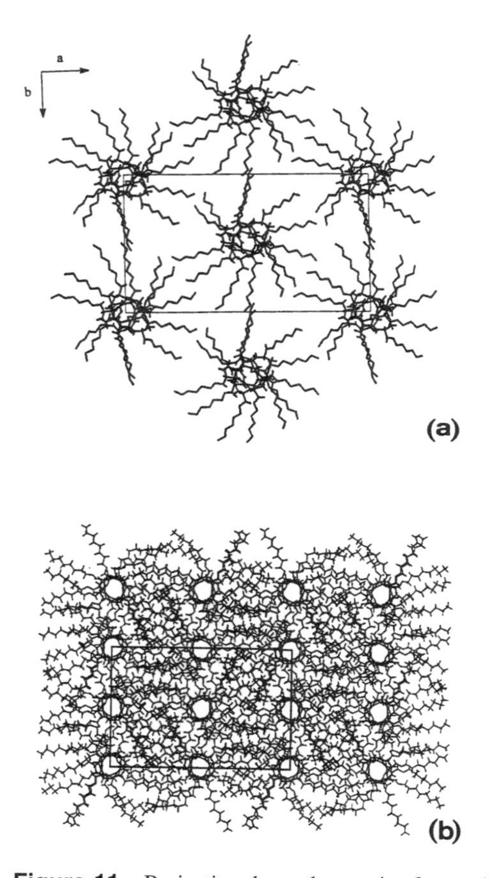

Figure 11 Projection down the c-axis of crystallized structures of medium size PαAA-n obtained by molecular mechanics calculations: (a) the hypothetically hexagonal structure of PαAA-6 (note that the helix is tilted about 30° with respect to the c-axis); (b) The crystal lattice model of PαAA-8 including 16 chains [note that overlapping of side chains is preferred along the direction normal to the layers (a-axis)]. (After Ref. 36.)

allow the polymer to crystallize in a three-dimensional lattice but too short to crystallize themselves in a separated phase.

The behavior of PαAA-8 was found to be remarkably different from that of PαAA-6 in spite of the fact that the alkyl sidechain is only two methylenes longer [36]. Casting from chloroform solutions produces a partially disordered structure (similar to that adopted by PαAA-6) that could be crystallized by annealing. The crystal structure consists of a layered arrangement of chains with an orthorhombic unit cell in the $P2_12_12$ space group. The intersheet distance is 18 Å and the distance separating the helices along the sheet is 12 Å. Model calculations showed that the sidechain trajectories in the interhelical space are almost fully extended and oriented normal to the helix sheet planes whereas those located within the sheets are folded. The degree of interdigitation into the molecular space of neighboring molecules greatly depends on the position of the residue along the 13/4 helix; whereas sidechains protruding normal to the sheet are intercalated by about 50% of their lengths; those coming out within the sheet plane show an almost negligible overlapping (Fig. 11b).

(c). Long Size Sidechain Systems. The combined DSC and X-ray analysis of PαAA-n with n ≥ 12 revealed the occurrence of two first-order transitions at temperatures T_1 and T_2 separating three structurally distinct phases [37]. Both transition temperatures and small-angle X-ray spacings characteristic of the phases decrease steadily with the number of methylenes in the sidechain (Fig. 12). Not only T_1 and T_2 values but also the enthalpy associated with transitions increases with the length of the alkyl sidechain. The low-temperature phase existing at $T < T_1$ is described as a layered structure with sidechains crystallized in a separated hexagonal lattice. The conformation of the sidechain in this phase is *all-trans* and the mainchain is in the familiar 13/4 helical conformation. The number of methylene units crystallized ranges from about 2 in PαAA-12 up to 12 in PαAA-22 representing in all cases a small fraction of the sidechains. The fact that sidechains prefer a hexagonal packing in the paraffinic crystallites over the monoclinic one usually adopted in poly(γ-glutamate)s is related to the restriction to chain mobility derived from the lacking of the flexible ethylene spacer linking the alkoxycarbonyl group to the mainchain. This feature, however, seems to have no significant influence on the molecular assembling of the helices.

T_1 is therefore attributed to the melting of the paraffinic phase composed of alkyl sidechains. The slight contraction of the intersheet distance observed for this transition is consistent with a small deviation of the C-C torsion angles from the *all-trans* conformation. Therefore the phase appearing above T_1 is regarded as a structure with the sidechains in a conformation not very different from that present at low temperatures. Solid-state NMR measurements gave strong support to these observations and corroborated that the α-helix conformation is retained by the polymer in this new phase. A detailed study carried out by DSC showed that crystallization of alkyl sidechains upon cooling from a temperature above T_1 does not obey the Avrami model [38], but proceeds instead at unexpectedly high initial rates. This fact can be explained by assuming that a very efficient self-nucleation is operating; such a mechanism is fully consistent with the arrangement accepted for the molten phase with sidechains located very near to their crystallized positions [39]. Further insight into the structure of the phase existing between T_1 and T_2 was provided by polarizing optical microscope observations. Strikingly beautiful colors changing from red to blue were displayed by

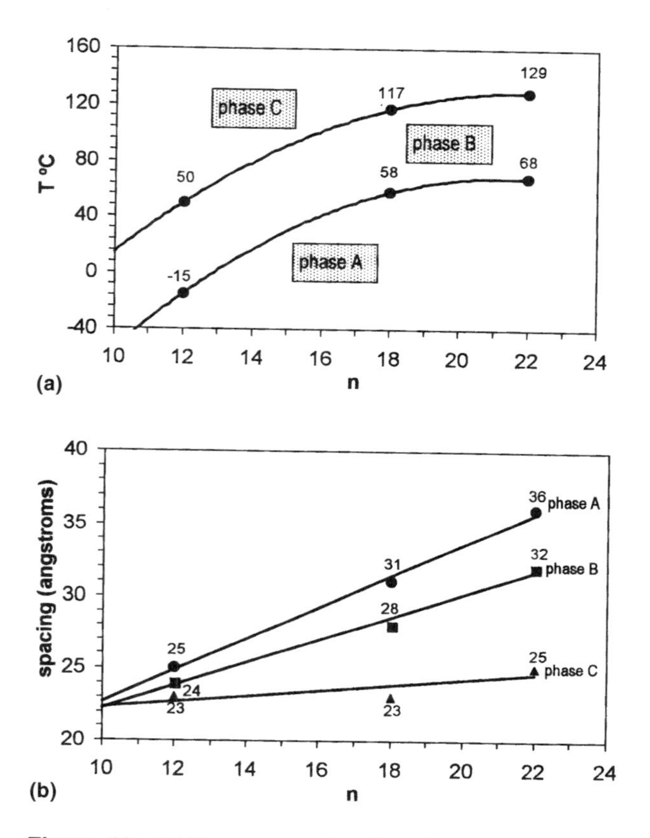

Figure 12 (a) Temperature domains of the three phases characterized in PαAA-n. (b) X-ray spacings of PαAA-n as a function of the number of carbon atoms contained in the alkyl sidechain. (Adapted from Ref. 37.)

uniaxially oriented films of PαAA-12 and PαAA-18 upon heating from T_1 up to T_2 to finally exhibit uniform white birefringence when the latter temperature was surpassed. Such a sequence of colors was found to reverse upon cooling. These effects are interpreted as due to selective reflection of circularly polarized light having a wavelength similar to the half-pitch of the supramolecular helical structure that gradually varies with temperatures [40].

The structural changes taking place at T_2 entail a considerable shortening in the periodicity of the lamellar structure. Differences with the low temperature phase ($T < T_1$) of about 11 Å for PαAA-18 and 8 Å for PαAA-16 are measured. Such differences imply that the layered structure is abandoned and helices are rearranged in a new phase characterized by an average interchain distance of

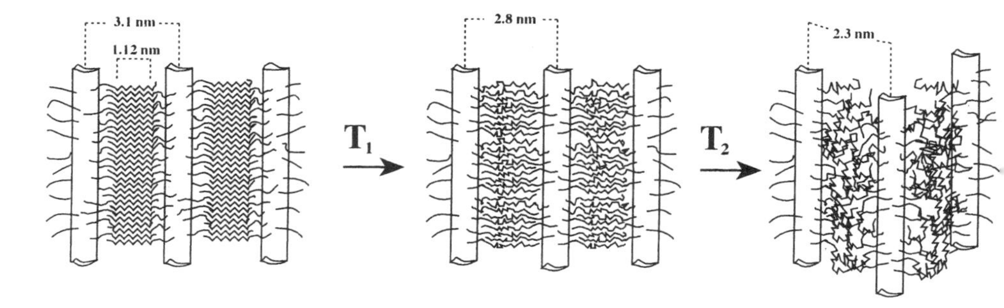

Figure 13 Schematic model illustrating the structural changes that take place in PαAA-n for n ≥ 12 upon heating. Both transitions are reversible with that occurring at low temperatures displaying a much faster switching. Distances are specified for the particular case of PαAA-18. (After Ref. 51.)

23–25 Å. In this high temperature, phase sidechains are probably in the random coil conformation allowing the mainchain helices to be loosely positioned in the space.

In summary, three structural phases differing in both the arrangement of the mainchain helical rods and the conformation of the sidechain are taken up by PαALA-n with n ≥ 12 as a function of temperature (Fig. 13). The phase existing between T_1 and T_2 is assumed to be cholesteric with the ability to crystallize upon cooling below T_1 and to convert into a nematic phase upon heating above T_2. Since the layered phase existing at low temperatures may be envisioned as a quenched smectic structure with helices immobilized by the crystallized sidechains, the structural transitions occurring in these systems appear to follow a smectic ↔ cholesteric ↔ nematic sequence.

B. Ionically Attached Long Alkyl Sidechain Polypeptides

Assemblies bonded by means of noncovalent interactions, such as hydrogen bonding or charge transfer, offer a number of advantages that make these systems particularly interesting as building blocks for generating supramolecular structures [41]. The preparation of such systems does not involve complicated chemical procedures, structure equilibrium may be realized, and rearrangement of the structure is feasible by adequate adjustment of the external conditions. Nevertheless it is remarkable that these assemblies are able to reproduce the mesophases known to exist in covalently linked comblike polymers. In particular, the superimposition of electrostatic free energy to soft and hard interactions characteristic of uncharged rodlike polymers gives rise to supramolecular arrangements of relevance to self-assembly of complex biological structures [42].

Formation of complexes from two polyelectrolytes of complementary charge has been recently reviewed [43]. Systems investigated so far include natural compounds such as polysaccharides as well as other synthetic pairs. Cell-like thin spherical membranes and other shapes displaying higher anysotropy can be also fabricated by selecting the adequate techniques. Nevertheless, the most studied self-assembling polymeric systems based on electrostatic interactions are complexes made of synthetic polyelectrolytes and oppositely charged low molecular weight amphiphilic molecules (see Chapter 11). The recent discovery of the ability of these systems to dissolve in organic solvents and spontaneously form well-defined supramolecular structures has motivated broad interest in this area. Today it is known that electrostatic complexes may generate stable assemblies with lamellar or cylindrical shape depending on the charge density of the polymer and the chemical nature of both the polymer and the surfactant [44].

Low molecular weight surfactants are known to form spherical micelles at the critical concentration. On the other hand complexes of different architecture (Fig. 14) are formed by coupling surfactant with polyelectrolytes according to the conformation assumed by the polymer [45]. Flexible polymer chains and surfactant molecules in solution are envisioned as discrete clusters of surfactant enclosed by polymer loops. In the case of rigid polymers the basic micelle is assumed to consist of a cylindrical assembly with ionically bound molecules aligned perpendicular to the axis of the polymer cylinder. These basic units may self-aggregate either longitudinally or side by side to yield nematic phases and gels. The assembly can be maintained in solution due to the presence of uncom-

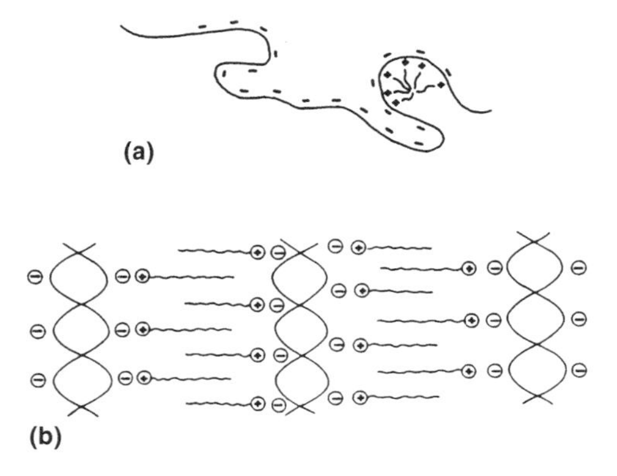

Figure 14 Scheme of polyelectrolyte–surfactant complexes in solution: (a) flexible polyelectrolyte chain; (b) rigid polyelectrolyte chain. (After Ref. 45.)

pensated charges at the cylinder skin. The regular distribution of rigid (polymer chain) and flexible (surfactant chain) components reproduces the arrangement found in both high-performance composites and biological complexes where highly oriented polypeptide or polysaccharide chains are embedded in an amorphous matrix of crosslinked proteins.

Complexes of α-helical polypeptides and oppositely charged low molecular weight surfactants constitute a novel class of comblike assemblies able to form lamellar structures in the solid state similar to those observed for covalently linked flexible-rigid polymers. Complexes of poly(γ,L-glutamic acid) [46,47] and poly(L-lysine) [48,49] with surfactant of opposite charge (Fig. 15) have been recently investigated by Tirrell's group. A brief account of the most remarkable features of the structures displayed by these complexes is given in the next two sections.

1. Poly(α,L-Glutamic Acid)-Alkyltrimethylammonium Cation Complexes

Stoichiometric ionic complexes of poly(γ,L-glutamic acid) and alkyltrimethylammonium cations with the long linear alkyl group being dodecyl, hexadecyl, and octadecyl were obtained as solid precipitates when equal amounts of aqueous solutions of sodium poly(γ,L-glutamate) and surfactant were mixed at room temperature at pH 8 [46,47]. The analysis of solid films of these complexes by FTIR and CD indicated that the polypeptide chain is predominantly in the α-helical conformation and low angle X-ray diffraction revealed that all the complexes adopt a lamellar structure made of alternating layers of polypeptide and surfactant (Fig. 16). The shorter surfactant alkyl chains containing 12 and 16 carbon atoms are disordered in the complex whereas the surfactant chain of 18 carbon atoms is crystallized in a hexagonal lattice.

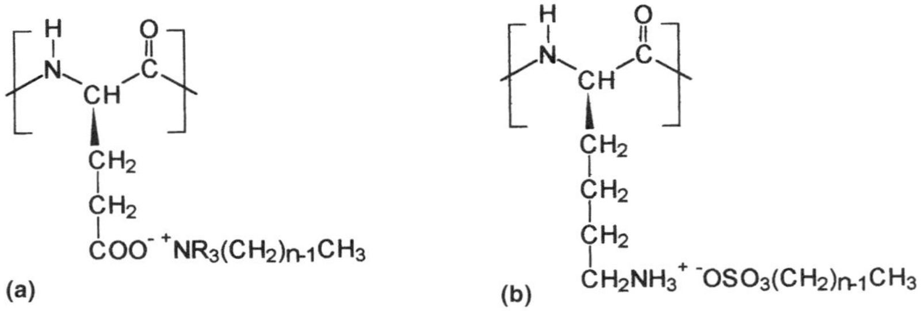

(a) $COO^- \; ^+NR_3(CH_2)_{n-1}CH_3$

(b) $CH_2NH_3^+ \; ^-OSO_3(CH_2)_{n-1}CH_3$

Figure 15 Chemical structures of (a) Poly(α,L-glutamic acid)-alkyltrimethylammonium cation complexes and (b) Poly(L-lysine)-alkyl sulfate anion complexes.

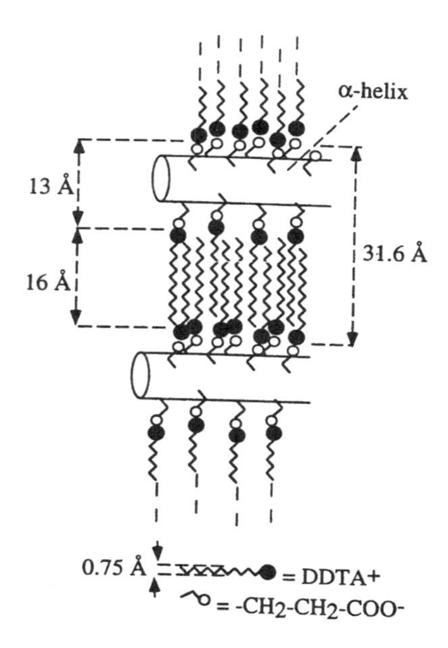

Figure 16 Scheme of the lamellar structure of the poly(γ,L-glutamic acid)-dodecyltrimethylammonium complexes viewed parallel to the layers and perpendicular to the α-helices axes. (After Ref. 46.)

A linear dependence of the long period of the lamellae on the number of carbon atoms in the surfactant chain was found. The slope of the line is about 1.3 Å per methylene indicating that surfactant chains are nearly fully extended, interdigitated, and oriented nearly normal to the lamellar surfaces. The paraffin crystallites of the octadecyltrimethylammonium complex melt at 48°C with an associated fusion heat of about 2.1 kcal · mol^{-1}. No other thermal transition was found to take place in the temperature range 10–170°C. The complexes do not show ordered melt and do not flow upon heating at temperatures below decomposition. They undergo irreversible changes in structure and properties upon storage, a process that is accompanied by generation of a small amount of crystalline surfactant. Such alteration is assumed to be due to unbinding of the alkyltrimethylammonium cations caused by uncontrolled hydrolysis.

2. Poly(L-Lysine)-Alkyl Sulfate Complexes

Stoichiometric complexes of poly(L-lysine) cations and dodecyl sulfate anions were investigated by Ponomarenko et al. [48,49]. The complexes are prepared by the same method used in the preparation of poly(γ,L-glutamic acid)-alkyltrimethylammonium complexes. The secondary structure adopted by the polypeptide in these complexes was found to be very susceptible to environmental conditions. In chloroform solution containing 1–2% (v/v) trifluoroacetic acid (TFA) the polypeptide chain is in the α-helical conformation whereas a disordered state is present at TFA concentrations above 5%. It was interesting to observe that a noticeable decrease in the ^{1}H spin-lattice relaxation time accompanied the helix-coil transition whereas no changes were detected in T_1 of protons attached to the surfactant counterparts. It was concluded therefore that in both states the surfactant alkyl chains are nearly extended and exposed to the solvent shielding both the polypeptide backbone and the joining ionic groups (Fig. 17).

In the solid state the poly(L-lysine)-dodecylsulfate complex is invariably organized in a lamellar structure consisting of layers of polypeptide chains alternating with double layers of surfactant molecules arranged tail to tail. However, both the secondary structure adopted by the polypeptide and the packing of the polymethylene chains in the paraffinic phase depend upon the history of the sample. In the powder isolated from synthesis, the structure of the complex is predominantly of β-sheet with sidechains in a liquidlike state characterized by only short-range order. On the contrary, films cast from a chloroform solution in which the polypeptide is in the disordered state exhibit an α-helix conformation of the mainchain with sidechains also uncrystallized but with the mobility more restricted than in the former case.

When a mixture of ethyl and octadecyl sulfates are mixed with poly(L-lysine) hydrobromide, the complex that is formed contains exclusively octadecyl sulfate anions. The selective binding of the long surfactant chains is attributed to the hydrophobic driving forces that are the determinant for micellization. Con-

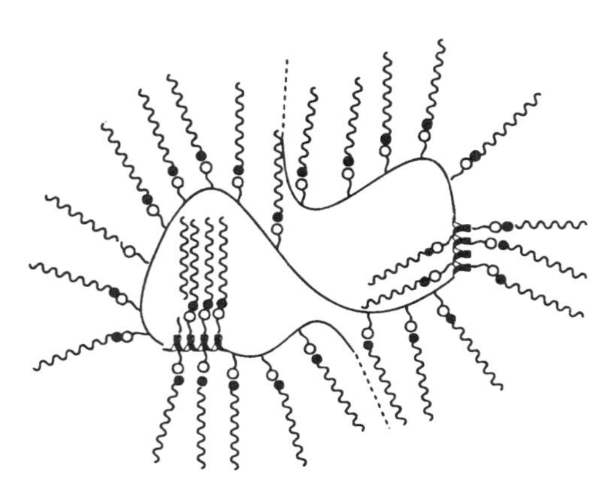

Figure 17 Schematic representation of the structure of the poly(L-lysine)-dodecyl sulfate complex with the polypeptide chain mainly in the coil conformation. (After Ref. 49.)

versely, complexes with the composition reflecting the composition of the mixture were obtained when two long chain surfactants were used. By this means it was feasible to prepare complexes of poly(L-lysine) with either octyl or octadecyl sulfates, and with mixtures of both surfactants for a variety of compositions. The method turns to be unique in providing ionically bonded comblike copolymers with a distribution predominantly in blocks. It seems that the hydrophobic self-association of the surfactants is combined with the ability of the electrostatically attached sidechains to move along the mainchain. The microstructure of these copolymers is therefore distinguished from covalently linked copoly(γ-alkyl-α,L-glutamate)s in which the attached sidechains invariably have a random distribution.

In agreement with other covalently linked systems previously discussed, the electrostatically bound supramolecular complexes of poly(L-lysine)-surfactant complexes adopt lamellar structures too, although the polypeptide is predominantly in the β-sheet conformation. A striking feature of these mixed systems is that the molecular organization in the supramolecular lamellar structure is governed not only by the surfactant chain length but also by the composition of the complex (Fig. 18). Thus the assembly generated by the poly(L-lysine)-octadecyl sulfate complex has sidechains crystallized in a hexagonal lattice with the chains interdigitated and oriented perpendicular to the sheet plane. Conversely, in the poly(L-lysine)-octyl sulfate complex, the chains are in the disordered state and arranged tail to tail. In complexes with mixed surfactants the lamellar organization approaches one or the other extreme according to composition. Assemblies containing more than 20% of octadecyl chains adopt a partially crystalline struc-

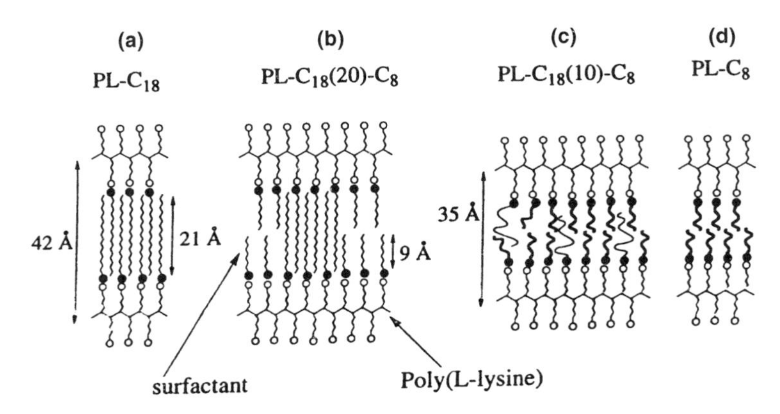

Figure 18 Arrangement of the surfactant molecules within the lamellae of complexes of poly(L-lysine)-alkyl sulfates containing mixed alkyl chains. Figures in brackets indicate the composition of the system. (After Ref. 49.)

ture similar to the poly(L-lysine)-octadecyl sulfate complex, whereas a disordered arrangement is preferred at lower contents.

III. BLOCK COPOLYMERS: ROD–COIL MAINCHAIN POLYMERS

Even though most experimental studies on the behavior of rod–coil block copolymers just occurred in the last two decades, previously a lot of theoretical work had been done for this interesting class of polymers. In one of the earliest theoretical works Semenov and Vasilenko introduced a microscopic model that accounts for the steric interactions among the rods, the stretching of the coils, and the unfavorable rod–coil interactions. They used the so-called strong segregation limit approach. Theories that deal with block copolymer phase behavior can be divided into two categories: strong segregation limit (SSL; Flory–Huggins parameter ($\chi N \gg 10$); and weak segregation limit (WSL, $\chi N \leq 10$). In the SSL theories well-developed microdomain structures are assumed to occur with relatively sharp interfaces as well as explicitly account for chain stretching. The WSL theories are premised on lower amplitude sinusoidal composition profiles and unperturbed Gaussian coils (neglected chain stretching) [50]. Assuming that the rods are strictly aligned in the direction normal to a lamellarlike assembly, Semenov and Vasilenko considered a nematic phase and a smectic-A lamellar phase where the rods remain perpendicular to the built sheets. The transition from a monolayer to a bilayer lamellar phase was also analyzed [51]. Semenov extended the work including smectic-C phases, with the possibility for the rods to tilt by an angle θ to the lamellar normal [52]. In a subsequent work multiblock linear macromolecules were considered. The authors constructed a theory for nematic–smectic and smectic-A-smectic-C transitions in the melt [53]. Williams and Fredrickson extended these calculations and incorporated nonlamellar phases where the rods form finite-sized disks (‘‘hockey pucks’’) surrounded by a corona of coil molecules. These pucks are predicted to pack and form a three-dimensional superstructure [54]. In the weak segregation limit spatially uniform planes of diblocks had been predicted by Schick and Holyst [55]. By varying the stiffness of the rod component the effect on the lamellar phases had been studied by Matsen [56]. Schick et al. introduced anisotropic interactions between the semiflexible blocks and further studied the effects of external fields [57]. They also studied spatially ordered phases in a rod–coil diblock copolymer melt by applying self-consistent field equations in the weak segregation limit or with a brushlike approximation in the strong segregation limit [58]. In a recent work Matsen and Barrett extended the Semenov–Vasilenko model by analyzing it with self-consistent field techniques for lamellar morphologies [59].

The phase behavior in solution of rod–coil polymers has been analyzed by Halperin [60,61] and by de Gennes and Raphael [62]. By incorporating enthalphic contributions from the solvent, ''tilted'' phases were predicted for copolymers in a selective solvent for the coil component by Halperin (see Chapter 3). Microdomains resembling ''plates,'' ''fences,'' and ''needles'' were instead considered by de Gennes. Sevick and Williams examined the morphology of rod–coil polymers grafted to repulsive surfaces by the end of the flexible segment in poor solvents. The predictions had to be altered in this case because the lack of flexibility of the rod component becomes more significant as the dimensionality is reduced from three (bulk) to two (surface). The micelles forming on a surface were predicted to be turnip and jellyfish-like [63].

Unfortunately not much work has yet been done on the solution properties of rod–coil systems in a concentration range spanning from micellar to liquid-crystal formation. In contrast to this many authors reported the bulk properties of their copolymer materials. Because of the complex nature of these systems it is not always easy to explain the observed bulk properties. The formation of the reported morphologies could certainly be better understood if there were a more thorough work on the solution properties of the related materials.

A. Helical-Rod Systems

1. Copolymers with Polypeptides

Polypeptides are classical examples of polymer molecules whose rigid regular conformation is stabilized by strong intramolecular hydrogen bonds. They can be found in a wide range of conformations from which the two major ones are a right-handed α-helix and a sheet structure. The α-helix is a rodlike structure in which the polypeptide mainchain is coiled and forms the inner part of the rod; the sidechains extend outward in a helical array. The helix is stabilized by hydrogen bonding between the NH and CO groups of the mainchain. The β-pleated sheet conformation differs markedly from the α-helix because it behaves as a sheet rather than a rod. Copolymers that incorporate polypeptide segments are not only attractive because of their rod–coil behavior but also as biocompatible and biodegradable materials which result from their natural or nature-identical component.

Gallot et al. reported on polybutadiene-block-poly(benzyl-L-glutamate) block copolymers that were synthesized by polymerization of the N-carboxyanhydrides of the corresponding aminoacids from aminoterminated polybutadiene [64]. (See Fig. 19 for an exemplary synthetic route.) From this polymer the authors also obtained polybutadiene-block-poly(N^5-hydroxypropylglutamine) by the reaction with 3-amino-1-propanol. By studying the resulting block copolymers by X-ray diffraction the authors found lamellar structures for the wide range

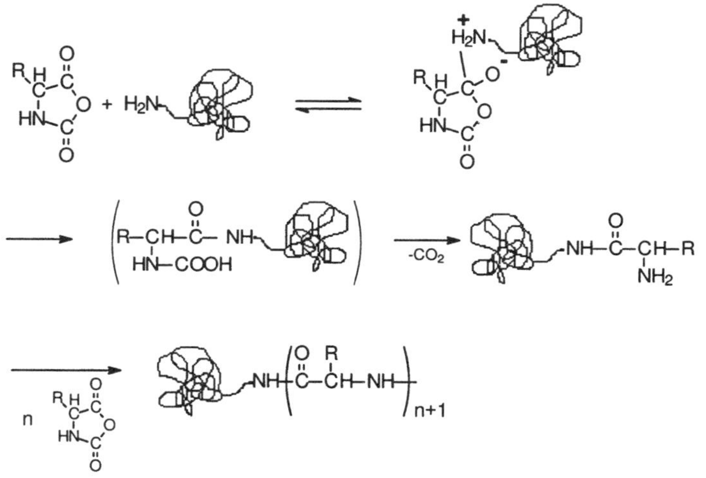

Figure 19 Exemplary synthetic scheme for the polymerization of N-carboxyanhydrides from aminofunctionalized polymers.

of peptide content of 20 to 75%. This result was also confirmed by studying film samples with electron microscopy. For that purpose mesomorphic gels were prepared in 2,3-dichloro-1-propene and then subjected to UV irradiation, ultramicrotomed, and stained by O_sO_4. The structural parameters obtained by both methods are in good agreement with one another. By analyzing the samples with X-rays in an angle domain corresponding to the periodicity between 3 and 50 Å the authors found that the polypeptide chains are in the α-conformation and packed in a hexagonal array. Furthermore they are folded into sheets showing a very uniform lamellar thickness even though the polypeptide blocks have a high polydispersity. The authors proposed a interdigitated (where the chains fold after crossing all the thickness of the polypeptide layer) and a bilayer arrangement (see Fig. 20) whereas the first assembly seems to be favored by energetical considerations.

Polystyrene-block-poly(L-lysine), polybutadiene-block-poly(L-lysine) [65], and block copolymers of polystyrene or polybutadiene with poly(carboxybenzoxy-L-lysine) [66] were also synthesized by Gallot et al. and evidence for lamellar assemblies was presented. Polysaccharide-block-polypeptide rod–coil systems were obtained by polymerizing the N-carboxyanhydrides of γ-benzyl-L-glutamate from the asparagine α-amino function of carbohydrate fractions of ovomucoid (a glycoprotein extracted from hen egg white). Lamellar assemblies in concentrated solutions and in the dry state were again observed [67]. The chain folding of the polypeptides in the lamellar structures was studied for these block

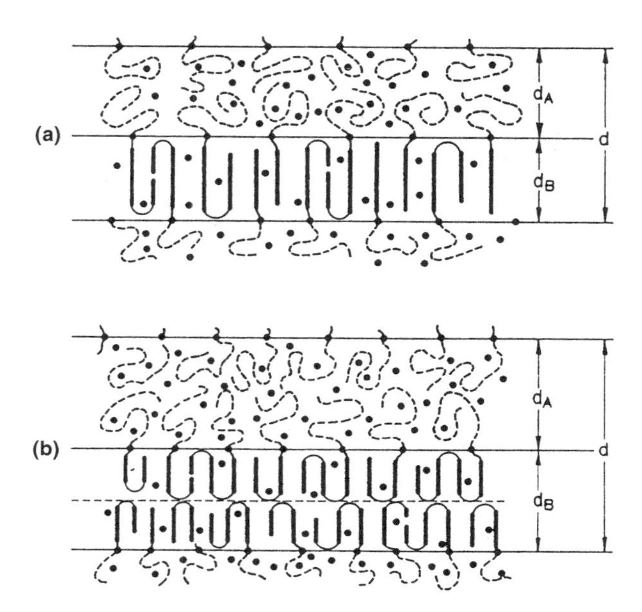

Figure 20 Schematic representation of the lamellar structure of block copolymers based on polypeptides as proposed by Gallot et al.: (a) the polypeptide chains fold after crossing all the thickness of the polypeptide layer and (b) the polypeptide chains fold after crossing only half the thickness of the layer. (Reprinted with permission from Ref. 64, Copyright 1976, Wiley-VCH.)

copolymers with X-ray diffraction and IR-spectroscopy. By studying systems with different vinyl polymers and different polypeptides in varying compositions they found that the number of folds depends upon the nature of the two blocks; for instance, in block copolymers with polystyrene blocks the number of folds of the polypeptide chains is greater than with polybutadiene blocks. Also poly-L-lysine seems to be more rigid than, for example, poly-γ-L-glutamate which is represented in the number of folds [68]. Gervais et al. used XPS spectroscopy to study the surface properties of polypeptide-block-polybutadiene-block-polypeptide triblock copolymers and found that the lamellar structures (which is as well observed by X-ray diffraction for these triblock systems) are perpendicular to the air–polymer interface. For more hydrophilic polypeptides they found that the surface is richer with polybutadiene whereas for hydrophobic polypeptides the surface resembled the block copolymer composition [69].

Poly(γ-benzyl-L-glutamate)-block-polyisoprene diblock copolymers were reported by Tirrell et al. [70]. The synthesis was based on coupling of carboxy-terminated polyisoprene and amino-terminated poly(γ-benzyl-L-glutamate) prepolymers. By dynamic light-scattering studies in dichloromethane they found that aggregates were observed in less than 24 h and remained stable over several days. In DMF two relaxation modes were observed that the authors assigned to translational diffusion of the micelles (slow mode) and to moleculary dispersed diblock molecules (fast mode). Since the radius of the sphere in DMF was found to be too large as to be assigned to a simple micelle, the authors proposed a multilayered complex sphere built by a simple sphere and successive diblock layer adsorption. By studying the system in a selective solvent for the rod segment (DMF) the observed micelles increased in size if the coil segment was extended. In contrast to this an extension of the coil segment does not influence the aggregate in dichloromethane.

Hayashi et al. [71] reported on A-B-A type block copolymers consisting of poly(γ-benzyl-L-glutamate) as the A component and polybutadiene as the B component prepared by polymerization of γ-benzyl-L-glutamate N-carboxyanhydrides from bisamino-terminated trans-polybutadiene. Circular dichroism spectroscopy in 1,2-dichloroethane and infrared spectroscopy in the solid state showed that the peptide chains form an α-helix. In the solid state the polybutadiene chains were in a random coil conformation forming cylindrical domains embedded in the polypeptide matrix phase, as shown by wide-angle X-ray diffraction and electron microscopy. The temperature dependence of the dynamic and loss moduli of the triblocks showed that the dynamic mechanical properties can be explained by the observed microseparated structure. By studying block copolymers with poly(γ-benzyl-DL-glutamate) the helical content of the D,L-copolymers in a helicogenic solvent and in solid state was estimated on the basis of ORD and IR spectra. The content of the right-handed α-helices and left-handed α-helices and random coil conformations of the polypeptide chains was determined. From wide-angle X-ray diffraction measurements it was shown that the hexagonal crystalline phase almost disappeared for D,L-copolymer membranes indicating a breakdown of the α-helical conformation in the mainchains. The morphological structure of these block copolymer membranes was studied with electron microscopy and the D,L-block copolymer membranes were found to build spherical structures when the volume fractions of the polybutadiene portion were rather high [72].

The authors also reported on A-B-A type block copolymers composed of poly(γ-benzyl-L-glutamate) as the A component and polyisoprene as the B component. By using wide-angle X-ray diffraction it could be shown that the block copolymers exhibit mesophase behavior in different solvents. In this case polyisoprene chains are in a random coil conformation and form domains embedded in the matrix phase consisting of poly(γ-benzyl-L-glutamate) chains in the α-helix conformation. Model analysis of the complex modulus of the membrane cast from solution suggested the occurrence of spherical and cylindrical domain structures in the membrane [73]. Morphological study of these triblock copolymers using pulsed nuclear magnetic resonance spectroscopy and electron microscopy revealed that the solid block copolymers are in microphase separated state. The NMR measurement was used to detect the interface of the block copolymer. Spin-lattice relaxation time T1 of the block copolymer revealed a domain size that is in good agreement with the result obtained from the electron microscopy [74].

The authors also analyzed the membrane surfaces of A-B-A triblock copolymers containing poly(ε-N-benzyloxycarbonyl-L-lysine) as the A component and polybutadiene as the B component. The surface characteristics of membranes built from the triblock were investigated by XPS, replication-electron microscopy, and wettability measurements. By investigating the surface with XPS measurements, the composition of the outermost surface proved to be quite different from the bulk composition; contact-angle measurements indicated the existence of an interfacial region between the α-helical polypeptide domains and the polybutadiene domains at the surfaces of the triblock membranes [75]. The authors found that the permeability of membranes prepared from their triblock systems was much higher than that of the polypeptide homopolymer and increased in proportion to the interfacial area between the domains. This suggests that residues near the end of the peptide chain and terminal residues of the diene block, which are located in the interfacial region between the domains, are responsible for water permeability, since the amine and carbonyl groups do not form intramolecular hydrogen bonds (see Fig. 21 for a schematic drawing of a membrane built from a diblock copolymer) [76]. The authors also took very great care studying the biocompatibility, biodegradability, and the membrane efficiency of their polymers [77–82].

Another type of triblock system was reported by Anderson et al. [83] who synthesized copolymers of the A-B-A type where A is poly(γ-benzyl-L-glutamate) and B is poly(butadiene/acrylonitrile). The authors studied their polymers cast from selective solvents by electron microscopy. In films prepared from dioxane (selective for the polypeptide block) they observed a lamellar structure whereas in films cast from chloroform (not preferential for either block) a nearly homogeneous phase mix became obvious. An interesting time-dependency of the morphologies was found in films processed from xylene: films from freshly prepared solutions showed a well-defined lamellar morphology whereas with films cast from solutions that had been stored for a minimum of two days a poorer domain formation occurred. The alternation of microphase structure may be due to gelation resulting in aggregation of the polypeptide helices.

2. Copolymers with Polyisocyanates, Polyisocyanide, and Polycarbodiimides

In contrast to polypeptides the conformational properties and chain rigidity of polyisocyanates, polyisocyanide, and polycarbodiimide is determined by their

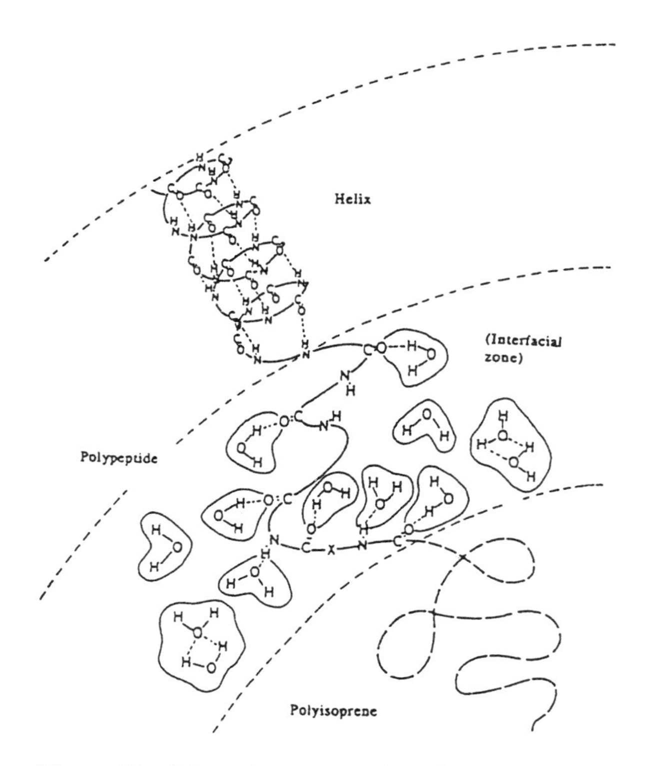

Figure 21 Schematic representation of a membrane formed by a polyisoprene-block-polypeptide diblock copolymer. In the interfacial zone (which is thought to be the main structural property that allows hydraulic permeability in this system) hydrogen-bonded water and clusters of water are drawn. (Reprinted from Ref. 78, Copyright 1996, with permission from Elsevier Science.)

primary chemical structure rather than by intramolecular hydrogen bonding. This results in a greater stability in the conformational characteristics in solution as compared to polypeptides. Furthermore it is possible to achieve high molecular weights in these systems and with certain synthetic methods (anionic synthesis, transition metal catalysis, etc.) even the polydispersity can be reduced, a reason that makes block copolymers based on these rigid segments very attractive.

Ober and Thomas et al. [84,85] reported on diblock copolymers containing polystyrene and polyhexylisocyanate, respectively, as the coil and the rodlike components (see Fig. 22a). The copolymers were synthesized anionically in a solvent mixture of THF and toluene and initiated by n-buthyllithium. In their first studies of this system the authors used a copolymer with a total molecular

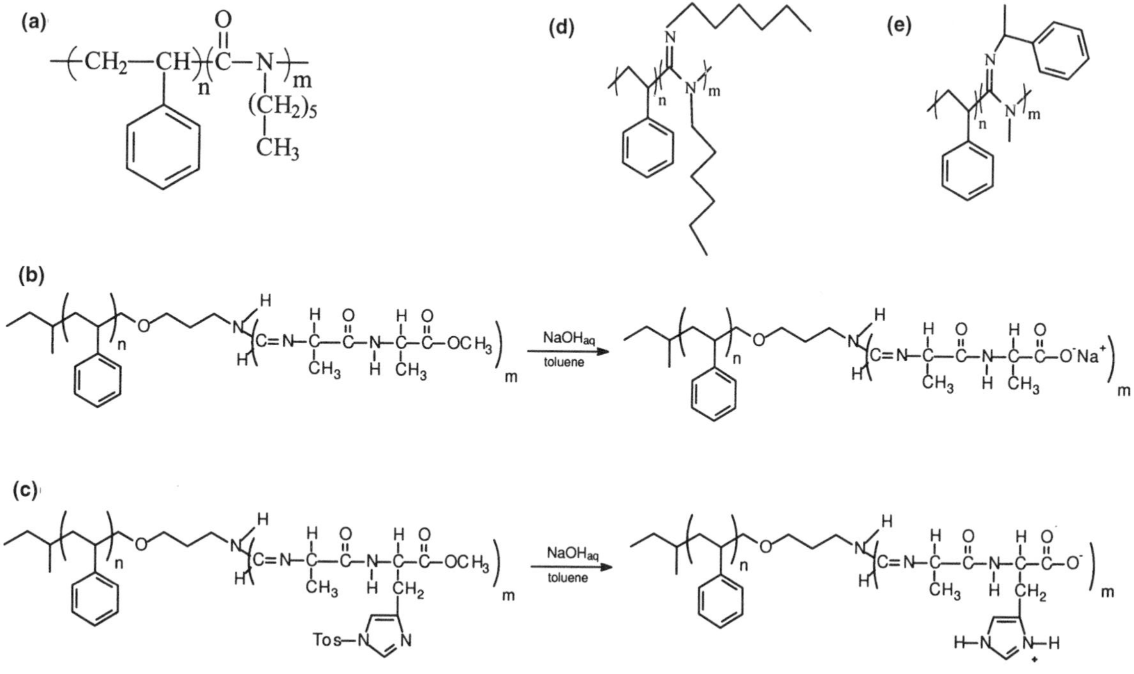

Figure 22 Chemical structures of (a) poly(hexylisocyanate); (d) poly(N,N′-di-*n*-hexylcarbodiimide); and (e) poly[(R/S)-N-(1-phenylethyl)-N′-methylcarbodiimide)]. (b) and (c) show the synthetic scheme of the removal of the ester functions of poly(isocyano-L-alanine-L-alanine) and poly(isocyano-L-alanine-L-histidine), respectively.

weight of 135.000 g/mol and a weight fraction of polystyrene of 0.2; the total polydispersity was 1.36. By studying concentrated (>15%) solutions of the block copolymer by optical polarized microscopy it became obvious that the solution contained liquid-crystalline regions. If the solution was sheared with a razor blade and allowed to relax a banded texture was seen with the observed strips perpendicular to the shearing direction. By studying the samples with transmission electron microscopy (thin films cast from toluene on carbon support; stained by RuO_4) a zigzag morphology was observed with a very high long-range order and a high degree of spatial correlation between adjacent layers (see Fig. 23a). The domain spacing (as monitored by TEM analysis) of the polystyrene-block was ~250 Å and of the polyhexylisocyanate domain ~1800 Å. A wide-angle electron diffraction pattern showed that the polyhexylisocyanate domains are crystalline showing a very high orientational order. The high number of reflections pointed out a high degree of crystallinity. For wide-angle X-ray scattering a polymorphism for the crystal data of the polyhexylisocyanate domain became obvious even though most of the WAXS data suggested hexagonal packing. This result was in accordance with the data obtained for pure polyhexylisocyanate. By analyzing the intensity for each layer in the electron diffraction pattern a 8_3 or 8_5 helix with a 1.95 Å translation and a rotation of 135° per monomeric unit along the c axis was suggested. The electron diffraction pattern also showed that the polyhexylisocyanate rods are oriented parallel to the pointing direction of the zigzags. The zigzag morphology is schematically shown in Fig. 23d either for a case where the rods are stacked in an interdigitated fashion or as a bilayer model, respectively. Calculation of the rod domain size for a copolymer with this composition suggests that the interdigitated model offers a better representation of the reported data. The authors also emphasized the tilting in the zigzag morphology in line with theoretical predictions of Halperin by judging the effects of the appropriate interfacial and deformational forces. The results they obtained are in good agreement with the observed morphology. The effect of changing the casting solvent on the morphology is quite dramatic. In the case of CCl_4 the authors observed fragmented polystyrene zigzags surrounded by a crystalline polyhexylisocyanate matrix with actually no long-range order. The morphology also revealed micellar-like assemblies having polystyrene as the core and polyhexylisocyanate as the corona (see Fig. 23c). By casting from $CHCl_3$ short zigzags were observable which showed a little more long-range order than in the CCl_4 case. This morphology also includes micellar polystyrene inclusions (see Fig. 23b).

The authors proposed a different mechanism for the formation of the above morphologies. During toluene casting the solution seems to separate into a solvent-rich phase and a nematic copolymer-rich phase. After further evaporation, a second transition to a smectic phase appears to occur and because toluene is a better solvent for the polystyrene block, the coil part can stretch and the zigzag morphology can build by tilted assembly of the rods and eventual ‘‘locking’’ of

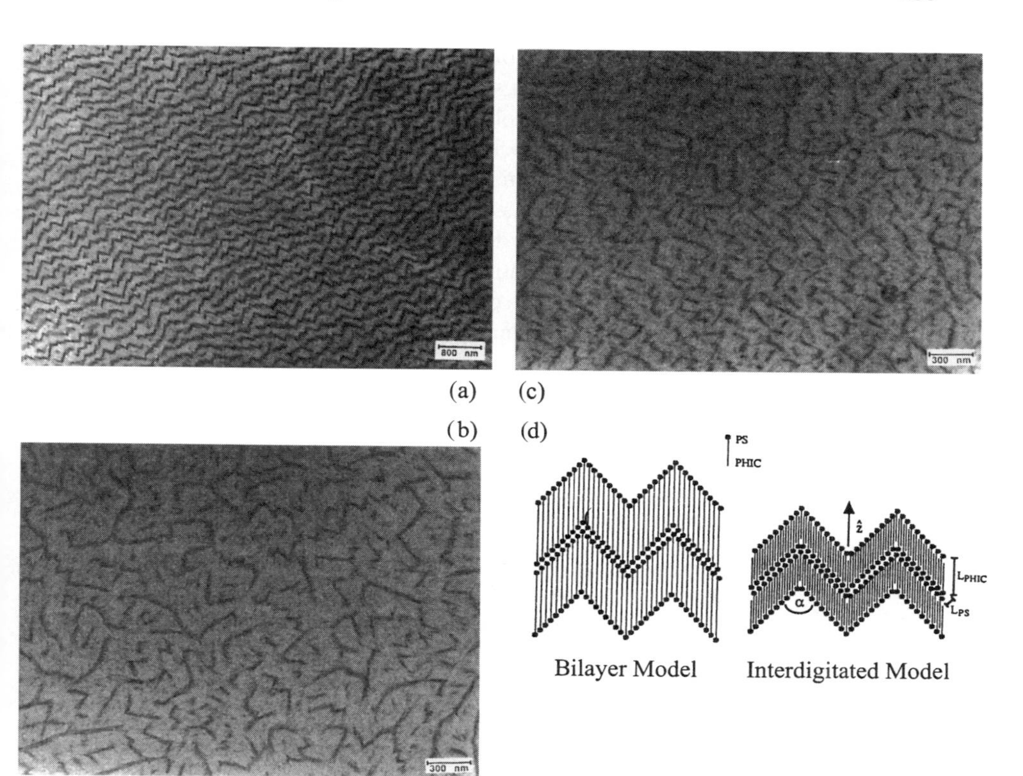

Figure 23 Bright field TEM images of a polyhexylisocyanate-block-polystyrene block copolymer with a molecular weight of 135.000 g/mol and a polystyrene weight fraction of 0.2 (thin films; stained with RuO_4): (a) cast from toluene; (b) caste from $CHCl_3$; and (c) CCl_4. Whereas (a) shows a zigzag morphology with very high long-range order, this order is lost for casting from $CHCl_3$ (b), and even more from CCl_4 (c). The proposed packing arrangements (for the zigzag morphology) either in an interdigitated or a bilayer fashion are schematically drawn in (d). (Reprinted with permission from Ref. 84. Copyright 1995, American Chemical Society.)

the structure by crystallization of the polyhexylisocyanate. The nonpolar CCl_4 is a better solvent for the polyhexylisocyanate and by evaporation the swelling of the polyhexylisocyanate mostly hinders the formation of a long-range structure. This also explains the micellar assembly of the polystyrene domains in this morphology. The morphology obtained by casting from $CHCl_3$ is an intermediate between that obtained from toluene and CCl_4 because $CHCl_3$ is not preferential to either block.

Ober and Thomas also reported on the effect of varying composition of the block copolymers. A wavy lamellar structure was observed for a copolymer with a molecular weight of 104.000 g/mol for the polystyrene block and of 73.000 g/mol for the polyhexylisocyanate block (Fig. 24a). The polyhexylisocyanate domain had an average width of ~60 nm. By calculating the predicted rod length for the appropriate molecular weight and assuming interdigitated stacking the authors predicted that the rods are tilted by approximately 60° with respect to the lamellar normal. Because of the relative high volume fraction of the polystyrene the system fails to show long-range order. Samples with a polystyrene block of 14.000 g/mol and a polyhexylisocyanate block of 36.000 g/mol form instead the already observed zigzag morphology. For this composition the authors proposed a bilayer arrangement of the rods because: (i) beam damage occurred from the middle of the rod domain (depolymerization from the chain ends); (ii) AFM studies showed a narrow dip in the middle of the polyhexylisocyanate domain;

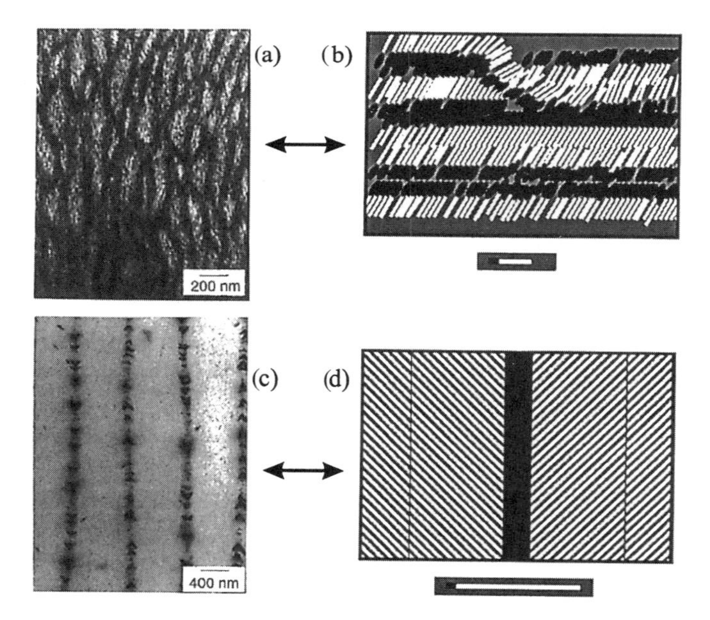

Figure 24 TEM micrograph of polyhexylisocyanate-block-polystyrene with the respective molecular weights (a) 73.000 and 104.000 g/mol; (c) 245.000 and 9.000 g/mol (thin film, cast from toluene and stained with RuO_4). The observed wavy lamellar morphology (a) can be explained by the schematic model in (b); the scheme (d) resembles the arrowhead morphology seen in (c). (Adapted from Ref. 85.)

and (iii) the domain spacing was in good accordance with the predicted rod length in a bilayer arrangement for the appropriate chain length.

For copolymers with even shorter polystyrene blocks another fascinating morphology was observed. In this morphology the polystyrene domains built arrowheads whose pointing direction flipped up and down with every other polystyrene-rich layer (see Fig. 24c). The morphology had an enormous long-range order over tens of micrometers. The electron diffraction pattern indicated that the orientation of the polyhexylisocyanate chain axis in adjacent layers alternated between 45° and −45° with respect to the layer normal. By comparing the predicted size of the polyhexylisocyanate domains with the one actually found in TEM, an interdigitated model seems to resemble a polymer system with a molecular weight of 386.000 g/mol for the polyhexylisocyanate block and of 7.100 g/mol for the polystyrene block best. A bilayer model offered instead a better representation for a copolymer system with a molecular weight of 245.000 g/mol for the polyhexylisocyanate block and of 9.300 g/mol for the polystyrene block.

Charged poly(styrene)-block-poly(isocyanodipeptides) were reported by Nolte et al. [86]. These block copolymers were synthesized by polymerization under catalysis of a Ni-complex with aminoterminated polystyrene using the isocyanides of dipeptides as monomers. Copolymers with different compositions of either isocyano-L-alanine-L-alanine (IAA) and isocyano-L-alanine-L-histidine (IAH) were prepared (see Fig. 22b, c). CD spectroscopy revealed a helical conformation for the isocyanide block [right-handed for poly(isocyano-L-alanine-L-alanine) and left-handed for poly(isocyano-L-alanine-L-histidine)]. The removal of the ester functions of the block copolymers resulted in negatively charged helical head-groups for the polystyrene-block-poly(isocyano-L-alanine-L-alanine) and zwitterionic head-groups for polystyrene-block-poly(isocyano-L-alanine-L-histidine) (see Fig. 22b, c). Because of the aggregation behavior of the amphiphilic blocks recording of CD spectra proved to be very difficult. To study these systems with electron microscopy the authors sonicated a 0.1 weight% dispersion in water (pH 7) for 1 hour at 70°C and after this buffered with sodium acetate to pH 5.6. Under these conditions [half of all acid functions protonated for polystyrene-block-poly(isocyano-L-alanine-L-alanine) and zwitterionic head-groups for polystyrene-block-poly(isocyano-L-alanine-L-histidine)] the block copolymers with a degree of polymerization of 40 for the polystyrene block and of 20 for the poly(isocyanide) block showed rodlike structures with a diameter of 12 nm (see Fig. 25a). These structures are thought to be micellar aggregates (core of polystyrene ~8 nm and corona of poly(isocyanide) ~2 nm) (see Fig. 25b). The rodlike structures assembled to a zigzag-like structure on mica and polyvinyl formaldehyde plastic plates as monitored by TEM and AFM. For longer poly(isocyanide) blocks (DP = 30) no clear morphologies could be seen; by making the poly(isocyanide) blocks shorter (DP ~ 10) the polystyrene-block-poly(isocyano-L-ala-

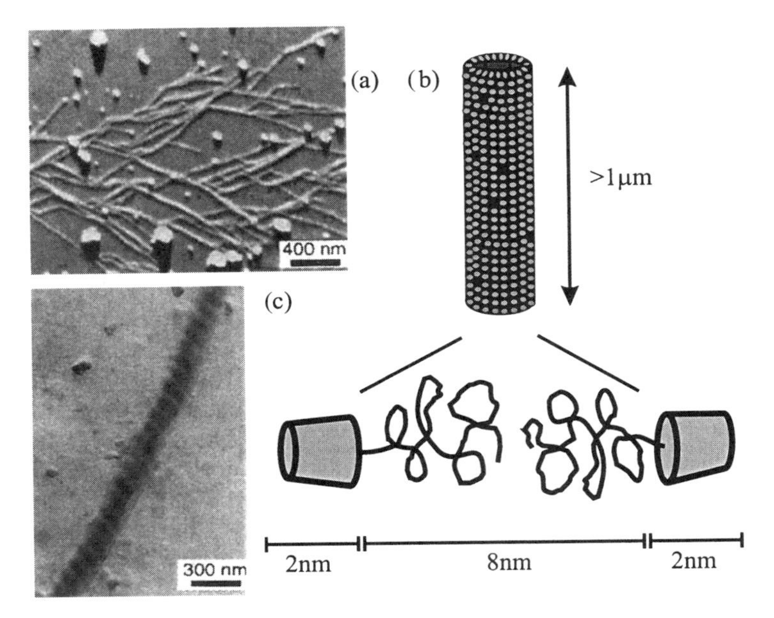

Figure 25 (a) AFM image of polystyrene-block-poly(isocyano-L-alanine-L-alanine) with a degree of polymerization of 40 for the polystyrene block and of 20 for the poly(isocyano-L-alanine-L-alanine) block. The observed rodlike structures can be explained by a schematic model as seen in (b). For a degree of polymerization of 40 for the polystyrene block and of 10 for the poly(isocyano-L-alanine-L-alanine) block a left-handed superhelix can be seen in TEM (c). [Both images in pH 5.6 sodium acetate buffer (0.2 mM)]. (Adapted from Ref. 86.)

nine-L-alanine) copolymer showed bilayer type structure in addition to collapsed vesicles, bilayer filaments, and left-handed superhelices (diameter 180 nm and pitch 110 nm) (see Fig. 25c). Contrary to this a polystyrene-block-poly(isocyano-L-alanine-L-histidine) with a degree of polymerization of 15 for the poly(isocyanide) blocks forms a right-handed superhelix with smaller dimensions (diameter 28 nm and pitch 19 nm) probably built of coiled rods.

Polybutadiene-block-polyisocyanide block copolymers synthesized by living transition metal catalysis were reported by Novak and Deming [87]. The catalyst used first polymerized butadiene and, after addition of methyl benzyl isocyanide, polymerized this second monomer without further including butadiene segments. A morphology based on spherical rod domains (on the order of 0.2 μm) within a polybutadiene matrix was observed using SEM. Block copolymers with a total molecular weight of 140.000 g/mol and a rod to coil ratio of

30%–70% were prepared. In a subsequent publication the synthesis of broken rods and also rod–coil–rod triblock copolymers by living bimetallic initiators was reported [88]. Novak and coworkers [89] also reported on polystyrene-block-poly(N,N′-di-n-hexylcarbodiimide) and polystyrene-block-poly[(R/S-N-(1-phenylethyl)-N′-methylcarbodiimide)] block copolymers (see Fig. 22d, e) which were synthesized by sequential anionic polymerization in benzene with sec-butyl lithium as initiator. The polymers were characterized by IR-spectroscopy and the composition of the diblocks was monitored by thermogravimetric analysis. In these polymers the rod segment is the major component (~83%). Samples were cast from different selective solvents and the microstructure (monitored by TEM) was found to be spherelike in hexane which is a good solvent for the polycarbodiimide and a nonsolvent for polystyrene. The morphology was found to be lamellar for benzene (a good solvent for polystyrene and θ-solvent for polycarbodiimide) and cyclohexane (a good solvent for polycarbodiimide and θ-solvent for polystyrene). In THF, which is a good solvent for both blocks, no clear morphology became obvious. The authors confirmed the microstructure produced from benzene solutions with atomic force microscopy.

B. Nonhelical Rod Systems

1. Copolymers with Rigid Aromatic Rod Segments

As already mentioned, the rigidity, for instance, in polypeptides is caused by the building of helical superstructures due to hydrogen bonding. Another reason for chain rigidity can be found in paralinked aromatic polyamides. The high resonance energy in the amide group leads to quasiconjugation and coplanarity in the structure that decreases the flexibility of the chains. Many of these structures exhibit a high equilibrium rigidity and can form nematic mesophases in concentrated solutions. The reason for the chain rigidity found in structures like poly(*para*-phenyleneethynylene) and poly(*para*-phenylene) is the symmetry of the bonds along the chain. The linear bonds of the phenylene units in poly(*para*-phenylene), for instance, only tolerate rotation of the repeating units around the polymer axis (for examples of some nonhelical rigid polymers refer to Fig. 26).

The work by Ciferri and coworkers [90–92] focused on the supramolecular organization of aromatic rod–coil copolyamides in diluted (isotropic) and moderately concentrated (lyotropic) phases. Their diblock copolymers were based on a rigid block of poly(p-benzamide) (PBA) having a DP about 100, and different comparable lengths of flexible blocks such as poly(m-phenylene isophtalamide) (MPD-I), poly(m-benzamide) (MBA) or poly(ethylene glycol) (PEG). The use of end-capped prepolymers and selective extraction techniques assured a strict two-block sequence, the absence of free homopolymers not strongly bound to

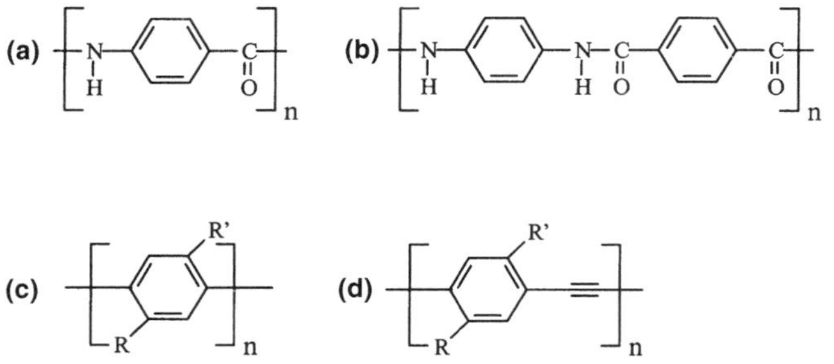

Figure 26 Chemical structures of rigid aromatic polymers: (a) poly(p-benzamide); (b) poly(p-phenyleneterephthalamide); (c) poly(p-phenylene); and (d) poly(p-phenyleneethynylene).

the copolymers, and a fractionation in terms of the rigid/flexible compositional distribution ratio.

From viscosity, light-scattering, and critical concentration data in H_2SO_4 and in N,N-dimethylacetamide (DMAc)/3%LiCl they concluded that in the latter solvent poly(p-benzamide) is not molecularly dispersed but rather occurs as supramolecular aggregates likely composed of seven polymer chains with a side-by-side shift of 1/4th of the molecular length. Since the critical concentration at which the mesophase appears for poly(p-benzamide)/poly(m-phenylene isophtalamide) and poly(p-benzamide)/poly(m-benzamide) copolymers in DMAc/3%LiCl was not much affected by the flexible blocks, they suggested that the organization within the mesophase could be represented by assemblies (Fig. 27) based on the above indicated aggregates of poly(p-benzamide) segments and on suitably oriented flexible blocks.

Solubility data (Fig. 28a), showing that the DMAc/3%LiCl solvent is a good solvent for both poly(p-benzamide) and poly(m-benzamide) [or poly(m-phenylene isophtalamide)] blocks, were interpreted to mean that in isotropic solutions a dispersion of assemblies such as that schematized in Fig. 9 occurs. Quite a different conclusion was reached in the case of the poly(p-benzamide)/poly(ethylene glycol) copolymer for which solubility data (Fig. 28b) showed a worsening of the solvent quality of DMAc toward poly(ethylene glycol) as the LiCl concentration was increased from 0.5 to 3.0%. A bilayer-like organization of the type envisioned by Halperin [61] was suggested. In the DMAc/0.5% LiCl solvent, the better solvated poly(ethylene glycol) segments screen from the solvent (Fig. 28b-left) relatively large clusters of poly(p-benzamide) aggregates. However, in the DMAc/3%LiCl solvent (Fig. 28b-right) the poly(p-benzamide) aggregates are exposed to the solvent, screening the less solvated flexible blocks.

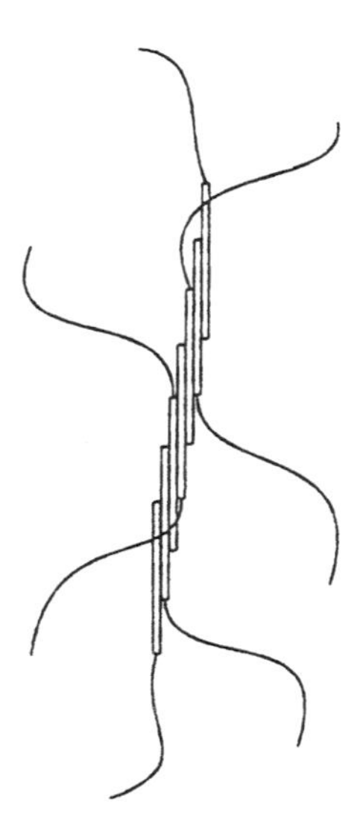

Figure 27 Schematic representation of a nematic aggregation of copolymer assemblies involving seven molecules each with a side-by-side shift of 1/4th of the length and alternating orientations. (Reprinted with permission from Ref. 90. Copyright 1997, American Chemical Society.)

The above structures were postulated to undergo a further level of organization upon increasing copolymer concentration until the formation of the liquid-crystalline and crystalline phases. However, no morphological studies have yet been reported. The copolymers of Ciferri and coworkers are of interest in the areas of mechanical and transport properties, and may represent a novel approach to the preparation of self-assembling high performance materials that avoids the current cumbersome route of engineered composites.

Jenekhe et al. [93–96] investigated the effect of supramolecular structure and morphology on the photophysical properties in polymer systems. For this purpose they used rod–coil di- and triblock copolymers and polymer blend systems.

The authors investigated the electronic energy transfer from a light-absorbing energy donating rod–coil system to a rigid-rod polymer (randomly dispersed in this polymer blend). (See Fig. 29a.) The rod–coil copolymer was varied by the number of methylene groups in the coil segments that resulted in a changed supramolecular structure and directly affected the Förster energy transfer efficiency (an increase for this efficiency for 12 methylene units as compared to 7 methylene units) [93].

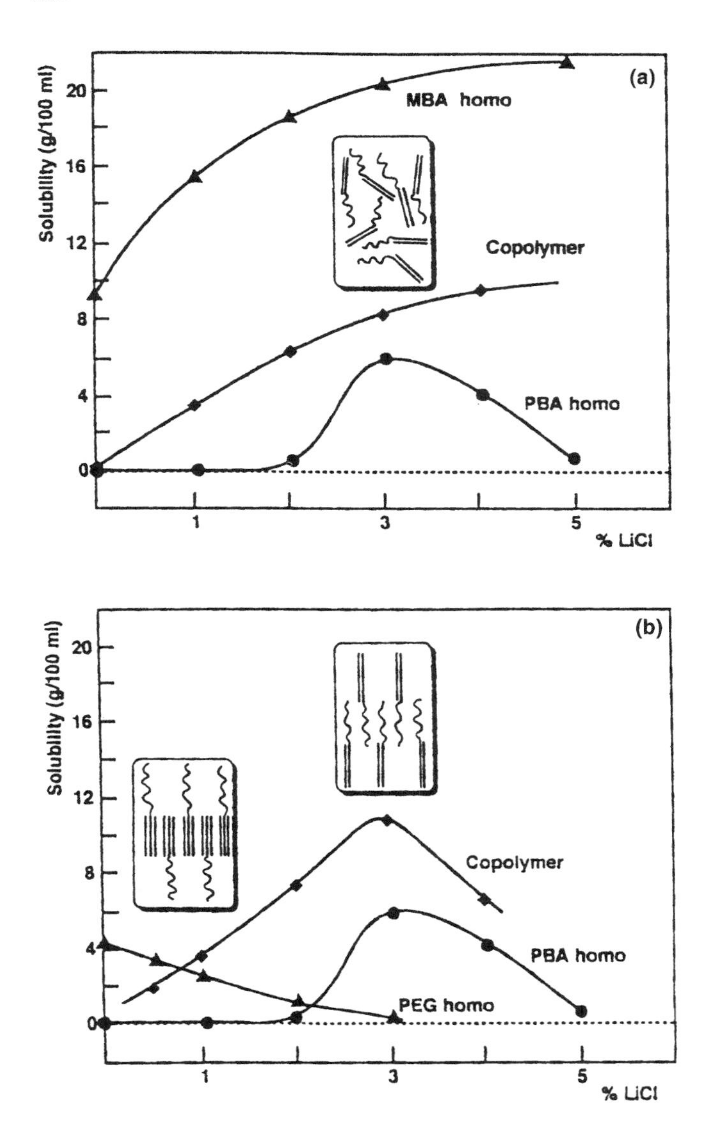

Figure 28 Solubility of copolymers and corresponding homopolymers in DMAc with varying LiCl concentrations: (a) Copolymer of poly(p-benzamide) and poly(m-benzamide); (b) copolymer of poly(p-benzamide) and poly(ethylene glycol). Possible supramolecular organizations in the isotropic solution are schematized in the insert. (Reprinted with permission from Ref. 92. Copyright 1999, Wiley-VCH.)

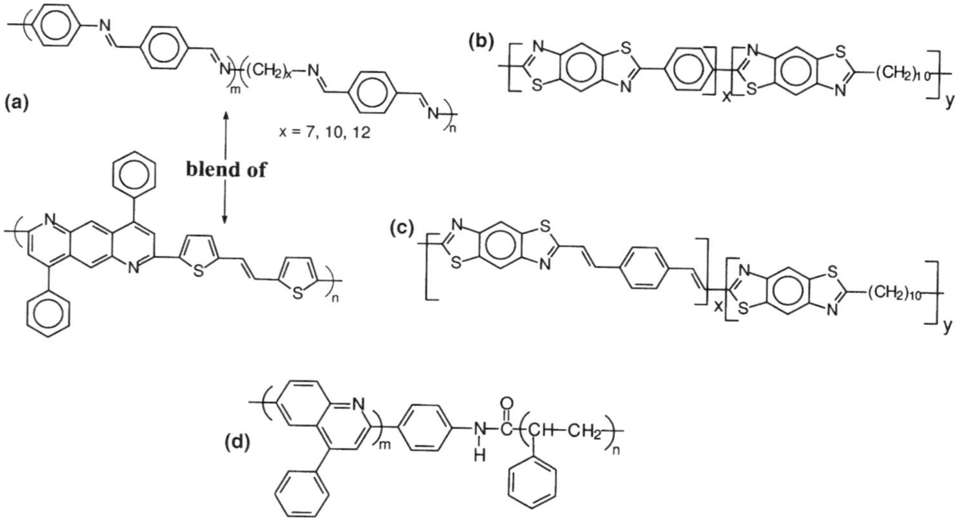

Figure 29 Chemical structures of (a) a blend of a light-absorbing energy donating rod–coil system and a rigid-rod homopolymer; (b) poly(1,4-phenylenebenzobisthiazole-co-decamethylenebenzobisthiazole); (c) poly((1,4-phenylenedivinylene)benzobisthiazole-co-decamethylenebenzobisthiazole); and (d) poly(phenylquinoline)-block-polystyrene.

For another type of rod–coil copolymers [94] the authors varied the fraction of the rod segment and analyzed the photoluminescence in this system. The random rod–coil system is built from the rigid segment poly(p-phenylenebenzobisthiazole) and the flexible poly(benzobisthiazoledecamethylene) (see Fig. 29b).

By studying the systems with X-ray diffraction and polarized optical microscopy the authors found that the morphology of their system was amorphous. A proposed picture of the morphology is seen in Fig. 30a. It became obvious that the photoluminescence emission changed with the composition from yellow (100% rod) to UV ($< 5\%$ rod). The corresponding photoluminescence spectra are seen in Fig. 30b.

For two series of rod–coil copolymers [95] poly(1,4-phenylenebenzobisthiazole-co-decamethylenebenzobisthiazole) (see Fig. 29b) and poly[(1,4-phenylenedivinylene)benzobisthiazole-co-decamethylenebenzobisthiazole] (see Fig. 29c) the authors also investigated the photophysical properties with varying composition. In these cases the photoluminescence quantum yield reached over seven fold higher values than for a pure conjugated homo polymer. Furthermore the emission color was tunable in the visible region by varying rod fraction.

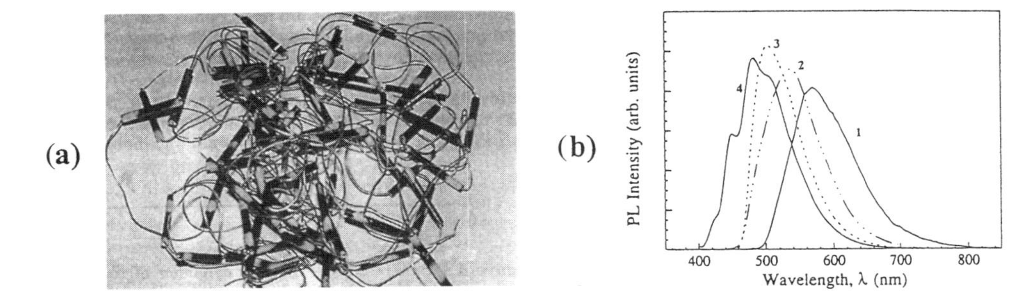

Figure 30 (a) Model of the supramolecular structure and morphology of the rod–coil copolymers by Jenekhe and coworkers. (b) The photoluminescence spectrum of poly(benzobisthiazole-1,4-phenylene) and copolymers: 1–100% PBZT, 2–40%; 3–20%; 4–5%. (Reprinted with permission from Ref. 94. Copyright 1994, American Chemical Society.)

A certain class of rod–coil copolymers studied by Jenekhe et al. (poly (phenylquinoline)-block- polystyrene) [96] (see Fig. 29d) proved to build fascinating supramolecular structures. For a block copolymer in which poly(phenylquinoline) had a degree of polymerization of 50 and the DP of polystyrene was 300 the authors observed micellar assemblies in the form of spheres, lamella, cylinders, and vesicles. (See Fig. 31.) The detected form of the supramolecular aggregates was dependent on the type of solvent or solvent mixture and the drying rate [the solvent mixtures were either trifluoroacetic acid (TFA) and dichloromethane or TFA and toluene (selective for the rod segment)]. Very rapid drying resulted in micellar assemblies with diameters of 0.5 to 10 μm whereas by slow evaporation of the solvent nonmicellar structures such as lamella, cylinders etc. became predominant [with diameters of 5–30 μm for lamellar aggregates, 0.5 to 1 μm for vesicles, and 1–3 μm for the cylinders (length of 5–25 μm)]. These huge assemblies cannot result from core-shell micelles as observed in other polymer systems because they are approximately 10 to 15 times larger than these micelles built by a rod–coil copolymer with the same composition. Therefore the authors proposed that these structures are built by hollow cavities. The covers of these cavities are highly ordered as they show crystalline features. By studying the structures with fluorescence microscopy it became obvious that the fluorescinating rod blocks are on the outermost surface of the observed assemblies and photoluminescence studies showed that the photophysical properties strongly depend on the supramolecular structure of the photoactive rod segments. (See Fig. 31.) The stability of these giant self-organized structures is thought to result from strong intermolecular hydrogen bonding caused by the linking amide unit in the poly(phenylquinoline)-block-polystyrene copolymer. (See Fig. 32.) It was also

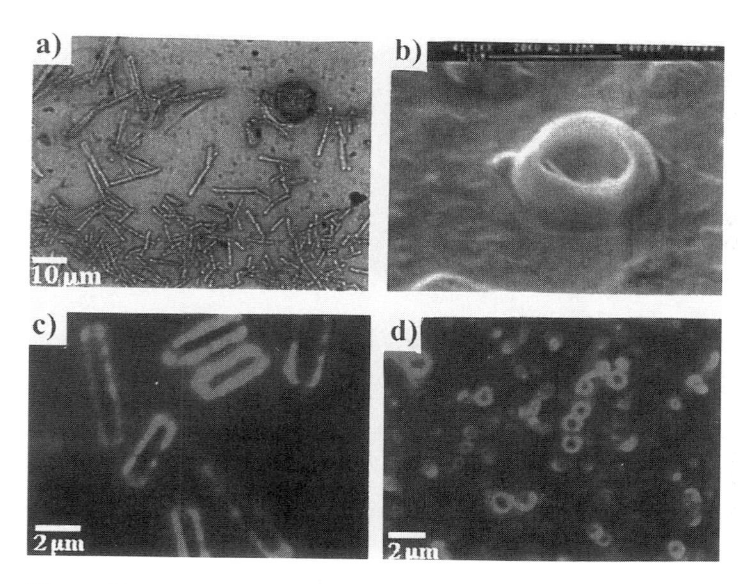

Figure 31 (a) Optical micrograph of cylinders built from poly(phenylquinoline)-block-polystyrene (9:1 TFA:DCM, 25°C); (b) scanning electron micrograph of giant vesicles built from the same block copolymer (1:1-1:4 TFA:DCM, 25°C). (c) and (d) show the fluorescence photomicrographs of the cylinders and the vesicles, respectively. (Adapted from Ref. 96.)

proposed that the stable aggregate structure results from efficient packing of the rod blocks.

The authors also found that their rod–coil systems proved to be feasible for encapsulating fullerens into the cavities. Compared to usual solvents for C_{60}, like dichloromethane or toluene, the solubility was enhanced by up to 300 times (in trifluoroacetic acid or mixtures of trifluoroacetic acid with toluene) when the molecules were encapsulated into rod–coil micelles. Such ''micro containers'' could also be used for a encapsulation process of other substances and so one can think of applications such as drug delivery, emulsions, etc.

Francois et al. [97,98] reported on the synthesis of polystyrene-block-poly (*para*-phenylene) block copolymers via the anionic polymerization to polystyrene-block-poly(1,3-cyclohexadiene) and subsequent aromatization. The micellar assemblies and molecular dispersed free chains could be separated by size exclusion chromatography and with this method the size of the spherical assemblies was found to be 30–50 Å for the poly(*para*-phenylene) core, and 3–15 nm for the full micelle. By evaporating the solvent from solutions in carbondisulphide in moist air, a hexagonal array of empty cells was observed from optical and

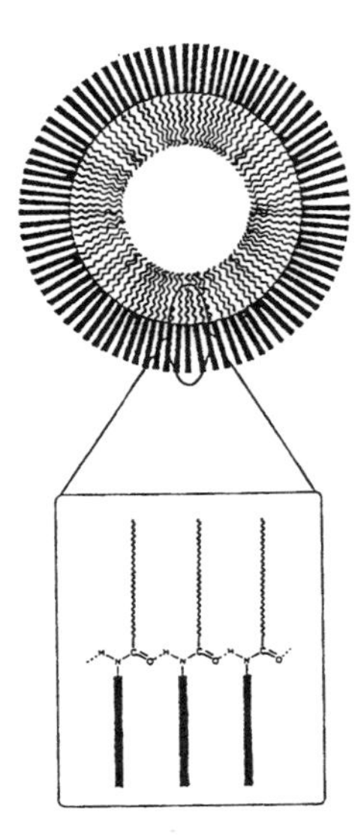

Figure 32 Schematic illustration of the self-assembly of poly(phenylquinoline)-block-polystyrene diblock copolymers into hollow aggregates. (Adapted from Ref. 96.)

electron microscopy. By removing the superficial skin of these cells an internal honeycomb-like structure became evident (see Fig. 33).

The size of the cell walls (as studied by AFM) was influenced by the length of the polystyrene block in the copolymer and for molecular masses larger than 50,000 the regularity in the structure gradually disappeared. The authors related the mechanism of formation of this structure to the classical ''phase inversion'' process for the production of polymeric membranes [99]. In a later work the authors reported on the possibility of monitoring this honeycomb structure in different polymeric systems including polystyrene-block-poly(*para*-phenylene), star branched polystyrenes, polystyrene-block-poly(3-hexylthiophene), polystyrenes with polar endgroups, and polymer blends [100]. Possible applications for such fascinating structures are in the area of polymeric membranes and optical devices.

Francois et al. also reported on the synthesis of polystyrene–polythiophene block and graft copolymers [101–103]. These materials incorporate the unique properties of polythiophenes with an easiness of solubilization and processing. The reported block copolymers showed nearly the same spectral characteristics as pure polythiophene and the authors proved that their block copolymers were still soluble after doping. On films of their copolymers subjected to a pyrolysis

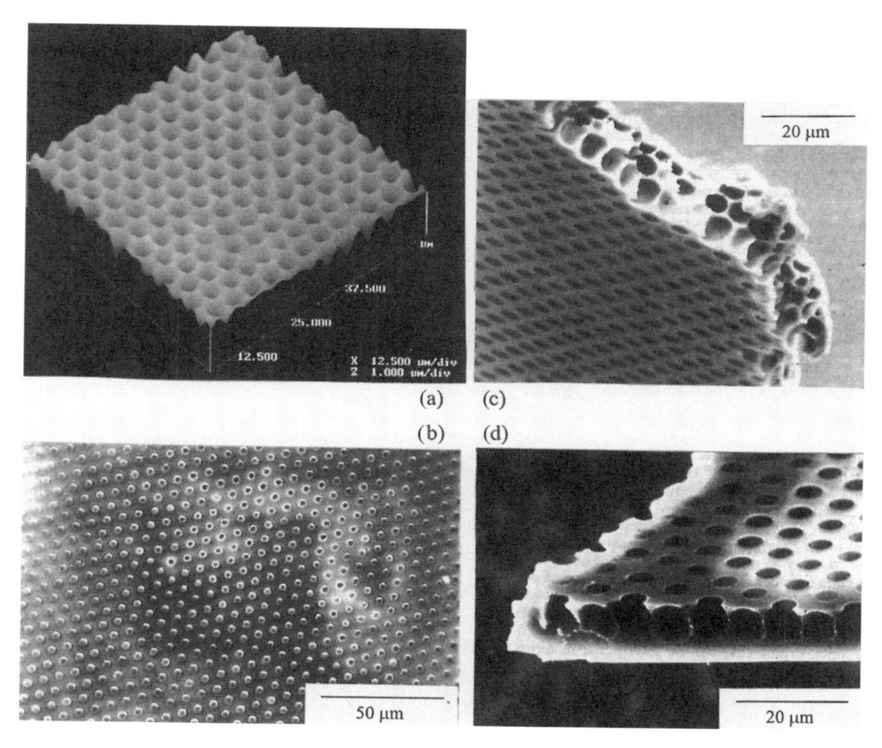

Figure 33 Scanning electron micrograph of polystyrene-block-poly(*para*-phenylene) (respective molecular weights 30.000 and 7.000 g/mol): (a) side view; (b) plan view; and (c) micrograph of a monolayer of pores in the copolymer film. Atomic force microscopy also manifests the honeycomb morphology (d). (Reprinted by permission from Ref. 99, copyright 1994, Macmillan Magazines Ltd.)

step they reported different morphologies (spheres and fibrils) depending on volume fraction and film thickness.

In a very recent publication Hempenius et al. [104] reported on polystyrene-oligothiophene-polystyrene triblock copolymers. Size-exclusion chromatography revealed the occurrence as self-assembled clustered as confirmed by scanning force microscopy of a dilute solution evaporated on mica. Incomplete covering of the mica surface allowed the observation of single droplets which can be attributed to single micelles (see Fig. 34a). The TEM image confirms the occurrence of micelles having a diameter of 12 nm corresponding to an aggregate of about 60 molecules, which is in line with results obtained from GPC (see Fig. 34b).

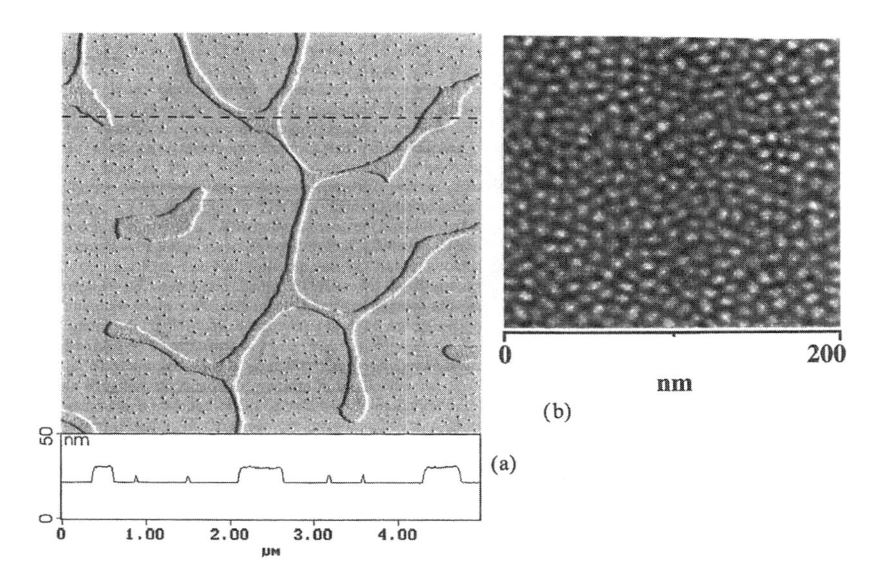

Figure 34 (a) Phase scanning force microscopy image of a polystyrene-block-oligothiophene-block-polystyrene triblock copolymer cast on mica from a solution in toluene (1 mg/ml). (b) Micelles built by the same triblock system as observed in a closed film by phase scanning force microscopy (after Fourier filtering). (Reprinted with permission from Ref. 104. Copyright 1998, American Chemical Society.)

Films of gold-stained micelles obtained by chemical oxidation of the cores with $HAuCl_4$ in toluene were monitored by TEM (the micellar cores appear as dark spots) and these doped micelles were still soluble. The electronic properties of the triblocks are in accordance with the ones of associated unsubstituted oligothiophenes.

In three mainly synthetic works Godt et al. and Müllen et al. reported on rod–coil systems having poly(*para*-phenyleneethynylene) or poly(*para*-phenylene) as the rod part. Müllen et al. [105] showed the synthesis of poly(*para*-phenyleneethynylene)-block-poly(ethyleneoxide) by condensation of monofunctionalized homopolymer blocks. Godt et al. [106] reported on the synthesis of poly(*para*-phenyleneethynylene)-block-polyisoprene and the transformation of such rod–coils into corresponding coil–rod–coil triblock copolymers. The synthesis of poly(*para*-phenylene) as the rod part and polystyrene or poly(ethyleneoxide) as the coil part by condensation of prepolymers was also reported by Müllen et al. [107].

2. Copolymers with Short Monodisperse Rod Segments

In this section the work on systems incorporating small organic molecules as the rigid segment is included even though these materials cannot be strictly considered as true block copolymers. Still these systems possess the major advantage of a monodisperse rod and therefore the resulting polymeric material has a very narrow polydispersity which is known to be very important for the results one can obtain on studying the material.

Stupp et al. reported on rod–coil systems containing a monodisperse rod part based on an azo dye bound to a rigid monomer [108–110]. For these systems the authors synthesized polyisoprenes anionically and terminated the living chains with carbon dioxide resulting in carboxylated polyisoprenes which were then coupled to the rigid block. (See Fig. 35a for the end structure.) Copolymers with varying volume fraction of the rod segment were prepared and the resulting morphological changes were monitored by TEM. For the copolymer with the highest rod volume fraction (f_{rod} = 0.36) the authors observed (after annealing a thin film at 140°C for 12 h) a morphology that seemed to consist of alternating strips of coils and rods (5–6 nm and 6–7 nm thickness, respectively; see Fig. 36a). Thicker films (cast from more concentrated solutions) showed a terracelike assembly. By studying these samples by electron tomography the authors found that the copolymers self-assembled into layered 2-D superlattices and ordered 3-D morphologies. For a rod volume fraction f_{rod} = 0.36 the same morphology as in the thin films was found for odd numbers of steps, but for even numbers of steps no contrast was observable. Slices parallel, perpendicular, and orthogonal to the film plane showed strip morphology built of discrete channellike objects in layers parallel to the surface (which are displaced to the strip above and below) with the long axis also parallel to the plane of the film. The loss in contrast for the even numbers of layers was attributed to the effectively uniform electron density across the plane of the sample. For odd numbers the alternating rod domains vary by one and a contrast is observable. The terraces in thick films are built of layers that contain the channellike objects. The strip morphology is schematically illustrated in Fig. 36b.

The authors correlated this morphology to cylinder phases in coil–coil diblock copolymers. However, the dimensions of the strips (6–7 nm) do not correspond to a radial arrangement of cylinders, but rather to interdigitated bilayers. For a rod volume fraction f_{rod} = 0.25 the rod domains organize into a superlattice with hexagonal order measuring approximately 7 nm in diameter and a domain spacing of 15 nm (see Fig. 37a). Studying the thick films with electron tomography, the layers showed a continuous network of small domains. Slices parallel, perpendicular, and orthogonal to the film plane showed that the rod domains were discrete objects with roughly the same dimensions in all directions. The aggregates are ordered in layers parallel to the surface. This morphology is sche-

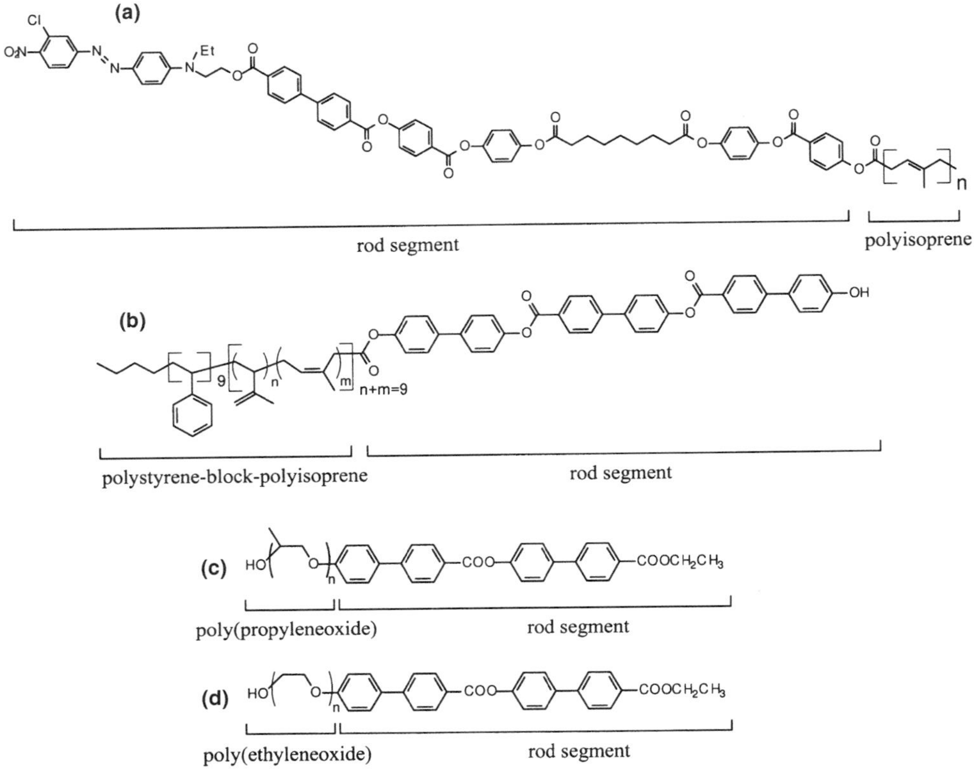

Figure 35 Chemical structures of rod–coil block copolymers with short, monodisperse rod segments: (a) block copolymer with polyisoprene and a rod segment built of an azo dye and a rigid monomer and (b) miniaturized triblocks from Stupp and coworkers; (c) and (d) show the chemical structures of block copolymers with either poly(propyleneoxide) or poly(ethyleneoxide) and a short rod segment as reported by Lee et al.

matically described in Fig. 37b. The sketch resembles a close-packed stacking of hexagonal superlattices.

An intermediate mix of the ribbon type and the discrete aggregate-like morphology was observed in thin films bearing a rod volume fraction $f_{rod} = 0.3$ (see Fig. 38a) which was as well found for up to three layers as monitored by electron tomography. However, by looking at more layers a network pattern based on hexagonal fused cells with a discrete rod aggregate as the nucleus was evident (see Fig. 38c). Cross-sectional slices showed that the hexagonal network is built by discrete rod aggregates having dimensions similar to those found for a rod

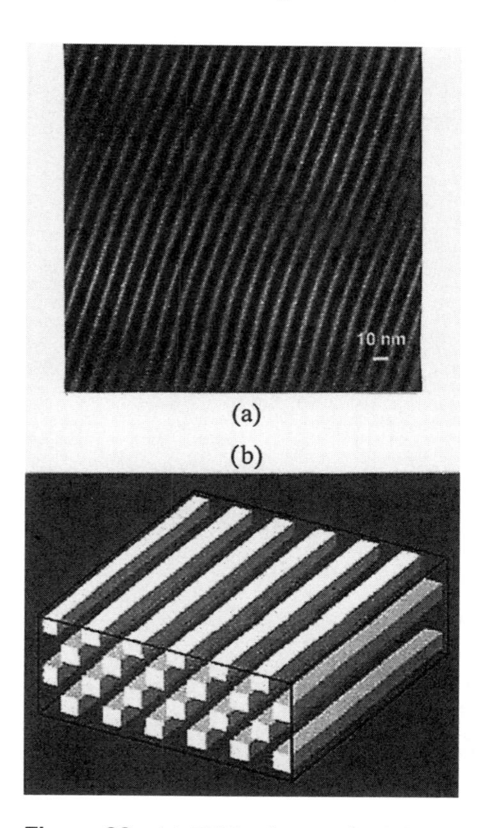

Figure 36 (a) TEM micrograph of the rod–coil block copolymers reported by Stupp et al. with a rod volume fraction of $f_{rod} = 0.36$ (thin film cast from cyclohexane, stained with OsO_4, and enhanced by translational Fourier filtering) and (b) a schematic model for the observed morphology of alternating strips. (Reprinted with permission from Refs. 109 and 110. Copyright 1994 and 1997, American Chemical Society.)

volume fraction of 0.25. Moreover, the ''nucleus'' (with higher contrast) is built by columns spawning the film thickness and discrete rod aggregates building the ''cell walls'' as seen in Fig. 38b. By looking at slices of the reconstructed volume the authors observed a 2-D hexagonal superlattice for each slice that is rotated by 30° with respect to the lattice above and below. The observed morphology was interpreted by a certain rotation of the hexagonal superlattices to each other and a certain distortion in the hexagonal packing of the rod domains. This is schematically shown in Fig. 38d.

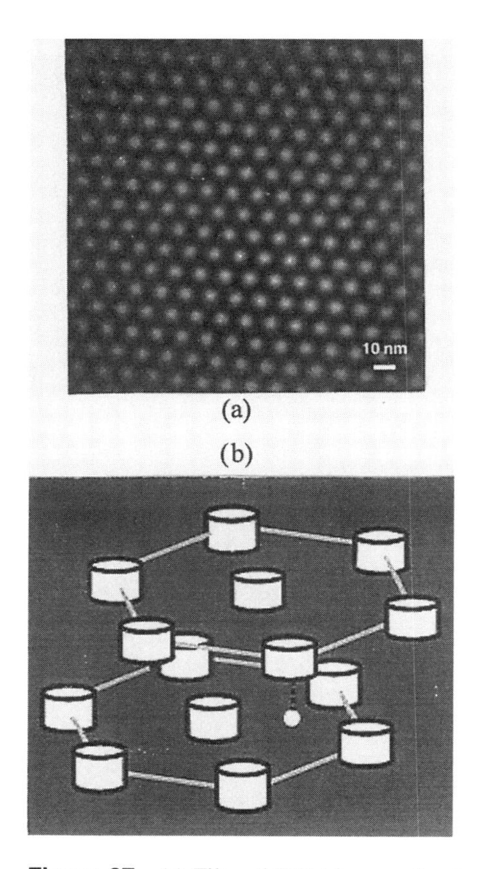

Figure 37 (a) Filtered TEM image of rod–coils with a rod volume fraction of f_{rod} = 0.25; (b) scheme for the hexagonal superlattice (built by rod aggregates). (Reprinted with permission from Refs. 109 and 110. Copyright 1994 and 1997, American Chemical Society.)

The effect of annealing was also studied by Stupp et al. While temperatures below 130°C showed no effect, increasing the temperature above 140°C caused the loss of long-range order. Because the crystal to liquid-crystal transition of the unconnected rod segment is around 140°C it was suggested that only at temperatures between 130°C and 140°C the molecules have the needed mobility to form the observed morphologies. For a copolymer with a rod volume fraction $f_{rod} = 0.19$ the sample loses the long-range order at 140°C and annealing at 100°C resulted in a microphase separated structure. This temperature effect on polymers which are more asymmetric in copolymer composition was interpreted as a hint

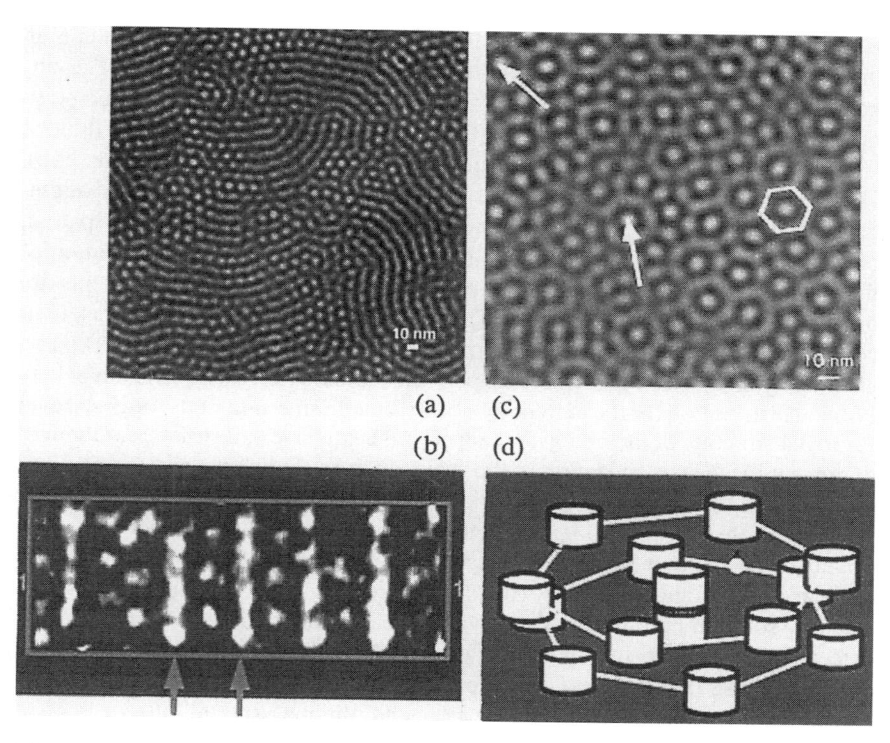

Figure 38 (a) Filtered TEM micrograph of a rod–coil block copolymer with a rod volume fraction of $f_{rod} = 0.3$. For three and more stacked layers (as monitored by electron tomography) the morphology (c) can be observed (the arrows point to the ''cell walls'' and the ''cell nuclei'' of the observed structure, respectively). A cross-sectional slice from the reconstructed volume (b) shows that the ''cell nuclei'' are formed by rod aggregates spawning the thickness of the film; the morphology is explained by the schematic drawing in (d) which shows the stacked and rotated hexagonal superlattices. (Reprinted with permission from Refs. 109 and 100. Copyright 1994 and 1997, American Chemical Society.)

that the loss of long-range order at higher temperatures is caused by an order–disorder transition. In the as cast state, samples with rod volume fractions $f_{rod} =$ 0.36 and $f_{rod} = 0.3$, respectively, showed ribbons and small aggregates of rod segments. The strips assemble parallel in some regions of the sample, but remain isolated in other parts and curve in the later regions. The thickness of ribbons was found to be 7 nm with length of up to 1 μm. For a nonannealed film of a copolymer with rod volume fraction $f_{rod} = 0.25$ only small aggregates were observed; for a rod volume fraction $f_{rod} = 0.19$ nanophase separation was detected.

Stupp et al. also reported on miniaturized triblock copolymers containing oligopolystyrene-block-oligoisoprene and a rigid structure [111–113]. The oligopolystyrene-block-oligoisoprene copolymer was synthesized via anionic polymerization and terminated by carbon dioxide. The carboxyfunctionalized diblock was afterwards coupled to the rod segment (biphenylester block) (see Fig. 35b). Thin films cast from chloroform (no staining) showed uniform nanosized aggregates in the transmission electron microscope. The electron diffraction pattern of the sample indicated crystalline regions in the morphological organization of the rod segments. The wide-angle X-ray pattern indicated that rod segments are oriented normal to the film plane. The authors emphasized the fact that their copolymers are composed of a uniform rod part (the biphenylester block) and a chemically diverse coil part (the oligopolystyrene-block-oligoisoprene diblock ⇒ Poisson distribution for the anionic synthesis; atactic structure of the oligostyrene; 1,4 to 3,4 structure of the oligoisoprene; etc.). The diverse coil structure is thought to prevent the 3-D organization of the structure even though the rod segment has a very strong tendency to aggregate via π–π overlap. The authors proposed (after intensive molecular modeling calculations) that these mini-copolymers build supramolecular units in the form of mushrooms containing 100 triblock units with a molar mass of about 200 kD which assemble in a ''cap to stem'' fashion. By studying solution-cast films in the TEM, a layered (lamellarlike) morphology was detected. This result supported the proposed mushroomlike assembly of the blocks because the dimensions of the bands can be correlated to either the extended rod (dark bands) or to the nonextended oligopolystyrene-block-oligoisoprene diblock (lighter band). The samples were found to be layered terracelike with a step-size under 100 Å, confirming that the sample is organized in monolayers rather than in bilayers. The structure is suggested to be formed by the mushrooms containing the miniblocks (see Fig. 39).

Lee et al. also reported on rod–coil systems with a small rod segment. Their copolymers consisted of poly(ethyleneoxide) or poly(propyleneoxide) and 4-[4′-oxy-4-biphenylcarbonyloxy]-4′-biphenylcarboxylate, synthesized by coupling of the rod segment to monotosylated poly(ethyleneoxide) or poly(propyleneoxide) (see Fig. 35c and d for the end structures). As alkali metal salts are selectively soluble in poly(ethyleneoxide) and poly(propyleneoxide) the authors also investigated the phase change of their copolymers by addition of $LiCF_3SO_3$. These polymers were characterized by differential scanning calorimetry, optical polarized microscopy, and X-ray scattering. First a rod–coil system with a poly (ethyleneoxide) block with a degree of polymerization of 12 was reported [114,115]. By complexation with 0.05–0.2 mol $LiCF_3SO_3$ per ethyleneoxide unit a smectic A phase was the high-temperature mesophase. For 0.2 mol salt per repeating unit an additional cubic phase was observed. For 0.25 mol salt per ethyleneoxide unit the smectic A phase disappeared and only a cubic mesophase remained. For 0.3 mol $LiCF_3SO_3$ a cylindrical micellar phase was the high-tem-

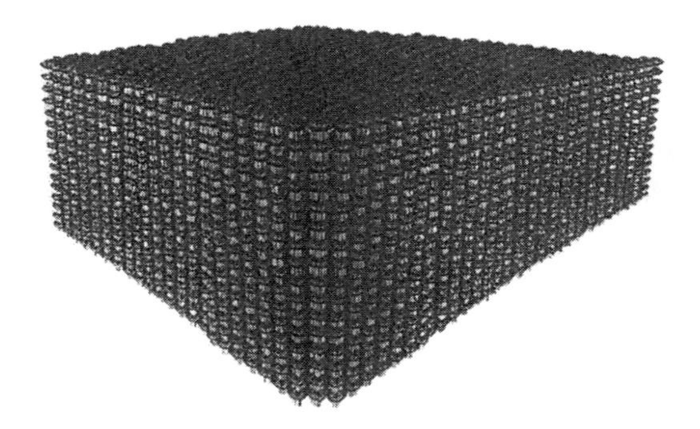

Figure 39 Schematic representation of the proposed structure built of the miniaturized triblocks reported by Stupp et al. The mushroom-shaped nanostructures are assembled in a three-dimensional structure in a ''cap to stem'' fashion. (Reprinted from Ref. 112. Copyright 1998, with permission from Elsevier Science.)

perature mesophase. For 0.4–0.7 mol salt per repeating unit the system only occurred in a cylindrical micellar mesophase. At higher salt concentrations the samples became amorphous. It was also noticed that increasing the salt concentration increased the mesomorphic–isotropic transition temperature and decreased the crystal melting temperature with the result of a higher thermal stability of the smectic phases. The authors also compared the above block copolymers based on poly(ethyleneoxide) with copolymers based on poly(propyleneoxide) having the same degree of polymerization (DP = 12). The latter copolymer did not exhibit a smectic A but showed a hexagonal columnar mesophase [116]. The differences between these similar systems are mostly caused by the steric hindrance of the additional methyl side-group in poly(propyleneoxide) and resulting different smectic ordering of the rods due to different coil-packing effects. Lee et al. also reported on blends of poly(ethyleneoxide) with their rod–coil block copolymers [117] and on coil–rod–coil triblock copolymers [with poly(propylenoxide) as the coil segment] and their complexes with $LiCF_3SO_3$ [118].

In a very thorough work on the influence of the length of the coil segment on the phase behavior the authors studied block copolymers with poly(propyleneoxide) with different degree of polymerization but the same rod segment as above [119,120]. Molecules with 7 and 8 propylene oxide units showed in DSC traces as in the polarization microscopy a crystalline phase that melted (coil segment) and recrystallized into a second crystal phase (rod segment) followed by a smectic C phase. The latter transformed to a smectic A phase on second heating. For block copolymers with 10 and 12 repeating units no birefringence between crossed polarizers was detected. DSC traces showed a melt transition and an additional phase transition accompanied by a decrease of viscosity suggesting a

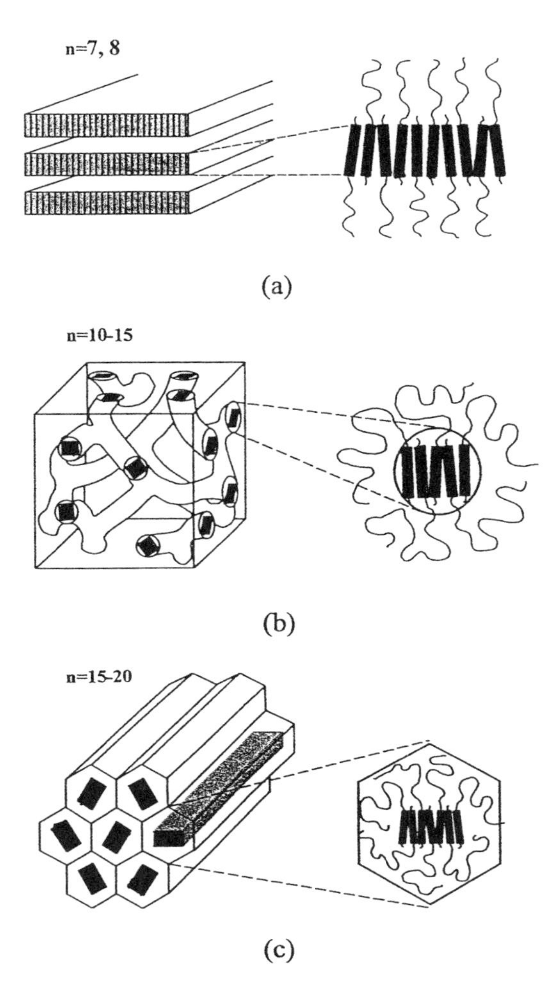

Figure 40 Schematic representation of self-assemblies in the rod–coil diblock copolymer system by Lee et al. (a) the lamellar smectic A, (b) the bicontinous cubic; and (c) the hexagonal columnar mesophase (n represents the degree of polymerization of the polyisoprene coil segment). (Reprinted with permission from Ref. 120. Copyright 1998, American Chemical Society.)

cubic mesophase. For 15 propylene oxide units a transition to an optical isotropic cubic mesophase occurred after crystalline melting followed at higher temperature by a hexagonal columnar mesophase which underwent isotropization at 46°C. Only a hexagonal columnar phase after crystalline melting was observable for block copolymers with 17 and 20 propylene oxide units. The authors also investigated the microstructure of the copolymers by X-ray diffraction either in the crystalline solid or in the liquid-crystalline phase. For molecules with 7 to 12 repeating units a lamellar structure with interdigitated rods in the crystalline state became obvious. With 15 to 17 repeating units the rods were tilted relative to the layer normal inside the lamellar structure. For 7 and 8 repeating units one could observe smectic C and smectic A phases; for 10 to 15 repeating units the system showed a bicontinuous cubic mesophase. For 15 to 20 repeating units a hexagonal columnar mesophase was observed (see Fig. 40).

This work shows very nicely the dependence of the microstructure on molecular composition as the mesophase shows a phase transition from lamellar to columnar phase with a bicontinuous cubic phase with Ia3d symmetry as the intermediate regime. Such behavior has already been observed within the phase diagram of certain coil–coil block copolymers (see Chapter 6).

In addition to the work on copolymers with uniform and monodisperse small rod molecules that is included in this chapter there is a lot of literature available on even smaller rodlike segments. A recent review was presented by Poser et al. [121].

Another class of rod–coil systems with helical rod segments is derived from amylose as one block. Most copolymer systems with amylose as one segment are based on an enzymatic grafting from polymerization with potato phosphorylase which was pioneered by Beate Pfannemüller. Linear, star, and comblike polymers carrying amylose chains were reported by Pfannemüller et al. [122–125]. Stadler and coworkers reported rod–coil systems composed of amylose blocks with polysiloxanes [126,127] (comblike structures) and polystyrene [128] (blocklike structures). Comblike structures based on polystyrene with amylose entities were synthesized by Kobayashi et al. [129].

ACKNOWLEDGMENTS

This chapter is dedicated to the memory of Professor Reimund Stadler whose sad and untimely death was a terrible loss for the scientific community. Katja Loos wishes to express a special ''thank you'' in particular to him for his constant encouragement and friendship. The authors are much indebted to Professor Alberto Ciferri for his advice and revision of the manuscript. Many thanks also to Volker Abetz, Carlos Alemán, Alexander Böker, Thomas Breiner, Jennifer David, Gerd Mannebach, Francisco López-Carrasquero, Antxon Martínez de Ilarduya-Saéz de Asteasu, Salvador León, Gabriele Rösner-Oliver, David Schlitzer, and David Zanuy for their hints, computer help, and proofreading. Sebastian Munoz-Guerra is grateful to DGICYT for grants PB-93-09660 and PB-96-0490 for financial support.

REFERENCES

1. VP Shibaev, M Palumbo, E Peggion. Biopolymers 14:73–81, 1975.
2. VP Shibaev, VV Chupov, VM Laktionov, NA Platé. Vysokomol Soedin B16:332–333, 1974.
3. VV Chupov, VP Shibaev, NA Platé. Vysokomol Soedin A21:218–228, 1979.
4. JM Fernández-Santín, J Aymamí, S Muñoz-Guerra, A Rodríguez-Galán, JA Subirana. Nature 311:53–54, 1984.
5. S Muñoz-Guerra, F López-Carrasquero, JM Fernández-Santín, JA Subirana. In: JC Salamone, ed. Encyclopedia of Polymeric Materials. 6:4694–4700, Boca Raton, FL: CRC, 1996.
6. NA Platé, VP Shibaev. J Polym Sci Macromol Rev 8:117–264, 1974.
7. A Turner-Jones. Makromol Chem 71:1–32, 1964.
8. N Morosoff, H Morawetz, B Post. J Am Chem Soc 87:3035–3040, 1965.
9. AZ Golik, AF Skyshevskii, II Adamenko. Zh Strukt Khim 8:1015–1020, 1967.
10. EF Jordan, DW Feldeisen, AN Wrigley. J Polym Sc A-1 9:1835–1852, 1971.
11. JC Smith, RW Woody. Biopolymers 12:2657, 1973.
12. DS Poché. Synthesis and Characterization of Linear and Star-Branched Poly(γ-stearyl-α,L-glutamate). PhD dissertation, Louisiana State University, 1990.
13. L Onsager. Ann NY Acad Sci 51:627–659, 1949.
14. WH Daly, II Negulescu, PS Russo, D Poche. In: P Shoever, AC Balazs, eds. Macromolecular Assemblies in Polymer Systems. ACS Symp Ser 493:292–299, Washington: Amer Chem Soc, 1992.
15. WH Daly, D Poché, I Negulescu. Prog Polym Sci 19:79–135, 1994.
16. EI Iizuka, K Abe, K Hanabusa, H Shirai. In: RM Ottenbrite, ed. Current Topics in Polymer Science. Munich: Hanser, 1987, pp 235–248.
17. A Thierry, A Skoulios. Mol Cryst Liq Cryst 41:125–128, 1978.
18. J Watanabe, H Ono, I Uematsu, A Abe. Macromolecules 18:2141–2148, 1985.
19. FJ Romero, JL Gómez, JM Barrales-Rienda. Macromolecules 27:5004–5015, 1994.
20. J Watanabe. Thermotropic Liquid Crystals in Polypeptides. Proceedings of the OUMS'93 on Ordering in Macromolecular Systems, Osaka, 1993, A Teramoto, M Kobayashi, T Norisuye, eds. Springer-Verlag, pp 99–113.
21. J Watanabe, H Ono. Macromolecules 19:1079–1083, 1986.
22. FJ Romero, JL Gómez, J Lloveras-Macia, S Muñoz-Guerra. Polymer 32:1642–1646, 1991.
23. J Watanabe, Y Takashina. Macromolecules 24:3423–3426, 1991.

24. D Frenkel. Liq Cryst 5:929–940, 1989.
25. M Hoshimo, H Nakano, H Kimura. J Phys Soc Jpn 46:1709, 1990.
26. J Watanabe, Y Takashina. Polym J 24:709–713, 1992.
27. J Watanabe, Y Fukuda, R Gehani, I Uetmasu. Macromolecules 17:1004–1009, 1982.
28. J Watanabe, M Gotoh, T Nagase. Macromolecules 20:298–304, 1987.
29. J Watanabe, T Nagase. Polymer J 19:781–784, 1987.
30. F López-Carrasquero, M García-Alvarez, S Muñoz-Guerra. Polymer 35:4502–4510, 1994.
31. M García-Alvarez, S León, C Alemán, JL Campos, S Muñoz-Guerra. Macromolecules 31:124–134, 1998.
32. A Martínez de Ilarduya, C Alemán, M García-Alvarez, F López-Carrasquero, S Muñoz-Guerra. Macromolecules 32:3257–3263, 1999.
33. JM Fernández-Santín, S Muñoz-Guerra, A Rodríguez-Galán, J Aymami, J Lloveras, JA Subirana. Macromolecules 20:62–68, 1987.
34. S Muñoz-Guerra, JM Fernández-Santín, C Alegre, JA Subirana. Macromolecules 22:1540–1545, 1989.
35. JM Montserrat, S Muñoz-Guerra, JA Subirana. Makromol Chem Macromol Symp 20/21:319–327, 1988.
36. JJ Navas, C Alemán, F López-Carrasquero, S Muñoz-Guerra. Polymer 38:3477–3484, 1997.
37. F López-Carrasquero, S Montserrat, A Martínez-Ilarduya, S Muñoz-Guerra. Macromolecules 28:5535–5546, 1995.
38. MJ Avrami. Phys Chem 7:1103–1112, 1939.
39. Y Calventus, P Colomer, J Malêk, S Montserrat, F López-Carrasquero, A Martínez-Ilarduya, S Muñoz-Guerra. Polymer 40:801–805, 1999.
40. F López-Carrasquero. Síntesis, Estructura y Propiedades de Poli(α-alquil-β,L-aspartatos). PhD dissertation, Universidad Politécnica de Catalunya, 1995.
41. JM Lehn. Angew Chem Int Ed Engl 29:1304–1319, 1990.
42. A Ciferri. Prog Polym Sci 20:1081–1120, 1995.
43. E Tsuchida. J Macromol Sci Pure Appl Chem A31:1–15, 1994.
44. M Antonietti, J Conrad. Angew Chem Int Ed Engl 33:1869–1870, 1994.
45. A Ciferri. Macromol Chem Phys 195:457–461, 1994.
46. EA Ponomarenko, AJ Waddon, KN Bakeev, DA Tirrell, WJ MacKnight. Macromolecules 29:4340–4345, 1996.
47. EA Ponomarenko, AJ Waddon, DA Tirrell, WJ MacKnight. Langmuir 12:2169–2172, 1996.
48. EA Ponomarenko, DA Tirrell, WJ MacKnight. Macromolecules 29:8751–8758, 1996.
49. EA Ponomarenko, DA Tirrell, WJ MacKnight. Macromolecules 31:1584–1589, 1998.
50. FS Bates. Science 251:898, 1991.
51. AN Semenov, SV Vasilenko. Sov Phys JETP 63(1):70, 1986.
52. AN Semenov. Mol Cryst Liq Cryst 209:191, 1991.
53. AN Semenov, AV Subbotin. Sov Phys JETP 74(4):690, 1992.
54. DRM Williams, GH Fredrickson. Macromolecules 25:3561, 1992.

55. R Holyst, MJ Schick. Chem Phys 96:727, 1992.
56. MWJ Matsen. Chem Phys 104:7758, 1996.
57. RR Netz, M Schick. Phys Rev Lett 77:302, 1996.
58. M Mueller, M Schick. Macromolecules 29:8900, 1996.
59. MW Matsen, CJ Barrett. Chem Phys 109:4108, 1998.
60. A Halperin. Europhys Lett 10:549, 1989.
61. A Halperin. Macromolecules 23:2724, 1990.
62. E Raphael, PG de Gennes. Makromol Symp 62:1, 1992.
63. EM Sevick, DRM Williams. Science 129/130:387, 1997.
64. B Perly, A Douy, B Gallot. Makromol Chem 177:2569, 1976.
65. J-P Billot, A Douy, B Gallot. Makromol Chem 177:1889, 1976.
66. J-P Billot, A Douy, B Gallot. Makromol Chem 178:1641, 1977.
67. A Douy, B Gallot. Makromol Chem 178:1595, 1977.
68. A Douy, B Gallot. Polymer 23:1039, 1982.
69. M Gervais, A Douy, B Gallot, R Erre. Polymer 29:1779, 1988.
70. D Vernino, D Tirrell, M Tirell. Polym Mater Sci Eng 71:496, 1994.
71. A Nakajima, T Hayashi, K Kugo, K Shinoda. Macromolecules 12:840, 1979.
72. T Hayashi, GW Chen, A Nakajima. Polym J (Tokyo) 16:739, 1984.
73. R Yoda, S Komatsuzaki, E Nakanishi, T Hayashi. Eur Polym J 31:335, 1995.
74. R Yoda, M Shimoda, S Komatsuzaki, T Hayashi, T Nishi. Eur Polym J 33:815, 1997.
75. K Kugo, Y Hata, T Hayashi, A Nakajima. Polym J (Tokyo) 14:401, 1982.
76. A Nakajima, K Kugo, T Hayashi. Polym J 11:995, 1979.
77. Y Yoshida, K Makino, T Ito, Y Yamakawa, T Hayashi. Eur Polym J 32:877, 1996.
78. R Yoda, S Komatsuzaki, T Hayashi. Eur Polym J 32:233, 1996.
79. R Yoda, S Komatsuzaki, T Hayashi. Biomaterials 16:1203, 1995.
80. R Yoda, S Komatsuzaki, E Nakanishi, H Kawaguchi, T Hayashi. Biomaterials 15: 944, 1994.
81. H Sato, A Nakajima, T Hayashi, GW Chen, YJ Noishiki. Biomed Mater Res 19: 1135, 1985.
82. GW Chen, T Hayashi, A Nakajima. Polym J (Tokyo) 16:805, 1984.
83. S Barenberg, JM Anderson, PH Geil. Int J Biol Macromol 3:82, 1981.
84. JT Chen, EL Thomas, CK Ober, SS Hwang. Macromolecules 28:1688, 1995.
85. JT Chen, EL Thomas, CK Ober, G-P Mao. Science 273:343, 1996.
86. JJLM Cornelissen, M Fischer, RJM Nolte. Science 280:1427, 1998.
87. TJ Deming, BM Novak. Macromolecules 24:5478, 1991.
88. BM Novak, TJ Deming. Macromol Symp 77:405, 1994.
89. JL David, SP Gido, BM Novak. Polymer Preprints 39/2:433, 1998.
90. P Cavalleri, A Ciferri, C Dell'Erba, M Novi, B Purevsuren. Macromolecules 30: 3513, 1997.
91. P Cavalleri, A Ciferri, C Dell'Erba, A Gabellini, M Novi Macromol Chem Phys 199:2087, 1998.
92. A Gabellini, M Novi, A Ciferri, C Dell'Erba. Acta Polym 50:127, 1999.
93. C-J Yang, SA Jenekhe. Supramol Sci 1:91, 1994.
94. SA Jenekhe, JA Osaheni. Chem Mater 6:1906, 1994.
95. JA Osaheni, SA Jenekhe. J Am Chem Soc 117:7389, 1995.

96. SA Jenekhe, XL Chen. Science 279:1903, 1998.
97. XF Zhong, B Francois. Makromol Chem 192:2277, 1991.
98. B Francois, XF Zhong. Synt Met 41–43:955, 1991.
99. G Widawski, M Rawiso, B Francois. Nature 369:387, 1994.
100. B Francois, O Pitois, J Francois. Adv Mater 7:1041, 1995.
101. T Olinga, B Francois. Makromol Chem Rapid Commun 12:575, 1991.
102. T Olinga, B Francois. J Chim Phys Phys Chim Biol 89:1079, 1992.
103. B Francois, T Olinga. Synth Met 57:3489, 1993.
104. MA Hempenius, BMW Langeveld-Voss, JAEH van Haare, RAJ Janssen, SS Sheiko, JP Spatz, M Moeller, EW Meijer. J Am Chem Soc 120:2798, 1998.
105. V Francke, HJ Raeder, Y Geerts, K Müllen. Macromol Rapid Commun 19:275, 1998.
106. H Kukula, U Ziener, M Schoeps, A Godt. Macromolecules 31:5160, 1998.
107. D Marsitzky, T Brand, Y Geerts, M Klapper, K Müllen. Macromol Rapid Commun 19:385, 1998.
108. LH Radzilowski, JL Wu, SI Stupp. Macromolecules 26:879, 1993.
109. LH Radzilowski, SI Stupp. Macromolecules 27:7747, 1994.
110. LH Radzilowski, BO Carragher, SI Stupp. Macromolecules 30:2110, 1997.
111. SI Stupp, V LeBonheur, K Walker, LS Li, KE Huggins, M Keser, A Amstutz. Science 276:384, 1997.
112. SI Stupp, M Keser, GN Tew. Polymer 39:4505, 1998.
113. GN Tew, L Li, SIJ Stupp. Am Chem Soc 120:5601, 1998.
114. M Lee, N-K Oh, H-K Lee, W-C Zin. Macromolecules 29:5567, 1996.
115. M Lee, N-K Oh, M-G Choi. Polymer Bulletin 37:511, 1996.
116. M Lee, N-K Oh, W-C Zin. Chem Commun 15:1787, 1996.
117. SH Ji, W-C Zin, N-K Oh, M Lee. Polymer 38:4377, 1997.
118. M Lee, B-K Cho. Chem Mater 10:1894, 1998.
119. M Lee, B-K Cho, H Kim, W-C Zin. Angew Chem 110:661–663, 1998.
120. M Lee, B-K Cho, H Kim, JY Yoon, WC Zin. J Am Chem Soc 120:9168, 1998.
121. S Poser, H Fischer, M Arnold. Prog Polym Sci 23:1337, 1998.
122. G Ziegast, B Pfannemüller. Makromol Chem Rapid Commun 5:373, 1984.
123. B Pfannemüller, T Dengler. Makromol Chem 189:1965, 1988.
124. WN Emmerling, B Pfannemüller. Colloid Polym Sci 261:677, 1983.
125. G Ziegast, B Pfannemüller. Carbohydr Res 160:185–204, 1987.
126. V von Braunmühl, G Jonas, R Stadler. Macromolecules 28:17, 1995.
127. V von Braunmühl, R Stadler. Macromol Symp 103:141, 1996.
128. K Loos, R Stadler. Macromolecules 30:7641, 1997.
129. K Kobayashi, S Kamiya, N Enomoto. Macromolecules 29:8670, 1996.

8

Polymers with Intertwined Superstructures and Interlocked Structures

Françisco M. Raymo and J. Fraser Stoddart
University of California, Los Angeles, California

I. MECHANICALLY INTERLOCKED MACROMOLECULES

Macromolecules incorporating repeating units connected by covalent bonds are widespread in nature [1]. Synthetic procedures for the construction of their artificial counterparts are well established [2]. Furthermore, the properties of these unnatural macromolecules are now rather well understood and, indeed, polymeric materials have found applications in numerous branches of science and technology [2]. In recent years, synthetic chemists have learned* how to introduce mechanical bonds (Fig. 1) into small molecules. Mechanically interlocked "rings," as well as "wheels" mechanically trapped onto "axles," can be constructed efficiently to afford molecular compounds, named catenanes and rotaxanes, respectively.† Metal coordination [18–32], donor/acceptor interactions [33–43], hydrogen bonds [44–64] and/or hydrophobic interactions [65–78] between appropriate components have all been employed to template‡ the formation of these exotic molecules. Making the transition from simple catenanes and rotaxanes to their macromolecular counterparts—namely, polycatenanes and polyrotaxanes,

---

* For books and reviews on mechanically interlocked molecules and macromolecules see Refs. 3–17.

† The term *catenane* derives from the Latin word *catena* meaning chain. The term *rotaxane* derives from the Latin words *rota* and *axis* meaning wheel and axle, respectively.

‡ For accounts and reviews on template-directed syntheses, see Refs. 79–90.

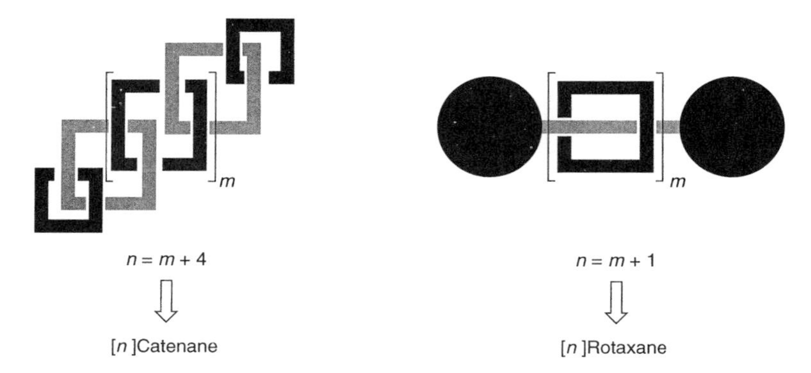

Figure 1 Schematic representation of an [*n*]catenane and an [*n*]rotaxane.

respectively—offers the possibility of generating a range of novel polymeric materials. Indeed, the fundamental difference between "conventional" macromolecules and these "unconventional" polymers (i.e., the presence of mechanical bonds) is expected to confer unusual properties upon them, which could have profound technological implications. Mechanically interlocked macromolecules can be regarded as the molecular-sized counterparts of macroscopic objects, such as abacuses, bearings, chains, and joints, which could become the components of some of the smallest possible devices sometime in the future.

II. POLYCATENANES

Despite some preliminary and highly speculative claims [91–97], the synthesis of a (*linear*) [*n*]catenane (Fig. 1) incorporating more than five mechanically interlocked rings has still to be achieved. A compound called Olympiadane [98–100] (i.e., a [5]catenane) has been isolated and fully characterized following a two-step procedure (Fig. 2) that relies upon the mutual recognition which exists among the components in contiguous rings. Reaction of **1** and **2** in the presence of the macrocyclic polyether **3** as a template leads to the [3]catenane **4** which was isolated in a yield of 10%, after counterion exchange. A small amount of a [2]catenane incorporating cyclobis(paraquat-4,4′-phenylene) and only one macrocyclic polyether component was also obtained. Each of the two macrocyclic polyether rings present in **4** is large enough to accommodate within its cavity at least one other bipyridinium unit belonging to a cyclophane. Thus, reaction of **5** and **6**, in the presence of the [3]catenane **4** as a template affords the [5]catenane **7**, with its five mechanically interlocked rings related to one another in a linear fashion,

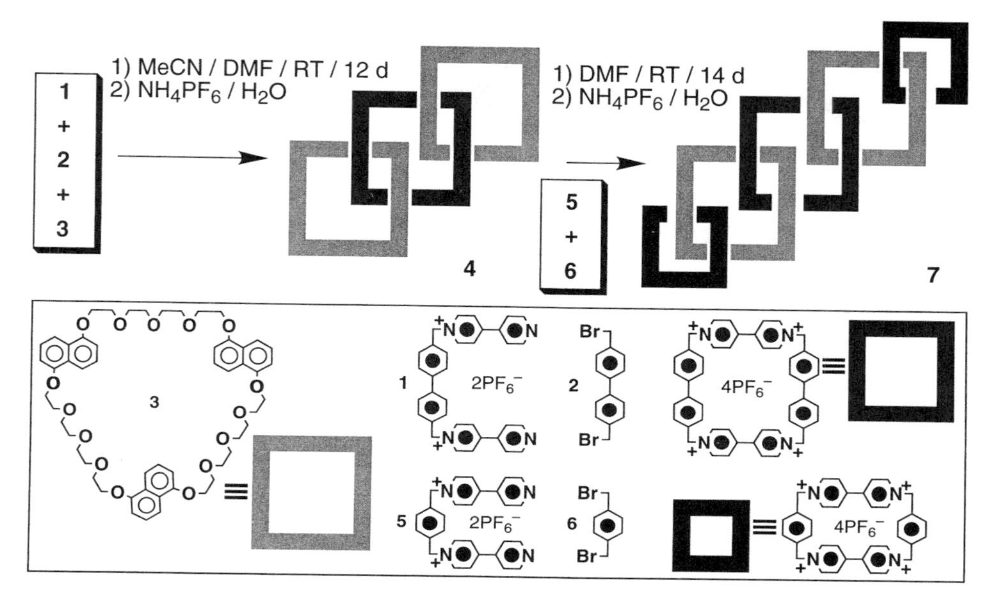

Figure 2 The two-step template-directed synthesis of the [5]catenane **7**.

in a yield of 18%. In addition, a [4]catenane, incorporating two different tetracationic cyclophanes and two identical macrocyclic polyether components, was also obtained in a yield of 51%. When the same reaction was carried out under ultrahigh pressure (12 kbar), the yield of the [5]catenane **7** rose to 30%. Furthermore, a [6]catenane and a [7]catenane were also obtained in yields of 28 and 26%, respectively. These higher *branched* catenanes incorporate one and two, respectively, additional cyclobis(paraquat-*p*-phenylene) cyclophane components. The [5]catenane **7** was characterized by liquid secondary ionization mass spectrometry (LSIMS) which revealed signals resulting from the loss of hexafluorophosphate counterions as well as from the loss of the component macrocycles. The mechanical-interlocking of the five ring components in this [5]catenane was demonstrated unequivocally by X-ray crystallography. Additionally, the X-ray structural analysis of **7** revealed that the $[\pi \cdots \pi]$ stacking interactions between the π-donors and π-acceptors are accompanied by [C—H $\cdots$ O] hydrogen bonds between selected polyether oxygen atoms and certain α-bipyridinium hydrogen atoms, as well as by [C—H $\cdots$ π] interactions between some of the 1,5-dioxynaphthalene hydrogen atoms and the phenylene rings in the tetracationic cyclophane components.

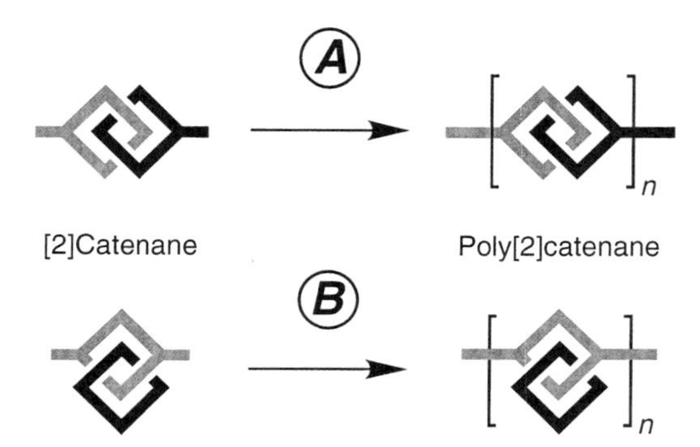

Figure 3 Synthetic strategies for the preparation of poly[2]catenanes.

Poly[2]catenanes (Fig. 3) are macromolecules composed of [2]catenane repeating units linked by covalent bonds. The syntheses of poly[2]catenanes have been realized according to the procedures A or B depending on whether the starting [2]catenane bears one reactive group on each of the two interlocked rings or two reactive groups on the same ring. [2]Catenanes, incorporating one reactive group attached to each of the two interlocked rings, have been obtained on account of the templating resulting from hydrogen-bonding interactions between the amide functions present in the two rings. Indeed, upon reaction [101] of the bisamine **8** with the bisacid chloride **9**, the [2]catenane **10**, which incorporates an aryl bromine substituent on each of its two ring components, self-assembles (Fig. 4) in a yield of 9%. Coupling of this bisfunctionalized [2]catenane in turn with the bisfunctionalized reagents **11** and **12** affords the poly[2]catenanes **13** and **14**, in yields of 84 and 99%, respectively. Gel permeation chromatographic (GPC) analysis of **13** and **14** revealed number-average molecular weights (M_n) of 3.0×10^3 and 3.3×10^3, respectively, with the respective weight-average molecular weights M_w being 3.6×10^3 and 5.0×10^3.

Comparison of the proton nuclear magnetic resonance (^{1}H-NMR) spectra, recorded in CD_2Cl_2 at various temperatures (−30 to 135°C), of the poly[2]catenanes **13** and **14**, with those of the corresponding monomeric [2]catenanes, shows that the signals are broadened significantly for the polymers. Presumably, the broad signals are a result of longer relaxation times imposed by the polymeric structure rather than of dynamic processes involving the relative movements of the interlocked components. Differential scanning calorimetry (DSC) revealed a glass transition at approximately 245°C, even though these polymers are only of moderate molecular sizes. A similar approach was employed [102] in the construction (Fig. 5) of the bisfunctionalized [2]catenane **17**. The bisamine **15** was reacted with the bisacid chloride **16** in the presence of Et_3N to afford the corre-

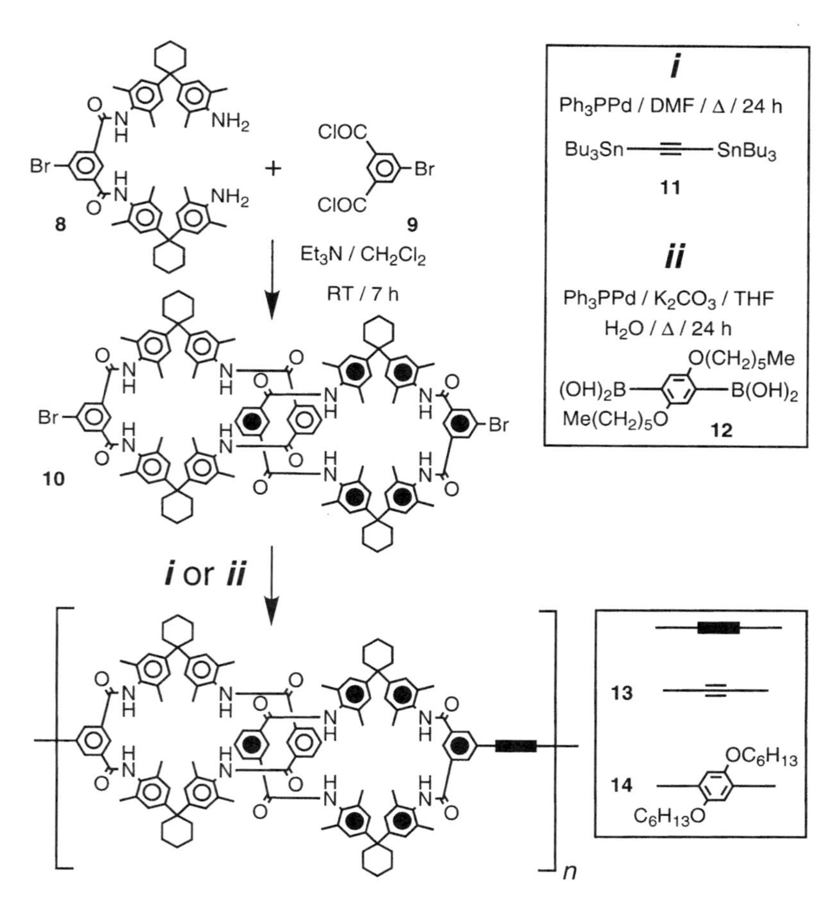

Figure 4 The hydrogen-bonding assisted template-directed synthesis of the [2]catenane **10** and its conversion into the poly[2]catenanes **13** and **14**.

sponding [2]catenane. Methylation of the amide groups, followed by the hydrolysis of the benzyloxy groups, afforded the [2]catenane **17** in a yield of 82%. Copolyesterification of this compound with the terephthalic acid derivative **18** under mild conditions afforded the poly[2]catenane **19**. Once again, the ^{1}H-NMR spectrum of the polymer exhibits significantly broader signals than that of the [2]catenane monomer. The formation of ester bonds was confirmed by infrared spectroscopy (IR) which reveals a band characteristic for the [C=O] group of the ester functions at 1754 cm^{-1}. The molecular weight distribution was investigated by GPC using polystyrene standards. These investigations revealed M_n and M_w val-

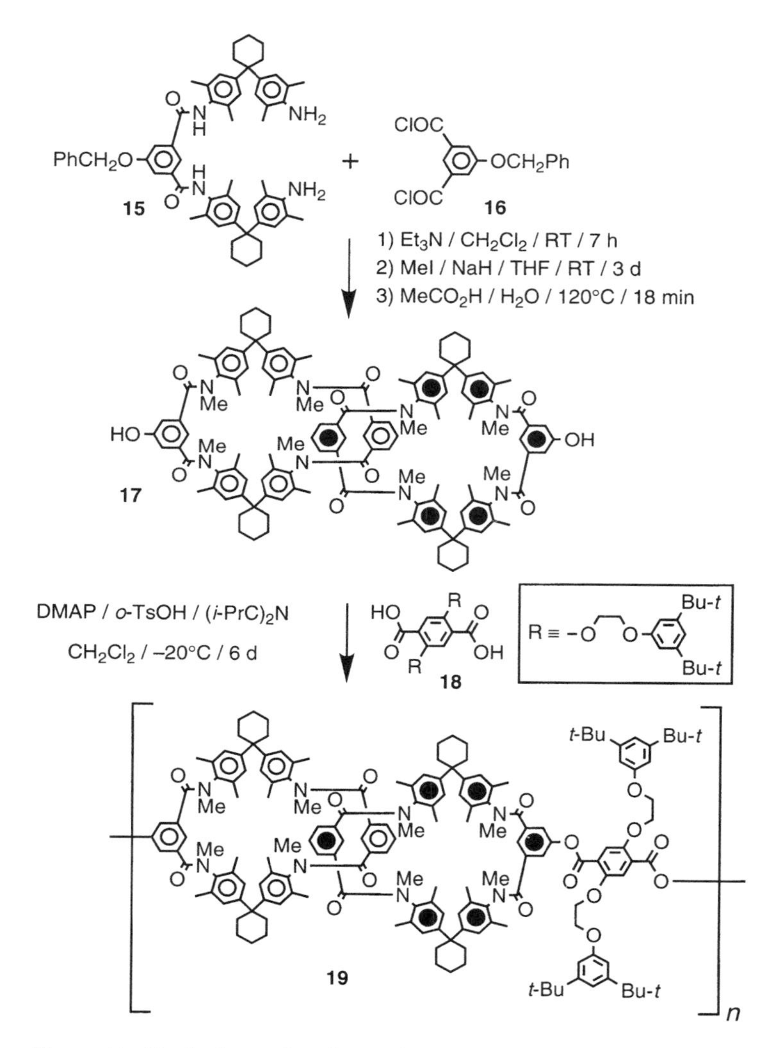

Figure 5 The hydrogen-bonding assisted template-directed synthesis of the [2]catenane **17** and its conversion into the poly[2]catenane **19**.

ues of 13.5 × 10^3 and 29.1 × 10^3, respectively, although the matrix-assisted laser desorption ionization time-of-flight (MALDI-TOF) mass spectrometric analysis of the poly[2]catenane **19** had indicated species with molecular weights of up to 52.0 × 10^3. Furthermore, DSC analysis of the phase transition behavior revealed a high glass transition temperature (T_g = 265°C), along with a very high thermal stability (380°C).

Threading of the phenanthroline-containing macrocyclic polyether **20** onto the acyclic phenanthroline-containing diphenol **21** occurs [103] in the presence of Cu^+ ions (Fig. 6). Reaction of the resulting complex **22** with the diiodide **23**, followed by reduction and demetalation, gave the [2]catenane **24**, carrying one hydroxymethyl group on each of its two ring components. Copolyesterification of this [2]catenane under mild conditions afforded (Fig. 7) the poly[2]catenane **25** with M_n and M_w values of 5.5 × 10^4 and 180 × 10^4, respectively. The metal-containing form of **24** was also copolymerized under identical conditions to yield a poly[2]catenate having values of 60.0 × 10^4 and 420 × 10^4 for M_n and M_w, respectively. Although the ^{1}H-NMR and ^{13}C-NMR spectra revealed resonances characteristic of the [2]catenane subunit as well as of the spacer unit, the signals were significantly broader than those observed in the spectra of the monomeric components. Thermogravimetric analysis of the metalated and demetalated forms of this poly[2]catenane revealed that the thermal stability is approximately 90°C higher in the absence of the metal ions. Similarly, the glass transition temperature of the demetalated form is higher than that of the metalated form by approximately 15°C. The bisfunctionalized [2]catenate **29** has been prepared [104] from the phenanthroline-containing compounds **26** and **27** (Fig. 8). These species form the complex **28** when mixed with $Cu(MeCN)_4BF_4$ in solution. Reaction of **28** with the appropriate diiodide gave the [2]catenate **29** which was (i) demetalated, (ii) deprotected, (iii) remetalated, (iv) copolymerized, and finally (v) demetalated to afford (Fig. 9) the poly[2]catenane **30**. The ^{1}H-NMR spectrum of this poly[2]-catenane recorded in $CDCl_3$ revealed very broad signals, presumably as a result of the restricted mobility of the interlocked components. The formation of amide bonds was confirmed by IR spectroscopy which indicated the presence of a band for the [C=O] groups at 1650 cm^{-1}. GPC analysis of **30** in DMF revealed M_n and M_w values of 8.1 × 10^5 and 33.2 × 10^5, respectively, for this polymer with numerous mechanical linkages.

Bisfunctionalized [2]catenanes have been also prepared by employing template-directed syntheses that involve the interaction of π-donors and π-acceptors. Reaction (Fig. 10) of the dibromide **31** with the dicationic salt **32** in the presence of either **33** or **34** as the macrocyclic polyether component afforded [105] the [2]catenanes **35** and **36**, respectively, after counterion exchange. The aromatic hydroxymethyl group located within the tetracationic cyclophane portion of the [2]catenane **35** was converted [106] into a chloromethyl group by treatment of **35** with 10 M HCl_{aq}. After counterion exchange, the chloromethyl group was

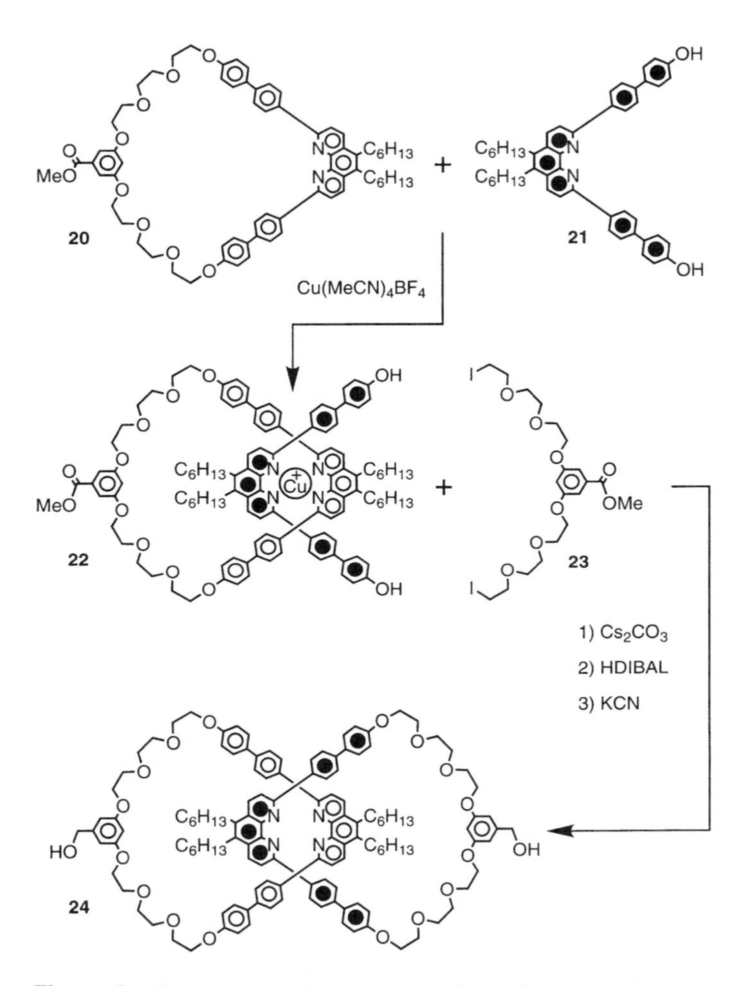

Figure 6 The metal-template synthesis of the [2]catenane **24**.

transformed into a bromomethyl group since it is reactive enough to enable in situ polyesterification in the presence of a base. Comparison of the ^{1}H-NMR spectrum of the resulting poly[2]catenane **37** with that of the monomeric [2]catenane **35** highlighted some significant differences. In particular, the methylene protons of the hydroxymethyl group of the [2]catenane give rise to a singlet. By contrast, the methylene protons of the ester groups of the poly[2]catenane appear as a multiplet. GPC analysis of the chloride salt of poly[2]catenane **37** revealed it to have an M_w value of 35.0×10^3. The [2]catenane **36** has a hydroxymethyl

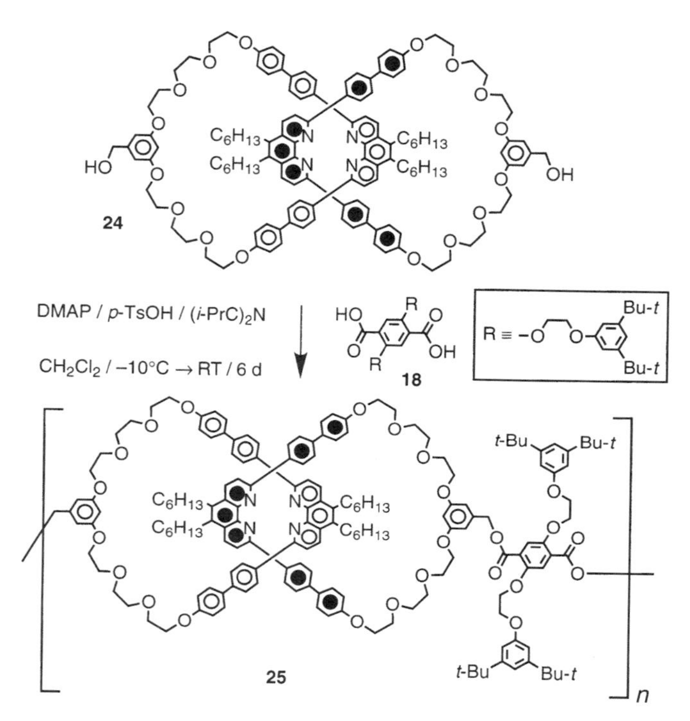

Figure 7 The synthesis of the poly[2]catenane **25**.

group attached to each of its two ring components. Its copolymerization with the bis-isocyanate **38** afforded [105] a poly[2]catenane **39** which can incorporate up to three different "bridging motifs." The diphenylmethane bridge can link together (i) two identical π-electron rich or (ii) two identical π-electron deficient macrocyclic components of two adjacent [2]catenane repeating units, or (iii) a π-electron rich to a π-electron deficient macrocyclic component. Presumably, the poly[2]catenane **39** is a mixture of constitutionally isomeric macromolecules. The ^{1}H- and ^{13}C-NMR spectra of this poly[2]catenane indicate the presence of signals characteristic of the [2]catenane subunit as well as of the bridging spacer unit. In addition, the IR spectrum of **39** showed bands at 3375 and 1734 cm^{-1} characteristic of the [N—H] and [C=O] groups, respectively, in the urethane linkages. GPC analysis of the chloride salt of the poly[2]catenane **39** revealed an M_w value of 26.5×10^3, suggesting that it incorporates 17 repeating units on average. The bisfunctionalized [2]catenane **42** incorporates two hydroxymethyl groups attached to the hydroquinone ring in its macrocyclic polyether salt component. This [2]catenane was synthesized [106] by reacting (Fig. 11) the dicationic

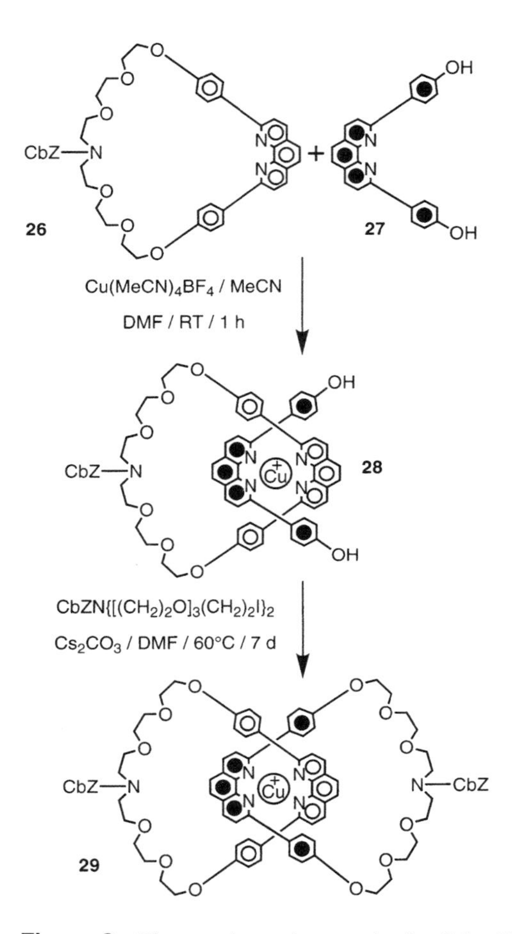

Figure 8 The metal-template synthesis of the [2]catenane **29**.

salt **32** with the dibromide **41** in the presence of the crown ether derivative **40**, followed by hydrolysis of the two acetyloxy groups, and counterion exchange. Copolymerization of the bisfunctionalized [2]catenane **42** with the bis-isocyananate **38** afforded the poly[2]catenane **43**. The IR spectrum of this poly[2]catenane confirmed the formation of urethane linkages, displaying two bands at 3375 and 1734 cm^{-1} for the [N—H] and [C═O] groups, respectively. GPC analysis of the chloride salt of **43** revealed a M_w value of 27.0×10^3, corresponding to a degree of polymerization of 20.

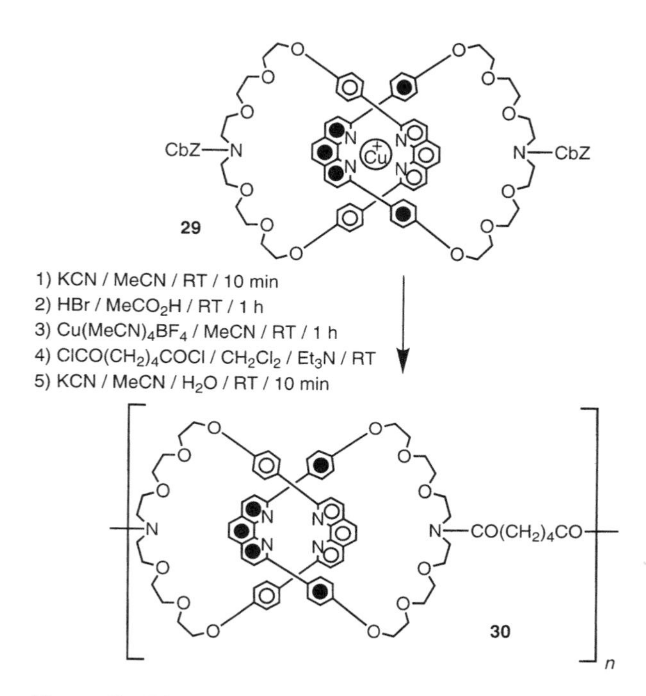

Figure 9 The synthesis of the poly[2]catenane **30**.

An alternative approach to mechanically interlocked macromolecules involves the copolymerization of bisfunctionalized bis[2]catenanes,* rather than of simple [2]catenanes carrying two functional groups. The bis[2]catenane **45**, which incorporates two hydroxymethyl groups, was obtained [106] by reacting (Fig. 12) the dibromide **31** with the dicationic salt **32** in the presence of bis(macrocyclic polyether) **44**. Copolymerization of the bis[2]catenane **44** and the bis-isocyanate **38** afforded the poly(bis[2]catenane) **46**. Similarly, the bis[2]catenane **48** has been constructed [106] by reacting (Fig. 13) the dibromide **31** with the dicationic salt **32** in the presence of bis(macrocyclic polyether) **47**. Copolymerization of the bis[2]catenane **48** with the bis-isocyanate **38** afforded the poly(bis[2]catenane) **49**. GPC analyses of the chloride salts of the poly(bis[2]catenane)s **46** and **49** revealed an M_w value of 45.0×10^3 for both polymers, indicating that they incorporate 15 repeating units on average. Their IR spectroscopic analysis

* For examples of bis[2]catenanes, see Refs. 107–112.

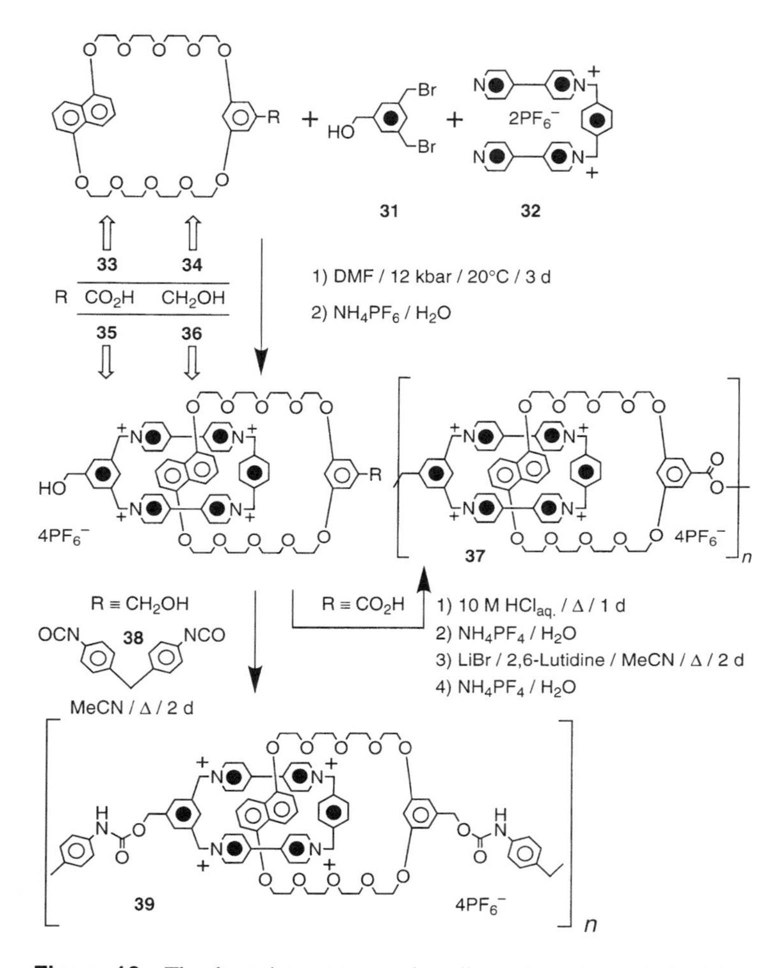

Figure 10 The donor/acceptor template-directed synthesis of the [2]catenanes **35** and **36** and their conversion into the poly[2]catenanes **37** and **39**, respectively.

demonstrated the formation of urethane linkages by showing the characteristic bands at 3375 and 1734 cm^{-1} in their spectra. The bis[2]catenane **51** has also been prepared [113] using a template-directed synthetic strategy (Fig. 14) relying upon interactions between π-donors and π-acceptors. Reaction of the dibromide **41** with the dicationic salt **50** in the presence of the bis(macrocyclic polyether) **44** afforded the bis[2]catenane **51**. This compound incorporates 2,2′-bipyridine units that are able, in principle, to bind transition metals. Indeed, on mixing the

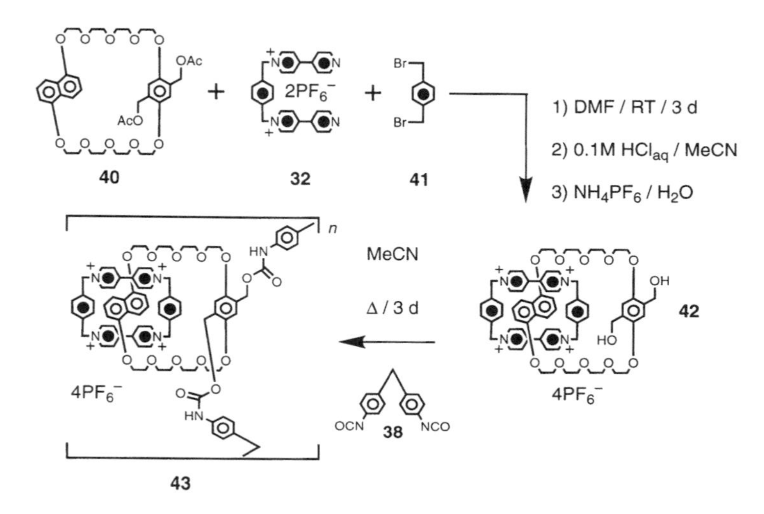

Figure 11 The donor/acceptor template-directed synthesis of the [2]catenane **42** and its conversion into the poly[2]catenane **43**.

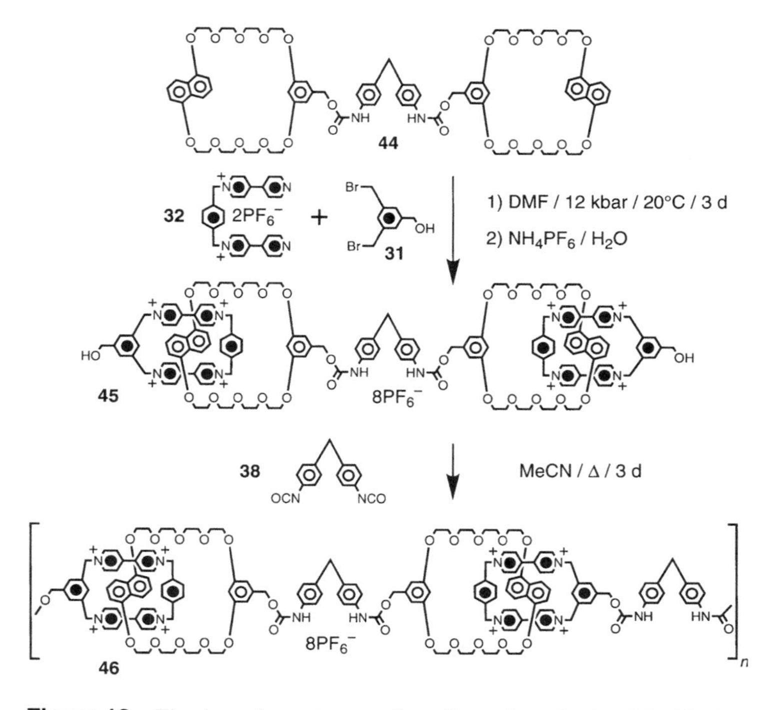

Figure 12 The donor/acceptor template-directed synthesis of the bis[2]catenane **45** and its conversion into the polybis[2]catenane **46**.

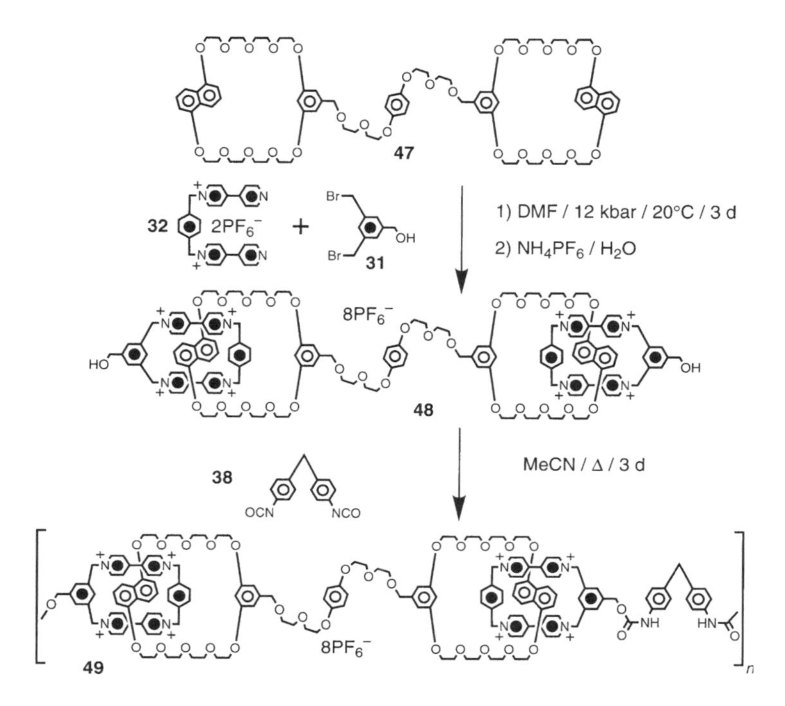

Figure 13 The donor/acceptor template-directed synthesis of the bis[2]catenane **48** and its conversion into the polybis[2]catenane **49**.

bis[2]catenane **51** with CF_3SO_3Ag, the poly(bis[2]catenane) **52** self-assembled spontaneously. Comparison of the ^{1}H-NMR spectra of the bis[2]catenane with that of the poly(bis[2]catenane) indicated the presence of significant differences which are particularly evident for the resonances associated with the 2,2′-bipyridine protons. GPC analysis of the chloride salt of the poly(bis[2]catenane) **52** revealed that it has an M_w value of 150.0×10^3. This particular poly(bis[2]catenane) is held together by a combination of covalent, mechanical, and coordinative bonds. Fascinating to say the least!

III. MAINCHAIN POLYROTAXANES

Mainchain polyrotaxanes can be synthesized (Fig. 15) from pseudopolyrotaxanes by attaching covalently bulky groups either at the termini or along the backbone of the acyclic component. Pseudopolyrotaxanes can be prepared following route C or D either by polymerizing *n* preformed pseudorotaxanes or by threading

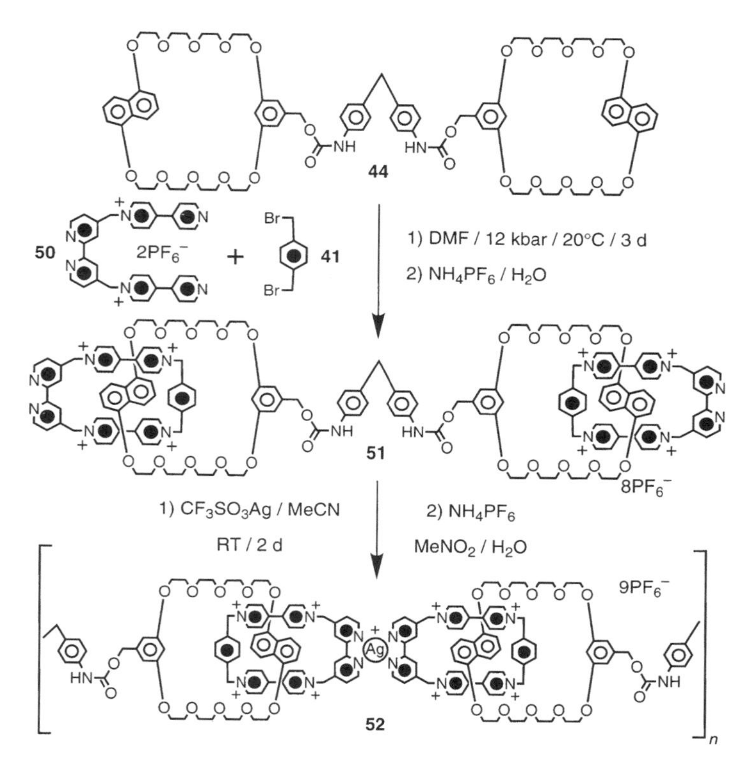

Figure 14 The donor/acceptor template-directed synthesis of the bis[2]catenane **51** and its conversion into the polybis[2]catenane **52**.

n rings onto a preformed acyclic polymer, respectively. The first examples of pseudopolyrotaxanes were prepared [114]* following route C by solution or interfacial copolymerization of diamines and diacid dichlorides in the presence of β-cyclodextrin (β-CD). A similar approach was employed [119] (Fig. 16) to incorporate α-cyclodextrin **55** into the polyrotaxane **56** which was obtained by reacting the diaminobenzidine **53** with the diol **54** in the presence of α-CD and $RuCl_2(PPh_3)_3$. Along its acyclic backbone, this polyrotaxane incorporates benzimidazole units that are too bulky to thread through the cavity of α-CD. As a result, each ring component is trapped mechanically between two benzimidazole units and encircles the aliphatic spacer that bridges the two bulky units. GPC analysis of the polyrotaxane **56** afforded M_n and M_w values of 4.0×10^3 and 7.2

* For other early examples of pseudopolyrotaxanes prepared following route C, see Refs. 115–118.

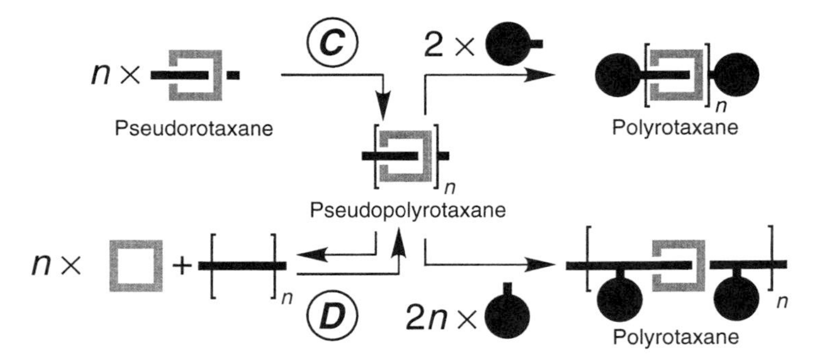

Figure 15 Strategies for the preparation of a pseudopolyrotaxane and its conversion into polyrotaxanes.

× 10^3, respectively, while ^{1}H-NMR spectroscopic studies showed a ring/repeating unit ratio of 0.163. Interestingly, the rigidity imposed on the polymeric backbone by the encircling α-CD rings imposes a 20°C increase in the T_g of the polyrotaxane relative to the T_g of the polymer in its free form.

Pseudopolyrotaxanes incorporating crown ethers have also been prepared following route C. For example, the reaction of ethylene oxide with the potassium salt of tetraethylene glycol [120,121], the free radical polymerization of acryloni-

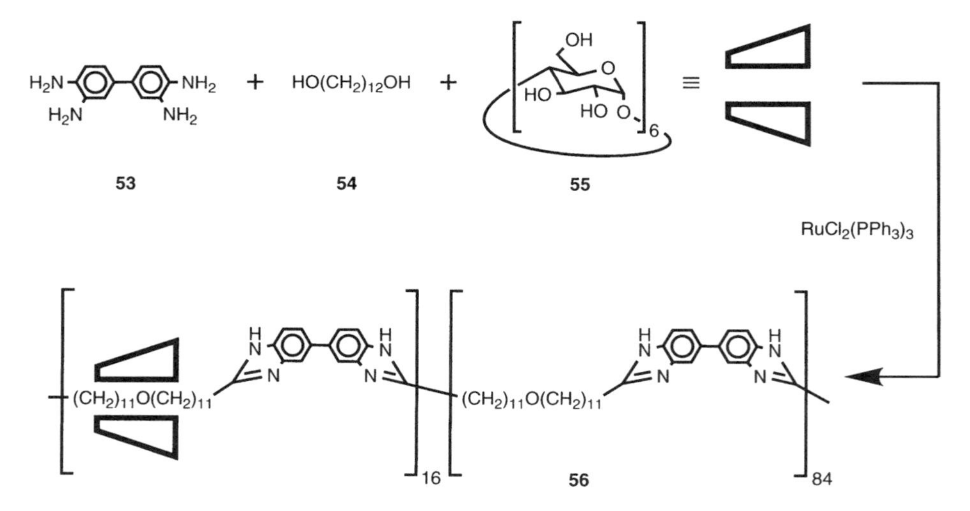

Figure 16 The synthesis of the polyrotaxane **56**.

trile or styrene [122–125], the polycondensation of diols and diacid dichlorides [126–130], and transesterification polymerizations [131,132] have all been carried out in the presence of a preformed crown ether, often used as the solvent, to afford pseudopolyrotaxanes or polyrotaxanes. However, it was found [126–130] that the threading efficiency increases significantly when bulky groups are introduced into the monomers. Thus, reaction (Fig. 17) of the tetraarylmethane-based diol **57** with the tetraarylmethane-based diacid dichlorides **58** in the presence of the crown ether **59** affords [129] the polyrotaxane **60** which was isolated pure following consecutive reprecipitations using THF/H_2O mixtures. In this polyrotaxane, each ring component is trapped mechanically between two tetraarylmethane groups encircling some of the aliphatic spacers that bridge pairs of bulky units. The ^{1}H-NMR spectra of the polyrotaxane **60** indicated the presence of signals for the cyclic and acyclic components and, from their relative intensities, a ring/repeating unit ratio of 0.172 was determined. GPC analysis of the polyrotaxane **60** revealed it to have M_n and M_w values of 25.2×10^3 and 87.3×10^3,

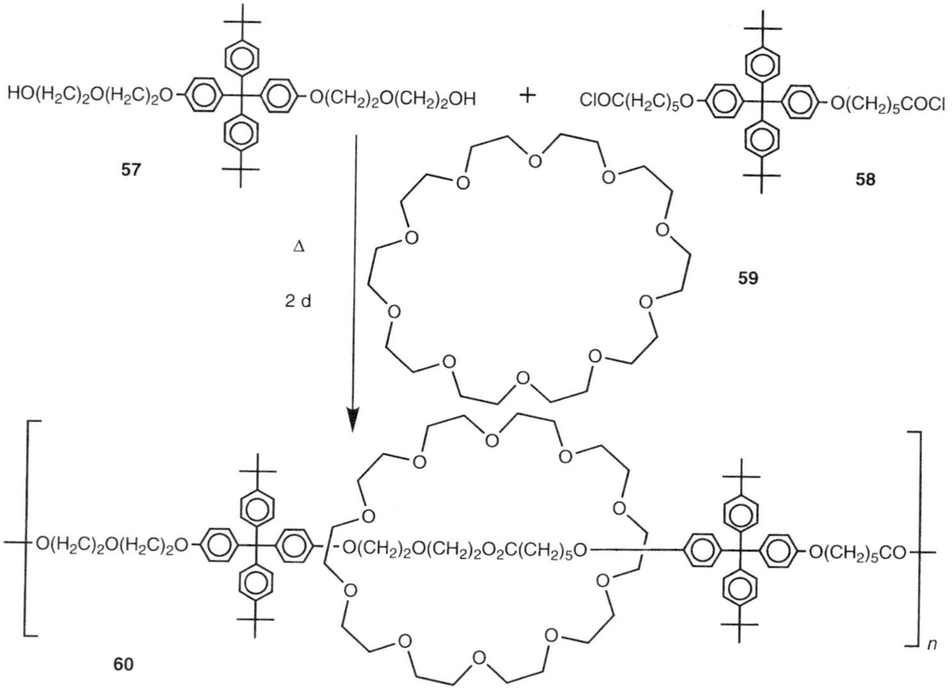

Figure 17 The synthesis of the polyrotaxane **60**.

respectively. Furthermore, the GPC traces did not contain the characteristic peak associated with the crown ether **59** in its free form, confirming that the acyclic and cyclic components of the polyrotaxane **60** are mechanically interlocked. A further increase in the threading efficiency was observed [133–135] when poly (urethane)s were employed as the acyclic components. Indeed, [N—H · · · O] hydrogen bonds [136] between the urethane hydrogen atoms of the polymer and the polyether oxygen atoms of the ring component are believed to assist the threading process. Thus, when diols are reacted with bis-isocyanates, pseudopolyrotaxanes incorporating up to 63% by mass of the ring component are obtained [133–135]. The polyrotaxane **62** was prepared [137,138] by reacting (Fig. 18) the tetraarylmethane-based diol **57** with the bis-isocyanate **61** in the presence of 30-crown-10 **(59)**. The absence of free 30-crown-10 was confirmed by GPC analysis which also revealed M_n and M_w values of 23.4×10^3 and 37.8×10^3, respectively. ^{1}H-NMR spectroscopic studies on this polyrotaxane indicated that it has ''solvent switchable'' microstructures. In $CDCl_3$, [N—H · · · O] hydrogen bonds

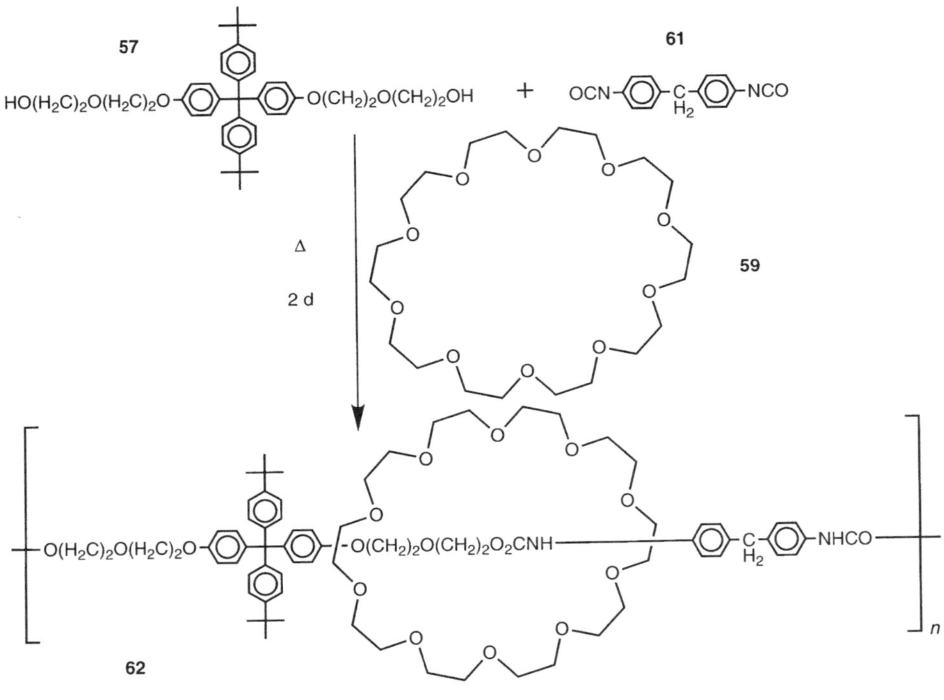

Figure 18 The synthesis of the polyrotaxane **62**.

between the urethane hydrogen atoms and the polyether oxygen atoms encourage the crown ether component to encircle the urethane units, as suggested by the upfield shift for the polyether protons that suffer shielding effects exerted by the diphenylmethane unit. However, the rings move toward the tetraarylmethane groups in $(CD_3)_2SO$ to allow for solvation of the urethane linkages by this highly polar solvating solvent.

Pseudopolyrotaxanes incorporating cycloalkanes [139], cyclodextrins [140–159], cyclourethanes [160–162], cyclophanes [163–165], and crown ethers [166–169] have all been prepared following route D. Hydrophobic interactions assist [170,171] the threading of α-CD (**55**) onto polyamine **63** in aqueous solution to yield (Fig. 19) pseudopolyrotaxanes. On threading, the specific viscosity (η_{sp}) increases gradually with time and reaches a constant value after 2 h when an equilibrium between complexed and uncomplexed species is reached. The opposite process was investigated by removing the macrocyclic component from the equilibrium by dialysis. The decreasing concentration of α-CD was monitored by optical rotation measurements. Complete dethreading was achieved after 15 h. Upon covalent attachment of nicotinoyl groups to some of the amine groups of this pseudopolyrotaxane, the threaded rings are trapped mechanically onto the

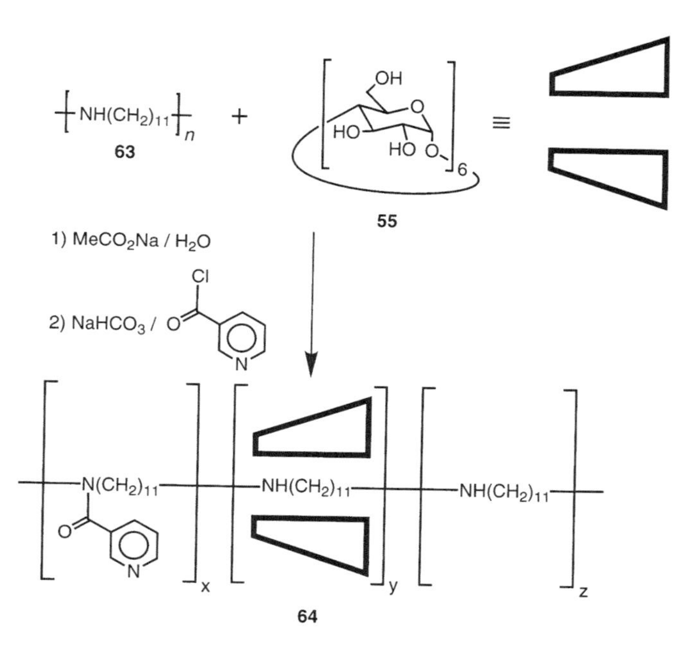

Figure 19 The synthesis of the polyrotaxane **64**.

polymeric backbone, affording the polyrotaxane **64**. From the relative intensities of the resonances in the ^{1}H-NMR spectrum of **64**, it was established that this polyrotaxane incorporates 10 mol% of threaded rings. A similar approach was employed in the supramolecularly assisted synthesis of a polyrotaxane incorporating two different types of macrocycles. When γ-CD (**65**) and β-CD (**66**) are combined (Fig. 20) with the stilbene-containing polymer **67** and the stilbene-containing "monomer" **68** in H_2O, the pseudopolyrotaxane **69** self-assembles [172]

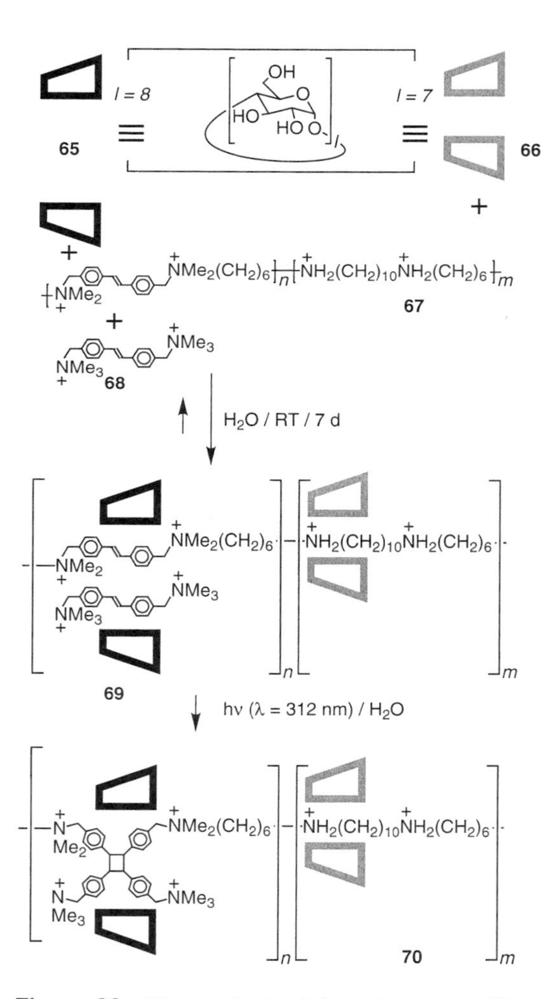

Figure 20 The synthesis of the polyrotaxane **70**.

spontaneously. On irradiation of an aqueous solution of **69** with ultraviolet (UV) light at 312 nm for 30 h tetraphenylenecyclobutane blocking groups are formed, yielding the polyrotaxane **70**. The formation of ''isolated'' phenylene rings was confirmed by the increased absorption at 250 nm. In addition, the signals for the [CH=CH] protons of the stilbene units disappear in the ^{1}H-NMR spectrum which also demonstrates that this polyrotaxane incorporates 20 mol% of threaded rings. GPC analysis did not show any free or easily removable CD ring, confirming that **70** is a polyrotaxane.

Threading of α-CD (**55**) onto the poly(ethylene glycol) bisamine **71**, followed by reaction (Fig. 21) with 2,4-dinitrofluorobenzene, affords [173–175] the polyrotaxane **72**. Mono- and bi-dimensional NMR spectroscopic analysis showed

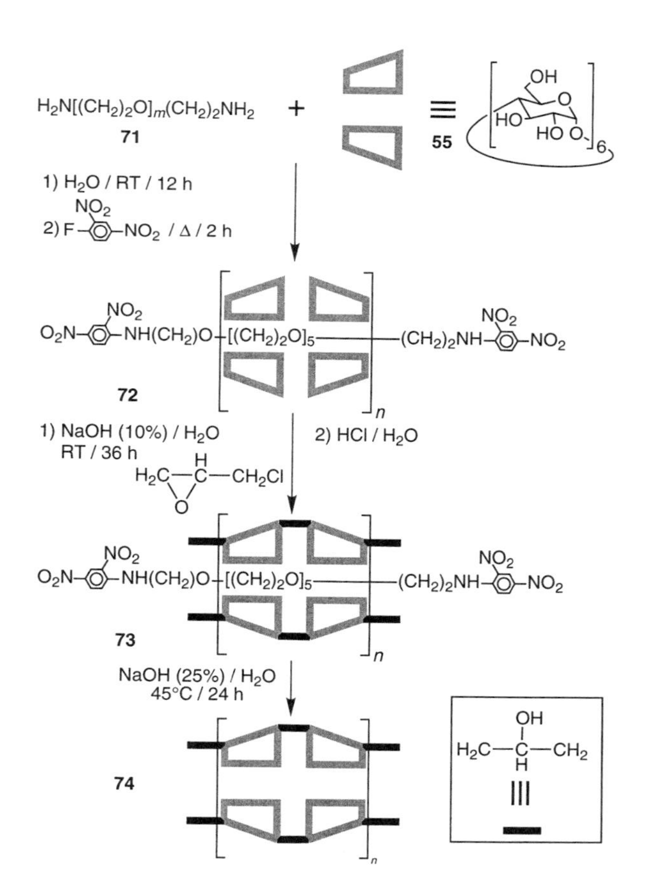

Figure 21 The template-directed synthesis of the cyclodextrin-based nanotube **74**.

signals for the ring, thread, and stopper components and confirmed that the poly (ethylene glycol) chain is inserted through the cavities of the rings. A combination of ^{1}H-NMR and UV spectroscopic and optical rotation data revealed that this polyrotaxane incorporates approximately one ring for every two bismethylene units, suggesting that the α-CD rings are close-packed from end to end along the polymer chain. On average, 15 to 20 α-CDs are incorporated into this polyrotaxane. However, a polyrotaxane incorporating as many as 37 α-CDs was obtained by the fractionation of the product. Treatment of **72** with epichlorohydrin resulted [176,177] in the covalent bridging of the threaded α-CD rings. Finally, cleavage of the terminal stoppers of **72** released the intact ''molecular tube'' **74**, composed of covalently bridged α-CD rings. GPC analysis of this novel material revealed that its average molecular weight is less than 20.0×10^3.

The formation of pseudopolyrotaxanes incorporating cyclobis(paraquat-*p*-phenylene) (**75**) and π-electron rich polyethers has been achieved [163–165] through a reliance on a combination of [C—H ··· O] hydrogen bonds, [π ···

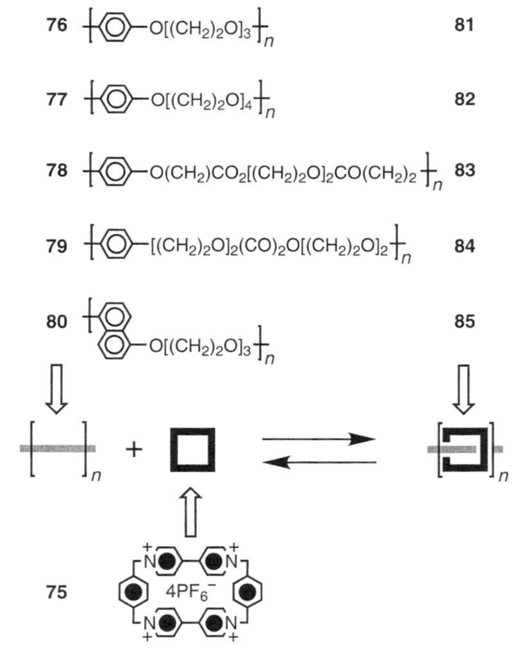

Figure 22 The self-assembly of pseudopolyrotaxanes incorporating the bipyridinium-based tetracationic cyclophane **75**.

π] stacking, and [C—H · · · π] interactions (see Fig. 22). The bipyridinium-based tetracationic cyclophane **75** binds [33–39] dioxyarene-based guests, producing complexes with pseudorotaxane geometries. When **75** is mixed with one of the dioxyarene-based polyethers **76–80** in solution, a purple color develops [178,179] as a result of the charge-transfer interactions between the π-electron rich and deficient recognition sites present in the thread and ring components of the pseudopolyrotaxanes **81–85**. Detailed ^{1}H-NMR spectroscopic investigation revealed that the amount of threaded rings is affected significantly (i) by the nature of the spacer unit separating the dioxyarene recognition sites and (ii) by the temperature.

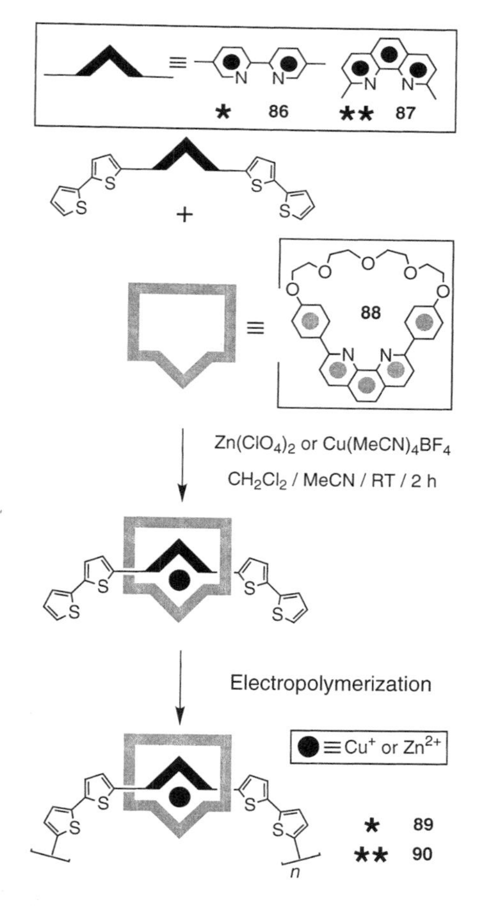

Figure 23 The synthesis of the metal-containing pseudopolyrotaxanes **89** and **90**.

In the case of pseudopolyrotaxane **79**, 94% of the π-electron rich units are encircled by a tetracationic cyclophane at −40°C.

Threading of the phenanthroline-based macrocycle **88** onto either the bipyridine- or the phenanthroline-based bis-thiophene derivatives **86** and **87**, respectively, occurs [180–182] (Fig. 23) spontaneously in the presence of metal ions (Zn^{2+} or Cu^{+}) which can be coordinated tetrahedrally. Electropolymerization of the resulting metal-containing pseudorotaxanes afforded the metal-containing pseudopolyrotaxanes **89** and **90**. The decomplexation/complexation of Zn^{2+} ions can be achieved reversibly in the case of **89**. By contrast, the reinsertion of Cu^{+} ions into **90** is only possible when the demetalation is performed in the presence of Li^{+} ions which prevent the collapse of the polymer by coordinating with the phenanthroline units. The redox and conducting properties of **89** are affected dramatically by the coordinated metal ions that produce charge localization and participate in conduction by means of redox processes.

IV. SIDECHAIN POLYROTAXANES

Sidechain polyrotaxanes incorporating CDs have been prepared [183–189] by reacting (Fig. 24) a preformed semirotaxane with a polymer having reactive groups on its sidechains. Upon mixing the methylated β-CD **91** with an acyclic compound bearing a trityl group at one end, the semirotaxane **92** precipitated [183] out and could be separated from its components by filtration. This semirotaxane was characterized by fast atom bombardment mass spectrometry (FABMS) which revealed peaks corresponding to the ‘‘molecular’’ ion of **92**.

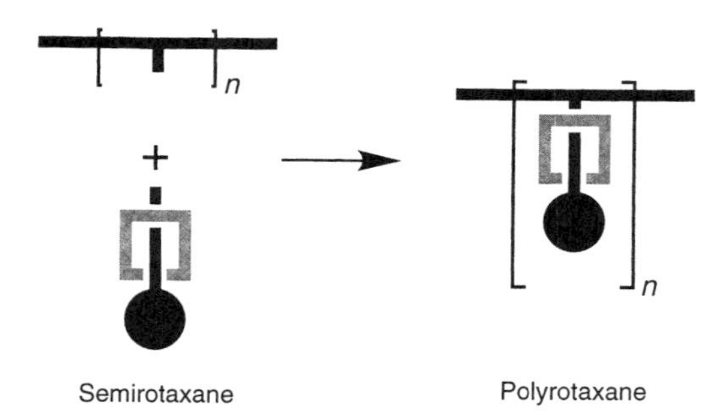

Figure 24 Synthetic strategy for the construction of sidechain polyrotaxanes.

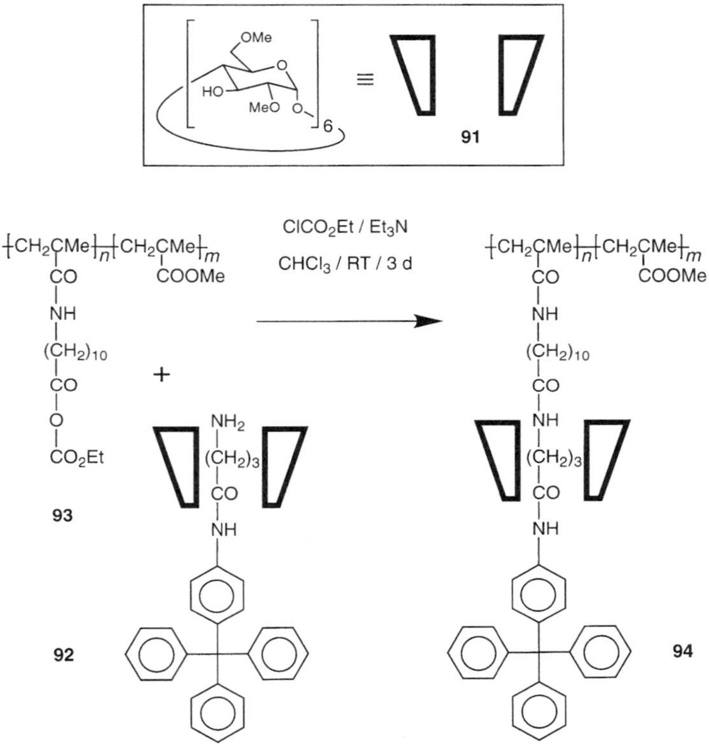

Figure 25 The synthesis of the polyrotaxane **94**.

Reaction of **92** with the preformed polymer **93** afforded (Fig. 25) the sidechain polyrotaxane **94**. The incorporation of β-CD rings into the polymer **94** was demonstrated by IR and ^{1}H-NMR spectroscopy: no free β-CD was detected by chromatography.

The presence of interlocked β-CD rings in the polyrotaxane **94** affects significantly its solubility and viscosity. For example, this polyrotaxane is soluble in Et_2O, whereas the parent polymer is not. Also, the viscosity of the polyrotaxane is lower than that of the parent polymer and shows a very different temperature dependence. A similar strategy was employed [190,191] to synthesize (Fig. 26) sidechain polyrotaxanes by connecting two semirotaxane subunits to the same sidechain. The semirotaxane **92** was reacted with the preformed polymer **95**, bearing two reactive groups per sidechain, to afford the polyrotaxane **96**. The

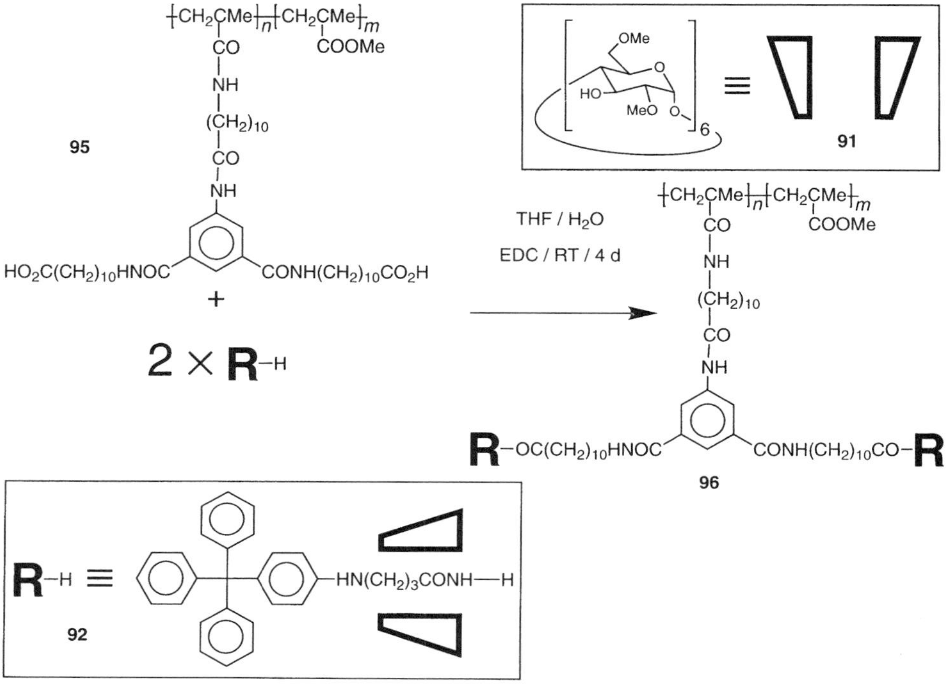

Figure 26 The synthesis of the polyrotaxane **96**.

^{1}H-NMR spectrum of the polyrotaxane **96** revealed signals broader than those observed for the parent polymer. These observations indicate that each CD ring moves rapidly on the ^{1}H-NMR timescale along the aliphatic chain at ambient temperature. However, the rings are located preferentially next to the terminal bulky groups at $-10°C$. DSC measurements showed an increase of the glass transition temperature of approximately 19°C for the polyrotaxane relative to the parent polymer. Another example of a sidechain polyrotaxane is illustrated in Fig. 27. In this instance, the semirotaxane **98**, incorporating the β-CD derivative **97**, was reacted [192] with the poly(benzimidazole) **99** in the presence of sodium hydride. The IR spectra of the resulting polyrotaxane **100** displayed the same bands that are observed for the parent polymer with the addition of a signal at 1040 cm^{-1} for the [C—O] groups of the CD rings. The amount of threaded macrocycles was determined by ^{1}H-NMR spectroscopy which indicated that 57% of the sidechains of the polyrotaxane **100** are encircled by β-CD rings. GPC

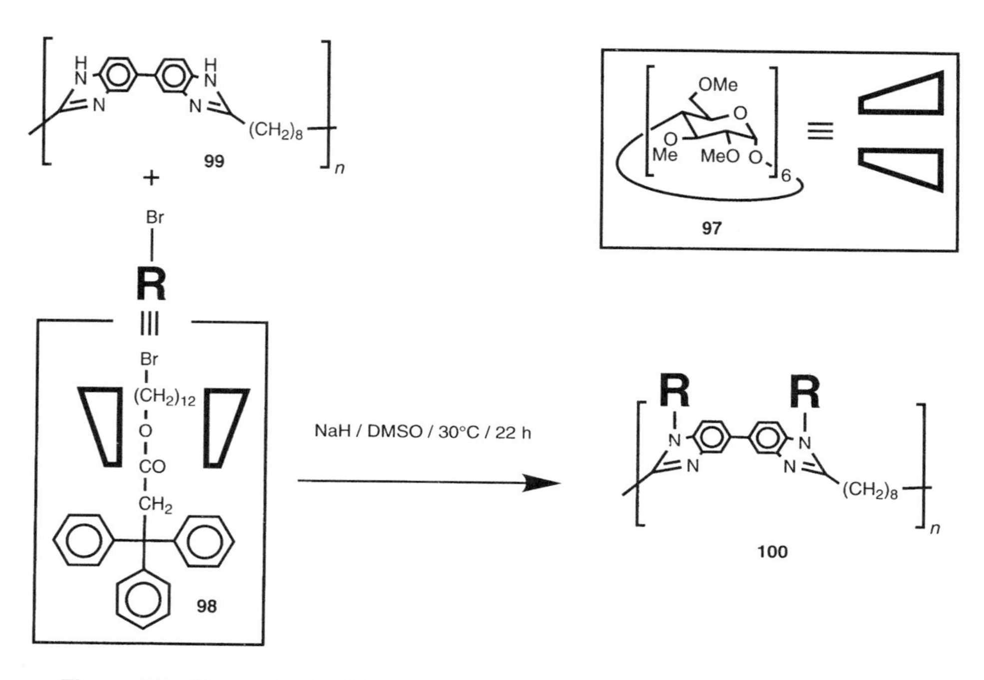

Figure 27 The synthesis of the polyrotaxane **100**.

analysis of this material revealed M_n and M_w values of 25.4×10^3 and 42.1×10^3, respectively.

An alternative approach to sidechain polyrotaxanes involves the threading of acyclic monomeric components through the cavities of large rings appended to a polymeric backbone. This strategy was employed [193–198] to generate (Fig. 28) sidechain pseudopolyrotaxanes. The polymers **101–103** were obtained by polymerizing preformed macrocyclic monomers. As a result of [C—H · · · O] hydrogen bonds and [$\pi \cdots \pi$] stacking interactions, insertion of the bipyridinium-based guests **104** and **105** inside the cavities of the appended rings of **101–103** affords sidechain pseudopolyrotaxanes. Interestingly, electrostatic perturbation on binding induces a significant decrease of carrier mobility and, as a result, conductivity in **101**. Similarly, a reduction of the fluorescence intensity of **102** and **103** is observed on binding. It is believed that the excitations diffuse along the polymer backbone and are quenched by the bipyridinium-based guests inserted through the appended macrocycles. Thus, in all instances, pseudopolyrotaxane formation can be easily detected by monitoring the change of the properties associated with the polymer backbone.

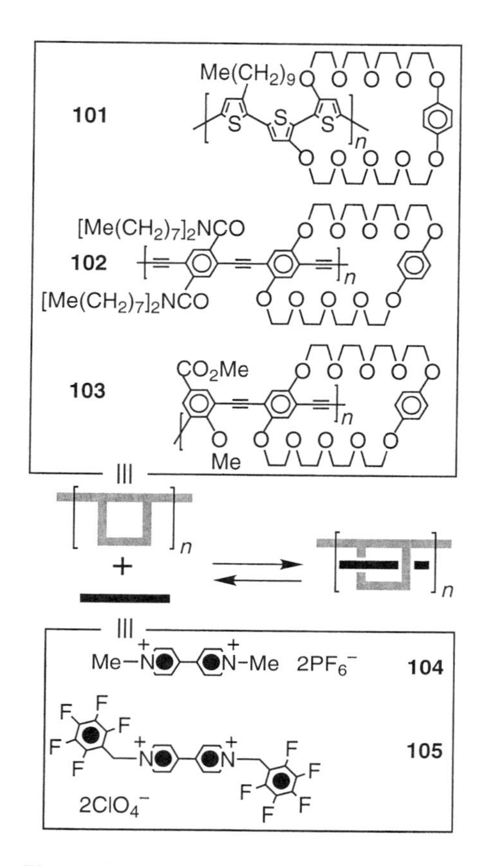

Figure 28 The self-assembly of pseudopolyrotaxanes incorporating π-electron rich and deficient components.

V. PROPERTIES

Since the synthetic strategies for the preparation of poly[2]catenanes have been developed only very recently [101–106,113], their properties are just beginning to be investigated. By contrast, the properties of pseudopolyrotaxanes and polyrotaxanes have been studied extensively and are understood in some considerable detail already. Indeed, it has been established that solubilities, phase transition behavior, hydrodynamic volumes, and the intrinsic viscosities of pseudopolyrotaxanes and polyrotaxanes are significantly different from those of their separate components. Solubilities can either be depressed or enhanced as a result of interlocking. For example, poly(ethylene glycol) and α-CD are soluble [147–159] in H_2O. By contrast, the pseudopolyrotaxane formed on threading poly(ethylene glycol) through many α-CD rings is not soluble in water. On the other hand, poly(acrylonitrile) is insoluble [168] in MeOH but it becomes soluble when encircled by 60-crown-20. In most instances, two distinct melting points, slightly lower than those of the separate acyclic and cyclic components, are associated with pseudopolyrotaxanes. This behavior is observed normally for those pseudopolyrotaxanes where the ring components are sufficiently mobile to aggregate, nucleate, and crystallize. Significant changes in the T_g values are observed when phase mixing of the mechanically interlocked components occurs. In polyurethane-based pseudopolyrotaxanes, the T_g corresponds [135] to a weighted average of those of the separate components, while the T_gs of sidechain polyrotaxanes tend [190] to be often higher than those of the parent components. Extensive GPC analyses have revealed [199] that hydrodynamic volumes and, as a result, intrinsic viscosities increase on mechanical interlocking. Indeed, the change of intrinsic viscosity on threading has been exploited [170] to monitor the formation of CD-containing pseudopolyrotaxanes.

VI. CONCLUSIONS

Polycatenanes and polyrotaxanes are macromolecules held together by a combination of covalent and mechanical bonds, in some instances aided and abetted by noncovalent bonding interactions. This ''unconventional'' combination of bonds imposes upon these polymeric systems unique topologies and unusual properties, making them particularly attractive synthetic targets. Not surprisingly, a number of synthetic strategies for the construction of mechanically interlocked macromolecules have been developed. Most of these synthetic approaches rely upon the assistance of noncovalent bonding interactions and borrow recognition motifs widely employed for the template-directed syntheses of monomeric catenanes and rotaxanes. Thus, poly[2]catenanes, as well as main- and sidechain polyrotaxanes, are now more than just simple chemical curiosities: they can be prepared easily and efficiently even in large quantities! Investigation of their physical properties has revealed that their rheological behavior can be affected drastically by the nature and frequency of interlocking structural elements, suggesting that mechanically bonded polymers might find some important applications in the near future.

REFERENCES

1. EA MacGregor, CT Greenwood. Polymers in Nature. New York: Wiley, 1980.
2. JC Salamone, ed. The Polymeric Materials Encyclopedia. Boca Raton, FL: CRC, 1996.
3. G Schill. Catenanes, Rotaxanes and Knots. New York: Academic, 1971.

4. DM Walba. Tetrahedron 41:3161–3212, 1985.
5. CO Dietrich-Buchecker, J-P Sauvage. Chem Rev 87:795–810, 1987.
6. CO Dietrich-Buchecker, J-P Sauvage. Bioorg Chem Front 2:195–248, 1991.
7. J-C Chambron, CO Dietrich-Buchecker, J-P Sauvage. Top Curr Chem 165:131–162, 1993.
8. HW Gibson, H Marand. Adv Mater 5:11–21, 1993.
9. HW Gibson, MC Bheda, PT Engen. Prog Polym Sci 19:843–945, 1994.
10. DB Amabilino, IW Parsons, JF Stoddart. Trends Polym Sci 2:146–152, 1994.
11. DB Amabilino, JF Stoddart. Chem Rev 95:2725–2828, 1995.
12. HW Gibson. Large Ring Molecules. JA Semlyen, ed. New York: Wiley, 1996, pp 191–202.
13. M Belohradsky, FM Raymo, JF Stoddart. Collect Czech Chem Commun 61:1–43, 1996.
14. FM Raymo, JF Stoddart. Trends Polym Sci 4:208–211, 1996.
15. M Belohradsky, FM Raymo, JF Stoddart. Collect Czech Chem Commun 62:527–557, 1997.
16. R Jäger, F Vögtle. Angew Chem Int Ed Engl 36:930–944, 1997.
17. J-P Sauvage, CO Dietrich-Buchecker, eds. Catenanes, Rotaxanes and Knots. Weinheim: VCH-Wiley, 1999.
18. J-P Sauvage. Acc Chem Res 23:319–327, 1990.
19. J-C Chambron, CO Dietrich-Buchecker, C Hemmert, AK Khemiss, D Mitchell, J-P Sauvage, J Weiss. Pure Appl Chem 62:1027–1034, 1990.
20. J-C Chambron, S Chardon-Noblat, A Harriman, V Heitz, J-P Sauvage. Pure Appl Chem 65:2343–2349, 1993.
21. J-C Chambron, CO Dietrich-Buchecker, J-F Nierengarten, J-P Sauvage. Pure Appl Chem 66:1543–1550, 1994.
22. JC Chambron, CO Dietrich-Buchecker, V Heitz, J-F Nierengarten, J-P Sauvage, C Pascard, J Guilhem. Pure Appl Chem 67:233–240, 1995.
23. J-C Chambron, CO Dietrich-Buchecker, J-P Sauvage. In: Comprehensive Supramolecular Chemistry. vol. 9. MW Hosseini, J-P Sauvage, eds. Oxford: Pergamon, 1996, pp 43–83.
24. F Bickelhaupt. J Organomet Chem 475:1–14, 1994.
25. M Fujita, K Ogura. Coord Chem Rev 148:249–264, 1996.
26. M Fujita. In: Comprehensive Supramoleuclar Chemistry, vol. 9. MW Hosseini, J-P Sauvage, eds. Oxford: Pergamon, 1996, pp. 253–282.
27. YM Jeon, D Whang, J Kim, K Kim. Chem Lett 503–504, 1996.
28. D Whang, YM Jeon, J Heo, K Kim. J Am Chem Soc 118:11333–11334, 1996.
29. D Whang, J Heo, CA Kim, K Kim. Chem Commun 2361–2362, 1997.
30. D Whang, K Kim. J Am Chem Soc 119:451–452, 1997.
31. D Whang, KM Park, J Heo, PR Ashton, K Kim. J Am Chem Soc 120:4899–4900, 1998.
32. SG Roh, KM Park, GJ Park, S Sakamoto, K Yamaguchi, K Kim. Angew Chem Int Ed 38:638–643, 1999.
33. DB Amabilino, JF Stoddart. Pure Appl Chem 65:2351–2359, 1993.
34. D Pasini, FM Raymo, JF Stoddart. Gazz Chim Ital 125: 431–435, 1995.
35. SJ Langford, JF Stoddart. Pure Appl Chem 68:1255–1260, 1996.

36. DB Amabilino, FM Raymo, JF Stoddart. In: MW Hosseini, J-P Sauvage, eds. Comprehensive Supramolecular Chemistry vol. 9. Oxford: Pergamon 1996, pp 85–130.
37. FM Raymo, JF Stoddart. Pure Appl Chem 69:1987–1997, 1997.
38. RE Gillard, FM Raymo, JF Stoddart. Chem Eur J 3:1933–1940, 1997.
39. FM Raymo, JF Stoddart. Chemtracts 11:491–511, 1998.
40. DG Hamilton, JKM Sanders, JE Davies, W Clegg, SJ Teat. Chem Commun 897–898, 1997.
41. AC Try, MM Harding, DG Hamilton, JKM Sanders. Chem Commun 723–724, 1998.
42. DG Hamilton, JE Davies, L Prodi, JKM Sanders. Chem Eur J 4:608–620, 1998.
43. DG Hamilton, N Feeder, L Prodi, SJ Teat, W Clegg, JKM Sanders. J Am Chem Soc 120:1096–1097, 1998.
44. CA Hunter, DH Purvis. Angew Chem Int Ed Engl 31:792–795, 1992.
45. CA Hunter. J Am Chem Soc 114:5303–5311, 1992.
46. CA Hunter. Chem Soc Rev 23:101–109, 1994.
47. FJ Carver, CA Hunter, RJ Shannon. J Chem Soc Chem Commun 1277–1280, 1994.
48. H Adams, FJ Carver, CA Hunter. J Chem Soc Chem Commun 809–810, 1995.
49. G Brodesser, R Güther, R Hoss, S Meier, S Ottens-Hildebrandt, J Schmitz, F Vögtle. Pure Appl Chem 65:2325–2328, 1993.
50. F Vögtle, T Dünnwald, T Schmidt. Acc Chem Res 29:451–460, 1996.
51. F Vögtle, R Jäger, M Händel, S Ottens-Hildebrandt. Pure Appl Chem 68:225–232, 1996.
52. AG Johnston, DA Leigh, RJ Pritchard, MD Deegan. Angew Chem Int Ed Engl 34: 1209–1212, 1995.
53. AG Johnston, DA Leigh, L Nezhat, JP Smart, MD Deegan. Angew Chem Int Ed Engl 34:1212–1216, 1995.
54. DA Leigh, K Moody, JP Smart, KJ Watson, AMZ Slawin. Angew Chem Int Ed Engl 35:306–310, 1996.
55. AG Johnston, DA Leigh, A Murphy, JP Smart, MD Deegan. J Am Chem Soc 118: 10662–10663, 1996.
56. DA Leigh, A Murphy, JP Smart, AMZ Slawin. Angew Chem Int Ed Engl 36:728–732, 1997.
57. AS Lane, DA Leigh, A Murphy. J Am Chem Soc 119: 11092–11093, 1997.
58. DA Leigh, A Murphy, JP Smart, MS Deleuze, F Zerbetto. J Am Chem Soc 120: 6458–6467, 1998.
59. AG Kolchinski, DH Busch, NW Alcock. J Chem Soc Chem Commun 1289–1291, 1995.
60. AG Kolchinski, NW Alcock, RA Roesner, DH Busch. Chem Commun 1437–1438, 1998.
61. PT Glink, C Schiavo, JF Stoddart, DJ Williams. Chem Commun 1483–1490, 1996.
62. PT Glink, JF Stoddart. Pure Appl Chem 70:419–424, 1998.
63. MCT Fyfe, JF Stoddart. Coord Chem Rev 183:139–155, 1999.
64. MCT Fyfe, JF Stoddart. Adv Supramol Chem 5:1–53, 1999.
65. JF Stoddart. Angew Chem Int Ed Engl 31:846–848, 1992.
66. H Ogino. New J Chem 17:683–688, 1993.
67. G Wenz, F Wolf; M Wagner, S Kubik. New J Chem 17:729–738, 1993.

68. R Isnin, AE Kaifer. Pure Appl Chem 65:495–498, 1993.
69. A Harada. Polym News 18:358–363, 1993.
70. A Harada, J Li, M Kamachi. Proc Jpn Acad 69:39–44, 1993.
71. G Wenz. Angew Chem Int Ed Engl 33:802–822, 1994.
72. A Harada. Coord Chem Rev 148:115–133, 1996.
73. A Harada. In: Large Ring Molecules. JA Semlyen, ed. New York: Wiley, 1996, pp 406–432.
74. A Harada. Supramol Sci 3:19–23, 1996.
75. A Harada. Adv Polym 133:142–191, 1997.
76. A Harada. Carbohydr Polym 34:183–188, 1997.
77. A Harada. Acta Polym 49:3–17, 1998.
78. SA Nepogodiev, JF Stoddart. Chem Rev 98:1959–1976, 1998.
79. DH Busch, NA Stephenson. Coord Chem Rev 100:119–154, 1990.
80. JS Lindsey. New J Chem 15:153–180, 1991.
81. GM Whitesides, JP Mathias, CT Seto. Science 254:1312–1319, 1991.
82. D Philp, JF Stoddart. Synlett 445–458, 1991.
83. DH Busch. J Inclusion Phenom 12:389–395, 1992.
84. S Anderson, HL Anderson, JKM Sanders. Acc Chem Res 26:469–475, 1993.
85. R Cacciapaglia, L Mandolini. Chem Soc Rev 22:221–231, 1993.
86. R Hoss, F Vögtle. Angew Chem Int Ed Engl 33:375–384, 1994.
87. JP Schneider, JW Kelly. Chem Rev 95:2169–2187, 1995.
88. D Philp, JF Stoddart. Angew Chem Int Ed Engl 35:1155–1196, 1996.
89. FM Raymo, JF Stoddart. Pure Appl Chem 68:313–322, 1996.
90. MCT Fyfe, JF Stoddart. Acc Chem Res 30:393–401, 1997.
91. G Karagounis, I Pandi-Agathokli. Prakt Akad Athenon 45:118–126, 1970.
92. G Karagounis, J Pandi-Agathokli, E Kondaraki. Chim Cronika 1:130–147, 1972.
93. G Karagounis, I Pandi-Agathokli, E Petassis, A Alexakis. Folia Bioch Biol Graeca 10:31–41, 1973.
94. G Karagounis, E Kontakari, E Petassis. Prakt Akad Athenon 49:118–126, 1973.
95. G Karagounis, I Pandi-Agathokli, E Kontakari, D Nikolelis. Prakt Akad Athenon 49:501–513, 1974.
96. G Karagounis, I Pandi-Agathokli, E Kontakari, D Nikoleis. IUPAC Colloid and Surface Science, International Conference, Selected Papers A 1:671–678, 1975.
97. G Karagounis, M Pandazi. Proceedings of the Fifth International Conference on Raman Spectroscopy, Freiburg im Breisgau, Freiburg, 1976, pp 72–73.
98. DB Amabilino, PR Ashton, AS Reder, N Spencer, JF Stoddart. Angew Chem Int Ed Engl 33:1286–1290, 1994.
99. DB Amabilino, PR Ashton, SE Boyd, JY Lee, S Menzer, JF Stoddart, DJ Williams. Angew Chem Int Ed Engl 36:2070–2072, 1997.
100. DB Amabilino, PR Ashton, SE Boyd, JY Lee, S Menzer, JF Stoddart, DJ Williams. J Am Chem Soc 120:4295–4307, 1998.
101. Y Geerts, D Muscat, K Müllen. Macromol Chem Phys 196:3425–3435, 1995.
102. D Muscat, A Witte, W Köhler, K Müllen, Y Geerts. Macromol Rapid Commun 18:233–241, 1997.
103. JL Weidmann, JM Kern, JP Sauvage, Y Geerts, D Muscat, K Müllen. Chem Commun 1243–1244, 1996.

104. S Shimada, K Ishiwara, N Tamaoki. Acta Chem Scand 52:374–376, 1998.
105. S Menzer, AJP White, DJ Williams, M Belohradsky, C Hamers, FM Raymo, AN Shipway, JF Stoddart. Macromolecules 31:295–307, 1998.
106. C Hamers, FM Raymo, JF Stoddart. Eur J Org Chem 2109–2117, 1998.
107. PR Ashton, AS Reder, N Spencer, JF Stoddart. J Am Chem Soc 115:5286–5287, 1993.
108. PR Ashton, JA Preece, JF Stoddart, MS Tolley. Synlett 789–792, 1994.
109. DB Amabilino, PR Ashton, JA Preece, JF Stoddart, MS Tolley. Am Chem Soc Div Polym Chem Polym Prepr 36(1):587–588, 1995.
110. PR Ashton, J Huff, IW Parsons, JA Preece, JF Stoddart, DJ Williams, AJP White, MS Tolley. Chem Eur J 2:123–136, 1996.
111. J Huff, JA Preece, JF Stoddart. Macromol Symp 102:1–8, 1996.
112. PR Ashton, T Horn, S Menzer, JA Preece, N Spencer, JF Stoddart, DJ Williams. Synthesis 480–488, 1997.
113. C Hamers, O Kocian, FM Raymo, JF Stoddart. Adv Mater 10:1366–1369, 1998.
114. N Ogata, K Sanui, J Wada. J Polym Sci Polym Lett Ed 14:459–462, 1976.
115. M Maciejewski, G Smets. Pr Nauk Inst Technol Organicz Tworz 16:57–69, 1975.
116. M Maciejewski, M Panasiewicz. J Macromol Sci Chem A12:701–718, 1978.
117. M Maciejewski. J Macromol Sci Chem A13:77–85, 1979.
118. M Maciejewski, A Gwizdowski, P Peczak, A Pietrzak. J Macromol Sci Chem A13: 87–109, 1979.
119. I Yamaguchi, K Osakada, T Yamamoto. J Am Chem Soc 118:1811–1812, 1996.
120. G Agam, D Graiver, A Zilkha. J Am Chem Soc 98:5206–5214, 1976.
121. G Agam, A Zilkha. J Am Chem Soc 98:5214–5216, 1976.
122. PT Engen, PR Lecavalier, HW Gibson. Am Chem Soc Div Polym Chem Polym Prepr 31(2):703–704, 1990.
123. HW Gibson, PT Engen, S-H Lee, S Liu, H Marand, MC Bheda. Am Chem Soc Div Polym Chem Polym Prepr 34(1):64–65, 1993.
124. HW Gibson, PT Engen. New J Chem 17:723–727, 1993.
125. SH Lee, PT Engen, HW Gibson. Macromolecules 30:337–343, 1997.
126. C Gong, HW Gibson. Macromolecules 29:7029–7033, 1996.
127. C Gong, HW Gibson. Macromol Chem Phys 198:2321–2332, 1997.
128. HW Gibson, S Liu, C Gong, Q Ji, E Joseph. Macromolecules 30:3712–3727, 1997.
129. C Gong, Q Ji, TE Glass, HW Gibson. Macromolecules 30:4807–4813, 1997.
130. C Gong, HW Gibson. Macromolecules 30:8524–8525, 1997.
131. C Wu, MC Bheda, C Lim, YX Shen, J Sze, HW Gibson. Polym Commun 32:204–207, 1991.
132. HW Gibson, S Liu, P Lecavalier, C Wu, YX Shen. J Am Chem Soc 117:852–874, 1995.
133. YX Shen, C Lim, HW Gibson. Am Chem Soc Div Polym Chem 32(1):166–167, 1991.
134. XY Shen, HW Gibson. Macromolecules 25:2058–2059, 1992.
135. XY Shen, D Xie, HW Gibson. J Am Chem Soc 116:537–548, 1994.
136. E Marand, Q Hu, HW Gibson, B Veystman. Macromolecules 29:2555–2562, 1996.
137. C Gong, HW Gibson. Angew Chem Int Ed Engl 36:2331–2333, 1997.
138. C Gong, TE Glass, HW Gibson. Macromolecules 31:308–313, 1998.

139. IT Harrison. J Chem Soc Chem Commun 384–385, 1977.
140. G Wenz, B Keller. Macromol Symp 87:11–16, 1994.
141. MB Steinbrunn, G Wenz. Angew Chem Int Ed Engl 35:2139–2141, 1996.
142. LP Meier, M Heule, WR Caseri, RA Shelden, UW Suter, G Wenz, B Keller. Macromolecules 29:718–723, 1996.
143. I Kräuter, W Herrmann, G Wenz. J Incl Phenom 25:93–96, 1996.
144. M Weickenmeier, G Wenz. Macromol Rapid Commun 18:1109–1115, 1997.
145. W Herrmann, B Keller, G Wenz. Macromolecules 30:4966–4972, 1997.
146. G Wenz, MB Steinbrunn, K Landfester. Tetrahedron 53:15575–15592, 1997.
147. A Harada, M Kamachi. J Chem Soc Chem Commun 1322–1323, 1990.
148. A Harada, M Kamachi. Macromolecules 23:2821–2823, 1990.
149. A Harada, J Li, M Kamachi. Chem Lett 237–240, 1993.
150. A Harada, J Li, S Suzuki, M Kamachi. Macromolecules 26:5267–5268, 1993.
151. A Harada, J Li, M Kamachi. Macromolecules 26:5698–5703, 1993.
152. J Li, A Harada, M Kamachi. Bull Chem Soc Jpn 67:2808–2818, 1994.
153. A Harada, J Li, M Kamachi. Macromolecules 27:4538–4543, 1994.
154. A Harada, J Li, M Kamachi. Nature 370:126–128, 1994.
155. J Li, A Harada, M Kamachi. Polym J 26:1019–1026, 1994.
156. A Harada, M Okada, J Li, M Kamachi. Macromolecules 28:8406–8411, 1995.
157. A Harada, S Suzuki, M Okada, M Kamachi. Macromolecules 29:5611–5614, 1996.
158. J Pozuelo, F Mendicuti, WL Mattice. Macromolecules 30:3685–3690, 1997.
159. A Harada, J Li, M Kamachi, Y Kitagawa, Y Katsube. Carbohydr Res 305:127–129, 1998.
160. TE Lipatova, LF Kosyanchuk, YP Gomza, VV Shilov, YS Lipatov. Dokl Akad Nauk SSSR Engl Trans 263:140–143, 1982.
161. TE Lipatova, LF Kosyanchuk, VV Shilov. J Macrol Sci Chem A22:361–372, 1985.
162. TE Lipatova, LF Kosyanchuk, VV Shilov, YP Gomza. Polym Sci USSR 27:622–629, 1985.
163. X Sun, DB Amabilino, IW Parsons, JF Stoddart. Am Chem Soc Div Polym Chem Polym Prepr 34(1):104–105, 1993.
164. X Sun, DB Amabilino, PR Ashton, IW Parsons, JF Stoddart, MS Tolley. Macromol Symp 77:191–207, 1994.
165. GJ Owen, P Hodge. Chem Commun 11–12, 1998.
166. PT Engen, PR Lecavalier, HW Gibson. Am Chem Soc Div Polym Chem Polym Prepr 31(2):703–704, 1990.
167. HW Gibson, PT Engen, S-H Lee, S Liu, H Marand, MC Bheda. Am Chem Soc Div Polym Chem Polym Prepr 34(1):64–65, 1993.
168. HW Gibson, PT Engen. New J Chem 17:723–727, 1993.
169. SH Lee, PT Engen, HW Gibson. Macromolecules 30:337–343, 1997.
170. G Wenz, B Keller. Angew Chem Int Ed Engl 31:197–199, 1992.
171. G Wenz, B Keller. Am Chem Soc Div Polym Chem Polym Prepr 34:62–63, 1993.
172. W Herrmann, M Schneider, G Wenz. Angew Chem Int Ed Engl 36:2511–2514, 1997.
173. A Harada, J Li, M Kamachi. Nature 356:325–327, 1992.
174. A Harada, J Li, T Nakamiysu, M Kamachi. J Org Chem 58:7524–7528, 1993.
175. A Harada, J Li, M Kamachi. J Am Chem Soc 116:3192–3196, 1994.

176. A Harada, J Li, M Kamachi. Nature 364:516–518, 1993.
177. A Harada. Am Chem Soc Div Polym Chem Polym Prepr 36(1):570–571, 1995.
178. PE Mason, IW Parsons, MS Tolley. Angew Chem Int Ed Engl 35:2238–2241, 1996.
179. PE Mason, IW Parsons, MS Tolley. Polymer 39:3981–3991, 1998.
180. SS Zhu, PJ Carroll, TM Swager. J Am Chem Soc 118:8713–8714, 1996.
181. SS Zhu, TM Swager. J Am Chem Soc 119:12568–12577, 1997.
182. PL Vidal, M Billon, B Divisia-Blohorn, G Bidan, JM Kern, JP Sauvage. Chem Commun 629–630, 1998.
183. M Born, H Ritter. Makromol Chem Rapid Commun 12:471–476, 1991.
184. M Born, T Koch, H Ritter. Acta Polym 45:68–73, 1994.
185. H Ritter. Macromol Symp 77:73–78, 1994.
186. M Born, T Koch, H Ritter. Macromol Chem Phys 196:1761–1767, 1995.
187. M Born, H Ritter. Macromol Rapid Commun 17:197–202, 1996.
188. O Noll, H Ritter. Macromol Rapid Commun 18:53–58, 1997.
189. O Noll, H Ritter. Macromol Chem Phys 199:791–794, 1998.
190. M Born, H Ritter. Angew Chem Int Ed Engl 34:309–311, 1995.
191. M Born, H Ritter. Adv Mater 8:149–151, 1996.
192. I Yamaguchi, K Osakada, T Yamamoto. Macromolecules 30:4288–4294, 1997.
193. MJ Marsella, PJ Carrol, TM Swager. J Am Chem Soc 116:9347–9348, 1994.
194. TM Swager, MJ Marsella, RJ Newland, Q Zhou. Am Chem Soc Div Polym Chem Polym Prepr 36(1):546–547, 1995.
195. Q Zhou, MR Ezer, TM Swager. Am Chem Soc Div Polym Chem Polym Prepr 36(1):607–608, 1995.
196. Q Zhou, TM Swager. J Am Chem Soc 117:7017–7018, 1995.
197. MJ Marsella, PJ Carroll, TM Swager. J Am Chem Soc 117:9832–9841, 1995.
198. Q Zhou, TM Swager. J Am Chem Soc 117:12593–12602, 1995.
199. HW Gibson, S Liu, YX Shen, M Bheda, SH Lee, F Wang. In: J Becher, K Schaumburg, eds. Molecular Engineering for Advanced Materials. Dordrecht: Kluwer, 1995, pp 41–58.

9

Dendrimeric Supramolecular and Supra*macro*molecular Assemblies

Donald A. Tomalia
University of Michigan, Ann Arbor, and Michigan Molecular Institute, Midland, Michigan

István Majoros
University of Michigan, Ann Arbor, Michigan

I. INTRODUCTION

> If all scientific knowledge were lost in a cataclysm, what single statement would preserve the most information for the next generations of creatures? How could we best pass on our understanding of the world? [I might propose:] ''All things are made of atoms—little particles that move around in perpetual motion, attracting each other when they are a little distance apart, but repelling upon being squeezed into one another.'' In that one sentence, you will see, there is an enormous amount of information about the world, if just a little imagination and thinking are applied.
>
> Richard P. Feynman [1]

In this simple quotation, Feynman has perhaps described Nature's ultimate example of a minimalist self-assembly. Most certainly this is not a molecular level self-assembly, but nonetheless atoms serve to remind us that self-organization of fundamental subatomic entities occurred to give us the most basic building blocks of the universe [2]. These self-assembly events were consummated some 10–13 billion years ago and marked a unique moment in time from which *first order* was forever derived from chaos. This was the genesis of the long, unrelenting evolutionary journey to more complex forms of natural matter.

The earliest events involved the assembly of subatomic particles into roughly spherical entities reminiscent of core-shell type architecture. First, lighter elements were formed followed by nuclear synthesis leading to the heavier elements. These discrete, quantized core-shell assemblies of electrons and nuclei were so precise, dependable, and indestructible in chemical reactions that they have functioned as the fundamental building blocks of the universe. Within these elements, Nature successfully organized nuclei and electrons to control atomic space at the subpicoscopic level (i.e., $<10^{-12}$ m) as a function of; size (atomic number), shape (bonding directionality), surface stickiness (valency), and flexibility (polarizability). These variables may, thus, be considered *critical atomic design parameters*—CADPs. This new order set the primordial stage for all evolutionary patterns that followed. These patterns seemed to follow the simple principle: ''*order begets order from chaos*'' [3–6].

The next phase in this evolutionary sequence involved the natural combination of these reactive elements to produce a bewildering array of simple molecular combinations derived from these core-shell atomic spheroids (i.e., NH_3, CH_4, urea, etc.) followed by the formation of more complex, but yet small molecules that included α-amino acids, nucleic acids, sugars, hydrocarbons, etc. Combinations and permutations of specific CADPs at the atomic level articulated molecular level architectures and incipient properties. One path led to *abiotic* molecular evolution (inorganic chemistry); whereas, the other initiated the *biotic* molecular evolution (organic chemistry), and ultimately life as we recognize it today.

The biotic molecular evolution was defined by the respective CADPs of the combined (atoms) required to produce this new molecular level order. It is now known that within this hierarchical level, new sizes, shapes, surface chemistries (functional groups/nonbonding interactions), flexibilities (conformations), and topologies (architectures) arise. These parameters may be visualized by the various shapes, valencies, and polarizabilities associated with the element carbon in its well-known sp, sp^2, or sp^3 hybridized states. We define these unique features as critical *small molecule design parameters*—CSMDPs. Molecular entities in this domain are generally less than 1000 atomic mass units, thus they occupy space of up to approximately 10 Å (1 nm) in diameter, when normalized as spheroids. They may be thought of as subnanoscale in dimension.

A. Suprachemical Categories and Dimensions

The rich patterns of electronegative and electropositive domains found in these small atom and molecule combinations allowed Nature to devise new rules and strategies for advancement to the next higher levels of ordered complexity by nonbonding interactions (suprachemistry) (e.g., *supramolecular*: ''higher in organization or more complex than a molecule; often: composed of many molecules)'' [7]. These strategies may be roughly categorized into several major types, namely:

Category I—Supra-atomic (Exo-): those assemblies involving small clusters of metal atoms or elements with subnano and nanoscale dimensions (e.g., quantum dots, etc.);

Category II—Supramolecular (Endo-): those assemblies leading to small and medium-sized supramolecular structure. These examples include primary convergent-type binding compounds (i.e., spheroidal guest–host structures, macrocyclic, carcerands, etc.);

Category III—Supramolecular (Exo-): those assemblies involving amphiphilic monomers that lead to medium-large supramolecular structures. These assemblies tend to function as transport entities, barriers, membranes, and container-type structures (i.e., micelles, liposomes, lipid bilayers, etc.); and

Category IV—Supramacromolecular (Exo-): those assemblies leading to precise, three-dimensional (3D) structure-controlled, noncovalently bound macromolecules. These supra*macro*molecular structures are derived from more complex, but precisely controlled macromolecular structures capable of information storage, expression, amplification, and use as functional/structural building blocks (e.g., protein folding, DNA-histone complexes, DNA expression, etc.).

Figure 1 illustrates these suprachemical categories as a function of dimensions. The objective of this account is to examine abiotic examples and parameters related to larger supramolecular and supra*macro*molecular dendritic structures analogous to those found in Categories II, III, and IV.

B. Progress in the Science of Abiotic Synthesis

Whereas Nature has been evolving the complexity of matter over the past 10 to 13 billion years, mankind formally began its journey directed at the ''*science of abiotic synthesis*'' only approximately 200 years ago. Beginning with Lavoisier's ''*atom hypothesis*,'' Dalton's ''*molecular hypothesis*'' followed by the initiation of ''*organic chemistry*'' with Wöhler's work, the progress of manmade synthetic evolution appears to be in its infancy compared to Nature's evolution [10,11].

Based on the various hybridization states of carbon and other elements in the periodic table, small molecule synthesis has led to at least four major architectural patterns. The major architectural classes may be visualized as described in

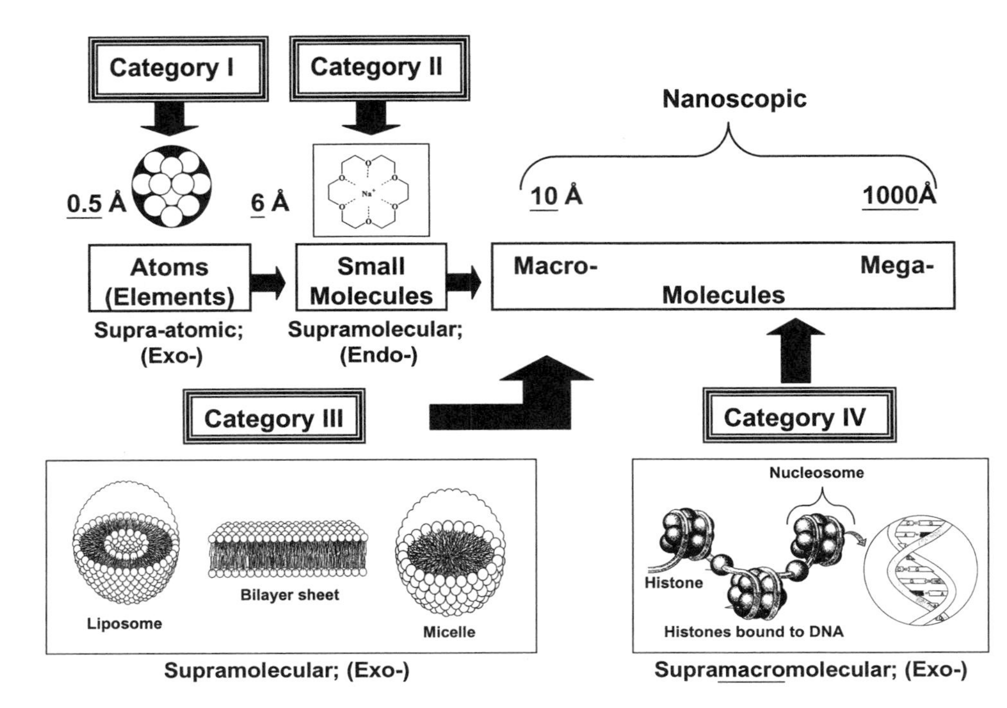

Figure 1 Suprachemistry categories/dimensions.

Fig. 2 and include; (I) Linear, (II) Bridged, (III) Branched, and (IV) Dendritic (cascade) architectures.

C. Progress in the Science of Abiotic Supramolecular Chemistry

> It seemed clear to me now, that the sodium (potassium) ion had fallen into the hole at the center of the molecule. C. J. Pedersen (Aldrichchim. Acta, 4, 1 (1971)).

This simple, but very bold statement made by C. J. Pedersen in the early 1970s literally ushered in the era of supramolecular chemistry based on further elaborations of these basic small molecule architectural classes. J. M. Lehn developed terms to describe two major categories of supramolecular receptors [12]. They are broadly defined as *endo-* and *exo-*receptors. The former present interactive sites that converge on a central locus leading to complexations that are cyclic or

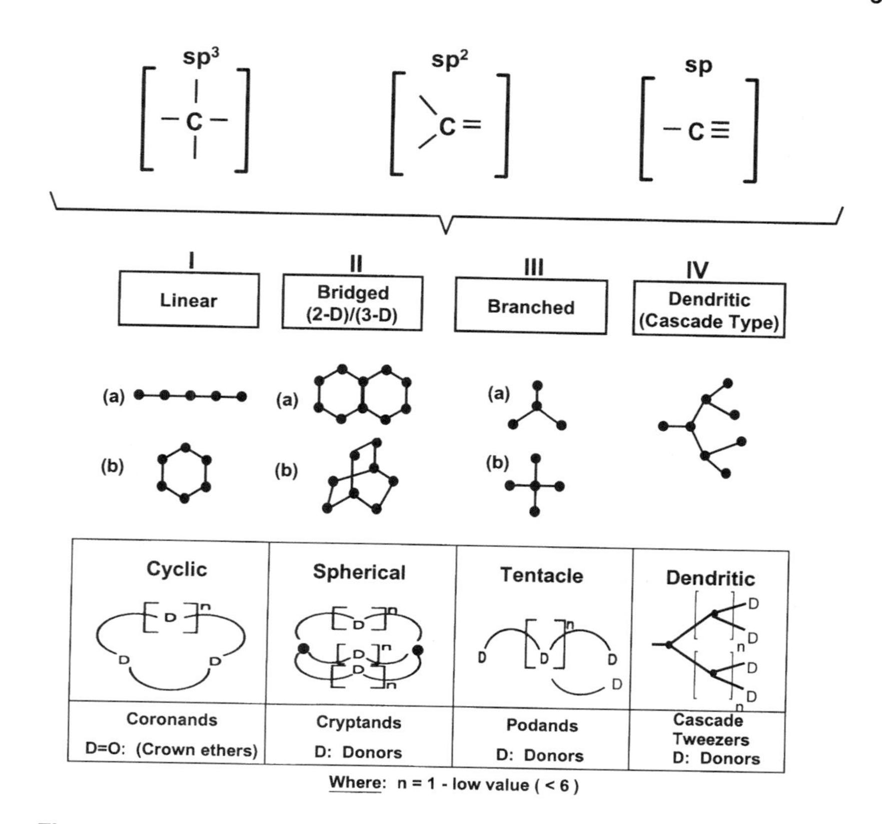

Figure 2 Major small molecule architectures defined as Classes I–IV with comparisons to known donor-type ligands.

severely bent molecules containing holes, clefts, or cavities. These neutral organic ligands are generally referred to as convergent binding molecules. They include small molecule *endo*-receptors such as crown ethers, cryptands, podands, and spherands. This work has been reviewed extensively by many workers [13–19]. *Exo*-complexation typically involves noncyclic molecules containing interactive sites that communicate outwardly. Considerably fewer examples of this type of complexation have been reported.

Recently, however, exciting new reports are appearing that describe the rapid self-assembly of highly ordered linear as well as two- and three-dimensional molecular structure [20]. These strategies usually involve *exo*-recognition among small components to produce tapes [21], squares [22–24], rosettes [25], and other interesting topologies [26–32].

The only topological type in this small molecule area that has not been exploited extensively as a neutral ligand in supramolecular chemistry was the Class IV dendritic (cascade)-type architecture. Vögtle et al. introduced this fourth major small molecule architecture in the late 1970s in a single isolated communication [33]. These low molecular weight dendritic (cascade) molecules served as precursors to macromolecules, which are now recognized as dendrons, dendrimers, and hyperbranched polymers [34,35]. Interest in the supramolecular aspects of dendritic systems has occurred only recently. As described earlier [36], they offer the potential for either *endo* or *exo* complexation and are the focus of this account.

It was only during the past several decades that substantial progress has been made in the use of amphiphilic reagents (monomers) in the construction of nonbonding macro-assemblies. These supramolecular constructs are of the Category II-type and usually lead to entities that are medium to large in dimensions. Amphiphilic reagents used for these constructions are generally derived from Class I (linear) or Class III (branched) small molecule architectures. They include a variety of surfactants, phospholipids, and amphiphilic oligomers as described in Fig. 1. This area has received considerable attention and has been extensively reviewed [37–39].

Many of the reactive linear and branched small molecule reagents were derived during the rich and prolific era of organic chemistry in the 19th century [11]. These developments provided many of the necessary reagents (monomers) upon which H. Staudinger based his work. With these building blocks, Staudinger initiated the synthetic macromolecular evolution some 65 years ago, as he introduced his ''macromolecular hypothesis'' [40]. This evolution has led to three major macromolecular architectures; namely linear, crosslinked (bridged), and branched types. These architectural classes parallel the major small molecule architectures (Figure 2) and are recognized as traditional polymers [41]. In all cases, these structures/architectures are produced by largely random, uncontrolled polymerization processes. These processes produce polydispersed (i.e., $M_w/M_n > 2-10$) products of many different molecular weights. In general, these are not structure-controlled macromolecular architectures such as one observes in biological systems. However, considerable recent progress has occurred in the areas of living-anionic [42], cationic [43], and radical polymerizations [44,45].

D. Structure-Controlled Macromolecules by Abiotic Synthesis

Structure-controlled abiotic synthesis of macromolecules that mimicked biological polymers was first reported by Merrifield nearly 30 years ago. Abiotic synthesis of poly(peptides) by use of solid phase synthesis was reported as early as 1963 [46]. This synthesis strategy was soon extended to the structure-controlled synthesis of poly(amides), poly(nucleotides), and poly(saccharides) [47]. Simply stated, the growing chain in all cases is covalently anchored with a cleavable linker to an insoluble substrate. Monomers are sequenced by means of protect/deprotection methods using linear genealogically synthesis schemes [48]. As early as 1979, we discovered simple synthetic strategies that have allowed us to produce structure-controlled macromolecules in ordinary laboratory glassware [49]. These strategies do not require biological components or immobilized substrate reactions. Utilizing traditional organic reagents and monomers such as ethylenediamine and alkyl acrylates, we are now able to routinely synthesize commercial quantities (kilograms) of controlled macromolecular structures with polydispersities of 1.0005–1.10. These new structures are referred to as *dendrons* or *dendrimers*.

Although the mimicry is less elegant and more minimalistic than that found in Nature, these synthetic strategies appear to mimic the four pervasive patterns devised by Nature for the structure control of natural macromolecules such as DNA, RNA, and proteins (Fig. 3).

1. Primary atomic (CADPs) or molecular (CMDPs) information is defined and stored in the initiator core or seed. This information which includes its size, shape, multiplicity (N_c), and chemistry (valency) is presented in an *exo-*

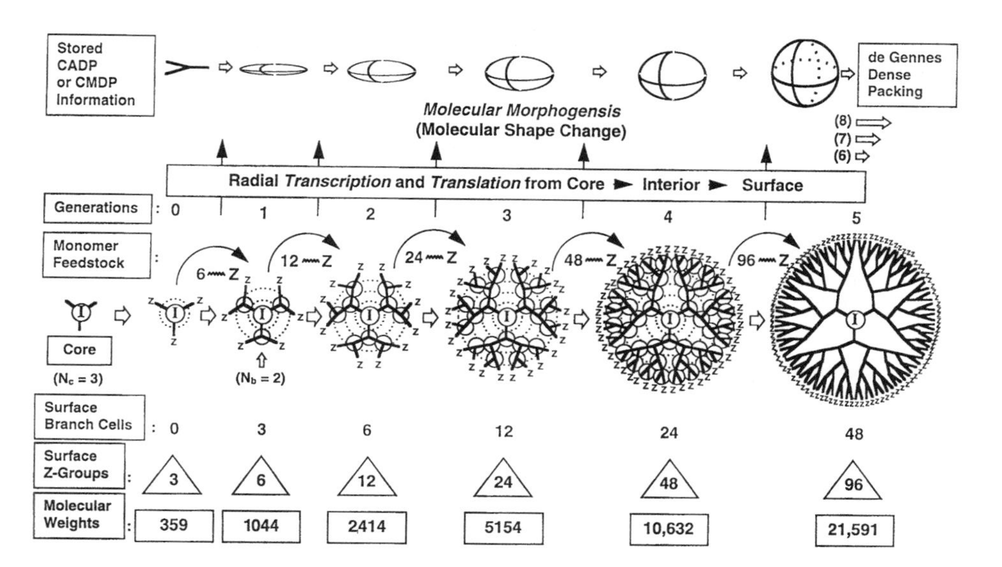

Figure 3 A dendrimer series with core multiplicity ($N_c = 3$) and branch cell multiplicity ($N_c = 2$) illustrating molecular shape changes, surface branch cells, surface Z-groups, and molecular weights as a function of generation.

fashion and communicated to a ''template polymerization region,'' namely, the terminal reactive groups.

2. Appropriate ''feedstock monomers'' such as acrylates, acrylonitrile, and other organic reagents are defined and adapted to various ''template polymerization'' schemes. This introduces geometrical amplification at the termini. These amplification values are defined by the multiplicity (N_b) of the branch junctures. Protect/deprotection procedures allow control of complementary chemical reactivities in the template polymerization region. The chemistry is designed to assure high-yield chemical bond formation at each iterative growth stage (generation = G) in an effort to avoid defects or digressions from geometrically ideal branch amplification.

Regioselective control of the template polymerization is obtained by transfer of genetic (i.e., CADP or CMDPs) information through the hierarchy of chemical bond connectivity involved in the dendrimer construction. This includes the following: [initiator core] (*DNA mimic*)—[*transcription*] $\rightarrow$ [interior branch cells] (*RNA mimic*)—[*translation*] $\rightarrow$ [surface branch cells] (*ribosome mimic*) $\rightarrow$ [terminal surface groups (Z)] (*template polymerization region*). In this minimalistic, abiotic core-shell construction, the architectural mimicry of a biological cell is readily apparent by the following comparison: [initiator core] $\cong$ [biotic cell nucleus], [interior branch cells] $\equiv$ [biotic cell cytoplasm], and [surface branch cells + terminal groups] $\equiv$ [biotic cell membrane].

Self replication of the primary genetic information (i.e., the initiator core) may occur according to a geometrically driven (2^G) amplification process. This specific self-replicating amplification process produces interior branch cells, surface branch cells, and surface functional groups, according to the geometrically progressive values illustrated. These familiar doubling values reflect the well-known (2^G) amplification values associated with biological cell division (mitosis) or DNA amplification (polymerase chain reactions, PCR). For this poly(amidoamine) PAMAM dendrimer family (Fig. 3), initiated from an ammonia core (N_c = 3) with a branch cell multiplicity (N_b = 2), the expected mass values of 359, 1044, 2414, 5154, 10,632 and 21,591 are obtained for generations 0–5, respectively. These values are verified routinely by electrospray or matrix-assisted, laser desorption mass spectroscopy (MALDI) methods. Polydispersity values (M_w/M_n) are obtained ranging from 1.000002–1.005 for this series. Over 50 different dendrimer families possessing compositionally different branch cells (i.e., carbon, nitrogen, silicon, sulfur, phosphorous, metals, etc.) and multiplicity values of N_c = 1–100 and N_b = 2–5 have been synthesized and characterized. Of course there may be some errors or defects (mutations) in these divergent dendrimer constructions just as there are well-known genetic defects and mutations in all biotic genealogically directed processes. This simply adds to the richness of the comparison between these abiotic (cells/organisms) and biotic (cells/organisms).

Biological cells may be thought of as core-shell-like microscale reactors designed to manufacture both structural and functional building blocks (proteins). Similarly, dendrimers may be thought of as nanoscale core-shell models with certain analogies to biological cells. In each case, core-shell characteristics and growth patterns are determined by a central library of information, which flows from the respective cores to the shell-like surfaces [48]. The major differences between the two systems are (a) scaling and (b) the mode of information transfer.

II. THE DENDRITIC STATE

Dendritic architecture has been widely recognized as a fourth major class of macromolecular architecture [50,51]. The signature for such a distinction is the unique repertoire of new properties manifested by this class of polymers. New properties and applications for this polymer class have been reviewed elsewhere [34,52]. Within the realm of macromolecular structure, dendritic architecture may be viewed as an intermediary architectural state that resides between linear (thermoplastic) structures and crosslinked (thermoset) systems [53]. As such, the dendritic state may be visualized as advancement from a lower order to a higher level of structural complexity [54,55]. Furthermore, recent developments in the synthetic control of macromolecular structure now suggest these transitions may involve various levels of structural control. In fact, these transformations may occur via statistical (A), semi-controlled (B), or Controlled (C) pathways. It is widely recognized that dendrons/dendrimers constitute a significant subclass of dendritic polymers and represent a unique combination of very high structural complexity, together with extraordinary structural control. The focus of this account is confined to the dendritic supramolecular and supra*macro*molecular aspects of assembling those entities indicated in the boxed area of Fig. 1. Furthermore, a cursory examination of the supramolecular properties of these dendritic structures is made as they relate to nanotechnology and issues of theoretical interest.

The assembly of reactive monomers [49], branch cells [14,52], or dendrons [56] around atomic or molecular cores to produce dendrimers according to divergent/convergent dendritic branching principles is well demonstrated [57]. Such systematic filling of space around cores with branch cells as a function of generational growth stages (branch cell shells) to give discrete quantized bundles of mass has been shown to be mathematically predictable [58]. Predicted theoretical molecular weights have been confirmed by mass spectroscopy [59,60] and other analytical methods [61]. In all cases their growth and amplification is driven by the general mathematical expressions shown in Fig. 4.

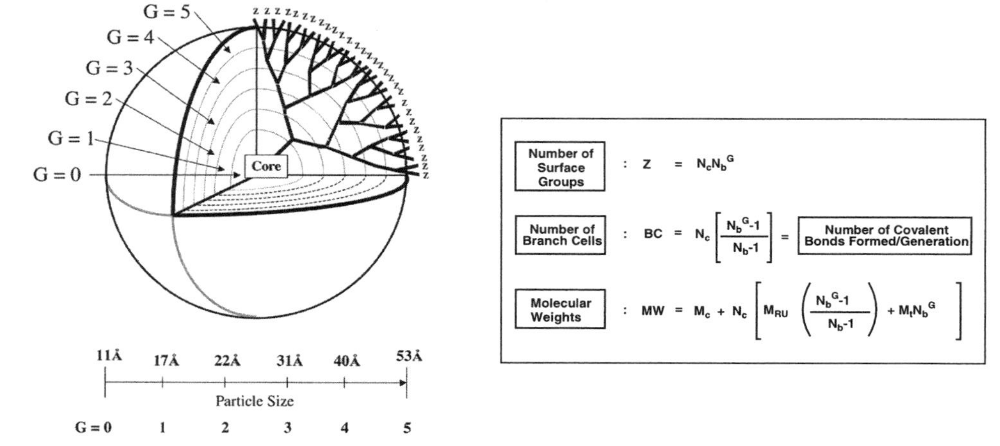

Figure 4 Core-shell dendrimer architecture with mathematics defining number of surface-groups (Z), number of branch cells (BC), theoretical molecular weights (MW), dimensions (Å) as a function of generation (G), branch cell multiplicity (N_b), core multiplicity (N_c), and core molecular weight (M_c).

A. A Comparison of Traditional Polymer Science with Dendritic Macromolecular Chemistry

It is appropriate to compare covalent bond formation in traditional polymer chemistry with that in dendritic macromolecular chemistry. This allows one to fully appreciate the implications and differences between the two areas in the context of supramolecular polymerization. Covalent synthesis in traditional polymer science has evolved around the use of reactive modules (AB-type monomers) that can be engaged in multiple covalent bond formation to produce single molecules. Such multiple bond formation is driven either by chain reaction or poly(condensation) schemes. Staudinger first introduced this paradigm in the 1930s by demonstrating that reactive monomers could be used to produce a statistical distribution of one-dimensional molecules with very high molecular weights (i.e., $>10^6$ daltons). These covalent synthesis protocols underpin the science of traditional polymerizations. As many as 10,000 or more covalent bonds can be formed in a single chain reaction of monomers. Although megamolecules with nanoscale dimensions may be attained, relatively little control can be exercised to precisely manage critical molecular design parameters such as: sizes, atom positions, covalent connectivity (i.e., other than linear topologies), or molecular shapes [43,44].

These polymerizations usually involve AB-type monomers based on substituted ethylenes, strained small ring compounds, or AB-type monomers that may undergo polycondensation reactions. The chain reactions may be initiated by free radical, anionic, or cationic initiators. Multiple covalent bonds are formed per chain sequence wherein the average lengths are determined by monomer to initiator ratios. Generally, polydispersed structures are obtained that are statistically controlled. All three classical polymer architectures (namely, Class I—linear, Class II—crosslinked (bridged), and Class III—branched topologies can be prepared by these methods, keeping in mind that simple introduction of covalent bridging bonds between polymer chains (Class I-type) is required to produce Class II crosslinked (thermoset) type systems [53].

In the case of dendron/dendrimer syntheses, one may view those processes leading to those structures as simply sequentially staged (generations), quantized polymerization events. Of course, these events involve the polymerization of AB_2 monomer units around a core to produce arrays of covalently bonded branch cells that may amplify up to the shell saturation limit as a function of generation.

Mathematically, the number of covalent bonds formed per generation (reaction step) in a dendron/dendrimer synthesis varies as a power function of the reaction steps (Fig. 5). This analysis shows that covalent bond amplification occurs in all dendritic strategies. This feature clearly differentiates dendritic processes from covalent bond synthesis found in traditional organic chemistry or polymer chemistry. Polymerization of AB_2 or AB_x monomers leading to hyperbranched systems also adheres approximately to these mathematics, however, in a more statistical fashion.

It is interesting to note that this same mathematical analysis may be used to predict the amplification of DNA by PCR methods or the proliferation of biological cells by mitosis as a function of generation. This comparison is described later.

B. Dendrimer Synthesis Strategies

Beginning in 1979 [35,49,62–67], certain major strategies have evolved for dendrimer synthesis. The first was the divergent method, wherein growth of a dendron (molecular tree) originates from a core site (root) (Fig. 6.) During the 1980s, virtually all dendritic polymers were produced by construction from the root of the molecular tree. This approach involved assembling monomeric modules in a radial, branch-upon-branch motif according to certain dendritic rules and principles [50]. This divergent approach is currently the preferred commercial route used by worldwide producers such as Dendritech (U.S.A), DSM (Netherlands), and Perstorp (Sweden).

A second method which was pioneered by Fréchet et al. in 1989 is referred to as the "convergent growth process" and proceeds in the opposite direction

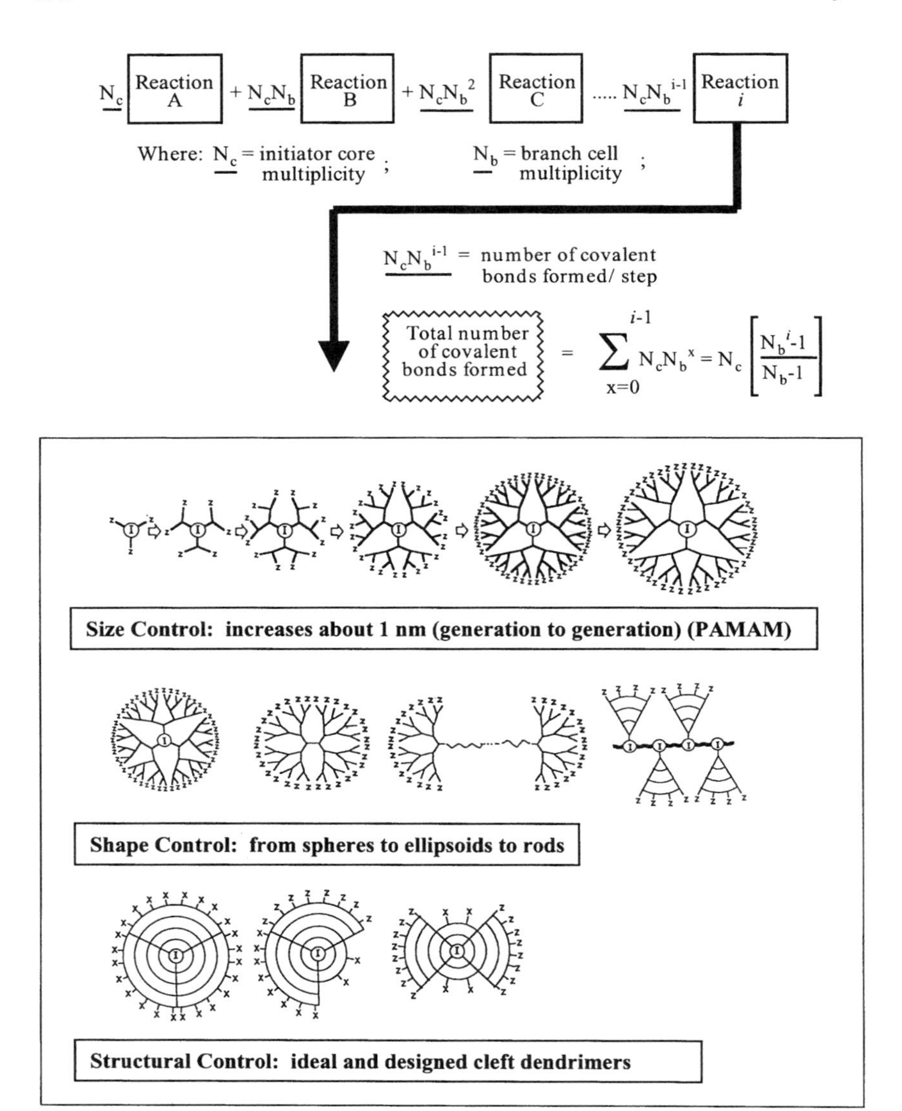

Figure 5 Dendritic macromolecular chemistry.

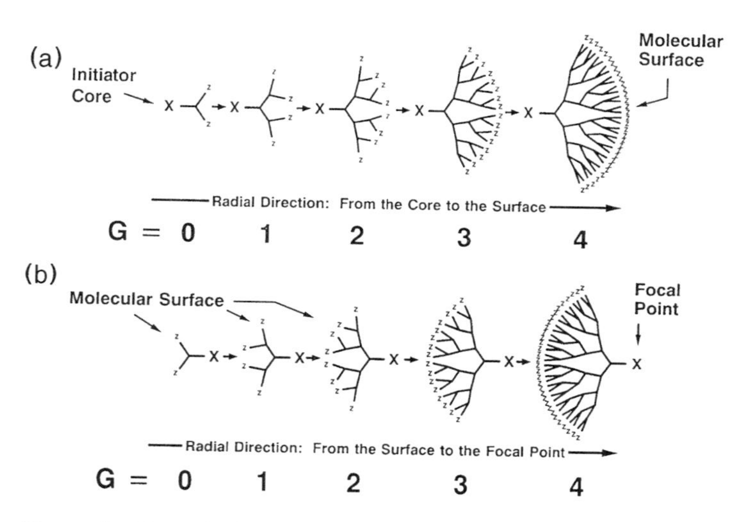

Figure 6 Two principal synthetic methods for building up hyperbranched dendritic macromolecules: (a) divergent and (b) convergent methods.

inward from what will become the dendrimer molecular surface to a focal point (root). In this matter, the latter results in formation of a single dendron, so that in order to obtain a multidendron structure one additional step is required (i.e., an anchoring reaction). In this reaction, several dendrons are connected via a covalently coupling reaction with a multifunctional core to yield the desired dendrimer product. Using these two major synthetic strategies, over 100 compositionally distinct dendrimer families have been synthesized and are reviewed elsewhere [14,35].

C. A Comparison of Divergent Abiotic Synthesis with the Biotic Strategy

Dendrimers may be thought of as nanoscale information processing devices. It is now well recognized that they possess the facility for transcribing and translating their core information into a wide variety of dimensions, shapes, and surfaces by organizing abiotic monomeric/branch cell reagents [48]. Furthermore, these transcription and translational events are accompanied by mathematically defined dendritic amplifications. The syntheses may be thought of as sequence staged, quantized polymerizations orchestrated to the amplification patterns dictated by the core (N_c) and branch cell (N_b) multiplicities. Presently, there appears to be very little limitation to the linear or branched polymerization units that may be used for these constructions. The limitations are determined primarily by the ability to control regiospecific reactivity of these construction units. Over 100 different types of polymerizable units have been used to make at least that many different compositional dendrimer families. These reactive monomeric reagents are processed by terminal group directed transcription into precisely defined peripheral assemblies. Such assemblies of polymerizable units organized in the terminal group region constitute well-defined shells (generations) [68]. These self-assembled shells are frozen genealogically by covalently fixing (translation of) these organized arrays.

Phenomenologically speaking, these molecular level events are reminiscent of those that occur during biotic protein synthesis. Within biotic cells, the objective is to provide a micron-sized compartment (reactor) for the synthesis and amplification of a requisite population of structural and functional proteins—the building blocks of life. These steps involve the dynamic transfer of information through space within the cell by specific carrier molecules (i.e., RNA, etc.). This journey begins with a supramolecular transfer of information in the nucleus (core) and terminates with a supramolecular-codon transfer of this information at the ribosome sites. These supramolecular events are covalently fixed as linear, genealogically directed sequences catalyzed by polymerases to produce the primary structure of proteins. Similarly, within the nanoscale environs of a dendrimer the information journey begins at the core of the dendrimer. Information such as shape, size, multiplicity, and directionality manifested at the dendrimer nucleus is transferred by means of presumed supramolecular events, followed by covalent bond formation at the terminal groups of the dendrimer construct. These informational parameters are processed by molecular level events that are dauntingly analogous to those that occur in biological cells. It should be apparent that the phenomena of transcription, translation, and amplification of this information has occurred unimolecularly within the space occupied by the dendrimer. Documentation of this information transfer is ultimately frozen and amplified at each generational level (G = 1–5) in the form of dendritic covalent connectivity.

Although these comparisons of information transfer phenomena have been made among vastly different dimensional scales the analogies appear to be remarkably consistent. Biotic cells manufacture precise protein building blocks either for structural or functional purposes. Similarly one may view the parallel products of dendrimeric growth as both structural (i.e., interior branch cells) and functional (i.e., surface branch cells) to the dendrimer. It should be noted that the final dendrimer products possess dimensions [i.e., poly(amidoamine) (PAMAMs)] that scale very closely to many important life-supporting protein building blocks, as illustrated in Fig. 48. However, the subtle shape and regiospecific chemistry control that Nature has mastered has yet to be attained with dendrimer constructs.

D. Supramolecular Aspects of the Classical Divergent Dendrimer Synthesis

1. The "All or Nothing" Observation

The divergent synthesis process has been the objective of much speculation and curiosity. Questions often asked are the following. Does the controlled generational growth of a dendrimer have any supramolecular characterizations? Is it an example of *exo*-molecular recognition and self-assembly, followed by covalent bond formation? The answer at this time is very likely. However, the evidence at this time is indirect and not unequivocal. Undoubtedly, the amphiphilic nature, the complementary shapes of the termini and the reagents, as well as the processing conditions will determine the degree of supramolecular character one might expect in these transformations.

First, the molecular recognition character at the terminal groups is largely determined by the complementary reactivities (communication) between the reagents and these terminal sites. The question remains—is there any evidence for enhanced or catalytic reactivity at these termini, which might suggest preorganization followed by covalent bond formation. The strongest evidence in support of this contention is the so-called ''all or nothing reactivity'' of dendrimer surfaces which has been observed in our laboratory, as well as Meijer's laboratory (Eindhoven University) with certain amphiphilic reagents. This observation is made routinely when allowing amphiphilic fatty acid chlorides to react with primary amine terminated dendrimers in the presence of an acid acceptor.

When using substoichiometric amounts of a fatty acid chloride in the presence of triethylamine with either amine terminated poly(amidoamine) PAMAM or poly(trimethyleneamine) POPAM-type dendrimers one observes only fully unreacted and fully reacted dendrimer products as illustrated in Fig. 7. This observation may be interpreted as a manifestation of regiospecific self-assembly, fol-

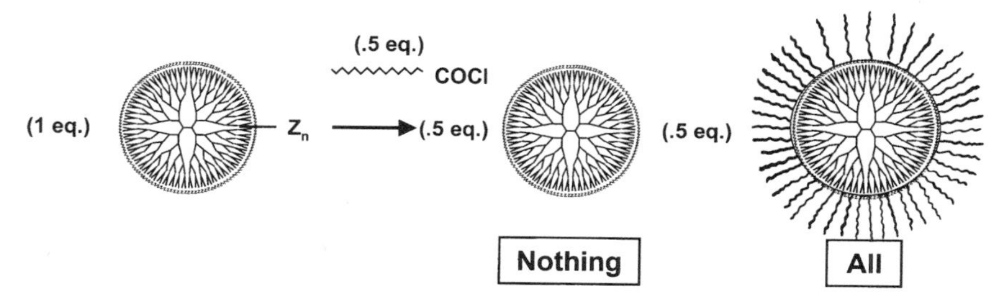

Figure 7 Reaction of substoichiometric amounts of acid chloride reagents with dendrimers to produce the ''all or nothing'' phenomena.

lowed by covalent amide bond formation, since the reactions are performed under kinetically favored mild conditions (i.e., 25–30°C). Alternatively, one might invoke some unidentified catalytic neighboring group effect. This catalytic effect may favor very rapid complete modification of those dendrimer surfaces that are initially substituted. However, very recent work by Froehling [222] supports the self-assembly hypothesis. When these same reactions were performed with fatty acids under thermodynamically driven (i.e., 130–150°C), azetroping conditions, a statistical distribution of substituents was observed in all cases. This is in complete contrast to the "all or nothing" reaction products, which are invariably obtained under milder, more kinetically driven conditions.

2. Sterically Induced Stoichiometry (SIS)

It was interesting to note that when these authors attempted to amidate higher generation poly(propyleneimine) (PPI) dendrimers under these same conditions, they observed regiospecific positional preference for single substitution at the available amine dyad termini. This observation is in complete agreement with the so-called "sterically induced stoichiometry" (SIS) hypothesis we have proposed previously [36,48,69]. This phenomenon is uniquely related to the size of a proposed surface modification reagent and the tethered congestion conditions that a dendrimer manifests at its surface. This becomes significant especially at higher generations. It was first predicted in 1983 [70], and occurs as a manifestation of de Gennes dense packing. As indicated in this experimental example, these congestion phenomena can affect dendrimer surface substituent patterns and reaction rates. We describe later how this congestion phenomenon literally determines the shell saturation levels for the supramolecular polymerization of dendrimers into core-shell tecto(dendrimers).

3. Dendrimer Structure Ideality—A Signature for Self-Assembly and for de Gennes Dense Packing

Further evidence to support the role of self-assembly in divergent dendrimer synthesis can be found in the construction of poly(amidoamine) PAMAM dendrimers. First, ideality and fidelity of structure is usually a signature of a self-assembly event. This is a universal observation throughout biological systems. In all cases, nearly ideal PAMAM structures are obtained *only* under mild conditions favoring such organizational events as a prerequisite to bond formation [71]. Attempts to impose more severe reaction conditions (i.e., reaction temperature >40°C dramatically reduces the ideality of dendrimer structures even in the early generations. Under mild, kinetically favorable reaction conditions, however, nearly ideal dendrimer structures are obtained up to the onset of de Gennes dense packing. Observation of ideal structure is consistent with amplification predictions based on core and branch cell multiplicities as well as observed mass defects by mass spectral analysis.

In order to better understand the parameters that cause sterically induced defects in dendrimer synthesis it is appropriate to review the contributing events. First, normal divergent growth occurs precisely as an ideal molecular structure consistent with molecular weights that predictably obey geometrical mathematics, as described earlier. It should be noted that the experimentally determined radius of a dendrimer molecule increases in a linear fashion as a function of generation, whereas the terminal groups amplify according to geometric progression laws. Therefore, ideal dendritic growth cannot extend indefinitely. Such a relationship produces a critical congestion state at some generational level. This creates a significant dilemma at a reacting dendrimer surface as a result of inadequate space to accommodate all of the mathematically required new monomer/branch cell units. This congested generational growth stage is referred to as the de Gennes dense packed state [70]. At this stage, the dendrimer surface has become so crowded with terminal functional groups that it is sterically prohibited from reacting completely to give ideal branching and dendrimer growth. The de Gennes dense packed state is the point in dendritic growth wherein the average free volume available to the reactive surface groups decreases below the molecular volume required for the transition state of the desired reaction to extend ideal branching growth to the next generation. Nevertheless, the onset of the de Gennes dense packed state in divergent synthesis does not preclude further dendritic growth beyond this point; however, it does mean that there will be notable mass defect digressions from ideality.

(a). Dramatic Changes in Dendrimer Container Properties Coincidental with de Gennes Dense Packing. We have previously described the dramatic influence that core multiplicity (N_c) and branch cell multiplicity (N_b) manifested on the onset of de Gennes dense packing [35,36]. As shown in Figs. 8a,b, it can be seen that for ammonia and ethylenediamine (EDA) core PAMAM dendrimers, respectively, nearly ideal molecular weight masses are observed by mass spectroscopy up to generations 5 and 4, respectively. It should be noted, a systematic pattern of mass defects is observed at a critical generation in each case, due to the sterically induced stoichiometry (SIS) induced by the de Gennes dense packing phenomenon [35,36]. For example, in the case of ammonia core ($N_c = 3$; $N_b = 2$) dendrimer mass defects are not observed until generation = 5 (Fig. 8a). At that generation, a gradual digression from theoretical masses occurs for generations =5–8, followed by a substantial break (i.e., $\Delta = 23\%$) between G = 8 and 9. This discontinuity in shell saturation is interpreted as a signature for de Gennes dense packing. It should be noted that shell saturation values continue to decline monotonically beyond this breakpoint down to a value of 35.7% of theoretical saturation at G = 12. A similar mass defect trend is noted for the EDA core,

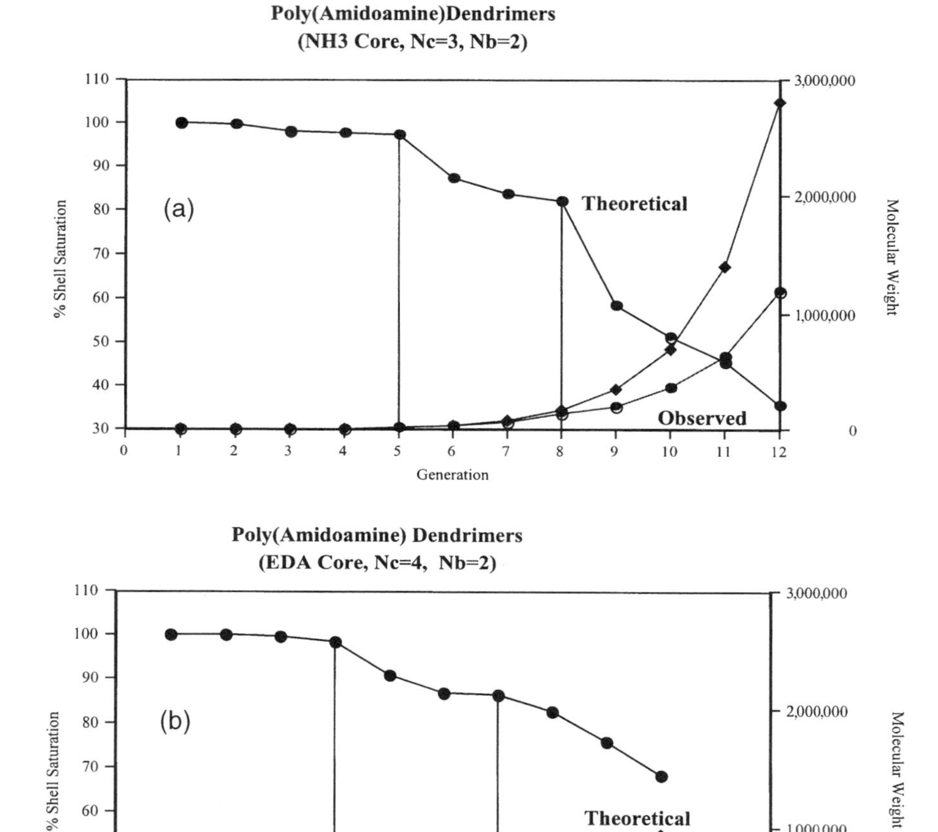

Figure 8 Shell saturation level (%), theoretical molecular weights, and observed molecular weights (mass spectroscopy) as a function of generation for (a) NH_3 core ($N_c = 3$, $N_b = 2$) and (b) EDA core ($N_c = 4$, $N_b = 2$) poly(amidoamine) PAMAM dendrimers.

PAMAM series ($N_c = 4$; $N_b = 2$); however, the shell saturation inflection point occurs at least one generation earlier (i.e., G = 4–7) (Fig. 8b) for the higher multiplicity EDA core. This suggests that the onset of de Gennes dense packing may be occurring between G = 7 and 8. These latter data are completely consistent with metal ion probe experiments [72]. It has been shown that the interior of hydroxyl terminated (EDA core) PAMAM dendrimers G = 1–6 is completely accessible to Cu^{++} hydrate. However, attempts to drive Cu^{++} .6 H_2O (metal ion hydrate) into the interior of G = 7–10 did not occur even under forcing conditions. Subsequent treatment of these respective Cu^{++} as a function of G = 1–10 (chelated) solutions with hydrogen sulfide manifested three different behaviors as a function of generation. Copious precipitates were obtained from Cu^{++}/G = 1–3. In contrast, completely soluble solutions were obtained for Cu^{++}/G = 4–6. Finally, a totally different precipitate was obtained for Cu^{++}/G = 7–10. TEM analysis of the three sets of metal/dendrimer combinations confirmed that G = 4–6 dendrimers had incarcerated copper sulfide and were functioning as *host container* molecules to the copper sulfide guest aggregation. Similar analysis of the last set indicated G = 7–10 were functioning as *surface scaffolding* with virtually no metal incorporation into the interior. This clearly defines a congestion periodicity pattern, as illustrated in Fig. 9, and reveals a unique pattern of dendrimer/metal ion relationships as a consequence of generational surface congestion. Obviously, these metal ion probe experiments are consistent with and support the mass spectral de Gennes dense packing signature hypothesis.

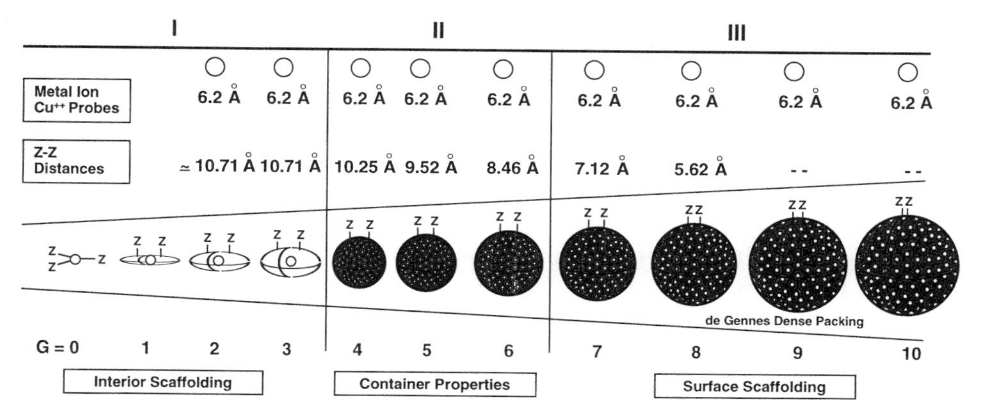

Figure 9 Periodic congestion patterns for a PAMAM EDA core ($N_c = 4$, $N_b = 2$) dendrimer series (G = 0–10) defining (I) interior scaffolding, (II) container-like properties, and (III) surface scaffolding properties when combined with an aqueous Cu^{+2} probe.

In summary, it is readily apparent that these abiotic, amplified genealogically directed polymerizations of AB_2 monomers have many things in common with biological systems. For example, certain biological processes, such as polymerase chain reactions PCR [73] or cell mitosis [8,9] may be thought of as analogous examples of amplification, but at much larger dimensional scales than are found in dendrimers. Most importantly, branch cell amplification and DNA amplification may be viewed as special examples of supramolecular directed polymerization leading in each case to very highly controlled macromolecular structures.

E. Strategies for the Supramolecular Assembly of Components to Produce Dendrimers

One of the major themes in biological systems is the noncovalent assembly of large structures from smaller components. Self-assembly is usually rapid, requires minimized energy for synthesis, and guarantees reproducible construction of complex products with high fidelity [8,9]. The presumed supramolecular construction of dendrimers from its constitutional components provides prime examples of abiotic supramolecular polymerizations to give well-defined structures with notable complexity. More recent developments in this area may be broadly categorized into those involving the supramolecular assembly of dendron components and those involving the assembly of subdendron units or branch cells.

1. Supramolecular Assembly of Dendrimers Based on Focal Point Functionalized Dendrons

This construction strategy evolved from the pioneering work of Zimmerman et al. reported in 1996 [74]. Their approach involved the hydrogen bond directed aggregation of Fréchet-type dendrons, which were terminated with two isophthalic diacid moieties at their focal point. In appropriate solvents, these assemblies were stable enough to be characterized by size exclusion chromotoraphy (SEC), laser light scattering (LLS), and vapor pressure osmometry (VPO). Dendrons as high as generation = 4 could be self-assembled by this process (Fig. 10).

Fréchet and coworkers [221] have recently described a similar self-assembly of benzyl ether dendrons, possessing carboxylic acid substituents at their focal points by metal–ligand coordination around a core of trivalent lanthanide metals (e.g., Er, Eu, or Tb). These self-assembled dendrimers were isolated by using ligand exchange reactions to produce structures derived from metal–ligand ionic interactions as shown in Fig. 11. As a consequence, these self-assembled dendrons served as a dendritic shell, which shielded the lanthanide atoms from one

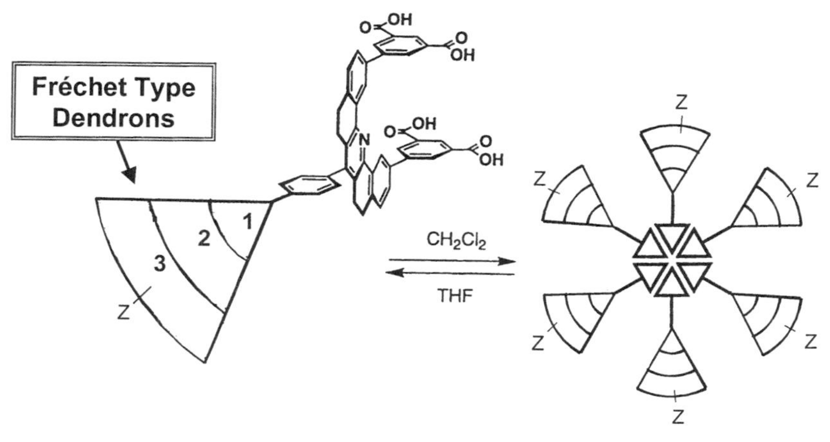

Figure 10 The self-assembly of a hexameric aggregate of wedgelike dendrons that are functionalized at their focal points by two isophthalic acid moieties. (From OA Matthews, Prog Polym Sci, 28, 1998. With permission from Elsevier Science.)

another. It was significant to note that these dendritic structures produced a substantial decrease in the rate of self-quenching (an energy transfer process) between metal atoms. This effect has been referred to as a *site isolation phenomenon.* The importance of this effect was clearly demonstrated as one progressed from lower to higher generations, wherein, the dendrimers exhibited vastly enhanced luminescence activity over the lower generation dendrimers.

2. Supramolecular Assembly of Dendrimers Based on the Assembly of Branch Cells

Numerous examples have been reported describing the use of hydrogen bonding between complementary groups to induce self-assembly [56,75]. Another approach is to use metal-induced coordination chemistry as the driving force in the assembly process. Metal-based coordination-driven methodology allows the rapid facile formation of discrete structures with well-defined shapes and sizes. Metal–ligand dative bonds are stronger than hydrogen bonds and have more directionality than other weak interactions such as π–π stacking, electrostatic hydrophobic/hydrophilic interactions, and even hydrogen bonding. One metal–ligand interaction can therefore replace several hydrogen bonds in the construction of supramolecular species. Perhaps the earliest work reported in this area was that described by Balzani et al. [76–82]. Referred to by Balzani as the ''complexes as ligands and complexes as metals'' strategy, the synthesis involved the divergent assembly of dendrimer structures as described in Fig. 12.

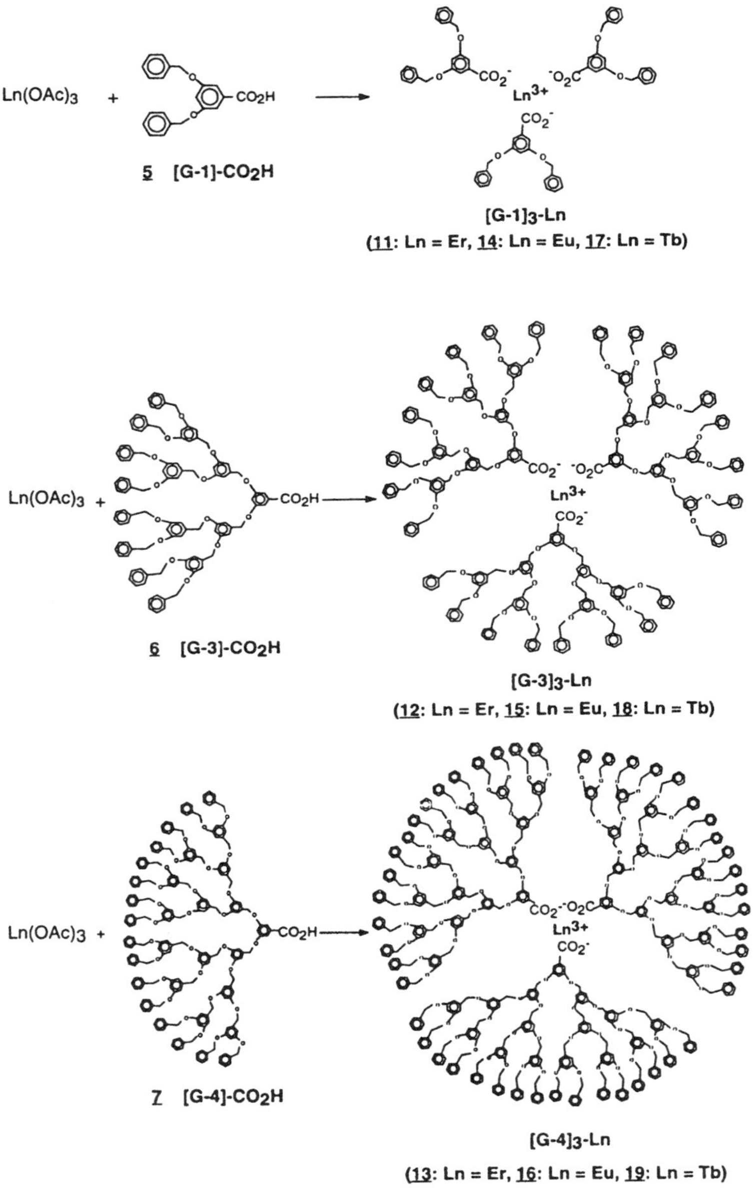

$Ln(OAc)_3$ + 5 [G-1]-CO_2H → [G-1]$_3$-Ln
(11: Ln = Er, 14: Ln = Eu, 17: Ln = Tb)

$Ln(OAc)_3$ + 6 [G-3]-CO_2H → [G-3]$_3$-Ln
(12: Ln = Er, 15: Ln = Eu, 18: Ln = Tb)

$Ln(OAc)_3$ + 7 [G-4]-CO_2H → [G-4]$_3$-Ln
(13: Ln = Er, 16: Ln = Eu, 19: Ln = Tb)

Figure 11 Self-assembly of focal point functionalized dendrons around metal cations according to Fréchet et al. (JMJ Fréchet et al. Chem Mater 10:287, 1998. Copyright 1998 American Chemical Society.)

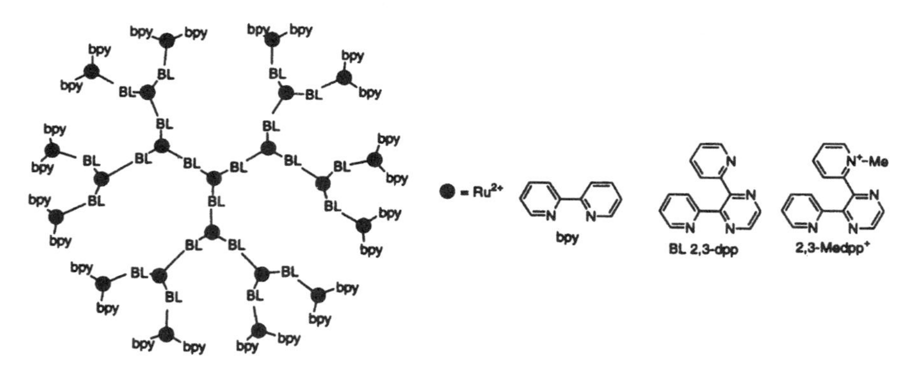

Figure 12 Dendritic ruthenium complex reported by Balzani and coworkers. (Courtesy Chem Rev 97:1706, 1997. Copyright 1997 American Chemical Society.)

In a similar fashion, Puddephatt and coworkers [83–85] described the use of platinum coordination chemistry involving a convergent strategy. By this method, they reported the synthesis of a generation = 4 dendrimer containing up to 28 coordination centers.

Perhaps the only true, reversible coordination, self-assembly for the synthesis of dendritic systems is that described in a series of papers by Reinhoudt and coworkers. They first reported the reversible assembly of hyperbranched spheres from palladium-substituted compounds by replacing a labile, coordinating nitrile ligand with a kinetically stable moiety located on an AB_2 monomer [86,87]. Subsequently, they used a combination of hydrogen bonding and noncovalent coordination chemistry to construct a unique rosette-type structure [88]. More recently, this group reported related chemistry that allows either a divergent or convergent approach [88] to dendrimer synthesis [89,90]. This strategy involves first the synthesis of three basic building units as described in Fig. 13. The controlled divergent assembly requires first the core unit containing three Pd—Cl pincer complexes, which are activated by removing chloride ion with $AgBF_4$. Subsequent addition of three equivalents of nitrile containing building block 2 yields a first generation metallo(dendrimer) possessing nitrile ligands coordinated to the palladium centers. Using this concept, they were able to synthesize metallodendrimers up to generation 5 [87]. They also described [90] the synthesis of building unit 3 with pyridine instead of nitrile ligand and were able to use this intermediate for the preparation of more stable metallodendrimers, due to the stronger coordination of pyridine ligands to the palladium centers.

The convergent route (Fig. 13c) begins with the synthesis of dendrons using building block 2 as a carrier unit and building block 3 as a building unit. The three dendrons were coupled to a trifunctional core 1 to form the dendrimer structure. Figure 13c illustrates a dendron (DG_2 and DG_3) constructed via controlled

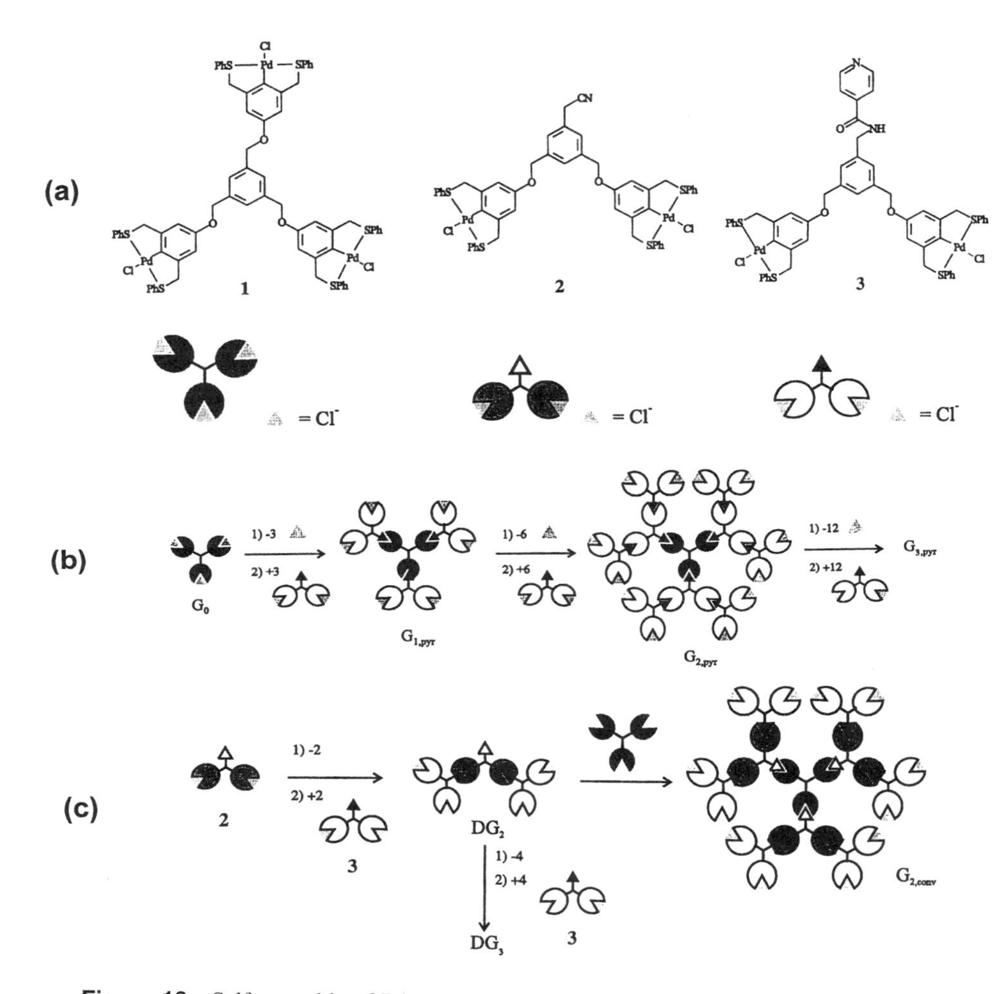

Figure 13 Self-assembly of Pd contained metallodendrimers using (a) building blocks 1–3 (b) divergent method, and (c) convergent method according to Reinhoudt et al. (Courtesy J Am Chem Soc 120:6242; 6243, 1998. Copyright 1998 American Chemical Society.)

convergent assembly. Activation of 2 with $AgBF_4$ and subsequent addition of two equivalents of 3 gave DG_2 after coordination of the pyridine ligands. Reiterating these steps led to higher generation metallodendrons. Evidence for the formation of all of these metallodendrimers was obtained by IR, H^1 NMR, ESMS, and MALDI-TOF mass spectrometry.

III. SUPRAMOLECULAR AND SUPRA*MACRO*MOLECULAR CHEMISTRY OF DENDRIMERS

In the very first papers published [62–67] it was apparent from electron microscopy data presented that dendritic architecture would offer rich possibilities in the area of supramolecular chemistry. It was virtually impossible to observe individual dendrimer modules due to their great propensity to form clusters or aggregates even when sampled from very dilute solutions. Similar observations were made by Newkome, as he published his early work on the related arborols [91]. In contrast, samples from more concentrated solutions displayed breathtaking dendritic arrays of microcrystallites upon drying [62,63]. Recent work by Amis et al. [92] using cryo-TEM and other methods has shown that dendrimers exhibit a great propensity to self-organize even in solution.

In the same early publications [62–67], the reported observation that copper sulfate could be solubilized in organic solvents with ester terminated PAMAM dendrimers to produce deep blue transparent chloroform solutions offered very early evidence for the unique *endo*-receptor (i.e., unimolecular inverse micellar) properties of these macromolecules. These properties are discussed in the following section.

A. The Dualistic Role of Dendrimers as Either *Endo*- or *Exo* Receptors

The field of dendritic supramolecular chemistry is young. As recently as three years ago fewer than a half dozen papers could be found on the subject [93]. Since that time, this field has expanded dramatically. As early as 1990, we commented extensively on the dualistic property of dendrimers [36]. At that time, it was noted that dendrimers could function as unimolecular *endo*-receptor-type ligands manifesting noncovalent chemistry reminiscent of traditional regular or inverse micelles or liposomes. Furthermore, it was noted that dendrimers also exhibited a very high propensity to cluster or complex in an *exo*-fashion with a wide variety of biological polymers (i.e., DNA proteins) or metals. In the following sections, we describe nonexhaustive samplings of recent work illustrating the supramolecular and supra*macro*molecular behavior of dendrimers that connects them to all three categories of suprachemical types observed in biological systems (Fig. 1).

B. Dendrimers as Unimolecular Nanoscale Cells or Container Molecules

1. Evolution of Abiotic Container Molecules—From Carcerands/Carceplexes to Dendrimers/Dendriplexes

The abiotic evolution of ''container-type'' molecules may be traced from the original synthesis of cubane [94], pentaprismane [95], and dodecahedrane [96].

These platonic solids are all closed surface compounds; however, their interiors are too small to host organic or inorganic compounds. It was not until 1983 that Cram et al. [97] reported a container-like hydrocarbon referred to as carcerand that was large enough to host simple organic compounds, inorganic ions, or gases [97]. This discovery has led to a vast array of small container-type molecules, which have been extensively reviewed [98]. Quite remarkably, this was approximately the same time that ester terminated PAMAM dendrimers were noted to dissolve and incarcerate copper salts to produce ''blue chloroform'' solutions due to their unimolecular, inverse micelle properties. This observation was publicly reported in 1984–1985 [62,63], during the same year that Smalley, Curl, and Kroto et al. described the first synthesis of buckminsterfullerene [99]. Of course, it is well known that ''bucky balls'' will host a variety of metals. Thus, it is apparent that the emergence of container molecules has followed the systematic enhancement of the organic host structure dimensions as illustrated in Fig. 14.

(a). Incarceration of Zero Valent Metals and Metal Compounds. Very recently, a novel and versatile method has been reported for the construction of stable, zero-valent metal quantum dots, using dendrimers as well-defined nanotemplate/containers. The concept involves the use of dendrimers as hosts to pre-organize small molecules or metal ions, followed by a simple in situ reaction that immobilizes and incarcerates these nanodomains (Fig. 15) [72,100]. The size, shape, size distribution, and surface functionality of the dendritic nanocomposites are determined and controlled by the architecture. Dendrimer-based nanocompos-

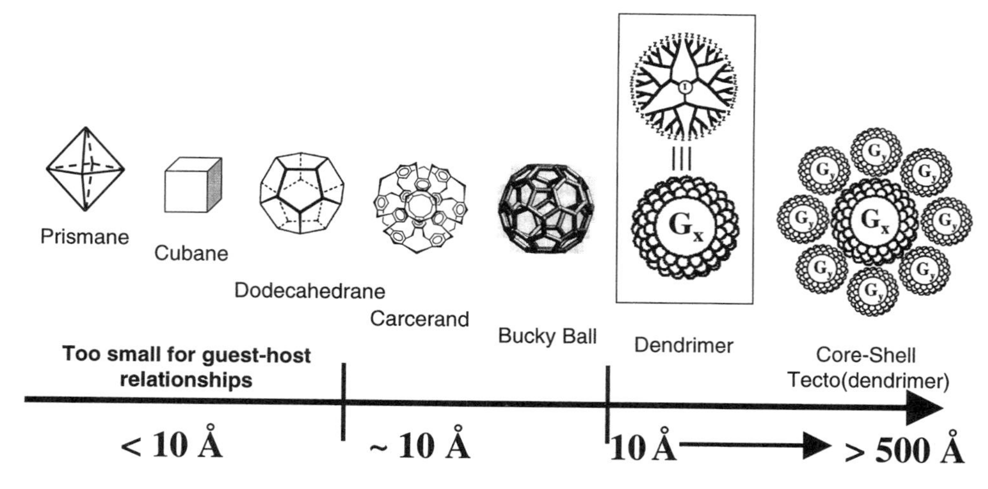

Figure 14 Evolution of abiotic container molecules as a function of dimensions and complexity.

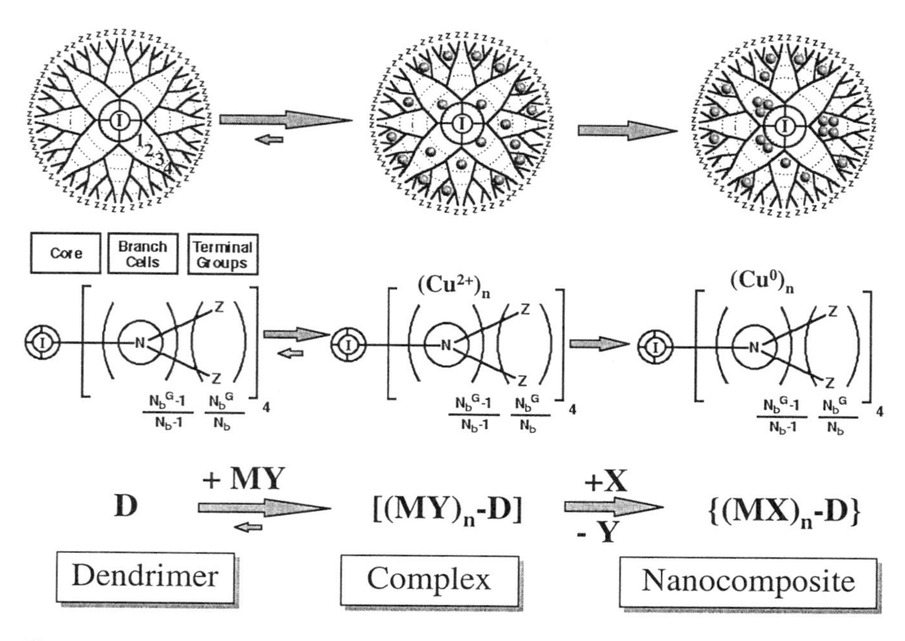

Figure 15 Construction of dendrimer nanocomposites by reactive encapsulation. Y (Cu^{+2}) denotes ligands after complex formation. X (Cu^0) represents zero valent metal incarcerated within the dendrimer interior after reduction. (Courtesy J Am Chem Soc 120:7355, 1998. Copyright American Chemical Society.)

ites display unique physical/chemical properties as a consequence of the atomic/molecular level interactions between the guest and host components within the dendrimer as well as the dendrimer interaction with various solvents and media.

Preparation of stable, zero valent metallic copper solutions were demonstrated in either water or methanol [72]. After complexation within various surface modified poly(amidoamine)(PAMAM) dendrimers, copper(II) ions were reduced to zero valent metallic copper thus providing a bronze, transparent dendrimer–metal nanocomposite soluble in water. Solubility of the metal domains is determined by the surface properties of the dendrimer host molecules; however, their solutions still display characteristic optical properties associated with metal domains. Both aqueous and methanolic solutions of copper clusters were stable for several months in the absence of oxygen. Similar work and results were also reported by Crooks et al. [101].

More recent work describing SAN characterization of copper sulfide PAMAM dendrimer nanocomposites [102] and extensions to silver and gold-based dendrimer nanocomposites have been reported [103]. This latter work confirms and demonstrates the periodic container properties described in Fig. 9.

(b). Incarceration of Organic Compounds, Unimolecular Encapsulation, the Dendritic Box. In another elegant study, Meijer and coworkers [104,105] skillfully enhanced an earlier concept for producing artificial cells by modifying dendrimer surfaces to induce ''unimolecular encapsulation'' behavior [36]. They referred to this new construction as the ''dendrimer box.'' Surface-modifying generation = 5, poly(propylene imine) (PPI) dendrimers [104] with 1-phenylalanine or other amino acids [106] induced dendrimer encapsulation by forming dense, hydrogen-bonded surface shells with almost solid-state character. Small guest molecules were captured in such dendrimer interiors and were unable to escape even after extensive dialysis [105]. The maximum number of entrapped guest molecules per dendrimer box was directly related to the shape and size of the guests, as well as to the number, shape, and size of the available internal dendrimer cavities. For example, 4 large guest molecules (e.g., Bengal Rose, Rhodamide B, or New Coccine) and 8–10 small guest molecules (e.g., p-nitrobenzoic acid, nitrophenol, etc.) could be simultaneously encapsulated within these PPI dendrimers containing 4 large and 12 smaller cavities (Fig. 16). Quite remarkably, this dendrimer box could also be opened to release either all or only some of the entrapped guest molecules [105]. For example, partial hydrolysis of the hydrogen-bonded shell liberated only small guest molecules, whereas total hydrolysis (with 12N HCl; 2 h at reflux) released all sizes of entrapped molecules.

2. Mimicry of Classical Regular Micelles

Based on qualitative evidence, Newkome et al. first hypothesized the analogy between dendrimers and regular micelles in 1985 [91]. Simultaneously, Tomalia

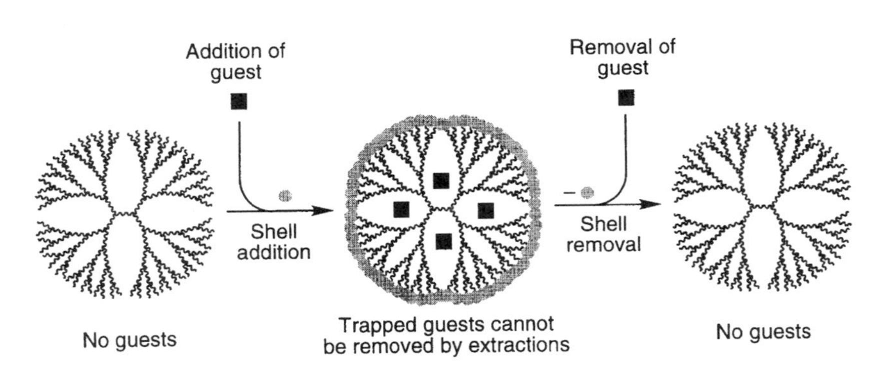

Figure 16 Principle of the dendritic box. (From OA Matthews, Prog Polym Sci, 28, 1998. With permission from Elsevier Science.)

et al. reported the direct observation of sodium carboxylated PAMAM dendrimers by electron microscopy in 1985 [62–67] and 1986 [107] which experimentally supported the fact that dendrimers clearly possessed topologies reminiscent of regular classical micelles. At that time, it was also noted from electron micrographs that a very high population of individual dendrimers possessed hollowness presumably due to the peripheral stacking association of terminal head-groups. This was experimentally confirmed [108] by noting the importance of branch cell symmetry as a requisite for interior void space. Such interior hollowness is observed in essentially all symmetrically branched dendrimers, but does not appear to exist in asymmetrically branched dendrimers such as those described by Denkewalter et al. [109–110]. Furthermore, experimentally determined hydrodynamic dimensions [108], shape confirmations, and comparisons of dendrimer termini (surface areas) as a function of generation with traditional micelle head-groups added further support to this hypothesis in 1985 to 1987 [107,108,113]. This unique dualist property of micelle topology and interior void space normally associated with liposomes was noted by Tomalia et al. in 1989 [114]. Subsequent nmr studies and computer-assisted simulations by Goddard et al. in 1989 [115], molecular inclusion work by Newkome et al. [116] in 1991, as well as extensive photochemical probe experiments by Turro et al. [117–120], have now unequivocally demonstrated the fact that symmetrically branched dendrimers may be viewed as unimolecular cells (nanoscale container molecules). Depending on the nature of their surface groups and interiors these dendrimers will manifest behavior reminiscent of either traditional regular or inverse micelles, however, with unique differences and advantages.

3. Mimicry of Classical Inverse Micelles

The first examples and observed dendritic inverse micelle properties were noted in the initial papers on poly(amidoamine) dendrimers published in 1984 to 1985 [62–67]. At that time, it was observed that methylene chloride or chloroform solutions of the methyl/alkyl ester modified dendrimers readily extracted copper ion (Cu^{+2}) from water into the organic phase. Beautiful ''blue chloroform'' solutions were obtained that were completely transparent and did not scatter light. It was assumed that the copper ions had been chelated into the interior and were being compatibilized by the more hydrophobic sheathing of the dendrimer surface groups. Variations of this work were both patented [121] and ultimately published [122], wherein PAMAM dendrimers were hydrophobically modified with alkyl epoxides and used to extract metal ions into toluene, styrene monomer, or a variety of other hydrophobic solvents.

Meanwhile, other examples of unimolecular dendritic, inverse micelles have been reported by Meijer et al. [123] and DeSimone et al. [124]. In the latter case, surface perfluorinated *dendri*-PPI dendrimers were used that have a high affinity for liquid CO_2. Thus, the hydrophilic dendrimer core provides a favorable environment for hydrophilic guests such as ionic methyl orange. Thus while the $-CO_2$-philic dendrimer shell allows the micelle to dissolve in CO_2, the ionic dye can be effectively transferred from a water layer into liquid CO_2.

C. Dendrimers as Nanoscale Amphiphiles

1. Dendritic Architectural and Compositional Copolymers

Perhaps one of the most elegant and complex examples of an *architectural copolymer* [125] that was constructed by both self-assembly and covalent synthesis is that reported by Stoddart et al. [126,127]. Stoddart and coworkers prepared a series of rotaxanes with dendritic stoppers using a ''threading approach'' (Fig. 17). Within this single rotaxane structure possessing dendritic stoppers one finds all three architectural types are represented. Although no amphiphilic properties were reported for this system, it has been shown that both compositional as well as architectural dendritic copolymers do manifest interesting amphiphilic properties leading to self-assembly processes.

It is very well known that the general area of traditional amphiphilic structure-property relationships can be broadly divided into two major types, namely, small molecule surfactants and amphiphilic block copolymers (see Chapter 3). Each of these classes has been the subject of extensive studies [37,128–131]. A detailed examination of the influence that increasing head/tail sizes and shapes have on the nature of aggregation has been limited to traditional structures. Such structures have included only low molecular weight surfactants, possessing

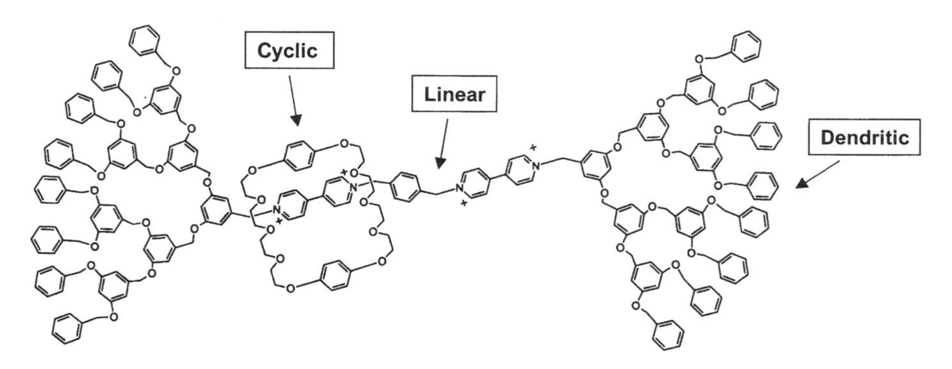

Figure 17 A rotaxane with dendritic stoppers. An example of a dendritic-linear-cyclic architectural copolymer. (Courtesy Chem Rev 1701, 1997. Copyright 1997 American Chemical Society.)

compact polar head-groups/tail, or classical linear, block copolymers (see Chapter 3).

With the advent of dendritic synthesis, the precise construction of nanoscale sizes and shapes has allowed the synthesis of new larger dimensioned amphiphilic structures. Perhaps the first well-documented example of dendrimers exhibiting *exo*-receptor, supramolecular behavior was that reported by Friberg et al. in 1988 [132]. It is well known that one may evolve a rich variety of lyotropic mesophases by merely coordinating the relative sizes of the amphiphilic components of a system (i.e., hydrophobic head-groups versus hydrophobic tail). This was accomplished by combining a $G = 2$, poly(ethyleneimine) (PEI) dendrimer (hydrophilic head-group) with octanoic acid (hydrophobic tail) to give a lamellar liquid-crystalline assembly. In effect, a nanoscale amphiphile was constructed in situ by the acid–amine reaction between octanoic acid and $G = 2$ *dendri*-PEI—$(NH)_{12}$ to produce an amphiphilic salt.

Additional examples of dendritic-type surfactants were reported in the early patent literature as hydrophobic core functionalized PAMAM dendrons [35,133]. For example, when the core was a 12 carbon chain, the first two generation PAMAM dendrons exhibited hydrocarbon solubility (TMF). However, beyond generation two, the amphiphilic dendrons were water-soluble. This demonstrates the effect one observes as the hydrophilic head-group is amplified through the HLB crossover point. The result is a dendron possessing a dominant hydrophilic head-group, which undoubtedly engulfs the tethered hydrocarbon core in aqueous medium. Many of these products were useful as amphiphilic reagents to produce novel Starburst® dendrimer-type micelles [133]. Among the many interesting properties exhibited by these micelles was the ability to sequester hardness ions (i.e., Mg^{+2}, Ca^{+2}, etc.) thus manifesting self-building surfactant properties within unimolecular structures.

Later extensions of this general concept were reported by Chapman, et al. [134] and Fréchet [52] wherein they reversed the amphiphilic components to produce a functionalized dendron possessing a hydrophobic head and a hydrophilic tail. The Chapman dendritic amphiphiles were derived from alkylene oxide tails and BOC terminated, Denkewalter-type dendritic heads. The Fréchet amphiphiles were obtained by attaching poly(ethylene oxide) to the focal point of a hydrophobic poly(ether) dendron as illustrated in Fig. 18 [135,136].

Nearly simultaneously, Fréchet et al. pioneered [52] the synthesis and development of a new type of amphiphilic dendrimer wherein the unimolecular architecture was differentiated. In these instances, Fréchet demonstrated that his poly(ether) dendrimers derived via convergent synthesis could be either homogeneously terminated or differentiated into hydrophobic and hydrophilic hemispheres, as illustrated in Fig. 19. These dendrimers could be oriented at an interface, as illustrated, or under the influence of an external stimulus. Indeed amphiphilic dendrimers form monolayers at the air–water interface and such den-

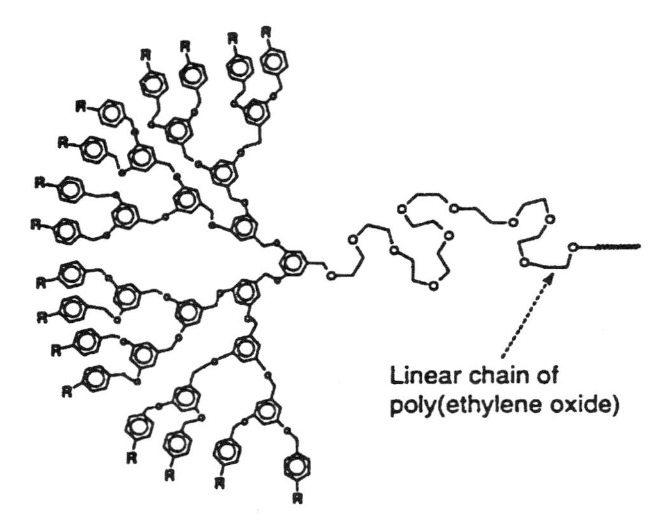

Figure 18 Hybrid dendritic linear polymer obtained by attaching poly(ethylene oxide) to the focal point of structure G-4.

drimers [137] are useful in forming interfacial liquid membranes or in stabilizing aqueous–organic emulsions. A dendrimer constructed from two segregated types of chain ends (i.e., half electron dominating and half electron withdrawing) can be oriented in an electrical field and has been shown to exhibit very large dipole moments [138].

One of the most comprehensive examinations of dendritic nanoscale amphiphiles involved structures that mimicked traditional surfactants. This work was reported in a thesis by van Hest and Meijer [93]. In this study, extensive hydrophobic tails were precisely constructed by the living anionic polymerization of styrene, followed by termination with a primary amine function. Dendronization of these amine functionalized poly(styrenes) produced focal point functionalized, hydrophobically substituted PPI dendrons, as illustrated in Fig. 20. More specifically, the living polystyrene anion is terminated with ethyleneoxide, thus providing a hydroxyl terminated poly(styrene) with molecular weights between 3000 and 5000 and a polydispersity <1.05. In a phase-transfer reaction the addition of acrylonitrile to the alcohol, followed by a (Raney Co/H_2) hydrogenation yields the primary amine terminated poly(styrene). Subsequent dendrimer construction is performed by the sequential Michael addition of acrylonitrile and heterogeneously catalyzed hydrogenation to form a variety of super amphiphiles with different sized nanoscale head-groups.

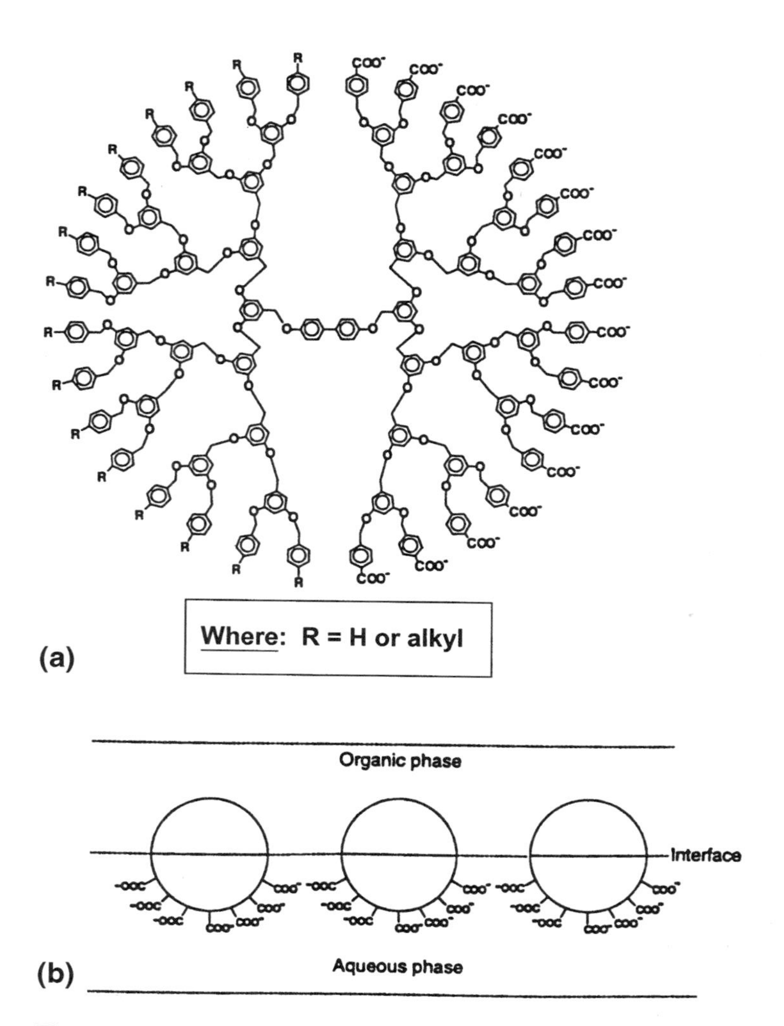

Figure 19 (a) Amphiphilic dendrimer obtained by hemispherical functionalization of half the terminal groups with carboxylic anions; (b) liquid membrane of amphiphilic dendrimer at the interface between water and an immiscible organic solvent.

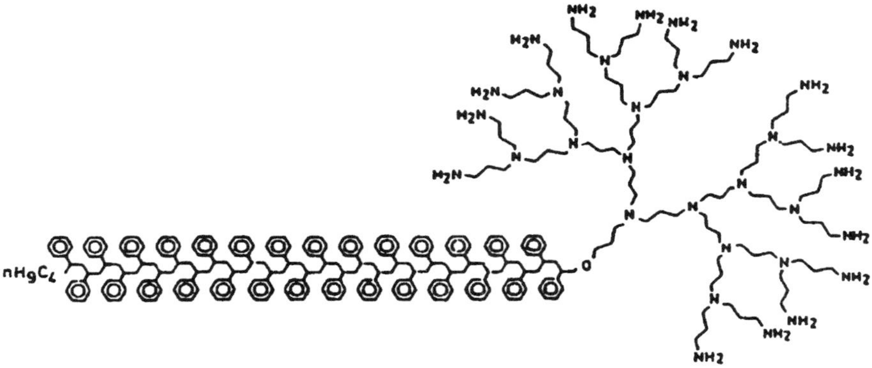

Figure 20 *Linear*-poly(styrene), focal point functionalized poly(propyleneimine) (PPI) dendrons; G = 4.0.

These new architectural copolymers derived from both linear and dendritic architectures exhibited very profound properties as nanoscale amphiphiles. In general, as one increases the head-group size from PS-*dendri*-PPI-$(NH_2)_4$ to PS-*dendri*-PPI-$(NH_2)_{32}$, the aggregation topologies change from planar bilayers to vesicles to rodlike micelles and finally to spherical micelles. Since only the head-group size and not the chemical composition of the amphiphilic structure is being changed, this study offers excellent proof for the validity of the Israelachvili et al. theory [131,139,140] of shape-dependent aggregation behavior.

Changing the head-group functionality from primary amine terminated to either carboxylic acid or quaternary amine terminated, produced amphiphilic behavior, which was very rational in the context of traditional theory. A major difference in all cases was the substantially larger dimensions of the aggregates compared to those obtained from traditional small molecule amphiphiles. In fact, one might visualize these dendritic amphiphiles as amplified forms of their traditional analogues. Aggregates and assemblies derived from these linear, dendritic architectural copolymers are roughly five times the size of those obtained from traditional surfactant molecules.

One of the earliest examples demonstrating the self-assembly of dendrimers based on inherent amphiphilic characters was that reported by Newkome et al. [141]. It was found that *dendri*-poly(amidoalcohols); G = 1–2 attached to various alkyne and hydrocarbon cores (Fig. 21) produced thermally reversible gels in aqueous solutions. It was postulated that formation of these rod-shaped aggregations was driven by a combination of hydrogen bonding of the hydroxyl terminal groups and the hydrophobic bonding of the core substituents in an orthogonal fashion. Various helical morphologies were proposed to account for the extraordi-

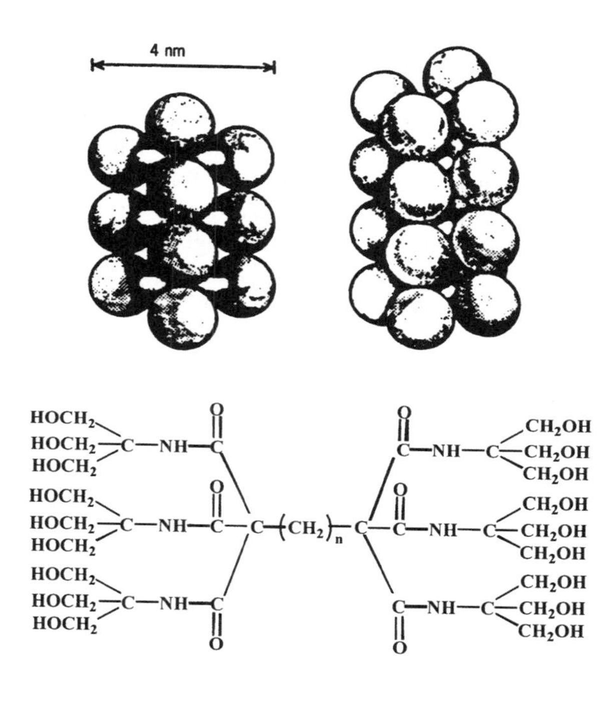

Figure 21 Molecular rods derived from micellarization of Starburst® poly(amidoalcohols).

narily large diameters (i.e., ≅ 600 Å) that were observed by electron microscopy [116]. It is quite likely that multiples of these rod aggregates may have self-assembled much as has been observed recently for covalently fixed, linear-dendritic architectural polymer rods reported recently by Tomalia et al. [125].

D. Dendrimers as Nanoscale Scaffolding

1. Dendritic Rods and Cylinders

(a). Congestion Induced Morphogenesis (Shape Control). Cylindrical rod-shaped dendrimer assemblies were first synthesized by Tomalia et al. as early as 1987 [35,36,142]. These structures represent some of the first examples of hybridized dendritic architecture. Since they possess a linear polymeric core and dendritic arms, they are called *architectural copolymers.* This work was recently

reported in detail [125]. The method involved the divergent dendronization of *linear* poly(ethyleneimine) (PEI) cores (see Scheme 1). These cores, which had DPs of ≅ 100–500, were dendronized by iterating with standard PAMAM chemistry from the active hydrogens on the backbone of the *linear* poly(ethyleneimine) (PEI). The first step involved the reaction of methyl acrylate followed by an excess of ethylenediamine to give first generation monodendrons along the PEI backbone. Reiteration of these steps led to the development of higher generations (G = 1–4). These final products were characterized by FTIR, C^{13} NMR, SEC, HPLC, MALDI-TOF MS, and transmission electron microscopy. Electron microscopy of sodium carboxylated forms of G = 1–3 revealed nondescript random coil topologies. However, as the G = 3 structure was advanced to the next generation (G = 4), a remarkable congestion-induced shape change occurred to produce rigid, rodlike cylinders as determined by electron microscopy. The rod diameters varied between 25–32 Å with lengths ranging between 500–3000 Å. Further-

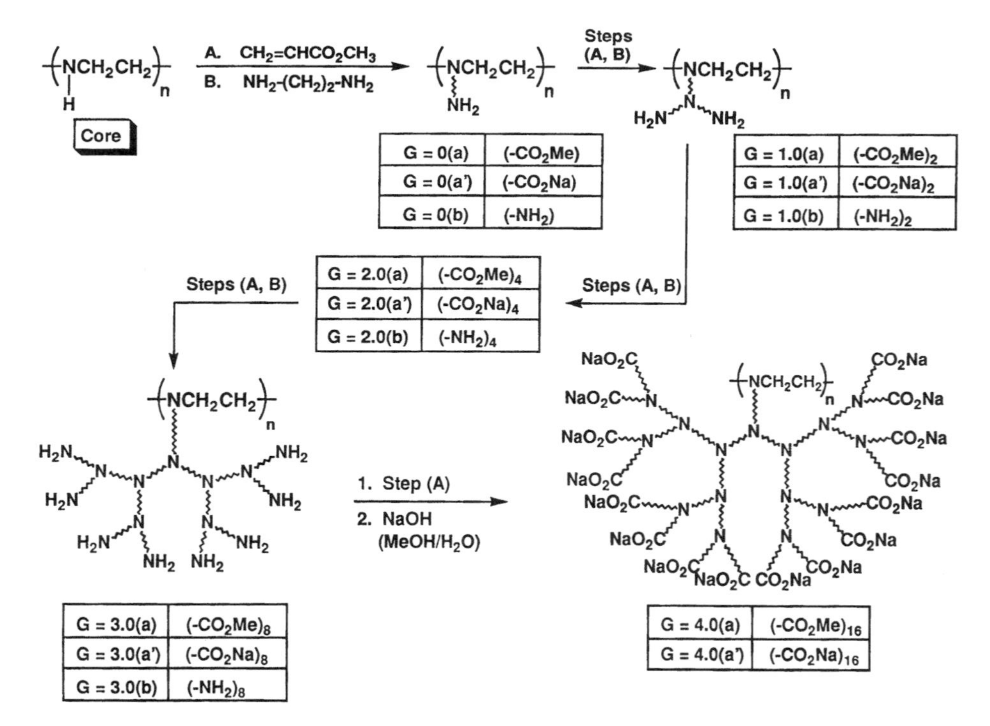

Scheme 1 Dendronization scheme for conversion of *linear* poly(ethyleneimine) cores to *dendri*-poly(amidoamine) hybrids. (Courtesy of J Am Chem Soc 1998, 120 (11), 2679. Copyright 1998 American Chemical Society.)

more, additional supra*macro*molecular assembly appeared to be occurring to give parallel clusters of the rods (Fig. 22).

As early as 1996, Schluter et al. [143] reported their first efforts to synthesize linear-dendritic architectural copolymers [144]. In general, their approaches involved some general strategies. The first method involved the synthesis of a linear core, followed by the coupling of preformed dendrons to produce the architectural copolymers. The second method involved the synthesis of dendrons possessing polymerizable functionality at their focal points. These dendritic macromonomers were then polymerized to give a polymeric mainchain with pendant dendrons. In each case, linear-dendritic architectural copolymers would be obtained. They would be expected to exhibit rodlike or cylindrical properties if the backbone core possessed a high degree of polymerization and the dendritic component were highly congested.

With the first method, Schluter et al. [145,146] utilized the Suzuki reaction to produce a poly (p-phenylene)-type backbone possessing reactive *ortho*-hydroxy-methyl substituents. These substituents were subsequently used to couple a variety of Fréchet-type monodendrons along the backbone as described in Fig. 23. In general, the lower generations (i.e., $G = 1$) coupled with linking efficiencies as high as 95%. However, coupling efficiencies for $G = 2$ or 3 decreased dramatically (i.e., 50–60%) unless appropriate 2-hydroxyethyl spacers were introduced to relieve steric problems [145].

The second approach involved the covalent attachment of Fréchet-type monodendrons (i.e., $G = 1$–3) to appropriate polymerizable reagents such as styrenes [143,146] or methacrylates [144] (Fig. 24). Suitable radical catalysts were used to polymerize these dendritic macromonomers into structures that were observed to be cylindrical rodlike architectures [144,146,147].

Percec and coworkers [148] utilized a similar strategy for the conversion of perfluorinated alkylene functionalized 3,4,5-trihydroxy benzoic acid-type dendrons into methyl methacrylate functionalized dendritic macromonomers. Characterization of the resulting linear-dendritic architectural copolymers involved DSC, X-ray diffraction, and thermal optical polarized microscopy. It was concluded that the self-assembly of the pendant dendritic mesogens forced the linear backbone into a tilted, helical ribbon-type structure. The self-assembly behavior was largely controlled by the multiplicity, composition, and molecular weights of the pendant dendritic mesogens.

Using a variation of this dendritic macromonomer method, Percec et al. [149,150] synthesized a variety of dendritic 7-oxanorbornene macromonomers and polymerized them with ROMP catalysts [150] as shown in Fig. 25, as well as with free radical or anionic catalysts. Yields were dependent upon the route used. Products were characterized by DSC, ^{13}C-NMR, and X-ray scattering. It was proposed that the resulting structures were mesogen-assembled supramolecular columns possessing single chain helicity.

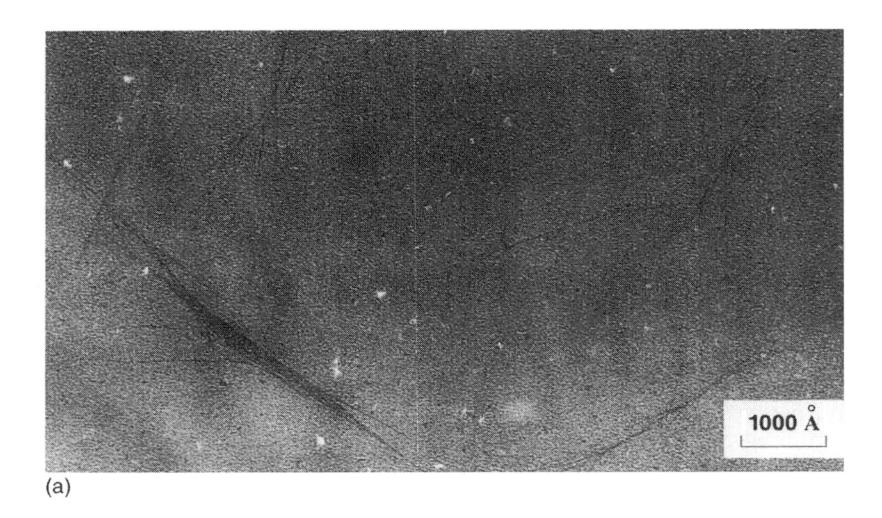

(a)

(b)

Figure 22 (a) Electron micrograph (TEM) of *linear* poly(ethyleneimine) (PEI) core; *dendri*-poly(amidoamine) PAMAM; G = 4(a); Z = $(\text{—}CO_2Na)_{16}$; N_c = 300–500 (Note: self-organization of dendrimer rods into parallel arrays). (b) Electron micrograph (TEM) of ammonia (NH_3) core; *dendri*-poly(amidoamine) PAMAM; G = 4(a); Z = $(\text{—}CO_2Na)_{48}$; N_c = 3 (Note: self-organization of dendrimer spheroids into clusters). (Courtesy J Am Chem Soc 120:2679, 1998. Copyright 1998 American Chemical Society.)

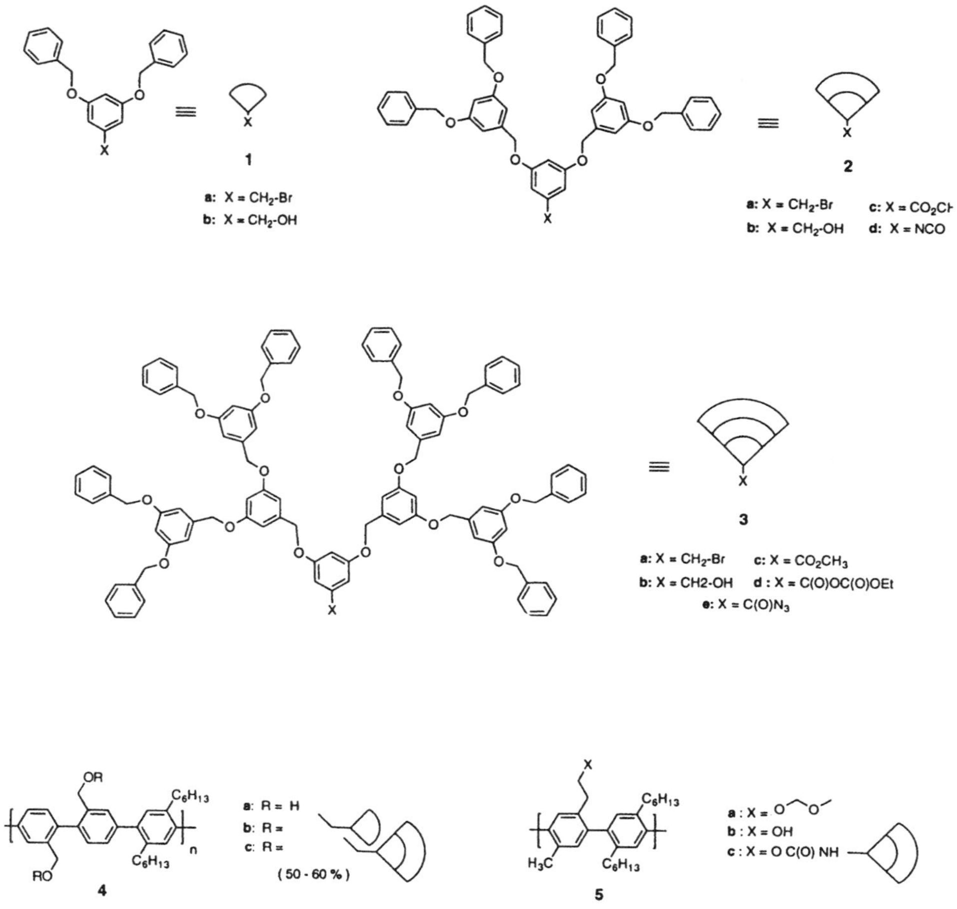

Figure 23 Fréchet-type monodendrons (Generations 1, 2, and 3) possessing various reactive groups at the focal point. Rod-shaped, dendritic-linear architectural copolymers obtained by coupling respective dendrons to linear poly(*para*-phenylene) (PPP) backbone. (Courtesy J Am Chem Soc 119:3297, 1997. Copyright 1997 American Chemical Society.)

(b). Dendrons as Mesogens—Dendromesogens. In addition to mainchain and sidechain liquid-crystalline polymers based on linear polymeric architecture, it was apparent as early as 1992 that new displays of mesogens presented on alternate architectures were possible. At that time, Ringsdorf et al. [151] and Percec et al. [152–155] pioneered extensive work in the area of dendritic mesogens. In general, this activity involved the examination of liquid-crystalline properties for

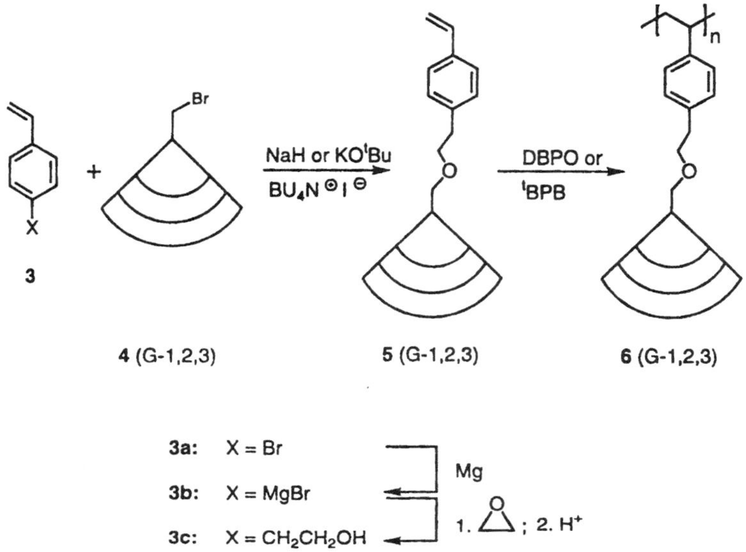

Figure 24 The synthetic sequence to macromonomers 5 (G-1,2,3) carrying dendritic fragments of the first (G-1), second (G-2), and third (G-3) generation and corresponding polymers 6 (G-1,2,3). (Courtesy Macromol Rapid Commun 17:519, 1996. With permission of Wiley-VCH Verlag GmbH, Germany.)

both hyperbranched and dendron-type architecture. For example, Percec et al. [156,157] synthesized low generation perfluorinated dendrons possessing a 15-crown-5-ether at their focal point. These unique structures were found to self-assemble into cylinders and ultimately into hexagonal columnar supramolecular arrays as illustrated in Fig. 26.

More recently, Lattermann et al. [158] described the attachment of mesogens to the terminal groups of poly(propyleneimine) dendrimers as a function of generation (i.e., G = 0–4). These new structures were referred to as *dendromesogens*. The authors noted that for the lower generations (i.e., G = 0–3), hexagonal columnar mesophases (Col_n) were observed. As the mesogen-induced congestion maximized at G = 4, mesomorphism disappeared and it was hypothesized that the dendromesogen transformed into a globular structure. These observations appear to parallel those of Tomalia et al. [125] and Percec et al. [159].

(c). Quasi-Equivalence of Dendritic Surfaces (Coats). An emerging direction in polymer chemistry is biomimicry of biological structures by using appropriate polymeric architectures, building blocks, and functionality. Quasi-equivalent building blocks or subunits are defined as chemically identical subunits, which may control their shape by switching between different conformational states

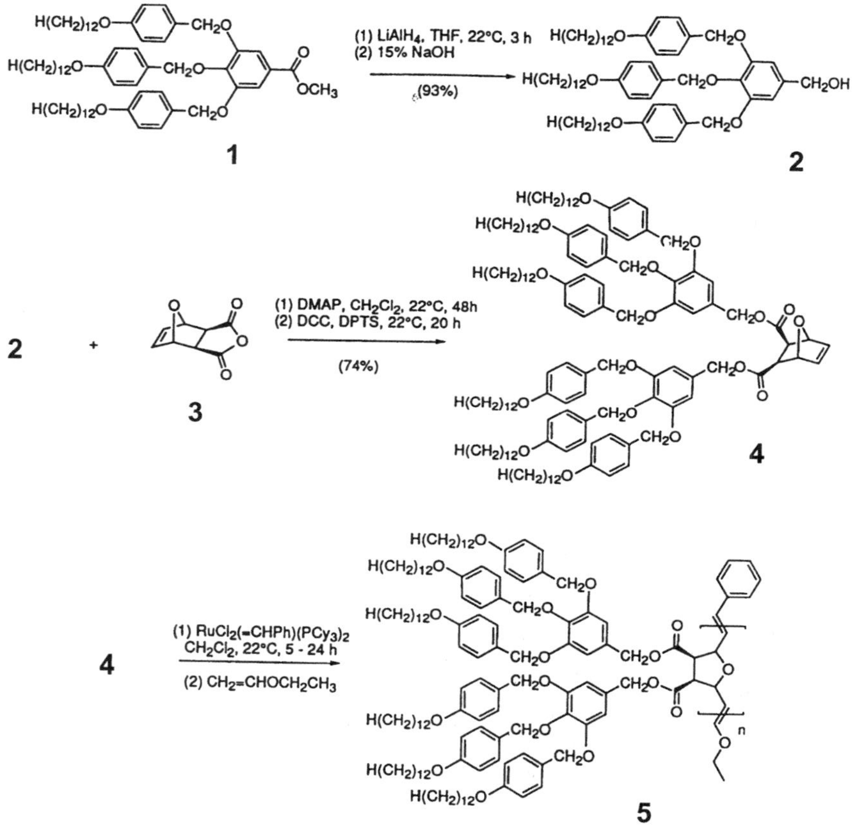

Figure 25 Synthesis of 7-oxanorbornene macromonomer derived from two (G-1) dendrons followed by ROMP polymerization into a linear dendritic hybrid architecture. (Courtesy of Macromolecules 1997, 30(19):5786. Copyright 1997 American Chemical Society.)

during the process of self-assembly [160,161]. Classic biological examples include the flat-tapered and conical protein subunits that are required for sheathing the nucleic acid component of a tobacco mosaic virus (TMV). By comparing very similar protein subunits it is apparent they will assume conical-type conformation to provide an appropriate sheathing for the protection of genetic material in an icosahedral virus, respectively [162]. In the former case, flat-tapered proteins self-assemble in the presence of a nucleic acid to generate rodlike viruses which have a helical symmetry. The cylindrical shape of this assembly induces a helical conformation to the nucleic acid (for example, a classic rodlike virus: TMV). On other hand, icosahedral viruses are approximated by a spherical shape

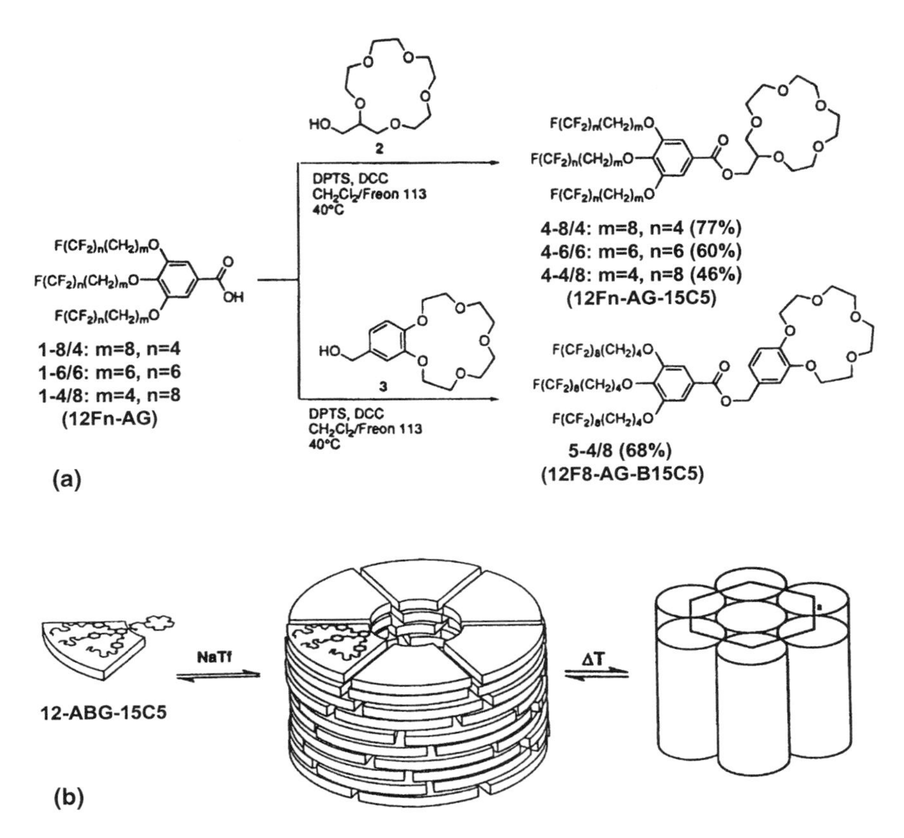

Figure 26 (a) Synthesis of monodendritic building unit with 15-crown-5 ether in the focal point; (b) self-assembly of cylindrical building blocks into a hexagonal columnar supramolecular architecture. (Courtesy J Am Chem Soc 118(41):9858, 1996. Copyright 1996 American Chemical Society.)

and are constructed from cone-shaped proteins. In the case of icosahedral viruses the nucleic acid adopts a random-coil conformation.

Using totally abiotic dendron subunits, Percec et al. [159,163] built structures that adapted either the shape of a rodlike virus with helical symmetry or an icosahedral virus with cone-shaped symmetry. Figure 27 outlines each of these viruses and their respective abiotic mimics.

In several seminal papers, Percec et al. [157,159] reported the demonstration of abiotic quasi-equivalency by controlling the degree of polymerization of various polymerizable dendrons described earlier. For example, 12 second-generation conical monodendrons (DP = 12) produced a spherical dendrimer

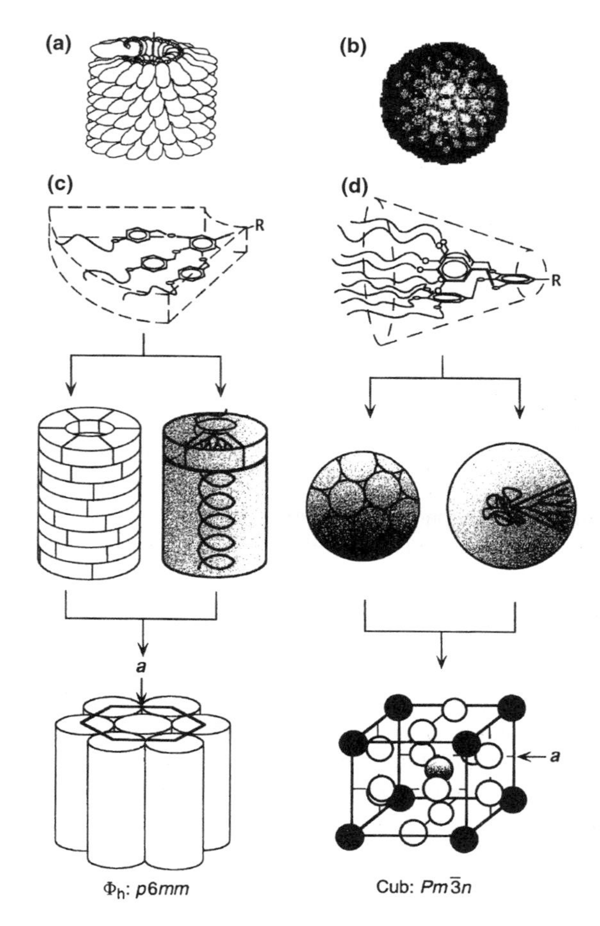

Figure 27 Quasi-equivalency of natural and synthetic supramolecular systems with cylindrical and spherical shapes: (a) tobacco mosaic virus, (b) icosahedral virus; synthetic analogue of (a) and (b) have been self-assembled from (c) tapered and (d) conical monodendrons. (Courtesy of Nature 1998, 391, 8, 161. Copyright 1998 Macmillan Magazines Limited.)

which results from self-organization into a cubic Pm3n three-dimensional (3-D) liquid-crystalline (LC) lattice. The attachment of methacrylate (12G2-AG-MA) or styrene (12G2-AG-S) monomer moieties to the dendron followed by radical initiation yields spherical (DP < 20) or cylindrical (DP > 20) polymers depending on the degree of polymerization (DP). The spherical polymers self-organize into the same cubic (Pm3n) 3-D lattice while the cylindrical ones self-assemble into a p6 mm columnar hexagonal (Fig. 28) (2-D liquid-crystal lattice). Furthermore, Percec found that the radical polymerization of the styrene or methylacrylic functionalized dendrons exhibited a very dramatic self-acceleration in polymerization kinetics. These kinetics are presumably enhanced by the unique self-assembly events that accompany these polymerizations [164]. As illustrated in Fig. 28, the principles for the design of such macromolecular structures are outlined.

2. Hypervalency/Hypercooperativity

Biological systems have long exploited the advantages of multivalent recognition events in the development of *exo*-supramolecular and supra*macro*molecular structures. Most notable is the power of multivalent hypercooperative binding associated with biological cell adhesion processes. It is now well recognized that this biological strategy is very versatile and ubiquitous. In the case of carbohydrate recognition at cell surfaces the preferred recognition mode is to involve many polyvalent soft recognition events that are geometrically optimized to provide hypercooperativity as opposed to single isolated events with harder recognition parameters. The power of this concept is illustrated in Fig. 29, eloquently described by Kiessling and Pohl [165], and further elaborated upon more recently by Whitesides et al. [166]. Simply stated, once a ligand has attached itself to a cell at one site, it suffers a smaller entropy loss by binding at neighboring sites. Mimicry of these biological adhesion parameters can be exquisitely modeled and tested with dendrimeric systems. Although a substantial number of linear, poly (valent) polymeric architectures [e.g., poly(acrylamides), etc.] have been tested with some success [167–170], efforts toward the use of dendrimer technology to create multivalent ligands for these purposes are in their infancy.

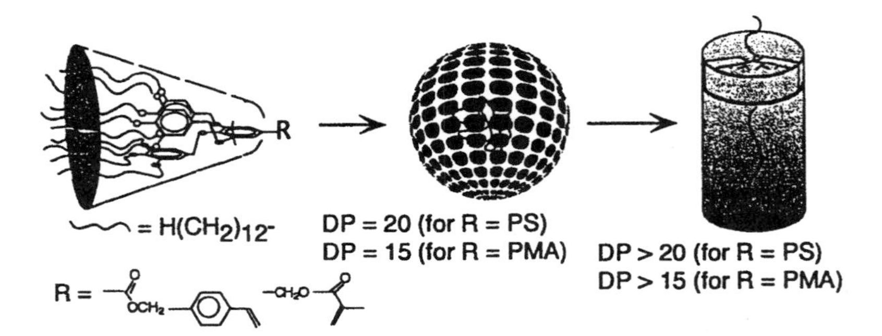

Figure 28 Scheme describing conversion of dendritic macromonomers to either spheroidal or cylindrical architectures depending on the degree of polymerization (DP). (Courtesy of J Am Chem Soc 1997, 119, 12978. Copyright 1997 American Chemical Society.)

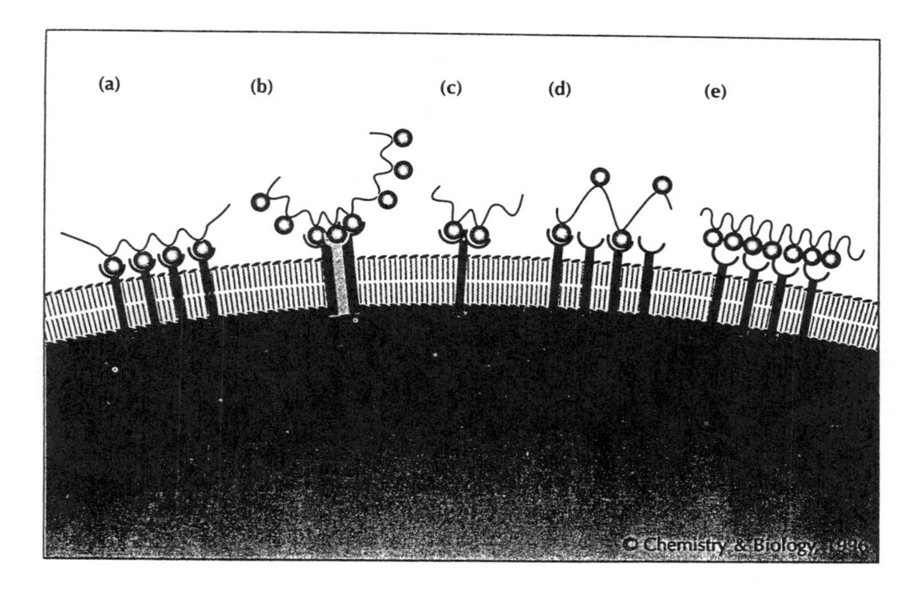

Figure 29 Specific recognition in multivalent interactions. Cells can use several strategies to bind to a multivalent ligand: (a) forming a cluster of many monovalent receptors on a small area of the cell surface; (b) using oligomeric receptors; or (c) using receptors with more than one saccharide-binding site. In all such systems, multivalent saccharide ligands bind more tightly to the cell than their monovalent counterparts. For a divalent ligand, the free energy of binding to a multivalent receptor array will be greater than the sum of the contributions of each individual site. This primarily results from the fact that once the ligand has attached itself to a cell by one site, it is closer to the second site and will suffer a smaller entropy loss by binding to it. Multivalent ligands with incompatible relative orientations (d) or spacing (e) of the saccharide units in the multivalent array will not bind tightly. (Courtesy of Chemistry & Biology 1996, 3(2):72. Copyright Current Biology Ltd., London.)

(a). Glyco(dendrimers). While searching for inhibitors of influenza virus hemagglutinin, Roy et al. [171–173] pioneered the synthesis and use of carbohydrate-substituted dendrimers, using solid-phase methodology to synthesize sialic acid decorated dendrimers. These dendrimers, containing 2, 4, 8, or 16 sialic acid residues, all inhibited the agglutination of erythrocytes by influenza virus (which is caused by hemagglutinin-mediated crosslinking of the erythrocytes) in the micromolar concentration range. (See Fig. 30.)

Very recent work in the Tomalia–Baker group [174] has shown that certain classes of dendritic architecture offer distinct advantages as scaffolding for presenting C-sialoside groups. For example, use of dendrigraft architecture [175] as

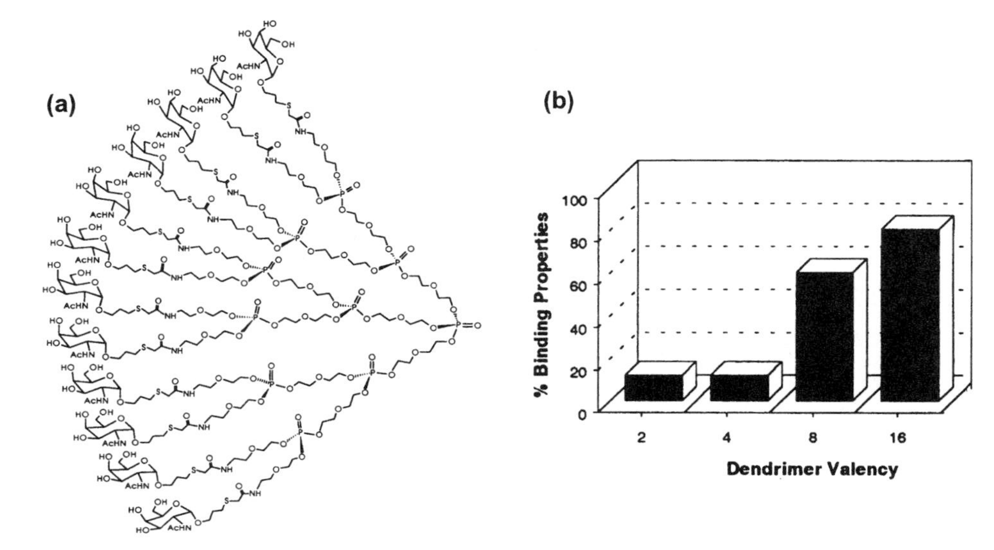

Figure 30 (a) Structure of phosphotriester GalNAc glycodendrimer. (b) Enzyme-linked lectin assays (ELLA) of L-lysine-based sialic acid dendrimers used as coating antigens in microtiter plates and detected with horseradish peroxidase labeled wheat germ agglutinin (HRPO-WGA). Binding activity as a function of dendrimer valency. (Courtesy of Polymer News, 1996, 21(7), 230. Copyright Gordon and Breach Publishers, Switzerland.)

scaffolding for the presentation of sialic acid groups was found to be more than 1000 times more effective than the corresponding linear architecture for the inhibition of influenza A viruses (see Fig. 31).

E. Dendrimers as Nanoscale Tectons (Modules)

1. Two-Dimensional Dendritic Assemblies

The self-assembly of dendrimers in two dimensions has been studied at both the air–water interface [i.e., Langmuir–Blodgett (LB) films] as well as at the air–bulk solid interface [i.e., self-assembled monolayer (SAM) films].

(a). At the Air–Water Interface [*LBs*]. Some of the earliest work was reported by Fréchet et al. in 1993 [176]. It involved the examination of LB films derived from the spreading of amphiphilic hydroxyl functionalized poly(arylether) dendrons as a function of generation level (i.e., G = 1–5) (Fig. 32). There was a strong dependence of the isotherm on molecular weight (generation level). The lower generations (i.e., G = 1–4) exhibited an increase in surface pressure through the liquid expanded phase (LE) followed by a peaked collapse transition, indicating a nucleation and growth event leading to a liquid condensed phase

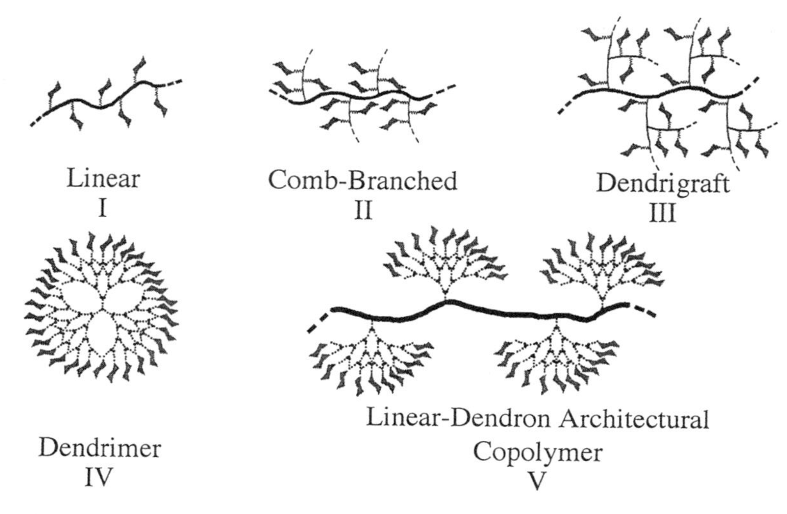

Figure 31 Model representations of sialic-acid-conjugated polymeric inhibitor subunits used to inhibit viral infection. Structures are not drawn to scale. (Courtesy of Bioconjugate Chemistry 1999, (10)2. Copyright 1999 American Chemical Society.)

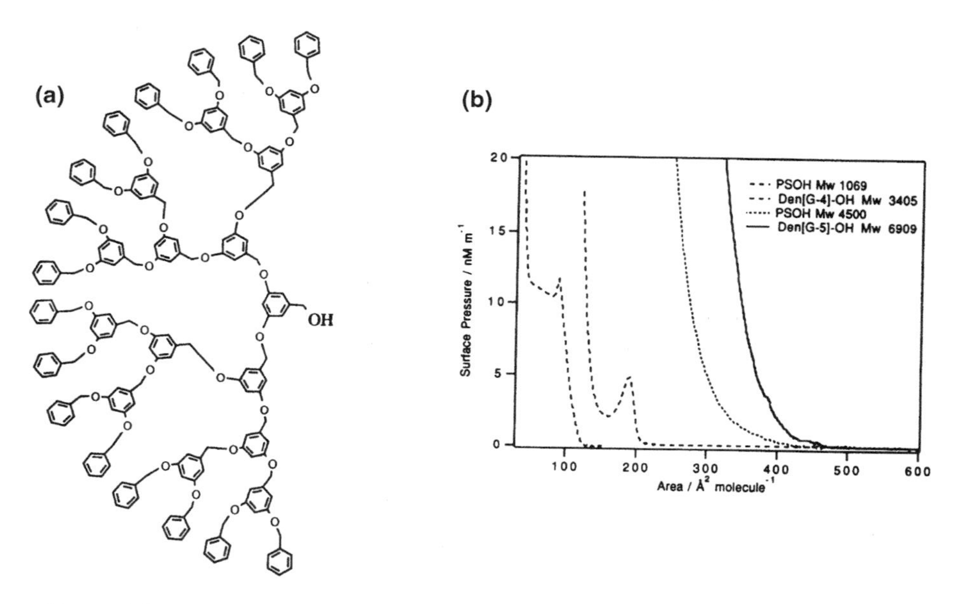

Figure 32 (a) Fourth generation [G=4] dendron, based on 3,5-dihydroxy benzyl alcohol core. (b) Isotherms of [G=4] and [G=5] dendrons showing differences due to compression rate and pause time before compression. (Courtesy of J Phys Chem 1993, 97, 294. Copyright 1993 American Chemical Society.)

(LC). A sharp transition in behavior occurred in progressing from G = 4 to 5. Advancement from G = 4 to 5 caused envelopment and isolation of the hydrophilic dendrimer focal point group from the water phase, thus proceeding directly to the condensed solid phase. In summary for generations 1–4, the dendrimers behave as classical surfactant molecules on a Langmuir trough. The isotherms of G = 5–6, however, manifest nonsurfactant behavior, once again reflecting the surface congestion-induced properties known for higher-generation dendrimers. Compression of the fourth generation poly(ether) dendrimer results in the formation of a stable bilayer. In this bilayer the dendrimers are compressed laterally in respect to the surface normal, producing an ellipsoid shape that is twice as high as it is broad (Fig. 33). Neutron-scattering studies on analogues with perdeuterated end-groups indicate that the terminal benzyl groups are located at the top of the lower layer [177]. More stable monolayers were formed when oligo(ethylene glycol) tails were used as core functionality [178,179].

Very interesting amphiphilic behavior at the air–water interface was observed by van Hest and Meijer [93] for hydrophobe (focal point) modified poly(propyleneimine) (PPI) dendrons [i.e., poly(styrene)-*dendri*-PPI $(NH_2)_n$, where n = 2,4,8]. These dendritic amphiphiles were essentially a reverse version of the Fréchet examples (see above). Only PS-*dendri*-PPI-$(NH_2)_n$ with n = 8 and 16 (i.e., G = 3 and 4) exhibited normal pressure-area isotherms. The lower generations (i.e., G = 1 and 2) all displayed isotherm curves indicating they transitioned directly to solid-state behavior.

Poly(propylene imine) dendrimers functionalized with hydrophobic alkyl chains (palmitoyl chains or alkyloxyazobenzene chains) assembled into stable monolayers at the air–water interface [122,180]. In the assemblies, the dendrimers adopted a cylindrical amphoteric shape in which the ellipsoidal dendritic moiety acted as a polar head-group and the alkyl chains arranged in a parallel fashion to form an apolar tail (Fig. 33b). This representation is based on the observation that the molecular area of a dendritic molecule increases linearly with the number of end-groups in this molecule.

Amphiphilic PAMAM dendrimers comparable in design to those reported

Figure 33 (a) Compressed dendrimer Langmuir bilayer; (b) dendrimer Langmuir monolayer.

for the poly(propylene imine) dendrimers have been studied at the air–water interface by Tomalia et al. [122]. The PAMAMs with aliphatic core groups of varying lengths (6, 8, 10, and 12 carbon atoms) also display the linear behavior between the molecular area at the compressed state and the number of end-groups per molecule. Tomalia et al. explain their findings in a model in which the lower generations are asymmetric like the poly(propylene imine) dendrimers, while the higher generations act as hydrophobic spheroids floating on the air–water interface. Since no indication for the latter behavior is found, it is proposed here that the amphiphilic PAMAM dendrimers of high generations when disposed on air–water interfaces, are also highly distorted with all aliphatic end-groups pointing upwards (Fig. 33b).

Most notable was the fact that it was shown that metal-loaded (Cu^{+2}) dendrimers can be readily organized into two-dimensional layers. Langmuir isotherms obtained for both unloaded and metal-loaded dendrimers in this series differed substantially from those observed by Fréchet. It was reassuring to note that radii measurements obtained from limiting area Langmuir–Blodgett film studies compared very favorably with radii determined by size exclusion chromatography measurements.

(b). At the Air–Bulk Phase Interface (SAMs). Very early observations [51,181] indicated the amine terminated PAMAM dendrimers exhibited tenacious adhesion to a variety of substrates. Early indications were that they formed SAMs on glass, silicon, or metal surfaces (i.e., gold, etc.). In a pioneering effort by Mansfield [182], it was predicted that dendrimers would exhibit various deformation modes on surfaces depending on generation and adsorption strength. This Monte Carlo simulation considered the adsorption of dendrimers on a surface at different interaction strengths. The calculations showed a flattening of the dendrimer shape with increasing adsorption strengths. As reflected in the ''phase'' diagram (Fig. 34), the mode of adsorption of the dendrimers is dependent on adsorption strength and on the generation number (higher-generation dendrimers have more interaction sites per molecule and, therefore, these dendrimers have a better chance to be adsorbed).

A wide variety of dendrimer adhesion modes have been defined experimentally that are beginning to fulfill the Mansfield predictions. They vary from self-assembled monolayers to multilayer assemblies as described in Fig. 35. Perhaps one of the first published works to clearly demonstrate the ability to construct mono/multilayers was reported by Regen and Watanabe [183]. They fabricated multilayers by repetitive activation with K_2PtCl_4 on a silicon wafer surface possessing primary amine groups, followed by deposition of PAMAM dendrimer, illustrated in Fig. 35(d). Reiteration of this sequence produced film thicknesses that were shown by elipsometry to increase linearly as a function of the number of cycles performed. After 12–16 cycles, multilayers with a thickness close to

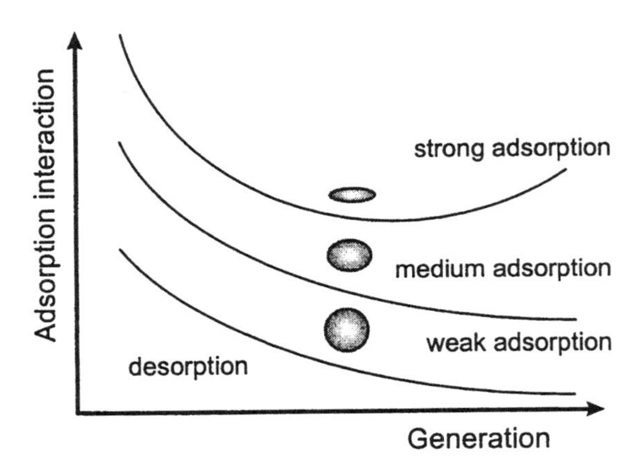

Figure 34 A "phase diagram" that shows how the shape of dendrimers in adsorbed monolayers depends on the strength of the adsorption interaction and the dendrimer generation. The data are based on calculations by Mansfield. (From Ref. 182.)

80 nm were obtained by using G = 8 or 6 PAMAM dendrimers, respectively. The elemental composition was confirmed by photoelectron spectroscopy (XPS) which both demonstrates the incorporation of PAMAM dendrimers as well as the necessity of the Pt^{+2} as a requisite component in the growth cycle.

An interesting type of dendrimer deformation has been reported by Crooks

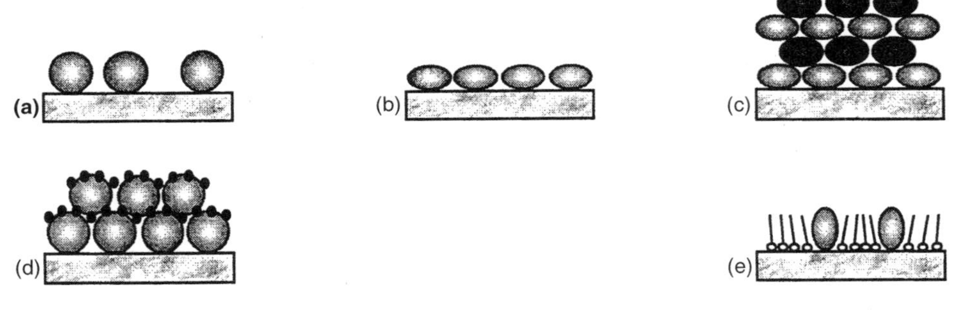

Figure 35 Schematic representation of the different modes of adsorption of dendrimers on surfaces: (a) adsorbed noninteracting dendrimers; (b) adsorbed dendrimers with surface-interacting end-groups; (c) interacting multilayer dendrimer films; (d) multilayer dendrimer films with ionic shielding; (e) mixed monolayer.

et al. [101,184,185]. Monolayers of PAMAM dendrimers adsorbed on a gold surface flatten due to multiple Au-amine interactions, but subsequent submission of alkanethiols to the surface results in a mixed monolayer in which the PAMAMs acquire a prolate configuration due to the shear exerted by the thiols (Fig. 35e). The shear originates from the stronger thiol–Au interaction as compared to the amine–Au interaction (see Chapter 10). If the adsorption time of the dendrimer monolayer is rather short (45 s instead of 20 h), exposure to hexadecanethiol results in piling up of the dendrimers to vacate the surface in favor of the thiols [101,184,185]. Eventually, this leads to complete desorption of the dendrimers from the surface.

Tsukruk et al. [186–188] reported on the alternating electrostatic layer-by-layer deposition of PAMAM dendrimer up to generation G = 10. This concept is illustrated in Fig. 35c. First an amine terminated dendrimer is adsorbed onto a SiO_2 surface followed by a carboxylic acid terminated dendrimer that deposits on top of a full generation. The cycle is repeated and multilayers are formed. They measured the layer thickness and found it to be below the theoretical one, which also supports the elastic deformation of dendrimer to obtain the most favorable energy balance. Molecular dendrimer dimensions follow scaling laws as a function of molecular weight of dendrimer with exponent of 0.27 (for spheres it is 1/3). Cast and spin-coated films exhibit large sensitivity toward conditions used during their formation. Thermal annealing can change dendrimer shape from oblate toward spherical. Change in surface characteristics (post- or pre-functional group modification) can largely influence the quality and properties of SAMs. Phase diagrams based on molecular modeling define the correlations among generation number, interaction strength, and shape (i.e., at strong interactions PAMAM G = 4 on Au dendrimer will be oblate on a surface, but at strong repulsion ($C_{16}SH$, PAMAM G = 8 on Au) dendrimer will be prolate. With weak interactions (G-0, G-2) dendrimer can retain sphere-like shapes.

Perhaps one of the most remarkable breakthroughs concerning dendrimeric SAMs is that reported recently by Fréchet et al. [189,190]. They have demonstrated that monolayers of dendritic polymers can be prepared by covalent attachment to a silicon wafer surface (Fig. 36a). These ultrathin polymer films can serve as effective resists for high-resolution lithography using the scanning probe microscope. These dendrimer films may be patterned using the SPL to create features with dimensions below 60 nm. Although very thin, the dendrimer films are resistant to an aqueous HF etch, allowing the production of a positive tone image as the patterned oxide relief features are selectively removed.

The patterned oxide relief features can be selectively removed under aqueous hydrofluoric acid (50:1, 60 sec) etching conditions resulting in a pattern transfer of raised oxide relief features into positive tone images. Figure 36b is a two-dimensional AFM image of the same patterns (from Fig. 36c) after etching

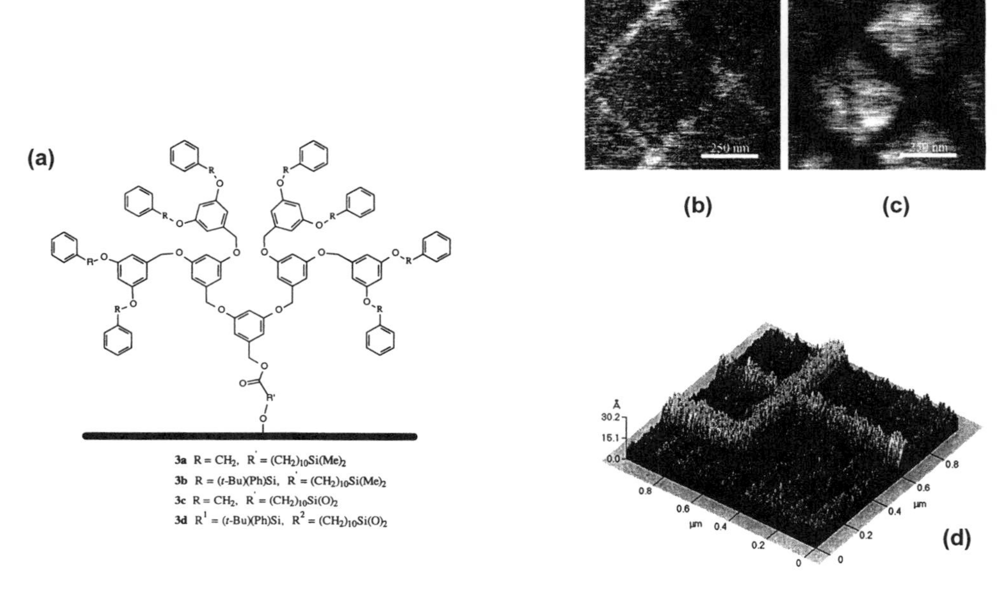

Figure 36 Covalent attachment of dendritic monolayers to a silicon wafer surface. (Courtesy of Cur Opinion in Colloid & Interface Sci, T Emrick, JMJ Fréchet, ''Self-Assembly of Dendritic Structures,'' 4, 1999 with permission from Elsevier Science.)

the wafer for 60 sec in 1M HF. The dark regions represent depressions approximately 2 nm deep into the silicon. Under these conditions, the dendrimer monolayer clearly resists the etch as evidenced from the lack of any line broadening or pitting in the unpatterned regions.

2. Three-Dimensional Dendritic Assemblies

(a). Statistical Structures. Earlier work by Tomalia et al. [55] has shown that dendrimers may be used as reactive (modules) building blocks to construct statistical three-dimensional covalent networks and gels. In contrast to traditional crosslinked networks [60], within the dendrimer networks one is able to observe conserved order in the form of unique topological features which have been seen in electron micrographs of the dendrimer system. Examples of linear, bridged, radially bridged, and macrocyclic bridged topologies were observed as shown in Fig. 37. Several of these topological types (i.e., linear and cyclic) are remarkably reminiscent of those that are obtained by noncovalent self-assembly of proteins.

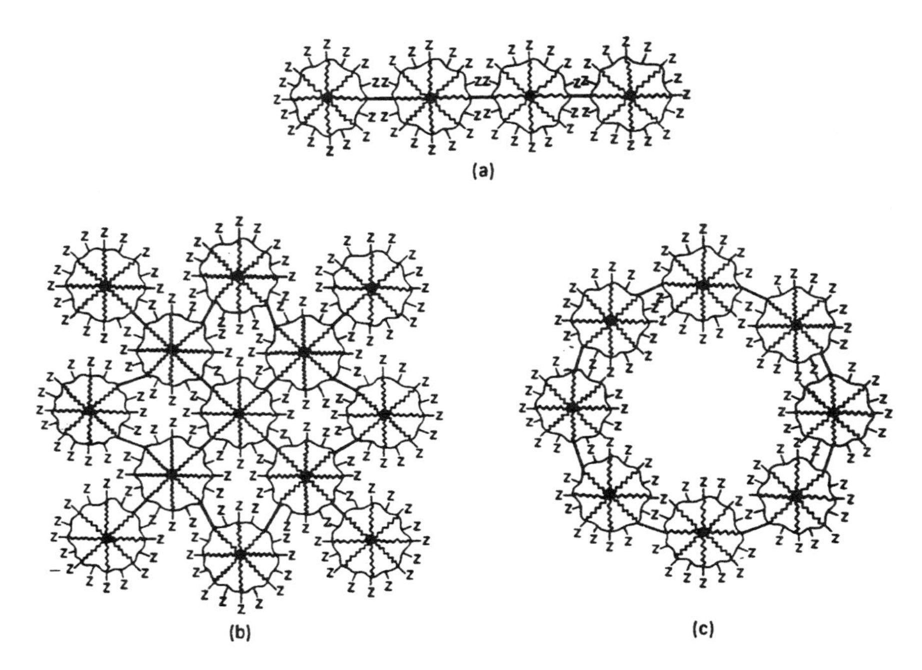

Figure 37 Bridged dendrimer types: (a) linear bridged, (b) radial bridged, (c) macrocyclic bridged.

However, virtually no reports have appeared concerning the self-assembly of dendrimers by noncovalent electrostatic methods until recently. Aida et al. [191] have reported the electrostatically directed assembly of prophyrin core dendrimers to produce large infinite aggregates as illustrated in Fig. 38. Energy transfer constants were obtained and used to determine distances between acceptors and donors as it was demonstrated that communication occurred between those domains.

Self-assembled, three-dimensional dendritic network structures possessing critical π-type surface groups have been reported to exhibit unique electrical conducting properties. Miller and coworkers [192] observed unusually high conductivities (i.e., 18 s/cm at 90% humidity) for generation = 3 PAMAM dendrimers surface-modified with cationically substituted naphthalene diimides (Fig. 39). In all cases, the conductivity was electronic and isotropic. Near-infrared spectra showed the formation of extensive π-stacking, which presumably favored electron hopping via a three-dimensional network.

Similarly, Wang and coworkers [193] reported that hyperbranched dendritic structures possessing poly(3-alkylthiophene) arms self-assemble into thin

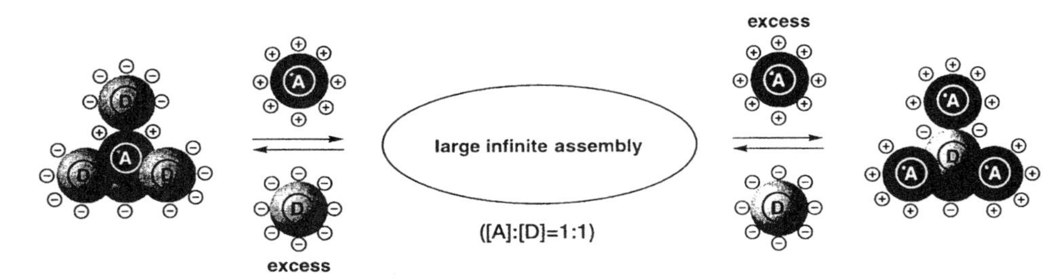

Figure 38 Schematic representation of the electrostatic interaction between negatively and positively charged dendrimer electrolytes. D represents a donor and A an acceptor in energy transfer. (Courtesy of Angew Chem Int Ed 1998, 37(11) 1533. Copyright Wiley-VCH Verlag GmbH, Weinheim, Germany.)

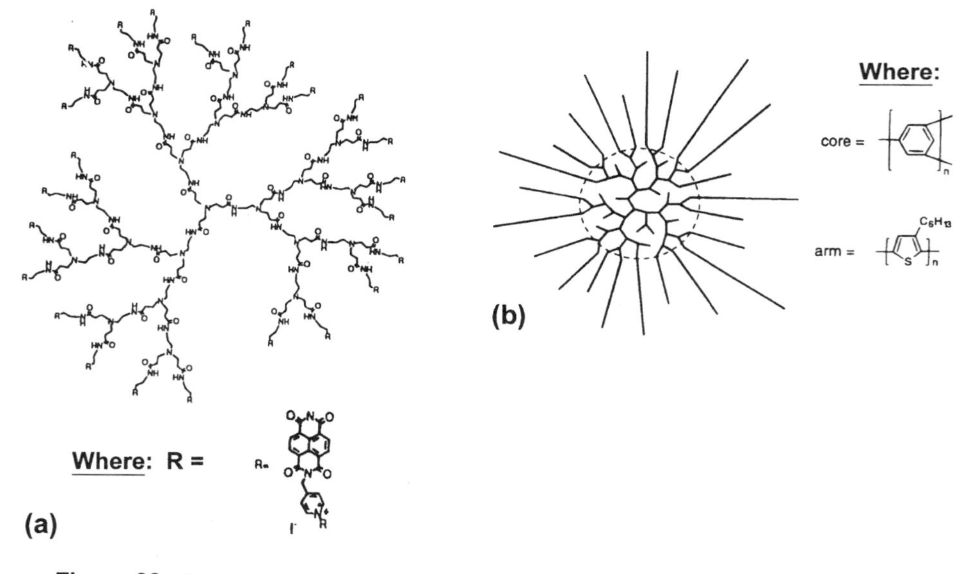

Figure 39 Examples of electrically conducting (a) dendrimeric and (b) hyperbranched star polymers. (Courtesy of J Am Chem Soc 1997, 119, 1006 and (b) 11106. Copyright 1997 American Chemical Society.)

films with morphological features, as well as electrical and optical properties that reveal a surprising degree of structural order. Typical conductivities varied between 42–65 s/cm.

(b). Structure Controlled Core-Shell Tecto(dendrimers). Core-shell architecture is a very recognizable concept in the lexicon of science. Beginning with the first observations by Galileo concerning the heliocentricity of the solar system [194] to the planetary model first proposed by Rutherford [1] and expounded upon by Bohr [195], such architecture has been broadly used to describe the influence that a central focal point component may exercise on its surrounding satellite components. Such has been the case recently at the subnanoscale level. Rebek et al. [196] have described the influence that a guest-molecule may have on self-assembling components that are directed by hydrogen-bonding preferences and the filling of space. It was shown that hydrogen-bonding preferences combined with spatial information such as molecular curvature can be used to self-assemble a single core-shell structure, as shown in Fig. 40 at the subnanoscopic level.

At the nanoscale-dimensional level it was shown by Hirsch et al. [197] that [60]-fullerenes could be used as a core tecton to construct a core-shell molecule with T_b-symmetrical C_{60} core and an extraordinarily high branching multiplicity of 12.

As described earlier, the assembly of reactive monomer [49] branch cells [14,52] around atomic or molecular cores to produce dendrimers according to divergent/convergent dendritic branching principles is well demonstrated [57]. The systematic filling of molecular space around dendrimer cores with monomer units or branch cells as a function of generational growth stages (dendrimer shells) to give discrete quantized bundles of mass has proven to be mathematically predictable [58]. This generational mass relationship has been demonstrated experimentally by mass spectroscopy [59,60] and other analytical methods [61,92]. These synthetic strategies have allowed the systematic control of molecu-

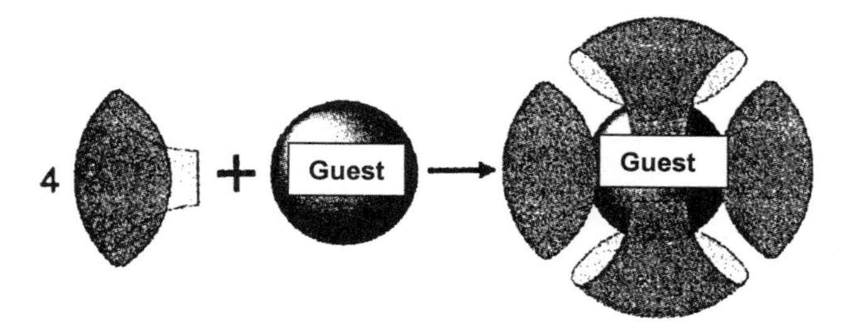

Figure 40 Self-assembly of a core-shell structure involving hydrogen bonding and complementary shape preferences according to Rebek et al., Ref. 196.

lar structure as a function of size [92], shape [125], and surface/interior functionality [36]. Such synthetic strategies have allowed construction of dendrimeric structures with dimensions that extend well into the lower nanoscale regions (i.e., 1–30 nm).

Since the very first reports on dendrimers in 1984 [62–67], we have proposed the use of these entities as fundamental building blocks for the construction of higher complexity structures on numerous occasions [35,58,198]. Early electron microscopy studies [55] and other analytical methods [61] indicated that the supra*macro*molecular assembly leading to formation of dimers, trimers, and other multimers of dendrimers occurred almost routinely; however, these were largely uncontrolled events.

Recent studies have shown that poly(amidoamine) PAMAM dendrimers are indeed very well-defined, systematically sized spheroids [92] as a function

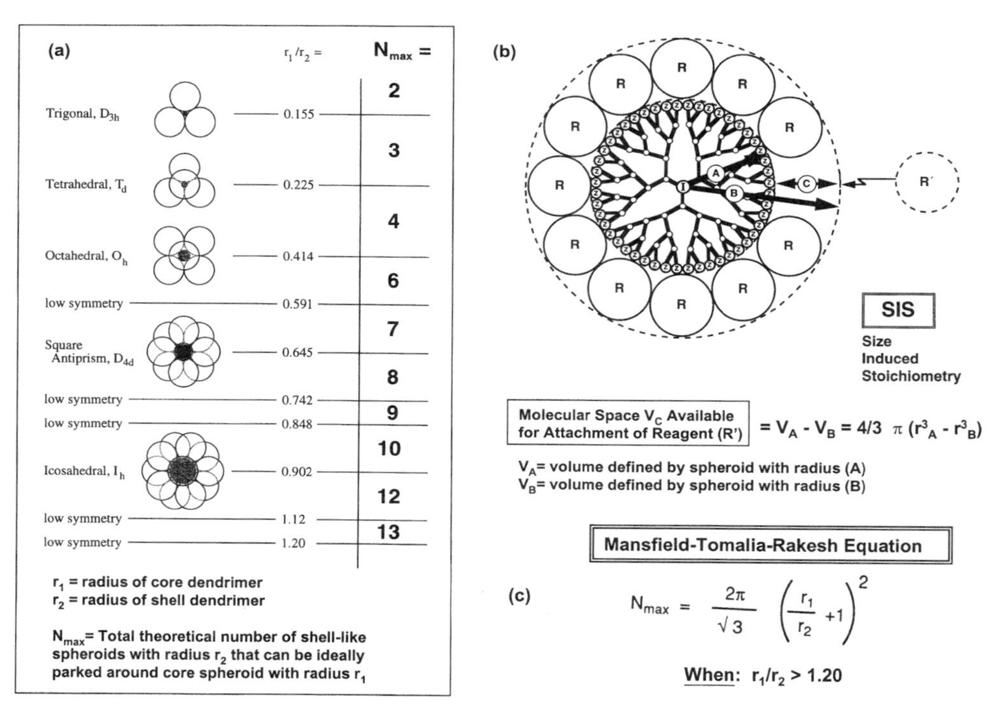

Figure 41 (a) Symmetrical properties for core-shell structures where $r_1/r_2 < 1.20$. (b) Sterically induced stoichiometry (SIS) based on respective radii (A) and (B) of dendrimers. (c) Mansfield–Tomalia–Rakesh equation for calculation of maximum shell filling when $r_1/r_2 > 1.20$.

of generation level. Furthermore, evidence has been obtained by SANS studies to show that these PAMAM dendrimers behave as originally described by de Gennes and Hervet [70]. The terminal groups remain largely at the periphery and are *exo*-presented with virtually no backfolding [68,199,200].

Anticipating the use of these nanoscale modules in a variety of construction operations we examined the random parking of spheres upon spheres [201]. From this study, it was rather surprising and pleasing to find that at low values of radii ratios (i.e., <1.2), absolutely beautiful symmetry properties appeared as illustrated in Fig. 41. However, at higher radii ratios, the mathematics resolved

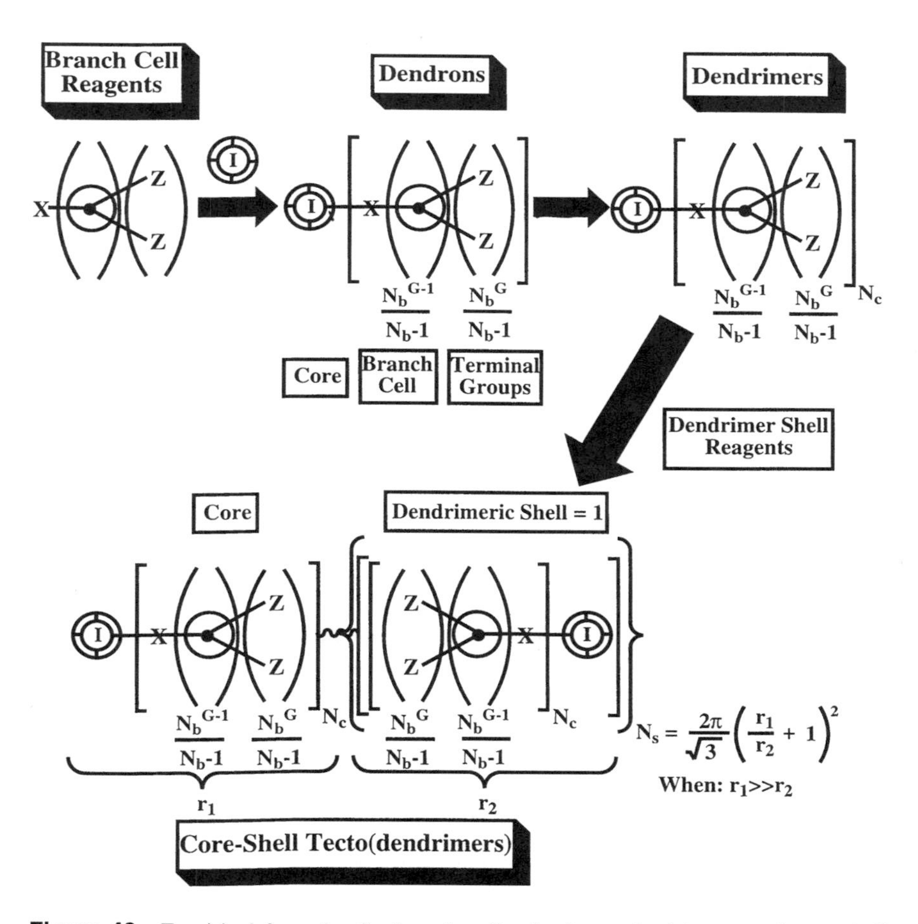

Figure 42 Empirical formulas for branch cells, dendrons, dendrimers, and core-shell tecto(dendrimers), where N_b = branch cell multiplicty; N_c = core multiplicty; G = generation; r_1 = radius of core dendrimer; r_2 = radius of shell dendrimer.

into the following Mansfield–Tomalia–Rakesh general expression. From this relationship, it is possible to calculate the number of spheroidal dendrimers one could possibly place in a shell around a core dendrimer as a function of their respective radii (Fig. 42).

Inspired by these derived values for shell filling around a central dendrimer core, we devised several synthetic approaches to test this hypothesis. The first method involved the direct covalent reaction of a dendrimer core with an excess of dendrimer shell reagent; referred to as the (a) Direct Covalent Method. The second method involved self-assembly by electrostatic neutralization of the dendrimer core with excess shell reagent to give the (b) Self-Assembly with Sequential Covalent Bond Formation Method. These strategies are described in Section IV. In each case, relatively monodispersed products are obtained. We call these new dendritic architectures *core-shell (tecto)dendrimers.*

IV. ASSEMBLY OF DENDRIMERS INTO PRECISE CORE-SHELL TECTO(DENDRIMERS)

A. Direct Covalent Method

This route produces partially filled shell structures and involves the reaction of a nucleophilic dendrimer core reagent with an excess of electrophilic dendrimer shell reagent as illustrated in Fig. 43 [54,202].

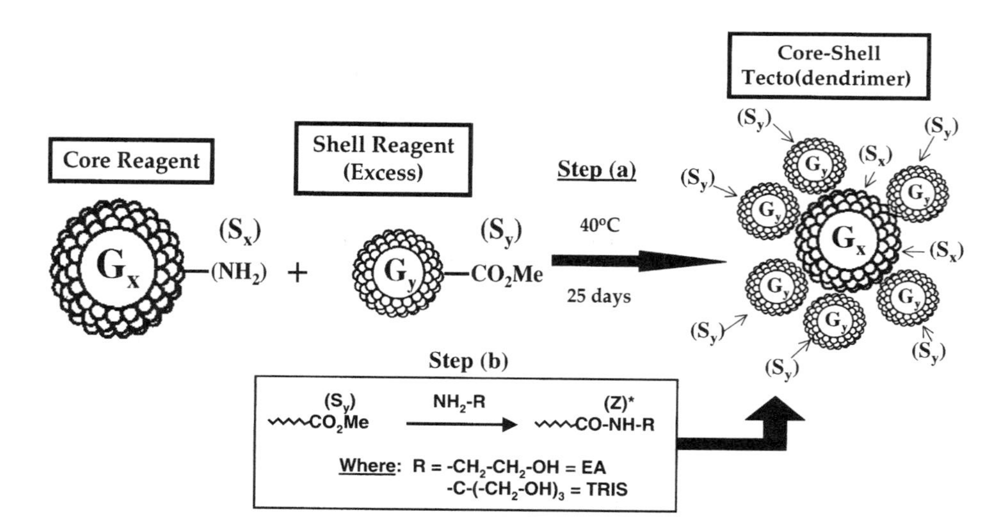

Figure 43 Synthetic scheme for core-shell tecto(dendrimers), where Z = surface functionality after reaction with R-type reagents.

Various poly(amidoamine) PAMAM dendrimer core reagents (i.e., either amine or ester functionalized) were allowed to react with an excess of appropriate PAMAM dendrimer shell reagents. The reactions were performed at 40°C in methanol and monitored by FTIR, ^{13}C NMR, SEC, and gel electrophoresis. Conversions in Step (a) (Fig. 43) were followed by the formation of shorter retention time, and higher molecular weight products using size exclusion chromatography (SEC). Additional evidence was gained by observing the loss of migratory band associated with the dendrimer core reagent present in the initial reaction mixture, accompanied by the formation of a higher molecular weight product, which displayed a much shorter migratory band position on the electrophoretic gel. In fact, the molecular weights of the resulting core-shell tecto(dendrimer) could be estimated by comparing the migratory time of the core-shell product (polyacrylamide) gel electrophoresis (PAGE) results (Table 1), with the migration distances of the PAMAM dendrimers (e.g., G = 2–10) used for their construction [61].

It was important to perform capping reactions on the surface of the resulting ester terminated core-shell products in order to pacify the highly reactive surfaces against further reaction. Preferred capping reagents were either 2-amino-ethanol or *tris*-hydroxymethyl aminomethane.

The capping reaction, Step (b) (Fig. 43), was monitored by following the disappearance of an ester band at 1734 cm^{-1}, using FTIR. Isolation and characterization of these products proved that they were indeed relatively mono(dispersed) spheroids as illustrated by AFM in Fig. 44. It was very important to perform the

Table 1 Analytical Evidence for Core-Shell Tecto(dendrimers)

$X;[(Y)(Z^*)]_n$	$G4[(G3);(EA)]_n$	$G5(G3);(TRIS)]_n$	$G6(G4);(TRIS)_n$	$G7(G5);(TRIS)_n$
Theoretical shell sat. levels(n)	9	15	15	15
Observed shell sat. levels(n*)	4	8–10	6–8	6
Percent theoretical shell sat. levels	44%	53–66%	40–53%	40%
MALDI-TOF-MS (MW):	56,496	120,026	227,606	288,970
PAGE (MW):	58,000	116,000	233,000	467,000
AFM: Observed dimensions:	25 × 0.38 nm (D,H)	33 × 0.53 nm (D,H)	38 × 0.63 nm (D,H)	43 × 1.1 nm (D,H)
CALC. (MW):	56,000	136,000	214,000	479,000

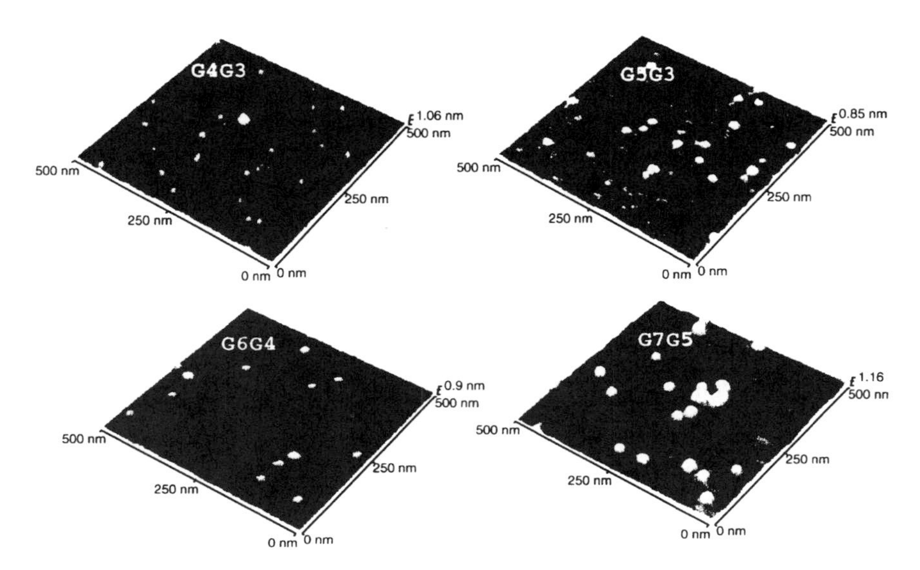

Figure 44 AFM images of core-shell tecto(dendrimer) PAMAMs. (Courtesy of PMSE, 1999, 80, 6. Copyright 1999 American Chemical Society.)

AFM analysis at very high dilution to avoid undesirable core-shell molecular clustering. In spite of these efforts, a small amount of clustering was still observed.

A distinct core-shell dimensional enhancement was observed as a function of the sum of the core-shell generation values used in the construction of the series (e.g., G4/G3, <G5/G3, <G6/G4, and <G7/G5). This is in sharp contrast to nondescript polydispersed dendrimer cluster/gel formation observed for 1 : 1 reaction ratios described in our earlier work [55].

Molecular weights for the final products were determined by MALDI-TOF-MS or PAGE. They were corroborated by calculated values from AFM dimension data (Table 1 and Fig. 44). These values were found to be in relatively good agreement within this series (Table 1).

Calculations based on these experimentally determined molecular weights allowed the estimation of shell filling levels for respective core-shell structures within this series. A comparison with mathematically predicted saturated shell structures reported earlier [201] indicates these core-shell structures are only partially filled (i.e., 40–66% of fully saturated values, see Table 1).

Functional groups differentiated clefts produced on the surfaces of these unique partially filled tecto(dendrimer) core-shell structures suggest rich possibilities for catalytic sites and other novel surface modifications that are presently under examination in our laboratory.

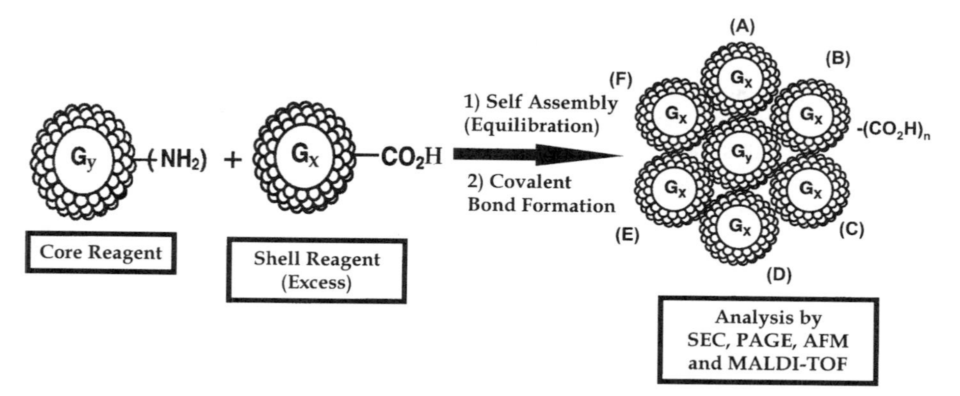

Figure 45 Self-assembly of core and shell dendrimers by charge neutralization (1) followed by covalent bond formation (2).

B. Self-Assembly with Sequential Covalent Bond Formation Method

The chemistry used in this approach involved the combination of an amine terminated dendrimer core with an excess of carboxylic acid terminated shell reagent dendrimer (Fig. 45) [203]. These two charge-differentiated species were allowed to equilibrate and self-assemble into the electrostatically driven core-shell tecto (dendrimer) architecture followed by covalent fixing of these charge-neutralized dendrimer contacts with carbodiimide reagents. Reactions were readily monitored

Table 2 Comparison of % Shell Filling as a Function of Core and Shell Dendrimers, Using the Supramolecularly Assisted Method Followed by Covalent Bond Formation

Core reagent	Shell reagent	Observed no. of parked dendrimers by mass spec.	Ideal no. of parked dendrimers[a]	Shell filling (%)
G5	G3)-COOH	10	12	83
G6	G3)-COOH	13	15	87
G7	G3)-COOH	15	19	79
G7	G5)-COOH	9	12	75

[a] Calculated using Mansfield–Tomalia–Rakesh equation for the maximum parking problem

$$N_{max} = \frac{2\pi}{\sqrt{3}}\left(\frac{r_1}{r_2} + 1\right)$$

Source: Ref. 201.

by SEC, gel electrophoresis, AFM, and MALDI-TOF mass spectroscopy. As might be expected, preliminary data show that the self-assembly method provides for more efficient parking of the dendrimer shell reagents around the core to yield very high saturation levels as shown in Table 2. Our present experimentation indicates that this method should allow the the assembly of additional shells in a very systematic fashion to produce precise nanostructures that transcend the entire nanoscale region (1–100 nm) [219].

V. OVERVIEW, PRESENT APPLICATIONS, DENDRITIC NANODEVICES

A. Overview of the Dendritic State

In summary, rheological (fluidity) investigations in our laboratories indicate that dendrimers behave like soft spherical bodies surrounded by relatively hard surface shells (i.e., like core-shell type entities) [204]. According to this emerging picture, the interiors of dendrimers may be deformable or rigid depending on the character of the monomers used in construction. The interiors may contain cavities, voids, and well-defined space capable of accommodating many small guest molecules such as solvents, dyes, and oligomers (intermediates between monomers and polymers). Guest molecules may enter or exit the dendrimer depending on their size and shape.

On the other hand, dendrimer surfaces appear to be impenetrable to other dendrimer molecules or natural and synthetic macromolecules, especially at higher generations. Dendrimers can thus be envisioned as unimolecular containers for encapsulations [36] or so-called dendrimer boxes [104]. Potential uses may include molecular delivery agents, transport vehicles, unimolecular micelles, molecule ball bearings, interphase catalysts, flow regulators in fluids, and highly monodispersed ‘‘dwarf latexes’’ for coatings, etc. If applications in these areas can be realized, the future of dendrimers and other polymers is very bright indeed.

Another fascinating area of dendrimer applications is based on their high surface functionality. No other class of synthetic or natural compounds contains so many reactive terminal groups per molecule as do the dendrimers. This provides major directions for possible exploitation. First, dendrimers can be modified in various ways with reagents of small molecular weight. It is thus possible to produce dendrimers with so-called *exo*-modified or differentiated surfaces. For example, attachment of catalytic or biological receptor sites suggests many possible applications. Furthermore, the dendrimer interiors may be modified in many yet specific ways. Interior differentiated dendrimers with different combinations of radial layers or segments may be prepared by using different dendrons or parts of dendrons in their construction. Incarceration of zero valent metals or their salts (i.e., iron, copper, silver, palladium, platinum, or cadmium sulfide, etc.) suggest many uses as catalysts, magnetic dendrimers, and quantum dots.

Equally significant, is the possibility of using the dendrimer surface reactivity to open a new branch of synthetic chemistry, namely, *nanoscopic chemistry.* This would involve using dendrimers as building blocks for the preparation of even larger compounds with nanoscopic and microscopic dimensions [58,205, 206]. Such megamolecules could result from reactions between dendrimers, as described in Section IV. B, or with appropriate biological macromolecules yielding covalently bonded nanoscopic structures with hybridized architectures (i.e., linear, branched, crosslinked, or dendritic types) [125]. Of course, this present account describes the numerous possibilities based upon various molecular recognition and noncovalent bonding processes, which include both supra- and supra*macro*-molecular interactions.

Perhaps most exciting is the emerging role that dendritic architecture is playing in the role of commodity polymers. Recent reports by Guan et al. [207] have shown that ethylene monomers polymerize to dendrigraft poly(ethylene) at low pressures. This occurs when using late transition metal or Brookhart-type catalyst (Fig. 46). Furthermore, these authors also stated that small amounts of dendrigraft poly(ethylene) architecture may be expected from analogous early transition metal–metallocene catalysts.

B. Dendritic Nanodevices

Abiotic–biotic hybrids composed of dendritic polymers and natural biopolymers have already found application as nanodevices. In many instances, these dendritic nanodevices are used in abiotic–biotic molecular recognition events involving suprachemistry. One prime example is the conjugation of poly(amidoamine) dendrimers to IgG antibodies for use in diagnostic immunoassay [208]. In that work, the architecturally precise dendrimers act as a replacement for a secondary antibody by spacing the primary antibody away from the solid phase. The replacement of a secondary binding antibody dramatically reduces lot rejections in manufacture and minimizes nonspecific interactions between analytes and the immobilized antibodies. In this nanodevice, the dendrimer acts both as an antibody replacement in contact with the solid phase as well as a macromolecular spacer to hold the antibody away from the solid phase.

Dendrimers have also found application as carriers of genetic materials into cells [209,210,220]. In this usage, the dendrimers can be thought of as a histone mimetic. With appropriately charged surface groups the polycationic dendrimers form a supra*macro*molecular complex with the polyanionic nucleic acid biopolymer. This dendrimeric nanodevice transports the genetic material across cell

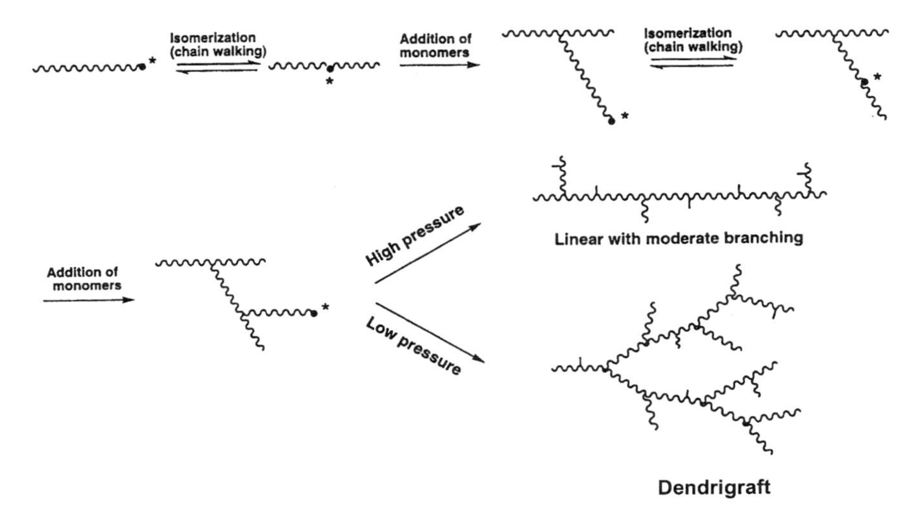

Figure 46 (a) Condensation polymerization of AB_2 monomers to give a hyperbranched polymer. Condensation of each monomer increases the active site from one to two. (b) ''Self-condensing'' polymerization of vinyl monomers to give a hyperbranched polymer. Addition of a new monomer increases the active site from one to two. (c) Proposed new approach to make hyperbranched polymer. The active site isomerizes to the internal backbone, and addition of monomers leads to branching.

membranes wherein the DNA is eventually incorporated into the cellular expression machinery to generate protein.

Dendrimer-based nanodevices have also found application in magnetic resonance imaging [211]. Dendrimer–chelate conjugates have been prepared that are extraordinarily robust complexes of gadolinium ions. The size of the dendrimer–gadolinium complexes allows sufficiently long residence times in the bloodstream for appropriate imaging studies. Furthermore, the nuclear relaxation parameters are dramatically enhanced when compared to conventional MRI contrast systems. Higher complexity conjugates have been prepared with target directors to introduce biotic molecular recognition functionality into these conjugates [211].

Dendritic polymers may also have application in the delivery of anticancer agents [212]. Dendrimers have been shown to selectively deliver a high payload of traditional chemotherapeutic agents, such as cisplatin, to a tumor [213]. Blood vessels in tumors have a higher permeability and poorer lymphatic drainage than vessels in normal tissue. This enhanced permeability and retention effect (EPR effect) allows selective delivery of a drug to a tumor, and has been demonstrated with nondendritic polymer–antitumor agent conjugates [213].

Dendrimer–saccharide conjugates are unique nanodevices in their ability to tightly bind lectins and other proteins with specific saccharide recognition capabilities [174]. The multivalency of dendrimer surfaces allows for a cooperative binding effect. Where single saccharide-receptor interactions are relatively weak, the multiplicity of poly(valent) interactions introduced by conjugating a

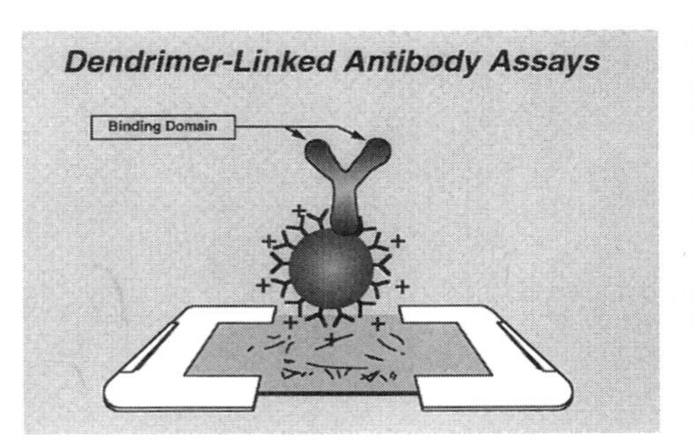

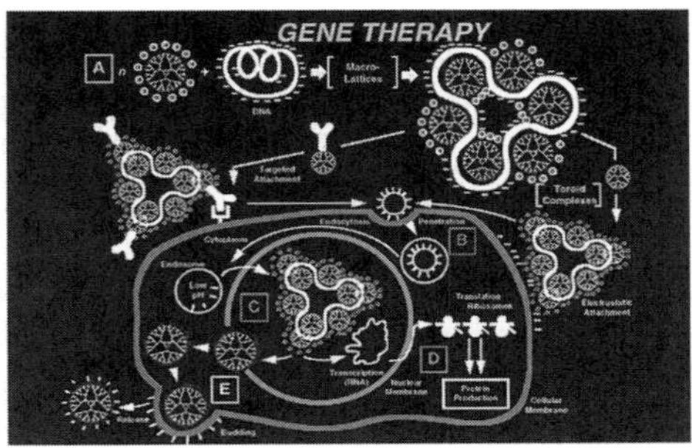

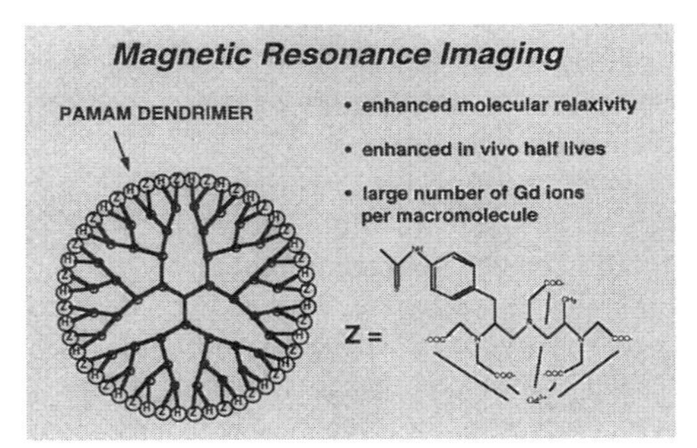

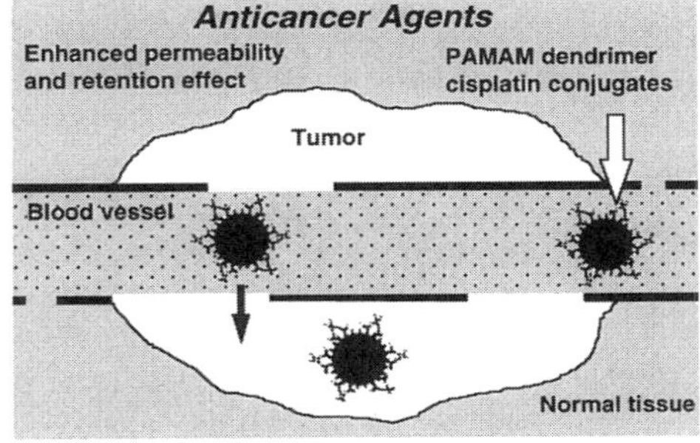

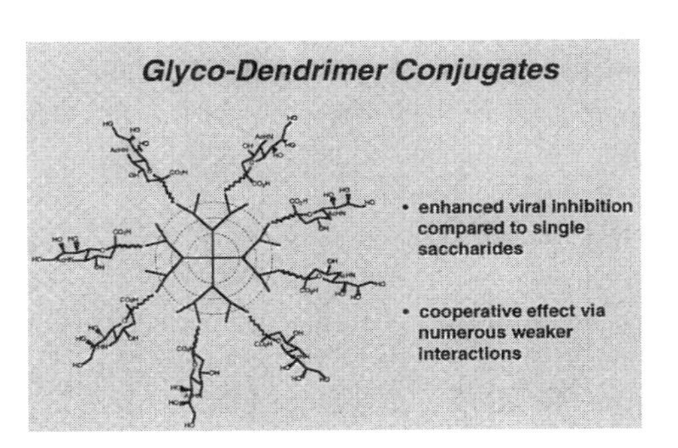

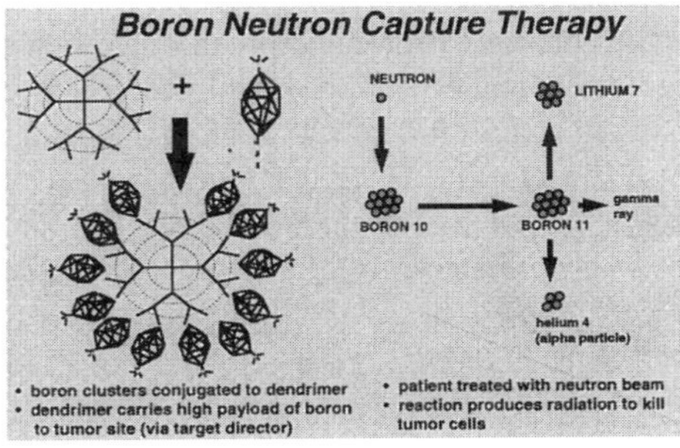

Figure 47 Some current nanobiological devices utilizing dendritic polymers.

large number of sugars onto a single dendrimer particle yields a cooperative unimolecular binding event that is thermodynamically much more favorable.

Dendritic polymers offer the potential to serve as amplifiers in a variety of applications. Researchers in boron neutron capture therapy have investigated the potential of dendrimers to deliver a large payload of boron to tumor sites by conjugating a large number of boron clusters to the surface of a dendrimer [214]. When irradiated with a neutron beam, the high cross-section of boron captures sufficient energy to produce a secondary radiation of sufficient energy to damage cells in the immediate vicinity of the boron. Again, conjugates with higher specificity can be prepared by introduction of target receptors [21].

Although many of the above examples are in the early stages of development, they have clearly demonstrated concepts that illustrate the current potential for dendrimers in a variety of nanodevices, many of them supramolecular or supra*macro*molecular in nature. Figure 47 is a pictorial summary of some of these nanobiological devices. Further information on the individual concepts may be found in the literature [34,215,216].

VI. THE FUTURE/CONCLUSIONS

> Nature is a very strange affair, and the strangeness already encountered by our friends the physicists are banalities compared to the queer things being glimpsed in biology, and the much queerer things that lie ahead.
>
> As these turn up . . . they will inevitably change the way the world looks. And when this happens, the view of life itself will also shift; old ideas will be set aside; the look of a tree will be a different look; the connectedness of all the parts of nature will become a reality for everyone, not just the mystics, to think about; painters will begin to paint differently; music will change from what it is to something new and unguessed at; poets will write stranger poems; and the culture will begin a new cycle of change.
>
> Lewis Thomas, 1985

This quotation conveys the daunting prophesies that may be expected from the convergence of supramolecular and supra*macro*molecular chemistry with nanoscale building blocks such as dendrimers or other nanoscale objects in the biological world.

It was both remarkable and surprising to find that many of these abiotic structure-controlled macromolecules (dendrimers, dendrigrafts, etc.) possess topologies, functions, and dimensions that scale very closely to a wide variety of important biological polymers and assemblies. Figure 48 compares poly(amidoamine) dendrimers as a function of generation with important biological structures that are both conserved and essential for life in the plant, as well as the animal, kingdom. History has shown that the introduction of traditional synthetic polymer

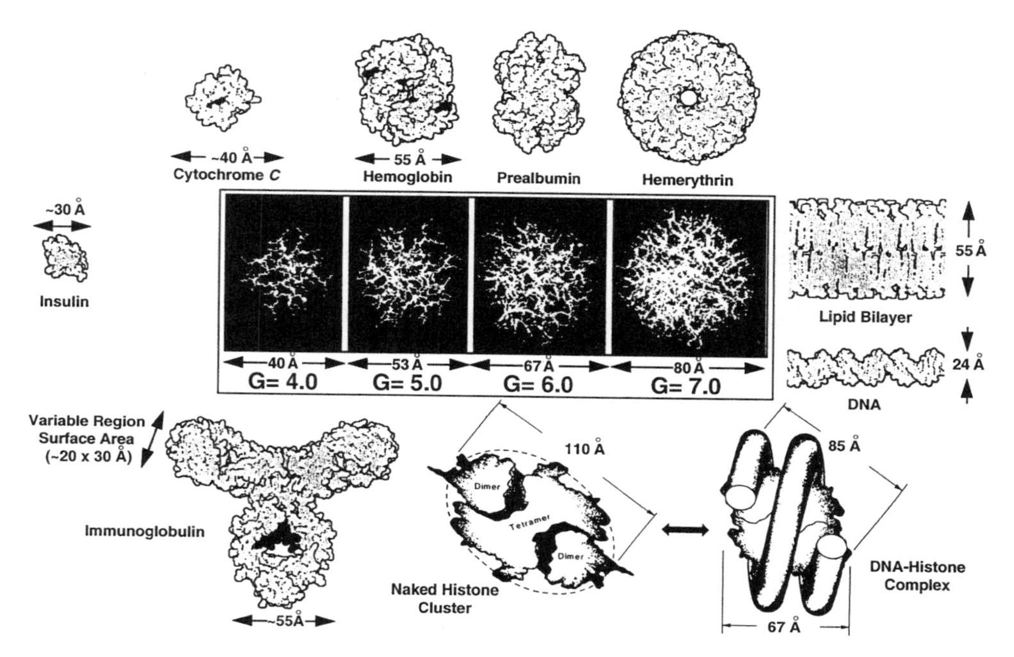

Figure 48 Scaled comparison of tridendron (NH_3 core) poly(amidoamine) dendrimers; generation = 4–7, sizes and shapes with various proteins. (Courtesy of J Mater Chem 1997, 7 (7), 1199. Copyright Royal Society of Chemistry.)

architectures (i.e., linear, crosslinked, etc.) by Staudinger, Carothers, Flory, and others [41,217] provided the basis for the replacement of many natural polymers (i.e., silk, rubber, cotton, etc.) accompanied by many improvements and advantages. With the present understanding of the covalent, supramolecular, and supra-*macro*molecular chemistry of both traditional and dendrimeric systems, it is now possible to visualize new architectural possibilities. This area will undoubtedly provide the enabling science for the replacement of many biological nanostructures such as histones, antibodies, hemoglobin, etc. [219]. In fact, recent developments are driving this new and emerging area of science, which is now being referred to as nanoscopic chemistry, biologic nanotechnology, or nanobiology. Briefly defined, it is the science of understanding covalent, structure-controlled synthesis, and characterization, combining rules and the supramolecular/supra-*macro*molecular dynamics of molecules and assemblies that manifest complexity and dimensions greater than traditional chemistry [20,218]. This new chemical science will undoubtedly evolve the necessary molecular level understanding to create requisite bridges to biology. These new connections to biology may lead one to expect many new paradigm shifts in the treatment and remediation of acute/chronic diseases, genetic defects, and enhancements of longevity (aging) leading to improvements in the human condition, as well as the biological environment in general.

Recent successes in the development of dendritic nanodevices reveal progress in this endeavor. It will not be surprising to see the evolution of synthetic immunosystems, gene expression protocol based on artificial histones, or the treatment of cancer with dendritic nanodevices in the foreseeable future.

As proposed earlier, dendritic biomimicry especially as it relates to dimensional size scaling, macromolecular structure control, shape/topological matching, genealogical sequencing, exponential amplification/polyvalency (hypercooperativity), site isolation, unimolecular encapsulation, and self-assembly are now possible with dendritic structures and architecture. In many cases, this mimicry has already been demonstrated in preliminary form (Fig. 47). Recent demonstration of these analogies not only has fulfilled some unexpected prophecies but also more importantly has set the stage (or the table) for an unimaginable banquet of new discoveries and properties. It is hoped that these endeavors will provide solutions to problems that presently plague and hamper the enhancement of the human condition.

Furthermore, we predict that the dendritic state will undoubtedly provide the quintessential scientific bridge by which abiotic molecular level scientists will be able to communicate and collaborate on such critical issues as disease control, increased agricultural production, and finally longevity, with enhanced quality of life. It is with these thoughts and premises that we should be very excited about the future of dendritic macromolecular technology and the role its supraproperties will provide as we enter the next millennium.

ACKNOWLEDGMENTS

The authors would like to thank the U.S. Army Research Laboratory (ARL/MMI Dendritic Polymer Center of Excellence), especially Dr. G. Hagnauer, and Edgewood Research, Dev. & Eng. Center (ERDEC), especially Dr. D. Durst, for financial support of this research and Ms. Linda S. Nixon for manuscript and graphics preparation.

REFERENCES

1. B Pullman. The Atom in the History of Human Thought. New York: Oxford University Press, 1998.
2. SE Mason. Chemical Evolution. Oxford: Clarendon, Oxford University Press, 1991.

3. M Eigen, W Gardiner, P Schuster, R Winkler-Oswatitsch. Evolution Now. New York: Freeman, 1982.
4. M Eigen. Naturwissenschaften 10:465, 1971.
5. H Kuhn, J Waser. Angew Chem Int Ed 20:500, 1981.
6. I Prigogine. Physics Today 25(12):38, 1972.
7. Webster's Third New International Dictionary. Springfield, MA: G & C Merriam, 1981.
8. B Alberts, D Bray, J Lewis, M Raff, K Roberts, JD Watson. Molecular Biology of the Cell. New York, London: Garland, 1994.
9. J Darnell, H Lodish, D Baltimore. Molecular Cell Biology. New York: Scientific American Books, 1986.
10. GF Joyce. Scientific American 90, 1992.
11. HW Salzburg. From Caveman to Chemist. Washington DC: American Chemical Society, 1991.
12. JM Lehn. Angew Chem Int Ed 27:89, 1988.
13. GW Gokel. Crown Ethers & Cryptands. Cambridge, UK: Royal Society of Chemistry, 1991.
14. GR Newkome, CN Moorfield, F Vogtle. Dendritic Molecules. Weinheim: VCH, 1996.
15. D Philip, JF Stoddart. Synlett 7:445, 1991.
16. PR Ashton, NS Isaacs, FH Kohnke, JP Mathias, JF Stoddart. Angew Chem Int Ed 28:1258, 1989.
17. PR Ashton, NS Isaacs, FH Kohnke, GS D'Alcontres, JF Stoddart. Angew Chem Int Ed 28:1261, 1989.
18. FH Kohnke, JP Mathias, JF Stoddart. Molecular Recognition: Chemical and Biochemical Problems. Cambridge, UK: Royal Society of Chemistry, 1989.
19. FH Kohnke, JP Mathias, JF Stoddard. Angew Chem Adv Mater 101:1129, 1989.
20. M Freemantle. Chem Eng News (April 19):51–58, 1999.
21. CM Drain, F Nifiatis, A Vasenko, JD Batteas. Angew Chem Int Ed 37:2344, 1998.
22. CM Drain, JMJ Lehn. J Chem Soc Chem Commun 2313, 1994.
23. RW Wagner, J Seth, SI Yang, D Kim, DF Bocian, D Holten, JS Lindsey. J Org Chem 63:5042, 1998.
24. RV Slone, JT Hupp. Inorg Chem 36:5422, 1997.
25. CM Drain, KS Russell, JM Lehn. Chem Commun 337, 1996.
26. E Alessio, M Macchi, S Heath, LG Marzilli. J Chem Soc Chem Commun 1411, 1996.
27. E Alessio, M Macchi, S Heath. J Chem Soc Chem Commun 36:5614, 1997.
28. Y Kobuke, H Miyaji. Bull Chem Soc Japan 69:3563, 1996.
29. RT Stibrany, J Vasudevan, S Knapp, JA Potenza, T Emge, HJ Schugar. J Am Chem Soc 118:3980, 1996.
30. H Tamiaki, T Miyatake, R Tanikaga, AR Holzwarth, K Schaffner. Angew Chem Int Ed 35:772, 1996.
31. DB Amabilino, CO Dietrich-Bucheker, J-P Sauvage. J Am Chem Soc 188:3285, 1996.
32. S Anderson, HL Anderson, JKM Sanders. Acc Chem Res 26:469, 1993.
33. E Buhleier, W Wehner, F Vögtle. Synthesis 155, 1978.

34. DA Tomalia, R Esfand. Chem Industry 11(2 June):416–420, 1997.
35. DA Tomalia, HD Durst. Genealogically Directed Synthesis: Starburst/Cascade Dendrimers and Hyperbranched Structures. In: E Weber ed. Topics in Current Chemistry vol. 165. Supramolecular Chemistry I—Directed Synthesis and Molecular Recognition, Berlin, Heidelberg: Springer-Verlag, 1993, pp 193–313.
36. DA Tomalia, AM Naylor, WA Goddard III. Angew Chem Int Ed Engl 29(2):138–175, 1990.
37. J Fendler. Membrane Mimetic Chemistry: Characterization and Applications of Micelles, Microemulsions, Monolayers, Bilayers, Vesicles, Host Guest Systems and Polyions. Chichester: Wiley Interscience, 1982.
38. C Tschierske. J Mater Chem 8(7):1485–1508, 1998.
39. C Tanford. The Hydrophobic Effect: Formation of Micelles and Biological Membranes. New York: Wiley, 1973.
40. Macromolekulare Chemie—Das Werk Hermann Staudingers in Seiner Heutigen Bedeutung. Munchen: Schnell und Steiner, 1967.
41. H Morawetz. Polymers: The Origin and Growth of a Science. Chichester, UK: Wiley, 1985.
42. HL Hsieh, RP Quirk. Anionic Polymerization: Principles and Practical Applications. New York: Marcel Dekker, 1996.
43. K Matyjaszewski. Cationic Polymerizations: Mechanisms, Synthesis, and Applications., New York: Marcel Dekker, 1996, pp 1–768.
44. K Matyjaszewski. Controlled Radical Polymerization, ACS Symposium Series 685, vol. 685. Washington, DC: ACS, 1997.
45. K Hatada, T Kitayama, O Vogl. Macromolecular Design of Polymeric Materials. New York: Marcel Dekker, 1997.
46. RB Merrifield. J Am Chem Soc 85:2149–2154, 1963.
47. RB Merrifield. Science 232:341–347, 1986.
48. MK Lothian-Tomalia, DM Hedstrand, DA Tomalia. Tetrahedron 53(45):15495–15513, 1997.
49. DA Tomalia. Scientific American 272(5):62–66, 1995.
50. DA Tomalia. Macromol Symp 101:243–255, 1996.
51. AK Naj. Persistent Inventor Markets a Molecule. In: The Wall Street Journal, New York, 1996, pp B1.
52. JMJ Fréchet. Science 263:1710–1715, 1994.
53. K Dusek. TRIP 5(8):268–274, 1997.
54. S Uppuluri, DR Swanson, HM Brothers II, LT Piehler, J Li, DJ Meier, GL Hagnauer, DA Tomalia. Poly Mater Sci Eng (ACS) 80:55–56, 1999.
55. DA Tomalia, DM Hedstrand, LR Wilson. Dendritic Polymers. In: Encyclopedia of Polymer Science and Engineering, vol. index volume. 2nd ed. New York: Wiley, 1990, pp 46–92.
56. F Zeng, SC Zimmerman. Chem Rev 97:1681–1712, 1997.
57. OA Matthews, AN Shipway, JF Stoddart. Prog Polym Sci 23:1–56, 1998.
58. DA Tomalia. Adv Mater 6(7/8):529–539, 1994.
59. GJ Kallos, DA Tomalia, DM Hedstrand, S Lewis, J Zhou. Rapid Commun Mass Spectrometry 5:383–386, 1991.
60. PR Dvornic, DA Tomalia. Macromol Symp 98:403–428, 1995.

61. HM Brothers II, LT Piehler, DA Tomalia. J Chromatography A 814:233–246, 1998.
62. DA Tomalia, H Baker, J Dewald, M Hall, G Kallos, S Martin, J Roeck, J Ryder, P Smith. Polym J (Tokyo) 17:117–132, 1985.
63. DA Tomalia, JR Dewald, MJ Hall, SJ Martin, PB Smith. Preprints of the First SPSJ International Polymer Conference, Society of Polymer Science, Kyoto, Japan, 65, 1984.
64. DA Tomalia. Akron Polymer Lecture Series, Akron, OH, April, 1984.
65. DA Tomalia. Flory–Pauling Macromolecular Conference—Frontiers in Synthetic Polymer Chemistry, Santa Barbara, CA, January, 1983.
66. DA Tomalia. Sixth Biennial Carl S. Marvel Symposium—Advances in Synthetic Polymer Chemistry, Tucson, AZ, March, 1985.
67. DA Tomalia. ACS Great Lakes/Central Regional Meeting, Kalamazoo, MI, May, 1984.
68. BJ Bauer, A Topp, TJ Prosa, EJ Amis, R Yin, D Qin, DA Tomalia. Polym Materials Sci Eng (ACS) 77:87–88, 1997.
69. AB Padias, HKH Jr. J Org Chem 52:5305–5312, 1987.
70. PG de Gennes, H Hervet. J Physique Lett 44:351–360, 1983.
71. AD Meltzer, DA Tirrell, AA Jones, PT Inglefield, DA Tomalia, DM Hedstrand. Macromolecules 25:4541–4548, 1992.
72. L Balogh, DA Tomalia. J Am Chem Soc 120:7355–7356, 1998.
73. KB Mullis, FA Faloona. Methods Enzymol 155:335, 1987.
74. SC Zimmerman, F Zeng, EC Reichert, SV Kolotuchin. Science 271:1095–1098, 1996.
75. M Conn, J Rebek, Jr. Chem Rev 97:1647–1668, 1997.
76. V Balzani, G Denti, S Serroni, S Campagna, V Ricevuto, A Juris. Proc Indian Acad Sci Chem Sci 105:421–434, 1993.
77. V Balzani, S Campagna, G Denti, A Juris, S Serroni, M Venturi. Coord Chem Rev 132:1–13, 1994.
78. V Balzani, S Campagna, G Denti, A Juris, S Serroni, M Venturi. Sol Energy Mater Sol Cells 38:159–173, 1995.
79. S Campagna, G Denti, S Serroni, A Juris, M Venturi, V Ricevuto, V Balzani. Chem Eur J 1:211–221, 1995.
80. A Juris, M Venturi, L Pontoni, IR Resino, V Balzani, S Serroni, S Campagna, G Denti. Can J Chem 1875–1882, 1995.
81. S Serroni, G Denti, S Campagna, A Juris, M Ciano, V Balzani. Angew Chem Int Ed Engl 31:1493–1495, 1992.
82. S Serroni, S Campagna, A Juris, M Venturi, V Balzani. Gazz Chim Ital 124:423–427, 1994.
83. S Achar, RJ Puddephatt. J Chem Soc Chem Commun 1895–1896, 1994.
84. S Achar, RJ Puddephatt. Angew Chem Int Ed Engl 33:847–849, 1994.
85. S Achar, JJ Vittal, RJ Puddephatt. Organometallics 15:43–50, 1996.
86. WTS Huck, FCJM van Veggel, BL Kropman, DHA Blank, EG Keim, MMA Smithers, DN Reinhoudt. J Am Chem Soc 117:8293–8294, 1995.
87. WTS Huck, FCJM van Veggel, DN Reinhoudt. Angew Chem Int Ed Engl 35:1213–1215, 1996.

88. WTS Huck, R Hulst, P Timmerman, FCJM van Veggel, DN Reinhoudt. Angew Chem Int Ed Engl 36:1006–1008, 1997.
89. WTS Huck, FCJM Van Veggel, DNJ Reinhoudt. J Mater Chem 7:1213–1219, 1997.
90. WTS Huck, LJ Prins, RH Fokkens, NMM Nibbering, FCJM Van Veggel, DN Reinhoudt. J Am Chem Soc 120:6240–6246, 1998.
91. GR Newkome, ZQ Yao, GR Baker, VK Gupta. J Org Chem 50:2003–2004, 1985.
92. CL Jackson, HD Chanzy, FP Booy, BJ Drake, DA Tomalia, BJ Bauer, EJ Amis. Macromolecules 31(18):6259–6265, 1998.
93. JCM van Hest. New Molecular Architectures Based on Dendrimers. PhD dissertation, Eindhoven University, Eindhoven, The Netherlands, 1996.
94. PE Eaton, TW Cole. J Am Chem Soc 86:962–964, 1964.
95. PE Eaton, YS Or, SJ Branka, BKR Shanker. Tetrahedron 42:1621, 1986.
96. LA Paquette, RJ Ternansky, DW Balogh, G Krentgen. J Am Chem Soc 105:5446, 1983.
97. DJ Cram. Science 219:1177, 1983.
98. DJ Cram, JM Cram. Container Molecules and Their Guests. Cambridge, UK: The Royal Society of Chemistry, 1994.
99. RF Curl, RE Smalley. Science 242:1017–1022, 1988.
100. L Balogh, DR Swanson, R Spindler, DA Tomalia. Poly Mater Sci Eng (ACS) 77: 118–119, 1997.
101. M Zhao, L Sun, RM Crooks. J Am Chem Soc 120(19):4877–4878, 1998.
102. NC Beck Tan, L Balogh, SF Trevino, DA Tomalia, JS Lin. Polymer 40:2537–2545, 1999.
103. L Balogh, R Valluzzi, KS Laverdure, SP Gido, GL Hagnauer, DA Tomalia. J Nanoparticle Res 00:1–16, 1999.
104. JFGA Jansen, EMM de Brabander-van den berg, EW Meijer. Science 266:1226–1229, 1994.
105. JFGA Jansen, EW Meijer, EMM de Brabander-van den Berg. J Am Chem Soc 117:4417–4418, 1995.
106. EMM de Brabander-van den Berg, A Nijenhuis, M Mure, J Keulen, R Reintjens, F Vandenbooren, B Bosman, R De Raat, T Frijns, S Wal. Macromol Symposia 77: 51–62, 1994.
107. DA Tomalia, H Baker, J Dewald, M Hall, G Kallos, S Martin, J Roeck, J Ryder, P Smith. Macromolecules 19:2466–2468, 1986.
108. DA Tomalia, M Hall, DM Hedstrand. J Am Chem Soc 109:1601–1603, 1987.
109. RG Denkewalter, JF Kole, WJ Lukasavage. U.S. Pat. 4410688. 1983.
110. RG Denkewalter, JF Kole, WJ Lukasavage. Chem Abstr 100:103907, 1984.
111. SM Aharoni, CR Crosby III, EK Walsh. Macromolecules 15:1093–1098, 1982.
112. SM Aharoni, NS Murthym. Polym Commun 24:132, 1983.
113. DA Tomalia, V Berry, M Hall, DM Hedstrand. Macromolecules 20:1164–1167, 1987.
114. DA Tomalia, DM Hedstrand, LR Wilson, DM Downing. Starburst Dendrimers: Size, Shape and Surface Control of Macromolecules. In: T Saegusa, T Higashimura, and A Abe, eds. Frontiers of Macromolecular Science, 32nd IUPAC Proceedings, Blackwell Publications, 1989, 207–212.

115. AM Naylor, WA Goddard III, GE Kiefer, DA Tomalia. J Am Chem Soc 111:2339–2341, 1989.
116. GR Newkome, CN Moorfield, GR Baker, AL Johnson, RK Behera. Angew Chem Int Ed 30(9):1176–1180, 1991.
117. MC Moreno-Bondi, G Orellana, NJ Turro, DA Tomalia. Macromolecules 23:910–912, 1990.
118. NJ Turro, JK Barton, DA Tomalia. Acc Chem Res 24(11):332–340, 1991.
119. KR Gopidas, AR Leheny, G Caminati, NJ Turro, DA Tomalia. J Am Chem Soc 113:7335–7342, 1991.
120. D Watkins, Y Sayed-Sweet, JW Klimash, NJ Turro, DA Tomalia. Langmuir 13: 3136–3141, 1997.
121. DM Hedstrand, BJ Helmer, DA Tomalia. U.S. Pat. 5,560,929. 1996.
122. Y Sayed-Sweet, DM Hedstrand, R Spindler, DA Tomalia. J Mater Chem 7(7): 1199–1205, 1997.
123. S Stevelmans, JCM van Hest, JFGA Jansen, DAFJ van Boxtel, EMM de Brabander-van den Berg, EW Meijer. J Am Chem Soc 118:7398–7399, 1996.
124. AI Cooper, JD Londono, G Wignall, JB McClain, ET Samulski, JS Lin, A Dobrynin, M Rubinstein, ALC Burke, JMJ Fréchet, JM DeSimone. Nature 389(September 25):368–371, 1997.
125. R Yin, Y Zhu, DA Tomalia. J Am Chem Soc 120:2678–2679, 1998.
126. DB Amabilino, PR Ashton, M Belohradsky, FM Raymo, JF Stoddart. J Chem Soc Chem Commun 751–753, 1995.
127. DB Amabilino, PR Ashton, V Balzani, CL Brown, A Credi, JMJ Fréchet, JW Leon, FM Raymo, N Spencer, JF Stoddart, M Venturi. J Am Chem Soc 118:12012–12020, 1996.
128. FM Menger, CA Littau. J Am Chem Soc 115:10083–10090, 1993.
129. Anionic Surfactants, Physical Chemistry of Surfactant Action. Surfactant Series vol. 11. New York: Marcel Dekker, 1981.
130. S Buckingham, C Garvey, G Warr. J Phys Chem 97:10236–10244, 1993.
131. J Israelachvili, D Mitchell, B Ninham. J Chem Soc Faraday Trans II 72:1525, 1976.
132. SE Friberg, M Podzimek, DA Tomalia, DM Hedstrand. Mol Cryst Liq Cryst 164: 157–165, 1988.
133. H Smith, DA Tomalia. U.S. Pat. 5,331,100. 1994.
134. T Chapman, G Hillyer, E Mahan, K Shaffer. J Am Chem Soc 116:11195–11196, 1994.
135. I Gitsov, KL Wooley, JMJ Fréchet. Angew Chem Int Ed Eng 31(9):1200–1202, 1992.
136. I Gitsov, KL Wooley, CJ Hawker, PT Ivanova, JMJ Fréchet. Macromolecules 26: 5621–5627, 1993.
137. CJ Hawker, KL Wooley, JMJ Fréchet. J Chem Soc Perkin Trans I:1287–1297, 1993.
138. KL Wooley, CJ Hawker, JMJ Fréchet. J Am Chem Soc 115:11496–11505, 1993.
139. JN Israelachvili, D Mitchell, B Ninham. Biophys Acta 470:158, 1977.
140. JN Israelachvili, S Marcelja, R Horn. Rev Biophys (13):121, 1980.
141. GR Newkome, GR Baker, S Arai, MJ Saunders, PS Russo, KJ Theriot, CN Moore-

field, LE Rogers, JE Miller, TR Lieux, ME Murray, B Philips, L Pascal. J Am Chem Soc 112:8458–8465, 1990.
142. DA Tomalia, P Kirchoff. U.S. Pat. 4,694,064. 1987.
143. I Neubert, E Amoulong-Kirstein, A-D Schluter. Macromol Rapid Commun 17: 517–527, 1996.
144. I Neubert, R Klopsch, W Claussen, A-D Schluter. Acta Polymer 47:455–459, 1996.
145. B Karakaya, W Claussen, K Gessler, W Saenger, A-D Schluter. J Am Chem Soc 119:3296–3301, 1997.
146. W Stocker, B Karakaya, LB Schurmann, PJ Rabe, A-D Schluter. J Am Chem Soc 120:7691–7695, 1998.
147. JMJ Fréchet, I Gitsov. Macromol Symp 98:441–465, 1995.
148. G Johansson, V Percec, G Ungar, JP Zhou. Macromolecules 29:646–660, 1996.
149. V Percec, D Schlueter, JC Ronda, G Johansson, G Ungar, JP Zhou. Macromolecules 29:1464–1472, 1996.
150. V Percec, D Schlueter. Macromolecules 30:5783–5790, 1997.
151. S Bauer, H Fischer, H Ringsdorf. Angew Chem Int Ed Engl 32:1589, 1993.
152. V Percec, P Chu, G Johansson, D Schlueter, JC Ronda, G Ungar. Polym Prepr 37: 68, 1996.
153. V Percec, M Kawasumi. Macromolecules 25:3843, 1992.
154. V Percec, P Chu, M Kawasumi. Macromolecules 27:4441, 1994.
155. V Percec, P Chu, G Ungar, J Zhou. J Am Chem Soc 117:11441, 1995.
156. V Percec, G Johansson, G Ungar, JP Zhou. J Am Chem Soc 118:9855–9866, 1996.
157. SD Hudson, H-T Jung, V Percec, W-D Cho, G Johansson, G Ungar, VSK Balagurusamy. Science 278:449–452, 1997.
158. JH Cameron, A Facher, G Lattermann, S Diele. Adv Mater 9(5):398–403, 1997.
159. V Percec, C-H Ahn, G Ungar, DJP Yeardly, M Moller. Nature 391:161–164, 1998.
160. DLD Gasper. Biophys J 32:103–138, 1980.
161. JD Watson. Molecular Biology of the Gene. W.A. Benjamin, Menlo Park, CA, 1976.
162. AJ Levine. Viruses. New York: W.H. Freeman, 1992.
163. V Percec, C-H Ahn, W-D Cho, G Johansson, D Schlueter. Macromol Symp 118: 33–43, 1997.
164. V Percec, C-H Ahn, B Barboiu. J Am Chem Soc 119:12978–12979, 1997.
165. LL Kiessling, NL Pohl. Chem Biol 3:71–77, 1996.
166. M Mammen, SK Choi, GM Whitesides. Angew Chem Int Ed Eng 37:2754–2794, 1998.
167. SK Choi, M Mammen, M Whitesides. Chem Biol 3:97–104, 1996.
168. KH Mortell, M Gingras, LL Kiessling. J Am Chem Soc 116:12053–12054, 1994.
169. C Fraser, RH Grubbs. Macromolecules 28:7248–7255, 1995.
170. KH Mortell, RV Weatherman, LL Kiessling. J Am Chem Soc 118:2297–2298, 1996.
171. R Roy. J Chem Soc Chem Comm 1869–1872, 1993.
172. R Roy, WKC Park, Q Wu, SN Wany. Tetrahedron 36:4377–4380, 1995.
173. R Roy. Polym News 21(7):226–232, 1996.
174. JD Reuter, A Myc, MM Hayes, Z Gan, R Roy, D Qin, R Yin, LT Piehler, R Esfand, DA Tomalia, J Baker, J.R. Bioconjugate Chem 10(2):271–278, 1999.

175. DA Tomalia, DM Hedstrand, MS Ferritto. Macromolecules 24:1435–1438, 1991.
176. PM Saville, JW White, CJ Hawker, KL Wooley, JMJ Fréchet. J Phys Chem 97: 293–294, 1993.
177. PM Saville, PA Reynolds, JW White, CJ Hawker, JMJ Fréchet, KL Wooley, J Pemford, JRP Webster. J Phys Chem 99:8283–8289, 1995.
178. JP Kampf, CW Frank, EE Malmstrom, CJ Hawker. Langmuir 15:227–233, 1999.
179. JP Kampf, CW Frank, EE Malmstrom, CJ Hawker. Science 282:1730–1733, 1999.
180. APHJ Schenning, C Ellisen-Roman, J-W Weener, MWPL Baars, SJ van der Gast, EJ Meier. J Am Chem Soc 120:8199–8208, 1998.
181. DA Tomalia, G Killat. U.S. Pat. 4,871,779. 1989.
182. ML Mansfield. Polymer 37:3835–3841, 1996.
183. S Watanabe, SL Regen. J Am Chem Soc 116:8855–8856, 1994.
184. M Zhao, H Tokuhisa, RM Crooks. Angew Chem Int Ed Engl 36:2596–2598, 1997.
185. A Hierlemann, JK Campbell, LA Baker, RM Crooks, AJ Ricco. J Am Chem Soc 120:5323–5324, 1998.
186. VV Tsukruk, F Rinderspacher, VN Bliznyuk. Langmuir 13:2171–2176, 1997.
187. VV Tsukruk. Adv Mater 10:253–257, 1998.
188. VN Bliznyuk, F Rinderspacher, VV Tsukruk. Polymer 39:5249–5252, 1998.
189. JMJ Fréchet, T Emrick. Curr Opin Colloid Interface Sci, 4:1999.
190. DC Tully, AR Trimble, JMJ Fréchet. Polym Prepr (ACS) 40(1):402–403, 1999.
191. N Tomioka, D Takasu, T Takahashi, T Aida. Angew Chem Int Ed 37(11):1531–1534, 1998.
192. LL Miller, RG Duan, DC Tulley, DA Tomalia. J Am Chem Soc 119(5):1005–1010, 1997.
193. F Wang, RD Rauh, TL Rose. J Am Chem Soc 119:11106–11107, 1997.
194. G Galileo. Dialogue Concerning the Two Chief World Systems: Ptolemaic and Copernican. Berkeley: University of California Press, 1967.
195. N Bohr. Nobel Laureate Lecture. 1922.
196. T Martin, U Obst, J Rebek, Jr. Science 281(September 18):1842–1845, 1998.
197. X Camps, H Schonberger, A Hirsch. Chem Eur J 3(4):561–567, 1997.
198. DA Tomalia, PR Dvornic. Dendritic Polymers, Divergent Synthesis (Starburst Polyamidoamine Dendrimers). In: JC Salamone, ed. Polymeric Materials Encyclopedia, vol. 3 (D–E). Boca Raton, FL: CRC Press, 1996.
199. A Topp, BJ Bauer, DA Tomalia, EJ Amis. Macromolecules 32:7232–7237, 1999.
200. A Topp, BJ Bauer, JW Klimash, R Spindler, DA Tomalia, EJ Amis. Macromolecules 32:7226–7231, 1999.
201. ML Mansfield, L Rakesh, DA Tomalia. J Chem Phys 105(8):3245–3249, 1996.
202. DA Tomalia, S Uppuluri, DR Swanson, HM Brothers II, LT Piehler, J Li, DJ Meier, GL Hagnauer, L Balogh. Dendritic Macromolecules: A Fourth Major Class of Polymer Architecture—New Properties Driven by Architecture. Boston: Materials Research Society, 1998.
203. S Uppuluri, DA Tomalia. Adv Matrls (in press).
204. S Uppuluri, SE Keinath, DA Tomalia, PR Dvornic. Macromolecules 31:4498–4510, 1998.
205. DA Tomalia. Dendrimers: Nanoscopic Modules for the Construction of Higher Ordered Complexity. In: J Michl ed. Modular Chemistry, Proceedings of the NATO

Advanced Research Workshop on Modular Chemistry, The Netherlands: Kluwer Academic 1997, 183–191.
206. PR Dvornic, DA Tomalia. Macromol Symp 88:123–148, 1994.
207. Z Guan, PM Cotts, EF McCord, SJ McLain. Science 283:2059–2062, 1999.
208. P Singh. Bioconjugate Chem 9:54–63, 1998.
209. A Bielinska, JF Kukowska-Latallo, J Johnson, DA Tomalia, J Baker, Jr. Nucleic Acids Res 24(11):2176–2182, 1996.
210. JF Kukowska-Latallo, AU Bielinska, J Johnson, R Spindler, DA Tomalia, J Baker, Jr. Proc Natl Acad Sci 93:4897–4902, 1996.
211. EC Wiener, MW Brechbiel, H Brothers, RL Magin, OA Gansow, DA Tomalia, PC Lauterbur. Magnetic Resonance Medicine 31(1):1–8, 1994.
212. DS Wilbur, PM Pathare, DK Hamlin, KR Buhler, RL Vesella. Bioconjugate 9: 813–825, 1998.
213. R Duncan. Polym Prepr (ACS) 40(1):285–286, 1999.
214. AH Soloway, W Tjarks, BA Barnum, F-G Rong, RF Barth, IM Codogni, JG Wilson. Chem Rev 98:1515–1562, 1998.
215. C Bierniaz. Dendrimers: Applications to Pharmaceutical and Medicinal Chemistry. In: Encycl of Pharm, Tech, vol. 18. New York: Marcel Dekker, 1998.
216. DA Tomalia, HM Brothers II. Regiospecific Conjugation to Dendritic Polymers to Produce Nanodevices. In: SC Lee and LM Savage, eds. Biological Molecules in Nanotechnology: Convergence of Biotechnology, Polymer Chemistry and Materials Science, vol. 1927. Southborough, MA: IBC, 1998.
217. ME Hermes. Enough for One Lifetime: Wallace Carothers, Inventor of Nylon. American Chemical Society and Chemical Heritage Foundation, 1996.
218. N Zimmerman, JS Moore, SC Zimmerman. Chem Industry (August 3):604–610, 1998.
219. M Freemantle. Chem Eng News (November 1), 77:27–35, 1999.
220. SK Ritter. Chem Eng News (November 8), 77:30–35, 1999.
221. M Kawa and JMJ Fréchet. Chem Mater 10:286–296, 1998.
222. PE Froehling, HAJ Linssen. Macromol Chem Phys 199:1691–1695, 1998.

10

Self-Assembled Monolayers (SAMs) and Synthesis of Planar Micro- and Nanostructures

Lin Yan
Bristol-Myers Squibb, Princeton, New Jersey

Wilhelm T. S. Huck and George M. Whitesides
Harvard University, Cambridge, Massachusetts

I. INTRODUCTION: SAMs AS TWO-DIMENSIONAL POLYMERS

A polymer, by conventional definition, is a macromolecule made up of multiple equivalents of one or more monomers linked together by *covalent bonds* (e.g., carbon–carbon, amide, ester, or ether bonds) [1]. These conventional polymers come in many configurations: for example, linear homopolymers, linear copolymers, block copolymers, crosslinked polymers, dendritic polymers, and others. The most common architecture for polymers is based on linear chains that may have other attached chains (branched, grafted, or crosslinked); that is, they are one-dimensional molecules. A few examples have been claimed as two-dimensional sheet polymers.*

A supramolecular polymer is a structure in which monomers are organized through *noncovalent interactions* (e.g., hydrogen bonds, electrostatic interactions, and van der Waals interactions) [4]. These less familiar types of polymers also exist in many forms. For example, molecular crystals are large collections of molecules arranged in a three-dimensional periodical lattice through noncovalent

* See Refs. 2 and 3 and references therein.

intermolecular interactions. Lipid bilayers are two-dimensional structures that exist in water, in which hydrocarbon tails aggregate to form a hydrophobic sheet in the form of a spherical shell, and polar or charged hydrophilic head-groups are exposed to water.

Self-assembled monolayers (SAMs) are highly ordered molecular assemblies that form spontaneously by chemisorption of functionalized molecules on surfaces, and organize themselves laterally, most commonly by van der Waals interactions between monomers [5]. We consider SAMs to be a type of two-dimensional polymer: they are, in a sense, a uniform supramolecular assembly of short hydrocarbon chains covalently grafted onto a macromolecular entity, that is, the surface. In SAMs, individual monomers (usually linear alkyl chains functionalized at one end or both) are not directly linked by covalent bonds to each other, but rather to a common substrate—a metal or a metal oxide surface. SAMs exist in a number of different types: homogeneous SAMs on planar and curved substrates, SAMs on metallic liquids, SAMs on nanoparticles, mixed SAMs, and two-dimensionally patterned SAMs. Table 1 compares some characteristics of SAMs and conventional polymers based on bonding and structural type.

Table 1 Comparison Between Conventional Polymers and SAMs, Considered as Two-Dimensional Polymers

		Conventional polymers	SAMs
Nature of bonding	Short range	Covalent bonding between adjacent monomers	Covalent bonding between head-groups and the substrate; van der Waals, H-bonding, ionic interactions between adjacent monomers
	Long range	van der Waals, H-bonding, ionic interactions between monomeric units proximate in space	No
Structural types		Homopolymers Copolymers (alternating, block, and random) Linear, branched, cross-linked, dendrimeric, etc.	Homogeneous SAMs Mixed SAMs Patterned SAMs
Conformational class		Extended Collapsed Random coiled	Crystalline Disordered Liquidlike

Figure 1 describes the formation of these two types of polymers schematically. A conventional linear polymer is formed by polymerization that links monomers through chemical reactions (e.g., free radical, ionic, and coordination addition, condensation, and ring opening reactions). Polymer chains are often conformationally disordered: in dilute solution, polymer chains are often coiled; in concentrated solution or in bulk, they are entangled. For SAMs, "polymerization" is a spontaneous process involving adsorption that connects monomers to a substrate, and self-organization that orders the system laterally through noncovalent intermolecular interactions. The strong chemical interaction between the head-group and the substrate renders the "shape" of a SAM two-dimensional, and its "size" the surface area of the substrate. The overall structure of SAMs is determined by the interaction of the head-group and the substrate, the lateral interaction between the neighboring monomers, and the structure of the constituent monomers. SAMs supported on metals or metallic oxides are not soluble, and thus provide no information about the behavior of two-dimensional soluble polymers. They are, however, excellent models for the surface chemistry of insoluble polymers, and among the motivations for the study of SAMs are to understand the physical-organic chemistry of polymer surfaces [6–8], and to develop

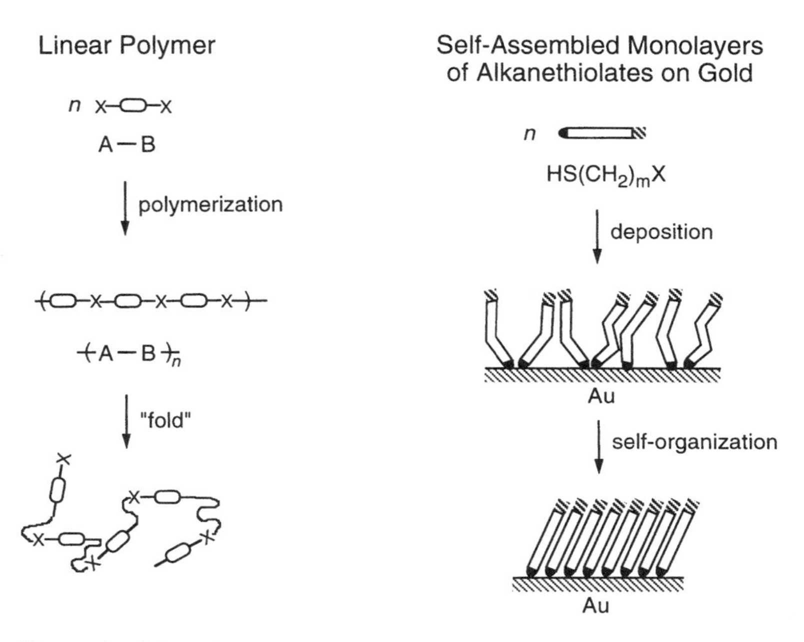

Figure 1 Schematic representation of the formation of conventional polymers and of self-assembled monolayers (SAMs).

methods that can be used to control interfacial properties of polymers at the molecular level [9].

This chapter considers SAMs as two-dimensional polymers, and describes the synthesis and structures of SAMs comprising one thiol, and mixed SAMs and patterned SAMs comprising more than one thiol (mainly on gold and silver). It reviews some recent studies of chemical transformations of terminal functional groups of SAMs after their assembly, and discusses two potentially useful chemical methods developed in our group for synthesis of mixed SAMs and patterned SAMs, and several of their applications.

II. SYSTEMS OF SAMs

The monomeric units of conventional polymers can be connected by different kinds of chemical bonds; correspondingly SAMs can have various chemical inter-

Table 2 Different Types of SAMs ($CH_3(CH_2)_nX$)

Head group (X)	Precursor	SAMs of	Substrate	Bonding
RS	RSH or $RS)_2$	Alkanethiolates	Au^{14}, Ag^{15}, Cu^{16}, Pd^{29}, Fe, $Fe_2O_3^{30}$, Hg^{31}, $GaAs^{32,33}$, InP^{34}	$RS^- \cdot M_n^{+n}$ or $RS)_2$; M^n
RSi(O)(O)(O)	$RSiCl_3$, $RSi(OCH_3)_3$, or $RSi(-OEt)_3$	Alkylsiloxanes	SiO_2, glass, $mica^{23,25}$, $Al_2O_3^{35}$, $Ga_2{}^{37}O_3^{32}$, $Au^{36,37}$	Polymeric siloxane
RA^-	RCO_2H	Acid-functionalized alkanes	$Al_2O_3^{35,38-42}$, In_2O_3/SnO_2^{43}, SiO_2^{41}, AgO, CuO^{40}	Acid-base
	RCONHOH		Au, Al_2O_3, ZrO_2, Fe_2O_3, TiO_2, AgO, CuO^{44}	
	RSO_2H		Au^{45}	
	RPO_3H_2		SiO_2, ZrO_2, Al_2O_3, TiO_2, $mica^{46-48}$	
RB	R_2S, R_3P	Base-functionalized alkanes	Au^{49-52}	Coordination
	RNC		Au^{53}, Pt^{50}	
R	RCH_3	Alkyl groups	Si, graphite $^{54-56}$	Covalent Si-C

actions between the head-group and the supporting substrate. A number of different types of SAMs have been explored; several of these systems have been reviewed [5,10–13]. Table 2 categorizes SAMs into groups based on the bonding between the head-group and the surface. The most widely studied systems have been SAMs formed by chemisorption of alkanethiols on gold [14], silver [15], or copper [16]. SAMs of phosphonates have been widely used to synthesize multilayer structures with application in nonlinear optic devices [17–19] and heterogeneous catalysis [18,19]. SAMs of siloxanes on glass and metallic oxides have been studied by Sagiv [22,23] and others [24–27], and widely used technologically in surface treatment [28]. SAMs covalently attached to these substrates provide a rugged system for various applications, but certain of these systems—especially those based on -$SiCl_3$ or -$Si(OEt)_3$ head-groups—can be difficult to synthesize, and the reactivity of these head-groups may be incompatible with other functional groups.

SAMs of alkanethiolates that present a wide range of functional groups on thin polycrystalline films of gold and silver are easy to prepare, and have been broadly applied in various fundamental and technological studies. They are well ordered, and the best characterized systems of organic monolayers presently known. In this chapter, we focus on them.

III. SYNTHESIS OF SAMs ON GOLD AND SILVER

The preparation of SAMs of alkanethiolates on gold and silver is straightforward. The metal substrates are prepared by evaporation of a thin layer of titanium or chromium (~1–5 nm; this layer of Ti or Cr promotes the adhesion of gold or silver to the supporting substrate) onto silica wafers, glass slides, or other flat surfaces, followed by deposition of gold or silver (~10–200 nm; in general, ≥40 nm is required to achieve a complete coverage of the substrate) [57]. SAMs of alkanethiolates (e.g., $X(CH_2)_nSH$, X is a terminal functional group) on gold and silver can be easily generated by immersing the metal substrate in 1–10 mM solutions of alkanethiols at room temperature; ethanol is commonly used as the solvent [Eq. (1)]; SAMs can also be generated using vapor phase deposition [58] or electrodeposition [59] of alkanethiols.

$$X(CH_2)_nSH + Au(0)_m \rightarrow X(CH_2)_nS^-Au(I) \cdot Au(0)_{m-1} + \frac{1}{2H_2} \quad (1)$$

Although formation of SAMs on gold is usually expressed as Eq. (1), the mechanistic details of this reaction remain incompletely understood. It is generally believed that the thiol group forms a thiolate ($RS^-Au(I)$) in its interaction with gold [13]. Some studies using grazing-angle X-ray diffraction have, however, been interpreted to suggest that the interaction of sulfur and gold involves a disulfide ($R_2S_2Au(0)$) [60]. Using molecular dynamics (MD), Gerdy and Goddard have calculated a hypothetical crystal structure for decyl disulfide on Au(111) surface, and found that the resulting structure was energetically stable and the X-ray diffraction pattern derived from such structure was indistinguishable from that observed experimentally [61]. Most of the theoretical studies have, however, been based on the assumption of interactions of thiolates and gold [62]. A conclusive description about the interaction between the sulfur and gold awaits additional experimental and theoretical studies [63]. The fate of the hydrogen atom of the thiol group has also not been resolved. Although there remain a number of uncertainties concerning the structure of the interface between SAMs and gold, most of the interest in SAMs has focused on the structure of the polymethylene $(CH_2)_n$ groups, and on the interaction of the tail groups with the solution; it is thus immaterial, to some extent, what the binding is between gold and surface.

The kinetics of formation of SAMs on gold has been studied using a range of methods: ellipsometry [64], contact angle [64], quartz crystal microbalance (QCM) [65–69], surface acoustic wave (SAW) [70], surface plasmon resonance (SPR) [71], optical second harmonic generation (SHG) [72], polarized infrared external reflectance spectroscopy (PIERS) [73], scattering Raman spectroscopy [74], and electrochemistry [75]. These studies provide a macroscopic picture of the processes that form SAMs: the growth rate is proportional to the number of unoccupied sites on gold, and can be described as a first-order Langmuir adsorption. Recent atomic force microscopy (AFM) [76] and scanning tunneling microscopy (STM) [77–79] studies depict a three-stage microscopic process:

1. Lattice gas phase—alkanethiols are confined on the surface and diffuse rapidly.
2. Low-density solid phase—molecular axes are aligned with the surface plane and the close pairing of thiol groups is maintained.
3. High-density pseudocrystalline solid phase—alkanethiols are closely packed and their axes are aligned with the surface normal.

Synthesis of SAMs is remarkably convenient: it requires only ambient conditions, and the substrate can be polycrystalline or even electroless gold and silver films [80,81]. Formation of SAMs of a single alkanethiol on gold is known to complete in a few minutes, and may occur in seconds during microcontact printing (μCP), a process in micropatterning [82,83]. SAMs on gold are one of the systems of SAMs most widely used.

IV. STRUCTURES OF SAMs ON GOLD AND SILVER

The molecular structures of SAMs have been studied extensively using various instrumental techniques: PIERS [84–87], X-ray photoelectron spectroscopy

(XPS) [64,88], AFM [89–91], grazing-angle X-ray diffraction [92], molecule beam diffraction [93,94], high-energy electron scattering [95], low-energy electron diffraction [57], electrochemistry [96,97], ellipsometry [85,98], and contact angle [87,98–100]. The packing of alkanethiolates on gold is influenced by the spacing of coordination sites and the interaction between the adjacent alkyl chains. Electron diffraction and low-energy helium beam diffraction studies suggest that the sulfur atoms are localized in threefold hollow sites of the Au(111) surface and form a commensurate triangular $\sqrt{3} \times \sqrt{3}R30°$ overlayer lattice (Fig. 2) [57,95,101]. In a SAM of *n*-alkanethiolates, the average cross-sectional area occupied by each thiolate is 21.4 Å^2; this value is larger than that of an alkane chain (18.4 Å^2). The alkyl chains adopt a largely *trans* conformation (for $n \geq 10$) and tilt ~30° with respect to the surface normal in order to maximize van der Waals interactions between adjacent polymethylene chains (the enthalpy of lateral interaction per CH_2 group is ~1.5 kcal/mol). Although the distance between nearest silver atoms (2.89 Å) on the Ag(111) surface is similar to that of gold (2.88 Å), in SAMs of alkanethiolates on silver, the sulfur atoms arrange themselves in a $\sqrt{7} \times \sqrt{7}R10.9°$ lattice and the alkyl chains are nearly perpendicu-

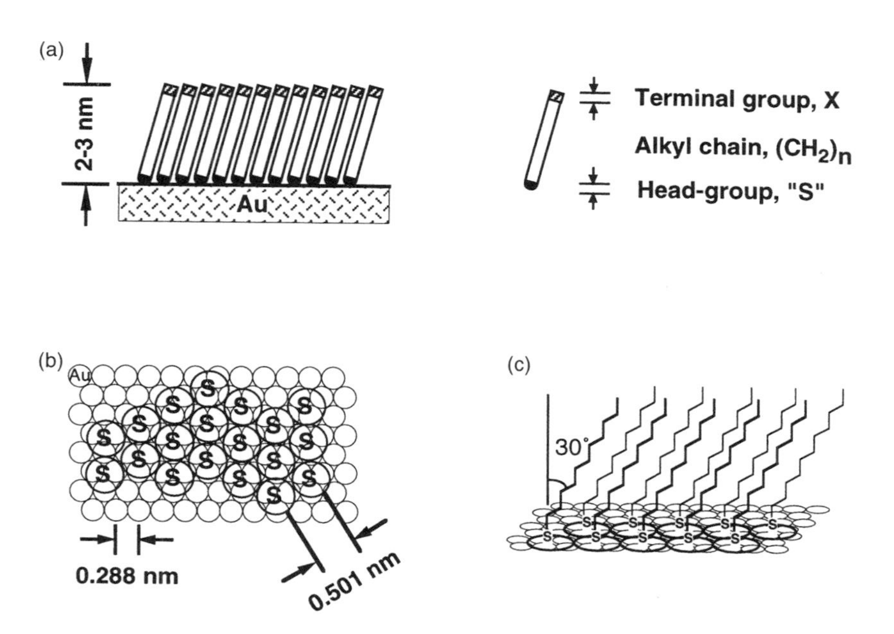

Figure 2 (a) A schematic representation of SAMs of *n*-alkanethiolates on gold. (b) The $\sqrt{3} \times \sqrt{3}R30°$ lattice of sulfur atoms on Au(111). (c) Alkyl chains adopt all *trans* conformation and tilt ~30° from the normal of the surface.

lar to the surface (tilt ~10° from the normal of the substrate) [102]. Using ab initio calculations, Ulman and coworkers suggest that the combination of the lateral discrimination of chemisorption potentials and unfavorable charge–charge interactions between both the thiolates and the underlying Au atoms in SAMs prevent the alkyl chains from packing as densely as those on silver [103]. SAMs formed on freshly prepared silver substrates generally have a lower population of gauche conformations than SAMs on gold, but silver is readily oxidized by oxygen in air and a thick film of silver oxide does not support a well-ordered SAM of alkanethiolates. In general, SAMs of alkanethiolates on freshly prepared silver should be considered more highly ordered than analogous structures on gold.

Although experimental studies have sketched a structural picture of SAMs on gold and silver, the details of this picture remain incompletely defined: among the remaining uncertainties are the exact position of sulfur atoms on gold, the bond angle of the metal-S-C group, the nature of the interaction of the chains with one another, and the nature of lateral movement. Theoretical studies can, in principle, contribute to understanding these issues, although SAMs represent very complex systems for computation. Most of the theoretical work has been carried out on SAMs on gold [62]. Ulman and coworkers have used a very simple model to simulate the thickness of the film, and the molecular orientation and packing of alkyl chains on gold [104]. To a first approximation, they first optimized the geometry of isolated molecules, and then constructed small hexagonal assemblies of these rigid molecules. Considering only van der Waals and electrostatic intermolecular interactions, they examined the interaction energy of a molecule with its neighbor as a function of tilt angles, and found that the calculated thickness and the tilt angles were the same as those established from experimental studies. Klein and coworkers have used MD to investigate the structure and dynamics of alkanethiols on gold [105–108]. They first pinned all the alkyl chains perpendicularly onto a well-defined triangular lattice with the nearest neighbor distance of 4.97 Å and then allowed the system to relax and to evolve into energetically minimal structures. Using both the united-atom model and the more realistic all-atom model, they found that most of the alkyl chains adopted *trans* conformations, and that the system had fewer gauche conformations when the metal-S-C bonds are colinear than 90°. They also found that the average conformation was temperature-dependent: SAMs were less ordered at high temperature, which observation agreed with the results derived from molecular beam studies. Siepmann and coworkers have used Monte Carlo (MC) methods to study the properties of SAMs on gold [109]. Grunze and coworkers have used stochastic global search to explore the configurational space of a SAM of octadecanethiol, $CH_3(CH_2)_{17}SH$, on gold [110]. They used four different force fields and found that several distinct monolayer structures could exist with energy difference less than 1 kcal/mol. Using ab initio calculation, Ulman and coworkers found that thiolates prefer the threefold hollow sites over the on-top sites, but Bishop and coworkers used MD and found that the energy difference within the surface corrugation potential is too small to pin sulfur atoms at any particular site [111].

The order of the terminal group and the top part of SAMs is determined not only by the sulfur atoms bound directly to the gold and the intermolecular interaction between the alkyl chains, but also by the size and geometry of the terminal group. STM studies show that SAMs of alkanethiolates on gold are heterogeneous and structurally complex. The alkyl chains form a ''superlattice'' at the surface of the monolayer of sulfur atoms, that is, a lattice with symmetry and dimensions different from that of the underlying hexagonal lattice formed by sulfur atoms. When alkanethiolates are terminated with end-groups other than the methyl group, the structure of the resulting SAM becomes less predictable. Nelles and coworkers have shown that the superlattice is dependent on the shape of the terminal groups: thiols having terminal groups with relatively spherical cross-sections form hexagonal lattices; thiols with more asymmetric cross-sections form centered rectangular lattices [112]. Sprik and coworkers have used STM and MD to study SAMs terminated with hydrophilic groups such as hydroxyl and amine groups, and found hydrogen bond induced reconstruction of the top layer and coadsorption of solvents [113].

For SAMs of thiols having tail groups more complicated than *n*-alkyl chains, the structures of SAMs depend on the size and geometry of these groups. Tao and others have shown that thiols derivatized with aromatic groups form well-ordered SAMs having a packing order different from *n*-alkanethiolates: the sulfur atoms form a $\sqrt{3} \times \sqrt{3}R30°$ lattice on gold, but the aromatic groups adopt a herringbone packing and are perpendicular to the surface [114–118]. SAMs of fluorinated alkanethiolates also pack differently [119,120]. The fluorinated alkyl groups have a van der Waals diameter of 5.6 Å (i.e., larger than that of the 5.0 Å diameter of normal alkyl groups). They form a 2×2 lattice and tilt $\sim 16°$ from the normal of surface.

Although SAMs are self-assembling systems and tend to reject defects, the presence of defects and pinholes is always observed [121–123]. A variety of factors influences formation and distribution of defects in a SAM, including the molecular structure of the surface, the length of the alkyl chain, and the conditions used to prepare SAMs. Grunze and coworkers have recently described a procedure in which a SAM of alkanethiolates on gold is treated with mercury vapor, then exposed again to a solution of alkanethiol. This procedure seems to heal defects in the SAM by increasing the density of alkyl chains and causing them to reorient to a tilt angle from the normal that resembles that characteristic of copper and silver [124].

V. FUNCTIONAL AND MIXED SAMs ON GOLD

SAMs presenting a variety of functional groups have been applied in a broad range of fundamental studies; representative areas include biocompatibility [125], wetting [126], adhesion [126,127], corrosion [14,128,129], and micro- and nanofabrication [130]. The strong chemo-specific interaction between the thiol group and gold (~24–40 kcal/mol) [12] gives SAMs high stability under mild conditions (room temperature) and allows SAMs to display a wide range of organic functionalities in high density ($\sim 2 \times 10^{14}$ molecules/cm^2) on the surface. Van der Waals interactions between adjacent polymethylene chains force alkanethiolates to pack at densities approaching those of crystalline poly(ethylene); these lateral interactions make SAMs impermeable to molecules in solution [131–134], and give them the electrical insulating properties similar to that of poly(ethylene) [135]. These characteristics allow the chemical and physical properties of the terminal functional groups that are exposed on the surface largely to determine the interfacial properties of a SAM. For example, oligo(ethylene glycol) and oligo(propylene sulfoxide) groups presented on SAMs of undecanethiolates on gold shelter the hydrophobic underlying polymethylene chains effectively from contacting proteins in solution and thus provide a surface that resists nonspecific adsorption of proteins [136–138]. These studies demonstrate that functional SAMs provide well-defined model systems to study surface phenomena.

SAMs that present a mixture of different functional groups–''mixed'' SAMs—provide desirable flexibility in design and synthesis of functional SAMs

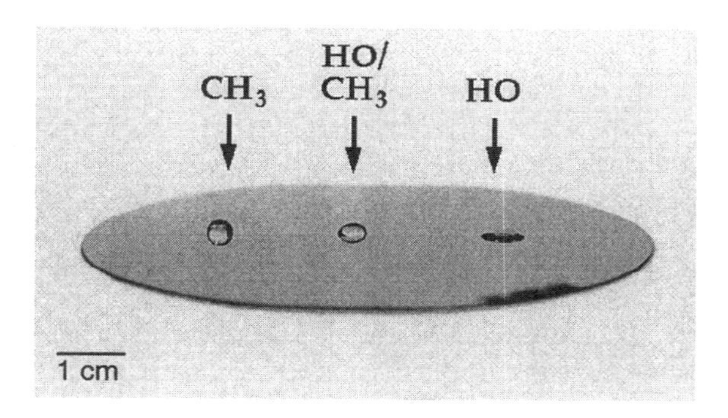

Figure 3 On this 4-inch gold-coated silicon wafer, three different kinds of SAMs were formed. The left part was covered with a SAM of $HS(CH_2)_{15}CH_3$ (a hydrophobic surface), the right part with a SAM of $HS(CH_2)_{11}OH$ (a hydrophilic surface), and the middle area with a mixed SAM of $HS(CH_2)_{15}CH_3$ and $HS(CH_2)_{11}OH$ (formed from a solution containing the two in a ~1:1 ratio). A droplet of water was placed on each of these regions. The shape of the droplet on the mixed SAM was intermediate between that on the hydrophobic and hydrophilic surfaces. This simple experiment shows that the terminal groups on the SAM can control interfacial properties (here, wetting) and also demonstrates the capability of mixed SAMs to tailor the surface properties by controlling the ratio of mixed functionalities on the surface.

that span a wider range of chemical and physical properties than do pure SAMs [21,139]. It becomes straightforward to tune continuously the interfacial properties of a SAM simply by varying the ratio of compositions of two thiols in the mixed SAM. Synthesis of mixed SAMs on gold is convenient: the gold-coated substrate is immersed in a solution containing two different thiols at a certain ratio, usually for 8 h; the resulting SAM contains a mixture of these two alkyl groups. Figure 3 shows the influence of a terminal group of SAMs on the shape of a droplet of water that contacts the surface. The wetting of water on the surface of a SAM having a ~1:1 mixture of HS $(CH_2)_{15}CH_3$ and $HS(CH_2)_{11}OH$ is intermediate between that of a SAM having $HS(CH_2)_{15}CH_3$ and that having a SAM of $HS(CH_2)_{11}OH$. The ratio of two components in the SAM is often different from that in solution [139]. Although the two organic groups of a mixed SAM are often well-mixed [140], thiols that have different properties can phase-separate into domains having only one type of thiols; whether this separation is kinetic or thermodynamic is not known [139,141]. AFM and STM studies show that the size of these phase-separated domains is ~50 nm [142–144]. The microscopic heterogeneity does not influence many of the macroscopic chemical and physical properties of the SAMs.

VI. CHEMICAL REACTIONS ON SAMs AFTER THEIR ASSEMBLY

Chemical transformation of terminal functional groups of SAMs *after* their assembly has recently attracted attention. There are several reasons to study chemical reactions on SAMs. They provide (i) controllable well-defined model systems with which to understand the influence of a surface on chemical reactions of functional groups; (ii) alternative methods to functionalize SAMs, to construct multilayers, and to attach molecules and biomolecules to surfaces; (iii) a possible basis for strategies for synthesis of combinatorial libraries of small molecules on a chip; and (iv) synthetic model systems that can be extended to chemical functionalization of polymer surfaces.

A variety of terminal functional groups and their chemical transformations on SAMs have been examined: for example, (i) olefins—oxidation [23,24,131,132], hydroboration, and halogenation [23,24]; (ii) amines—silylation [145,146], coupling with carboxylic acids [22,146], and condensation with aldehydes [22,147]; (iii) hydroxyl groups—reactions with anhydrides [148,149], isocyanates [150], epichlorohydrin [151], and chlorosilanes [152]; (iv) carboxylic acids—formation of acyl chlorides [153], mixed anhydrides [154], and activated esters [148,155]; (v) carboxylic esters—reduction and hydrolysis [156]; (vi) thiols and sulfides—oxidation to generate disulfides [157–159] and sulfoxides [160]; and (vii) aldehydes—condensation with active amines [161]. Nucleophilic displacement on SAMs has also been investigated mainly on SAMs of alkylsiloxanes on Si/SiO_2 [146,162–164]. These studies have shown that many organic reactions that work well in solution are difficult to apply to transformations at the surface, because the surface is a sterically hindered environment, and backside reactions (e.g., the S_N2 reaction) and reactions with large transition states (e.g., esterification, saponification, Diels–Alder reaction, and others) often proceed slowly. At present, only few synthetic methods are used, but those are capable of introducing a wide range of functional groups onto the surface.

The difficulty of developing useful chemical transformations on the surface is substantially compounded by lack of efficient techniques to identify products and to establish their yields after each reaction. PIERS and XPS are two particularly useful methods to characterize chemical transformations on the surface. PIERS provides direct evidence of transformation of an infrared-active functional group; XPS furnishes evidence for the presence or absence of an element characteristic of the functional group being introduced or eliminated, and is often used to estimate qualitatively the yield of chemical transformation. Other methods, such as ellipsometry, contact angle, secondary ion mass spectroscopy (SIMS) [165], and AFM [166] also provide complementary and valuable support for characterization of chemical reactions on SAMs. Establishing unambiguously the products and yield of a chemical reaction on the surface often requires a combination of information from several techniques.

We have developed two convenient chemical methods that may have general utility for rapid synthesis of functional SAMs. The first procedure has three steps (Fig. 4) [167]: preparation of a well-ordered homogeneous SAM of 16-mercaptohexadecanoic acid on gold; conversion of the terminal carboxylic acid groups into interchain carboxylic anhydrides by reaction with trifluoroacetic anhydride; and reaction of the interchain carboxylic anhydride with an alkylamine to give a mixed SAM presenting carboxylic acids and *N*-alkyl amides on its surface. Figure 5 summarizes the characterization of the product of transformation of terminal carboxylic acids to interchain carboxylic anhydrides, and of anhydrides to a ~1:1 mixture of acids and amides. SAMs of 16-mercaptohexadecanoic acid show C=O stretches at 1744 and 1720 cm^{-1} that are characteristic of terminal carboxylic acid groups presented on a SAM [87]. Treatment of the SAM with trifluoroacetic anhydride gives a SAM having C=O stretches at 1826 and 1752 cm^{-1}; these frequencies are characteristic of a carboxylic anhydride group [168]. XPS studies of the resulting SAM show no fluorine. The combination of these results indicates that the carboxylic acid group is converted into an interchain carboxylic anhydride group by trifluoroacetic anhydride, rather than to a mixed trifluoroacetic carboxylic anhydride. After reaction of the interchain carboxylic anhydride with *n*-undecylamine (taken as a representative *n*-alkylamine), the C=O stretches of the anhydride disappear completely, and two new absorption bands appear at 1742 and 1563 cm^{-1}; these bands are assigned as the C=O stretch of a carboxylic acid and an amide II band, respectively. XPS also

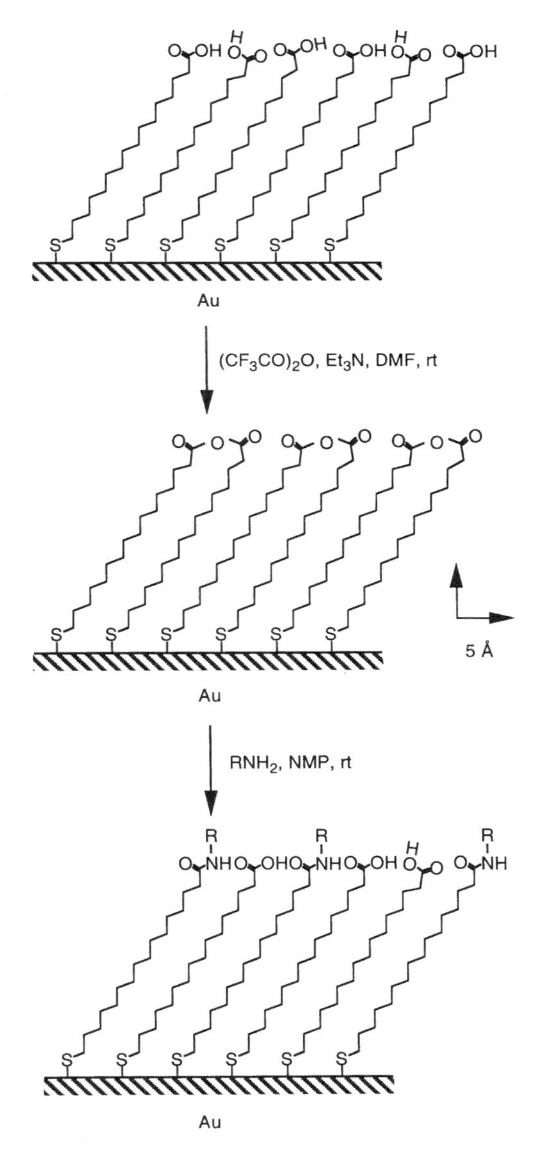

Figure 4 Schematic representation of formation of the interchain carboxylic anhydride and reaction of the anhydride with an alkylamine. The alkyl chains of the original SAM of the carboxylic acid in these SAMs are in *trans* conformation; the methylene groups near the functional groups may adopt gauche conformation and be less ordered. Carboxylic acids in the SAMs of the carboxylic acid, and in the mixed SAMs of amides and carboxylic acids, are hydrogen bonded to neighboring polar groups. The interchain carboxylic anhydrides orient largely parallel to the surface normal. (From Ref. 167.)

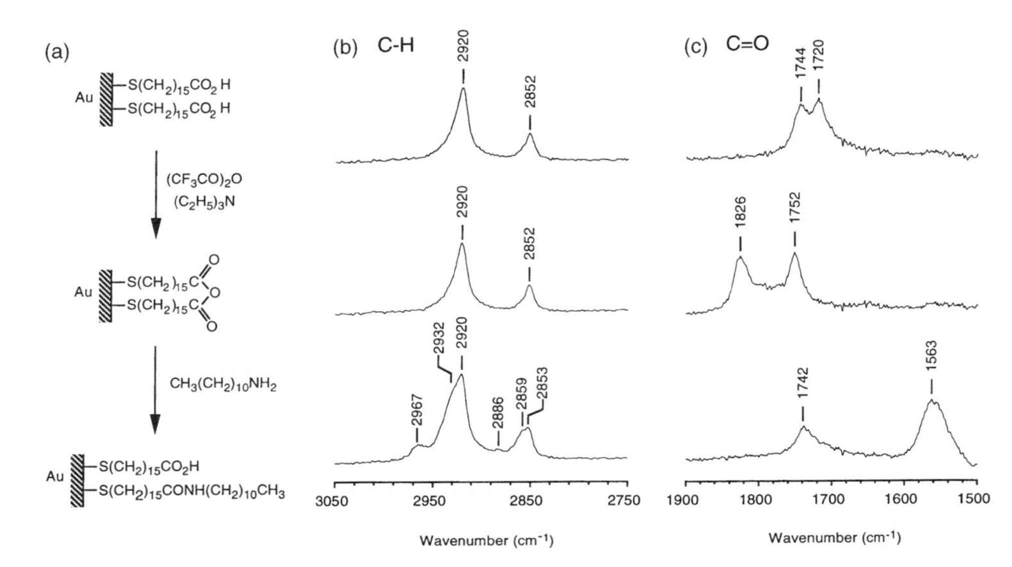

Figure 5 Comparison of PIERS spectra of the SAMs of the carboxylic acid, the interchain carboxylic anhydride, and a mixture of carboxylic acids and *n*-undecylamides on gold: (a) schematic representation of formation of the interchain carboxylic anhydride and the reaction of the anhydride and *n*-undecylamine; (b) the PIERS spectra of these SAMs in the C—H stretching region; (c) the PIERS spectra of these SAMs in the C=O stretching region. (From Ref. 167.)

indicates the presence of nitrogen in the resulting sample. The combination of these results suggests the coupling of amines to the SAM. PIERS further shows that there are two low-frequency shoulder peaks at 2932 and 2859 cm^{-1} in the resulting sample, which are the methylene stretches of the *n*-undecyl chain; these peak positions suggest that the alkyl chains of the original SAM remain in a *trans* conformation, but that the new alkyl chains are disordered and contain more gauche conformations [167]. These transformations on the surface occur rapidly, and in close to quantitative yield. This new chemical method using interchain carboxylic anhydride as a reactive intermediate has allowed for rapid introduction of many functionalities into SAMs, for example, *n*-alkyl groups [167], perfluorinated *n*-alkyl groups [169], peptides [170], charged groups (sulfonate and guanidine groups) [169], and polymers containing amine groups [e.g., poly(ethylene imine)] [171]; all that is required is a molecule containing that functionality and also an active amine group. This method provides access to SAMs that can be inconvenient or impossible to prepare using the older methods.

Figure 6 compares the second method—the common intermediate method—with the older but more commonly used method. This method also has

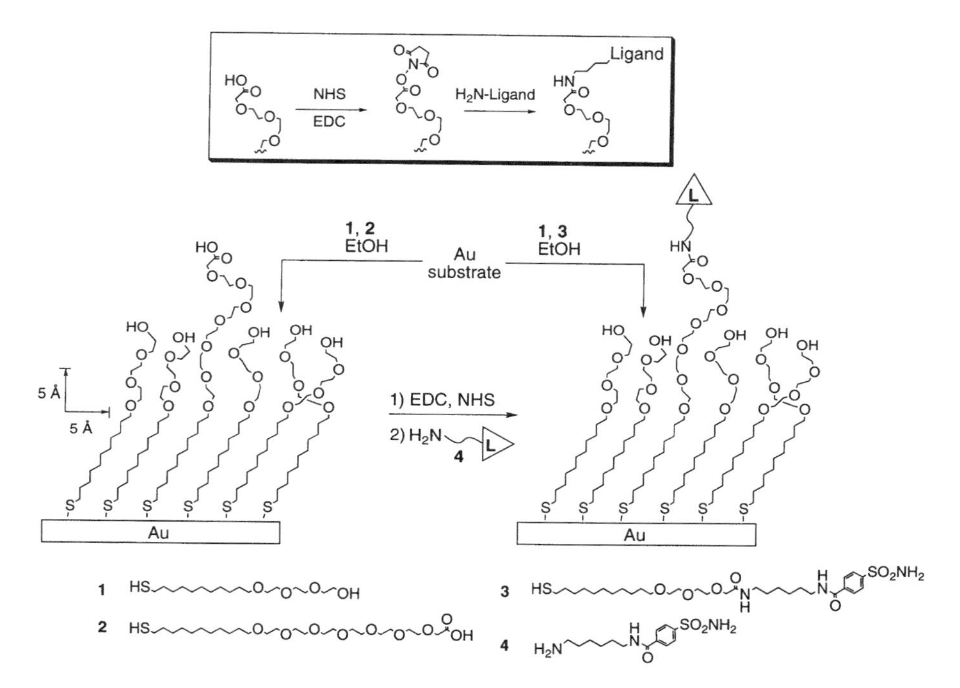

Figure 6 Schematic comparison of the common intermediate method and the older method involving the synthesis of ligand-terminated alkanethiols for preparation of SAMs presenting ligands. In the common intermediate method, a SAM bearing carboxylic acid groups is formed by immersing gold substrates in a mixture of alkanethiols **1** and **2**. This mixed SAM, after activation with NHS and EDC, presents an active NHS ester group on the surface that serves as a common intermediate for the attachment of different ligands by amide bond formation. The upper panel illustrates the chemical transformations involving carboxylic acid groups: (i) activation of carboxylic acid groups with NHS and EDC to generate active NHS esters, and (ii) displacement of the NHS group with an amino group on the ligand (a benzenesulfonamide-containing amine **4** as a representative ligand) or ϵ-amino groups of lysine residues of proteins to form an amide bond. The polymethylene chains of the alkanethiols in the SAMs are drawn in all-*trans* conformation; this conformation has been observed in SAMs of long chain alkanethiols on gold. The oligo(ethylene glycol) groups are depicted with little or no ordering; the detailed conformation in these SAMs has not been firmly established. (From Ref. 172.)

three steps [172]: formation of a mixed SAM of alkanethiolates on gold derived from the tri(ethylene glycol) ($(EG)_3OH$) terminated thiol ($HS(CH_2)_{11}(OCH_2CH_2)_3OH$; **1**) and the hexa(ethylene glycol)-carboxylic acid ($(EG)_6CO_2H$) terminated thiol ($HS(CH_2)_{11}(OCH_2CH_2)_6OCH_2CO_2H$; **2**); generation of activated *N*-hydroxylsuccinimidyl (NHS) esters of thiol **2**; and reaction with proteins, peptides, or small molecules containing active amine groups. These reactions are characterized using PIERS and ellipsometry. Figure 7 shows the PIERS spectra of SAMs of **1**, of **2**, of an authentic thiol ($HS(CH_2)_{11}(OCH_2CH_2)_6OCH_2CONH(CH_2)_6NHCOC_6H_4SO_2NH_2$; **3**), of a mixed SAM comprising **1** and **2**, and of the products of the subsequent reactions. Upon treatment of the mixed SAM

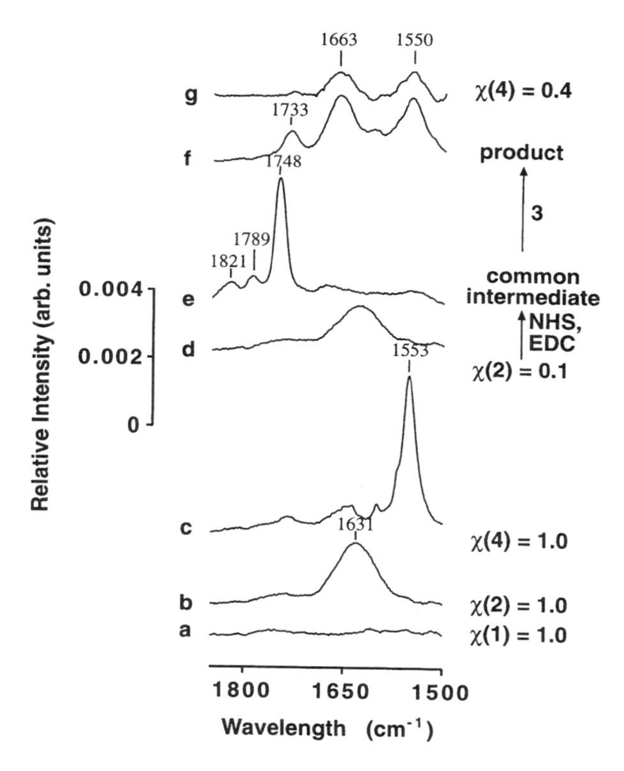

Figure 7 PIERS data for: (i) a homogeneous SAM of **1**; (ii) a homogeneous SAM of **2**; (iii) a homogeneous SAM of **3**; (iv) a mixed SAM comprising **1** and **2** with $\chi(\mathbf{2}) = 0.10$; (v) a mixed SAM comprising **1** and **2** with $\chi(\mathbf{2}) = 0.10$ after activation with NHS and EDC in H_2O; (vi) a mixed SAM comprising **1** and **2** with $\chi(\mathbf{2}) = 0.10$ after treatment with NHS and EDC followed by reaction with **4**; and (vii) a mixed SAM comprising **1** and **3** with $\chi(\mathbf{3}) = 0.40$. (From Ref. 172.)

with *N*-hydroxylsuccinimide (NHS) and 1-ethyl-3-(3-dimethylaminopropyl)carbodiimide (EDC), the appearance of bands diagnostic for the NHS ester at 1789 cm^{-1} (symmetric stretch of imide C=O groups) and 1821 cm^{-1} (C=O stretch of the activated ester carbonyl group) indicates the formation of active NHS ester groups on the surface. The complete disappearance of the band at 1631 cm^{-1}, which is assigned as the C=O stretch of a carboxylic acid, suggests near quantitative conversion of carboxylic acid groups to NHS esters. After treatment with a benzenesulfonamide-containing amine ($NH_2(CH_2)_6NHCOC_6H_4SO_2NH_2$; **4**), PIERS shows the appearance of bands at 1550 and 1660 cm^{-1} (characteristic of N—H bending modes) and a weak band at 1733 cm^{-1} (C=O stretch of carboxylic acid groups). Comparison by PIERS with an authentic SAM of **3** established that benzenesulfonamide ligands were covalently attached to the mixed SAM through amide bonds. Bovine carbonic anhydrase II (CA) can recognize and reversibly bind to the immobilized benzenesulfonamide ligand on the surface [172]. We have also used this procedure to attach proteins to SAMs [172].

VII. PATTERNING OF SAMs ON GOLD IN THE PLANE OF THE MONOLAYER

Patterning SAMs in the plane of the monolayer is useful in determining the two-dimensional distribution of chemical and physical properties on a surface. There are several methods available for generation of patterned SAMs. Microcontact printing is a convenient technique that ''stamps'' a pattern of SAM directly on a surface (Fig. 8) [82,83]. In μCP, an elastomeric poly(dimethylsiloxane) (PDMS) stamp—fabricated by casting and curing PDMS against masters that present patterned photoresist on silicon wafers—is wetted or inked with an alkanethiol and brought into contact with the gold-coated substrate; SAMs form on the areas that contacted the stamp. SAMs presenting different functional groups can be subsequently formed on the uncontacted areas, either by μCP or by immersion in a solution of another thiol. The edge resolution of the patterns resulting from μCP is ~50 nm [173]. AFM studies show that the structural order characterizing alkanethiolates in SAMs formed by μCP is the same as that deriving from immersion of the gold substrate in a solution of thiol [174]. μCP offers a convenient, low cost, flexible, and nonphotolithographic method to pattern SAMs on large areas [175] and curved substrates [176], and also to pattern SAMs of other systems: alkanethiolates on coinage metals [134], alkylsiloxanes on Si/SiO_2 and glass [177,178], and colloids [179,180] and proteins on various substrates [181,182].

Combination of μCP and chemical reaction on a reactive SAM, for example, a SAM presenting interchain carboxylic anhydride groups, can simplify and extend μCP [169]. Figure 9 shows an example: a PDMS stamp with protruding

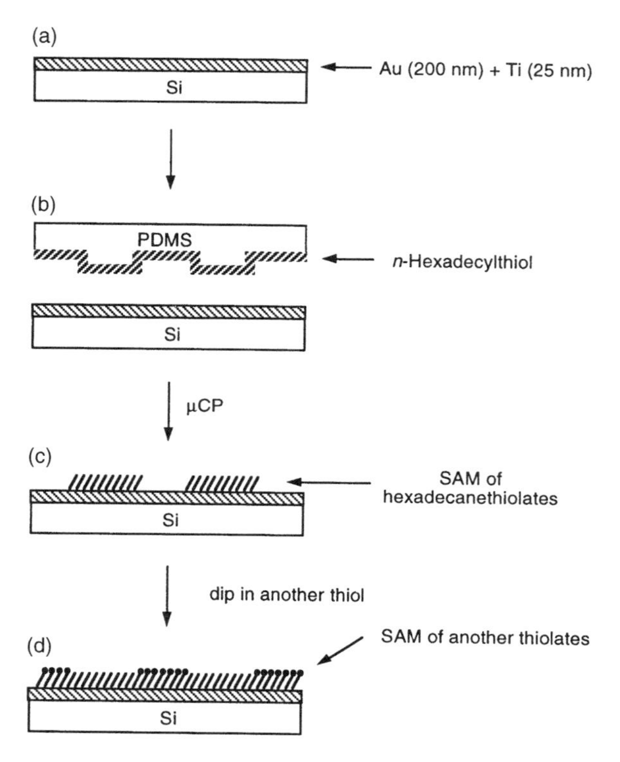

Figure 8 Schematic illustration for μCP of *n*-hexadecanethiol on gold followed by dipping into a solution containing another thiol.

features (squares having ~10 μm on a side) prints *n*-hexadecylamine on the reactive SAM to give a SAM comprising a~1:1 mixture of *N*-alkyl amides and carboxylic acids. The remaining anhydride groups in the uncontacted regions are allowed to react with another amine, $CF_3(CF_2)_6CH_2NH_2$ to give a patterned SAM having regions presenting *N*-hexadecyl amides and fluorinated *N*-alkyl amides. SEM images (Fig. 10) indicate that the edge resolution of these squares is at submicron scale (≤100 nm). The high contrast and uniformity in the SEM and SIMS images (Fig. 10) suggest that μCP delivered a well-defined pattern of *n*-hexadecylamine to the reactive SAM on both gold and silver. The key chemical reaction—the reaction of amine and surface anhydride—proceeds rapidly and in close to quantitative yield under ambient experimental conditions, with good edge

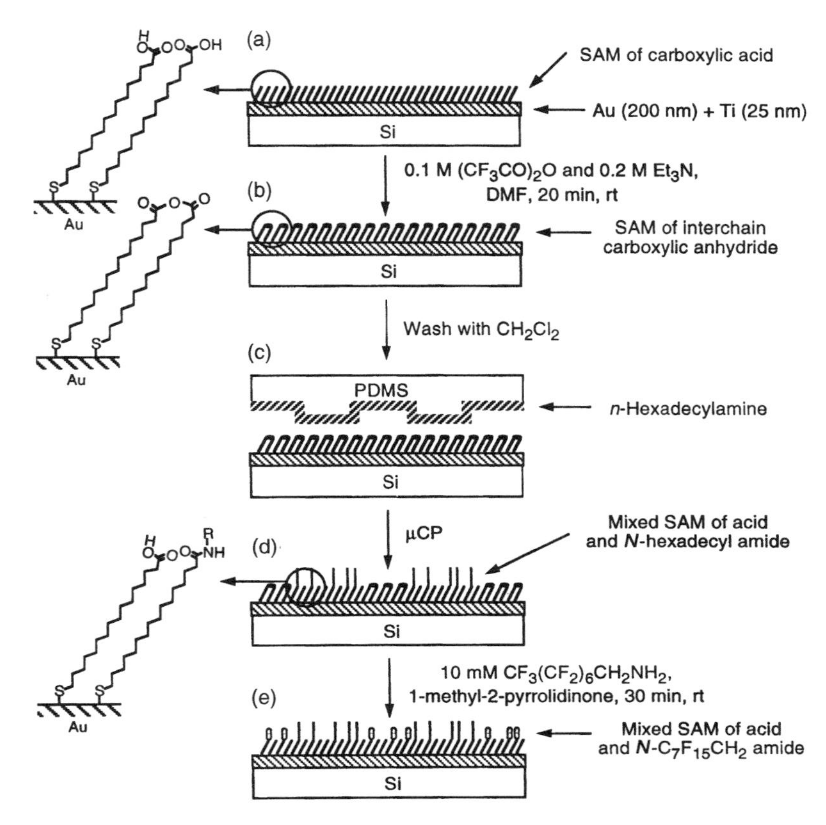

Figure 9 Schematic outline of the procedure for patterning a SAM that presents two different *N*-alkyl amides using μCP and a chemical reaction. The diagram represents the composition of the SAM but not the conformation of the groups in it. (From Ref. 169.)

definition. This method provides a straightforward route to patterned SAMs that present a variety of functional groups.

VIII. APPLICATIONS

SAMs of alkanethiolates on gold provide excellent model systems for studies on interfacial phenomena (e.g., wetting, adhesion, lubrication, corrosion, nucleation, protein adsorption, cell attachment, and sensing). These subjects have been re-

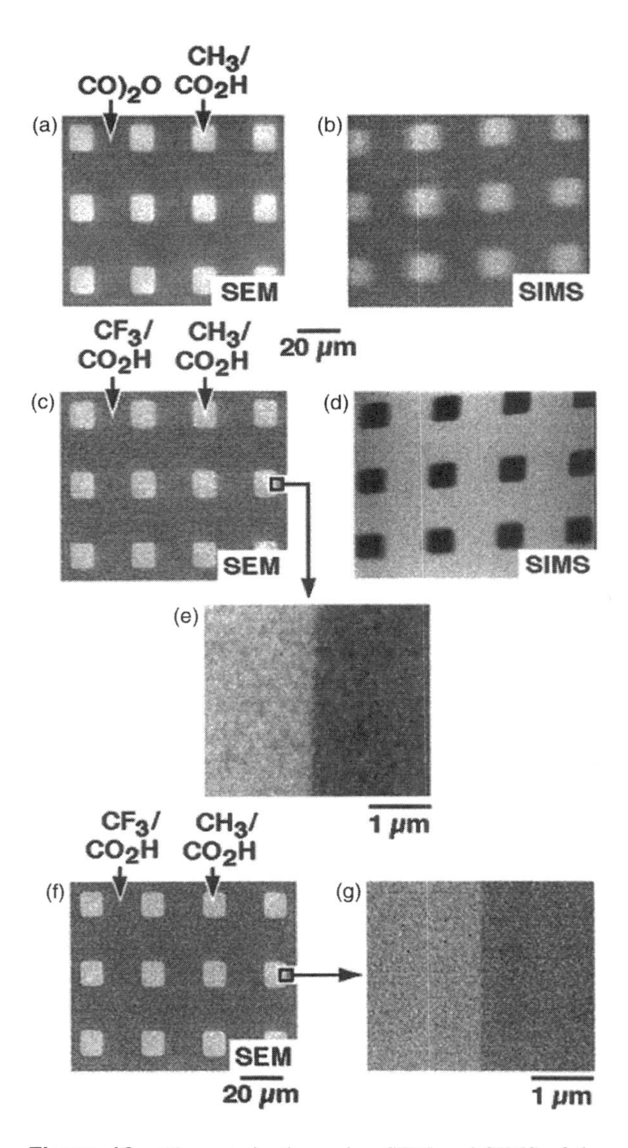

Figure 10 Characterization using SEM and SIMS of the patterned SAMs on Au (a)–(e) and on Ag (f) and (g) generated by μCP of *n*-hexadecylamine on the reactive SAM followed by reaction with $CF_3(CF_2)_6CH_2NH_2$. The light areas in the SEM images were the regions contacted by the stamp. The light squares in SIMS image (b) contained nitrogen while the dark regions did not; the light regions in SIMS image (d) had fluorine while the dark squares did not. The patterns in the SIMS images were distorted because the sample holder was slightly tilted during acquisition of these images. (From Ref. 169.)

viewed previously [125,183–185]. Here we focus on applications that involve using chemical synthesis of functional SAMs after their assembly.

A. Patterning Thin Films of Polymer

Patterned thin films of polymers have many applications e.g., in preventing etching [177], in molecular electronics [186–188], in optical devices [189,190], in biological [191] and chemical sensors [154,192], and in tissue engineering [193]. Thin films of polymers that have reactive functional groups present a surface that can be further modified by chemical reactions [194,195]. There are several methods available for attaching polymers to SAMs: electrostatic adsorption of polyelectrolytes to an oppositely charged surface [196,197], chemisorption of polymers containing reactive groups to a surface [198,199], and covalent attachment of polymers to reactive SAMs [151,154,200,201]. There are presently only a few methods available for patterning thin films of polymers on SAMs; these include procedures based on photolithography [200,201], templating the deposition of polymers using patterned SAMs [197,202,203], and templating phase-separation in diblock copolymers [204,205]. Patterned thin films of polymers attached covalently to the surface are more stable than are ones only physically adsorbed. Photochemical pattern transfer offers only limited control over the surface chemistry, the properties, and the structure of the modified surfaces.

Combination of μCP and chemical modification of the reactive SAM presenting interchain carboxylic anhydride groups provides a convenient method for patterning thin films of amine-containing polymers having submicron-scale edge resolution on the surface [171]. Figure 11 describes this approach. A reactive SAM presenting interchain carboxylic anhydride groups is prepared using trifluoroacetic anhydride (Fig. 4). A PDMS stamp with protruding squares (~10 μm on a side) on its surface is oxidized for ~10 sec with an oxygen plasma. The oxidized PDMS stamp is immediately inked with a 1 wt% solution of poly(ethylene imine) (PEI) in 2-propanol and placed in contact with the substrate. The anhydride groups in the regions that contacted the PDMS stamp react with the amine groups of PEI. Removal of the stamp and hydrolysis of the remaining anhydride groups with aqueous base (pH = 10, 5 min) give a surface patterned with PEI. All these procedures are carried out under ambient conditions; the entire process—from the readily available SAMs of 16-mercaptohexadecanoic acid to the final patterned PEI films—can be completed in less than one hour.

The AFM images acquired in contact mode show that μCP delivered a well-defined pattern of PEI to the reactive SAM (Fig. 12). The resulting thin films of PEI are nearly continuous, but their surfaces are not smooth at the nanometer scale (Fig. 12b). The roughness of these films is controlled in part by the surface topology of the polycrystalline gold substrate, and probably also by the presence of gel or dust particles in the PEI. Line analysis indicates that the aver-

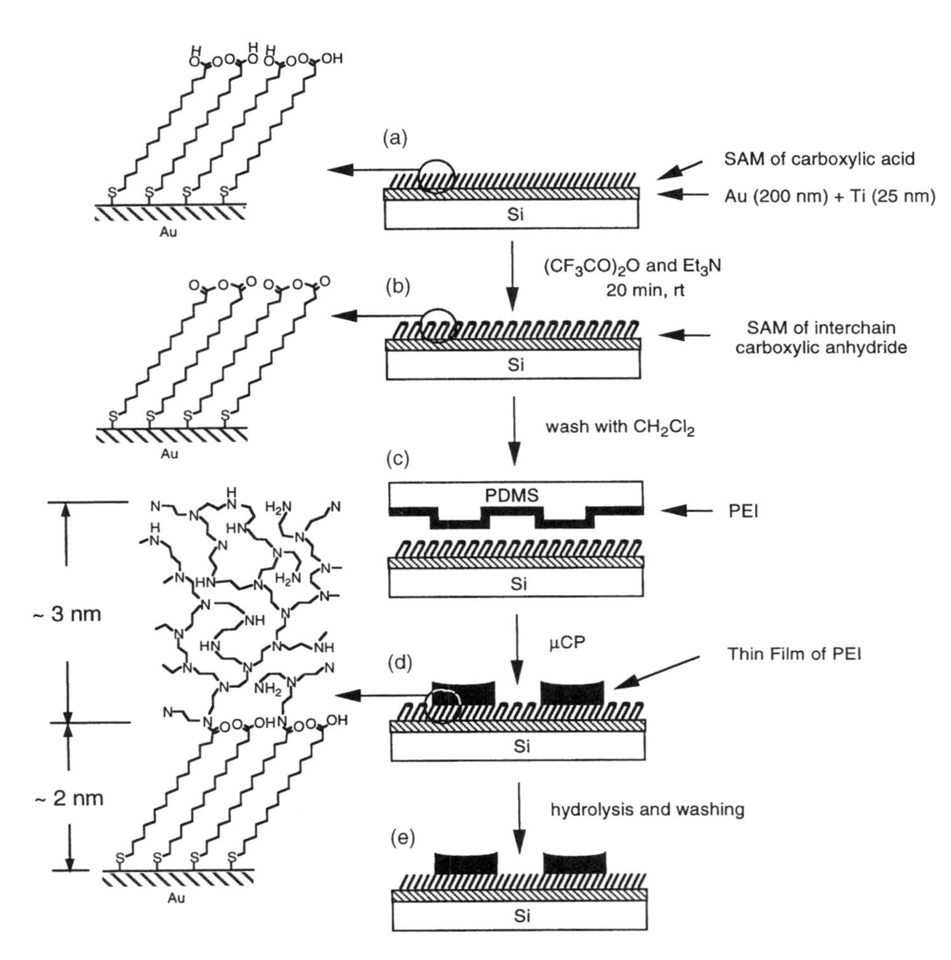

Figure 11 Schematic description of the procedure for patterning thin films of PEI on the surface of a SAM using μCP and a chemical reaction. The scheme suggests the composition of the SAM, but not the conformation of the groups in it; it also makes no attempt to represent either the conformation of the polymer or the distribution of functional groups on the polymer backbone.

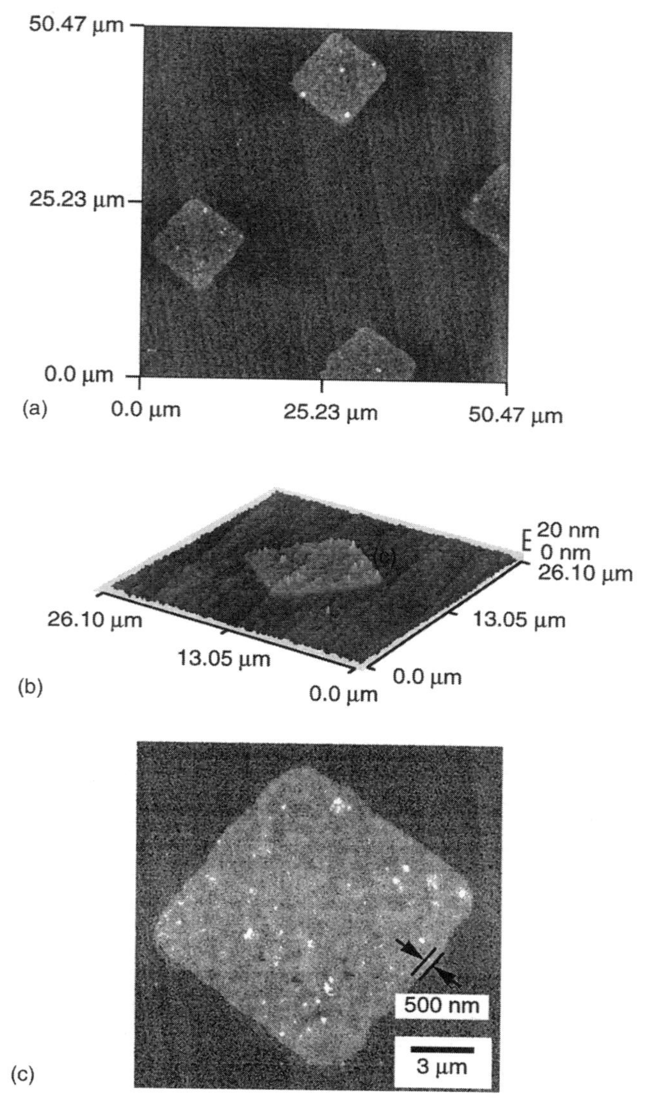

Figure 12 Contact mode AFM characterized the patterned thin films of PEI generated by μCP of PEI on the reactive SAM; these images show a sample patterned by μCP, followed by hydrolysis of the remaining, unreacted interchain carboxylic anhydrides with aqueous base. The light squares in the AFM images (a)–(c) were thin films of PEI on the regions contacted by the PDMS stamp. The AFM image (c) shows that the PEI film was separated by a well-defined boundary (with roughness <500 nm) from the regions presenting carboxylic acid groups. The lines in the images were artifacts generated by the instrument. (From Ref. 171.)

age thickness of the patterned thin films is ~3 nm. Figure 12c suggests that the edge resolution of these squares is at the submicron scale (<500 nm); this value is larger than that obtained in μCP of *n*-hexadecylamine on the reactive SAM (<100 nm) and of alkanethiolates on gold (<50 nm). Because PEI is a hydrophilic polymer, it is essential to make the hydrophobic PDMS stamp hydrophilic using oxygen plasma prior to inking in order to form continuous, patterned thin films of PEI on the surface [206]. PIERS studies further show that PEI is covalently linked to the SAM by amide bonds and that the PEI films are thus more stable under both acidic and basic conditions than are polymers physically adsorbed on SAMs of carboxylic acids.

The covalently attached PEI films make a large number of reactive amine groups available for further chemical modification of the surface. These amine groups can react with other functional groups (e.g., acyl chlorides and carboxylic anhydrides) to introduce different organic functionalities into the surface, and to attach polymers that have such organic functional groups. We have shown that the amine groups of the attached PEI film can react with perfluorooctanoyl chloride, palmitoyl chloride, palmitic anhydride, and poly(styrene-*alt*-maleic anhydride) [171].

B. Facile Preparation of SAMs That Present Mercaptan, Charged, and Polar Groups

Chemical reaction provides a straightforward method for the preparation of SAMs that present a variety of functional groups—especially polar, charged, or structurally complex groups, such as peptides, polymers, and oligosaccharides—that are difficult to prepare using deposition of thiols terminated at these groups [169].

Figure 13 shows a patterned SAM presenting methyl and thiol groups, and the subsequent assembly of Au nanoparticles in the regions presenting thiol groups. The patterned SAM is generated by μCP of *n*-hexadecylamine on the reactive SAM followed by reaction with cysteamine ($HSCH_2CH_2NH_2$). Patterned SAMs presenting thiol groups are difficult to prepare by conventional μCP. The resulting patterned substrate is then immersed in an aqueous suspension of Au nanoparticles stabilized with citrate anions. SEM images show that the Au nanoparticles assemble predominately in the thiol-presenting regions, and the width of the border separating the region having adsorbed nanoparticles from that having none is <100 nm.

We have prepared SAMs presenting sulfonates and guanidines by allowing the reactive SAM to react with 3-amino-1-propanesulfonic acid ($H_2N(CH_2)_3SO_3H$) and agmatine sulfate ($H_2N(CH_2)_4NHC({=}NH)NH_2{\cdot}H_2SO_4$). For the SAM presenting sulfonates, XPS shows a signal at 168.5 eV; we assign this peak to S(2p) of a sulfonate, and the advancing and receding contact angles

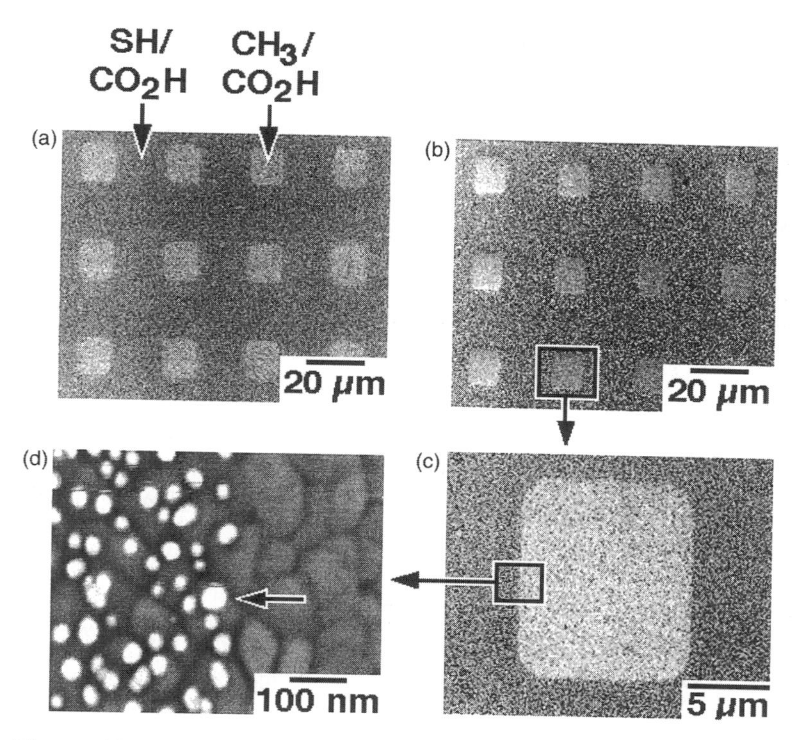

Figure 13 SEM images of (a) a patterned SAM presenting methyl groups (white squares) and thiol groups; (b), (c), and (d) patterned deposition of Au nanoparticles (white dots pointed by an arrow in (d) in the regions that present thiol groups. The background texture in (d) is the "islands" that are formed on evaporating gold under the conditions we used. The mean diameter of the Au nanoparticles was ~20 nm. (From Ref. 169.)

of water are both less than 10° (the lowest value we can measure). For the SAM presenting guanidines, XPS shows an N(1 s) signal at 400.4 eV, the advancing contact angle of water is 47°, and the receding contact angle 30°.

C. Wetting

Previous extensive studies of carboxylic acid functionalized poly(ethylene) films (PE—CO_2H) [207–209], of SAMs terminated in ionizable acids and bases [100], of mixed SAMs of carboxylic acid- and methyl-terminated alkanethiolates [210], and of SAMs of dialkyl sulfides on gold [211] have established the utility of contact angle titration in characterizing the wetting properties of interfaces. We have determined the advancing contact angle of water θ_a as a function of pH for several mixed monolayers obtained by allowing the interchain anhydride to react with homologous *n*-alkylamines (*n*-$C_nH_{2n+1}NH_2$, n = 0, 1, 4, 6, 11, and 18) (Fig. 14). In these mixed SAMs, the polar carboxylic acid groups are buried beneath hydrocarbon layers of different thickness.

The values of θ_a for the SAMs derivatized with long *n*-alkylamines (n = 11 and 18) do not change with pH. These alkyl groups form thick hydrophobic films (ca. 10 and 15 Å, respectively) that prevent water from contacting the buried carboxylic acid groups. The values of θ_a change with pH for mixed SAMs derivatized with short *n*-alkylamines (n = 1, 4, and 6). There are significant features of these data. First, the titration curves do not reach a plateau at high pH; the same behavior is observed for the mixed SAMs of carboxylic acid- and methyl-terminated alkanethiolates on gold prepared from mixtures of $HS(CH_2)_{10}COOH$ and $HS(CH_2)_{10}CH_3$ [211]. By contrast, the carboxylic acid functionalized material obtained by oxidizing poly(ethylene) (PE—CO_2H) does achieve plateau values at high pH [207]. Second, the onset points of ionization are approximately pH

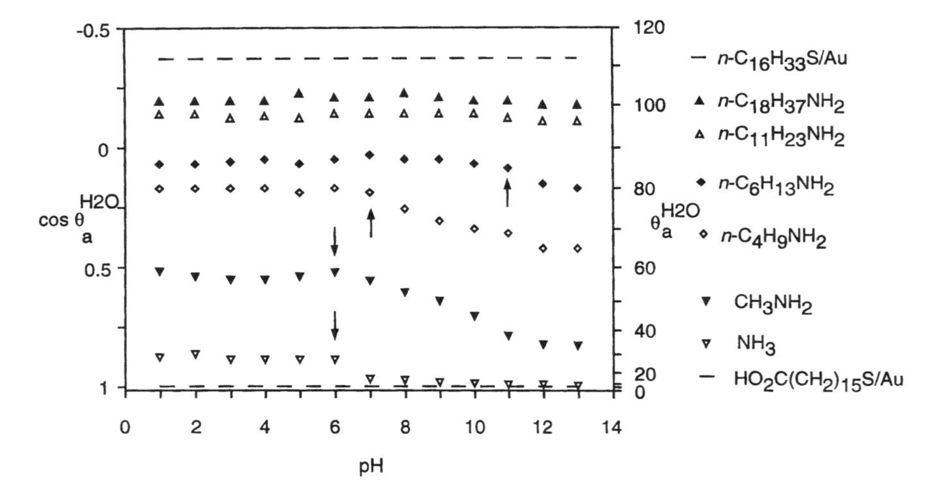

Figure 14 Dependence of the advancing contact angle (θ_a) of buffered aqueous solutions of different values of pH on the SAMs comprising a mixture of carboxylic acids and amides, generated by reactions of the interchain anhydride and alkylamines (*n*-$C_nH_{2n+1}NH_2$, n = 0, 1, 4, 6, 11, and 18). The curves are labeled by the respective alkylamines on the right side of the plot. Two dashed lines on the top and the bottom of the plot are reference data for θ_a of $CH_3(CH_2)_{15}S/Au$ and $HO_2C(CH_2)_{15}S/Au$, respectively, as indicated. Arrows indicate the onset points of ionization of carboxylic acid groups.

11 for the *n*-hexylamine-modified SAM and approximately pH 7 for the *n*-butylamine-modified SAM. The values of θ_a also change with pH for mixed SAMs comprising carboxylic acids and *N*-methyl amide and primary amides. These titration curves are very similar to that of the mixed SAM of *N*-butyl amides and carboxylic acids. Both the onset points of ionization are approximately at pH 6. Such shifts from the values expected on the basis of titration in aqueous solution are similar to those observed with mixed SAMs terminated with carboxylic acid and methyl groups [211].

Contact angle titrations of these *n*-alkylamine-modified SAMs suggest that these systems are more like mixed SAMs of methyl- and carboxylic acid-terminated alkanethiolates on gold than they are like PE—CO_2H.

D. Patterning Ligands on Reactive SAMs

Patterning ligands on surfaces has several applications: for example, in biosensors using patterned proteins and cells, and in diagnostic tools using patterned DNA fragments, antibodies, and antigens. The techniques used for patterning ligands are often based on photolithographical procedures: these methods may be incompatible with many types of ligands.

Combination of μCP and chemical reaction on a reactive SAM presenting a mixture of active NHS ester groups and oligo(ethylene glycol) groups provides a convenient, inexpensive, and versatile method for patterning ligands on surfaces that can be recognized and bound specifically by biomacromolecules [212]. Figure 15 describes a procedure that has several steps: (i) formation of mixed SAMs presenting thiol **1** and **2**; (ii) activation by immersion in a solution of EDC (0.1 M) and pentafluorophenol (0.2 M) for 10 min; and (iii) printing a biotin-containing amine to the reactive SAMs by bringing a freshly oxidized PDMS stamp, inked with the ligand, in contact with the substrate for 5 min. The formation of patterned SAMs presenting biotin ligands was imaged by fluorescence microscopy of substrates that were incubated in a solution of fluorescently labeled antibiotin antibody (Fig. 16a). The patterns were also detected using a different approach, in which the substrate was incubated sequentially in solutions of streptavidin, biotin-conjugate protein G, fluorescently labeled goat antirabbit IgG, and imaged by fluorescence microscopy (Fig. 16b). The smallest features resolved in images obtained by these methods were squares with a 5 μm side. The high contrast between the fluorescent regions and nonfluorescent ones indicates that the PDMS stamp delivered biotin ligands to the SAM and that they were bound specifically by its binding proteins. The coupling yields were estimated using SPR to be ~75–90% of that obtained by immersion. It was also found that oxidation of the PDMS stamp prior to inking was critical for good coupling yields.

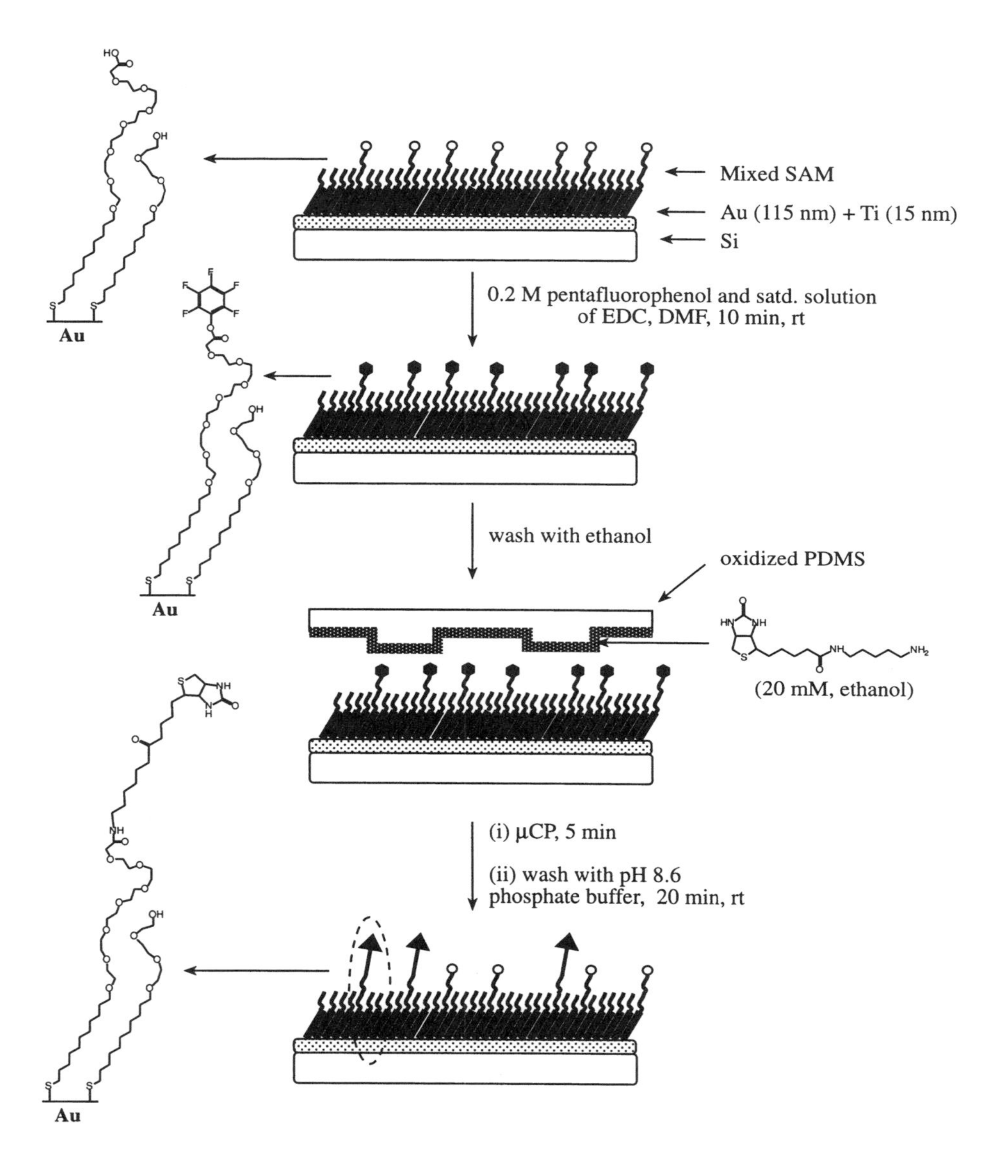

Figure 15 Schematic representation of the procedure used for patterning biotin ligands onto SAMs consisting of activated carboxylic esters. (From Ref. 212.)

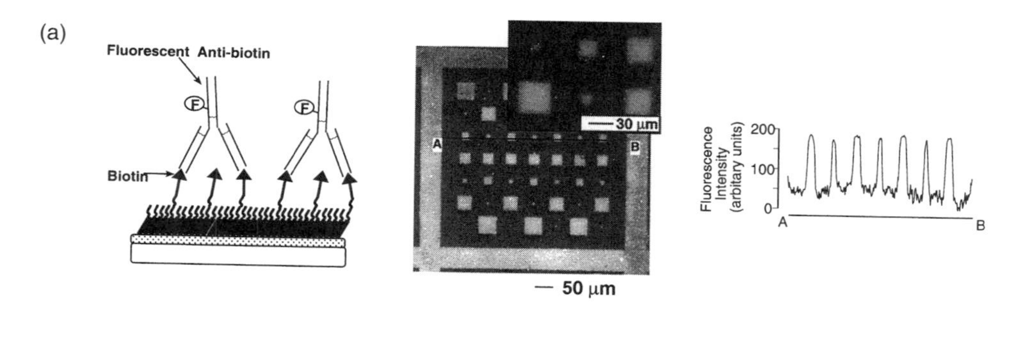

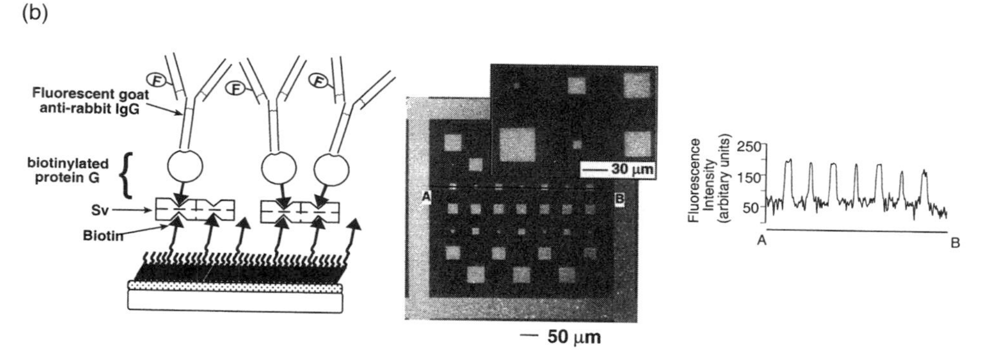

Figure 16 Fluorescence microscopy images of patterned SAMs having biotin groups (χ ~0.02) and schematic representations of the surfaces during fluorescence detection. The intensity of fluorescence in regions having biotin groups and not having ones across the line (AB) was analyzed using the NIH image software. (a) Images of fluorescently labeled antibiotin bound to patterned SAMs presenting biotin groups. (b) Images of fluorescently labeled goat anti rabbit IgG bound to SAMs presenting biotin groups that were sequentially incubated in solutions of streptavidin and biotin-conjugated protein G. (From Ref. 212.)

IX. CONCLUSIONS

SAMs can be considered as a form of insoluble, two-dimensional, and grafted polymers: that is, the substrate surface is the backbone, and the attached alkanethiolate (RS^-) groups are the sidechains. The backbone already exists before the SAM is formed: the preparation of SAMs is more like postpolymerization modification than it is like polymerization. The packing of the alkyl chains is such that the alkyl chains are highly ordered and crystallize. In SAMs of *n*-alkanethiolates on gold, the alkyl chains adopt an all-*trans* conformation and tilt ~30° from the normal of the substrate to maximize intermolecular interactions between adjacent chains. There is no comparable phenomenon to this tight packing of sidechains in conventional polymers. The high density of sidechains leads to anomalous reactivities of functional groups incorporated in the sidechains. SAMs provide a sterically highly congested environment. Even terminal functional groups are sterically hindered; their chemical transformations on the surface are often slower than they would be in solution. Because terminal groups are embedded in a sea of hydrophobic polymethylene chains, they are difficult to ionize. On the other hand, the high density of functional groups can result in favorable neighboring and chelating interactions.

The formation of SAMs on gold and silver is simple and convenient. The strong chemoselective interactions between the head-groups and the substrate allow SAMs to present a wide range of functional groups on the surface. Mixed and patterned SAMs deliver flexible control over the lateral distribution of chemical functionalities. Both the interchain carboxylic anhydride and the activated carboxylic esters strategy minimize the amount of orgainic synthesis required for the preparation of functionalized alkanethiols to prepare SAMs that present a wide range of chemical functionalities. They are straightforward methods and could find many applications.

ACKNOWLEDGMENTS

This work was supported by DARPA and the ONR. Lin Yan thanks Ned Bowden for technical assistance in obtaining Fig. 3.

REFERENCES

1. P Flory. Principles of Polymer Chemistry. Ithaca, NY: Cornell University Press, 1953.
2. J-M Lehn. Supramolecular Chemistry, Concepts and Perspectives. Weinheim: VCH, 1995.
3. SI Stupp, S Son, HC Lin, LS Li. Science 259:59–63, 1993.
4. H Rehage, M Veyssié. Angew Chem Int Ed Engl 29:439–448, 1990.
5. A Ulman. Introduction to Thin Organic Films: From Langmuir–Blodgett to Self-Assembly. Boston: Academic, 1991.
6. CD Bain, GM Whitesides. Angew Chem Int Ed Engl 28:506–516, 1989.
7. GM Whitesides. Chimia 44:310–311, 1990.
8. GM Whitesides, CB Gorman, eds. Self-Assembled Monolayers: Models for Organic Surface Chemistry. Handbook of Surface Imaging and Visualization. Boca Raton, FL: CRC, 1995, pp 713–733.

9. GS Ferguson, MK Chaudhury, HA Biebuyck, GM Whitesides. Macromolecules 26:5870–5875, 1993.
10. LH Dubois, RG Nuzzo. Annu Rev Phys Chem 43:437–463, 1992.
11. S Yitzchaik, T Marks. Acc Chem Res 29:197–202, 1996.
12. AR Bishop, RG Nuzzo. Curr Opin Coll Interf Sci 1:127–136, 1996.
13. A Ulman. Chem Rev 96:1533–1554, 1996.
14. RG Nuzzo, BR Zegarski, LH Dubois. J Am Chem Soc 109:733–740, 1987.
15. PE Laibinis, MA Fox, JP Folkers, GM Whitesides. Langmuir 7:3167–3173, 1991.
16. PE Laibinis, GM Whitesides. J Am Chem Soc 114:9022–9027, 1992.
17. H Lee, LJ Kepley, H-G Hong, TE Mallouk. J Am Chem Soc 110:618–620, 1988.
18. D Li, MA Ratner, TJ Marks, C Zhang, J Yang, GK Wong. J Am Chem Soc 112: 7389–7390, 1990.
19. ME Thompson. Chem Mater 6:1168–1175, 1994.
20. SMC Neiva, JAV Santos, JC Moreira, Y Gushikem, H Vargas, DW Franco. Langmuir 9:2982–2985, 1993.
21. MGL Petrucci, AK Kakkar. J Chem Soc Chem Commun 1577–1578, 1995.
22. J Sagiv. Isr J Chem 18:339–345, 1979.
23. J Sagiv. J Am Chem Soc 102:92–98, 1980.
24. I Haller. J Am Chem Soc 100:8050–8055, 1978.
25. SR Wasserman, GM Whitesides, IM Tidswell, BM Ocko, PS Pershan, JD Axe. J Am Chem Soc 111:5852–5861, 1989.
26. SR Wasserman, Y-T Tao, GM Whitesides. Langmuir 5:1074–1087, 1989.
27. T Nakagawa, K Ogawa, T Kurumizawa. Langmuir 10:525–529, 1994.
28. B Buszewski, RM Gadzata-Kopciuch, M Markuszewski, R Kaliszan. Anal Chem 69:3277–3284, 1997.
29. TR Lee, PE Laibinis, JP Folkers, GM Whitesides. Pure Appl Chem 63:821–828, 1991.
30. G Kataby, T Prozorov, Y Koltypin, H Cohen, CN Sukenik, A Ulman, A Gedanken. Langmuir 13:6151–6158, 1997.
31. KJ Stevenson, M Mitchell, HS White. J Phys Chem B 102:1235–1240, 1998.
32. CW Sheen, J-X Shi, J Martensson, AN Parikh, DL Allara. J Am Chem Soc 114: 1514–1515, 1992.
33. JF Dorsten, JE Maslar, PW Bohn. Appl Phys Lett 66:1755–1757, 1995.
34. Y Gu, Z Lin, RA Butera, VS Smentkowski, DH Waldeck. Langmuir 11:1849–1851, 1995.
35. DL Allara, RG Nuzzo. Langmuir 1:45–52, 1985.
36. HO Finklea, LR Robinson, A Blackburn, B Richter. Langmuir 2:239–244, 1986.
37. DL Allara, AN Parikh, F Rondelez. Langmuir 11:2357–2360, 1995.
38. DL Allara, RG Nuzzo. Langmuir 1:52–66, 1985.
39. PE Laibinis, JJ Hickman, MS Wrighton, GM Whitesides. Science 245:845–847, 1989.
40. Y-T Tao, GD Hietpas, DL Allara. J Am Chem Soc 118:6724–6735, 1996.
41. SH Chen, CW Frank. Langmuir 5:978–987, 1989.
42. YG Aronoff, B Chen, G Lu, C Seto, J Schwartz, SL Bernasek. J Am Chem Soc 119:259–262, 1997.
43. TJ Gardner, CD Frisbie, MS Wrighton. J Am Chem Soc 117:6927–6933, 1995.

44. JP Folkers, CB Gorman, PE Laibinis, S Buchholz, GM Whitesides, RG Nuzzo. Langmuir 11:813–824, 1995.
45. JE Chadwick, DC Myles, RL Garrell. J Am Chem Soc 115:10364–10365, 1993.
46. W Gao, L Dickinson, C Grozinger, FG Morin, L Reven. Langmuir 12:6429–6435, 1996.
47. JT Woodward, A Ulman, DK Schwartz. Langmuir 12:3626–3629, 1996.
48. L Bertilsson, K Potje-Kamloth, H-D Liess. J Phys Chem B 102:1260–1269, 1998.
49. CD Bain, HA Biebuyck, GM Whitesides. Langmuir 5:723–727, 1989.
50. JJ Hickman, PE Laibinis, DI Auerbach, C Zou, TJ Gardner, GM Whitesides, MS Wrighton. Langmuir 8:357–359, 1992.
51. K Uvdal, I Persson, B Liedberg. Langmuir 11:1252–1256, 1995.
52. HA Biebuyck, CD Bain, GM Whitesides. Langmuir 10:1825–1831, 1994.
53. AC Ontko, RJ Angelici. Langmuir 14:3071–3078, 1998.
54. MR Linford, CED Chidsey. J Am Chem Soc 115:12631–12632, 1993.
55. A Bansal, X Li, I Lauermann, NS Lewis. J Am Chem Soc 118:7225–7226, 1996.
56. JM Buriak, MJ Allen. J Am Chem Soc 120:1339–1340, 1998.
57. P DiMilla, JP Folkers, HA Biebuyck, R Harter, G Lopez, GM Whitesides. J Am Chem Soc 116:2225–2226, 1994.
58. LH Dubois, R Zegarski, RG Nuzzo. J Chem Phys 98:678–688, 1993.
59. DE Weisshaar, BD Lamp, MD Porter. J Am Chem Soc 114:5860–5862, 1992.
60. P Fenter, A Eberhardt, P Eisenberger. Science 266:1216–1218, 1994.
61. JJ Gerdy, WA Goodard. J Am Chem Soc 118:3233–3236, 1996.
62. JI Siepmann, IR McDonald. Thin Films 24:205–226, 1998.
63. T Sawaguchi, F Mizutani, I Taniguchi. Langmuir 14:3565–3569, 1998.
64. CD Bain, EB Troughton, Y-T Tao, J Evall, GM Whitesides, RG Nuzzo. J Am Chem Soc 111:321–335, 1989.
65. K Shimazu, I Yag, Y Sato, K Uosaki. Langmuir 8:1385–1387, 1992.
66. DS Karpovich, GJ Blanchard. Langmuir 10:3315, 1994.
67. TW Schneider, DA Buttry. J Am Chem Soc 115:12391, 1993.
68. Y-T Kim, RI McCarley, AJ Bard. Langmuir 9:1941, 1993.
69. W Pan, CJ Durning, NJ Turro. Langmuir 12:4469–4473, 1996.
70. RC Thomas, L Sun, RM Crooks, AJ Ricco. Langmuir 7:620, 1991.
71. TT Ehler, N Malmberg, LJ Noe. J Phys Chem B 101:1268, 1997.
72. M Buck, F Eisert, J Fischer, M Grunze, F Traeger. Appl Phys A A53:552–6, 1991.
73. RH Terrill, TA Tanzer, PW Bohn. Langmuir 14:845–854, 1998.
74. MA Bryant, JE Pemberton. J Am Chem Soc 113:8284–8293, 1991.
75. CA Widrig, C Chung, MD Porter. J Electroanal Chem 310:335–359, 1991.
76. K Hu, AJ Bard. Langmuir 14:4790–4794, 1998.
77. GE Poirier, ED Pylant. Science 272:1145–1148, 1996.
78. R Yamada, K Uosaki. Langmuir 14:855–861, 1998.
79. GE Poirier. Langmuir 15:1167–1175, 1999.
80. Y Xia, N Venkateswaran, D Qin, J Tien, GM Whitesides. Langmuir 14:363–371, 1998.
81. Z Hou, NL Abbott, P Stroeve. Langmuir 14:3287–3297, 1998.
82. A Kumar, NA Abbott, E Kim, HA Biebuyck, GM Whitesides. Acc Chem Res 28: 219–226, 1995.

83. J Tien, Y Xia, GM Whitesides. Thin Films 24:227–253, 1998.
84. RG Nuzzo, FA Fusco, DL Allara. J Am Chem Soc 109:2358–2368, 1987.
85. MD Porter, TB Bright, DL Allara, CED Chidsey. J Am Chem Soc 109:3559–3568, 1987.
86. AN Parikh, DL Allara. J Chem Phys 96:927–945, 1992.
87. RG Nuzzo, LH Dubois, DL Allara. J Am Chem Soc 112:558–569, 1990.
88. C Zubraegel, C Deuper, F Schneider, M Neumann, M Grunze, A Schertel, C Woell. Chem Phys Lett 238:308–312, 1995.
89. CA Widrig, CA Alves, MD Porter. J Am Chem Soc 113:2805–2810, 1991.
90. CA Alves, EL Smith, MD Porter. J Am Chem Soc 114:1222–1227, 1992.
91. GE Poirier, MJ Tarlov. Langmuir 10:2853–2856, 1994.
92. P Fenter, P Eisenberger. Phys Rev Lett 70:2447, 1993.
93. CED Chidsey, G-Y Liu, P Rowntree, G Scoles. J Chem Phys 91:4421–4423, 1989.
94. N Camillone III, CED Chidsey, G-Y Liu, TM Putvinski, G Scoles. J Chem Phys 94:8493–8502, 1991.
95. L Strong, GM Whitesides. Langmuir 4:546–558, 1988.
96. JJ Hickman, D Ofer, C Zou, MS Wrighton, PE Laibinis, GM Whitesides. J Am Chem Soc 113:1128–1132, 1991.
97. DM Collard, MA Fox. Langmuir 7:1192–1197, 1991.
98. CD Bain, EB Troughton, Y-T Tao, J Evall, GM Whitesides, RG Nuzzo. J Am Chem Soc 111:321–335, 1989.
99. LH Dubois, BR Zegarski, RG Nuzzo. J Am Chem Soc 112:570–579, 1990.
100. TR Lee, RI Carey, HA Biebuyck, GM Whitesides. Langmuir 10:741–749, 1994.
101. CED Chidsey, DN Loiacono. Langmuir 6:709, 1990.
102. PE Laibinis, GM Whitesides, DL Allara, Y-T Tao, AN Parikh, RG Nuzzo. J Am Chem Soc 113:7152–7167, 1991.
103. H Sellers, A Ulman, Y Shnidman, JE Eilers. J Am Chem Soc 115:9389–9401, 1993.
104. A Ulman, JE Eilers, N Tillman. Langmuir 5:1147–1152, 1989.
105. J Hautman, ML Klein. J Chem Phys 91:4994–5001, 1989.
106. J Hautman, ML Klein. J Chem Phys 93:7483–7492, 1990.
107. J Hautman, JP Bareman, W Mar, ML Klein. J Chem Soc Faraday Trans 87:2031–2037, 1991.
108. W Mar, ML Klein. Langmuir 10:188–196, 1994.
109. JI Siepmann, IR McDonald. Molec Phys 79:457–473, 1993.
110. AJ Pertsin, M Grunze. Langmuir 10:3668–3674, 1994.
111. KM Beardmore, JD Kress, N Gronbech-Jensen, AR Bishop. Chem Phys Lett 286: 40–45, 1998.
112. G Nelles, H Schonherr, M Jaschke, H Wolf, M Schaub, J Kuther, W Tremel, E Bamberg, H Ringsdorf, H-J Butt. Langmuir 14:808–815, 1998.
113. M Sprik, E Delamarche, B Michel, U Rothlisberger, ML Klein, H Wolf, H Ringsdorf. Langmuir 10:4116–4130, 1994.
114. Y-T Tao, M-T Lee, S-C Chang. J Am Chem Soc 115:9547–9555, 1993.
115. E Sabatani, J Cohen-Boulakia, M Bruening, I Rubinstein. Langmuir 9:2974–2981, 1993.

116. JM Tour, LI Jones, DL Pearson, JJS Lamba, TP Burgin, GM Whitesides, DL Allara, AN Parikh, SV Atre. J Am Chem Soc 117:9529–9534, 1995.
117. A-A Dhirani, RW Zehner, RP Hsung, P Guyot-Sionnest, LR Sita. J Am Chem Soc 118:3319–3320, 1996.
118. T-W Li, I Chao, Y-T Tao. J Phys Chem B 102:2935–2946, 1998.
119. CA Alves, MD Porter. Langmuir 9:3507–3512, 1993.
120. G-Y Liu, P Fenter, CED Chidsey, DF Ogletree, P Eisenberger, M Salmeron. J Chem Phys 101:4301–4306, 1994.
121. K Edinger, A Golzhauser, K Demota, C Woll, M Grunze. Langmuir 9:4–8, 1993.
122. C Schonenberger, JAM Sondag-Huethorst, J Jorritsma, LGJ Fokkink. Langmuir 10:611–614, 1994.
123. PG van Patten, JD Noll, ML Myrick. J Phys Chem B 101:7874–7875, 1997.
124. J Thome, M Himmelhaus, M Zharnikov, M Grunze. Langmuir 14:7435–7449, 1998.
125. M Mrksich, GM Whitesides. Ann Rev Biophys Biomol Struct 25:55–78, 1996.
126. JP Folkers, PE Laibinis, GM Whitesides. J Adhesion Sci Tech 6:1397–1410, 1992.
127. MK Chaudhury, GM Whitesides. Science 255:1230–1232, 1992.
128. GK Jennings, JC Munro, T-H Yong, PE Laibinis. Langmuir 14:6130–6139, 1998.
129. FP Zamborini, RM Crooks. Langmuir 14:3279–3286, 1998.
130. Y Xia, GM Whitesides. Angew Chem Int Ed Engl 37:550–575, 1998.
131. R Maoz, J Sagiv. Langmuir 3:1034–1044, 1987.
132. R Maoz, J Sagiv. Langmuir 3:1045–1051, 1987.
133. N Abbott, A Kumar, GM Whitesides. Chem Mater 6:596–602, 1994.
134. Y Xia, X-M Zhao, E Kim, GM Whitesides. Chem Mater 7:2332–2337, 1995.
135. MA Rampi, OJA Schueller, GM Whitesides. Appl Phys Lett 72:1781–1783, 1998.
136. KL Prime, GM Whitesides. Science 252:1164–1167, 1991.
137. KL Prime, GM Whitesides. J Am Chem Soc 115:10714–10721, 1993.
138. L Deng, M Mrksich, GM Whitesides. J Am Chem Soc 118:5136–5137, 1996.
139. CD Bain, GM Whitesides. J Am Chem Soc 110:6560–6561, 1988.
140. L Bertilsson, B Liedberg. Langmuir 9:141–149, 1993.
141. JP Folkers, PE Laibinis, GM Whitesides, J Deutch. J Phys Chem 98:563–571, 1994.
142. Y Li, J Huang, RTJ McIver, JC Hemminger. J Am Chem Soc 114:2428–2432, 1992.
143. Y Sato, R Yamada, F Mizutani, K Uosaki. Chem Lett 987–988, 1997.
144. WA Hayes, H Kim, X Yue, SS Perry, C Shannon. Langmuir 13:2511–2518, 1997.
145. DG Kurth, T Bein. Angew Chem Int Ed Engl 31:336–338, 1992.
146. DG Kurth, T Bein. Langmuir 9:2965–2973, 1993.
147. JH Moon, JW Shin, SY Kim, JW Park. Langmuir 12:4621–4624, 1996.
148. DA Hutt, GJ Leggett. Langmuir 13:2740–2748, 1997.
149. S Pan, DG Castner, BD Ratner. Langmuir 14:3545–3550, 1998.
150. H-J Himmel, K Weiss, B Jager, O Dannenberger, M Grunze, C Woll. Langmuir 13:4943–4947, 1997.
151. S Lofas, B Johnsson. J Chem Soc Chem Commun 1526–1528, 1990.
152. A Ulman, N Tillman. Langmuir 5:1418–1420, 1989.
153. RV Duevel, RM Corn. Anal Chem 64:337–342, 1992.

154. M Wells, RM Crooks. J Am Chem Soc 118:3988–3989, 1996.
155. GJ Leggett, CJ Roberts, PM Williams, MC Davies, DE Jackson, SJB Tendler. Langmuir 9:2356–2362, 1993.
156. J Wang, JR Kenseth, VW Hones, J-B Green, MT McDermott, MD Porter. J Am Chem Soc 119:12796–12799, 1997.
157. YW Lee, J Reed-Mundell, CN Sukenik, JE Zull. Langmuir 9:3009–3014, 1993.
158. SM Amador, JM Pachence, R Fischetti, JPJ McCauley, ABI Smith, JK Blasie. Langmuir 9:812–817, 1993.
159. P Kohli, KK Taylor, JJ Harris, GJ Blanchard. J Am Chem Soc 120:11962–11968, 1998.
160. N Tillman, A Ulman, JF Elman. Langmuir 5:1020–1026, 1989.
161. RCJ Horton, TM Herne, DC Myles. J Am Chem Soc 119:12980–12981, 1997.
162. N Balachander, CN Sukenik. Langmuir 6:1621–1627, 1990.
163. TS Koloski, CS Dulcey, QJ Haralson, JM Calvert. Langmuir 10:3122–3133, 1994.
164. GE Fryxell, PC Rieke, LL Wood, MH Engelhard, RE Williford, GL Graff, AA Campbell, RJ Wiacek, L Lee, A Halverson. Langmuir 12:5064–5075, 1996.
165. A Benninghoven, B Hagenhoff, E Niehuis. Anal Chem 65:630A–640A, 1993.
166. S Akari, D Horn, H Keller, W Schrepp. Adv Mater 7:549–551, 1995.
167. L Yan, C Marzolin, A Terfort, GM Whitesides. Langmuir 13:6704–6712, 1997.
168. RG Cooks. Chem Ind 142, 1955.
169. L Yan, X-M Zhao, GM Whitesides. J Am Chem Soc 120:6179–6180, 1998.
170. J Rao, L Yan, B Xu, GM Whitesides. J Am Chem Soc 121:2629–2630, 1999.
171. L Yan, WTS Huck, X-M Zhao, GM Whitesides. Langmuir 15:1208–1214, 1999.
172. J Lahiri, L Isaacs, J Tien, GM Whitesides. Anal Chem 71:777–790, 1999.
173. HA Biebuyck, GM Whitesides. Langmuir 10:4581–4587, 1994.
174. NB Larsen, H Biebuyck, E Delamarche, B Michel. J Am Chem Soc 119:3017–3026, 1997.
175. Y Xia, D Qin, GM Whitesides. Adv Mater 8:1015–1017, 1996.
176. R Jackman, J Wilbur, GM Whitesides. Science 269:664–666, 1995.
177. Y Xia, M Mrksich, E Kim, GM Whitesides. J Am Chem Soc 117:9576–9577, 1995.
178. NL Jeon, K Finnie, K Branshaw, RG Nuzzo. Langmuir 13:3382–3391, 1997.
179. PC Hidber, W Helbig, E Kim, GM Whitesides. Langmuir 12:1375–1380, 1996.
180. PC Hidber, PF Nealey, W Helbig, GM Whitesides. Langmuir 12:5209–5215, 1996.
181. A Bernard, E Delamarche, H Schmid, B Michel, HR Bosshard, H Biebuyck. Langmuir 14:2225–2229, 1998.
182. CD James, RC Davis, L Kam, HG Craighead, M Isaacson, JN Turner, W Shain. Langmuir 14:741–744, 1998.
183. GM Whitesides, PE Laibinis. Langmuir 6:87–96, 1990.
184. KR Stewart, GM Whitesides, HP Godfried, IF Silvera. Rev Sci Instr 57:1381–1383, 1986.
185. M Mrksich, GM Whitesides, ed. Using Self-Assembled Monolayers That Present Oligo(ethylene glycol) Groups to Control the Interactions of Proteins with Surfaces. Poly(ethylene glycol) Chemistry and Biological Applications, vol. 680. Washington, DC: American Chemical Society, 1997, pp 361–373.

186. M Nishizawa, M Shibuya, T Sawaguchi, T Matsue, I Uchida. J Phys Chem 95: 9042–9044, 1991.
187. PL Burn, A Kraft, DR Baigent, DDC Bradley, AR Brown, RH Friend, RW Gymer, AB Holmes, RW Jackson. J Am Chem Soc 115:10117–10124, 1993.
188. L Dai, HJ Griesser, X Hong, AWH Mau, TH Spurling, Y Yang, JW White. Macromolecules 29:282–287, 1996.
189. PL Burn, AB Holmes, A Kraft, DDC Bradley, AR Brown, RH Friend, RW Gymer. Nature 356:47–49, 1992.
190. BG Healey, SE Foran, DR Walt. Science 269:1078–1080, 1995.
191. W Knoll, M Matsuzawa, A Offenhausser, J Ruhe. Isr J Chem 36:357–369, 1996.
192. Y Liu, M Zhao, DE Bergbreiter, RM Crooks. J Am Chem Soc 119:8720–8721, 1997.
193. R Langer, JP Vacanti. Science 260:920–926, 1993.
194. SA Sukhishvili, S Granick. Langmuir 13:4935–4938, 1997.
195. Y Liu, ML Bruening, DE Bergbreiter, RM Crooks. Angew Chem Int Ed Engl 36: 2114–2116, 1997.
196. G Decher. Science 277:1232–1237, 1997.
197. PT Hammond, GM Whitesides. Macromolecules 28:7569–7571, 1995.
198. JM Stouffer, TJ McCarthy. Macromolecules 21:1204–1208, 1988.
199. TJ Lenk, VM Hallmark, JF Rabolt, L Haussling, H Ringsdorf. Macromolecules 26:1230–1237, 1993.
200. L Rozsnyai, MS Wrighton. J Am Chem Soc 116:5993–5994, 1994.
201. G Mao, DG Castner, DW Grainger. Chem Mater 9:1741–1750, 1997.
202. SL Clark, MF Montague, PT Hammond. Macromolecules 30:7237–7244, 1997.
203. E Kim, GM Whitesides, LK Lee, SP Smith, M Prentiss. Adv Mater 8:139–142, 1996.
204. A Karim, JF Douglas, BP Lee, SC Glotzer, JA Rogers, RJ Jackman, EJ Aims, GM Whitesides. Phys Rev E: Stat Phys, Plasmas, Fluids, Relat Interdiscip Top 57: R6273–R6276, 1998.
205. M Boltau, S Walheim, M Jurgen, G Krausch, U Steiner. Nature 391:877–879, 1998.
206. S Zhang, L Yan, M Altman, M Lässle, H Nugent, F Frankel, DA Lauffenburger, GM Whitesides, A Rich. Biomaterials 20:1213–1220, 1999.
207. SR Holmes-Farley, RH Reamey, TJ McCarthy, J Deutch, GM Whitesides. Langmuir 1:725–740, 1985.
208. SR Holmes-Farley, GM Whitesides. Langmuir 2:62–76, 1986.
209. SR Holmes-Farley, C Bain, GM Whitesides. Langmuir 4:921–937, 1988.
210. CD Bain, GM Whitesides. J Am Chem Soc 111:7164–7175, 1989.
211. CD Bain, GM Whitesides. Langmuir 5:1370–1378, 1989.
212. J Lahiri, E Ostuni, GM Whitesides. Langmuir 15:2055–2060, 1999.

11

Architecture and Applications of Films Based on Surfactants and Polymers

Masatsugu Shimomura
Hokkaido University, Sapporo, Japan

I. INTRODUCTION

A large interest is growing in a biomimetic approach for molecular engineering of advanced materials for molecular electronics, photonics devices, sensor devices, and so on. Biological membranes are basic functional "molecular devices" in biological systems and multicomponent complex systems mainly consisting of lipid bilayers and biological macromolecules. The lipid bilayer is a nanoscopic two-dimensional layered supramolecular assembly and a structural element of biomembranes. In turn, biomembranes are elemental structural components of mesoscopic submicrometer-size subcellular apparatuses, i.e., organellas. Biological cells are micrometer-size assemblies of the organellas, and basal building units of biological tissues and living organs at the macroscopic scale. Structural hierarchical assembling of biomolecules from nanoscopic to macroscopic scale in biological systems is a vital and constitutional base of living organs.

Surfactant bilayers are nanoscopic layered supramolecular structures composed of synthetic surfactant molecules such as the double-chain ammonium amphiphile [1] which is spontaneously assembled in water. A large variety of bilayer-forming amphiphiles that are not directly related to biolipids has been synthesized [2–5]. Structural characteristics of surfactant bilayers such as two-dimensional molecular ordering, thermal phase transition, and controllable two-dimensional molecular distribution, are in common with those of biological membranes. Immobilization of aqueous surfactant bilayers with preservation of the structural characteristics is imperatively required if the surfactant bilayers are utilized as functional advanced materials. Bilayer-forming amphiphiles can form self-standing thin films having multilayered lamella structure by simple casting on solid surfaces [6]. Composites with polymer compounds result in improved mechanical stability and processability of immobilized bilayer films [7].

Another method for the fabrication of thin organic film is the Langmuir–Blodgett (LB) technique that is one of the best known methods for the layer-by-layer architecture of two-dimensional molecular nanostructures with single or multicomponent systems [8–11].* From the viewpoint of materials fabrication, practical durability (e.g., mechanical, thermal, and chemical stability) of the LB films is indispensably required. A large number of polymeric LB films have been prepared [12,13] to improve the stability of organic thin films. Alternating multilayers of polymer and amphiphile with stoichiometric ion pairing can be deposited onto solid substrates by the conventional LB technique [14].

The polyion complex technique [15] was proposed as a convenient preparative method to immobilize water-soluble, bilayer-forming amphiphiles and polymers as composite thin films. A water-insoluble polyion complex is precipitated when the aqueous solution of the charged bilayer membrane is mixed with a water solution of the countercharged polyelectrolyte. The polyion complex monolayer also can be formed in situ when the charged amphiphiles are directly spread on the surface of the aqueous polyelectrolyte solution. The principle of the polyion complex LB technique has been expanded into a general and simple alternative deposition method of charged polymers by Decher [16] (this volume chapter 12).

The first half of this chapter (Sections II–IV) briefly describes molecular design, fabrication, and application of two-dimensional supramolecular nanostructures, i.e., immoblized bilayer films, polymeric LB films, and alternatively deposted polymer multilayers, which are the best-suited engineered assemblies employed in the first step of the biomimetic approach for material design and fabrication.

The second step of the biomimetic approach to material engineering is the organization of the nanoscopic assemblies up to mesoscopic superstructured organizations. Large numbers of mesoscopic organizations of polymer assemblies have been found as micro- and macrophase separation structures of block copolymers and polymer blends, respectively [17] (this volume, chapters 6 and 7). The main driving force for mesoscopic structuring in polymer films is the incompatibility of each block of the copolymers or of blended polymers. The surface energy of substrates is another factor affecting the microstructuring when the polymer

* Special issues for Langmuir–Blodgett films are listed in Ref. 11.

films are thinning in the mesoscopic scale (<100 nm). Recently Steiner and coworkers demonstrated controlled mesoscopic structuring of polymer blends in a cast thin film [18], where the microphase separation was induced by a prepatterned substrate surface modified by the microcontact printing technique proposed by Xia and Whitesides [19]. Microcontact printing is an efficient method for pattern transfer when using poly(dimethylsiloxane) (PDMS) as an elastomeric stamp, and self-assembled monolayers of thiole derivatives (SAMs) as ink materials, respectively. Patterned modification of the substrate surface with SAMs can regulate the local surface energy and then polymer wettability of the substrate for polymer casting.

Aksay and coworkers [20] produced mesoscopic patterning of oriented nanostructured silica thin films polymerized by a surfactant-templated sol-gel technique [21] in combination with a micromolding technique, which is another microfabrication technique without photolithography proposed by Xia and Whitesides [19]. A network pattern of microcapillaries (submicrometer scale) was transferred to an elastomeric PDMS stamp as a microreplica molding. An aqueous mixture of tetraethoxysilane and a cationic surfactant (CTAC: cetyltrimethylammonium chloride) was introduced into the microcapillaries. After hydrolysis of tetraethoxysilane at the cationic interface of the tubular surfactant assemblies, a mesoscopic supramolecular structure hierarchically constituted from hexagonally packed nanoscopic tubules of silica was formed in the microscopic capillary.

The hierarchical structuring of supramolecular assembling observed in biological systems is often dynamic self-organization processes with a large energy consumption. As a general physical phenomenon, we know another type of dynamic self-organization: the so-called *dissipative structure* which is generated under chemical or physical conditions far from equilibrium. Many dynamic structures of self-organization with various spatial scales, so-called spatiotemporal patterns, are known to be formed in the dissipative processes from mesoscopic size of submicrons to macroscopic size. Several types of regular patterns (spirals in the Belousov–Zhabotinsky reaction systems, honeycomb and stripes in Rayleigh–Bénard convections, striped Turing patterns in polymer gel reactors [22], fingering patterns at the liquid interface [23], granular flow [24], polymer transport [25], and hexagons in oscillating granular layers [26] are reported to be formed during dissipative processes. While we do not claim that the mechanism of formation of dissipative structures is the mechanism actually controlling the hierarchical structuring of biological systems, we present here a study of these dynamical structures for model systems and point out their possible applications.

The dissipative structure is essentially a dynamic structure requiring continual energy dissipation, but dynamic structures can be frozen as stationary stable structures. The casting process of polymer solutions on solid surfaces is complex enough to form dissipative structures. Thus, some spatiotemporal polymer patterns are expected to be immobilized after rapid solvent evaporation. Section V of this chapter describes the freezing of dissipative structures generated in the casting process of polymer solutions.

II. CRYSTAL ENGINEERING OF IMMOBILIZED BILAYER MEMBRANES

A. Design of Molecular Orientation in Two-Dimensional Supramolecular Nanostructure

Bilayer membranes are two-dimensional supramolecular assemblies that are spontaneously formed in water and are representatively known as biomembranes and liposomes in biological science. Since Kunitake's report of the formation of the bilayer membrane from a simple double longchain ammonium salt **1** (see Scheme 1) in 1977 [1], a large number of synthetic amphiphiles have been reported as bilayer-forming surfactant molecules. Singlechain surfactants having aromatic π-electron groups, e.g., biphenyl [27], anthracene [28], carbazole [29], etc. were found to form bilayer membranes and the relation between chemical structure and aggregation morphology was systematically investigated [30,31].

For the use of the nanostructured molecular assemblies as novel functional materials, the aqueous solutions of bilayer membranes need to be immobilized

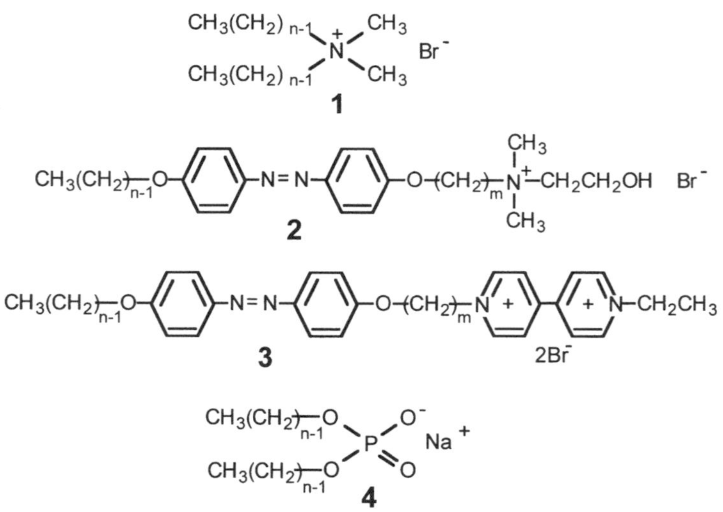

Scheme 1

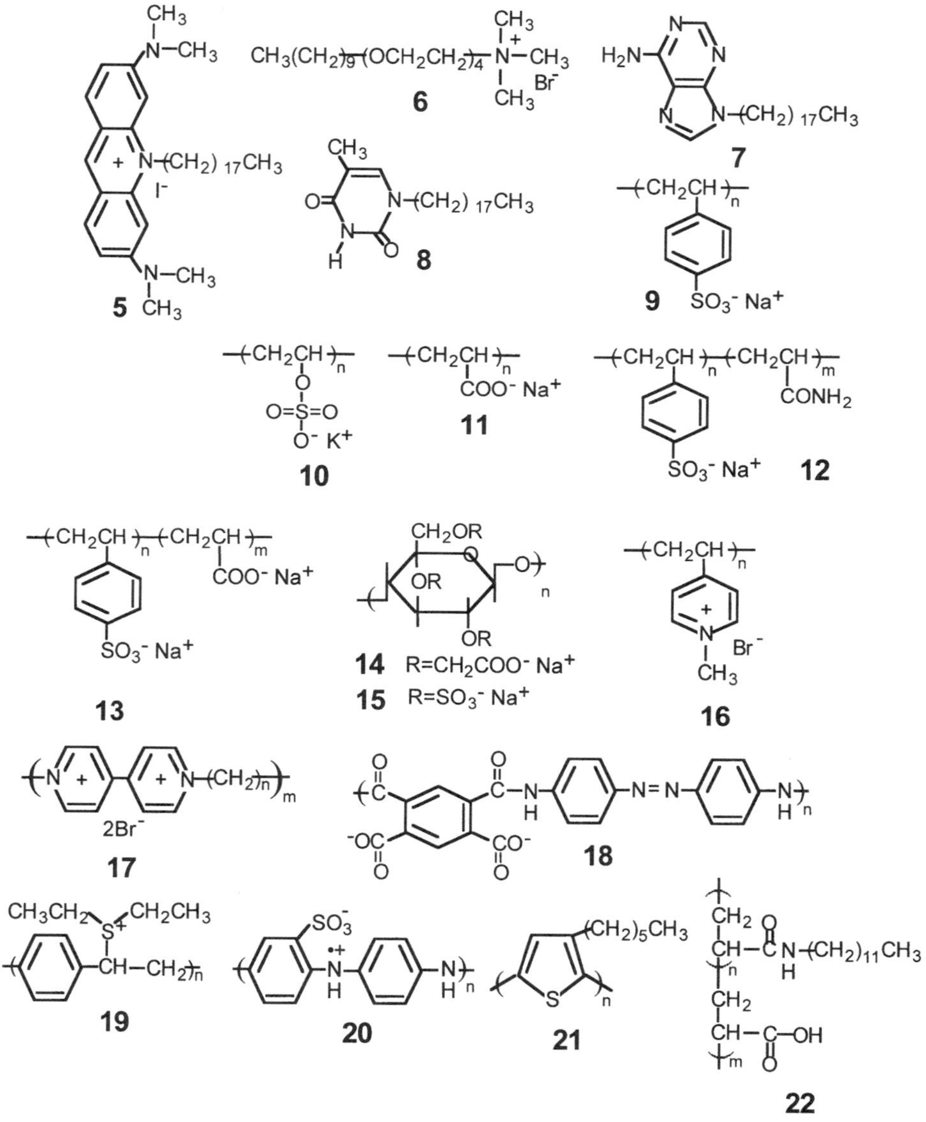

14 $R = CH_2COO^- Na^+$

15 $R = SO_3^- Na^+$

Scheme 1 (continued)

in the macroscopic solid state for easy handling. The casting method is a simple way to immobilize bilayer membranes on a solid support by solvent evaporation from an aqueous solution. As shown in Fig. 1a, upon casting from the aqueous bilayer membrane of the azobenzene amphiphile **2**, a self-standing transparent macroscopic film of several micrometers thickness is easily peeled off the solid substrate [32]. The scanning electron micrograph (SEM) shown in Fig. 1c and X-ray diffraction studies strongly indicate that the cast film is composed of highly oriented and multiply stacked lamellae of bilayer membranes parallel to the film surface. The repeating period calculated from the X-ray diffraction patterns is consistent with the bilayer thickness. The X-ray diffraction patterns of the cast film are found to be almost the same as those of the single crystals whose structures have been completely determined by Okuyama and Shimomura et al. [33,34].

A strong intermolecular interaction among π-electron chromophores is observed as Davydov splitting in the absorption spectrum. Kasha proposed a ''molecular exciton theory'' for the strong π–π interaction in molecular aggregates, e.g., molecular crystals, and the spectral shift due to the Davydov splitting can be described as a function of chromophore orientation [35,36]. Due to the two-

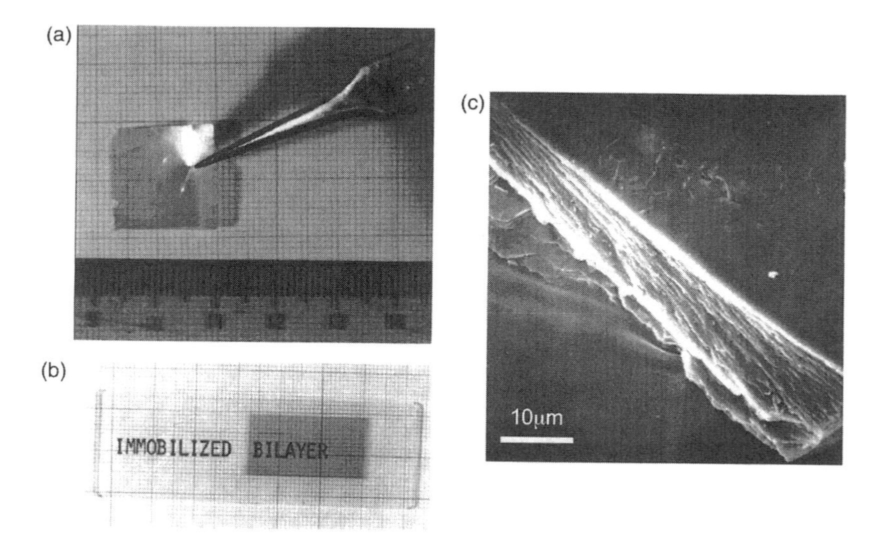

Figure 1 Cast films of azobenzene bilayer membranes from water solutions: (a) a self-standing film peeled off a casting substrate; (b) an optical transparent film prepared on a quartz plate; (c) SEM image of film cross-section.

dimensional stacking of the aromatic moiety in the aqueous bilayer assemblies, the absorption spectrum of the azobenzene chromophore is strongly affected by the intermolecular π–π interaction [37,38]. The azobenzene chromophore has two strong absorption bands attributed to π–π^* electronic transitions in the UV-visible region and a broad weak band of n–π^* transitions at around 450 nm. The two π–π^* bands located around 250 nm and 355 nm (for the molecularly dispersed chromophore in ethanol) are attributed to a transition dipole moment along the short and long axes of the azobenzene chromophore, respectively.

An extensive spectral variation is observed for the long axis transition when the azobenzene amphiphiles **2** are dissolved in water as bilayer membranes. As a result of a systematic investigation we have determined the relation between the chemical structure and absorption spectrum [38]. The absorption maximum is strongly dependent on the alkyl chain length. Homologous series of $m = 5$ show a large bathochromic shift to around 400 nm. Hypsochromic shift to 300 nm is found when the difference of the two alkyl chains ($m - n$) is larger than two (Fig. 2a). The Davydov splitting in the absorption spectrum of the azobenzene bilayer membranes strongly suggests the formation of Frenkel excitons [36] which are attributed to a strong intermolecular coupling among the excited states of the aromatic π-electron groups. According to the semiquantitative calculation of molecular exciton theory of Kasha [35], the blue and the red shifts relative to the isolated azobenzene chromophore are ascribed to side-by-side and head-to-tail chromophore orientation in the bilayer membrane, respectively. An X-ray structural analysis of the single crystal confirms that the spectral estimation of the molecular packing is in fact correct (Fig. 2b) [33,34].

Optically transparent films were prepared on quartz plates by simple casting of the aqueous bilayer solutions (Fig. 1b). The absorption spectrum of the cast film is found to be strongly affected by the chemical structure of the amphiphile. From systematic investigation of spectral measurement and X-ray structural analysis of solvent cast films of 43 azobenzene amphiphiles, it can be concluded that the alkyl chain length of the bilayer-forming amphiphile is a definitive determining factor for the molecular orientation in the cast film [39]. Based on the systematic X-ray structural analysis of the single crystals of the azobenzene amphiphiles, the orientational variety in the cast film can be described as diversity of molecular packing in the two-dimensional bilayer assemblies [40]. Since the cross-sectional area of the hydrophilic head (A_1) of the azobenzene amphiphile is about twice as large as that of the aromatic moiety (A_2), two types of the most densely molecular packing models are plausible for bilayer assemblage: tilt molecular orientation with tilting angle α satisfying $A_2 = A_1 \sin \alpha$, and interdigitated orientation of two molecules to balance $A_1 = 2A_2$. Head-to-tail and side-by-side chromophore orientation are, respectively, accomplished in these two molecular packing models.

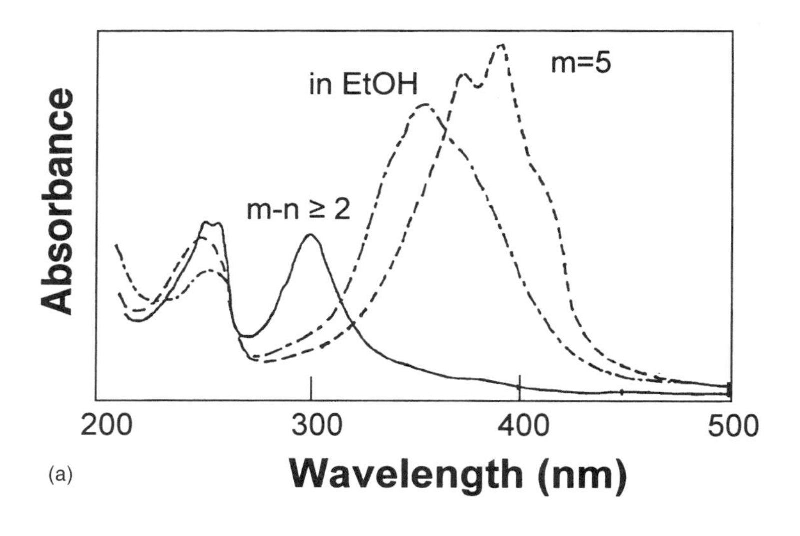

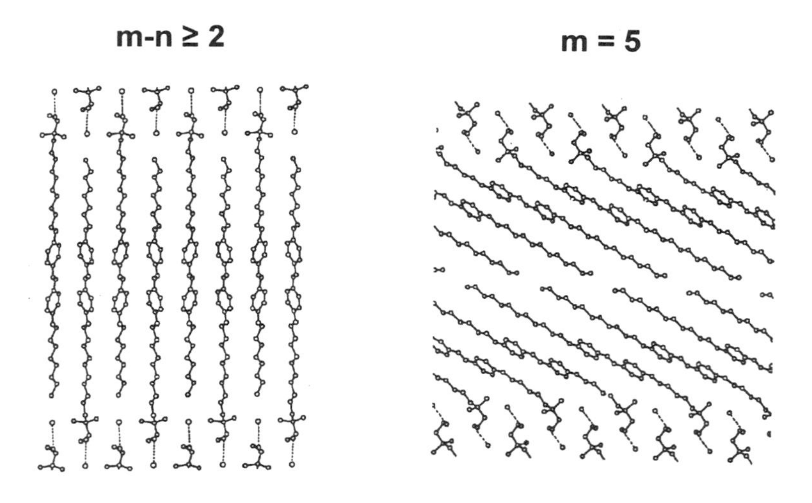

Figure 2 Davydov splitting and crystal structures of azobenzene **2** bilayer assemblies. (a) UV-vis absorption spectra of aqueous bilayer solutions and ethanol solution. Similar spectral splitting is observed in cast films. (b) Schematic illustration of molecular orientation in single crystals.

B. Spectral Switching of Bilayer Films Based on Structural Polymorphism

Reversible transformation between two molecular packing modes in a cast bilayer film was observed as a peculiar solid–solid phase transition with a large thermal hysteresis when the azobenzene bilayer membrane **2** ($n = 8$, $m = 10$) was cast from water solutions [41]. The cast film showed two endothermic peaks in differential scanning calorimetry at 115°C and 177°C, corresponding to a solid-to-liquid crystal and a liquid crystal-to-isotropic phase transition, respectively. Absorption maximum moved from 300 nm, typical for the side-by-side chromophore orientation, to 360 nm on heating above the first phase transition temperature. On cooling from the liquid-crystalline state to room temperature, a new absorption spectrum similar to the head-to-tail chromophore orientation was observed (Fig. 3a). Spectral shift from 370 nm to 360 nm occurred at 60°C in the second heating. The X-ray diffraction patterns of two room-temperature films, states (1) and (3) in Fig. 3, were completely different. An interesting finding is an isothermal phase transition, coupled with spectral shift, induced by moistening of the cast film. The spectral state of the annealed film (370 nm) was very stable in dry atmospheric conditions but immediately shifted to 300 nm in humid conditions at room temperature. The absorption spectrum and X-ray diffraction pattern of the moistened film were identical to those of the as-cast film. Relaxation from the metastable solid state with the head-to-tail molecular packing [state (3)] to the stable solid state with the side-by-side orientation [state (1)] is accelerated by the moisture treatment. Water molecules bound to the hydrophilic ammonium head appear to act as plasticizers of the cast film. Repeated cycles of the thermal transition and of the isothermal moisture treatment produced reversible switching of the spectral shift and molecular packing (Fig. 3b) [42].

A large spectral change similar to that of the moisture-induced isothermal phase transition was found when the annealed film was photoirradiated at room temperature in a dry atmospheric condition [43]. A large absorption band located at 370 nm attributed to a π–π^* transition of the *trans*-isomer decreased with UV light irradiation concurrently with an increase of 300 nm absorption, whose spectral shape was quite different from that of the *cis*-isomer (λmax: 320 nm). The absorption band at 450 nm attributed to the n–π^* transition of the *cis*-isomer increased during the initial stage of the photoirradiation and then decreased gradually upon further irradiation. Eventually the absorption spectrum of the UV-irradiated film was identical to that of the moisture-treated film.

Generally in an ethanol solution, visible light irradiation (n–π^* excitation) accelerates a reverse reaction from *cis* to *trans*. Alternating irradiation by UV and visible light in solution gives a reversible periodic response of the absorption spectrum. In the cast film, the 370 nm absorption was gradually reduced by the alternating UV and visible irradiation.

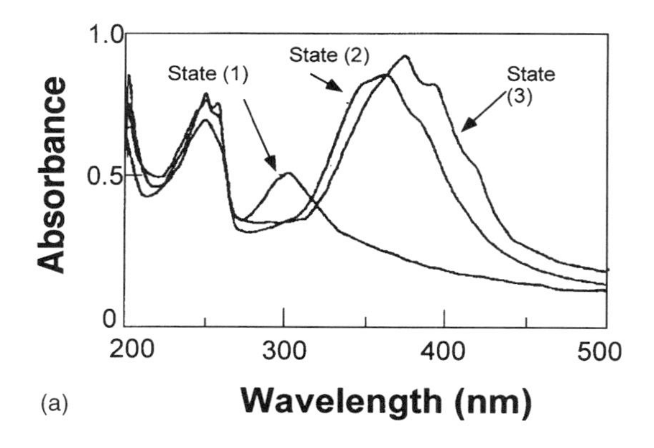

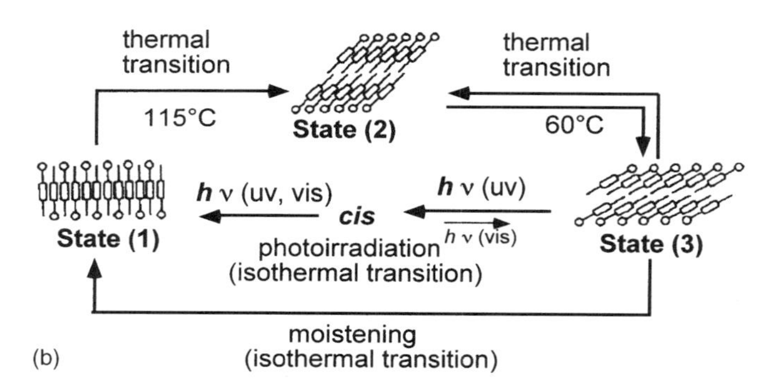

Figure 3 Spectral switching of the cast azobenzene bilayer film by thermal and isothermal phase transition based on structural polymorphism of two-dimensional molecular packing. (a) Spectral state (1) of as-cast film moves to state (2) upon heating above the solid-to-liquid crystal phase transition (115°C). Cooling to room temperature gives state (3) and state (2) is regenerated in the second heating (60°C). State (1) appears again by an isothermal moisture treatment of state (3) as well as by UV-light irradiation of state (3). (b) Schematic illustration of structural change.

The spectral change due to the UV light irradiation is ascribable to the isothermal phase transition triggered by the photochemical isomerization reaction. The *trans*-isomer in the metastable solid phase with the head-to-tail chromophore orientation is converted to the *cis*-isomer by the UV light irradiation. The backward reaction from *cis* to *trans* can also partly proceed during the UV light irradiation. When the cast film is irradiated below the solid-to-liquid crystal phase transition temperature, the *trans*-isomer is regenerated by the backward reaction from the stable solid state with the side-by-side chromophore orientation showing the 300 nm absorption maximum. Due to large steric hindrance in the well-packed side-by-side chromophore orientation the *trans*-isomer is never converted to the *cis*-isomer on prolonged UV-light irradiation.

Reversible spectral switching produced by the alternating combination of heating and UV-irradiation, as well as by the coupling of heating and moistening, can be successfully repeated many times. Since the spectral change induced by the peculiar phase transition based on the polymorphism of the cast bilayer film is in a ''one-way'' direction, the nanostructured bilayer assembly is applicable to an erasable optical memory material with spectral bistability. These results are successful examples of ''crystal engineering'' that enable structural prediction and functional design of two-dimensional molecular assemblies to be used potentially as novel engineered materials.

III. PREPARATION OF POLYMERIC LAYER-BY-LAYER FILMS BY POLYION COMPLEX TECHNIQUE

A. Immobilization of Bilayer Assemblies as Polymer Composites

Polymerization of the bilayer films is another way of immobilization for material fabrication. Free-standing films are easily prepared when cast films of bilayer membranes containing suitable reactive groups are polymerized. A free-standing film prepared by photo polymerization of cast films of diacetylene amphiphiles was reported by O'Brien and Kuo [44].

Blending with polymers is another way of preparing polymer film. Bilayer membranes can be immobilized as polymer composites by each of the following methods.

1. Porous polymers, e.g., a membrane filter or a nylon capsule [45], are soaked in organic solvents containing bilayer-forming amphiphiles and then dried.
2. Bilayer-forming amphiphiles are cast with hydrophobic polymers, e.g., poly(vinylchloride), from organic solvents. Large clusters of the bilayer membrane are formed as phase separated micro domains in the polymer matrix [46].
3. Bilayer membranes are cast with water-soluble polymers from an aqueous solution. Poly(vinylalcohol) is a suitable inert matrix for supporting bilayer membranes [47]. The water-solubility of the composite film with poly(vinylalcohol) can be lowered by coating with cellulose acetate [48] and crosslinking by irradiation [49].

Polyion complex technique is a unique method of supramolecular polymerization of bilayer membranes without conventional polymerization procedures [14,50]. Water-insoluble polyion complexes are precipitated when the aqueous solution of the charged bilayer membrane is mixed with a water solution of the countercharged polyelectrolyte. The polyion complexes can be formed as thin films by usual casting from organic solutions. The fundamental bilayer structure and characteristics are essentially maintained in the immobilized cast films of the polyion complexes. X-ray diffraction of the solvent cast polyion complex films reveals the layered structure with repeating spacing corresponding to the bilayer thickness [50–52].

The advantage of the polyion complex technique is in the combination of the characteristics of polymers and amphiphiles. Chemical structure of polyelectrolytes used as counterions are selected not only from synthetic vinyl polymer derivatives such as poly(styrene sulfonate) **9**, poly(vinyl sulfate) **10**, poly(acrylic acid) **11**, etc., but also from biopolymers such as DNA, polyamino acids, and polysaccharides. The bilayer characteristics occasionally can be affected by electrostatic interaction with polymers. It is known that the bilayer characteristics are often disturbed when the bilayer membrane is chemically or supramolecularly linked to some polymers. Ringsdorf and coworkers showed that introduction of the hydrophilic spacer groups into the polymer mainchain effectively reduces the steric disturbance of the polymer chain on ordered molecular packing of bilayer assemblies [53]. Increased flexibility of the polymer chain of the copolymer might allow the higher ordering of the alkyl chain packing of the bilayer-forming amphiphiles [54].

The absorption maximum of the cast film of the azobenzene bilayer membrane **2** ($n = 8$, $m = 10$) complexed with poly(styrene sulfonate) is located at 323 nm. The slight red shift from the original cast film (300 nm) without polyelectrolytes suggests that the chromophore packing in the polyion complex film is loosened due to the steric influence of the polymer mainchain. The long spacing of the complex film estimated from X-ray diffraction data is slightly larger than the original film but the layer correlation is poor. The thermal phase transition coupled with spectral change was also found in cast films of the polyion complex. The absorption spectrum of the annealed complex film (λmax : 347 nm) is different from that of the original film (360 nm; see Fig. 3a). When acrylarmide groups are introduced as uncharged spacer segments into the mainchain of poly(styrene sulfonate), i.e., copolymer **12**, the complex film shows a spectral resemblance to the original film [55]. The layer correlation of the complex film with the copolymer **12** is better than that of the complex film with the homopolymer **9**.

The peculiar isothermal phase transition with a large spectral shift induced by the moisture treatment and the photochemical isomerization is observed in the cast films of the polyion complexes, too. Stability of the annealed state against humidity and the speed of the photoinduced phase transition are extremely dependent on the chemical structure of the polyanions [43]. These findings indicate that the chemical structure of the polymeric counterpart can be one of the indispensable tuning factors of the molecular assemblage in the nanostructured molecular systems. Molecular ordering and spectral properties could be regulated by pH change when the azobenzene bilayer membranes are complexed with poly(acrylic acid) or copolymers of styrene sulfonate and acrylic acid **13** [7].

B. Langmuir–Blodgett Films Prepared by Polyion Complex Technique

The polyion complexes of bilayer-forming amphiphiles and polyelectrolytes can form the regularly ordered two-dimensional layered structures at the air–water interface as well as in the solvent cast films. The polyion complex monolayer can be formed directly on the surface of the aqueous polyelectrolyte solution when the bilayer-forming amphiphiles are spread from organic solution [15].

Due to the high water solubility of a divalent cationic head-group, a viologen amphiphile **3** (n = 12, m = 5) [56] could not form a stable monolayer on a pure water subphase. A monolayer is stabilized when an anionic polyelectrolyte is added to the water subphase. Pressure-area (π–A) isotherms of the complex monolayers are shown in Fig. 4. The shape of the pressure-area isotherms is strongly dependent upon the chemical structure of the anionic polyelectrolyte. A condensed monolayer was formed over a highly diluted solution (ca. 0.06 mM)

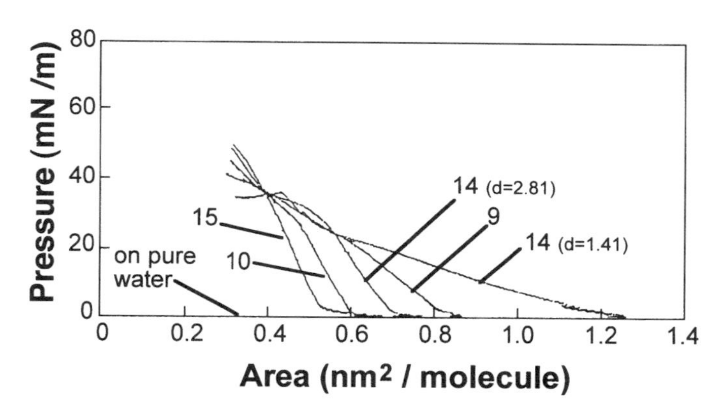

Figure 4 Pressure-area isotherms of polyion complex monolayers of a viologen amphiphile **3** with the anionic polyelectrolyte at 20°C. d is the degree of substitution of the carboxymethyl group.

of potassium poly(vinyl sulfate). Poly(styrene sulfonate) in the subphase caused the monolayer to expand more. The π–A isotherm on carboxymethylcellulose **14** shown in Fig. 4 was strongly dependent on the degree of substitution (d) of carboxymethyl group per one glucose unit. The π–A isotherm of the monolayer complexed with a higher substituted polymer (d = 2.81) was more condensed than that of lower substituted polymers (d = 1.41). Highly sulfated cellulose polymer **15** also gave a condensed monolayer. Molecular orientation in the polyion complex monolayer is strongly affected by the charge density of the polyelectrolyte [57–59].

Alternatively deposited multilayers of polymer and amphiphile can be prepared over solid substrates by the conventional LB technique [15,60]. The deposition pattern of monolayer over substrates is strongly dependent not only upon the deposition pressure but also upon the chemical structure of the polymer. Three types of deposition are found when a solid substrate is immersed in and pulled up from the water subphase by the vertical dipping method. The deposition pattern known as Y-type deposition is obtained when the monolayer is deposited during both down and up strokes of the substrate. Z- and X-type depositions are obtained when the monolayer is deposited only during the up and down strokes, respectively. Complex monolayers of the viologen amphiphile with polymer **14** (d = 1.41) were deposited as a Y-type LB film at 20 mNm^{-1} and as a Z-type film at 10 mNm^{-1}. The thermal stability of the polyion complex LB film is larger than that of the LB film formed from the corresponding amphiphile without polymer [61]. The X-ray diffraction from (1,0,0) layers of the LB film of an anionic amphiphile **4** (n = 16) disappeared above 80°C, in good agreement with the bulk melting point of the amphiphile. Layered structure of the complex LB film with a cationic polyelectrolyte **16** was stable up to 160°C.

C. Stoichiometric Ion Pairing in Polyion Complex Assemblies

Stoichiometric ion pairing between the charged amphiphile and polyelectrolyte at the air–water interface was observed by thickness measurement using the surface plasmon polariton (SPP), resonance experiment [62], and reflection spectroscopy [63]. Adsorption kinetics of polymer **9** to the surface monolayer of amphiphile **1** (n = 18), was directly observed as SPP resonance. The reflectivity sharply decreases at the reflection angle where the SPP at the interface between the metal (silver) and the monolayer is excited by P-polarized light of the He–Ne laser. The shift of the resonance angle is described as a product of changes of thickness and refraction indices of the thin layer film at the interface. Therefore, the adsorption process can be detected as a change of the thickness of the adsorbed polymer. Miyano et al. [62] found that the adsorption proceeds in three steps. Polymers begin to adsorb to the monolayer, forming a uniform sheet with thickness of

0.6 nm that corresponds to the thickness of the poly(styrene sulfonate) chain. Subsequently the polymers attach to the surface already covered with polymer sheet and then the thickness increases to 1.8 nm at the final stage. Only the first layer of the adsorbed polymer is transferred with the amphiphile monolayer because the elemental analysis performed by X-ray photoelectron spectroscopy (XPS) clearly shows the stoichiometric ion pairing in the deposited polyion complex films prepared by the conventional LB technique [15].

Stoichiometric ion pairing leads to an ordering of the polymer at the two-dimensional charged interface of the amphiphile bilayer or monolayer. Peculiar odd–even effects of the alkyl chain length were found in the cast polyion complex film of the anionic amphiphile **4** (n = 12) and cationic ionen polymers having a viologen group **17** [64]. The viologen groups are covalently connected through methylene segments. Strongly colored cation radicals were formed by a chemical or an electrochemical reduction of the polyion complex films. Figure 5a shows the electrochemical generation of viologen radicals in the complex film of the

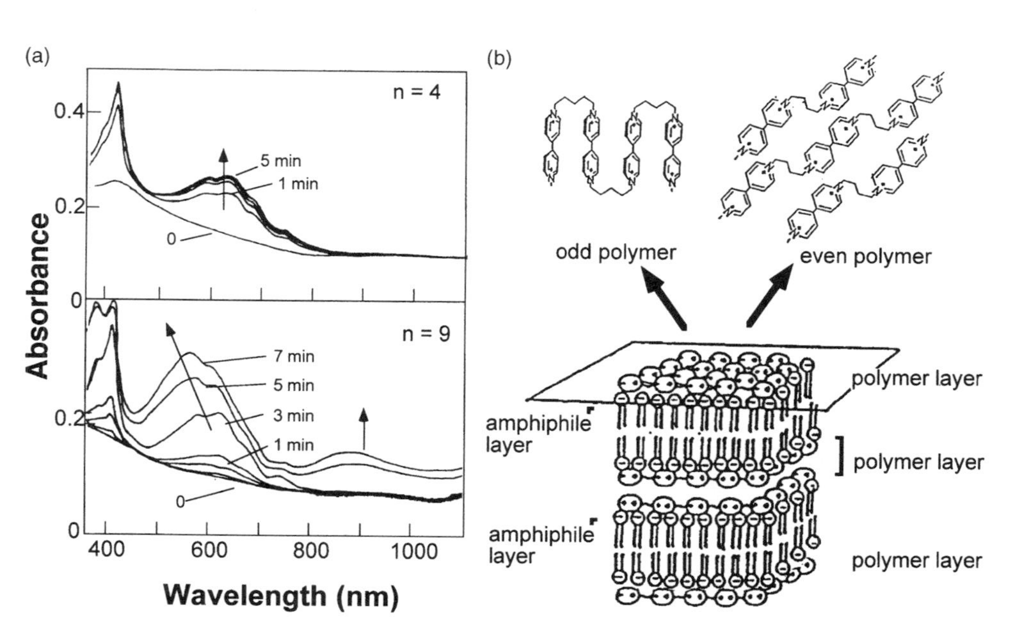

Figure 5 Electrochemical cation radical formation in the polyion complex films of the viologen polymers **17** and anionic amphiphile **4** cast on ITO electrodes: (a) spectral changes of even and odd polymers during electrochemical reduction; (b) schematic illustrations of the organization of the polymer chain intercalated into the amphiphile bilayers. Upper figure describes in-plane view of the polymer layer.

even polymer **17** (n = 4) immobilized on an ITO electrode at the applied voltage of −0.6 V vs. SCE (saturated calomel electrode). In contrast, the spectral change of the odd polymer is not as simple. At the beginning of the reduction, the absorption maximum was located at 610 nm. During the course of reduction, the absorption maximum shifted to 560 nm and a broad absorption at 900 nm appeared indicating the dimer radical formation. The formal redox potential for the first redox reaction of viologen, measured by a cyclic voltammetry (CV) of the complex film cast on a graphite electrode, also showed an odd–even effect. Diffusion coefficients of the electrochemical electron transfer in the odd polymers were smaller than those in the even polymers, suggesting that the dimer radical in the odd polymer acts as a trapping site of the hopping electrons. If a gauche-$(trans)_{n-3}$-gauche conformation of the methylene chain segment is fixed at the anionic charged interface, an intramolecular association of the neighboring viologen groups within the linear polymer is possible in the odd polymer, but is suppressed in the even polymer (Fig. 5b).

D. Potential Applications of Engineered Polymer Assemblies Prepared by Polyion Complex Technique

The polyion complex LB film of polyamid acid **18** and amphiphilic amines is a precursor of a polyimide LB film. Elimination of the amphiphiles concomitant with imidization reaction of the polyamid acid occurs in gently heated acetic acid and pyridine solution [65]. Pinhole-free polyimide films with a monolayer thickness of approximately 0.4 nm were applicable as insulating layers of MIM (metal–insulator–metal) devices [66], rubbing-free aligning layers of liquid crystal cells [67], thermally stable polymer thin films for electronic applications [68], and electroluminescence devices [69]. Thermal elimination of sulfonium groups from polyion complex LB film of anionic amphiphile **4** (n = 12) and sulfonium polymer **19**, gave poly (*p*-phenylenevinylene) thin film [70,71] and the third harmonic generation (THG) of the film was reported by Saito et al. [72]. Electronically conducting polymers having charged moiety can be prepared as layered thin films by the polyion complex LB technique. Rubner et al. have prepared the polyion complex monolayer of stearylamine and the π-conjugated polymer sulfonated polyanilin **20** [73]. Photochemical isomerization of azobenzene chromophores can be applicable for molecular switching or information storage devices. Complex formation with polyelectrolytes at the interface can provide free volume available to the *trans-cis* isomerization of azobenzene moiety [74,75]. Stroeve and his group deposited polyion complex monolayers of arachidic acid and poly(ethyleneimine) onto a microporous support membrane for preparing gas-separation films [76].

IV. TWO-DIMENSIONAL DNA ASSEMBLIES AS NOVEL FUNCTIONAL MATERIALS

Double-helical DNA is a supramolecular architecture composed of complementary base-pairings of adenine–thymine and cytosine–guanine based on specific hydrogen bonding. Stacking interaction of π-electrons at the ground states of the base-pairs is observed as a spectral hypochromic effect (reduction of the molecular extinction coefficient) in the UV-vis absorption spectrum. The hypochromism is ascribed to weak intermolecular interaction, since the average distance between two base-pairs is about 0.34 nm in B-form DNA. Recently some articles have claimed that DNA can act as a π-electron medium for the photoinduced electron transfer because of its close stacking of base-pairs [77–79]. The mechanism for the DNA-mediated long-range electron transfer is, however, now under heated discussion [80,81].

Aiming toward the functional applications of DNA as novel molecular devices, polynucleic acids have been immobilized as two-dimensional molecular assemblies by means of the specific intermolecular interaction at the air–water interface [82,83] as well as by the polyion complex technique [84]. We describe here the effect of DNA on photoinduced electron transfer in polyion complex LB monolayers deposited on an ITO electrode [85], or cast in polyion complex films [86].

A. DNA Monolayers Complexed with Cationic Amphiphiles at the Air–Water Interface

The polyion complex technique is a useful method for assembling water-soluble polyelectrolytes on opposite charged surfaces of monolayers at the air–water interface. DNA was assembled with countercharged amphiphiles and fixed as a cast-oriented film on solid substrates by Okahata et al. [84]. DNA molecules are expected to be arranged in an orderly manner maintaining the double-helical structure as counterparts of cationic monolayers at the air–water interface (Fig. 6). Some cationic dyes, so-called intercalators, can be incorporated into base-pairs of the double-stranded DNA with high affinities. A new type of amphiphilic intercalator, 10-octadecyl acridine orange iodide **5**, which can form a monolayer at the air–water interface was synthesized to confirm the double-helical structure of DNA complexed with the monolayer at the air–water interface. The amphiphilic intercalator was prepared by a quaternarization reaction of acridine orange with alkyliodide [82].

The interaction between cationic amphiphiles and anionic polymers at the air–water interface is reflected in the π–A isotherms of the monolayer. The π–A isotherm of the amphiphilic intercalator on a DNA subphase was found to be

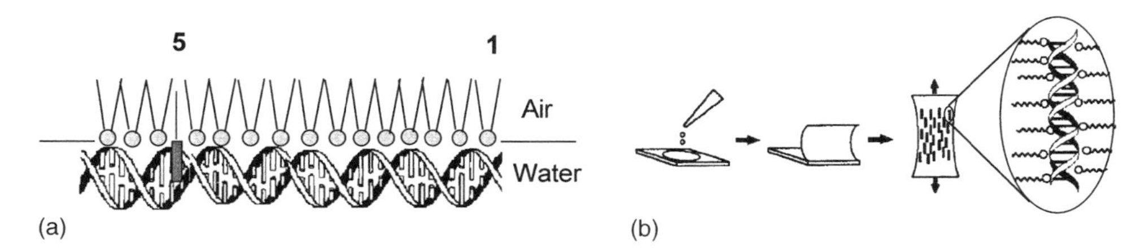

Figure 6 Schematic illustrations of DNA-amphiphile complexes (a) at the air–water interface; (b) in the stretched cast film.

more expanded and with a higher collapse pressure than that on a pure water subphase. Due to the hydrophobic microenvironment effect of the stacked base-pairs in the double-helical DNA, fluorescence intensity of the intercalator was remarkably enhanced. Green fluorescence (535 nm) attributed to monomeric acridine fluorophore was strongly enhanced when the intercalator monolayer was spread on the DNA subphase. Thus the DNA molecule complexed with the cationic monolayer kept the double-helical structure.

B. DNA Enhanced Electron Transfer from Acridine Monolayer to ITO Electrode

We have investigated the effect of DNA on the fluorescence lifetime of a binary component monolayer of **5** and **1** ($n = 18$). Fluorescence emission spectrum and lifetime were measured by a fiber optics spectrometer and a pico-second pulse laser system with a streak camera through a fluorescence microscope equipped with a Langmuir trough. Due to the intercalation of the acridine moiety into the base stacking, fluorescence intensity and lifetime measured on the DNA subphase are larger and longer than those on the subphase of polymer **14**, which was used as a control polymer without nucleo base. Strong fluorescence quenching was, however, found when polyG·polyC was dissolved in a water subphase. Static quenching of acridine fluorescence, probably due to the electron transfer from guanine base, occurred because the fluorescence lifetime was not as reduced as the fluorescence intensity.

The photoinduced electron transfer to an ITO electrode was measured by an electrochemical method. The binary component monolayers complexed with anionic polymers were deposited on ITO electrodes, and the electrode was soaked in an aqueous 0.1 M KCl solution containing EDTA as a sacrificial electron donor. As shown in Fig. 7, the polyion complex monolayer deposited on the ITO electrode generated photocurrent with visible light irradiation (>450 nm). Anodic photosensitized current of the amphiphilic acridine dye **5** is enhanced by complex

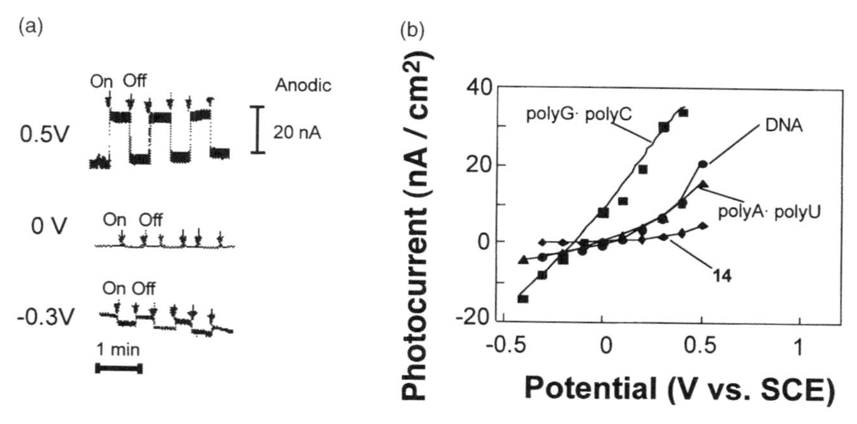

Figure 7 Photocurrent generation of polyion complexed DNA monolayers deposited on ITO electrodes: (a) photoswitching of current generation at 0.5, 0, and −0.3 V vs. SCE by visible light irradiation of the DNA complex; (b) polymer dependence of I–V curves of complex monolayers. Polymer **14** is used as control.

formation with polynucleic acids. A long-lived excited state of the acridine monolayer complexed with DNA can transfer its photoexcited electron more effectively than that complexed with polymer **14**. PolyG · polyC, whose guanine base is assumed to be oxidized by the excited state of the acridine moiety, is most effective for photocurrent generation (Fig. 7b). If the one-dimensional stacking of the guanine bases along the polymer backbone can transport a hole (the cation radical of guanine which is generated after the oxidative electron transfer between the excited state of the acridine donor) the double-stranded guanine polynucleotide could act as a π-electron medium.

C. Anisotropic Electric Conductivity in Cast Polyion Complex Film

A self-standing polyion complex film of DNA and the singlechain cationic amphiphile **6** was stretched two to three times in length by hand at room temperature. An X-ray diffraction and a polarized optical experiment strongly suggest that the DNA molecules are aligned with a distance of 4.1 nm along the stretching direction (Fig. 6) [83]. The stacking period of the base-pairs in the stretched film is similar to that of the B-form DNA in a water solution. The direct current conductivity of the stretched film put on a comb-shaped electrode plate was measured at room temperature under vacuo (0.1 mmHg). A large ohmic current was gener-

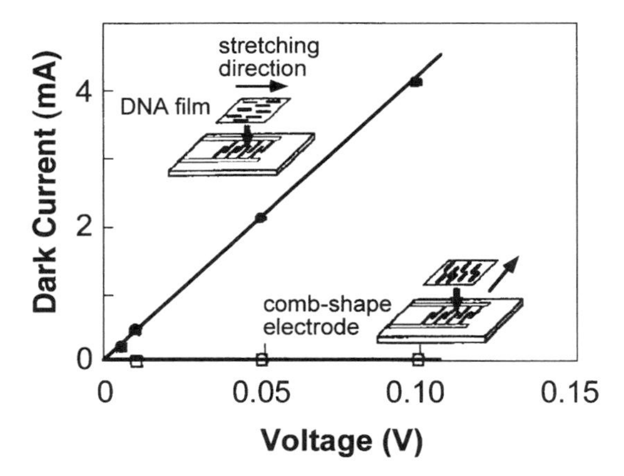

Figure 8 Anisotropic electric conduction in the stretched DNA film complexed with cationic amphiphiles. A schematic illustration of the stretched film is shown in Fig. 6(b).

ated with increasing applied potential. Figure 8 clearly shows a large anisotropy of the dark current along the stretched direction. A large anisotropic response was also found in the photocurrent generation when a small amount of acridine orange was doped into the polyion complex film and photoirradiated by visible light.

D. Formation of DNA Mimetics at the Air–Water Interface Based on Hydrogen Bonding Interaction

Complementary base-pairings of adenine–thymine and cytosine–guanine with specific hydrogen bonding are essential for double-helix formation of DNA. Many efforts have been made to prepare artificial base-pairs at the air–water interface [87]. To simplify mechanism discussions of the photoinduced electron transfer in the stacked base pairs we have attempted cutting stacked base-pairs out of DNA helices. We have already succeeded in preparing two-dimensional DNA mimetics by using molecular-recognition-directed self-assembly of octadesyl cytosine and guanine nucleosides based on specific hydrogen bonding at the air–water interface [88]. From an in situ fluorescence imaging experiment [89,90] of the surface monolayer, we have concluded that our DNA mimetics were composed of stacked cytosine–guanine base-pairs because the amphiphilic acridine intercalator **5** was incorporated into the crystalline domains of the nucleobase monolayer.

Octadecyladenine **7** and octadecylthymine **8** were prepared by alkylation of adenine and thymine with octadecyliodide in the presence of sodium hydride in dry dimethylformamide. Figure 9a shows the π–A isotherms of **7** and **8** on a pure water subphase. Two amphiphiles were premixed at various molar ratios in chloroform solution. In the case of a single-component monolayer of **7** or **8**, the π–A isotherms indicate formation of expanded monolayers. On the other hand, the isotherms indicate formation of a stable and condensed monolayer by mixing. A phase-diagram of the molecular area and surface pressure shows that a 1 : 1 mixture gives the minimum area. The molecular area of the 1 : 1 mixture, estimated to be 0.35 nm^2/molecule from the isotherm, is consistent with the averaged edge area of purine and pymidine rings. Thickness of the mixed monolayer determined by an X-ray diffraction experiment (2.73 nm) is consistent with the Watson–Crick base pairing at the air–water interface (Fig. 9c).

The triple helix is another type of superstructure of DNA [91]. A-T-T (A-T-U) and G-C-C base-trimers are formed by combination of Watson–Crick and Hoogsteen-type hydrogen bonding. Base-trimers are formed in the DNA-mimetic system at the air–water interface between the Watson–Crick-type monolayer and

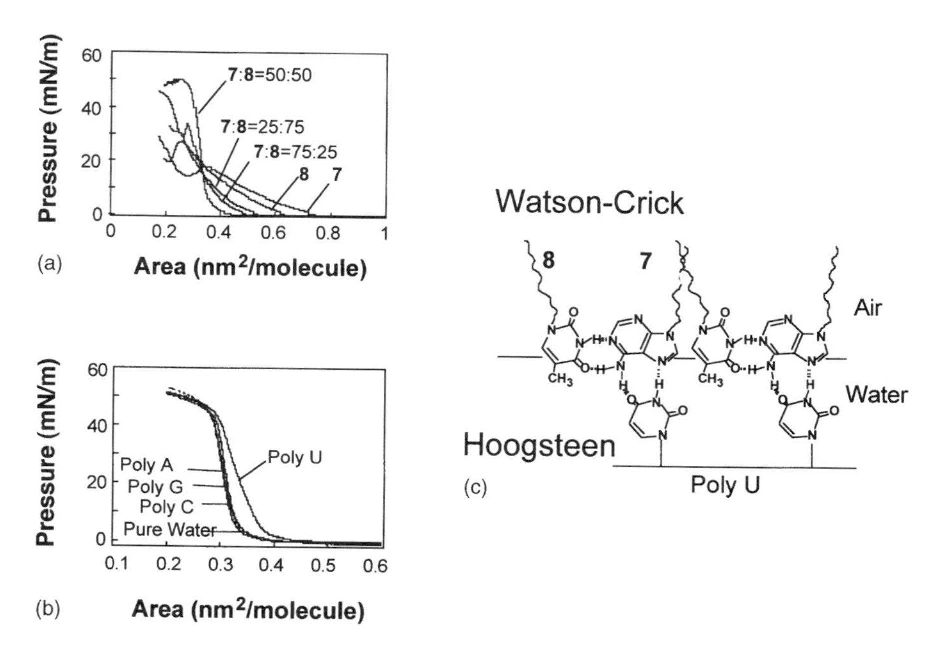

Figure 9 Pressure-area isotherms of nucleobase amphiphile monolayers: (a) mixed monolayers spread on pure water subphase; (b) 1:1 mixture spread on various polynucleotide subphases; (c) schematic illustration of triplex formation at the air–water interface.

a single-strand polynucleotide. Figure 9b shows π–A isotherms of the 1:1 mixture monolayer on various polynucleotide subphases. It clearly indicates that only single-stranded poly U, which is known to form A-T-U base-triplex by a Hoogsteen-type hydrogen bond, can change the isotherm. A new type of layer-by-layer hydrogen-bonding assemblies of polynucleic acid can be constructed when the base-triplex monolayer schematically shown in Fig. 9c is deposited by the conventional LB technique.

V. MESOSCOPIC PATTERN FORMATION BY DISSIPATIVE STRUCTURES

A. Mesoscopic Pattern Formation in Simply Cast Polyion Complex: Under Dry Conditions

In order to prepare macroscopically homogeneous films suitable for the optical experiments described in the previous section, fairly concentrated chloroform solutions of polyion complexes (>1 wt%) are normally employed for casting. As shown in Fig. 1c, film thickness is in the micrometer range. It is well known that owing to dewetting of polymer from substrates [92], homogeneous thin polymer films (thickness <100 nm) hardly can be prepared by simple solvent casting [93,94]. After dewetting on substrates thin polymer films are often ruptured to form irregular network patterns similar to a Voroni pattern [95]. Homogeneity of cast polymer films is strongly affected by many physical and chemical factors, e.g., concentration of the casting solution, temperature, solvent type, atmospheric condition of casting chamber, surface properties of casting substrate, etc.

An in situ observation of the casting and dewetting processes from highly diluted polymer solutions (a few milligrams per liter) by fluorescence microscopy was carried out in a glove box under controlled atmospheric humidity [96]. To visualize the evaporation process by fluorescence images a small amount of an amphiphilic cationic fluorescence probe, octadecyl rhodamin B, was added as a counterion of polymer **9** as well as the bilayer-forming amphiphile **1** (n = 18). To visualize the convection flow in the thin liquid film of the polymer solution, fine talc particles (ca. 10 μm in diameter) were also added. Several microliters of the polymer solution were spread homogeneously as a thin liquid film (ca. 1 cm^2 area) on the freshly cleaved mica surfaces.

Mainly due to solvent evaporation and surface tension, the solution front is always directed toward the center of the liquid film. At the first stage of evaporation the solution front recedes smoothly; then after a short while it starts receding intermittently, with jump. The jumping instability with a stick and slip motion of the receding front is ascribable to the local gelation effect of a polymer at the three-phase line (liquid–substrate–air boundary) where polymer concentration is assumed to be higher than the bulk polymer solution. The local gelation prevents the front from receding. After deposition of the condensed polymer on the solid surface the front can recede again until the next gelation. The condensed polymer regions left behind the jumping front are observed as periodic lines perpendicular to the receding direction.

Condensation of the polymer at the three-phase line can be observed directly by a fluorescence microscope. The solution front is clearly observed as a bright red boundary line between the dark substrate and the bulk solution. An interesting finding is a periodic polymer condensation along the three-phase line that is observed as fingering instability similar to the ''tear of Wine'' [97] (Fig. 10a). The fingering inhomogeneity of the polymer concentration at the three-phase line can be attributed to the Marangoni effect [97] which is visualized as fast lateral migration and coarsening of the fingers along the solution front. The fingers often start to grow up as lines perpendicular to the three-phase line. Figure

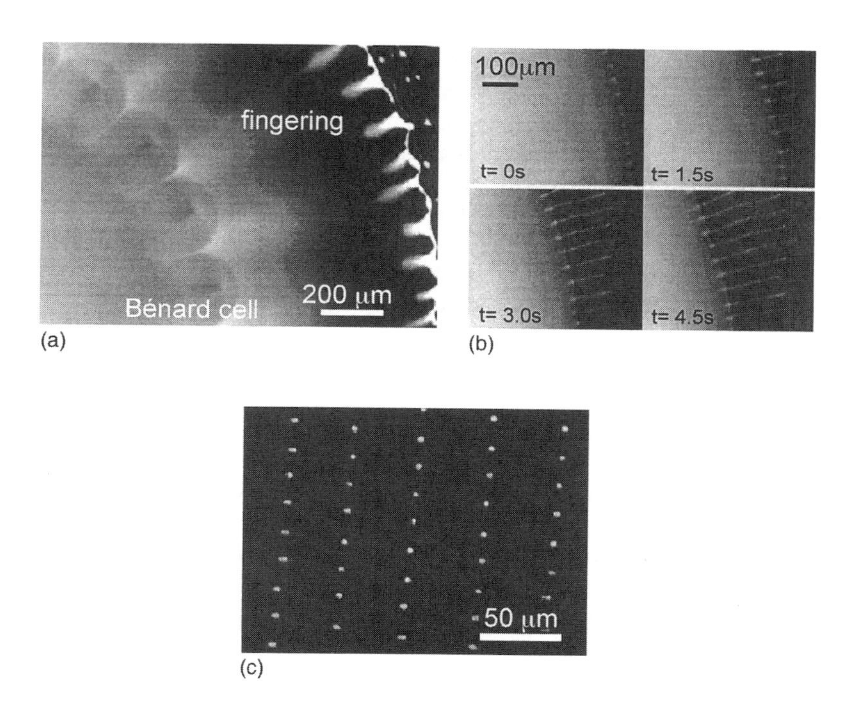

Figure 10 Dissipative structures formed in casting polymer solution: (a) Bénard cells and fingering instability formed in chloroform solution of polyion complex of **1** and **11**; (b) snapshots of stripe formation from the fingering instability at the three-phase line synchronized with receding front; (c) regularly arrayed polymer dots formed by dewetting in the stripes.

10b shows snapshots of the line formation from the periodically generated fingers. The fingers are straightened as regular stripes concomitant with the smooth receding of the solution front.

The stripes grow to several millimeters in length and the line width of each stripe measured by an atomic force microscope is several hundred nanometers. A minimum value of thickness, approximately 3 nm, corresponds to the bilayer thickness of the polyion complex deduced from X-ray diffraction. Dimensions of the mesoscopic stripe patterns are altered by casting conditions. The larger the concentration and molecular weight of polymers, the larger are the dimensions of the stripe. Regularly arranged polymer dots shown in Fig. 10c are occasionally formed when dewetting of the polymer occurs in the regularly arrayed stripe lines.

As shown in Fig. 10a several circular domains that resemble Rayleigh–Bénard convection cells are formed in the central part of the polymer solution. Local convection in the circular domains is observed as very fast movement of the talc particles which are trapped and turning around in the individual cells. At the first stage of casting the lifetime of the convection cells is very short, in the subsecond range, and new cells are continuously supplied from the center of the casting solution. With increasing viscosity of the solution due to solvent evaporation the lifetimes of cells are prolonged. A bright stream connecting the cell and the fingers reveals another convectional flow. The talc experiment clearly shows the radial convection between the three-phase line and the solution center. The particles keep going to the solution front and back to the cell boundary. The periodic fingering instability at the three-phase line is assumed to be generated by the radial convectional streams originated from the regular arrangement of the Bénard-type convection cells.

B. Mesoscopic Pattern Formation in Simply Cast Polyion Complex: Under Wet Conditions

In the final stage of casting, macroscopically homogeneous and optically transparent polymer films, so-called cast films, are formed under the dry atmospheric condition when polymer concentration is high enough to prevent dewetting. However, under humid casting conditions, the final films are neither homogeneous nor optically transparent. Figure 11 shows a fluorescence and an atomic force micrograph of a cast film prepared under high humidity. The cast film has a regular honeycomb pattern with a size of a few micrometers per each cell.

Based on the in situ microscopic observation of the casting process under high humidity, the formation mechanism of the mesoscopic honeycomb structure is schematically summarized in Fig. 12 [98]. After spreading the chloroform solution of the polyion complex on the substrate, the solvent starts to evaporate. Due

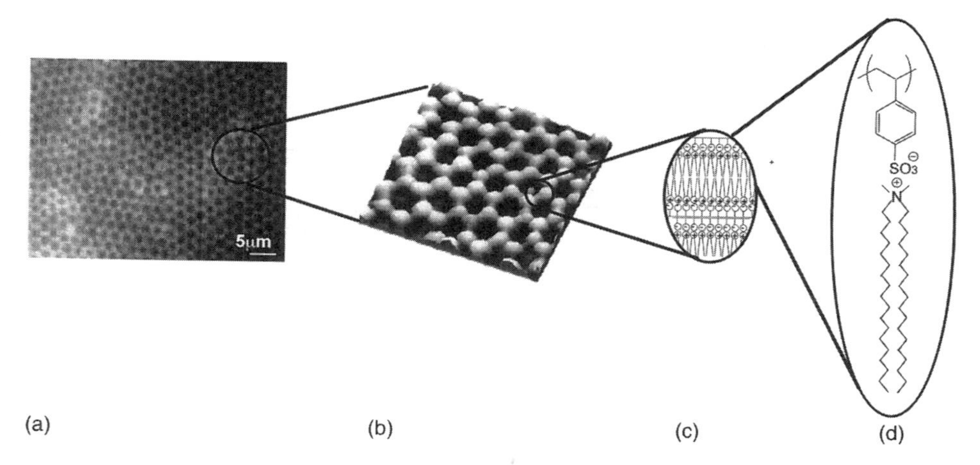

Figure 11 Honeycomb-patterned film prepared under high atmospheric humidity: (a) fluorescence micrograph of honeycomb-patterned polyion complex film; (b) atomic force micrograph; (c) schematic illustration of nanoscopic layer-by-layer assembly; (d) molecular structure of polyion complex.

to the heat of evaporation micron-size water droplets are condensed onto the solution surface. Hexagonally packed droplets are transported to the three-phase line by the convectional flow or the capillary force generated at the solution front. Since the surface tension between water and chloroform is reduced by the polyion complex of the bilayer-forming surfactant, the water droplets are stabilized against fusional coalescence. Upon evaporation of chloroform the solution front moves over the hexagonal array of water droplets. After water evaporation, regularly ordered honeycomb patterns of polymer assemblies are formed on the substrates [98,99]. Concentration and relative humidity of the casting condition are definitively controlling factors of the honeycomb size. The polymer rim surrounding the honeycomb hole becomes thinner with decreasing concentration and the hole size becomes larger with increasing humidity.

Bragg peaks with 3.6 nm spacing, that is, well corresponding to the bilayer thickness, observed in the X-ray diffraction pattern of the honeycomb film strongly suggest that the mesoscopic honeycomb patterns are constituted from nanostructured lamella assemblies of the polyion complexes. UV-visible absorption spectra of the patterned films (honeycomb, dots, and stripes) prepared from the polyion complexes of the azobenzene amphiphiles are identical to those of the macroscopically homogeneous cast films described in the previous section. It is concluded that the dissipative structures (Bénard cell, fingering instability)

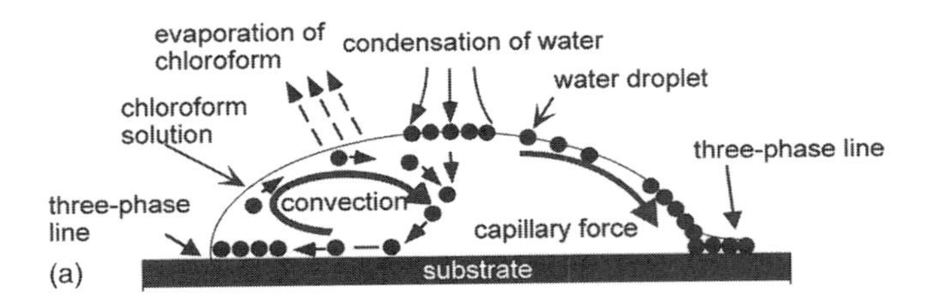

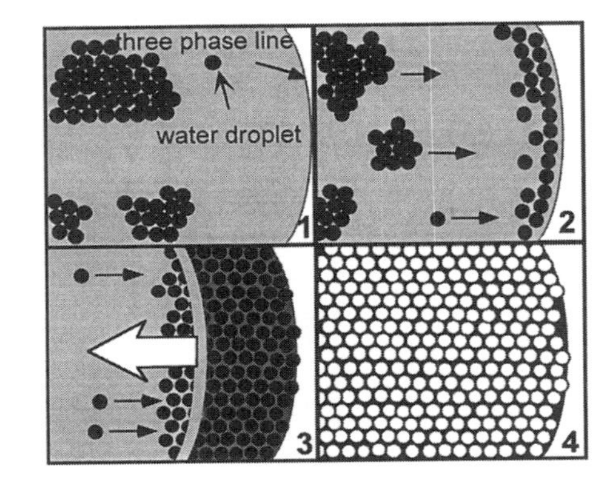

Figure 12 Schematic illustrations of the formation mechanism of the honeycomb pattern: (a) cross-sectional view of the casting solution; (b) idealized snapshots of a top view.

generated in casting solutions are frozen on solid surfaces as the mesoscopic patterns representing hierarchical structures of the nanoscopic bilayer assemblies (Fig. 11).

C. Photoconductivity Measurement of Single Supramolecular Assembly of Polythiophene

Polyhexylthiophene **21**, one of well-known conducting polymers, can form mesoscopic two-dimensional patterns on solid surfaces by also using the simple casting method. The mesoscopic stripe prepared from chloroform solution is expected to be a supramolecular cord because several hundred polymer chains are tied up in a bundle. To measure the photoconductivity of the single bundle one side of the stripe is contacted with an Ag-paste electrode and another side is contacted

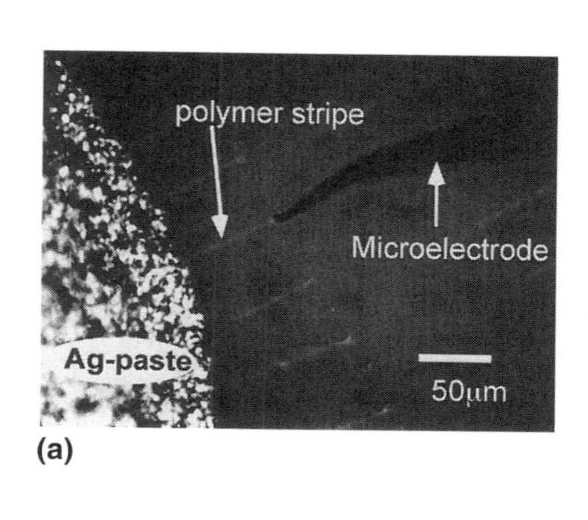

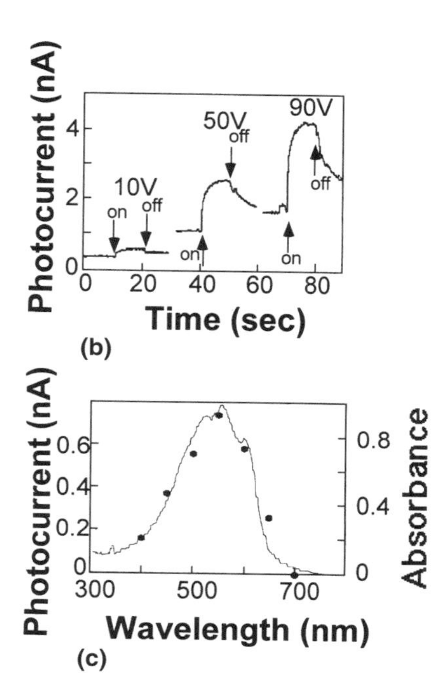

Figure 13 Photocurrent measurement of a polymer bundle prepared on a mica surface: (a) optical micrograph of the experimental setup for photoconduction measurement of a single stripe of polymer **21**; (b) apparent photocurrent generation at various applied voltages irradiated with green light (510 ~ 540 nm); (c) photocurrent action spectrum of a single polymer bundle.

with a microelectrode with the help of a micromanipulator (Fig. 13a). When the line was irradiated with 510 ~ 550 nm light, photocurrent was generated [100]. Figure 13b shows the action spectrum of the photocurrent of a single stripe line. If the polymer chain is spontaneously oriented along the long axis of the finger growth by the mesoscopic cavity effect, size effects of the mesoscopic line width on the polythiophene conductivity are expected.

D. Preparation of Two-Dimensionally Ordered Gold Pattern by Using Mesoscopic Polymer Patterns as Templates

A copolymer of acrylic acid and N-dodecyl acrylamide [13,101] **22** is surface active enough to stabilize water droplets formed on the cast polymer solution.

A regularly ordered honeycomb pattern of the acrylamide polymer was formed on an ITO electrode. After ion coating this structure with a thin gold film, the organic polymer part was dissolved in chloroform and then the substrate was washed by sonication. An optical microscopic observation indicates that the gold film is deposited into the center voids of the honeycomb because the light transmission through the honeycomb pattern in the micrographs turned completely reverse after gold deposition.

If the polymer is not completely removed, the residual polymer is expected to act as an insulating cover layer of the ITO substrate. Figure 14a shows an atomic force image of the polymer–gold hybrid structure after the incomplete removal of the polymer. Mapping of electric conduction measured by a modified AFM apparatus clearly shows that micrometer-scale conducting gold dots are arrayed in an orderly fashion in the insulating copolymer thin film matrix (Fig. 14b). By using the self-organized honeycomb patterns of the amphiphilic copolymer as a template, micropatterns of gold dots can be fabricated without photolithography [100].

We are currently investigating the ability of the honeycomb-patterned film of polymer **22** and its derivatives as artificial extracellular matrices for biological cell culture [102].

E. Immobilization of Dissipative Structure Based on DNA

By using the simple casting method we prepared mesoscopic stripe patterns of a DNA–amphiphile complex cast from chloroform solutions on a mica substrate

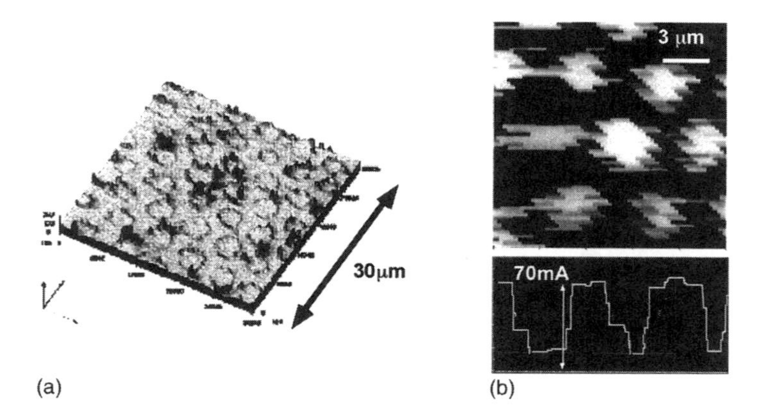

Figure 14 Two-dimensionally oriented gold dot arrays templated by the polymer honeycomb pattern: (a) AFM image of polymer–gold hybrid; (b) electrical current mapping and profile of gold dot prepared on an ITO substrate by a modified AFM apparatus.

[103]. Complexation with amphiphilic counterions was indispensable to the solubilization of DNA in chloroform. Here we show other methods of DNA patterning from water solutions without using any amphiphilic counterion. DNA patterns were prepared from water solutions by the following methods: (1) an aqueous DNA solution mixed with an aqueous $CaCl_2$ solution was cast onto a freshly cleaved mica surface and dried in air at room temperature; and (2) an aqueous DNA solution (containing rhodamine-B as a fluorescence probe) mixed with/without an aqueous alginic acid solution was dropped onto a fleshly cleaved mica surface and dried by heating to approximately 80°C. An AFM imaging of a DNA–Ca^{2+} complex film, prepared by method (1), in propanole shows formation of network structures of DNA (Fig. 15a). Periodicity is about 2~3 nm and almost comparable to the diameter of DNA. Figure 15b shows a fluorescence image of DNA film prepared by method (2). Mesoscopic lines parallel to the receding direction of the solvent were observed in the cast film. When alginic acid was added in the aqueous DNA solution, more regularly aligned stripe patterns were obtained (Fig. 15c). Height of those lines was below 10 nm.

VI. CONCLUDING REMARKS

The layer-by-layer assemblies of amphiphiles, polymers, low molecular weight compounds, and inorganic compounds are the structural elements of novel engineered materials. Casting is the simplest manner of material fabrication. Details of the organization in the cast film, e.g., orientation, film thickness, etc., are, however, not under precise control. Spatiotemporal dissipative structures occurring in casting solutions are frozen on solid surfaces as regular mesoscopic polymer assemblies with the hierarchical structures of the nanoscopic bilayer assemblies. The latter finding is likely applicable as a novel general method of mesoscopic structure formation without lithographic procedures because the formation of dissipative structures is a general physical phenomenon for any polymeric material. To fabricate the mesoscopic structure with good reproducibility, we are now preparing regular stripes under the controlled continuous front receding [104] under specified conditions (dipping speed, temperature, solvent evaporation, etc.). Conductivity measurements of patterned mesoscopic DNA structures, and single molecule detection of DNA embedded in the mesoscopic polymer pattern are now in progress.

The polyion complex LB technique is the most precise fabrication method of layer-by-layer polymer assemblies because the lateral molecular ordering and layer structure (X-, Y-, and Z-type deposition and heterogeneous deposition) of the polymer films can be controlled. The molecular assembly in polyion complex monolayers can be switched by pH change [105] or photochemical reaction [106]. Film uniformity of the polyion complex monolayer was improved by in situ ther-

(a)

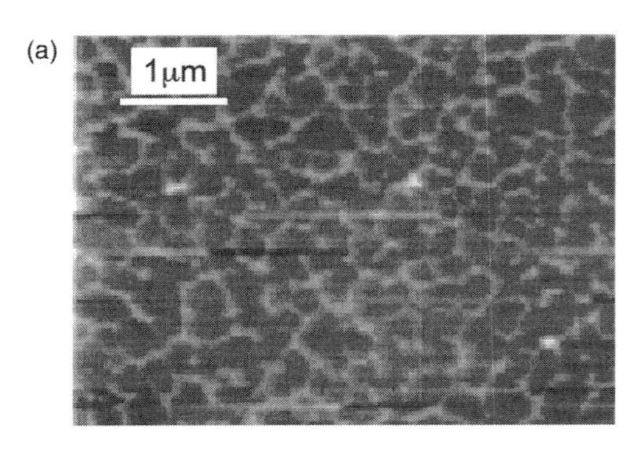

(b)

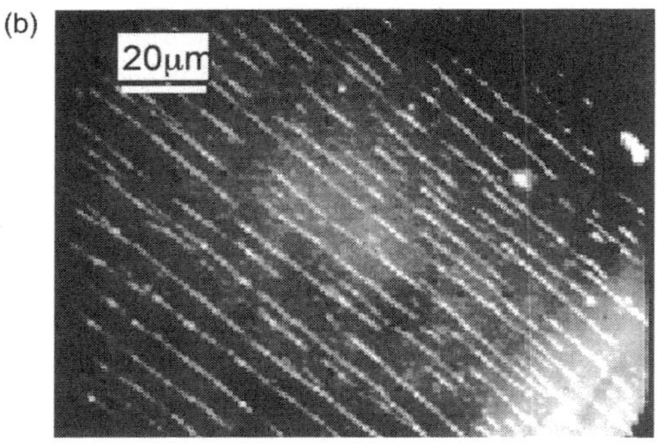

(c)

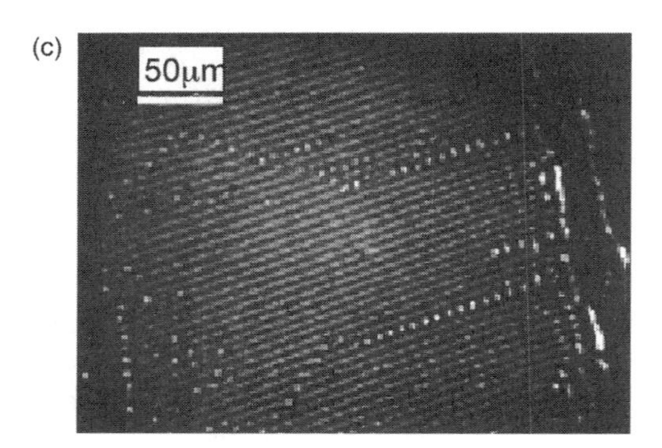

Figure 15 Mesoscopic pattern of DNA assemblies: (a) AFM image of DNA–Ca^{2+} complex film prepared by method (1) (see text); (b) fluorescence image of DNA stripe cast on mica from water solution; (c) fluorescence image of DNA–alginic acid complex cast from water solution.

mal treatment [107,108]. The crystallization process at the air–water interface can be monitored by in situ imaging by a fluorescence microscope [90] or a Brewster angle microscope [109,110]. Since the production of the LB films has been simplified by Albrecht et al. [111], the polyion complex LB films are applicable for molecular electronics and photonics devices, as well as sensor devices based on the engineered assemblies with layer-by-layer structures.

REFERENCES

1. T Kunitake, Y Okahata. J Am Chem Soc 99:3860–3861, 1977.
2. T Kunitake. Angew Chem Int Ed Engle 31:709–726, 1992.
3. T Kunitake. In: Comprehensive Supramolecular Chemistry vol. 9. Oxford: Pergamon, 1996, pp 351–406.
4. J Fendler. Membrane Mimetic Chemistry. New York: Wiley, 1982.
5. JH Fuhrhop, J Köning. Membranes and Molecular Assemblies. Cambridge: Cambridge University Press, 1994.
6. T Kunitake, M Shimomura, A Harada, K Okuyama, T Kajiyama, M Takayanagi. Thin Solid Films 121:L89–L91, 1984.
7. M Shimomura. Prog Polym Sci 18:295–339, 1993.
8. G Roberts. Langmuir–Blodgett Films. New York: Plenum, 1990.
9. A Ulman. An Introduction to Ultrathin Organic Films from Langmuir–Blodgett to Self-Assembly. New York: Academic, 1991.
10. MC Petty. Langmuir–Blodgett Films. An Introduction. Cambridge: Cambridge University Press, 1996.
11. Thin Solid Films 68:1980; 99:1983; 132–134:1985; 159–160:1988; 178–180: 1989; 210–211:1992; 242–244:1994; 284–285:1996; 327–329:1998.
12. H Ringsdorf, B Schlarb, J Venzmer. Angew Chem Int Ed Engl 27:113–158, 1988.
13. T Miyashita. Prog Polym Sci 18:263–294, 1993.
14. M Shimomura, T Kunitake. Thin Solid Films 132:243–248, 1985.
15. T Kunitake, A Tsuge, N Nakashima. Chem Lett 1783–1786, 1984.
16. G Decher. Science 277:1232–1237, 1997.
17. H Hasagawa, T Hashimoto. In Comprehensive Polymer Science. London: Pergamon, 1996.
18. M Böltau, S Walheim, J Mlynek, G Krausch, U Steiner. Nature 391:877–879, 1998.
19. Y Xia, GM Whitesides. Angew Chem Int Ed 37:550–575, 1998.
20. M Trau, N Yao, E Kim, Y Xia, GM Whitesides, IA Aksay. Nature 390:674–676, 1997.
21. IA Aksay, M Trau, S Manne, I Honma, N Yao, L Zhou, P Fenter, PM Eisenberger, SM Gruner. Science 273:892–898, 1996.
22. Q Quyang, H Swinney. Nature 352:610–612, 1991.
23. AM Cazabat, F Heslot, SM Troian, P Carles. Nature 346:824–826, 1990.
24. O Pauliquen, J Delour, SB Savage. Nature 386:816–817, 1997.

25. BN Preston, TC Laurent, WD Comper, GJ Checkley. Nature 287:499–503, 1980.
26. F Melo, PB Umbanhowar, HL Swinney. Phys Rev Lett 75:3838–3841, 1995.
27. Y Okahata, T Kunitake. Ber Bunsenges Phys Chem 84:550–556, 1980.
28. M Shimomura, H Hashimoto, T Kunitake. Chem Lett 1285–1288, 1982.
29. T Kunitake, M Shimomura, Y Hashiguchi, T Kawanaka. J Chem Soc Chem Commun 833–835, 1985.
30. T Kunitake, Y Okahata. J Am Chem Soc 102:549–553, 1980.
31. T Kunitake, Y Okahata, M Shimomura, S Yasunami, K Takarabe. J Am Chem Soc 103:5401–5413, 1981.
32. T Kunitake, M Shimomura, A Harada, K Okuyama, T Kajiyama, M Takayanagi. Thin Solid Films 121:L89–91, 1984.
33. K Okuyama, C Mizuguchi, G Xu, M Shimomura. Bull Chem Soc Jpn 62:3211–3215, 1989.
34. G Xu, K Okuyama, M Shimomura. Mol Cryst Liq Cryst 213:105–115, 1992.
35. M Kasha. Spectroscopy of the Excited State. New York: Plenum, 1976.
36. JD Wright. Molecular Crystals. New York: Cambridge University Press, 1987.
37. M Shimomura, T Kunitake. Chem Lett 1169–1172, 1981.
38. M Shimomura, R Ando, T Kunitake. Ber Bunsenges Phys Chem 87:1134–1143, 1983.
39. M Shimomura, S Aiba, N Tajima, N Inoue, K Okuyama. Langmuir, 11:969–976, 1995.
40. K Okuyama, M Shimomura. New Functionality Materials, vol. C, Synthesis Process and Control of Functionality Materials. Tokyo: Elsevier Science, 1993.
41. K Okuyama, M Ikeda, S Yokoyama, Y Ochiai, Y Hamada, M Shimomura. Chem Lett 1013–1016, 1988.
42. M Shimomura, Y Hamada, N Tajima, K Okuyama. J Chem Soc Chem Commun 232–234, 1989.
43. M Shimomura, N Tajima, K Kasuga. J Photopolym Sci Tech 4:267–270, 1991.
44. T Kuo, DF O'Brien. J Am Chem Soc 110:7571–7573, 1988.
45. Y Okahata. Acc Chem Res 19:57–63, 1986.
46. T Kajiyama, A Kumano, M Takayanagi, Y Okahata, T Kunitake. Chem Lett 645–648, 1979.
47. S Hayashida, H Sato, S Sugawara. Chem Lett 625–626, 1983.
48. M Shimomura, T Kunitake. Polym J 16:187–190, 1984.
49. N Higashi, T Kunitake. Polym J 16:583–585, 1984.
50. M Shimomura, K Utsugi, J Horikoshi, K Okuyama, O Hatozaki, N Oyama. Langmuir 7:760–765, 1991.
51. A Takahara, N Morotomi, S Hiraoka, N Higashi, T Kunitake, T Kajiyama. Macromolecules 22:617–622, 1989.
52. Y Okahata, G Enna. J Phys Chem 92:4546–4551, 1988.
53. A Laschewsky, H Ringsdorf, G Schmidt, J Schneider. J Am Chem Soc 109:788–796, 1987.
54. J Schneider, H Ringsdorf, JF Rabolt. Macromolecules 22:205–210, 1989.
55. K Okuyama, M Shimomura. New Developments in Construction and Function of Organic Thin Films. Tokyo: Elsevier Science, 1996.

56. M Shimomura, S Aiba, S Oguma, M Oguchi, M Matsute, H Shimada, R Kajiwara, H Emori, K Yoshiwara, K Okuyama, T Miyashita, A Watanabe, M Matsuda. Supramol Sci 1:33–38, 1994.
57. M Shimomura, T Tsukada, K Kasuga. J Chem Soc Chem Commun 845–846, 1991.
58. M Shimomura, K Kasuga, T Tsukada. Thin Solid Films 210/211:375–377, 1992.
59. J Engelking, H Menzel. Thin Solid Films 327–329:90–95, 1998.
60. M Shimomura, Y Hamada, T Onosato. Thin Solid Films 160:287–297, 1988.
61. C Erdelen, A Laschewsky, H Ringsdorf, J Schneider, A Schuster. Thin Solid Films 180:153–166, 1989.
62. K Miyano, K Asano, M Shimomura. Langmuir 7:444–445, 1991.
63. K Asano, K Miyano, H Ui, M Shimomura, Y Ohta. Langmuir 9:3587–3593, 1993.
64. M Shimomura, K Utsugi, J Horikoshi, K Okuyama, O Hatozaki, N Oyama. Langmuir 7:760–765, 1991.
65. M Suzuki, M Kakmoto, T Konshi, Y Imai, M Iwamoto, T Himno. Chem Lett 395–398, 1986.
66. M Iwamoto, X Xu. Thin Solid Films 284–285:936–938, 1996.
67. A Baba, F Kaneko, K Shinbo, K Kato, S Kobayashi. Thin Solid Films 327–329: 353–356, 1998.
68. MP Srinivasan, FJ Jing. Thin Solid Films 327–329:127–130, 1998.
69. A Wu, T Fujuwara, M Jikei, M Kakimoto, Y Imai, T Kubota, M Iwamoto. Thin Solid Films 327–329:901–903, 1998.
70. Y Nishikata, M Kakimoto, Y Imai. J Chem Soc Chem Commun 1040–1041, 1988.
71. M Era, H Shinozaki, S Tokito, T Tsutsui, S Saito. Chem Lett 1097–2001, 1988.
72. K Kamiyama, M Era, T Tsutsui, S Saito. Jap J Appl Phys 29:L840–L842, 1990.
73. JH Cheung, E Punkka, M Rikukawa, RB Rosner, AT Royappa, MF Rubner. Thin Solid Films 210–211:246–249, 1992.
74. H Tachibana, R Azumi, M Tanaka, M Matsumoto, S Sako, H Sakai, M Abe, Y Kondo, N Yoshino. Thin Solid Films 284–285:73–75, 1996.
75. N Nishiyama, M Fujihira. Chem Lett 1257–1260, 1988.
76. PJ Bruinsma, P Stroeve, CL Hoffmann, JF Rabolt. Thin Solid Films 284–285: 713–717, 1996.
77. CJ Murphy, MR Arkin, Y Jenkins, ND Ghatlia, SH Bossmann, NJ Turro, JK Barton. Science 262:1025–1029, 1993.
78. TJ Meade, JF Kayyem. Angew Chem Int Ed Engl 34:352–354, 1995.
79. FD Lewis, T Wu, Y Zhang, RL Letsinger, SR Greenfield, MR Wasielwski. Science 277:673–676, 1997.
80. DN Beratan, S Priyadarshy, SM Risser. Chem Biol 4:3–8, 1997.
81. S Priyadarshy, SM Risser, DN Beratan. J Phys Chem 100:17678–17682, 1996.
82. K Ijiro, M Shimomura, M Tanaka, H Nakamura, K Hasebe. Thin Solid Films 284–285:780–783, 1996.
83. K Tanaka, Y Okahata. J Am Chem Soc 118:10679–10683, 1996.
84. Y Okahata, K Ijiro, Y Matsuzaki. Langmuir 9:19–21, 1993.
85. M Shimomura, J Matsumoto, F Nakamura, T Ikeda, T Fukazawa, K Hasebe, T Sawadaishi, O Karthaus, K Ijiro. Polym J 31:1115–1120, 1999.
86. Y Okahata, T Kobayashi, K Tanaka, M Shimomura. J Am Chem Soc 120:6165–6166, 1998.

87. K Ariga, T Kunitake. Acc Chem Res 31:371–378, 1998.
88. M Shimomura, F Nakamura, K Ijiro, H Taketeuna, M Tanaka, H Nakamura, K Hasebe. J Am Chem Soc 119:2341–2342, 1997.
89. M Lösche, E Sackmann, H Möhwald. Ber Bunsenges Phys Chem 87:848–852, 1983.
90. M Shimomura, K Fujii, T Shimamura, M Oguchi, E Shinohara, Y Nagata, M Matsubara, K Koshiishi. Thin Solid Films 210–211:98–100, 1992.
91. W Saenger. Principles of Nucleic Acid Structure. New York: Springer-Verlag, 1984.
92. PG de Gennes. Soft Interfaces, New York: Cambridge University Press, 1997.
93. G Reiter. Langmuir 9:1344–1351, 1993.
94. A Faldi, RJ Composto, KI Winey. Langmuir 11:4855–4861, 1995.
95. TG Stange, DF Evans, WA Hendrickson. Langmuir 13:4459–4465, 1997.
96. N Maruyama, T Koito, T Sawadaishi, O Karthaus, K Ijiro, N Nishi, S Tokura, S Nishimura, M Shimomura. Supramol Sci 5:331–336, 1998.
97. R Vuilleumier, V Ego, L Neltner, AM Cazaba. Langmuir 11:4117–4121, 1995.
98. N Maruyama, T Koito, J Nishida, S Nishimura, X Cieren, O Karthaus, M Shimomura. Thin Solid Films 327–329:854–856, 1998.
99. G Widawski, M Rawiso, B François. Nature 369:387–389, 1994.
100a. M Shimomura, T Koito, N Maruyama, K Arai, J Nishida, L Gråsjö, O Karthaus, K Ijiro. Mol Cryst Liq Cryst 322:305–312, 1998.
100b. O Karthaus, T Koito, M Shimomura. Mater Sci Eng C 1999 (in press).
101. T Miyashita, M Nakaya, A Aoki. Thin Solid Films 327–329:833–836, 1998.
102. T Nishikawa, J Nishida, R Ookura, S Nishimura, S Wada, T Karino, M Shimomura. Supramol Sci 1999 (in press).
103. M Shimomura, O Karthaus, N Maruyama, K Ijiro, T Sawadaishi, S Tokura, N Nishi. Rep Prog Polym Phys Jpn 40:523–524, 1997.
104. D Bensimon, AJ Simon, V Croquette, A Bensimon. Phys Rev Lett 265:4754–4757, 1995.
105. M Shimomura, S Aiba. J Chem Soc Jap Chem Indust Chem 905–916, 1993.
106. M Matsumoto, D Miyazaki, M Tanaka, R Azumi, E Manda, Y Kondo, N Yoshino, H Tachibana. J Am Chem Soc 120:1479–1484, 1998.
107. M Shimomura, K Fujii, P Karg, W Frey, E Sackmann, P Meller, H Ringsdorf. Jap J Appl Phys 27:L1761–L1763, 1988.
108. M Shimomura, M Oguchi, K Kasuga, K Fujii, E Shinohara, S Kondo, N Tajima, K Koshiishi. Thin Solid Films 243:358–360, 1994.
109. D Hönig, D Möbius. J Phys Chem 95:4590–4592, 1991.
110. S Hénon, J Meunier. Rev Sci Instrum 62:936–939, 1991.
111. O Albrecht, H Matsuda, K Eguchi, T Nakagaki. Thin Solid Films 338:252–268, 1999.

12
Supramolecular Polyelectrolyte Assemblies

Xavier Arys, Alain M. Jonas, André Laschewsky, and Roger Legras
Université Catholique de Louvain, Louvain-La-Neuve, Belgium

I. INTRODUCTION

In recent years, much progress has been achieved in the preparation and characterization of organic and hybrid organic–inorganic ultrathin multilayers. The growing interest for these systems, both from the fundamental and from the applied side, is partly due to the unusual physical properties of nanostructured materials, and to the potential applications resulting from these properties, in particular in the field of integrated molecular optics and electronics. Size quantization, for instance, occurs when electron-hole pairs are confined in domains whose dimensions are comparable to the wavelength of the de Broglie electron, and the mean free path of excitons [1]. This in turn enables the control of the emission color of light-emitting diodes (LEDs) [2]. Another reason for the current growing interest in organic multilayer films is the similarity among the methods for obtaining such molecular assemblies and the principles governing the self-organization of organic molecules in natural systems [3,4]. Being able to control the spatial arrangement of biological molecules on a nanometer scale opens the way to numerous applications, among which molecular recognition [5] and multistep catalysis [6] can be cited. At the frontier between biology and physics, molecular bioelectronics is also gaining in interest [7]. An example of development in this field is the fabrication of systems that mimic photosynthesis, the goal being to design artificial systems for the efficient conversion of solar energy into chemical or electrical energy [8–10].

The eldest technique for the fabrication of multilayer films was invented in the 1920s by Langmuir and Blodgett [11–13]. A renaissance of this technique occurred in the 1960s under the impetus of Kuhn and coworkers [14]. Significant progress in the understanding of monolayer assemblies has been gained since the late 1970s [15], in part due to the development of new characterization techniques. In order to build up Langmuir–Blodgett (LB) films, amphiphilic molecules are spread on water and compressed until they form a solidlike two-dimensional phase. A film of these molecules is then transferred onto a solid surface by dipping a substrate through the air–water interface. Multilayers are realized by repetitive dipping. LB films are highly ordered and have controlled uniform thickness. However, the requirements for substrates are stringent: they must be smooth, homogeneous [16], and have a regular shape. Furthermore, expensive equipment is required, and the coating of areas larger than 10 cm^2 is demanding. Even more, LB-multilayers are metastable, and as a result have a limited stability against solvents or thermal treatments, although strategies have been designed to overcome these problems [17–19]. Finally, defects formed in a given layer are difficult to cover up by subsequent layers. Although possible, multilayer heterostructures, i.e., composed of layers of different amphiphiles, or noncentrosymmetric multilayers, are not easily realized. Noncentrosymmetric assemblies are necessary for second-order nonlinear optical (NLO), piezo- and pyro-electric effects.

Another approach to assemble layered structures was presented in the 1980s by Netzer and Sagiv [20]. Their strategy was based on a two-step sequence involving the chemisorption of a monolayer followed by chemical activation of the exposed surface, in order to provide polar adsorption sites for the anchoring of the next monolayer. The reagents used had a trichlorosilane head-group for the chemisorption step, and a reactive functionalized end-group for the creation of a hydroxylated surface after the chemical activation step. Examples of such activation reactions are the conversion of double bonds through hydroboration-oxidation [21], $LiAlH_4$ reduction of a surface ester group [22], or photolysis [23]. These covalently linked multilayers are highly ordered, and substrates of any size or shape can be coated by this technique. They are stable against solvents or thermal treatments. Noncentrosymmetric assemblies are easily obtained [24]. Nevertheless, the high specificity of the chemisorption step imposes major requirements on the chemical nature and homogeneity of the substrate; this may limit the practical potential of this technique. The optimum buildup requires a nearly quantitative surface activation step; otherwise defects will appear and grow with layer number. This prerequisite often limits the number of layers to small values. Notwithstanding, it has been reported that, in some instances, thick films with up to 500 layers could be built [22]. Finally, trichlorosilane derivatives are hard to handle due to their moisture sensitivity. Replacing trichlorosilane derivatives by methoxy- or ethoxysilanes allows minimizing this problem, at the expense of a much slower reaction rate [22].

Concomitant with the development of the trichlorosilane technique, a method for the chemisorption of monolayers of disulfides on gold was developed by Nuzzo and Allara [25]. It has been rapidly generalized to the adsorption of monolayers of alkanethiols on noble metal surfaces [22,26]. Using ω-mercaptoalkanoic acid, and converting the acid surface to a copper salt on which another thiol monolayer can adsorb, Ulman and coworkers showed that this chemical self-assembling technique is also useful for multilayer buildup [27,28]. Alternatively α,ω-dithiols and colloids, like CdS nanoparticles, may be used [29].

In 1988, Mallouk and coworkers showed that multilayer films could be prepared simply by sequential complexation of Zr^{4+} and α,ω-bisphosphonic acid [30]. The technique has been extended to the complexation of organophosphonates with a variety of other metals [31,32], and even to completely different metal–ligand systems [33–38]. The multilayers grown by this technique are well ordered, robust, and relatively easy to prepare. Substrates of any size or shape can be covered, and even noncentrosymmetric multilayers may be obtained [39]. Based on very specific interactions, the technique is, however, restricted to a narrow class of chemical compounds.

In 1966, Iler presented a technique for building films of controlled uniform thickness by the alternate adsorption of positively and negatively charged colloidal particles [40]. Although a few other singular attempts have been reported afterwards [41–43], the development of the electrostatic self-assembly (ESA) technique is mainly due to the work of Decher and coworkers. In their first article, Decher and Hong used anionic and cationic bipolar amphiphiles containing rigid biphenyl cores [44]. The adsorption of such molecules leads to a surface charge reversal, and multilayer assemblies of both compounds can thus be obtained by alternatively dipping a substrate in solutions of the anionic and cationic amphiphiles. In the same year, they published an article extending the method to polyelectrolytes [45]. The crucial factor for successful deposition is the surface charge reversal upon adsorption, which can usually be obtained by a proper choice of deposition parameters. In the meanwhile, many charge-bearing molecules or particles have been deposited by this technique, and numerous examples of successful deposition can be found in the literature. Just to name a few, films have been prepared by ESA from proteins [46], DNA [47], dyes [48], inorganic platelets [49], latex particles [50], dendrimers [51], and even viruses [52]. The thickness of the coatings can be adjusted by changing processing parameters, such as ionic strength, pH, or the nature of the solvent. Substrates of any size or shape are suited, if they have a charged surface. As methods exist to bring charge to virtually any type of organic or inorganic surface, there is practically no limitation to the nature of the substrate. The versatility of the technique has triggered an abundant and rapidly increasing literature, scattered in myriad journals, so that it is opportune to update the existing reviews of the field [53–59].

This success of ESA is probably also linked to the ease with which stable films with precision in the range of a few Ångstrøms can be built by this method, without expensive equipment. Accordingly, this technique can be referred to as a ''molecular beaker epitaxy'' [60]. There seems to be no limitation to the maximum number of layers that can be deposited, and films with up to 1000 layers have been realized [61]. This is partly due to the self-healing properties of the multilayer buildup, when polyelectrolytes are used. Furthermore, the process is relatively fast, can be easily automated [62–64]*, and is a priori environment friendly, since water is usually used as the solvent. Finally, aperiodic multilayers are easily prepared [65]. Among the drawbacks of the technique, the structure of the films is in general fuzzy [53], meaning that there is normally an important interpenetration between neighboring layers. The magnitude of the interpenetration is of the order of the layer thickness [66]. Thus, the term ''layer pair'' should preferably be used instead of ''bilayer'' to denote a positive/negative pair of molecules, since a structural subunit composed of well-defined bilayers often does not exist [67]. Although some exceptions have been mentioned, films made by ESA are usually centrosymmetric. This limits the scope of potential applications. A possible way to generate noncentrosymmetric coatings and to obtain films made up of only one type of polyelectrolyte is to combine the ESA technique with a chemical activation step [68].

Subsequently, several authors have shown that interactions other than the electrostatic ones between oppositely charged molecules could be used to prepare multilayer assemblies (Fig. 1). For example, Rubner and Stockton have demonstrated the successful growth of films made of polyaniline (PAn) alternating with nonionic water-soluble polymers [69]. Four different nonionic water-soluble polymers were used: poly(vinyl pyrrolidone) (PVP), poly(vinyl alcohol) (PVA), poly(acrylamide) (PAAm), and poly(ethylene oxide) (PEO). The authors presumed that the multilayer buildup is due to hydrogen bonding, supporting this hypothesis by infrared (IR) spectroscopy showing the formation of such bonds in the case of the PAn/PVP system. Films have also been prepared with PVP and poly(acrylic acid) (PAA) [70,71], for which again hydrogen-bonding interactions were identified by IR spectroscopy. Interestingly, the strong interaction between

Figure 1 Examples of polymer pairs suited for layer-by-layer assembly (from top to bottom): by electrostatic interactions: poly(styrene sulfonate)/poly(allylamin hydrochloride); by H-bonding: polyaniline/poly(vinylpyrrolidone) (from Ref. 69); by H-bonding: poly(4-vinyl pyridine)/poly(acrylic acid) (from Ref. 71); by electron-donor–electron-acceptor interaction: poly(2-(9-carbazolyl)ethyl methacrylate)/poly(2-(3,5-dinitrobenzoyl)oxyethyl methacrylate) (from Ref. 75); by specific interaction: biotinylated poly(lysine)/streptavidin (from Ref. 79).

* Our group has such an automated system at its disposal, which was purchased from Riegler & Kirstein GmbH.

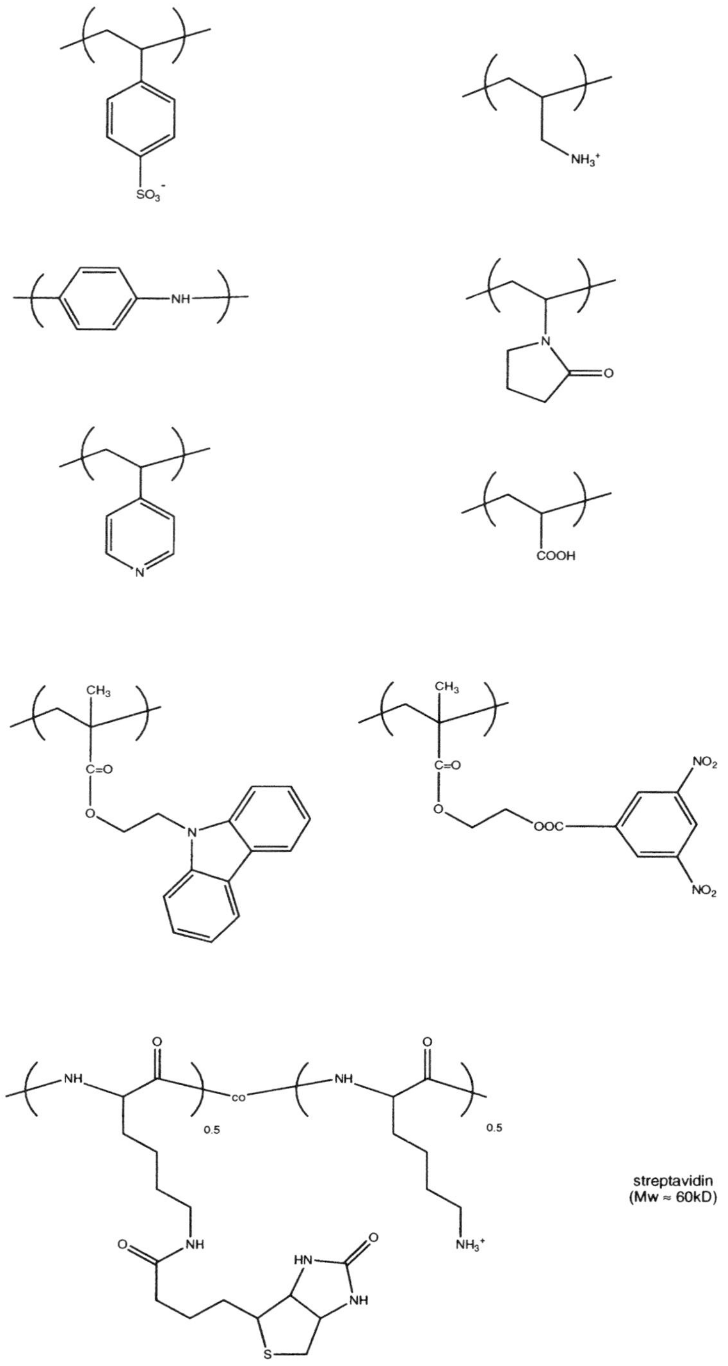

SO_3^-

NH_3^+

NH

N

O

N

COOH

CH_3

C=O

O

N

CH_3

C=O

O

OOC

NO_2

NO_2

O

NH

co

0.5

O

NH

0.5

O

NH

HN

O

NH

S

NH_3^+

streptavidin
(Mw ≈ 60kD)

both polymers led to the formation of a precipitate upon mixing of both solutions. On the other hand, combinations involving polymers with hydrogen-bonding capabilities such as PVP/PVA, PAAm/PVP, PAAm/PVA, and PAAm/PEO were attempted for multilayer buildup, however, without success [72]. Multilayer buildup based on hydrogen-bonding interactions was also combined with ESA to produce multilayers [73].

Another strategy for multilayer assemblies of polymers was demonstrated by Shimazaki et al. [74,75], who used the sequential adsorption of polymers having, respectively, electron-donating and electron-accepting pendant groups.

Finally, we cite as examples of multilayers based on interactions other than electrostatic ones, protein multicomponent films, based on specific interactions between these biological molecules or between these biological molecules and polyelectrolytes. For example, immunoglobulin G can be assembled with the anionic poly(styrene sulfonate) at pH values above and below its isoelectric point, thus implying that this process is not electrostatically driven [76]. Also, layer-by-layer deposition of avidin and biotin-labeled poly(amine)s leads to the formation of multilayer assemblies through avidin–biotin complexation [46,77–79], even in the case of electrostatic repulsion arising from the net positive charges of avidin (isoelectric point at pH 9.0–10) and poly(amine)s [77,78]. The assembly of glucose oxidase multilayers through use of bispecific antibody interlayer [80], and multilayer buildup of concanavalin A with glycogen [81] are based on the occurrence of specific interactions as well. By contrast, attempts to build multilayers of poly(uridylic acid) and poly(adenylic acid) were unsuccessful, probably because the electrostatic repulsion between these polyanions dominates the binding energy arising from pairing the nuclear bases [82].

Interestingly, ESA bears much formal similarity with the successive ionic layer adsorption and reaction (SILAR) method originally developed in 1985 by Nicolau [83] and Lindroos et al. [84]. This technique allows us to obtain thin inorganic films of controlled thickness based on successive adsorption from aqueous solutions of small ions on selected substrates.

II. A CASE STUDY: PAH/PSS MULTILAYERS

As an introductory example, we summarize the literature on the buildup of multilayers made of poly(allylamine hydrochloride) (PAH, Fig. 2) and poly(styrene sulfonate, Fig. 2) (PSS). This polyelectrolyte pair is among the first systems employed for ESA, and has been up to now the most thoroughly studied system [53,66,76,85–117]. The role of this pair of polyelectrolytes as a model system is somewhat unfortunate because PAH is a weak polyelectrolyte. Indeed, the dependence of its charge on the experimental conditions such as pH renders the comparison and interpretation of the results obtained by different authors diffi-

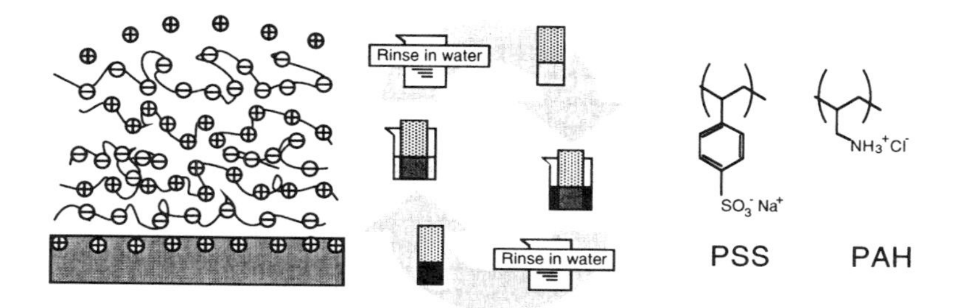

Figure 2 Left: schematic molecular representation of polyelectrolyte multilayers. Center: layer-by-layer electrostatic self-assembly: films are alternately dipped in a polycation and a polyanion solution. A rinsing step follows every adsorption step. Right: chemical structure of standard polyelectrolytes PAH and PSS.

cult. Another complication may result from the hydrogen-bonding abilities of the primary amine groups of PAH. Fortunately, no such complications are expected with the strong polyelectrolyte PSS.

The principle of the layer-by-layer growth is illustrated in Fig. 2. A substrate is alternately dipped in dilute solutions of PSS and PAH. If the surface charge of the substrate is positive, the cycle is initiated by dipping the substrate in the PSS solution. Conversely, if the surface charge of the substrate is negative, the first dipping occurs in the PAH solution. Note that two polyelectrolyte layers, a polyanionic and a polycationic one, are deposited per dipping cycle. The samples are rinsed after each immersion in a polyelectrolyte solution. Without the rinsing steps, an adhering layer of solution would be left on the surface of the substrate, leading to coprecipitation upon immersion in the following polyelectrolyte solution, and eventually to the incorporation of precipitated particles in the films [44]. Drying is sometimes applied after each rinsing step.

As presented in the introduction, the nature of the substrate used for the above-described electrostatic layer-by-layer deposition is not crucial, as long as it is charged. For example, PAH/PSS multilayers have been prepared on plasma-treated glass [85,86], as cleaved mica [87], silicon functionalized by 3-aminopropyldimethoxysilane [88], surface-oxidized poly(4-methyl-1-pentene) [89], chemically modified poly(chlorotrifluoroethylene) (PCTFE) [90,91], plasma-modified poly(tetrafluoroethylene) (PTFE) [92,93], chemically modified as well as untreated poly(ethylene terephtalate) (PET) [94], gold surfaces modified with mercaptopropionic acid [76], and many others. PAH/PSS multilayers have also been grown on chemically modified polystyrene [95] and melamine formaldehyde latex particles [96,97]. Multilayer assembly on latex particles was either accomplished by adsorption from solutions of high polyelectrolyte concentration with intermediate centrifugation cycles in water as a washing procedure, or by adsorption without centrifugation, but at polyelectrolyte concentrations just beyond the onset of ζ-potential saturation (see below).

The deposition process has been followed by UV spectroscopy [88] and quartz crystal microbalance (QCM) [76,98], indicating, for carefully selected process parameters, a linear growth of the amount of material deposited with the dipping cycle. Ellipsometry [99], X-ray reflectometry (XRR) [88], and surface plasmon resonance [76] showed a concomitant increase in film thickness, with 1 to 10 nm deposited per dipping cycle, depending on the selected set of processing parameters. The growth of PAH/PSS multilayers on colloidal particles could be followed by dynamic light scattering, single particle light scattering, and fluorescence intensity measurements [95,96]. There seems to be no limit to the maximum number of PAH/PSS layer pairs that can be deposited, and assemblies with up to 200 layers were reported [87].

Although multilayers can be grown on many different supports, some substrates cause reduced growth steps during the first dipping cycles [76,87]. Similarly, the roughness of the first deposited polyelectrolyte layers is close to that of the substrate, but further dipping cycles result in a smoother film surface, at least for carefully chosen processing parameters [87]. In the case of PAH/PSS multilayers, there is no consensus in the literature as to whether the substrate influences the layer-by-layer deposition process of PAH/PSS multilayers beyond the onset of regular film growth as a function of the number of dipping cycles (see also the section on polyelectrolyte adsorption and multilayer buildup).

X-ray photoelectron spectroscopy (XPS) has been used to measure the nitrogen/sulfur ratio of the PAH/PSS multilayers, as a function of the number of deposited layers [89,93,94]. This ratio shows a pronounced odd–even trend with the number of deposited layers, being higher when PAH is the outermost layer, and lower when PSS is the outermost layer. This trend is maintained even for high numbers of layers, and reveals an effective stratification of the multilayers. Contact angle measurements [89,93,94] also show a pronounced odd–even trend in the measured angles.

If the films prepared by ESA have a well-defined multilayered structure, reflectometry techniques, like XRR and neutron reflectometry (NR), should allow the determination of the periodically varying refractive index profile along the surface normal [118], due to the well-defined supramolecular structure. Nevertheless, Bragg peaks are observed neither by X-ray reflectometry, nor by neutron reflectometry (when all PSS layers are deuterated) [100,101]. In the latter case at least, the reason for the absence of Bragg peaks cannot be ascribed to the lack of contrast between hydrogenated PAH and deuterated PSS layers, since neutron scattering lengths for hydrogen and deuterium are drastically different [119]. However, when deuterated PSS layers are separated by one or more nondeuterated PSS layers, Bragg reflections are seen by neutron reflectometry [53,66,

100,101]. From these measurements, the interfacial width (Gaussian width) between neighboring polymer layers was determined to be ~1.9 nm and ~3 nm, respectively, for films prepared from 2 M NaCl [101] and 3 M NaCl [66]. This important interfacial width is responsible for the absence of Bragg peaks in XRR measurements and in NR measurements when all layers are deuterated (see Fig. 3). The interfacial width must be considered as originating from both the interfacial roughness (or interfacial waviness) and the polymer interdiffusion. Assuming that the interfacial roughness between two neighboring layers is equal to the roughness of the film–air interface, the interdiffusion of two successive polyelectrolyte layers was estimated to be ~1.2 nm for films deposited from 2 M NaCl. Off-specular reflectivity measurements would enable us to disclose the relative contributions of interfacial waviness and polymer interdiffusion to the interfacial width [120]. Contact angle measurements [89,93,94] support the hypothesis of strong interpenetration of neighboring layers. Indeed, these data indicate that wettability is controlled by the outermost layer when the layers are sufficiently thick, and by at least the two final layers when the layers are thinner, suggesting strong interdiffusion of the layers.

The thickness of the PAH and PSS layers have been determined by NR to be respectively ~2 nm and ~3.5 nm for films deposited from 2 M NaCl and to be, respectively, ~1.8 nm and ~5.8 nm for films prepared from 3 M NaCl [66,101]. Determining a stoichiometry from these results is not straightforward, but it seems that the PAH:PSS ratio is ≥ 1. The stoichiometry was determined from XPS measurements to give an ammonium to sulfate ion ratio ranging from ~1/1 to ~2/1, apparently depending on the nature of the substrate [89,93,94]. We stress that these ratios relate the number of PAH to PSS repeat units, and not directly the number of cationic to anionic groups, which depends on the degree of protonation of PAH.

If the complexation of the anionic and cationic groups is not stoichiometric, some extrinsic charge compensation should occur, and the presence of counterions in the multilayers is thus expected. Unfortunately, NR measurements are not able to determine whether inorganic salts are included in the deposited films [101]. Most of the time their presence also could not be unambiguously determined by XPS [89,93,94], due to problems of sensibility and possible removal of HCl during the measurement, which is performed under ultrahigh vacuum. However, 0.8% molar fraction of Na has been reported by an author [102]. This number does not exactly correspond to the molar fraction of Na inside the multilayers, since XPS measurements probe the inner layers and the outer layer (this latter layer must contain counterions). Furthermore, XPS is more sensitive to the outer layer than to the inner layers. Measurements of the electrical properties of the multilayers also suggest the presence of mobile ions in the films [102]. Ions can also be trapped in the multilayers, especially in the case of deposition from solutions of high ionic strength. The presence of $MnCl_2$ has indeed been

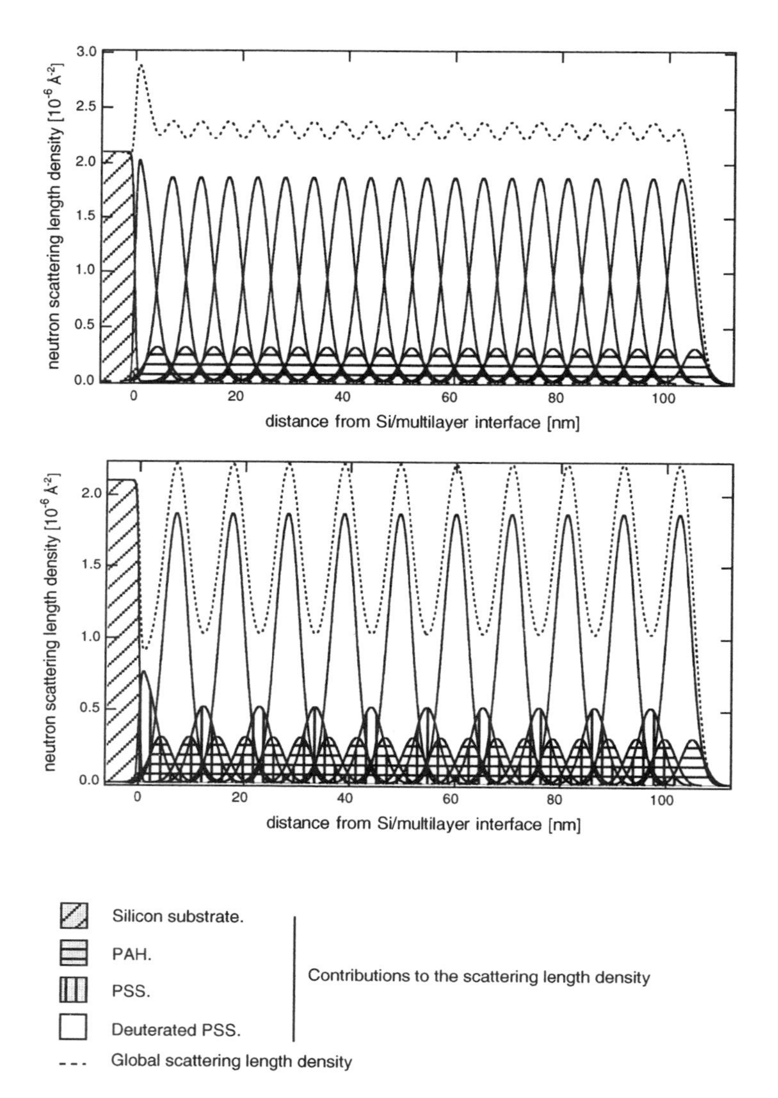

Figure 3 Top: neutron scattering length density profile for a Si/{PSS-deuterated/PAH}$_{20}$ multilayer. Bottom: neutron scattering length density profile for a Si/{PSS-deuterated/PAH/PSS/PAH}$_{10}$ multilayer. No Bragg peaks are observed by neutron reflectometry for the film with all PSS layers deuterated, because oscillations in the scattering length density are too weak. The scattering density profile corresponding to the film with every second PSS layer deuterated shows marked oscillations; accordingly, Bragg peaks are experimentally observed. (Data adapted from Ref. 66.)

detected by XPS when this salt was added to the PSS solution used for the multilayer buildup [94].

It has been calculated from a detailed analysis of NR data that much water is present in these films [101]: about six water molecules are bound per PSS monomer, and one water molecule per PAH monomer, so that water molecules occupy 40% of the film volume. Dehydration of the film leads to shrinking, but a significant part of the volume previously occupied by water seems to remain unfilled by the polyelectrolyte chains.

Of key importance for the electrostatic layer-by-layer deposition process is the occurrence of reproducible surface charge reversal upon polyelectrolyte adsorption, whatever the number of dipping cycles. For PSS concentrations larger than approximately 5 mg/L, it has been shown by surface force measurements. (SFA) that the charge of a dioctadecyldimethylammonium-covered mica surface was reversed upon PSS adsorption [103]. The thickness of the monolayer was determined to be in the range of 1 to 3 nm, depending on the molar mass. The adsorption of PAH on pure mica from 0.01 monomol/L solutions containing 1 mM NaCl has also been studied [104]. Adsorbed layers were ~0.7 nm thick in air. In water, mobile chains extended as far as 2.5 to 4 nm from the surface. Surface charge reversal occurred, as could be demonstrated by measuring the interaction between the polycation monolayer and a bare mica surface. A surface potential of ~34 mV has been estimated by fitting the experimental data. Similarly, surface charge reversal upon PSS adsorption on the PAH-covered mica surface could be evidenced by measuring the interaction between the polycation monolayer and a substrate covered with a polycation/polyanion bilayer. The surface potential was estimated to range from −77 to −98 mV, depending on the NaCl concentration. Approaching two bilayer-covered PSS-terminated substrates at sufficiently small distances reveals an attractive segment–segment interaction, most likely resulting from ion pair formation between anionic and cationic monomer units. This suggests an intermingling of the layers, with polycation and polyanion chains dangling in water. In agreement with the preceding results, a higher charge density when PSS is the outer layer than when it is PAH has been suggested by more pronounced shifts in the apparent pKa value for the titration of dyes inserted in PAH/PSS multilayers when PSS is the outer layer than when it is PAH [105].

PAH/PSS multilayers deposited on latex particles offer the possibility of electrophoretic mobility measurements. The electrophoretic mobilities were measured as a function of pH and ionic strength (electrophoretic fingerprinting) [106], showing that, for pH values below approximately 8, the surface appears to be positively charged upon PAH adsorption. At higher pH the negative charge of the PAH-covered polystyrene latex results from deprotonation of PAH, allowing approximately 1/3 of the original negative surface charge of the latex to be seen. A very pronounced negative surface charge is observed following PSS adsorption for (nearly) all the pH range investigated ($3 \leq pH \leq 10$). At high pH an apparent increase of the PSS charge of approximately one third of the value seen at lower pH is observed, due to deprotonation of the PAH at high pH and subsequent freeing of some of the PSS-charged groups. The influence of the next PAH layer is fully consistent with the previous results. The thickness of the top hairy layer was determined to be of the order of 1 nm, and counterion adsorption to the charged groups of the top layer could be observed. pK values for PAH were estimated to be in the range between 9 and 10, and to lay between 3.5 and 4.5 for PSS (which is higher than usually accepted). pK values for PAH are in agreement with values reported by other authors [121,122]. The multilayer film growth was followed by electrophoresis, for films grown from PSS and PAH solutions containing 0.5 M NaCl [95–97]. The measured mobilities were converted into a ζ-potential using the Smoluchowski relation. The ζ-potential showed a pronounced odd–even trend with the number of deposited layers. Irrespective of the number of dipping cycles, ζ-potential values for latex particles having PAH or PSS as the outermost layer were, respectively, $\sim$40 mV and ~ -40 mV.

Finally, studies of the Förster resonance energy transfer between 6-carboxyfluorescein (6-CF) and rhodamine B-labeled melamine formaldehyde particles covered by PAH/PSS multilayers showed that 6-CF interacts with PAH when the outer polyelectrolyte layer is PAH [97]: upon increasing the 6-CF concentration, an increase of the fluorescence intensity is observed only above a critical concentration, assumed to correspond to the saturation of the charged groups of PAH by the anionic 6-CF. This phenomenon has been used to titrate the number of amino sites of PAH not interacting with PSS in the polyelectrolyte multilayer film; it has been estimated as 1.6 per nm^2.

Considering the values of ζ-potential presented by different authors, a wide scatter is observed in the values reported, even in the values reported for PAH and PSS only. This is partly due to the fact that the multilayers were prepared under different conditions, and in part to difficulties linked to the techniques used for determining ζ-potentials. Nevertheless, charge reversal upon adsorption has been clearly demonstrated.

Obviously, many parameters can be ajdusted in the layer-by-layer deposition process: dipping time, polyelectrolyte concentration, pH, ionic strength, nature of the solvent, molar mass of the polyelectrolyte, presence or absence of a drying step, etc. In the following, we review the influence of some of these parameters on the growth of PAH/PSS multilayers.

A proper choice of the dipping time presupposes a study of the adsorption kinetics. In situ measurements of the adsorption of PAH (from a 0.01 M solution with 0.5 M $MnCl_2$ added) on a PSS-coated optical waveguide showed that the adsorption is practically completed after 3 minutes [107]. It was shown that the kinetics of adsorption are not transport limited. Similar results hold for subsequent polyelectrolyte adsorption. Ellipsometric measurements of the thickness of adsorbed PSS and PAH layers during film assembly and in situ QCM measurements confirms that the adsorption proceeds mainly during the first minutes [87,98]. Ex situ measurements by XPS of the kinetics of adsorption of PAH or PSS on PAH/PSS multilayers deposited on polymer surfaces [89,93,94] also indicated that the adsorption time is shorter than 10 min. Nevertheless, the adsorption of the first PAH layer on nonchemically treated PET and of the first PSS layer on chemically modified PTFE took, respectively, 30 and 20 min. The adsorption of a PSS layer on a silanized silicon wafer has been followed ex situ by atomic force microscopy, by interrupting the adsorption after given periods of time [108,109]. The formation of isolated islands is observed in the initial stage of PSS deposition, especially on defect sites (scratches, microparticles, etc.). For deposition times from 1 to 5 minutes, a sharp increase of surface coverage by random islands is observed. After 10 minutes, the surface exhibits a very homogeneous morphology with a roughness as low as 2 nm claimed. These results indicate that the adsorption of PSS is a two-stage process: macromolecular chains are anchored to the surface by some segments during the short initial stage, and then relax to dense packing during the long second stage. In agreement with the above-mentioned in situ measurements, this implies that diffusion-limited adsorption mechanisms are not adequate to explain the experimental results. Adsorption of a second PAH layer on top of the PSS monolayer follows similar tendencies, although a complete PAH film is formed faster [108,109]. Other authors have reported that PAH/PSS multilayers could only be assembled if the duration of each adsorption cycle were not shorter than 10 minutes [87], and that films assembled with dipping times close to this limit showed pronounced roughness. Accordingly, dipping times of 10 minutes or more are used in the literature.

Films grown from PSS (0.5 M $MnCl_2$) and PAH (2 M NaCl) with PSS concentrations of 0.1, 0.02, 0.004 mole per liter showed the same growth step [87], suggesting a weak influence of polyelectrolyte concentration on the deposition process. The effect of PSS concentration has also been studied by monitoring the UV-vis absorbance of PAH/PSS multilayers with 8 polyelectrolyte layers built from 5×10^{-4} M PAH solution and PSS solutions of various concentrations [102]. Changing the PSS concentration from 4×10^{-6} to 1×10^{-3} M increases the adsorbed amount only by a factor of 3 (Fig. 4). Nevertheless, as mentioned before, a lower limit to the PSS concentration must exist, since surface charge overcompensation has to occur for successful multilayer buildup [103]. The UV absorption was reported to be very weak for the multilayers assembled from 5×10^{-7} M PSS solutions (no salt added), and XRR showed that films could not be grown from 0.001 M PSS (0.5 M $MnCl_2$) solutions [87].

Changing the ionic strength also offers the possibility to fine-tune the film thickness of the PAH/PSS multilayers (Fig. 5). The addition of electrolytes enables the adjustment of the average thickness of each oppositely charged layer

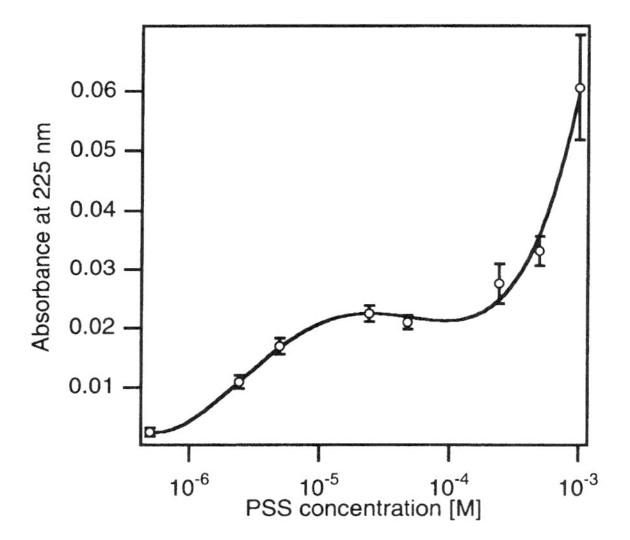

Figure 4 Dependence of the absorbance at 225 nm of ESA PAH/PSS multilayers on the concentration of the PSS solutions. The PAH concentration is 5×10^{-4} M. (Data taken from Ref. 102.)

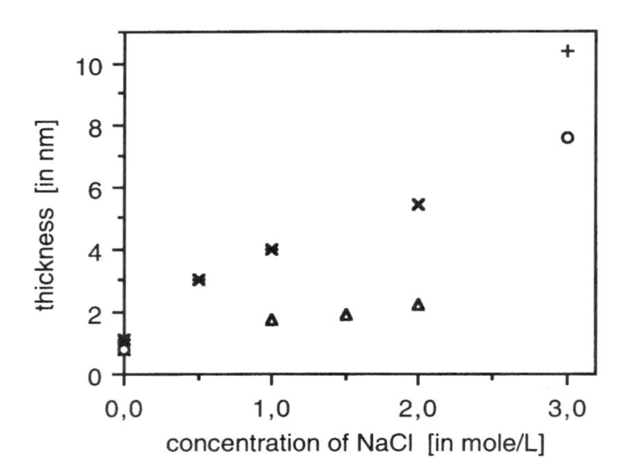

Figure 5 Evolution of the thickness of a layer pair of PAH and PSS in polyelectrolyte multilayers, made from solutions with increasing amounts of NaCl added. (△): NaCl added to the PSS solutions only. (+), (×), (*), (○): NaCl added to both PSS and PAH solutions. [Data taken from Ref. 110 (△); Ref. 66 (×); Ref. 101 (○); Ref. 100 (+); Ref. 105, 115 (*).] Thicknesses are determined by XRR. Concentration of the polyelectrolyte solutions: 0.003 M or 0.01 M (repeat unit). Solutions contain 0, 0.003 M, or 0.01 M HCl.

pair, with a precision as good as 0.05 nm having been claimed [110]. The available range goes from 1.09 nm (no salt added) to 10.4 nm (3 M NaCl) [100]. The structure of the PAH/PSS multilayers [101] and, in particular, the morphology of the film surface [111] and the interpenetration of the polyelectrolyte layers [110], are also affected by the ionic strength. Screening of electrostatic interactions is responsible for the observed increase of the film thickness with increasing salt concentration, due to a more coiled conformation of the polyelectrolytes. Interestingly, this fine-tuning of the layer thickness can be used for the construction of complex film architectures, by stacking layers of different thicknesses of PAH and PSS [87]. This is possible because the swelling of the polyelectrolyte multilayers when immersed in solutions of different ionic strengths is only on the order of 10 to 15% [112], whereas the influence of the ionic strength on the thickness increment covers a much wider range.

Deposition with various water–glycerol mixtures showed a twofold increase in thickness while increasing the glycerol fraction from 20 to 50% [87]. Clearly, changing the solvent offers new processing potentials, although this has not been much explored so far.

The effect of the molar mass of PSS is small: increasing its degree of polymerization from 900 to 5000 changed the overall thickness by less than 6% [101].

As mentioned above, no Bragg peak is found when PAH/PSS multilayers are characterized by XRR. Nevertheless, when a drying step was inserted every 2, 3, or 4 dipping cycles, Bragg peaks corresponding to these periods were reported [113]. The peaks sharpen when the number of adsorption/drying cycles is increased. The introduction of a drying step after every dipping cycle does not generate Bragg peaks. No effect on the growth step per bilayer is detected. These results have been rationalized by suggesting that the drying step induces denser packing of the outer polyelectrolyte layer, reducing the mutual penetration of chain loops and ends between neighboring layers. Annealing at 90°C did not suppress the supramolecular structure, but a prolonged exposure to water (four days) did [113].

In another experiment [87], a sample with PSS as the outer layer has been dried, and then dipped again for some time in the PSS solution. This drying/adsorption cycle has been repeated three more times. As thickness measurements showed, only upon a total 120 min adsorption did the process attain its saturation. Altogether, the PSS thickness reached 10.4 nm, although for undried samples adsorption did saturate after 10 min to a thickness of 4 nm. Surprisingly, this stepwise treatment also had no influence on the subsequent growth step.

Finally, we stress the stability of the PAH/PSS multilayers, which do not redissolve in pure water (which is very important for the rinsing step of the deposition process) or in solutions of high ionic strength [112]. The thickness of films immersed 10 days in pure water or submitted to water vapor for 2 months present unchanged thickness [116]. Furthermore, films heated for more than 1 week to

200°C do not show any noticeable deterioration, but release some water, which leads to some shrinking of the films. After dehydration of the films over P_2O_5, recovery of water is almost complete (95% of the water molecules return into the film structure) [101]. However, PAH/PSS multilayers delaminate at pH 10, presumably due to deprotonation of the amine groups [117].

III. POLYELECTROLYTE ADSORPTION AND MULTILAYER BUILDUP

That polyelectrolyte multilayer buildup is based on electrostatic interactions is intuitively a very reasonable assumption. Taking into account the weak steric requirements of ionic bonds, this assumption explains why so many polycation–polyanion pairs can lead to successful multilayer buildup, with only some rare unsuccessful cases reported (see application range). It also explains why attempts at film deposition based on the alternate adsorption of polyelectrolytes of identical charge were doomed to failure [123,124]. Another implication is that a given protein can be used for multilayer formation either as a polycation or a polyanion by adsorbing it at a pH respectively under and above its isoelectric point. Hemoglobin, for example, is assembled as a positively charged entity at pH 4.5 (with PSS), and as a negatively charged entity at pH 9.2 [with poly(ethylenimine) (PEI)] (Fig. 6a) [124]. Electrostatic interactions not only control the deposition process, but also determine the stability of polyelectrolyte multilayers: inverting the charge of one of the polyelectrolytes by changing the pH in the film [125], or suppressing the charge of one of the constituent polyelectrolytes by thermochemistry [126] or by changing the pH [117,127] leads to destruction of the multilayer assembly. The strong influence of the ionic strength on the multilayer buildup is also indicative of the major role played by electrostatic interactions.

In the following, we discuss to what extent theories on the adsorption of polyelectrolytes on solid substrates, based on electrostatic interactions, can provide insight into the mechanisms governing polyelectrolyte multilayer buildup. Indeed, the ensemble consisting of a substrate and the polyelectrolyte multilayer covering it may be seen as a substrate for another polyelectrolyte adsorption, although the complexation in the loose outer layer, with chains dangling around, must be somewhat different from the adsorption on a rigid surface. Accordingly, a part of this section is devoted to phenomena particular to the adsorption at this swollen interface, or occurring over several dipping cycles. Another reason to present the theories of the adsorption of polyelectrolytes on solid substrates is that this adsorption is the initial step for polyelectrolyte multilayer buildup. In the last part of this section, the effects of different parameters influencing the layer-by-layer deposition are examined.

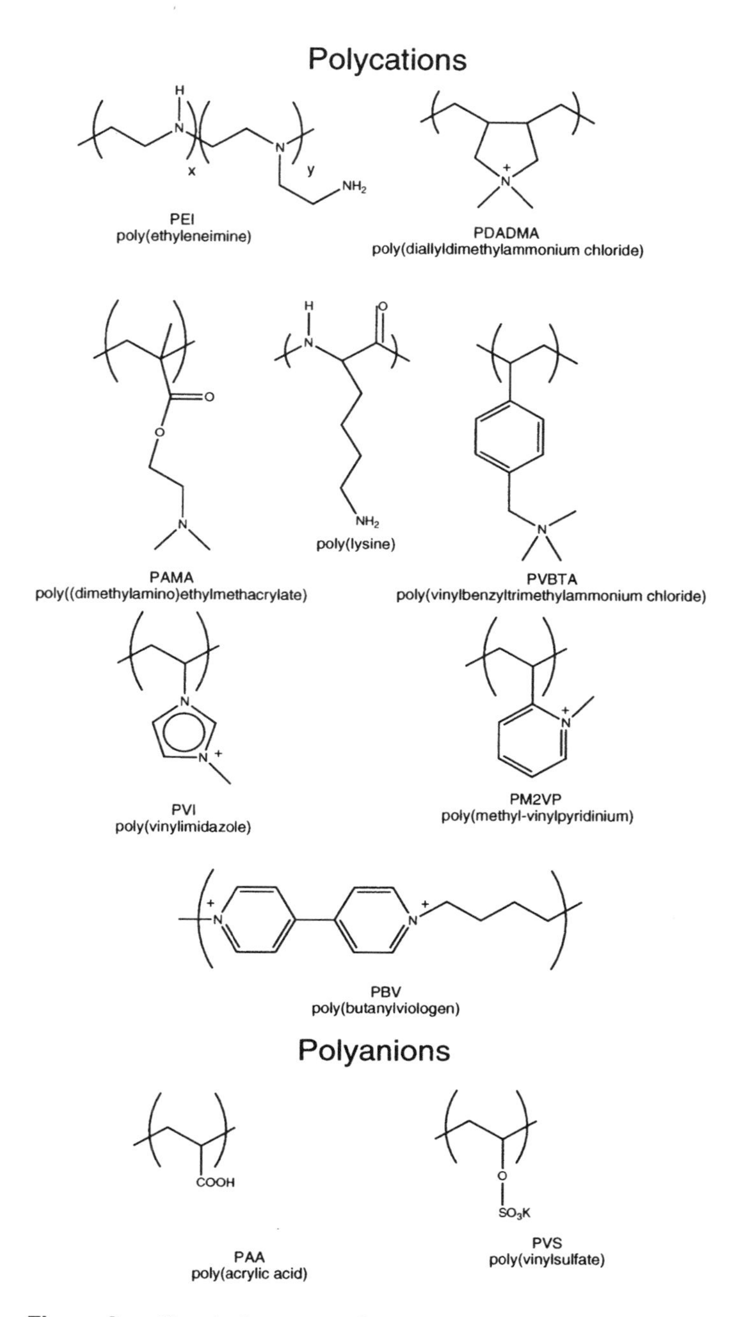

Figure 6a Chemical structure of some polyelectrolytes that have been used for electrostatic self-assembly.

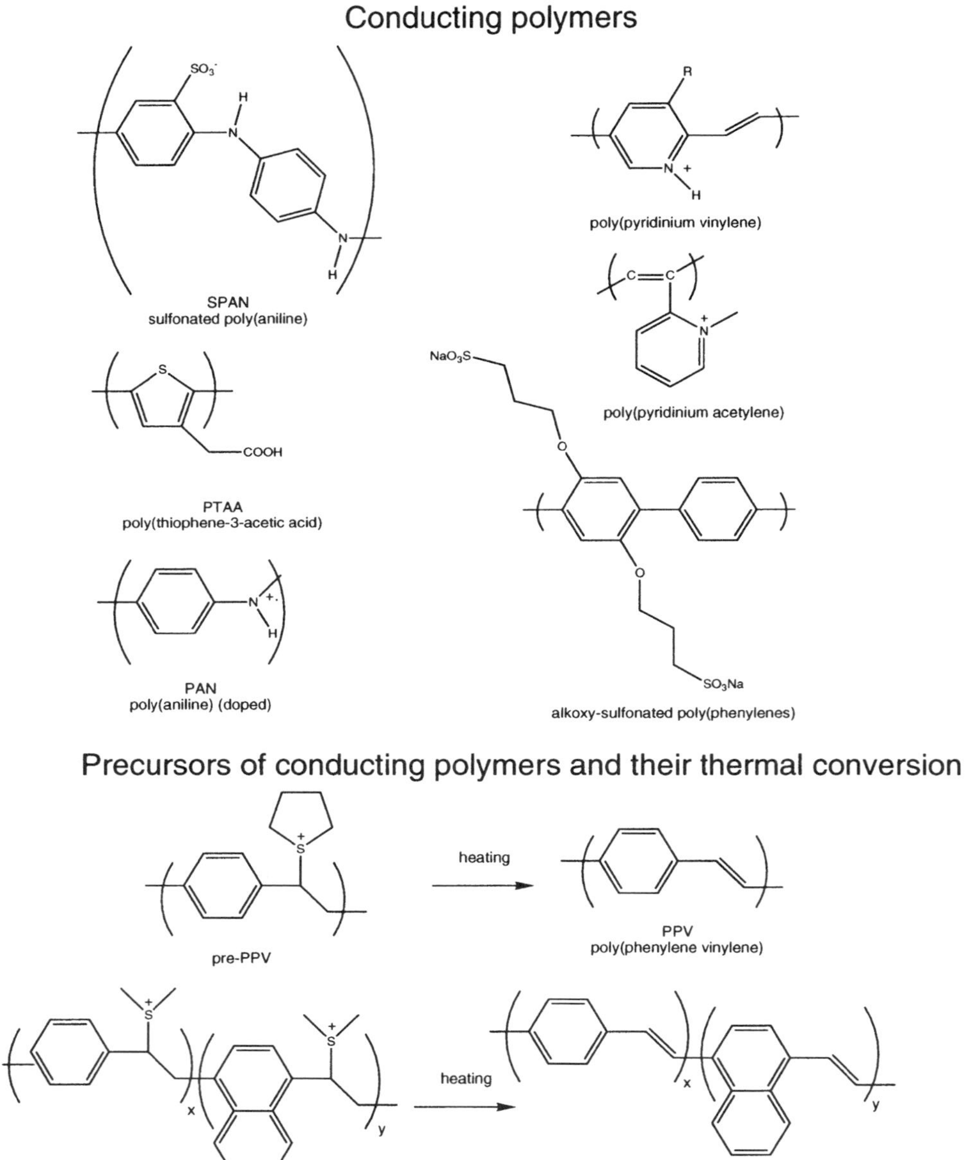

Conducting polymers

SPAN
sulfonated poly(aniline)

poly(pyridinium vinylene)

poly(pyridinium acetylene)

PTAA
poly(thiophene-3-acetic acid)

PAN
poly(aniline) (doped)

alkoxy-sulfonated poly(phenylenes)

Precursors of conducting polymers and their thermal conversion

pre-PPV

heating

PPV
poly(phenylene vinylene)

heating

copoly-(1,4-PV 1,4-NV)
poly(phenylenevinylene naphtylene vinylene)

Figure 6b Top: chemical structure of some conducting polyelectrolytes that have been used for electrostatic self-assembly. Bottom: chemical structure of some precursors of conducting polyelectrolytes that have been used for electrostatic self-assembly.

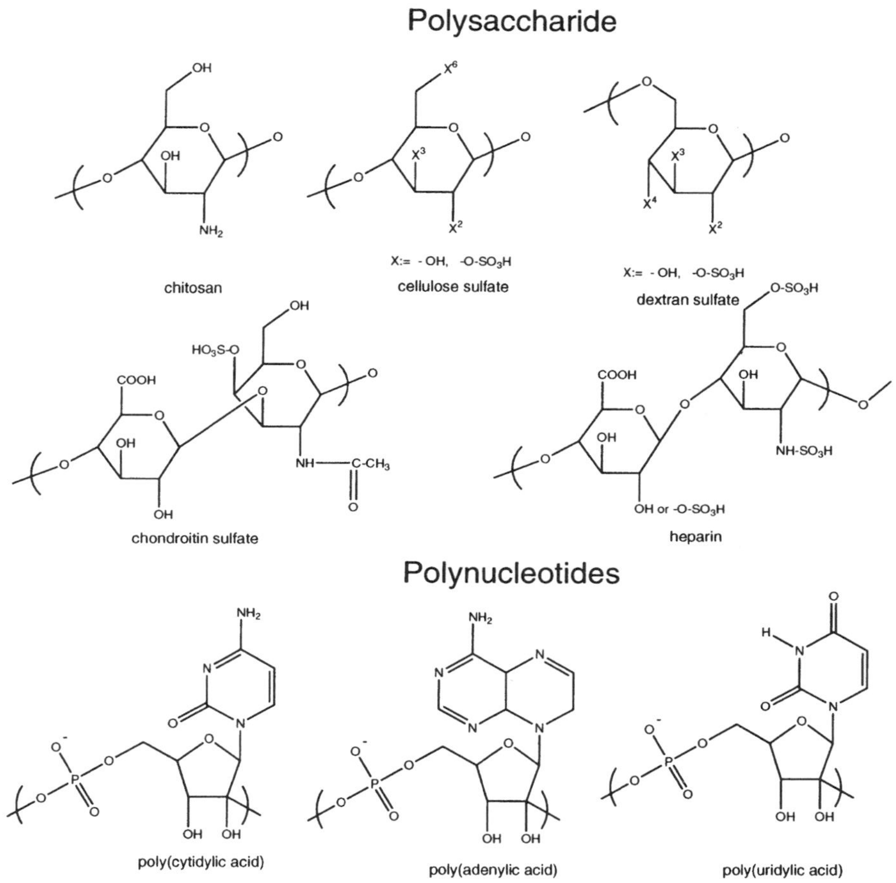

Figure 6c Top: chemical structure of some polysaccharides that have been used for electrostatic self-assembly. Bottom: chemical structure of some polynucleotides that have been used for electrostatic self-assembly.

The simplest adsorption theory of polyelectrolytes on an oppositely charged surface is based on an ion-exchange model [128]: the polyelectrolyte competes with the counterions to pair with the charged sites on the surface. In the pure electrosorption model, repulsion between the charged segments of the polyelectrolyte opposes accumulation in the surface region, but the segment–surface interaction promotes adsorption. The polyelectrolyte loses entropy upon adsorption, but more entropy is gained by the liberated counterions, resulting in a total gain of entropy. Since surface–counterions and polyelectrolyte–counterions interactions are replaced by surface–polyelectrolyte and anion–cation interactions, the enthalpic change should be small, and polyelectrolyte adsorption is thus mainly an entropy-driven phenomenon. Long chains should thus replace short ones. The entropy loss experienced by a stiffer polyelectrolyte is smaller, and it can be anticipated that the adsorbed amount is more important for such polymers [129]. As a consequence of intrachain repulsion, highly charged polyelectrolytes adsorb with a flat conformation, forming rather thin adsorbed layers. By contrast, the balance between entropic and enthalpic energy will lead to the formation of many more tails and loops in the case of polyelectrolytes with moderate segment charge. Indeed, such conformations are entropically favored, and, due to the long-range character of electrostatic interactions, adsorption energy is also gained for segments not attached to the surface [130]. The behavior of weak polyelectrolytes is more complex: their degree of ionization depends on the local pH of the solutions, which, close to the surface, is different from that in the bulk of the solution [131]. Aside from the pure electrosorption, specific interactions between the polyelectrolyte and the surface may also play a role in the adsorption behavior of the polyelectrolytes. In this respect, the effect of ionic strength is instructive. At high ionic strength, the segment–segment interaction is screened, and the polyelectrolyte behaves like an uncharged polymer: it adopts conformations with loops and tails, and the adsorbed amount is expected to increase (screening-enhanced regime). On the other side, the electrostatic attraction by the surface is also reduced, which can lead to a decreased adsorption (screening-reduced regime) or even to complete desorption of the polyelectrolyte chains: a fully screened polyelectrolyte can only adsorb if there is an attractive nonelectrostatic interaction with the surface. The balance between electrostatic and nonelectrostatic interactions determines the final behavior of the system. An increase of the adsorbed amount is usually observed for strong polyelectrolytes. In the screening-reduced regime, a decrease in adsorbed amount does not necessarily result in a decrease in film thickness [132]. Another complication is the commonly encountered reduced solubility of screened polyelectrolytes, which may lead to the precipitation of the polyelectrolyte at very high ionic strengths. Counterions can also displace the adsorbed polyelectrolyte if they have a specific affinity for the surface, and if their concentration is high enough. Experimentally, the ion-exchange model has been qualitatively verified: the adsorbed amount vs. segment charge, surface charge, or ionic strength follows the expected trends. Arguments coming from the ion-exchange theory are often used as a qualitative explanation of the effect of different parameters on the multilayer growth.

The ion-exchange model nevertheless has shortcomings. This theory based on a mean surface potential cannot explain why certain polycations adsorb on net positively charged TiO_2, where positive and negative charges coexist [133]. Theories taking into account the heterogeneity of surfaces have been developed

[134–137]. Similarly, it has been observed that amphoteric molecules (i.e., bearing positive and negative charges) are able to adsorb on surfaces with a charge opposite to their net charge [138,139]. Also, correct matching of the average distance between the charges of the polyelectrolyte plays a role [140]. It is thus the topological possibility of forming ion pairs rather than the net surface charge that determines adsorption.

A more fundamental shortcoming of the ion-exchange model is the assumption of a thermodynamic equilibrium. Even if only the strong and numerous electrostatic interactions with the charged surface are considered, adsorbed polyelectrolytes are expected to have a small mobility. Photobleaching experiments showed indeed that the diffusion coefficient for PAH adsorbed on mica is almost zero [141] (although movements of the whole molecule could be driven by capillarity); plastically deformed PSS/PAH bilayers did not anneal [141]. Polyelectrolyte adsorption may thus be expected to be an irreversible process. This problem has been tackled by different experiments; literature data indicate that polyelectrolyte adsorption on solid substrates is not a very reversible process [142]. Reversibility is enhanced in the presence of salt, and for low molar mass polyelectrolytes. As could be expected, irreversibility is also observed during the multilayer buildup by the layer-by-layer ESA. We have already seen that PAH/PSS multilayers remained unchanged when immersed in water (i.e., irreversibility upon dilution), and did not dissolve even at high ionic strength. Similarly, multilayers of poly-(L-lysine) (Fig. 6a) and copper phthalocyanine tetrasulfonic acid resist 10 min immersion in 1 M NaCl [48]. A PSS/poly(vinylbenzyltrimethylammonium chloride) (PVBTA) (Fig. 6a) bilayer immersed in pure water does not show desorption either [143]. The PVBTA layer is not displaced by the addition of 0.2 M YCl_3, although the highly charged Y^{3+} competes very strongly for the negatively charged sulfonate groups. Self-exchange experiments give similar results: the radiolabeled PSS situated in the outer layer of the PSS/PVBTA bilayer is not exchanged by a large excess of unlabeled PSS of the same molar mass. Radiolabeled poly(N-methyl-2-vinylpyridinium) (PM2VP) was used to build up multilayers with PSS. PM2VP-capped multilayers were not exchanged by poly(diallyldimethylammonium chloride) (PDADMA) (Fig. 6a) for exposure times up to 2 h. Partial exchange of only about 20% was observed after several days [126]. In contrast the PAH outer layer of a PAH/PSS multilayer was at least partially exchanged by fluorescence-labeled PAH [104], but, in this later case, 1 M NaCl was added to the polyelectrolyte solutions, which probably increased adsorption reversibility, as just shown above. Experiments testing the reversibility upon pH changes have also been carried out [144]. In these experiments, multilayers of PSS and a weak polycation, poly(dimethylaminoethylmethacrylate) (PAMA) (Fig. 6a) are built up at pH 8. The pH is then decreased to 4, increasing the charge of PAMA by protonation. According to ion-exchange theories, PAMA desorption should occur. Nevertheless, no desorption occurs when

PSS is the outer layer, proving that PAMA is effectively trapped inside the multilayer, and only limited desorption occurs when PAMA forms the outermost layer. The desorbed molecules are probably loosely attached PAMA molecules in the outer layer. Cassagneau et al. [145] mentioned that poly(pyrrole)/PSS and poly(pyrrole)/α-ZrP bilayers are so stable that even sonication in ethanol cannot destroy these assemblies. It is questionable whether an equilibrium theory can be used to study systems like polyelectrolyte multilayers exhibiting such pronounced adsorption hysteresis. Before concluding this discussion about the irreversibility, we want to stress that the aforementioned lack of mobility of the polyelectrolyte in the layers does not preclude local movements of parts of the polyelectrolyte chains, and especially of the pendant group. Rearrangement of pendant chromophores trapped in the multilayers has indeed been revealed by SHG measurements [146,147], and by the dependence of spectral shifts (metachromic effect) on the number of dipping cycles in UV-measurements [140,148,149]. In this context, it is probably important to remember that water molecules are thought to occupy 40% of the volume in the PAH/PSS multilayers [101].

Another feature unexplained by the ion-exchange theory is the often-found charge overcompensation upon polyelectrolyte adsorption. In the case of pure physisorption, ion-exchange theory predicts only a minute overcompensation, due to a balance between entropic and electrostatic energy. Notwithstanding, nonelectrostatic attractive interactions may lead to some supplementary overcompensation. It is nevertheless hard to believe that nonelectrostatic interactions are present in all the observed cases of overcompensation, considering the huge number of different substrates and polyelectrolyte pairs concerned. Furthermore, it has been shown that quaternized poly(vinylpyridine) (PM2VP) (Fig. 6a) has no specific interactions with TiO_2, by measuring the adsorbed amount as a function of ionic strength [133], but very stable multilayers could be built on this substrate by alternatively adsorbing PM2VP and PSS [144]. The successful multilayer buildup is indirect evidence of overcompensation upon adsorption of PM2VP on TiO_2. An explanation of the overcompensation is best sought in kinetic phenomena. Adsorption is often observed to proceed in steps: a first, rapid one taking place within minutes, and a second, slow one that can be as long as several hours or days, due to reconformation of the adsorbed layer [133]. If reconformation is slower than adsorption, charge overcompensation is expected: during adsorption, polyelectrolyte chains anchor by only some of their charged groups to the surface, and, before they have time to reconform in order to occupy the neighboring charged sites of the surface, these latter sites are occupied by other polyelectrolyte chains. The attached polymer molecules progressively create a surplus of charge that leads to an electrostatic barrier which prevents the attachment of other polyelectrolyte chains by repelling them: the phenomenon is self-regulating. Solutions of higher ionic strength shield the electrostatic interactions better, leading to a lowering of the activation energy necessary for getting over the electrostatic barrier. Furthermore, increasing the ionic strength fastens the reconformation steps, by decreasing the strength of the ion pairs. The abovementioned irreversibility corroborates this model of kinetically hindered equilibrium.

Charge overcompensation is theoretically a prerequisite for multilayer buildup. Every example of successful multilayer buildup is thus an indirect proof of charge reversal upon polyelectrolyte adsorption. By contrast, measurements of the ζ-potential, and of its sign reversal during multilayer buildup, are direct proofs of charge overcompensation. In addition to the already mentioned case of PAH/PSS multilayers [95,96,97,104,106], there have been only a few such measurements. Our group has presented ζ-potential values calculated from streaming potential measurements for planar substrates covered by multilayers made of poly(vinylsulfate)(PVS) (Fig. 6a) and an ionene [68]. From streaming potential measurements again, the pH dependence of the ζ-potential of different multilayers assembled by ESA has been reported by Schwarz et al. [150]. Electrophoretic measurements have been used to determine the ζ-potential of multilayer-coated polystyrene latex particles [95,96]; alternatively positive and negative potentials were observed, respectively, when polycation and polyanion formed the outer layer, for the PSS/PDADMA and deoxyribonucleic acid (DNA)/PDADMA layer pairs. The observed reproducible values of the ζ-potential from one dipping cycle to the next are consistent with the usually observed linear increase in film thickness during polyelectrolyte multilayer buildup. Linear increase of the adsorbed amount also requires an approximately constant roughness at the multilayer surface. Hoogeveen et al. [144] studied the ζ-potential of poly(vinylimidazole) (PVI)/poly(acrylic acid) (PAA) (Fig. 6a) multilayers on silica particles as a function of d_q, the degree of quaternization of PVI. They showed that alternatively positive and negative potentials are obtained, respectively, when PVI and PAA are the outer layers if $d_q > \sim 0.1$. The higher the d_q, the higher the absolute values of the measured ζ-potential are. Interestingly, no multilayer growth is observed when $d_q = 0$, and stable multilayer formation occurs only when $d_q > 0.18$. This demonstrates experimentally a direct link between polyelectrolyte charge, charge overcompensation, and multilayer buildup. Schlenoff et al. [126] have used radiolabeled counteranions to determine surface excess charge in PSS/PDADMA multilayers. They found 45 and 46% of uncompensated charges on the surface of multilayers capped, respectively, with the polycation and the polyanion.

The picture that emerges is that of a multilayer whose topmost adsorbed layer contains trains anchoring the polyelectrolyte to the multilayer, and dangling loops or tails. These loops and tails are responsible for the charge overcompensation which, in turn, determines the amount adsorbed in the next step. The adsorbed amount depends thus on the nature of the underlying layer. For example, QCM monitoring of the deposition of PSS shows the adsorbed amount depending on the nature of the underlying polycation [87] or protein layer [124]. Also, we observed that the thickness of xAyAxAy multilayers (where A denotes a polyanion, x and y denote polycations) was significantly different from the thickness of yAxAyAx multilayers although they both contain the same number of layers of A, x, and y (see Fig. 7) [151]. Similar observations have been made with other systems too [140,147,148,152,153], demonstrating that the thickness of the layers in multilayers built by ESA depends on the chosen polycation–polyanion pair. Frequently in the literature authors assume, however, that the polyelectrolyte layer thickness is an intrinsic property of the polyelectrolyte, and subtract the polyelectrolyte layer thickness measured on other systems from their own measured growth increment, in order to calculate the contribution of the other component of their multilayer to the overall growth step.

According to the above discussion, the conclusion of such reasoning must be taken with much care, as it may at most be only a crude approximation. In agreement with the ion-exchange theory, if the loops and tails determine the adsorbed amount, this amount should diminish with increasing polyelectrolyte charge density and decreasing surface charge density. This has been nicely illustrated in a study of the buildup of a PAA/PAH multilayer as a function of the pH of the polyelectrolyte solutions used for the assembly [154] (Fig. 8). In the pH range used in this study, PAH is fully protonated. Thus, with PAA as the outer layer, the pH controls the surface charge, since PAA is a weak polyelectrolyte, without changing the linear charge density of the adsorbing PAH. It is found that the average PAH layer thickness increases with increasing surface charge density (increasing the pH of the PAH dipping solution). When PAH is the outer layer, the pH determines the linear charge density of the adsorbing PAA, without changing the surface charge density. It was found that the average PAA layer thickness increases with decreasing linear charge density (decreasing the pH of the PAA dipping solution). Absolute values of the linear charge density, however, may be difficult to determine accurately, and nonelectrostatic interactions may play a role when the linear charge density becomes weak.

We have just seen that the formation of loops and tails seems to determine the adsorbed amount by dipping cycle. The possibilities for loops and tails formation are different in the vicinity of a solid surface or in the loose outer layer of polyelectrolyte multilayers. An effect of the substrate on the first deposited layers is thus to be expected, causing a nonlinear growth of the multilayer thickness for the first layers versus number of dipping cycles. Such nonlinear growths have indeed been often observed, for example, in Refs. 76 and 155 to 158. In agreement with the above, Lösche et al. [101] pointed out that linear increase of the PAH/PSS multilayer thickness is observed to start for a thickness that coincides with the length scale of interdiffusion between these polyelectrolytes, suggesting a causal relationship. Interestingly, looking at the adsorption of different polyelectrolytes on the same surface, nonlinear growth of the initial layers may occur for only some of them [156,159,160]. In some cases, nonlinear growth may also be

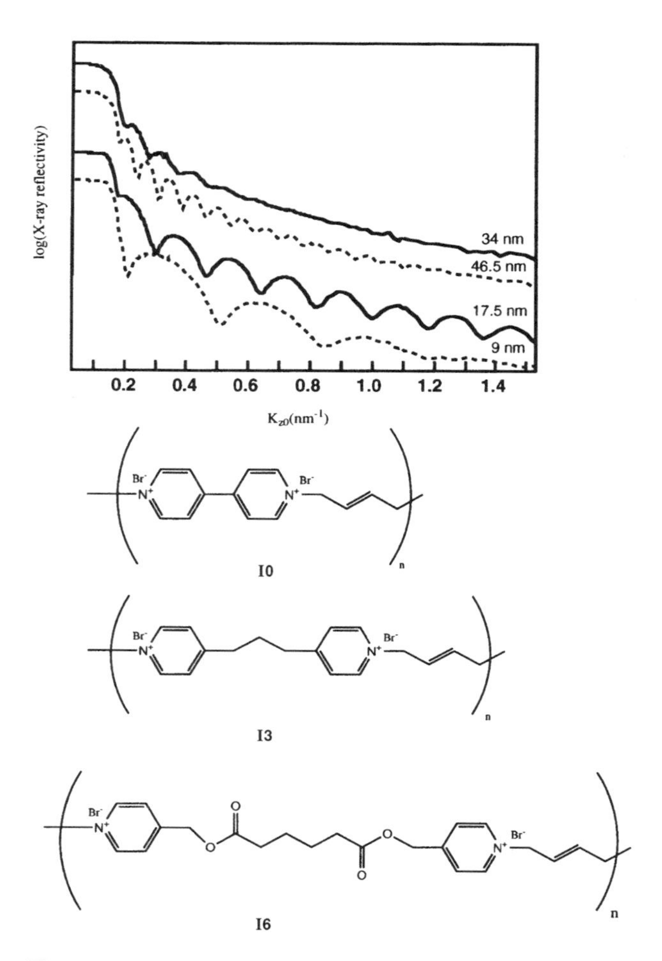

Figure 7 X-ray reflectivity versus K_{z0}, the component of the wavevector of the incident photon perpendicular to the interfaces for different ESA multilayers containing aromatic ionenes (I0, I3, I6, below) and poly(vinylsulfate) (PVS). The curves have been shifted vertically for clarity. From top to bottom, the structure of the different samples is: Si/I3/PVS/I0/PVS/I3/PVS/I0, Si/I0/PVS/I3/PVS/I0/PVS/I3, Si/I0/PVS/I6/PVS/I0/PVS/I6, SI/I6/PVS/I0/PVS/I6/PVS/I0. The thickness of these multilayers in nm is reported in the figure. (From Ref. 151.)

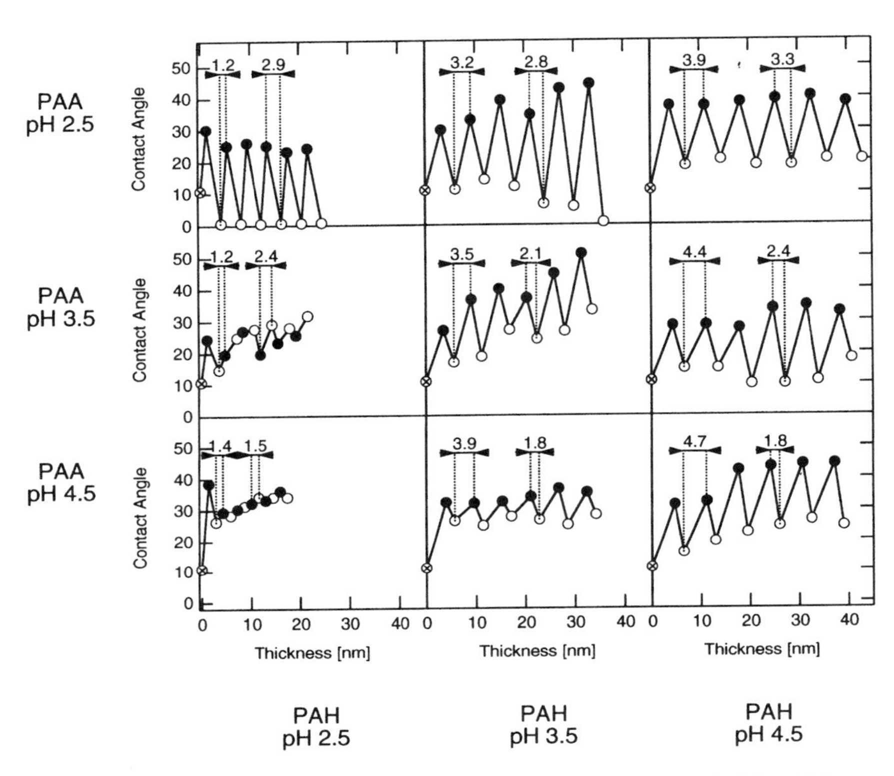

Figure 8 Contact angle vs. thickness in nm for multilayers composed of 1 to 12 layers. The pH of the PAA and PAH dipping solutions are indicated, respectively, to the left and under the graphs. Open circles: multilayers with PAA as outermost layer. Filled circles: multilayers with PAH as outermost layer. The average incremental contribution of PAH and PAA to the multilayer thickness are reported in nm in the figure; left: contribution of PAH; right: contribution of PAA. (Data taken from Ref. 154.)

attributed to a lack of affinity between the polyelectrolyte and the substrate [158]. In this case, the multilayer growth begins by adsorption at imperfections, forming islands that finally merge together with an increasing number of deposition cycles. Linear growth usually occurs after less than 5 dipping cycles, but in some instances up to 20 dipping cycles are needed [155]. Once the regime of linear growth has been reached, we may wonder whether the growth increment is the same for multilayers deposited on different substrates, if they are made of the same given polyelectrolyte pair. A growth increment independent of the substrate would strongly suggest that the structure and properties of the outer layer converge towards a state independent of the substrate. Unfortunately, this question is still open, and cases for [6,86,123,147,158,161–164] and against [89,93,94, 165,166] this view have been reported, although there is a vast majority of ''for'' cases. As a result of the outpacing of ''for'' cases, polyelectrolyte multilayers have even been proposed as decoupling layers for the adsorption of proteins on a variety of substrates [167]. Finally, we stress that the influence of the substrate on polyelectrolyte multilayer buildup may also be detected by measuring properties other than the growth step per dipping cycle. In some instances, these methods are even more sensitive, revealing the effect of the solid surface although a linear growth of the film thickness versus number of dipping cycles is observed. Examples are the formation of aggregates resulting in a shift of the maximum absorbance in the UV-vis measurements [140,148,168] or a decrease in the overall degree of orientation of NLO chromophores, as evidenced by SHG experiments [147,169–171]. An example from our group is presented in Fig. 9. Shifts in electroluminescence and photoluminescence with decreasing number of layers have also been reported, but they are due rather to a confinement of electron and holes than to the presence of the substrate [172–174].

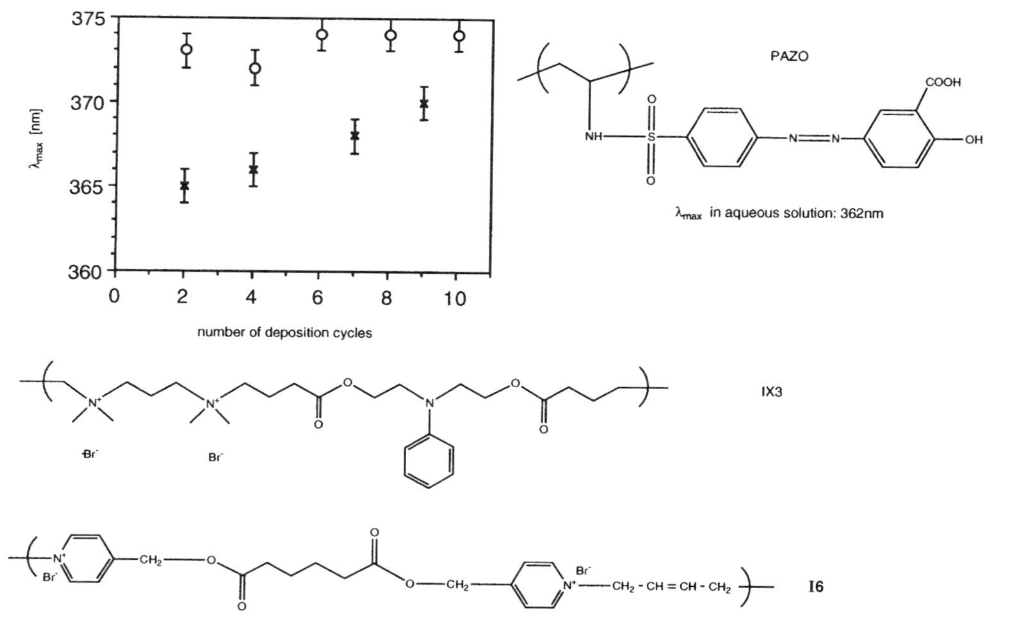

Figure 9 Shift of the absorbance maximum of the colored polyanion PAZO in growing multilayers made by ESA with different polycation partners (○ = IX3 and × = I6). (Data taken from Ref. 140.)

Since the multilayer buildup by the ESA does not proceed under thermodynamic equilibrium conditions, it is questionable whether the approximately 1:1 charge stoichiometry predicted by the ion-exchange theory is found in the multilayers. Indeed, it is known from electrochemical manipulations after multilayer construction that a certain amount of extrinsic charge compensation does not lead to multilayer destruction [126]. Furthermore, it is also conceivable that some buried charges of the outer layer may not be accessible for polyelectrolyte complexation in the next adsorption step. In fact evidence for both stoichiometric [67,126,144,175] and nonstoichiometric [89,93,94,144] multilayers can be found in the literature, although only few data are available on this topic. It would be interesting to establish the minimum ion pair density required for the stability of the multilayer.

A related problem that may occur during multilayer buildup is the desorption of some of the polyelectrolyte from the outer layer caused by the adsorbing polyelectrolyte, in order to form a soluble complex. This problem is easily detected when using strongly colored weakly charged ionenes [176] (Fig. 10), and has been observed for other systems as well by in situ measurements, or simply from a smaller thickness of the film after an adsorption cycle [98,114,144,177]. Analysis by high-performance liquid chromatography of PSS solutions, which were used for PAH/PSS multilayer buildup, also revealed the presence of a small amount of PAH/PSS complex in the solution [95]. Serizawa et al. could circumvent the desorption problem by adsorbing from a solvent in which the complex was insoluble [177]. The problem of desorption is particularly acute in the case of multilayers containing small charged molecules, although it can be somewhat controlled by a proper setting of the ionic strength [178].

An interesting phenomenon arising from the use of polyelectrolytes for the buildup of multilayers is the observed self-healing capability of the layer-by-layer deposition process. An example of this is the restoration of a smooth film surface after the addition of a much rougher layer, such as a virus [52] or globular protein layer [179]. The buildup of composite films made of montmorillonite (a clay platelet) and PDADMA leads to a surface presenting large pits, up to 700 nm diameter and 30 nm depth [180]. These pits also were smoothed during the layer-by-layer deposition; 188 nm diameter and 14 nm deep pits were covered after only one deposition cycle. These self-healing capabilities may be a drawback for patterning [181], but enable the buildup of polyelectrolyte multilayer on weakly charged surfaces [158,182]. In the latter case, adsorption occurs at isolated imperfections, forming islands that later grow vertically and laterally until they coalesce. The ability to bridge the gaps has been shown to be directly related to the molar mass of the polyelectrolyte [182]. We believe that the increase of the self-healing capabilities with increasing molar mass is a general trend.

Many parameters may influence the deposition process of polyelectrolyte multilayers: the linear charge density, flexibility, molar mass and polydispersity

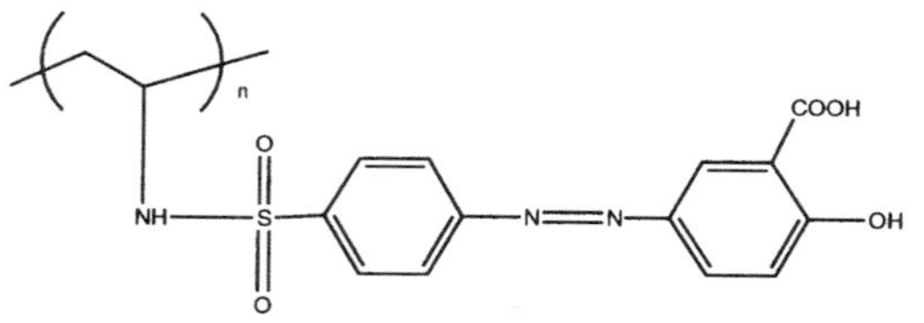

PAZO [113,140,146,148,156,168]

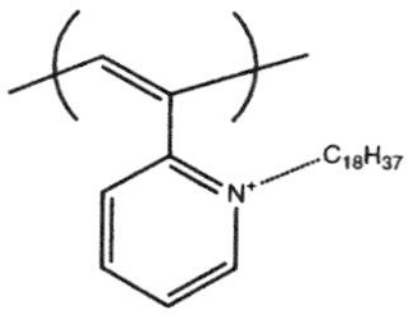

P-2PA [248]

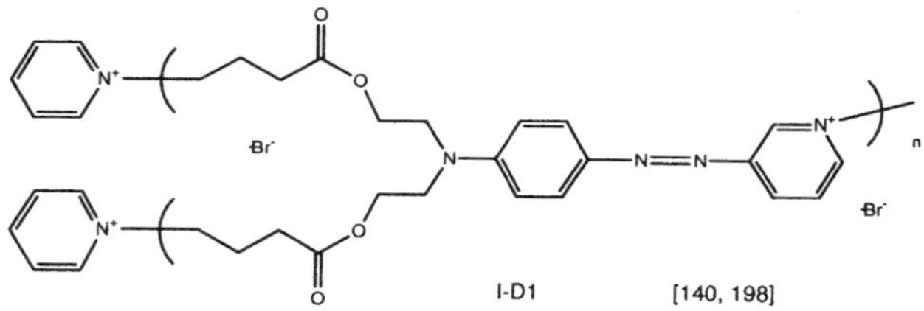

I-D1 [140, 198]

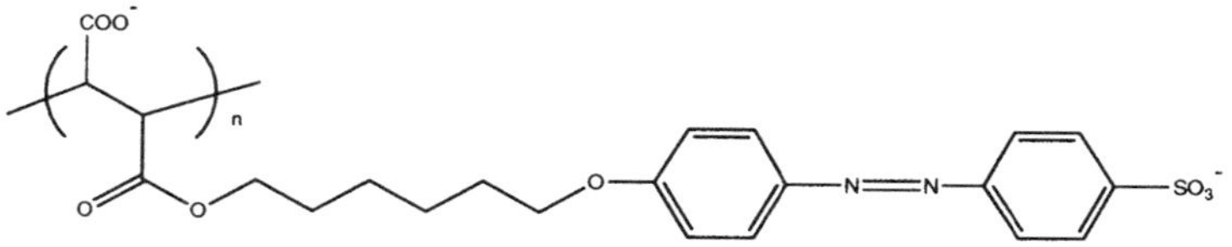

A-D1 [223]

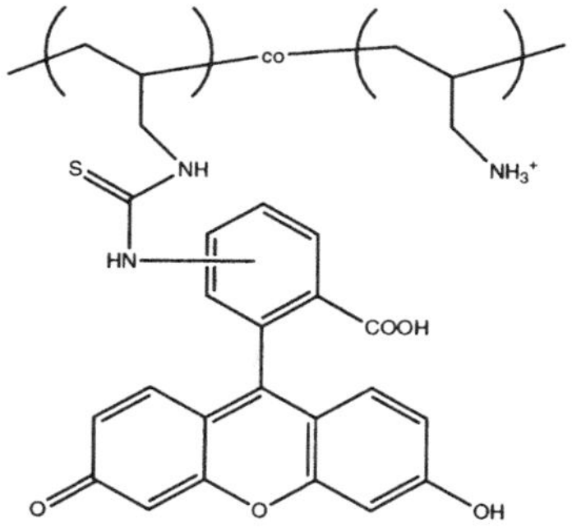

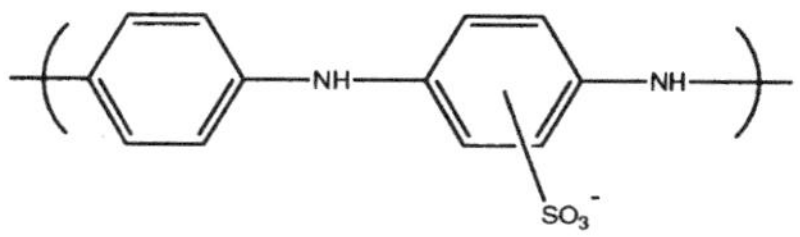

PAni-S [161,162,166,174,299]

PAH-F

[104,105,114,115,265]

Figure 10 Examples of colored polyelectrolytes employed in ESA multilayers.

of the polyelectrolytes, the existence of specific interactions, the matching of the charge–charge distance, the substrate charge and its distribution on the surface, the ionic strength, pH, temperature, and polyelectrolyte concentration of the dipping solutions, the nature of the solvent [177], the adsorption time, the presence or not of a drying step, or even the application of an external potential [180,183].

Concerning the linear charge density of the polyelectrolyte, we have already cited the work of Yoo et al. [154] and of Hoogeveen et al. [144]. The latter author showed the existence of a critical linear charge density for the successful buildup of polyelectrolyte multilayers, depending slightly on the ionic strength of the solutions. Our group has confirmed these results in the case of ionenes [157,184]; a critical charge density close to that found by Hoogeveen et al. was established. Also, liquid-crystalline (LC) polymers containing one charge per 80 carbon atoms could be used successfully for multilayer buildup, but not LC polymers with an ionic group content of one charge per 176 carbon atoms [185]. As the charge density is decreased, other interactions may come into play. This in turn may render the determination of a critical charge density difficult. Interestingly, decreasing the charge density may then lead to even better assemblies [140]. Further studies are clearly required to clarify this notion.

Aside from the already mentioned case of PAH/PSS multilayers, the effect of molar mass has also been studied in the case of polyaniline/PSS multilayers [123] (Fig. 6b); no dependence on the molar mass of PSS was found when it was varied from 5,000 to 1,000,000. Very weak dependence on the molar mass of the polycation has been observed in the case of PAH/bolaamphiphiles multilayers [186]. Interestingly, even poly(thiophene-3-acetic acid) oligomers (Fig. 6b) of only 8 to 12 repeat units may be useful in assembling multilayers [161].

As mentioned earlier, an increase in ionic strength generally leads to an increased thickness in the case of the adsorption of strong polyelectrolytes. Accordingly, an increased film thickness is generally observed in the case of polyelectrolyte multilayers prepared from solutions of high ionic strength [62,67, 110,126,153,172,184,187–189] (see also Fig. 5). The thickness can be changed by as much as one order of magnitude. An increase in ionic strength may also increase [62,188,189] or reduce the film roughness [190].

Similar to what is observed in the case of PAH/PSS multilayers, a slight increase of the adsorbed amount is observed for higher polyelectrolyte concentrations [123,143,161,162,191], leading to an increase of the growth step per dipping cycle [123,162,192].

Numerous kinetic investigations have been reported in the literature, aiming at a determination of an appropriate dipping time. Rapid adsorption usually takes place in the first few minutes [107,140,146,159,170,193–195], sometimes followed by a slower adsorption step as adsorption saturation sets in [132,156,162, 187,196,197] (Fig. 11). Nevertheless, adsorption may sometimes take more than one hour [198–200]. Kinetics have been shown to depend on the substrate [163],

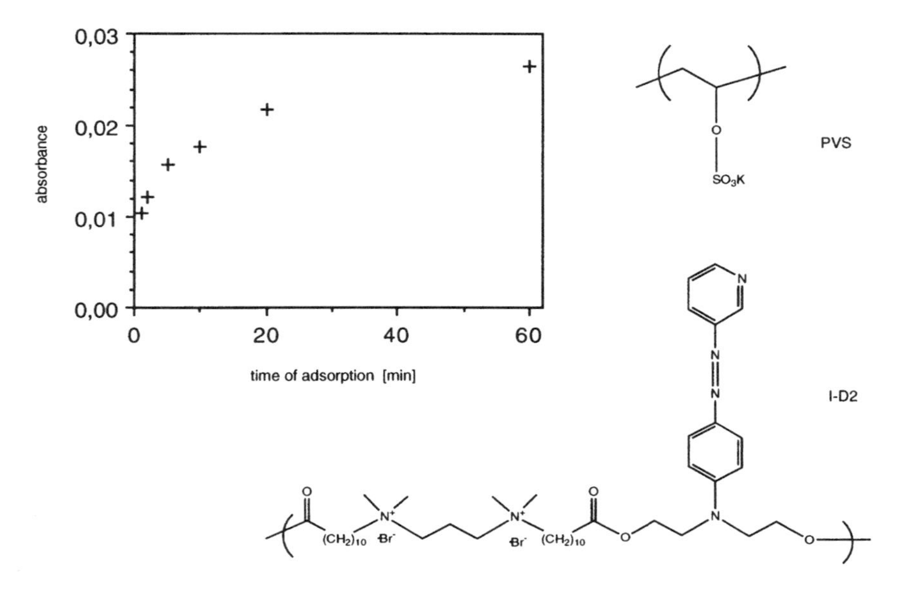

Figure 11 Kinetics of the adsorption of colored ionene I-D2 on a layer of polyvinylsulfate, as followed by vis-spectroscopy. (Data taken from Ref. 140.)

the ionic strength of the solution [187], and the polyelectrolyte concentration [123,143,161,162]. Other parameters, such as the molar mass, have also been shown to play a role in the case of the adsorption on a solid surface [133]. Since most of the physisorbed amount is deposited at the very beginning of the adsorption process, it is tempting to build multilayers with adsorption times shorter than the time required for saturation of the adsorption. Multilayers can indeed be successfully assembled for dipping times shorter than the time required for saturation of the adsorption [51,140,188], and, in some instances, dipping times shorter than five seconds have been shown to be convenient [49,158,201]. Considering the amount of kinetic studies and the number of publications in the field of ESA, it must be recognized that dipping time is usually arbitrarily selected; this is not a very satisfactory situation, as the optimal dipping time is most certainly strongly system-dependent.

As is the case for PAH/PSS multilayers, the presence of an apparently insignificant drying step may have different effects on the deposition process. It has indeed been mentioned that sample drying speeds the relaxation of the polyelectrolyte chains [201], increases the adsorbed amount [127,148,202], and, especially when performed by spinning the samples, results in a lower roughness of the films [153,203].

Studying the kinetics of adsorption of poly(o-methoxyaniline), it has been found that the adsorbed amount is notably increased when adsorption is interrupted several times for drying and measuring the sample [163]. Furthermore, the time required for reaching the saturation of the adsorption is enhanced. This is similar to what we mentioned previously in the case of the adsorption of PSS [87]. We believe this is a general phenomenon, and we must thus warn against kinetic measurements obtained by repeatedly dipping and drying the same sample: the time required for reaching saturation may be systematically overestimated.

IV. SUITABLE SUBSTRATES

Most ESA multilayers have been built on glass, silicon, quartz, and ITO (indium-tin-oxide), since these substrates are particularly well suited for a number of experimental techniques. Another interesting point is that their surface physicochemistry is well studied, and convenient procedures can thus be found in the literature to clean and charge their surfaces. When hydroxyl groups are present, a correct choice of the pH is sufficient to generate a useful surface charge. Other examples of procedures to generate charges on these substrates are: bolaform amphiphile adsorption, plasma deposition, or silanization (see Section I). A number of polyelectrolytes, such as PEI or PDADMA [183], may even adsorb on the bare surface of these and other substrates, and thus be used as a precursor layer for further multilayer assembly.

A number of other substrates have also been used for the ESA. Among the metals, gold has been widely used. Its surface is easily charged by the chemisorption of alkanethiols (see Section I). Successful multilayer assembly on bare gold has also been reported [76]. Examples of other metals that have been successfully used are platinum [204], silver [124], and aluminum [205]. For the latter, the surface must be assumed to consist of aluminum oxide.

A whole range of polymer substrates has also been used. Aside from the polymers already mentioned in the PSS/PAH case study, poly(propylene) (PP), poly(methylmethacrylate) (PMMA) [206,198], poly(ethylene) (PE), polyurethane [207], and poly(vinyl chloride) (PVC) [208] have also been used, just to cite a few among the most important industrially. Successful multilayer assembly on polymeric substrates is obtained either by chemically modifying the polymer surface, or by functionalization of the polyelectrolyte so that nonelectrostatic interactions may occur.

Substrates covered by LB films have also been used [209]; as many different substrates are suitable for LB films, LB film deposition can be seen as a versatile way of bringing charges to these surfaces prior to ESA.

Substrates need not be planar for ESA, as exemplified by the already mentioned buildup of PAH/PSS multilayers on latex particles, and by the buildup of multilayers on optic fibers [210]. Substrates even need not be solid, as multilayers have been grown on alginate gel beads [211].

Although successful multilayer buildup has been demonstrated on many different substrates, very little is known on the adhesion of these multilayers to their supports (see Section VIII). Especially when adsorption of the first layer is only due to specific interactions between the polyelectrolyte and the substrate, good adhesion is questionable [91,94,198].

V. APPLICATION RANGE

The adsorption of cationic bipolar amphiphiles has been shown to reverse the surface charge of a mica substrate, and to form well-ordered monolayers [212,213]. Successful charge reversal is obtained only with a careful molecular design of the bolaamphiphiles, which must be long enough to avoid flat adsorption on the surface and favor lateral interactions, and short enough to avoid adsorption as a loop [44,214]. The problem of loop formation vanishes if a rigid core, like a biphenyl, azobenzene, or diacetylene unit, is inserted in the middle of the molecule, and consequently all the bolaamphiphiles presented in the following contain such rigid units. Nevertheless, successful self-assembly of a monolayer of a flexible amphiphile has also been reported [215]. Alternately adsorbing cationic and anionic bolaform amphiphiles leads to the formation of multilayers whose thickness depends linearly on the number of dipping cycles [44]. Multipolar molecules have also been used for the assembly of multilayers, offering the advantage of an increased tolerance against the formation of small defects [216,217]. It has been shown that composite multilayers made of bolaform amphiphiles and inorganic nanoparticles [209,218,219] or enzymes [220,221] could be successfully assembled. Nevertheless, so far most studies on multilayers containing bolaamphiphiles and, more generally, multipolar dye molecules concerned composite assemblies with polyelectrolytes [48,149,155,164, 178,197,222–225]. A number of studies have been driven by potential applications arising from the insertion of dye molecules in a multilayer. For example, electroluminescent [152,226–228], photoisomerizable [63,229,232], SHG-active [233,234], or pH sensitive dyes [149,235] have been assembled in multilayers with suitable polyelectrolytes. The possibility of manipulating several dyes in a multilayer opens the door to other applications, such as the fabrication of tunable color filters [228]. The benefits of the association with polyelectrolytes appear clearly when looking at the case of composite assemblies with a pH-sensitive dye, congo red. Indeed, unlike dye monolayers on glass, the multilayer can be cycled from very low pH to very high pH many times without noticeable dye desorption [235]. A number of studies have also been devoted to the photopolymerization of acetylene and cinnamoyl-containing bolaamphiphile layers [186,

212,213,236–238], which may increase the chemical, thermal, and mechanical stability of the multilayers containing these molecules. The successful photopolymerization implies the existence of well-ordered amphiphilic domains in these layers, even when sandwiched between polyelectrolyte layers. Such well-ordered layers might be thought to have interesting barrier properties [186]. We have already mentioned the importance of lateral interactions for the successful formation of bipolar amphiphile monolayers. These interactions lead sometimes to aggregation in solutions. Amphiphiles with only one end charged may also lead to such aggregates in solution, forming charged bilayer membranes. Multilayers have been made by the ESA technique by adsorbing alternately such positively and negatively charged membranes, or by adsorbing such membranes alternatively with polyelectrolytes [239].

Polyelectrolyte–polyelectrolyte multilayers have even attracted more interest, partly due to their self-healing capabilities, leading to a reliable and easy buildup, and to the wide variety of useful polyelectrolytes, enabling tailoring of the properties of the multilayers. Many different polyelectrolytes have been used [58,140,240–244], among which some are commercially available, such as PSS, PVS, PEI, PAH, PDADMA, and PAA, for example. LC polyelectrolytes [140,185, 245], block copolymers [246–248], and dendrimers [51,249,250] are readily assembled in multilayer structures. Reproducible deposition of PEI/Cu^{2+} complex in alternation with PSS to produce multilayers has also been demonstrated [243]. A number of conducting polymers, such as poly(thiophene acetic acid), sulfonated poly(aniline), poly(pyridinium acetylene) (Fig. 6b), bearing charged groups, have successfully been used for the electrostatic layer-by-layer self-assembly [162,248,251]. The partial doping of conjugated polymers that have no charged groups in their chemical structure, such as polyaniline, leads to the presence of delocalized charges along the polymer backbone. This delocalized charge has been shown to be suitable for multilayer buildup by the ESA technique [123,192]. Alternate dipping of a substrate into a chemically active aqueous solution of an in situ polymerized polymer, such as poly(pyrrole) or poly(aniline), and a solution of a polyanion also leads to satisfactory multilayer assembly [252].

The biochemistry of living organisms relies heavily on macromolecules, not only for controlling the cell metabolism, but also as structural elements. Most of these macromolecules can be classified as polysaccharides, nucleic acids, and proteins. Polysaccharides are rather rigid long chains built from one or two monomeric units, and are mainly used by nature as building blocks or for food storage. Some polysaccharides are made of monomers bearing chemical groups such as carboxylic acids or amines, making them suitable candidates for the layer-by-layer electrostatic multilayer buildup. Multilayers containing chitosan [127,177, 187], chondroitin sulfate [187], dextran sulfate [125,208], cellulose sulfate [153], and heparin [175,207,208] (Fig. 6c) have indeed been successfully assembled. Adsorption of some of these polysaccharides may enhance surface biocompatibility, a property needed for cell culture and human implants, for example. Since the long-term stability of these films in physiological solutions or in blood plasma is of importance, the presence of covalent bonds between the layers would be an advantage. Such bonds have been created by posttreatment of the multilayers, or from adsorption with a reactive polyelectrolyte [125,207,208].

Ribonucleic and desoxyribonucleic acids, which are used by the cell to store and transmit information, have covalently linked backbones made of alternating pentoses and highly negatively charged phosphates. They can thus be used as polyanions for the layer-by-layer deposition process. Whether DNA conserves its double-helical structure is of importance for potential applications such as the express-diagnostics of a virus, or for the control of environmental pollution with carcinogenic molecules [253]. For example, the conservation of the double helix is of importance for the biosensing of a number of drugs from the alkaloid and antibiotic classes that intercalate in this double helix [254]. A conformation transition inside the multilayers may arise from temperature [47] or pH changes [255]. Homopolynucleotides such as polycytidylic [255], polyuridylic, and polyadenylic acid (Fig. 6c) [82] have also been deposited.

Proteins, which are used for structural or for metabolic purposes by the cell, are made of amino acids. Some of them bear a negatively charged lateral chain, such as aspartic or glutamic acid, while others have a positively charged lateral chain, such as lysine, arginine, or histidine. Most of the proteins are thus amphoteric, and may be globally negatively or positively charged when used, respectively, at a pH above or under their isoelectric point. Accordingly, composite protein-polyelectrolytes [81,114,124,125,159,179,256] and protein-DNA [196] multilayers have been prepared by the layer-by-layer deposition technique. Potential applications for such films range from nonthrombogenic surfaces [207] to biosensing [257,258], through the immobilization of enzymes for biocatalysis [6]. It is of importance for these applications that the proteins are not denatured by the deposition process in order to remain biologically active. In this respect, ESA, which does not involve any enzyme modification or covalent bonding should preserve in most cases the functional characteristics of the proteins [81, 124,159]. In biosensing, it may be hoped that multilayer buildup allows for a tunable sensitivity through a control of the number of layers, while the selectivity can be modified by an appropriate choice of the biospecific biomolecule [220, 259,260]. Protein multicomponent films are also promising for sequential enzyme reactions through vectorial transfer of chemicals, electrons, or energy [5].

Supramolecular biological assemblies, like virus [52] or membrane fragments [261,262] containing bacteriorhodopsin have also been inserted in multilayers built by ESA.

The insertion of inorganic nanoparticles or platelets with defined chemical and physical properties at precise locations in a multilayer is of interest for fundamental research, as well as for a number of potential applications. ESA may be useful in assembling such multilayers if suspensions of charged inorganics can be prepared [1,40]. For example, suspensions of delaminated clay platelets, which are intrinsically charged, have been successfully used to assemble multilayers. Both the positively charged hydrotalcite [245] and the negatively charged hectorite [49,263] and montmorillonite [81,124,147,180,193,245] clay platelets can be utilized for the buildup. AFM and X-ray diffraction measurements have shown that the platelets lie flat on the substrate, as could be expected. Interestingly, it has been shown that, in some instances, more than one layer of platelets may be deposited [49,180]; charge balance considerations [126] seem to be able to explain this behavior. This is further corroborated by similar phenomena observed during adsorption of purple membrane fragments [261]. Due in part to the high aspect ratio of the clay platelets, hybrid polymer–clay assemblies possess a set of unique mechanical, electrical, and gas permeation properties, and as a result have been used in gas permeation membranes [182] or as insulating layers [183]. Platelets may also be useful, simply to prevent interpenetration of adjacent layers. Exfoliated suspensions of other intrinsically charged inorganic sheets, such as zirconium phosphate (α-ZrP) [194,264–267], MoS_2 [268], SnS_2 [268], or graphite oxide [269–271] have also been demonstrated for the electrostatic self-assembly. Note that after multilayer assembly graphite oxide can be reduced to graphite whose conductive and magnetoresistive properties are interesting.

The insertion of clusters and nanoparticles in multilayers may be useful too. Stable colloidal dispersions of negatively charged silica (at pH 10) [272], for example, have been successfully used for ESA. Various polyoxometalates [60,165,273–275], as well as gold [276–278] and platinum nanoparticles [279], were incorporated by this process in multilayer assemblies. A composite assembly containing polyelectrolytes and a membrane bilayer with a layer of gold particles on top were reported [280]. Many different semiconductor nanoparticles have been employed for the layer-by-layer deposition process, for example, cationic TiO_2 nanoparticles (isoelectric point corresponds to pH = 4.5–6.8) [204,281], but also PbI_2 [209,218], PbS [204], CdS [204,282], CdSe [145,283], and ZnS [284] and ''coupled'' TiO_2/PbS [285,286] nanoparticles. The considerable interest for these semiconducting systems arises in part from their special properties caused by quantum-size and surface effects, and the potential applications taking advantage of these effects. Preparation of multilayers containing these nanoparticles is not straightforward. Much attention has to be paid to the process parameters influencing the preparation of the colloidal suspension. Correct preparation must lead to a narrow size distribution of the particles and a correct surface charge, sufficiently important to allow for self-assembly but sufficiently small to prevent desorption during the rinsing step. Addition of stabilizers may be needed to prevent coagulation of the nanoparticles. As a last example of insertion of nanoscale particles, we cite multilayer magnetic thin films made by the consecutive adsorption of Fe_3O_4 particles stabilized by PDADMA and a polyamic acid salt [287].

A number of other systems have been shown to be suitable for multilayer buildup by the ESA, such as charged latex nanospheres [50,200,288–290] or metallosupramolecular complexes [291–293]. ESA has also been combined with Langmuir–Blodgett transfer [294–297], and even superlattice films have been produced.

Figure 12 presents a schematic of a virtual assembly in which all the kinds of molecules that have been shown to be usable for the ESA would be combined. It may seem from the preceding that any molecule or nano-object bearing charges is readily deposited by ESA. Although this is probably close to reality, we temper this conclusion here by a few remarks and some examples to the contrary [47,124]. First, we have already mentioned the existence of a critical charge density for the ESA process, and this may explain some of the unsuccessful attempts with polyelectrolytes [182] and other charged entities [145,220,261]. Second, in the case of weak polyelectrolytes, the pK values of the polycation and polyanion

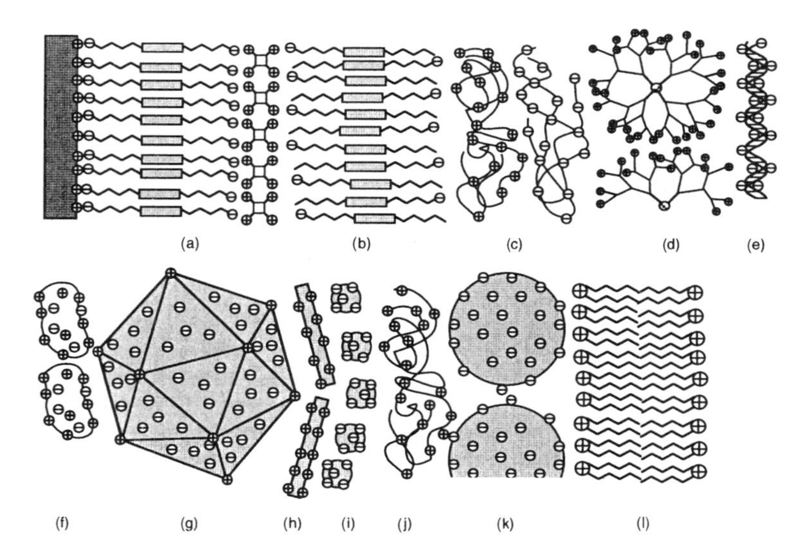

Figure 12 Scheme of the structure of a virtual multilayered heterostructure made from all kinds of molecules or particles that have been shown to be usable for ESA. Components: (a) boladications and other multipolar molecules; (b) lipid layers; (c,j) polyelectrolytes; (d) dendrimers; (e) DNA; (f) proteins; (g) viruses; (h) inorganic sheets; (i) inorganic nanoparticles; (k) latex nanospheres; (l) LB interlayers. (Adapted from Ref. 87.)

(or the isoelectric point in the case of proteins) must fit in order for both polymers to be sufficiently charged and thus adsorb in a consecutive fashion; this also may explain some unsuccessful depositions [166]. The inhomogeneous repartition of the charges on the surface of proteins complicates their behavior, and has been claimed to be responsible for some multilayer deposition failure [114]. Finally, in the case of small molecules or nanoparticles, redissolution during rinsing or formation of a soluble complex [258] during adsorption may also be a major concern, as discussed above.

VI. STRUCTURE OF THE MULTILAYERS

As shown in the case study on PAH/PSS multilayers, the interdiffusion between layers is so important in this system that no Bragg peaks are observed by XRR measurements, nor by NR measurements (when every PSS layer is deuterated) (Fig. 3). However, it has been shown that these films are stratified. Is this combination of stratification and important interdiffusion a general feature of polyelectrolyte multilayers?

Stratification of PAA/PAH multilayers is suggested by contact angle measurements showing odd–even trends as a function of the number of deposited layers [154] (Fig. 8). Also, stratification has been proved in the case of multilayers containing more than two polyelectrolytes: $\{ABCB\}_n$, $\{(AB)_n\ (AC)\}_m$, $\{(AB)_n\ (AC)(AB)_m(AC)\}_p$ films have been assembled which give rise to Bragg peaks detected by XRR [65,298] and NR [203,298]. By selective deuteration at varying intervals along the film normal, the stratification of PAH/sulfonated polyaniline multilayers also has been demonstrated to be preserved at least for the deposition of 40 bilayers [299].

By contrast, neither X-ray nor neutron [300] reflectivity measurements on multilayers assembled from only one polyanion/polycation pair have shown Bragg peaks so far, except for one type of specific system, discussed later in this section. This may be due either to a lack of contrast between the polyelectrolytes constituting the multilayer or to a (too) large interfacial width. In our opinion, it is mainly the important interfacial width that is responsible for the absence of Bragg peaks. Indeed, as shown above, Bragg peaks have been observed in the case of multilayers made of more than two polyelectrolytes, showing that usually polyanions and polycations have sufficiently different electron densities to give rise to Bragg peaks. Indeed, even a drying step applied at periodic intervals during multilayer buildup has been claimed to induce sufficient contrast to give rise to Bragg peaks [65]. Finally, important interfacial widths have been demonstrated for a number of systems [67,154,203,298,299].

As discussed in the PAH/PSS case study, both the interfacial roughness and the polymer interdiffusion contribute to the observed interfacial width. The reported interfacial roughnesses represent thus an upper limit to the extent of the interdiffusion of polyelectrolytes in neighboring layers. Strong interdiffusion between polyelectrolyte layers has been shown in the case of multilayers containing poly(butanylviologen) (PBV) (Fig. 6a) [67]: changing the distance between the redox active layers by interposing nonelectrochemically active layer pairs, it was determined that PBV was spread over a distance of at least 2.5 layer pairs. Covering a layer of partially biotinylated poly(L-lysine) by an increasing number of PSS and nonbiotinylated poly(L-lysine) layers, and looking at the amount of streptavidin that was still able to adsorb by avidin–biotin complexation, the interdiffusion of the partially biotinylated poly(L-lysine) in at least four neighboring polyelectrolyte layers has been demonstrated [79]. Strong interdiffusion of the polyelectrolytes in the neighboring layers has also been shown in the case of PAA/PAH multilayers, by performing contact angle measurements as a function of polycation and polyanion layer thickness [154]. This has been corroborated by a simple surface dying technique, using methylene blue as dye [154]. Strong interpenetration of the layers in ESA multilayers is also supported by nonradiative energy transfer measurements [202].

Interdiffusion seems to be less important for the rigid conjugated polyelectrolytes, but large interfacial widths are nevertheless observed [203,298,299]. For example, PAH/sulfonated polyaniline (Fig. 6b) multilayers exhibit interfacial width in the nanometer range, with an internal organization decaying monotonically away from the substrate. This suggests that the interfacial width is primarily due to the accumulation of defects as the multilayers are assembled, i.e., is primarily due to interfacial roughness. Neutron reflectivity measurements on multilayers containing poly(phenylenevinylene) (PPV) have shown that the interfacial width, although again in the nanometer range, was sufficiently small to maintain the PPV layers well separated from each other [203,298]. Confinement of electron-hole pairs in these layers is thus expected. PPV, which is prepared by thermal conversion of a charged precursor, is immiscible with PSS; phase separation is likely to occur during curing, participating in a further decrease of the interdiffusion of the rigid PPV in the PSS layer [101]. By contrast, photoluminescent shifts roughly proportional to $1/d^2$, where d is the thickness of the assembly, were observed for multilayers containing another conjugated copolymer [172] and thin polyelectrolyte insulating layers ($\sim$ 7Å), as expected for confined photogenerated electron-hole pairs in an infinite square potential well (the whole film). Nevertheless, the photoluminescence of films with 40 Å thick insulating layers is independent of the multilayer thickness, suggesting a confinement of electron-hole pairs in the conducting layers, and thus an interdiffusion of the conducting polymer smaller than the insulating spacer layer thickness.

Measurements in our group are in sharp contrast with the preceding results showing high internal roughness and/or strong interdiffusion of adjacent layers. X-ray reflectivity measurements on multilayers containing some particular ionenes reveal indeed the presence of Bragg peaks [140,157,243,244,301]. In some instances, Bragg peaks up to the fourth order were observed [140] (Fig. 13). This undoubtedly points to a small interdiffusion of the layers, coupled with a small waviness of the internal interfaces. The reason for this good ordering in the films is probably to be discovered in the low linear charge density of the ionenes used, which let other nonelectrostatic interactions come into play. Interestingly, the same lamellar structure with a repetition distance of about 2.4 nm has been found both in these multilayers and in the polyelectrolyte complexes obtained by mixing the polyanion solution with the ionene solutions. However, such a relationship between the structure of polyelectrolyte multilayers and the corresponding polyelectrolyte complexes is certainly not an absolute rule [301].

Interdiffusion and, to a certain extent, interface waviness can be suppressed by using more rigid ionic blocks for the multilayer assembly. As a consequence, Bragg peaks may appear in X-ray measurements of some multilayers containing platelets [49,180,194,263,269], LB [295], or bolaamphiphile [220,223] interlayers. Bragg peaks may also be seen in the case of multilayers containing colloidal nanoparticles, but in this case, the spacer layer should be thick enough to avoid the overlap of the layers containing the nanoparticles [276]. Bragg peaks have also been claimed for samples made of bolaamphiphiles and colloidal nanoparticles [218]. The successful photocrosslinking of some multilayers containing

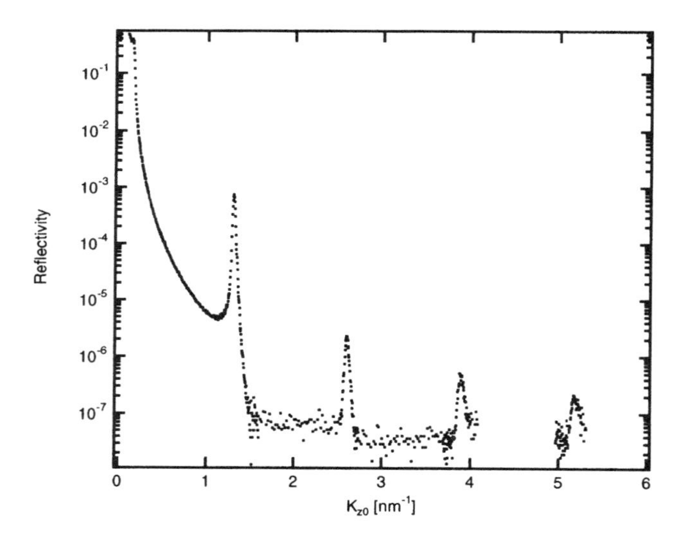

Figure 13 X-ray reflectivity vs. K_{z0}, the component of the wavevector of the incident photon perpendicular to the interfaces, for an ESA multilayer containing 20 bilayers of I-D2 and PVS deposited on silicon. (Data taken from Ref. 140.)

bolaamphiphiles implies good packing of these molecules in the multilayers, as seen above. An example of a multilayer made of bolaamphiphiles and multipolar molecules sufficiently well ordered to give rise to Bragg peaks has also been reported [216]. Note, however, that even for systems discussed in this section, the appearance of Bragg peaks is exceptional.

The formation of aggregates [48,63,140,148,149,168,216,232,233] and the preferential orientation of molecules or molecular fragments has been investigated by many different techniques including SHG [61,146,147,169,170,261,171, 302] (Fig. 14), spectroscopic techniques [161,224,303–305], and ellipsometry [99]. Naturally, the results depend heavily on the system under investigation.

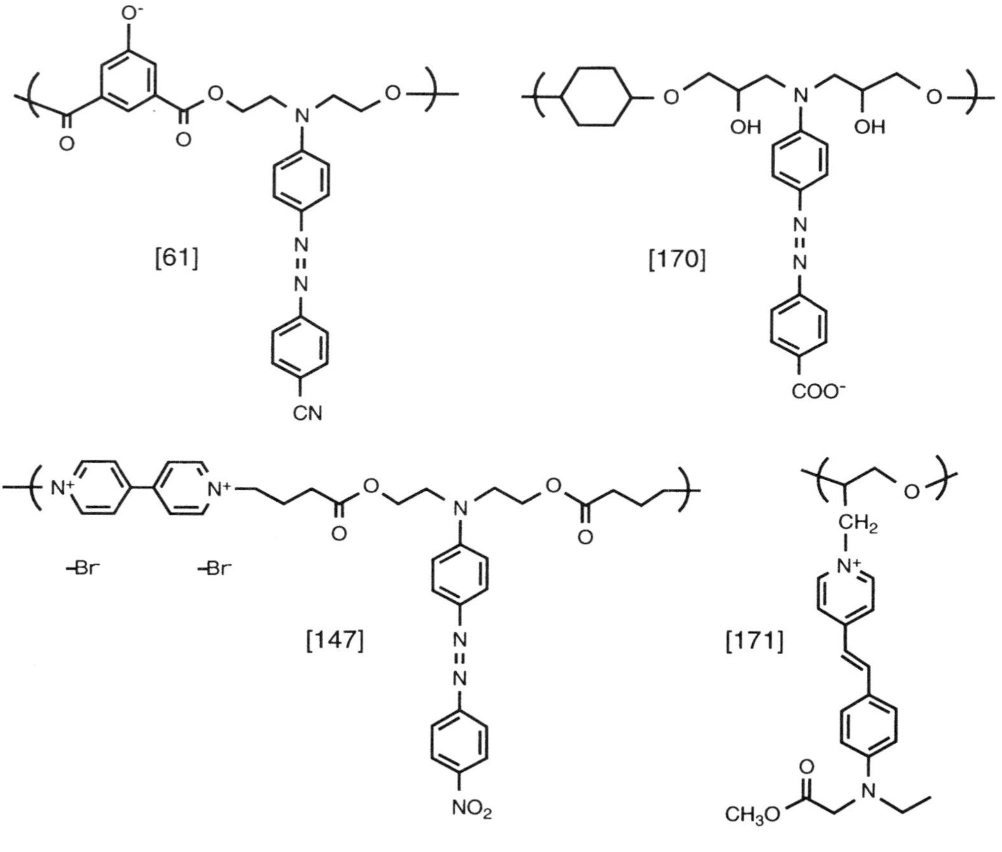

Figure 14 Examples of functional polyions employed in ESA multilayers to study second-order nonlinear optical effects.

Note that in one instance, orientation in the dipping direction has been reported [48].

The outer layer of the polyelectrolyte multilayers is expected to be significantly different from the bulk of the film, but not much is known about this particular layer. X-ray and neutron reflectometry, as well as AFM measurements, have been used to measure the roughness of this layer, which ranges from molecularly flat [109,179] to tens of Ångstrøms [203], depending on the system under investigation. Diffusion experiments point to a looser molecular packing in the outer layer, at least for PAH/PSS multilayers [114,115]. As expected, more counterions are found in the outer layer than in the bulk of the film [126].

Of prime importance for potential applications in integrated optics or electronics is the possibility of patterning the polyelectrolyte multilayers with micron resolution. Microcontact printing [181,305–308], ion beam lithography [309], and photostructuration [46] have been used to this end. The presence of salt in the polyion solutions affects both the selectivity of the deposition and the surface roughness of the film, and can even lead to a ''negative'' patterning when combined with a drying step [181,189,306,308].

VII. CHEMICAL MODIFICATIONS OF THE DEPOSITED FILMS

Chemical modifications of the layers after deposition open new opportunities to further tune the properties of the multilayers. The crosslinking of protein multilayers by different chemical reactions [125,207,208], and the photocrosslinking of unsaturated bonds has already been cited [186,212,213,236–238]. Other examples of crosslinking reactions can be found in the literature devoted to ESA [117,198,310]. Crosslinking may be used, for instance, to improve the chemical [117,125,198,207,213,310], mechanical [198], and thermal [236] resistance of the films, reduce the interdiffusion between neighboring layers, or change the barrier properties of the multilayers [186].

Some interesting molecules cannot be incorporated in multilayers by ESA, simply because they are not charged. Chemistry is very useful here also, since it provides a way around this problem: a charged precursor of the molecule is used for the assembly, and is then converted into the desired molecule. Some examples are the reduction of graphite oxide in graphite, the thermal conversion of PPV-precursor [153,298,311–313] or of precursors of other conjugated polymers [73,172,173] repeat units (Fig. 6b), and the use of polyimide precursors [62]. Remember, however, that if no specific interactions come into play, the stability of the multilayers may be a concern during these chemical reactions, since these reactions suppress the charges that link the neighboring layers [126].

Electrochemistry in multilayers [165,196,217,256,259,269], and particularly in multilayers containing viologen moieties [67,126,314,315], has been extensively studied, partly because of the potential applications in electrocatalysis, sensing, or electrochromism, but also because of the wealth of fundamental information that it may provide. Electrochemical methods also provide the opportunity to control the ionic content of the films [126,315].

Another benefit of chemistry results from its combination with ESA in building up thin organic films. Our group has worked on a new route for the development of organic multilayers [127,157,244,316,317], consisting of the electrostatic adsorption of a polyelectrolyte, followed by a chemical reversal of its charge (Fig. 15). The process can be cycled as many times as desired. Note that, in the case of one such activation reaction, a negative ζ-potential has been measured, although the total ratio of cationic to anionic groups could not be more than 2:1, even after 100% conversion [68]. This strongly suggests a noncentrosymmetric type of ordering, as confirmed by SHG measurements. The NLO properties of these films have prompted the extension of this type of reaction to obtain noncentrosymmetric ordering of functional dyes in normal ESA multilayers: in the classical ESA of a polyanion with a polycation, a step consisting of the chemical grafting of an uncharged NLO-active dye is inserted after every polycation adsorption step [242].

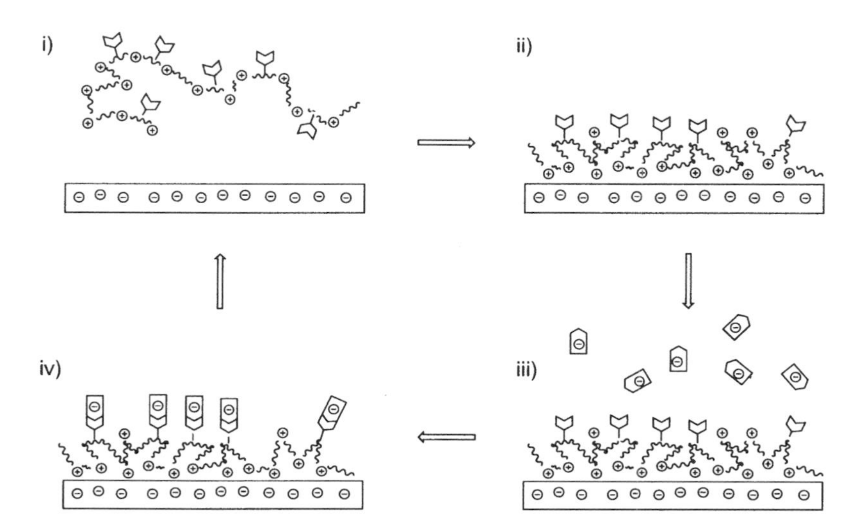

Figure 15 Schematic drawing of the multilayer preparation by combined ESA and chemical activation. (Adapted from Ref. 68.)

VIII. PHYSICAL PROPERTIES

In the following, some physical properties of the polyelectrolyte multilayers are presented. The number of studies focusing on these issues is limited, and often system dependent. Although we must thus warn against excessive generalization, these results may be useful as guidelines for further work in this field.

In sharp contrast to most of the LB films, the multilayers made by ESA usually show good ageing properties [114,203,300], although they also are not in thermodynamical equilibrium. This is mostly due to the numerous ion pairs, which lead to a reduced mobility of the polyelectrolyte chains and thus to the observed irreversibility or reduced reversibility of the adsorption process.

Numerous ionic bonds are also present in the polyelectrolyte complexes obtained by mixing solutions of polycations and polyanions of high charge density. These bonds are responsible for the formation of compact precipitates that are usually insoluble even at high salt concentrations, but can be solubilized in ternary solvent mixtures [318]. Accordingly, good solvent resistance of multilayers made of strong polyelectrolytes is usually reported [62,167,316], even at high ionic strength [112], but dissolution in a ternary mixture may occur [310]. Multilayers made of weak polyelectrolytes may dissolve at a pH that reduces or inverts their charge [117,125,127,207]. As already mentioned, solvent resistance may be improved by crosslinking [117,125,198,207,208,213,310].

The ion pairing is also most probably responsible for the good thermal behavior reported [46,65,188], at least as long as no degradation or conformational transition occurs [47]. Crosslinking can also be used in an attempt to further increase the thermal stability [236]. Nevertheless, heating may lead to local mobility of the chains or pendant groups; rearrangement of chromophores trapped in the multilayers above a (system-dependent) critical temperature has been reported [146,168,171]. This local mobility is also probably responsible for the sometimes-mentioned increase of internal organization upon annealing of the multilayers [295,298].

So far, investigations of the mechanical properties of the multilayers assembled by ESA have been limited to tape peel tests [40,91,94,161,198,205,217]. Failure occurring in the layer, at the interface with the substrate, or in the tape have been reported, depending on the system under investigation. Again, crosslinking may be useful to improve the mechanical properties of the multilayers [198].

In contrast, many more studies have been devoted to the diffusion of small molecules in the multilayers, because of its importance in applications such as permeation membranes or biosensing. Both IR spectroscopy [319] and fluorescence measurements [320] have shown the diffusion of protons in the multilayers, and thus an influence of the pH of the outer solution even far inside the films.

The influence of water on the thickness of the multilayers is also well documented [112,321]. Diffusion of radiolabeled salt ions has also been measured [126,315]. Voltamperometry showed that PAH/PAA films had little effect on the diffusion of $Fe(CN)_6^{3-}$, but that PAH/PSS films could hinder its transport [117]. 6-carboxy-fluorescein [97], acridine orange [82], daunomycin [254], 2′–3′ cyclic adenosine monophosphate [255], bisulfite [316], and different diazonium salts [148,316] have been shown to permeate deeply in multilayers built by ESA. Immunoglobulin G (IgG) could permeate or not in a superlattice made of anti-IgG layers and PAH/PSS spacer layers, depending on the thickness of the spacer layer. The diffusion constants of rhodamine and of 2,2,6,6,-tetramethyl-4-piperidinol-1-oxide (TEMPOL) in PAH/PSS multilayers have been quantified [114,115].

Aside from the numerous studies over electrochemistry in the multilayers, and over multilayers containing conducting polymers, some more useful information concerning the electrical properties of multilayers may be found in the literature. Shinbo et al. have studied the electrical properties of PAH/PSS multilayers, and showed that the conductivity was due to hopping conduction, probably due to the migration of mobile ions [102]. As can be expected, humidity greatly influences the electrical properties. ESA has also been used to assemble a metal–insulator–gold nanocluster–insulator–metal heterostructure that allows the observation of single electron charging effects at ambient temperature [322].

IX. POTENTIAL APPLICATIONS

The huge number of organic and inorganic molecules that are suitable for the ESA, combined with the simplicity of this deposition technique paves the way for numerous potential applications. However, due to the relatively recent development of the technique, the number of patents in the field is still rather limited [323–334], and, in the best case, only feasibility studies have been performed to date. The transfer of these technologies to industry will require the fulfillment of numerous other requirements, such as cost effectiveness, stability of the devices, etc. In the following, some of the potential applications proposed in the literature are reviewed.

To begin, multilayers can be used to modify surfaces, and in particular their biocompatibility (e.g., for implants) [175,187,207,335] or their wettability (e.g., for antifogging mirrors or eyeglasses) [154]. Obviously, only a limited number of layers are needed for these applications; the fact that polyelectrolyte adsorbs readily on various substrates is here a clear advantage.

Multilayers may also be used for their permeation properties. Accordingly, membranes covered by multilayers have been employed for gas permeability measurements and for pervaporation studies [89,182,186,336–338]. These measurements showed, for example, O_2/H_2O, CO_2/O_2, or toluene–heptane selectivity. The permeation properties of polyelectrolyte multilayers are also important when they are used for the encapsulation of enzymes [211] or living cells [339,340]. The deposition of polyelectrolyte multilayers or of hybrid polyelectrolyte/inorganic multilayers on latex particles, and the subsequent dissolution or calcination of the latex beads leads to the fabrication of hollow spheres [95–97,341–343]. Potential applications of such hollow spheres are numerous, for example, for the controlled release and targeting of drugs.

Polyelectrolyte multilayers are able to very effectively immobilize various molecules such as enzymes, for example. These multilayers have been shown to exhibit useful biocatalytic (i.e., enzymatic) [220,221] activities; hybrid multilayers containing polyoxometalate are promising for catalytic applications [165]. ESA multilayers have also been used for molecular recognition [199,344] or, more specifically, as biosensors [76,167,253–255,345,346]. Electrocatalytic [196,256,259] and electrosensing [257] capabilities of multilayers deposited on electrodes have also been demonstrated. As already mentioned, the possibility of putting different enzymes at specific locations in the multilayer has been employed to design systems for multistep catalysis [6] (Fig. 16).

Applications in the domain of data storage have also been envisioned. Multilayers with alternating layers of magnetic nanoparticles and polyimide have already been prepared, yielding an average magnetic flux per bilayer of 850 nT at a distance of 2 cm [287]. Optical data storage based on the Weigert effect, which consists of birefringence or dichroism induced by irradiation with polarized light, can be envisioned in the case of multilayers containing azobenzene functionalized polymers [230,231]. Indeed, these derivatives are known to undergo *cis-trans* and *trans-cis* photoisomerization upon adequate light irradiation. Multilayers containing azobenzene-functionalized polymers have also been shown to be useful for the photofabrication or photoerasure of surface relief gratings [347], and as command surfaces for LC display devices [63,232].

As stated in the application range section, electroactive, conducting, semiconducting, and insulating layers can be inserted in the multilayers assembled by ESA. Furthermore, we mentioned the possibilities of patterning the deposition of the multilayers. Various applications are thus expected in the electronic and optoelectronic fields. Thin conducting layers may find applications in transparent electrodes, antistatic coatings, or electromagnetic interference (EMI) shielding, for example [69]. Such thin conducting layers are also useful for the fabrication of organic LEDs. Potential advantages of organic LEDs are the possibility of fabrication of large area devices, and their relatively low cost [311]. Nevertheless, their long-term stability is a concern in some instances [283], and the processability of some electroactive polymers is challenging [174]. ESA enables an easy assembly of complex heterostructures containing very thin layers of these elec-

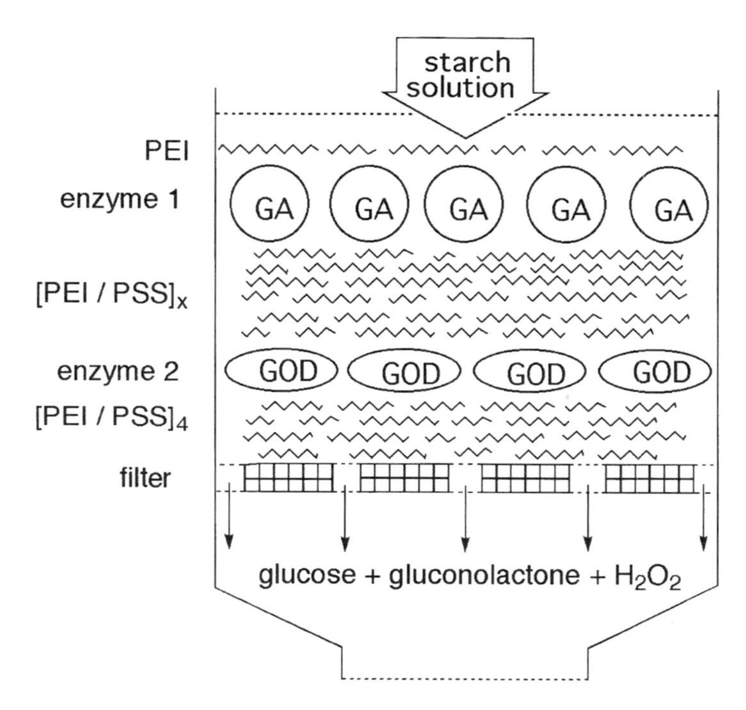

Figure 16 Scheme of the sequential enzymatic conversion of starch by glucoamylase (GA) and by glucose oxidase (GOD) in multilayers prepared by ESA. (Adapted from Ref. 6.)

troactive polymers, with a level of control higher than what would be possible by spin-coating, and more easily than by the LB technique [174]. As a consequence, organic LEDs have been prepared by the ESA; the quality of these films is of at least the same quality as spin-cast films [54]. Precursors of poly(phenylene vinylene) [311–313] and of copolymers (phenylenevinylene-naphtylenevinylene) [73,172,173], as well as poly(pyridinium vinylene) [166,174,348] and alkoxy-sulfonated poly(phenylenes) [160,349] (Fig. 6b), are examples of electroactive polymers that have been used for the ESA of LEDs. Devices have also been prepared from an electroluminescent ruthenium complex, either incorporated in the main chain of a polyester, or used as such [152,226,228,235,350]. Multilayers made of conducting polymers and nanoparticles are also potentially interesting as electroluminescent devices [283,351]. Indeed, their emission wavelength can be controlled by the size of the nanoparticles, while the transport properties of electrons and holes can be adjusted by a correct choice of the polymer. Furthermore, CdSe/PPV devices showed longer lifetimes than devices prepared from only one of these components. LEDs based on a conjugated bipyridinium have also been reported in the literature [227]. The behavior and performances of the LEDs have been shown to depend on the nature of the counterpolyanion [312]. By combining chemical tuning with size effects, the emission can be finely tuned, and may cover the entire visible range [172]. The photocurrent generation has been measured in different systems containing colloidal nanoparticles [183, 204,286]. In this respect, ''coupled'' type nanoparticles are particularly interesting, since charge injection from one semiconductor into another can lead to efficient and longer charge separation by minimizing the electron-hole recombination pathway [286]. Such systems are thus anticipated to have potential applications in solar energy conversion. Solar energy conversion by biomimetic light-harvesting systems would be very interesting, but it requires the possibility of controlling the supramolecular arrangement of the assemblies [8]. The versatility of ESA that enables an easy juxtaposition of electron donors and acceptors at controlled distances for the photoinduced electron transfer, would be an interesting way towards such systems [264,265]. Multilayers containing bacteriorhodopsin also show interesting photoelectric properties, in particular, the generation of a differential photocurrent. This, together with the long-term stability and photochromic properties of bacteriorhodopsin, leads to potential applications ranging from imaging devices to light-sensitive alarm devices through devices for motion and direction detection [261,262].

A number of other applications have been proposed, including SHG (see Section VI), electrochromism (see Section VII), high density rechargeable lithium-ion batteries [270,271], tunable color filters [228], or pH indicators [235]. A pH-sensitive optrode has also been proposed [210], but first attempts were unsuccessful due to the absence of response of the pH-sensitive dye used (phenol red) to the pH of the surrounding solution, when trapped in the multilayer.

Despite the wide range of possible applications of the proposed ESA multilayers, we somewhat temper here the widespread enthusiasm. It should be stressed, for example, that these films are sensitive to atmospheric humidity and other environmental conditions [112,321]. The environment-dependent swelling of the multilayers can be prejudicial for some applications. Among other potentially unwanted aspects of these multilayers, we can also cite the ionic conductivity due to small ions trapped in the films, although this has not been much studied so far [102]. Therefore, much work will be needed in the future to establish the practical scopes and limits of the ESA method.

X. CONCLUSIONS

In comparison with other techniques for the preparation of thin organic multilayers, ESA presents a range of advantages, among which the most prominent is probably its versatility. Indeed, using this method, a broad range of organic and inorganic molecules can be assembled at a nanometer scale. Such control of the vertical stacking of molecules, complemented with patterning capabilities, will certainly lead to a broad range of applications in integrated optics and electronics as well as in biotechnology. A limitation of this deposition technique is the often-encountered lack of internal organization of the ESA multilayers.

Much progress in the understanding of the physical processes underlying the ESA has been gained over these last years, and the influence of the main processing parameters is now clearly established. Nevertheless, the optimization of the multilayer buildup remains essentially an experimental trial-and-error process, and nonequilibrium theories taking into account all observed phenomena are still lacking. Consequently, the topic will remain an active field of research in the coming years.

We therefore think the future is bright for ESA and many exciting developments are still to come.

ACKNOWLEDGMENTS

The authors would like to thank B. Laguitton, A. Delcorte, P. Bertrand, and G. Decher for stimulating discussions. This work was supported by the Fonds National de la Recherche Scientifique F.N.R.S. and by the DG Recherche Scientifique of the French Community of Belgium (convention 94/99–173).

REFERENCES

1. JH Fendler. Chem Mater 8:1616, 1996.
2. A Dodabalapur. Solid State Commun 102:259, 1997.
3. S Mann. Nature 365:499, 1993.
4. H Ringsdorf, B Schlarb, J Venzmer. Angew Chem Int Ed Engl 27:113, 1988.
5. T Kunitake. Thin Solid Films 284:9, 1996.
6. M Onda, Y Lvov, K Ariga, T Kunitake. J Ferment Bioeng 82:502, 1996.
7. C Nicolini. Thin Solid Films 284:1, 1996.
8. H Byrd, EP Suponeva, AB Bocarsly, ME Thompson. Nature 380:610, 1996.
9. A Hagfeldt, M Grätzel. Chem Rev 95:49, 1995.
10. V Bach, D Lupo, P Comte, JE Moser, F Weissörtel, J Salbeck, H Spreitzer, M Grätzel. Nature 395:583, 1998.
11. I Langmuir. Trans Faraday Soc 15:62, 1920.
12. KB Blodgett. J Am Chem Soc 57:1007, 1935.
13. KB Blodgett, I Langmuir. Phys Rev 51:964, 1937.
14. H Kuhn, D Möbius, H Bücher. In: A Weissberger, BW Rossiter, eds. Physical Methods of Chemistry. Vol 1, Part 3B. New York: Wiley, 1972, pp 577.

15. JD Swalen. J Mol Electron 2:155, 1986.
16. O Albrecht, A Laschewsky, H Ringsdorf. Macromolecules 17:937, 1984.
17. VV Arslanov. Adv Colloid Interf Sci 40:307, 1992.
18. T Miyashita. Prog Polym Sci 18:263, 1993.
19. F Embs, D Funhoff, A Laschewsky, U Licht, H Ohst, W Prass, H Ringsdorf, G Wegner, R Wehrmann. Adv Mater 3:25, 1991.
20. L Netzer, J Sagiv. J Am Chem Soc 105:674, 1983.
21. R Maoz, L Netzer, J Gun, J Sagiv. J Chim Phys 85:1059, 1988.
22. A Ulman. Chem Rev 96:1533, 1996.
23. RJ Collins, IT Bae, DA Scherson, CN Sukenik. Langmuir 12:5509, 1996.
24. TJ Marks, MA Ratner. Angew Chem Int Ed Engl 34:155, 1995.
25. RG Nuzzo, DL Allara. J Am Chem Soc 105:4481, 1983.
26. LH Dubois, RG Nuzzo. Annu Rev Phys Chem 43:437, 1992.
27. SD Evans, A Ulman, KE Goppert-Berarducci, LJ Gerenser. J Am Chem Soc 113: 5866, 1991.
28. TL Freeman, SD Evans, A Ulman. Langmuir 11:4411, 1995.
29. K Hu, M Brust, AJ Bard. Chem Mater 10:1160, 1998.
30. H Lee, LJ Kepley, H-G Hong, TE Mallouk. J Am Chem Soc 110:618, 1988.
31. DL Feldheim, TE Mallouk. Chem Commun 2591, 1996.
32. AC Zeppenfeld, SL Fiddler, WK Ham, BJ Klopfenstein, CJ Page. J Am Chem Soc 116:9158, 1994.
33. MA Ansell, AC Zeppenfeld, K Yoshimoto, EB Cogan, CJ Page. Chem Mater 8: 591, 1996.
34. XQ Zhang, HM Wu, ZH Lu, XZ You. Thin Solid Films 284/285:224, 1996.
35. H Xiong, M Cheng, Z Zhou, X Zhang, JC Shen. Adv Mater 10:529, 1998.
36. I Ichinose, T Kawakami, T Kunitake. Adv Mater 10:535, 1998.
37. DM Sarno, D Grosfeld, J Snyder, B Jiang, WE Jones. Polym Prepr Am Chem Soc Polym Chem Div 39(2):1101, 1998.
38. DL Thomsen III, F Papadimitrakopoulos. Macromol Symp 125:143, 1997.
39. HE Katz, G Scheller, TM Putvinski, ML Schilling, WL Wilson, CED Chidsey. Science 254:1485, 1991.
40. RK Iler. J Colloid Interface Sci 21:569, 1966.
41. C-G Gölander, H Arwin, JC Eriksson, I Lundstrom, R Larsson. Colloids Surf 5: 1, 1982.
42. P Fromherz. In: W Baumeister, W Vogell, eds. Electron Microscopy at Molecular Dimensions. Berlin: Springer, 1980, pp 338–349.
43. RH Tredgold, CS Winter, ZI El-Badawy. Electron Lett 21:554, 1985.
44. G Decher, JD Hong. Makromol Chem Macromol Symp 46:321, 1991.
45. G Decher, JD Hong. Ber Bunsenges Phys Chem 95:1430, 1991.
46. JD Hong, K Lowack, J Schmitt, G Decher. Prog Colloid Polym Sci 93:98, 1993.
47. Y Lvov, G Decher, G Sukhorukov. Macromolecules 26:5396, 1993.
48. TM Cooper, AL Campbell, RL Crane. Langmuir 11:2713, 1995.
49. ER Kleinfeld, GS Ferguson. Science 265:370, 1994.
50. VN Bliznyuk, VV Tsukruk. Polym Prepr Am Chem Soc Div Polym Chem 38(1): 963, 1997.
51. S Watanabe, SL Regen. J Am Chem Soc 116:8855, 1994.

52. Y Lvov, H Haas, G Decher, H Möhwald, A Mikhailov, B Mtchedlishvily, E Morgunova, B Vainshtein. Langmuir 10:4232, 1994.
53. G Decher. Science 277:1232, 1997.
54. G Decher, M Eckle, J Schmitt, B Struth. Curr Opin Colloid Interface Sci 3:32, 1998.
55. A Laschewsky. Eur Chem Chronicle 2:13, 1997.
56. W Knoll. Curr Opin Colloid Interface Sci 1:137, 1996.
57. G Decher. In: JC Salamone, ed. The Polymeric Materials Encyclopedia: Vol 6: Synthesis, Properties, and Applications. Boca Raton, FL: CRC, 1996, pp 4540–4546.
58. G Decher. In: J-P Sauvage, MW Hosseini, eds. Comprehensive Supramolecular Chemistry. Oxford: Pergamon, 1996, pp 507–528.
59. M Sano, Y Lvov, T Kunitake. Ann Rev Mater Sci 26:153, 1996.
60. SW Keller, H-N Kim, TE Mallouk. J Am Chem Soc 116:8817, 1994.
61. KM Lenahan, YX Wang, YJ Liu, RO Claus, JR Heflin, D Marciu, C Figura. Adv Mater 10:853, 1998.
62. JW Baur, P Besson, SA O'Connor, MF Rubner. Mater Res Soc Proceedings 413: 583, 1996.
63. RC Advincula, D Roitman, C Frank, W Knoll, A Baba, F Kaneko. Polym Prepr Am Chem Soc Dir Polym Chem 40(1):467, 1999.
64. Riegler & Kirstein GmbH, Wiesbaden, Germany.
65. G Decher, Y Lvov, J Schmitt. Thin Solid Films 244:772, 1994.
66. J Schmitt, T Grünewald, G Decher, PS Pershan, K Kjaer, M Lösche. Macromolecules 26:7058, 1993.
67. D Laurent, JB Schlenoff. Langmuir 13:1552, 1997.
68. A Laschewsky, B Mayer, E Wischerhoff, X Arys, P Bertrand, A Delcorte, A Jonas. Thin Solid Films 284/285:334, 1996.
69. WB Stockton, MF Rubner. Mater Res Soc Proceedings 369:587, 1994.
70. L Wang, Z Wang, X Zhang, J Shen, L Chi, H Fuchs. Macromol Rapid Commun 18:509, 1997.
71. L Wang, Y Fu, Z Wang, Y Fan, X Zhang. Langmuir 15:1360, 1999.
72. WB Stockton, MF Rubner. Macromolecules 30:2717, 1997.
73. H Hong, D Davidov, M Tarabia, H Chayet, I Benjamin, EZ Faraggi, Y Avny, R Neumann. Synth Met 85:1265, 1997.
74. Y Shimazaki, M Mitsuishi, S Ito, M Yamamoto. Langmuir 13:1385, 1997.
75. Y Shimazaki, M Mitsuishi, S Ito, M Yamamoto. Langmuir 14:2768, 1998.
76. F Caruso, K Niikura, DN Furlong, Y Okahata. Langmuir 13:3422, 1997.
77. J-I Anzai, Y Kobayashi, N Nakamura, M Nishimura, T Hoshi. Langmuir 15:221, 1999.
78. J-I Anzai, M Nishimura. J Chem Soc Perkin Trans 2:1887, 1997.
79. T Cassier, K Lowack, G Decher. Supramol Sci 5:309, 1998.
80. C Bourdillon, C Demaille, J Moiroux, J-M Saveant. J Am Chem Soc 116:10328, 1994.
81. Y Lvov, K Ariga, I Ichinose, T Kunitake. Thin Solid Films 285:797, 1996.
82. GB Sukhorukov, H Möhwald, G Decher, YM Lvov. Thin Solid Films 284/285: 220, 1996.

83. YF Nicolau. Appl Surf Sci 22/23:1061, 1985.
84. S Lindroos, T Kanniainen, M Leskelä. Appl Surf Sci 75:70, 1994.
85. Y Lvov, G Decher, H Haas, H Möhwald, A Kalachev. Physica B 198:89, 1994.
86. Y Lvov, H Haas, G Decher, H Möhwald, A Kalachev. J Phys Chem 97:12835, 1993.
87. YM Lvov, G Decher. Crystallogr Rep 39:628, 1994.
88. G Decher, JD Hong, J Schmitt. Thin Solid Films 210/211:831, 1992.
89. J-M Leväsalmi, TJ McCarthy. Macromolecules 30:1752, 1997.
90. V Phuvanartnuruks, TJ McCarthy. Polym Prepr Am Chem Soc Div Polym Chem 38(1):961, 1997.
91. V Phuvanartnuruks, TJ McCarthy. Macromolecules 31:1906, 1998.
92. MC Hsieh, RJ Farris, TJ McCarthy. Polym Prepr Am Chem Soc Div Polym Chem 38(2):670, 1997.
93. MC Hsieh, RJ Farris, TJ McCarthy. Macromolecules 30:8453, 1997.
94. W Chen, TJ McCarthy. Macromolecules 30:78, 1997.
95. GB Sukhorukov, E Donath, H Lichtenfeld, E Knippel, M Knippel, A Budde, H Möhwald. Colloids Surf A 137:253, 1998.
96. GB Sukhorukov, E Donath, S Davis, H Lichtenfeld, F Caruso, VI Popov, H Möhwald. Polym Adv Technol 9:759, 1998.
97. F Caruso, E Donath, H Möhwald. J Phys Chem B 102:2011, 1998.
98. A Baba, F Kaneko, RC Advincula. Polym Prepr Am Chem Soc Dir Polym Chem 40(1):488, 1999.
99. A Tronin, Y Lvov, C Nicolini. Colloid Polym Sci 272:1317, 1994.
100. D Korneev, Y Lvov, G Decher, J Schmitt, S Yaradaikin. Physica B 213/214:954, 1995.
101. M Lösche, J Schmitt, G Decher, WG Bouwman, K Kjaer. Macromolecules 31: 8893, 1998.
102. K Shinbo, K Suzuki, K Kato, F Kaneko, S Kobayashi. Thin Solid Films 327/329: 209, 1998.
103. P Berndt, K Kurihara, T Kunitake. Langmuir 8:2486, 1992.
104. K Lowack, CA Helm. Macromolecules 31:823, 1998.
105. R von Klitzing, H Möhwald. Langmuir 11:3554, 1995.
106. E Donath, D Walther, VN Shilov, E Knippel, A Budde, K Lowack, CA Helm, H Möhwald. Langmuir 13:5294, 1997.
107. JJ Ramsden, YM Lvov, G Decher. Thin Solid Films 254:246, 1995.
108. VN Bliznyuk, DW Visser, VV Tsukruk, AL Campbell, T Bunning, WW Adams. Polym Prepr Am Chem Soc Dir Polym Chem 37(2):608, 1996.
109. VV Tsukruk, VN Bliznyuk, D Visser, AL Campbell, TJ Bunning, WW Adams. Macromolecules 30:6615, 1997.
110. G Decher, J Schmitt. Prog Colloid Polym Sci 89:160, 1992.
111. VV Belyaev, AL Tolstikhina, ND Stepina, RL Kayushina. Crystallogr Rep 43:124, 1998.
112. GB Sukhorukov, J Schmitt, G Decher. Ber Bunsenges Phys Chem 100:948, 1996.
113. G Decher, Y Lvov, J Schmitt. Thin Solid Films 244:772, 1994.
114. R von Klitzing, H Möhwald. Thin Solid Films 284/285:352, 1996.
115. R von Klitzing, H Möhwald. Macromolecules 29:6901, 1996.

116. R Kayushina, Y Lvov, N Stepina, V Belayev, Y Khurgin. Thin Solid Films 284/285:246, 1996.
117. ML Bruening, JJ Harris, PM DeRose. Polym Prepr Am Chem Soc Dir Polym Chem 40(1):451, 1999.
118. TP Russell. Mat Sci Rep 5:171, 1990.
119. JS Higgins, HC Benoît. Polymers and Neutron Scattering. Oxford: Clarendon, 1994.
120. VW Stone, AM Jonas, X Arys, R Legras. (To be published).
121. EM Arnett. Prog Phys Org Chem 1:223, 1963.
122. HC Brown, DH McDaniel, O Häfliger. In: EA Braude, FC Nachod, eds. Determination of Organic Structures by Physical Methods. New York: Academic, 1955, pp 567.
123. JH Cheung, WB Stockton, MF Rubner. Macromolecules 30:2712, 1997.
124. Y Lvov, K Ariga, I Ichinose, T Kunitake. J Am Chem Soc 117:6117, 1995.
125. E Brynda, M Houska. Macromol Rapid Commun 19:173, 1998.
126. JB Schlenoff, H Ly, M Li. J Am Chem Soc 120:7626, 1998.
127. M Koetse, A Laschewsky, T Verbiest. Mat Sci Eng C382: in press, 1999.
128. GJ Fleer, MA Cohen Stuart, JMHM Scheutjens, T Cosgrove, B Vincent. Polymers at Interfaces. London: Chapman & Hall, 1993.
129. P Linse. Macromolecules 29:326, 1996.
130. HGM van de Steeg, MA Cohen Stuart, A de Keizer, BH Bijsterbosch. Langmuir 8:2538, 1992.
131. MR Böhmer, OA Evers, JMHM Scheutjens. Macromolecules 23:2288, 1990.
132. OJ Rojas, PM Claesson, D Muller, RD Neuman. J Colloid Interface Sci 205:77, 1998.
133. NG Hoogeveen, MA Cohen Stuart, GJ Fleer. J Colloid Interface Sci 182:133, 1996.
134. D Andelman, J-F Joanny. Macromolecules 24:6040, 1991.
135. D Andelman, J-F Joanny. J Phys II 3:121, 1993.
136. M Muthukumar. Curr Opin Colloid Interface Sci 3:48, 1998.
137. M Muthukumar. J Chem Phys 103:4723, 1995.
138. KW Mattison, PL Dubin, IJ Brittain. J Phys Chem B 102:3830, 1998.
139. S Neyret, L Ouali, F Candau, E Pefferkorn. J Colloid Interface Sci 176:86, 1995.
140. P Fischer, A Laschewsky, E Wischerhoff, X Arys, A Jonas, R Legras. Macromol Symp 137:1, 1995.
141. K Lowack, CA Helm. Macromolecules 28:2912, 1995.
142. NG Hoogeveen, MA Cohen Stuart, GJ Fleer. J Colloid Interface Sci 182:146, 1996.
143. JB Schlenoff, M Li. Ber Bunsenges Phys Chem 100:943, 1996.
144. NG Hoogeveen, MA Cohen Stuart, GJ Fleer. Langmuir 12:3675, 1996.
145. T Cassagneau, TE Mallouk, JH Fendler. J Am Chem Soc 120:7848, 1998.
146. Y Lvov, S Yamada, T Kunitake. Thin Solid Films 300:107, 1997.
147. A Laschewsky, E Wischerhoff, M Kauranen, A Persoons. Macromolecules 30:8304, 1997.
148. D Cochin, A Laschewsky. Macromol Chem Phys 200:609, 1999.
149. A Laschewsky, B Mayer, E Wischerhoff, X Arys, A Jonas. Ber Bunsenges Phys Chem 100:1033, 1996.
150. S Schwarz, K-J Eichhorn, E Wischerhoff, A Laschewsky. Colloids Surf 1999 (accepted for publication).

151. B Laguitton, X Arys, AM Jonas, A. Laschewsky. Unpublished results, 1997.
152. A Wu, J Lee, MF Rubner. Thin Solid Films 327/329:663, 1998.
153. B Lehr, M Seufert, G Wenz, G Decher. Supramol Sci 2:199, 1995.
154. D Yoo, SS Shiratori, MF Rubner. Macromolecules 31:4309, 1998.
155. M Lütt, MR Fitzsimmons, DQ Li. J Phys Chem B B102:400, 1998.
156. R Advincula, E Aust, W Meyer, W Knoll. Langmuir 12:3536, 1996.
157. X Arys, AM Jonas, B Laguitton, R Legras, A Laschewsky, E Wischerhoff. Prog Org Coat 34:108, 1998.
158. ER Kleinfeld, GS Ferguson. Chem Mater 8:1575, 1996.
159. Y Lvov, K Ariga, T Kunitake. Chem Lett 2323, 1994.
160. S Kim, J Jackiw, E Robinson, KS Schanze, JR Reynolds, J Baur, MF Rubner, D Boils. Macromolecules 31:964, 1998.
161. M Ferreira, MF Rubner. Macromolecules 28:7107, 1995.
162. M Ferreira, JH Cheung, MF Rubner. Thin Solid Films 244:806, 1994.
163. M Raposo, RS Pontes, LHC Mattoso, ON Oliveira. Macromolecules 30:6095, 1997.
164. F Saremi, G Lange, B Tieke. Adv Mater 8:923, 1996.
165. I Moriguchi, JH Fendler. Chem Mater 10:2205, 1998.
166. J Tian, CC Wu, ME Thompson, JC Sturm, RA Register. Chem Mater 7:2190, 1995.
167. A Diederich, M Lösche. Adv Biophys 34:205, 1997.
168. S Dante, R Advincula, CW Frank, P Stroeve. Langmuir 15:193, 1999.
169. S Yamada, A Harada, T Matsuo, S Ohno, I Ichinose, T Kunitake. Japan J Appl Phys 36:L1110, 1997.
170. S Balasubramanian, XG Wang, HC Wang, K Yang, J Kumar, SK Tripathy, L Li. Chem Mater 10:1554, 1998.
171. MJ Roberts, GA Lindsay, WN Herman, KJ Wynne. J Am Chem Soc 120:11202, 1998.
172. H Hong, M Tarabia, H Chayet, S Davidov, EZ Faraggi, Y Avny, R Neumann, S Kirstein. J Appl Phys 79:3082, 1996.
173. H Hong, D Davidov, Y Avny, H Chayet, EZ Faraggin, R Neumann. Adv Mater 7:846, 1995.
174. M Onoda, H Nakayama, T Yamaue, K Tada, K Yoshino. Jpn J Appl Phys Part 1 36:5322, 1997.
175. M Houska, E Brynda. J Colloid Interface Sci 188:243, 1997.
176. X Arys, AM Jonas, A Laschewsky, R Legras. Unpublished results, 1998.
177. T Serizawa, H Goto, A Kishida, T Endo, M Akashi. J Polym Sci: Part A: Polym Chem 36:801, 1999.
178. MR Linford, M Auch, H Möhwald. J Am Chem Soc 120:178, 1998.
179. M Onda, Y Lvov, K Ariga, T Kunitake. Jpn J Appl Phys Lett 36:L1608, 1997.
180. NA Kotov, T Haraszti, L Turi, G Zavala, RE Geer, I Dékány, JH Fendler. J Am Chem Soc 119:6821, 1997.
181. SL Clark, PT Hammond. Adv Mater 10:1515, 1998.
182. NA Kotov, S Magonov, E Tropsha. Chem Mater 10:886, 1998.
183. JJ Fendler. Stud Surf Sci Catal 103:261, 1997.
184. X Arys, AM Jonas, B Laguitton, A Laschewsky, R Legras, E Wischerhoff. Thin Solid Films 329:734, 1998.
185. M Paßmann, G Wilbert, D Cochin, R Zentel. Macromol Chem Phys 199:179, 1998.

186. F van Ackern, L Krasemann, B Tieke. Thin Solid Films 327/329:762, 1998.
187. Y Lvov, M Onda, K Ariga, T Kunitake. J Biomater Sci Polym Ed 9:345, 1998.
188. Y Lvov, G Decher, H Möhwald. Langmuir 9:481, 1993.
189. SL Clark, MF Montague, PT Hammond. Polym Prepr Am Chem Soc Div Polym Chem 38(1):967, 1997.
190. G Mao, Y-H Tsao, M Tirrell, HT Davis, V Hessel, H Ringsdorf. Langmuir 11: 942, 1995.
191. LA Godinez, R Castro, AE Kaifer. Langmuir 12:5087, 1996.
192. JH Cheung, AF Fou, MF Rubner. Thin Solid Films 244:985, 1994.
193. Y Lvov, K Ariga, I Ichinose, T Kunitake. Langmuir 12:3038, 1996.
194. HN Kim, SW Keller, TE Mallouk, J Schmitt, G Decher. Chem Mater 9:1414, 1997.
195. W Knoll. Ann Rev Phys Chem: 569, 1998.
196. YM Lvov, Z Lu, JB Schenkman, X Zu, JF Rusling. J Am Chem Soc 120:4073, 1998.
197. K Ariga, Y Lvov, T Kunitake. J Am Chem Soc 119:2224, 1997.
198. A Laschewsky, E Wischerhoff, P Bertrand, A Delcorte. Macromol Chem Phys 198: 3239, 1997.
199. A Laschewsky, E Wischerhoff, S Denzinger, H Ringsdorf, A Delcorte, P Bertrand. Chem Eur J 3:34, 1997.
200. T Serizawa, H Takeshita, M Akashi. Langmuir 14:4088, 1998.
201. YM Lvov, JF Rusling, DL Thomsen, F Papadimitrakopoulos, T Kawakami, T Kunitake. Chem Commun 1229, 1998.
202. D Cochin, J-L Habib Jiwan, A Laschewsky, E Wischerhoff, M van der Auweraer, E Rousseau. Unpublished 1999.
203. H Hong, R Steitz, S Kirstein, D Davidov. Adv Mater 10:1104, 1998.
204. NA Kotov, I Dekany, JH Fendler. J Phys Chem 99:13065, 1995.
205. G Kim, RJ Farris, TJ McCarthy. Polym Prepr Am Chem Soc Div Polym Chem 38(2):672, 1997.
206. A Delcorte, P Bertrand, E Wischerhoff, A Laschewsky. Langmuir 13:5125, 1997.
207. E Brynda, M Houska. J Colloid Interface Sci 183:18, 1996.
208. H Kim, MW Urban. Langmuir 14:7235, 1998.
209. M Gao, X Zhang, B Yang, J Shen. J Chem Soc Chem Commun 2229, 1994.
210. W Fabianowski, M Roszko, W Brodzińska. Thin Solid Films 327–329:743, 1998.
211. P Rilling, T Walter, R Pommersheim, W Vogt. J Membr Sci 129:283, 1997.
212. G Mao, Y-H Tsao, M Tirrell, HT Davis, V Hessel, H Ringsdorf. Langmuir 9:3461, 1993.
213. G Mao, Y-H Tsao, M Tirrell, HT Davis, V Hessel, H Ringsdorf. Langmuir 11: 942, 1995.
214. G Mao, Y-H Tsao, M Tirrell, HT Davis, V Hessel, J van Esch, H Ringsdorf. Langmuir 10:4174, 1994.
215. B Sellergren, A Swietlow, T Arnebrant, K Unger. Anal Chem 68:402, 1996.
216. X Zhang, M Gao, X Kong, Y Sun, J Shen. J Chem Soc Chem Commun 1055, 1994.
217. K Araki, MJ Wagner, MS Wrighton. Langmuir 12:5393, 1996.
218. JS Do, TH Ha, JD Hong, K Kim. Bull Korean Chem Soc 19:257, 1998.
219. M Gao, Mi Gao, X Zhang, Y Yang, B Yang, J Shen. J Chem Soc Chem Commun 2777, 1994.

220. W Kong, X Zhang, ML Gao, H Zhou, W Li, JC Shen. Macromol Rapid Commun 15:405, 1994.
221. W Kong, LP Wang, ML Gao, H Zhou, X Zhang, W Li, JC Shen. J Chem Soc Chem Commun 1297, 1994.
222. G Decher, JD Hong. Ber Bunsenges Phys Chem 95:1430, 1991.
223. M Gao, X Kong, X Zhang, J Shen. Thin Solid Films 244:815, 1994.
224. TM Cooper, AL Campbell, RL Crane. Polym Prep Am Chem Soc Div Polym Chem 36(1):377, 1995.
225. DQ Li, M Lütt, MR Fitzsimmons, R Synowicki, ME Hawley, GW Brown. J Am Chem Soc 120:8797, 1998.
226. JK Lee, DS Yoo, ES Handy, MF Rubner. Appl Phys Lett 69:1686, 1996.
227. HP Zheng, RF Zhang, Y Wu, JC Shen. Chem Lett 909, 1998.
228. D Yoo, A Wu, J Lee, MF Rubner. Synth Met 85:1425, 1997.
229. I Ichinose, H Tagawa, S Mizuki, Y Lvov, T Kunitake. Langmuir 14:187, 1998.
230. J-D Hong, E-S Park, A-L Park. Bull Korean Chem Soc 19:1156, 1998.
231. F Saremi, B Tieke. Adv Mater 10:388, 1998.
232. R Advincula, E Fells, N Jones, J Guzman, A Baba, F Kaneko. Polym Prepr 40: 443, 1999.
233. S Yamada, A Harada, T Matsuo, S Ohno, I Ichinose, T Kunitake. Japan J Appl Phys 36:L1110, 1997.
234. H Fukumoto, Y Yonezawa. Thin Solid Films 327/329:748, 1998.
235. D Yoo, J-K Lee, MF Rubner. Mater Res Soc Proceedings 413:395, 1996.
236. F Saremi, E Maassen, B Tieke, G Jordan, W Rammensee. Langmuir 11:1068, 1995.
237. F Saremi, B Tieke, G Jordan, W Rammensee. Supramol Sci 4:471, 1997.
238. F Saremi, B Tieke. Adv Mater 7:378, 1995.
239. I Ichinose, K Fujiyoshi, S Mizuki, Y Lvov, T Kunitake. Chem Lett 257, 1996.
240. TS Lee, J Kim, J Kumar. Macromol Chem Phys 199:1445, 1998.
241. KS Alva, J Kumar, KA Marx, SK Tripathy. Macromolecules 30:4024, 1997.
242. V Charlier, A Laschewsky, B Mayer, E Wischerhoff. Macromol Symp 126:105, 1997.
243. DM Kaschak, JT Lean, CC Waraksa, GB Saupe, H Usami, TE Mallouk. J Am Chem Soc 121:3435, 1999.
244. A Delcorte, P Bertrand, X Arys, A Jonas, E Wischerhoff, B Mayer, A Laschewsky. Surf Sci 366:149, 1996.
245. D Cochin, M Paßmann, G Wilbert, R Zentel, E Wischerhoff, A Laschewsky. Macromolecules 30:4775, 1997.
246. L Balogh, L Samuelson, KS Alva, A Blumstein. J Polym Sci Part A Polym Chem 36:703, 1998.
247. L Balogh, L Samuelson, KS Alva, A Blumstein. Macromolecules 29:4180, 1996.
248. A Blumstein, L Samuelson. Adv Mater 10:173, 1998.
249. VV Tsukruk. Adv Mater 10:253, 1998.
250. VV Tsukruk, F Rinderspacher, VN Bliznyuk. Langmuir 13:2171, 1997.
251. J Tian, ME Thompson, C-C Wu, JC Sturm, RA Register, MJ Marsella, TM Swager. Polym Prepr 35(2):761, 1994.
252. AC Fou, MF Rubner. Macromolecules 28:7115, 1995.
253. GB Sukhorukov, MM Montrel, AI Petrov, LI Shabarchina, BI Sukhorukov. Biosensors Bioelectronics 11:913, 1996.

254. MM Montrel, GB Sukhorukov, AI Petrov, LI Shabarchina, BI Sukhorukov. Sensors Actuators B 42:225, 1997.
255. MM Montrel, GB Sukhorukov, LI Shabarchina, NV Apolonnik, BI Sukhorukov. Mat Sci Eng C 5:275, 1998.
256. Y Sun, J Sun, X Zhang, C Sun, Y Wang, J Shen. Thin Solid Films 327–329:730, 1998.
257. SF Hou, HQ Fang, HY Chen. Anal Lett 30:1631, 1997.
258. F Caruso, DN Furlong, K Ariga, I Ichinose, T Kunitake. Langmuir 14:4559, 1998.
259. J Hodak, R Etchenique, EJ Calvo, K Singhal, PN Bartlett. Langmuir 13:2708, 1997.
260. F Caruso, K Niikura, DN Furlong, Y Okahata. Langmuir 13:3427, 1997.
261. J-A He, L Samuelson, L Li, J Kumar, SK Tripathy. Langmuir 14:1674, 1998.
262. J-A He, L Samuelson, L Li, J Kumar, SK Tripathy. J Phys Chem B 102:7067, 1998.
263. GS Ferguson, ER Kleinfeld. Adv Mater 7:414, 1995.
264. SW Keller, SA Johnson, ES Brigham, EH Yonemoto, TE Mallouk. J Am Chem Soc 117:12879, 1995.
265. DM Kaschak, TE Mallouk. J Am Chem Soc 118:4222, 1996.
266. HG Hong. Bull Korean Chem Soc 16:1145, 1995.
267. J Kerimo, DM Adams, PF Barabara, DM Kaschak, TE Mallouk. J Phys Chem B 102:9451, 1998.
268. PJ Ollivier, NI Kovtyukhova, SW Keller, TE Mallouk. Chem Commun 15:1563, 1998.
269. NA Kotov, I Dékany, JH Fendler. Adv Mater 8:637, 1996.
270. T Cassagneau, JH Fendler. Adv Mater 10:877, 1998.
271. JH Fendler. Croatica Chemica Acta 71:1127, 1998.
272. K Ariga, Y Lvov, M Onda, I Ichinose, T Kunitake. Chem Lett 2:125, 1997.
273. F Caruso, DG Kurth, D Volkmer, MJ Koop, A Müller. Langmuir 14:3462, 1998.
274. I Ichinose, H Tagawa, S Mizuki, Y Lvov, T Kunitake. Langmuir 14:187, 1998.
275. D Ingersoll, PJ Kulesza, LR Faulkner. J Electrochem Soc 141:140, 1994.
276. J Schmitt, G Decher, WJ Dressick, SL Brandow, RE Geer, R Shashidhar, JM Calvert. Adv Mater 9:61, 1997.
277. W Schrof, S Rozouvan, E Van Keuren, D Horn, J Schmitt, G Decher. Adv Mater 3:338, 1998.
278. Y Liu, Y Wang, RO Claus. Chem Phys Lett 298:315, 1998.
279. Y Liu, RO Claus. J Appl Phys 85:419, 1999.
280. T Yonezawa, S-Y Onoue, T Kunitake. Adv Mater 10:414, 1998.
281. Y Liu, A Wang, R Claus. J Phys Chem B 101:1385, 1997.
282. MY Gao, X Zhang, B Yang, F Li, JC Shen. Thin Solid Films 284:242, 1996.
283. M Gao, B Richter, S Kirstein, H Möhwald. J Phys Chem B 102:4096, 1998.
284. J Sun, E Hao, Y Sun, X Zhang, B Yang, S Zou, J Shen, S Wang. Thin Solid Films 327/329:528, 1998.
285. Y Sun, E Hao, X Zhang, B Yang, M Gao, J Shen. Chem Commun 2381, 1996.
286. Y Sun, E Hao, X Zhang, B Yang, J Shen, L Chi, H Fuchs. Langmuir 13:5168, 1997.
287. Y Liu, A Wang, RO Claus. Appl Phys Lett 71:2265, 1997.
288. MR Talingting, YM Ma, C Simmons, S. Webber. Polym Prepr Am Chem Soc Div Polym Chem 40(2):1090, 1999.

289. K-U Fulda, D Piecha, B Tieke, H Yarmohammadipour. Prog Colloid Polym Sci 101:178, 1996.
290. K-U Fulda, A Kampes, L Krasemann, B Tieke. Thin Solid Films 327–329:752, 1998.
291. T Salditt, Q An, A Plech, C Eschbaumer, US Schubert. Chem Commun 2731, 1998.
292. US Schubert, C Eschbaumer, Q An, T Salditt. Polym Prepr Am Chem Soc Div Polym Chem (40)1:414, 1999.
293. M Schütte, DG Kurth, MR Linford, H Cölfen, H Möhwald. Angew Chem Int End 37:2891, 1998.
294. B Lindholm-Sethson. Langmuir 12:3305, 1996.
295. R Advincula, W Knoll. Coll Surf A123/124:443, 1997.
296. Y Lvov, F Eßler, G Decher. J Phys Chem 97:13773, 1993.
297. H Bock, RC Advincula, EF Aust, J Käshammer, WH Meyer, S Mittler-Neher, C Fiorini, J-M Nunzi, W Knoll. J Nonlinear Opt Phys Mater 7:385, 1998.
298. M Tarabia, H Hong, D Davidov, S Kirstein, R Steitz, R Neumann, Y Avny. J Appl Phys 83:725, 1998.
299. GJ Kellogg, AM Mayes, WB Stockton, M Ferreira, MF Rubner, SK Satija. Langmuir 12:5109, 1996.
300. R Bijlsma, AA van Well, MA Cohen Stuart. Physica B 234/236:254, 1997.
301. X Arys. Understanding Ordering in Polyelectrolyte Multilayers: Effect of the Chemical Architecture of the Polycation. PhD dissertation, Université catholique de Louvain-La-Neuve, Belgium, 2000.
302. S Balasubramanian, X Wang, HC Wang, L Li, DJ Sandman, J Kumar, SK Tripathy, MF Rubner. Polym Prepr Am Chem Soc Div Polym Chem 38(2):502, 1997.
303. K Yang, S Balasubramanian, XG Wang, J Kumar, S Tripathy. Appl Phys Lett 73: 3345, 1998.
304. X Wang, S Balasubramanian, L Li, X Jiang, DJ Sandman, MF Rubner, J Kumar, SK Tripathy. Macromol Rapid Commun 18:451, 1997.
305. VG Gregoriou, R Hapanowicz, SL Clark, PT Hammond. Appl Spectrosc 51:470, 1997.
306. SL Clark, M Montague, PT Hammond. Supramol Sci 4:141, 1997.
307. PT Hammond, GM Whitesides. Macromolecules 28:7569, 1995.
308. SL Clark, MF Montague, PT Hammond. Macromolecules 30:7237, 1997.
309. A Delcorte. Static Secondary Ion Mass Spectrometry of Thin Organic Layers. PhD dissertation, Université Catholique de Louvain-La-Neuve, Belgium, 1999.
310. JQ Sun, T Wu, YP Sun, ZQ Wang, X Zhang, JC Shen, WX Cao. Chem Commun 1853, 1998.
311. O Onitsuka, AC Fou, M Ferreira, BR Hsieh, MF Rubner. J Appl Phys 80:4067, 1996.
312. AC Fou, O Onitsuka, M Ferreira, MF Rubner, BR Hsieh. J Appl Phys 79:7501, 1996.
313. M Onoda, K Yoshino. Jpn J Appl Phys 34:260, 1995.
314. J Stepp, JB Schlenoff. J Electrochem Soc 144:L155, 1997.
315. JB Schlenoff, D Laurent, H Ly, J Stepp. Adv Mater 10:347, 1998.
316. M Koetse, A Laschewsky, B Mayer, O Rolland, E Wischerhoff. Macromolecules 31:9316, 1998.
317. A Laschewsky, B Mayer, E Wischerhoff, X Arys, A Jonas, M Kauranen, A Persoons. Angew Chem Int Ed Engl 36:2788, 1997.
318. J Smid, D Fish. In: HF Mark, NM Bikales, CG Overberger, G Menges, eds. Encyclopedia of Polymer Science & Engineering, vol 11. New York: Wiley, 1988, p 720.

319. M Müller, T Rieser, K Lunkwitz, S Berwald, J Meier-Haack, D Jehnichen. Macromol Rapid Commun 19:333, 1998.
320. R von Klitzing, H Möhwald. Langmuir 11:3554, 1995.
321. GB Sukhorukov, G Decher, J Schmitt. Personal communication, 1997.
322. DL Feldheim, KC Grabar, MJ Natan, TE Mallouk. J Am Chem Soc 118:7640, 1996.
323. G Decher, J-D Hong. Eur Patent 00 472 990 A2, 1992.
324. G Decher, J-D Hong. US Patent 00 5 208 111 A, 1993.
325. MF Rubner, JH Cheung. Int Patent 95/02251, 1995.
326. MF Rubner, JH-W Cheung. US Patent 00 5 536 573 A, 1996.
327. MF Rubner, JH Cheung. US Patent 00 5 518 767 A, 1996.
328. M-S Sheu, I-H Loh. US Patent 00 5 700 559 A, 1997.
329. M-S Sheu, I-H Loh. US Patent 00 5 837 377 A, 1998.
330. M-S Sheu, I-H Loh. US Patent 00 5 807 636 A, 1998.
331. GS Ferguson, ER Kleinfeld. US Patent 5 716 709 A, 1998.
332. H-U Siegmund, L Heiliger, B van Lent, A Becker. Eur Patent 00 561 239 A1, 1993.
333. H-U Siegmund, L Heiliger, B van Lent, A Becker. Int Patent 96/18498, 1996.
334. H-U Siegmund, L Heiliger, B van Lent, A Becker. US Patent 00 5 711 915 A, 1998.
335. DL Elbert, CB Herbert, JA Hubbell. Langmuir 15:5355, 1999.
336. P Stroeve, V Vasquez, MAN Coelho, JF Rabolt. Thin Solid Films 284:708, 1996.
337. P Zhou, L Samuelson, KS Alva, C-C Chen, RB Blumstein, A Blumstein. Macromolecules 30:1577, 1997.
338. L Krasemann, B Tieke. J Membr Sci 150:23, 1998.
339. B Jacob, J Schrezenmeir, R Pommersheim, W Walter. Diabetologia 36:A189, 1993.
340. B Jacob, A Gaumann, R Pommersheim, W Vogt, J Schrezenmeir. Diabetologia 37: A216, 1994.
341. E Donath, GB Sukhorukov, F Caruso, SA Davis, H Möhwald. Angew Chem Int Ed Engl 37:2202, 1998.
342. F Caruso, H Lichtenfeld, M Giersig, H Möhwald. J Am Chem Soc 120:8523, 1998.
343. F Caruso, RA Caruso, H Möhwald. Science 282:1111, 1998.
344. X Yang, S Johnson, J Shi, T Holesinger, B Swanson. Sensors Actuators B 45:87, 1997.
345. G Decher, B Lehr, K Lowack, Y Lvov, J Schmitt. Biosensors Bioelectronics 9: 677, 1994.
346. G Decher, F Eßler, J-D Hong, K Lowack, J Schmitt, Y Lvov. Polym Prepr Am Chem Soc Div Polym Chem 34(1):745, 1993.
347. NK Viswanathan, S Balasubramanian, L Li, J Kumar, SK Tripathy. J Phys Chem B 102:6064, 1998.
348. J Tian, C-C Wu, ME Thompson, JC Sturm, RA Register, MJ Marsella, TM Swager. Adv Mater 7:395, 1995.
349. JW Baur, S Kim, PB Balanda, JR Reynolds, MF Rubner. Adv Mater 10:1452, 1998.
350. J-K Lee, D Yoo, MF Rubner. Chem Mater 9:1710, 1997.
351. H Mattoussi, LH Radzilowski, BO Dabbousi, EL Thomas, MG Bawendi, MF Rubner. J Appl Phys 83:7965, 1998.

13
Functional Polymer Brushes

Jürgen Rühe and Wolfgang Knoll
Max-Planck-Institute for Polymer Research, Mainz, Germany

I. INTRODUCTION

Thin organic or polymeric layers on solid substrates play a key role in many processes aimed at modifying surface properties. The general idea is to optimize the bulk properties of a device or system component independently from its surface properties.

Despite the enormous practical importance, little is known about the fundamentals of what actually determines the surface properties of the thin and ultrathin organic/polymeric films used as functional coatings. Many physical and chemical factors have been identified but the overall picture is very complex and hence we are far from a coherent understanding.

A currently very fashionable recipe for the preparation of well-controlled surface layers is the use of molecules with a reactive head-group amenable to undergo a covalent bond with a complementary chemical site on the surface of the substrate to be modified. This leads to the formation of so-called self-assembled monolayers SAM [1] with structural properties similar to Langmuir layers prepared from amphiphiles at the water–air interface [2]. Examples are silanes on oxide surfaces or thiols, and monosulfides or disulfides on noble metal surfaces (see also Chapter 7).

This way, surface coatings can be obtained with an extreme degree of positional and orientational order, and in some cases even crystalline packing has been observed. If molecules are assembled that carry at their tail end a specific chemical moiety or a biochemically active group one can obtain, as a result, a 2-D arrangement of these functionalities. Examples are numerous: end-groups that can dissociate, such as COO^-, SO^-_3, NH^+_3, etc and generate a pH-dependent surface charge layer [3]; methyl end-groups or fluorocarbon segments at the end of the assembling units can convert a hydrophilic surface into a highly water-repellent hydrophobic one; through the introduction of ''ligands'' as recognition sites in bioaffinity assays one generates surfaces, e.g., on top of the transducer of a biosensor, that very specifically bind proteins from solution [4].

Several applications of the SAMs have been demonstrated rather convincingly, e.g., as surface modifiers in corrosion protection or to manipulate the electron transfer rates from proteins to electrode surfaces [5], as alignment layers for liquid-crystalline samples [6], or for the induction of the nucleation and growth of inorganic crystals from supersaturated salt solutions [7].

In some of these applications the intrinsic limitations of this strictly 2-D arrangement of the functional groups become obvious: there is an upper limit in the surface density of the functionality given by the surface area cross-section of the assembled unit, and the arrangement of the individual functional unit at such high packing densities in some cases leads to a mutual blocking or, at least, to a limited accessibility.

II. POLYMERS AT SOLID SURFACES

One obvious solution to the above problem is the extension into the third dimension, i.e., the use of polymers carrying the functional groups along the chain, thus generating higher cross-sectional densities of these groups simultaneously guaranteeing good accessibility.

Under certain (limited) circumstances a mere physisorption of the functional polymers might work, in particular, if only transient surface coatings are employed. For example, the spin-coating of a clean Si-wafer by a photoresist is sufficient to prepare a photoreactive layer that can be used to generate a pattern by mask-illumination, etch this into the Si to laterally structure the substrate, and then dissolve the polymer away again.

However, in many other situations, the polymer coating has to withstand exposure to solvents, good solvents in most cases. For example, the binding reactions between polymer- and hence substrate-bound ligands and proteins from solution in a bioassay require the extended exposure of the physisorbed polymer to an aqueous phase. This almost inevitably leads to dissolution, desorption, displacement, or delamination.

One strategy to improve the adhesion between a substrate and a functional coating is the establishment of a covalent chemical bond between the polymer molecules, of which the functional layer consists, and the surface of the substrate. Most approaches in that direction use polymers carrying an ''anchor'' group either as an end-group or in a sidechain, which can be reacted with appropriate sites at the substrate surface yielding surface-attached monolayers (''grafting to'') [8–12]. Such a chemisorption process is in some aspects closely related to the formation of self-assembled monolayers of low molecular weight compounds mentioned above.

Some rather strict limitations apply for structures, which can be realized following such a ''grafting-to'' strategy. First, the use of reactive anchor groups for the surface-attachment of the polymers imposes some rather strict limitations on the choice of functional groups available for incorporation into the polymer. The functional groups of the polymer can compete with the anchor moieties for surface sites (competitive adsorption) (Fig. 1). Especially if functional polymers containing highly polar or charged groups are employed, adsorption of the functional groups to the surface can be very strong and compete very effectively with the chemisorption process. Such a competition between anchor and functional group has been observed, for example, in the case of the attachment of a low molecular weight alkoxysilane containing amine groups to a silicon oxide surface [13–15]. In this system interactions between the (basic) amine groups of the silane and the (acidic) silanol groups of the silicon oxide substrate can strongly compete with the chemisorption reaction of the alkoxysilyl moiety with the substrate silanol groups. As a result, a layer was obtained where both strong physisorption due to acid-base interactions and chemisorption occurred.

On the other hand, to obtain a fast surface attachment reaction with a high number of chains covalently bound to the substrate, rather reactive anchor groups

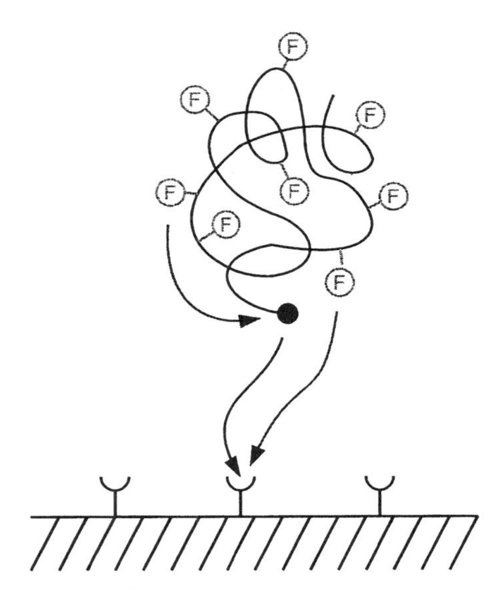

Figure 1 Schematic depiction of the chemisorption of polymer molecules with functional groups (F) to solid surfaces through the anchor group (black dot).

are required, which do not tolerate the coexistence of a large number of functional groups in the polymer (Fig. 1). For example, if a chlorosilyl group is chosen as an anchor group for the attachment to a silicon oxide surface, this choice prevents the incorporation of all functional groups into the polymer that contain amine-, hydroxyl-, or carboxylic-acid moieties. It should be noted, however, that the choice of a less reactive anchor group will not at all solve the problem. The ratio between the undesired side reaction, the reaction of the functional group with the anchor group, which leads to loss of anchor moieties, and the desired reaction with a group at the surface of the substrate, which results in a successful chemisorption reaction, depends solely on the ratio of the nucleophilicity/electrophilicity of the functional groups in the polymer and the surface groups. This situation is represented by the arrows (solid line) in Fig. 1. For a given anchor group the only important factor to be considered is whether the groups at the surface are sufficiently nucleophilic or electrophilic to guarantee a reaction that is faster than all side reactions. Only such systems will allow an efficient immobilization process and a clearcut structure of the resulting monolayer. Surface groups, however, are almost by definition not highly reactive due to the close proximity of the groups in the surface layer.

Another complication inherent in "grafting-to" processes is an intrinsic limitation of the film thickness and accordingly the number of functional groups per surface area that can be obtained by such an approach. Films generated by chemisorption from solution are usually limited to (dry) thicknesses of 1 to 5 nm. As soon as the surface becomes significantly covered with attached chains, the polymer concentration at the interface becomes larger than the concentration of the macromolecules in solution. Additional chains, trying to reach the surface, have to diffuse against this concentration gradient that increases with the amount of attached polymer. This diffusion barrier makes the immobilization at the surface more and more unfavorable (Fig. 2). Thus, the rate of the attachment reaction levels off quickly and further polymer is linked to the substrate only at a very low rate due to this kinetic hindrance. Indeed, it has been shown both theoretically [16,17] and experimentally [8–12] that once the surface-attached coils overlap, the attachment of further polymer molecules takes place only on a logarithmic time scale. Films generated by this technique are accordingly intrinsically limited concerning the film thickness for all practical experimental time scales. Additionally it should be noted that, even if this kinetic limitation is somehow circumvented, the attachment of chains to a strongly covered surface becomes unfavorable also for thermodynamic reasons [16,17]. At high grafting densities the surface-attached polymer chains are in a rather stretched conformation due to the presence of strong segment–segment interactions [8]. A chain, which is now becoming attached to the surface, has to change from a coil conformation in solution to a stretched (brushlike) conformation at the surface. The entropy loss during this process is only compensated by the establishment of one chemical

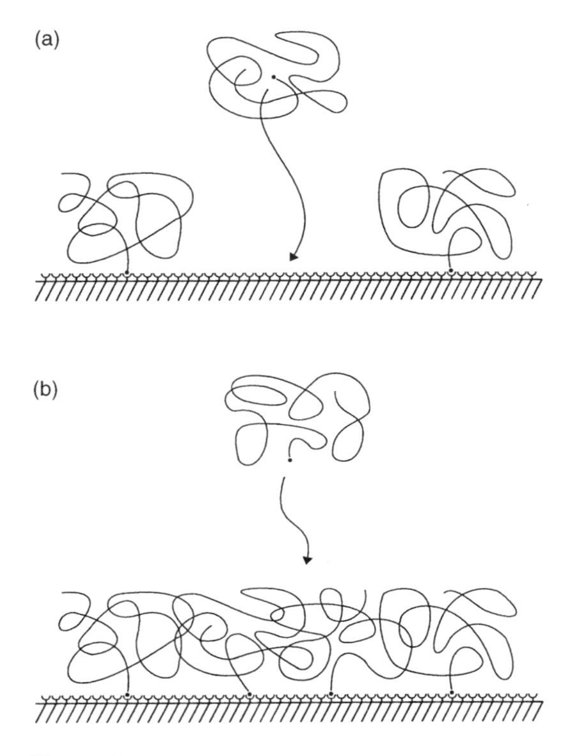

Figure 2 Attachment of polymer molecules to solid surfaces: (a) low surface coverage (distance between anchor points is much larger than the dimensions of the polymer molecules); (b) significant coverage of the surfaces (polymer molecules attached to the surface overlap).

bond, namely, the one connecting the polymer to the surface. The higher the graft density of the chains at the surface is, the stronger will be the entropy penalty.

These limitations of ''grafting-to'' processes can be overcome if the polymer molecules are directly generated at the surface of the substrate using surface-attached initiator molecules [18–20]. If the polymer is ''grown'' at the surface no significant diffusion barrier exists as only low molecular weight compounds—the monomers—have to reach the growing chain ends during monolayer formation. Thus, the kinetic barrier preventing the formation of thick monolayers of preformed polymers can be circumvented.

Most systems described in the literature make use of immobilized azo compounds as the initiating species [18–20]. These monolayers are in many cases created from self-assembled monolayers of silanes that are then further modified in various steps to yield the actual monolayer of the initiator [18–22]. This approach suffers from major disadvantages: (1) the various reaction steps on the surfaces are not quantitative and, hence, the graft densities of the initiator monolayer and, subsequently, of the polymer layer will be low and not easy to reproduce; and (2) incomplete conversions and side reactions may result in a surface with an unknown number and quantity of different structures. Some of these species may strongly influence the subsequent polymerization process, making it rather difficult to understand the mechanism of this reaction.

To avoid these problems we follow a ''grafting-from'' concept which is schematically depicted in Fig. 3 [23–25]. In a first step the initiator carrying an appropriate anchor group is attached to the surface via a covalent bond (formation of a self-assembled monolayer of the initiator). As the initiator is attached to the

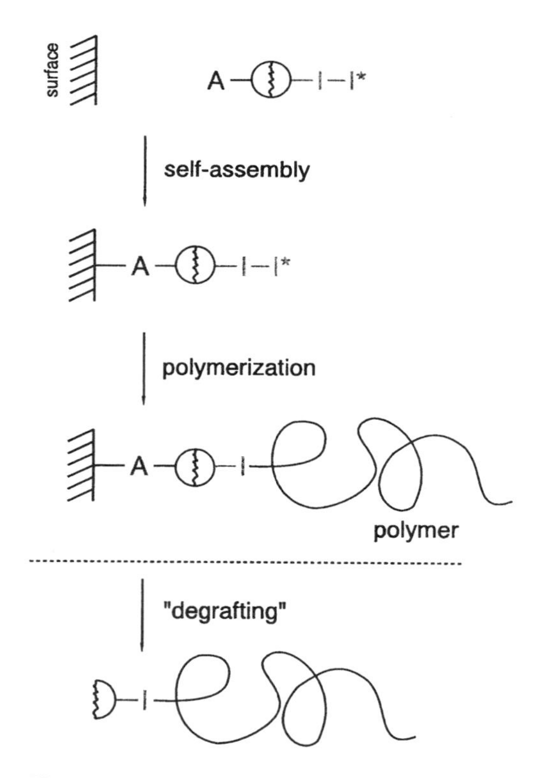

Figure 3 Concept for the growing of polymer molecules by using self-assembled monolayers of initiators.

surface by a one-step reaction such a procedure allows a high and reproducible graft density of the initiator. In a second step the polymerization reaction is carried out starting from the surface-attached initiator. This approach can be used for all polymerization reactions following a chain mechanism such as cationic, anionic, and radical chain polymerization. However, the latter is a most suited one for the preparation of functional monolayers as radical chain processes tolerate the presence of many different functional groups in the monomer. In the work reviewed below we used an azo initiator which is a close analogue to azobisisobutyro nitrile AIBN, the most common radical chain initiator. We attached it to silicon and other oxide surfaces using a monochlorosilyl group (see Fig. 4). The initiator can be activated either by heat or light. The latter allows for the photolithographic micropatterning [26,27] of the surface-attached polymer monolayers and for the generation of multifunctional patterns at selected areas of the substrate.

In the context of the main thrust of this book, i.e., supramolecular polymers, it should be noted that it is the mere presence of the interface that leads to the unique organization and hence physical properties of polymer brushes: the chains, grown at high surface densities, all end-grafted by one of their two chain ends to the solid support, avoid crowding by stretching away from the interface. At equilibrium the reduction in interaction energy compensates the concomitant

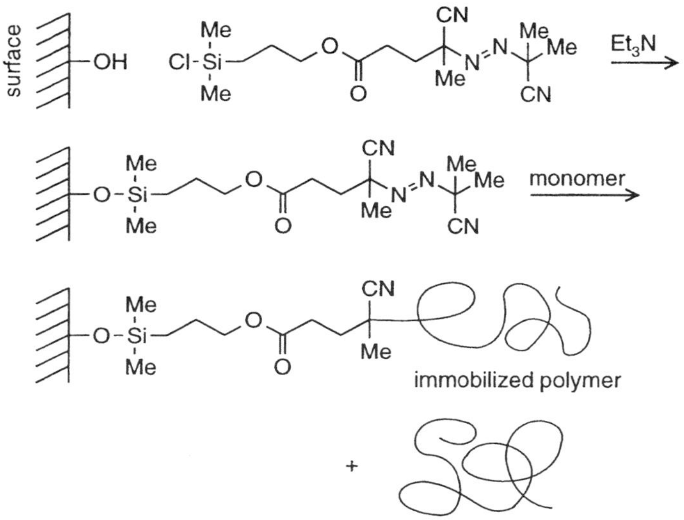

Figure 4 Surface initiated radical chain polymerization at SiO_x surfaces using self-assembled monolayers of azoinitator similar to AIBN with a chlorosilane anchor group.

loss in conformational entropy. The individual chain this way attains a configuration that can be far away from the Gaussian coil in solution. The assembly of chains forced by the surface attachment and noncovalent interactions constitute a supramolecular system of unique physical, e.g., mechanical or tribological, properties.

In order to evaluate the chemical structure, the molecular weights and molecular weight distributions of the polymers formed during the surface polymerization process, it is desirable to detach the surface-attached macromolecules from the substrate after generation of the polymer monolayer and characterize them by using common techniques of polymer analysis. This can be achieved by introducing a third, cleavable group to the structure of the initiator [23–25]. This group can now act as a chemical''break-seal''for the connection between the polymer chain and the surface. The only prerequisite for the successful detachment of the polymer is that the break-seal is accessible during the cleavage process; i.e., the polymer has to be in a strongly swollen state during the cleaving reaction. In the case described here we used an ester group as the break-seal and cleaved the polymer from the surface through an acid-catalyzed transesterification reaction.

III. CONTROL OF THICKNESS OF POLYMER MONOLAYERS

An important prerequisite for the preparation of tailormade functional polymer monolayers is the possibility of adjusting the mass of the surface-attached material and accordingly the layer thickness over a wide range. In the system described here the (dry) film thickness d can be adjusted both by controlling the number of surface-attached chains per area, or the graft density Γ, and the length of the attached polymer chains, i.e., the number averaged molecular weight $\overline{M}_n$ (Fig. 5):

$$d = \frac{\Gamma \, \overline{M}_n}{\rho} \tag{1}$$

ρ denotes here the density of the polymer in the monolayer. The graft density of the chains can be influenced either by directly controlling (reducing) the graft density of the initiator or by adjusting the conversion of the initiator. The first case can be achieved by diluting the initiator with an inert compound, which is unable to start a polymerization reaction, or by statistically deactivating some fraction of the initiator species [28]. Both dilution and deactivation lead to a decrease of the surface concentration of active initiator species.

When the graft density of the polymer is controlled by adjusting the initiator conversion the statistical nature of the initiation process is to be considered. In

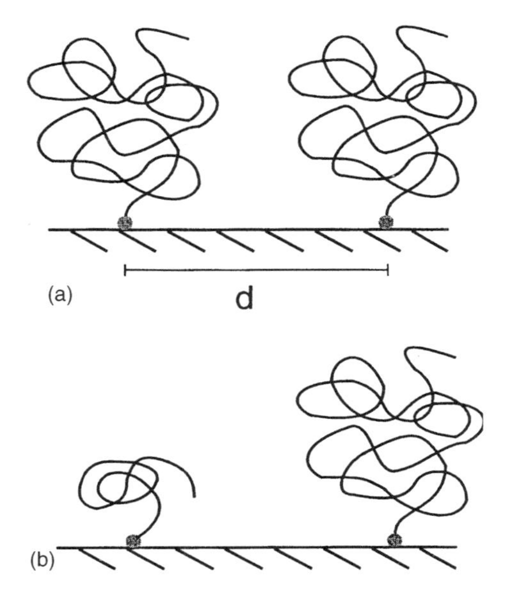

Figure 5 Control of the film thickness of surface-attached polymer monolayers: (a) control of the graft density; (b) control of the molecular weight of the polymer.

fact, as in any radical chain polymerization reaction, not all initiator molecules are activated at once and start to grow a polymer chain. For example, for the initiator shown in Fig. 4 half of the initiator molecules become activated within 20 hours if the reaction temperature is set to 60°C. Compared to this the growth of the individual chains is orders of magnitude faster. This allows control of the graft density of the polymer molecules at the surface simply by selecting the polymerization time t:

$$\Gamma = f[I_0](1 - e^{-kt}) \tag{2}$$

f represents here the radical efficiency factor for the given initiator/solvent/monomer system, $[I_0]$ is the graft density of the initiator prior to the start of the polymerization reaction and k the rate constant of the initiator decomposition. For long polymerization times (larger than one-half lifetime of the initiator) it has to be considered that the radical efficiency decreases with increasing conversion, and is accordingly also a function of the polymerization time. The thickness of the surface-attached layers can thus be predicted from the above, independently measured parameters.

An example for the control of the thickness of the surface-bound monolayers by adjustment of the polymerization time is given in Fig. 6. Some typical

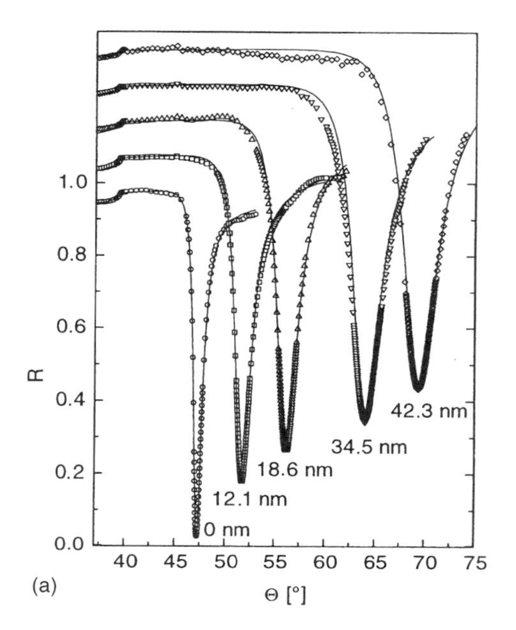

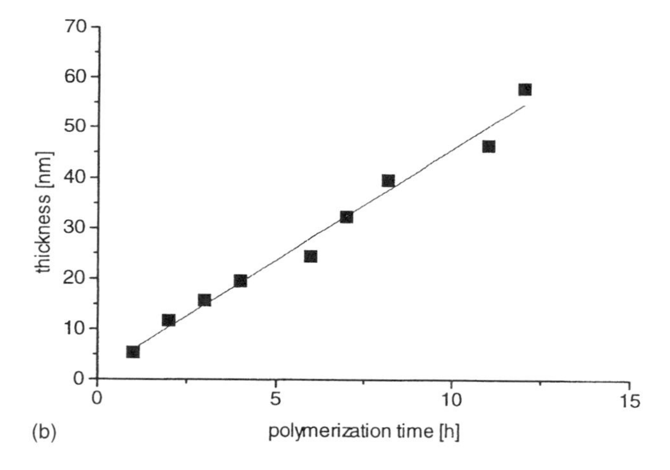

Figure 6 (a) SPR reflectivity curves of polystyrene monolayers as a function of polymerization time (initiator conversion); substrate (glass, 50 nm Ag, 30 nm SiO_x), T = 60°C styrene: toluene 1:1, t = 0, 2, 4, 6, and 7 hours; the reflectivity curves are shifted by a constant additive factor for clarity; (b) layer thickness of the obtained monolayers. (From Ref. 29.)

examples of reflectivity curves obtained in surface plasmon resonance (SPR) experiments [29] are shown for a set of polystyrene monolayers attached to glass/silver/SiO_x substrates (Fig. 6a). During the preparation of these samples all polymerization parameters were kept constant, except the polymerization time which was systematically varied. It can be clearly seen, that with increasing polymerization time the resonance minimum shifts to higher angles, indicating an increase in the film thickness. The thickness of the obtained monolayers calculated by using the Fresnel equations as a function of the polymerization time is shown in Fig. 6b. A simple box model was assumed for the model fit calculations and the refractive index of bulk polystyrene was used. No saturation limit of the layer thickness is visible due to the fact that within the time-frame of the experiments less than half of the initiator has decomposed. Each datapoint shown in this plot was measured on a different sample. The close agreement with the results of the experiments and the theoretically expected kinetics (solid line in Fig. 6b) demonstrate that the monolayer generation can be carried out in a well-reproducible fashion. It should be emphasized that all samples were extracted for 15 to 48 h with boiling toluene, which is a good solvent for polystyrene, in a continuous extraction setup, so that no physisorbed polymer remained in the film. Physisorbed, i.e., nonattached polymer, can originate from the second radical generated during the initiator decomposition, which is not attached to the surface and gives rise to the growth of free polymer in solution (cf. Fig. 4). Further sources could be free polymer originating from chain transfer reactions to solvent or monomer and/or from thermally initiated polymerization. It has been shown [28] that such extraction procedures are sufficient to remove all polymer from the film that is not covalently linked to the substrate.

An important parameter for any application using surface-attached polymer monolayers is the homogeneity of the layers. To obtain a more quantitative and detailed impression of the film morphology AFM measurements were performed. Figure 7 shows an AFM image taken from a $5 \times 5\ \mu m^2$ large area of a polystyrene film attached to a silicon wafer. The conditions for the preparation of the sample were very similar to the ones in the SPR experiments depicted in Fig. 6. The average film thickness of the sample shown is approximately 200 nm. It should be emphasized, that the scale in the z-direction of the image is only 5 nm and that all height values are within this range. When a profile along an arbitrary chosen line is taken (Fig. 7b), it can be seen that the strongest height differences (peak to valley value) observed along this line are only roughly 2.7 nm. The rms (root mean square) roughness of this section of the film was about 0.5 nm which is essentially the same value as that of the silicon substrate prior to any surface modification. Both values show that the film is very homogeneous and smooth on this length scale.

As pointed out above, the layer thickness is not only a function of the number of chains attached per surface area, but also a function of the molecular

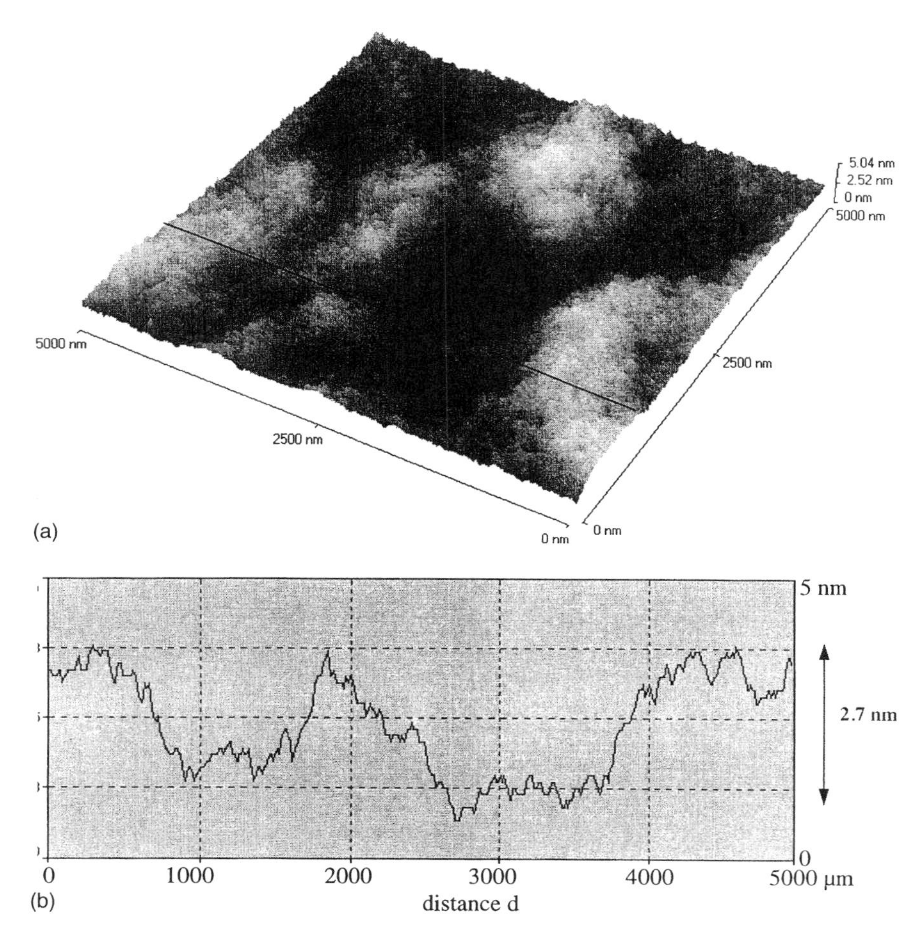

Figure 7 (a) AFM image of a 200 nm thick polystyrene layer attached to a silicon wafer; the layer has been prepared by the ''grafting-from'' technique; (b) height profile along a line shown in (a).

weight of the surface-bound polymer. Thus all parameters that influence the molecular weight of the polymer during a free radical chain polymerization reaction can be used to control the film thickness. Examples are monomer concentration, reaction temperature during the polymerization process, addition of transfer agents, and the addition of ''free'' initiator (e.g., AIBN in solution [30]). It is beyond the scope of this review to discuss all the different possibilities for con-

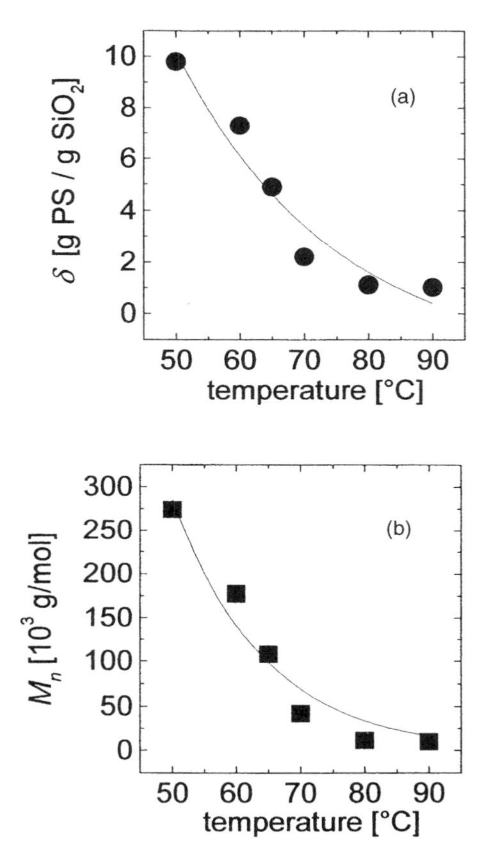

Figure 8 (a) Graft density δ and (b) averaged molecular weight M_n of the covalently attached polymers as a function of temperature during the polymerization reaction; I(Azo) = 1.82 mol/m^2 [M_0] = 1.74 mol 1^{-1}.

trolling the molecular weight of the surface-attached chains. Only the influence of the polymerization temperature on the film thickness of the monolayer is discussed.

In Fig. 8a the mass of polystyrene monolayers on high surface area silica gels is shown as a function of the temperature during the polymerization. All experiments were carried out in such a way that the initiator conversion, and accordingly, the graft density of the surface-attached chains were constant. Higher temperatures T increase the number of simultaneously growing chains, and increase the probability for deactivation, hence decreasing the degree of polymerization $\overline{P}_n$ of the surface-attached chains according to:

$$\ln \overline{P}_n = \text{const} + \frac{E_{A,\overline{P}_n}}{RT} \tag{3}$$

Here $E_{A,\overline{P}_n}$ represents the overall activation energy for the polymerization process and R the universal gas constant. The molecular weights of the degrafted polymer chains are shown in Fig. 8b).

The influence of the monomer concentration on the film thickness of the surface-attached monolayer is shown in Fig. 9a using poly(methylmethacrylate) PMMA monolayers on glass substrates as an example. With increasing monomer concentration the molecular weight and the film thickness increase. Slight deviations from a linear relationship between monomer concentration and film thickness can be attributed to radical transfer to solvent. As an example for the good optical quality of the monolayers, a waveguide spectrum [29] of an approximately 1700 nm thick PMMA monolayer (100 vol% sample in Fig. 9a) is shown in Fig. 9b.

By adjusting the two sets of parameters (graft density and molecular weight), polymers having molecular weights of $\overline{M}_n > 10^6$ g/mol can be attached to surfaces with average distances between the anchor points of less than 3 nm resulting in surface-attached monolayers with thicknesses exceeding 2 μm (solvent-free state). To illustrate the meaning of these values a schematic representation of such a monolayer is shown in Fig. 10. The distances between the anchor groups of the different surface-attached species are drawn to scale. The different symbols depict the anchoring sites of the different species immobilized onto the substrate (however, the size of the symbols is not related to the dimensions of the molecules). The schematic representation refers to a polymerization time, at which about 30% of the azo molecules have been thermally decomposed. The other 70% are still unchanged. Most of the generated radicals were unable to escape the solvent cage and did not start a polymerization reaction. However, about 30 to 40% of the decomposed initiator molecules, described by the radical efficiency factor in Eq. (2), successfully initiated a surface polymerization process. The surface-attached polymer chains are also represented by their respective anchor points. Under the conditions of the specific experiment the molecular weight of the polymer was $\overline{M}_w = 300.000$ g/mol. Such a molecule has a radius of gyration of approximately 21.5 nm according to light scattering measurements. The projection of the equivalent sphere representing the polymer molecule is also shown in Fig. 10. Already at this low conversion roughly 170 other polymer molecules compete for the same space at the surface of the substrate. At higher conversion and/or with higher molecular weights of the attached chains, as for example described in Fig. 9, even more polymer chains compete for the available space, leading to situations where roughly 2000 polymer molecules are attached

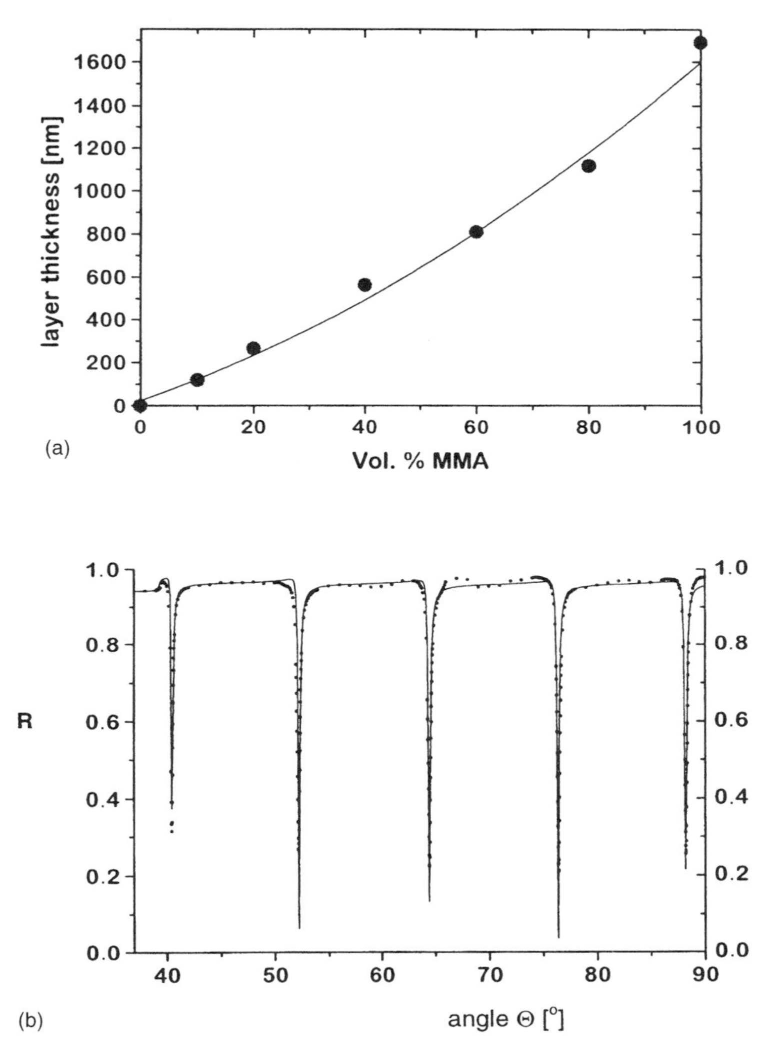

Figure 9 (a) Film thickness of PMMA monolayers as a function of monomer concentration during the polymerization reaction (temperature 60°C, t = 18 h); all samples have been extracted with toluene after stopping of the polymerization reaction for 20 to 48 hours; (b) waveguide spectrum of a 1690 nm thick PMMA layer; solid line is a calculation according to Fresnel equations.

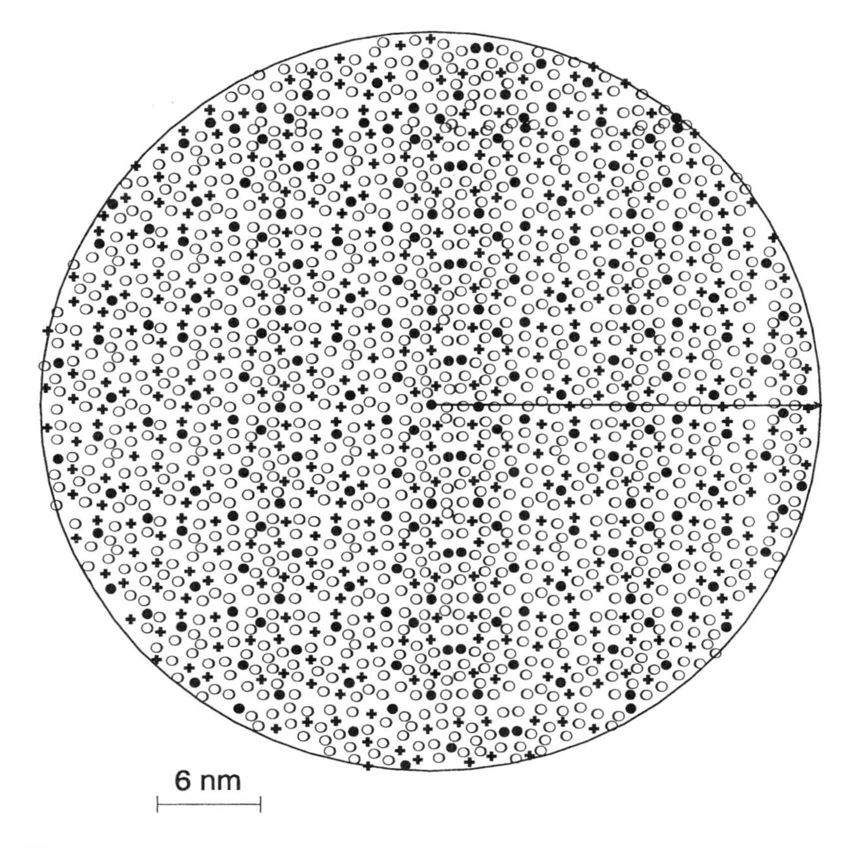

Figure 10 Schematic illustration of the surface coverage of an SiO_x surface using a self-assembled monolayer of the azo initiator shown in Fig. 4 after 10 hours of polymerization at 60°C (to scale): ○ = still unreacted initiator molecules; + = decomposed initiator molecules, which did not start a polymerization reaction; ● = surface-attached polymer chains. The symbols represent the anchoring sites of the molecules and are not related to their size. The circle depicts the size of the projection of one polymer molecule $\langle R_G \rangle$ in solution with $M_W = 290{,}000$ g mol^{-1}.

to within the projection of an equivalent sphere representing the radius of gyration of a free polymer chain!

IV. POLYMER BRUSHES FROM FUNCTIONAL MONOMERS

In the previous section only polymers without functional groups were discussed. However, the concept of using self-assembled initiator monolayers and growing the polymer directly at the surface of the substrate can be applied for a large number of different monomers. Some representative examples of polymer monolayers, which have been realized following this ‘‘grafting from’’ approach are depicted in Fig. 11. In principle, all monomers that can be transformed into high molecular weight polymers by free radical chain polymerization reactions can be used.

If it is desired to synthesize functional monolayers containing groups which either have a strong tendency for chain transfer or contain very valuable groups, it is impractical to introduce the moieties directly into the monomer as in both cases only rather low film thickness will be obtained. In the first case this is due to termination of the growth of polymer chains. Any chain transfer to monomer or solvent leads only to the formation of additional polymer molecules in solution, which is washed away after completion of the reaction. For a surface polymerization reaction every transfer reaction is in fact a termination reaction.

If on the other hand monomers, which contain very complicated and expensive functional groups, are used, it will be difficult to perform polymerization reactions with high monomer concentrations if large substrates are to be modified. Such processes would require unreasonably large amounts of monomer. In such cases one would try to dilute the monomer by addition of large quantities of solvent. However, a reduction of the monomer concentration will lead to a corresponding decrease in the molecular weight and, accordingly, film thickness of the functional monolayer. In both cases it is advantageous to introduce the functional groups after the polymerization step via a polymer-analogous reaction.

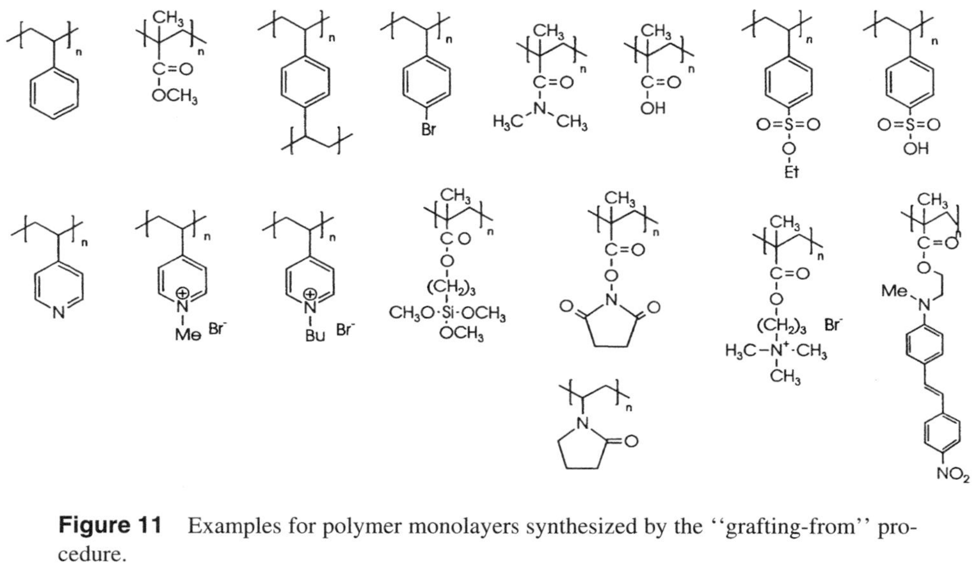

Figure 11 Examples for polymer monolayers synthesized by the ‘‘grafting-from’’ procedure.

In the following we describe two selected systems where the polymer monolayers carry functional groups—polyelectrolyte brushes and sidechain liquid crystal brushes.

A. Polyelectrolyte Brushes

The physical properties of polyelectrolyte (PEL) molecules in contact with the surface of a solid substrate are very different from those of uncharged polymers. When polymer molecules carrying charges are brought into contact with a surface, the conformation of the polymer molecules in the monolayer and, accordingly, the structure and the physical properties of the layers are dominated by electrostatic interactions [8]. The interaction with the substrate is determined by the charge density of the polymer [31], the sign and density of the surface charge [32], and the presence of low molecular weight salt in the surrounding electrolyte [33–35]. Usually polyelectrolytes adsorb only in small amounts from solution onto solid surfaces as the electrostatic repulsion between the charged segments opposes the accumulation of polymer chains at the surface. The only way around this is to screen the charges of the chains in contact with the surface through addition of large amounts of salt. The salt addition suppresses the electrostatic interactions and the polymer chains adsorb at the surface in a very similar way to that of a neutral polymer molecule. However, regardless of whether salt is added during the adsorption of polyelectrolytes, the amount of polymer which is adsorbed to a solid surface is typically between 0.1 and 5 mg/m^2 [8].

As the interaction between physisorbed polyelectrolyte molecules with the surfaces of the substrates is governed by electrostatics, polyelectrolyte layers, which are only physically attached, present an intrinsic problem concerning the stability of the layers in different environments. The polyelectrolyte monolayers can be completely or partially removed from the surface either by changing the sign of the surface charge (which leads to an electrostatic repulsion between the polymer segments and the substrate) [32] or by displacement through of a competing compound [36,37]. The situation is completely different if a chemical bond is generated between the polyelectrolytes and the surface. The amount of polymer at the surface is in this case independent of the sign and density of the surface charge of the substrate and the polymer can be removed from the surface only if the bond is cleaved.

To avoid the limitations of ''grafting-to'' procedures we have generated polyelectrolyte molecules by growing the chains at the surface of the substrate as described above for neutral polymers. Either an ionic monomer such as styrene sulfonate is polymerized directly [38] or the PEL monolayer generation is performed in a two-step process [39]. First, a neutral brush is grown at the surface, which is then converted into the PEL brush in a polymer-analogous reaction. In this case it is very important that the degree of functionalization is close to $f = 1$, i.e., that all repeat units are transformed into the charged species. Otherwise a mixture of unreacted hydrophobic and of charged hydrophilic units is obtained, causing phase separation inside the layer and accordingly making the understanding of the obtained structures very difficult.

Poly-4-vinylpyridine (PVP) monolayers were generated at the surface of the substrate as shown in Fig. 12 [39]. At first the azo initiator was self-assembled on planar silicon oxide substrates. In the next step the polymer monolayer was generated by thermal activation of the attached initiator and radical (in situ) chain polymerization of 4-vinylpyridine directly at the surface. The polymerization re-

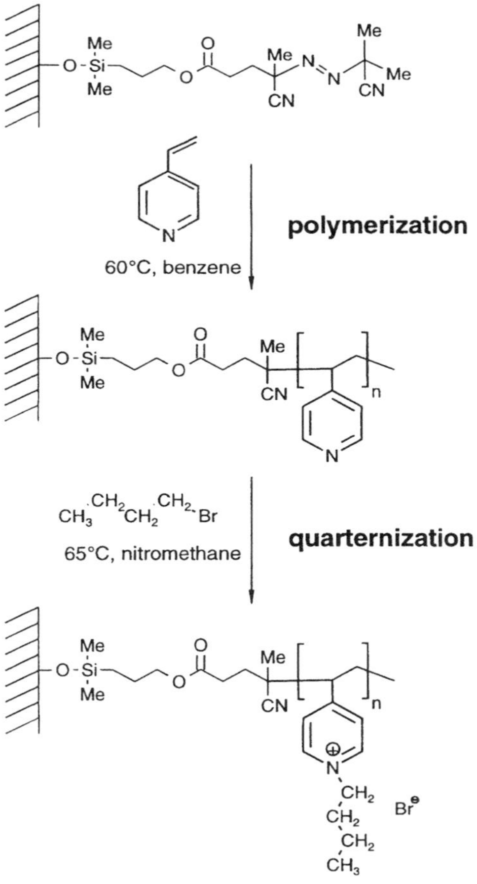

Figure 12 Reaction scheme for the preparation of monolayers of poly(4-vinylpyridine) brushes.

actions were carried out in benzene solution in a way similar to that described by Luskin [40] and by Boyes and Strauss [41] for the radical chain polymerization of PVP in solution.

As pointed out above, the thickness of the attached polymer monolayers is a function of the number of polymer chains attached per surface area (graft density) and of the molecular weight of the attached polymer chains. The layer thicknesses of the surface-attached monolayers are measured by waveguide spectroscopy. Three typical reflectivity curves of monolayers attached to SiO_x surfaces are shown in Fig. 13. It can be seen clearly that with increasing polymerization time the resonance minimum of the first waveguide mode shifts to higher angles of incidence and additional waveguide modes can be excited. The thickness of the obtained PVP layers as a function of the polymerization time is shown in Fig. 14. Thickness and refractive index were calculated from the reflectivity

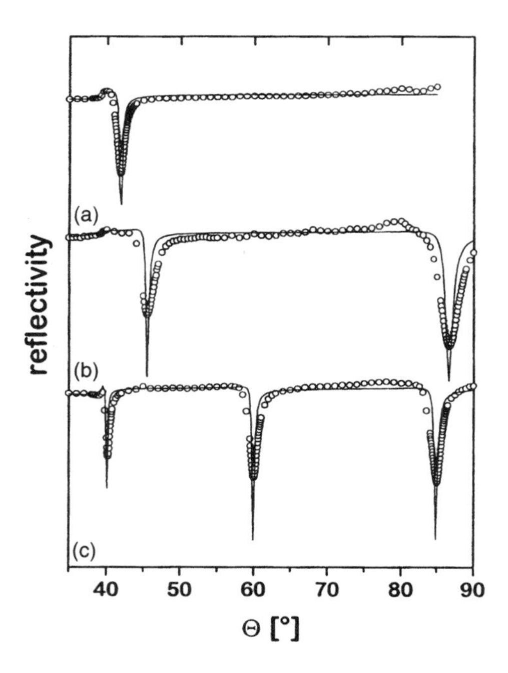

Figure 13 Waveguide-spectra (p-polarization) of (a) 230 nm, (b) 520 nm, and (c) 1030 nm thick PVP layers on glass/silver (50 nm)/SiO_x (30 nm) substrates. The layers were prepared on the substrates by radical polymerization of 4-vinylpyridine in benzene (50 mol%) at 60°C for 4.5, 6, and 24 hours, respectively. After polymerization the substrates were extracted for 15 hours in methanol. The solid lines are the calculations according to the Fresnel equations.

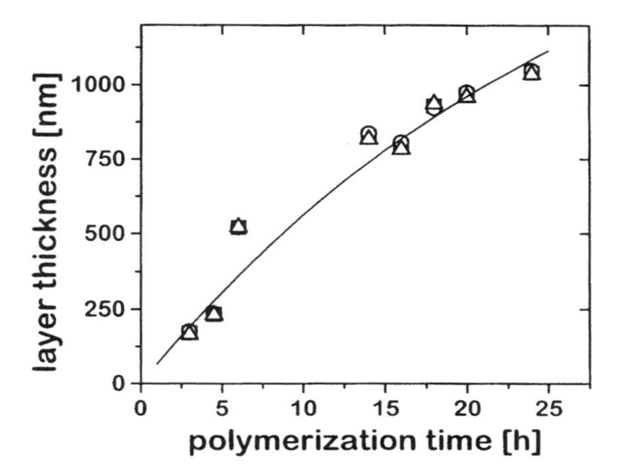

Figure 14 Thickness of PVP monolayers as a function of polymerization time, measured by waveguide spectroscopy. The polymerization reactions were carried out at 60°C in benzene using a monomer concentration of 50 mol%. After the polymerizations all substrates were extracted for 15 hours with methanol. The solid line represents the graft density of the polymer molecules calculated from the decomposition kinetics and radical efficiency of the initiator.

curves using the Fresnel formalism [29]. The different symbols represent the results of measurements on different spots of the same sample.

The solid line shown in the graph is not a fit to the datapoints, but represents calculated values based on the kinetics of the azo decomposition, the molecular weight of the surface-attached chains, and the radical efficiency factor. All parameters are essentially the same as for the polymerization of styrene with the same surface-attached initiator described above.

Quarternization of the PVP monolayer with n-butylbromide or methyliodide in nitromethane [42] leads to monolayers of positively charged n-alkyl PVP molecules. As by this approach rather thick polymer films can be obtained, the reaction can be followed by simple transmission FT-IR spectroscopy. Figure 15a shows FTIR spectra of two approximately 35 nm thick PVP layers attached to both surfaces of a silicon wafer after different reaction times. The conversion of the quarternization reaction can be calculated from the integrated intensities of the absorption maxima at 1638 cm^{-1} and at 1597 cm^{-1}. From the fact that no significant absorption at 1600 cm^{-1} can be detected after roughly 80 hours of reaction time it can be concluded that the reaction can be carried out to almost quantitative conversion. In Fig. 15b the conversion of the quarternization reaction is shown as a function of the reaction time. The results of measurements from

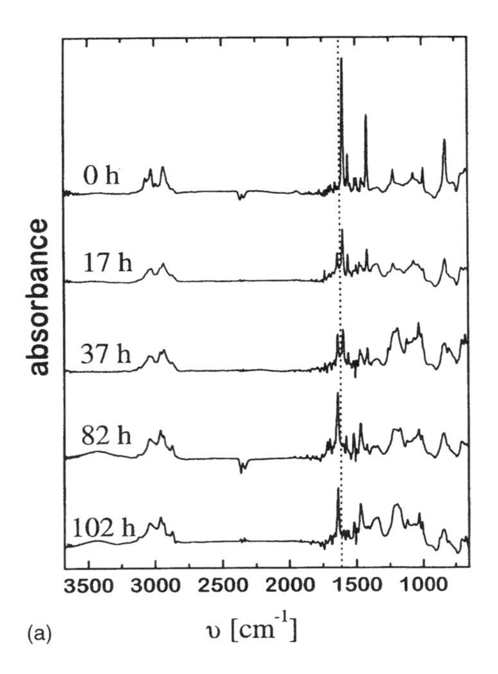

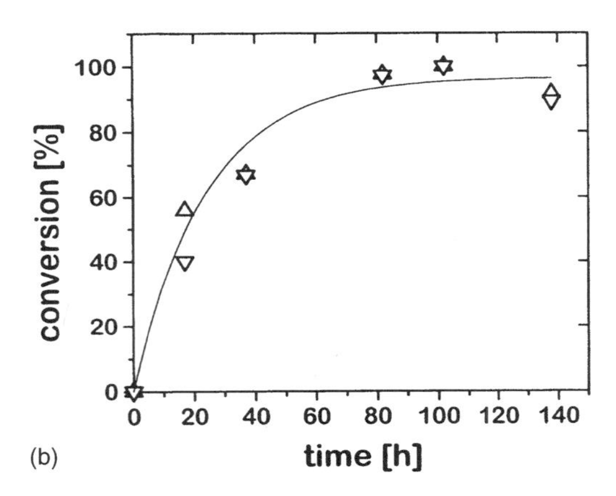

Figure 15 (a) FTIR-spectra of a 2 × 35 nm thick PVP monolayer attached to both sides of a silicon wafer. The spectra were measured during the quarternization of a covalently attached PVP monolayer at given reaction times. The quarternization was carried out with 0.5 M n-butylbromide in nitromethane at 65°C, (b) conversion of the quarternization reaction as a function of the reaction time of the layers in (a). Conversions were calculated by measuring the integrated intensity of the maximum of the IR-band from PVP at 1600 cm^{-1} (△) and from BuPVP at 1640 cm^{-1} (▽).

Fuoss and Strauss on the kinetics of the quarternization of PVP in solution are included for comparison in Fig. 15b [42]. The close agreement between the kinetics of the quarternization of the surface-attached monolayers and the rate of the same reaction in solution demonstrate that the covalent attachment of the chains to the surface has no significant influence on the polymer-analogous reaction.

The thickness of the attached layer increases strongly as a result of the quarternization of the polymer. As an example the waveguide spectra of one PVP sample before and after quarternization with methyl iodide are shown in Fig. 16. The thickness of the neutral PVP monolayer was 490 nm, the thickness of the MePVP layer was 870 nm. This thickness increase is caused by the increase of the molecular weight of the repeat unit M from 106 g/mol (VP) to 245 g/mol (BuPVP) or 246 g/mol (MePVP). However, for the calculation of the expected increase of the thickness it has to be considered that the MePVP is under ambient conditions in a rather strongly swollen state due to the presence of air moisture. The dry film thickness d after quarternization, determined in a moisture-free environment, was 741 nm. In addition, the density ρ of the polymer in the monolayer increases during the quarternization reaction due to the presence of the iodide counterions. The degree of functionalization ζ of the layer can be calculated according to

$$\zeta = \frac{d_{MePVP}}{d_{PVP}} \frac{M_{PVP}}{M_{MePVP}} \frac{\rho_{MePVP}}{\rho_{PVP}} \tag{4}$$

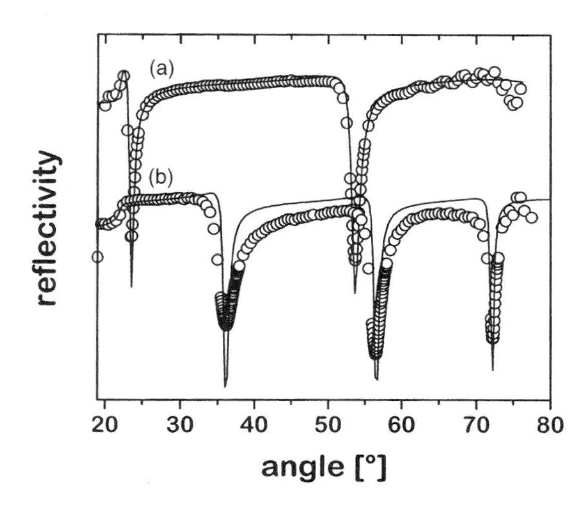

Figure 16 Waveguide-spectra (p-polymerization) of (a) a 490 nm thick PVP brush on a LaSFN9/Au/SiO_x substrate. The brush was prepared as mentioned in Fig. 4; (b) the same sample at a constant relative humidity of 70% after quarternization with methyliodide: thickness: 870 nm.

The thickness increases of the polymer layer during the quarternization reaction or during moisture exposure can serve as examples that polymer layers prepared by the ''grafting-from'' approach show a strong response towards low molecular weight compounds in their environment due to the large number of reactive sites present in the polymer layer. As the layers can be swollen in appropriate solvents and polymer-analogous reactions can be carried out on the tethered molecules, it can be envisioned that the grafted monolayers can be used to quantify low molecular compounds present in a surrounding analyte. Whereas thickness changes due to chemical reactions in self-assembled low molecular weight monolayers are typically only on the order of a few Ångstrøms, films such as the ones described here can show a much stronger response to the presence of low molecular weight compounds. Thickness increases of dozens or even hundreds of nanometers caused by reaction with a low molecular mass present in the analyte can easily be achieved. This might render the ''grafted-from'' polymer layers interesting components in sensor applications.

B. Brushes with Mesogenic Side-Groups

A second example for the synthesis of a functional surface-attached polymer brush is the preparation of monolayers of a liquid-crystalline polymer (LCP) with mesogenic units in the sidechain [43]. Such a system could be of interest for the preparation of alignment layers for liquid-crystal displays (LCD) [44,45]. Alignment layers are key components for the production of LCDs [46–50]. As the orientation of the nematic director of the LC in contact with a surface is energetically degenerate, domains are formed in the LC layer that lower the contrast of the display. Thin films of polymers such as rubbed polyimide film (''alignment layers'') are frequently used to align the liquid-crystalline molecules in order to form extended monodomains.

The purpose of a LC-brush at the surface of a LC cell would be to induce a certain orientation of the nematic director in the adjacent bulk liquid-crystal phase (Fig. 17). Interestingly, a tilted orientation of the whole liquid-crystalline system may result due to a competition between the orienting action of the stretched polymer chains and the preferred orientation of the LC induced by the surface anchoring to the substrate. This is important because for optimum operation of a LCD a tilt of the LC director with respect to the aligning surface is beneficial. In super-twisted nematic (STN) displays, for example, pretilt angles of more than 10° are required. An alignment mechanism based on polymer brushes requires, however, strong swelling of the brush in the bulk nematic medium (Fig. 17). For sidechain LCPs it is possible to use low molecular weight nematics which are miscible with the polymer.

We have shown [43], that it is possible to synthesize sidechain LCP brushes by thermally initiated polymerization reactions using the surface-attached azo-

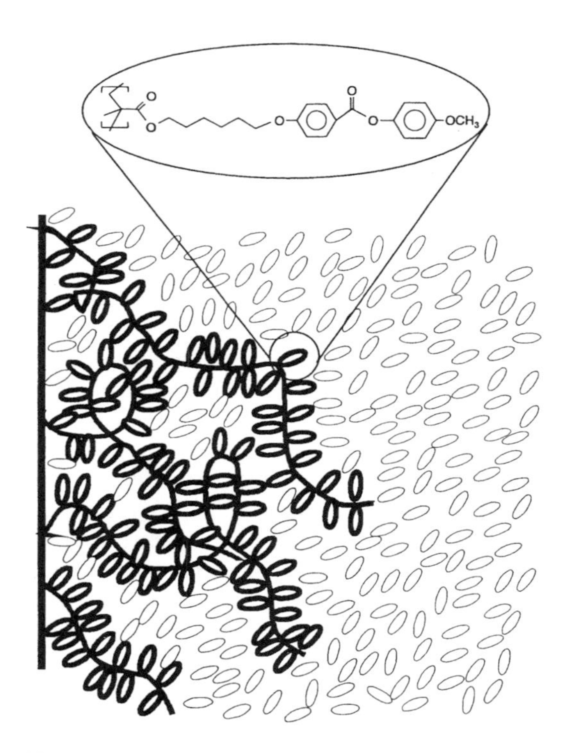

Figure 17 Schematic depiction of a LC brush swollen in a low molecular weight nematic. Competing orienting actions from the brush and the bare surface may result in a tilted alignment. The chemical structure of the sidechain mesogen used in this study is indicated in the inset.

initiator already discussed above. The structure of the polymer monolayer is shown in Fig. 17. By adjusting monomer concentration and polymerization time, monolayers of LC polymers with thicknesses ranging from a few nanometers to more than 200 nm can be obtained.

Directly after preparation the LCP brushes are found to be isotropic. Apparently the mesogenic units have been quenched into an isotropic glass by the rapid drying process. However, a nematic texture gradually appears when the samples are annealed at the glass transition temperature of the LCP ($T_g \sim 50°C$) or slightly above. The monolayers become isotropic at $T_{N,I} \sim 113°C$. Although LC brushes are monolayers in the strict sense of the definition (one single layer of molecules attached to the surface of the substrate), the layers are thick enough to allow for the direct optical observation of textures.

The textures of the brushes observed in the polarizing microscope show domains that are much larger than the film thickness. Even for brushes as thin as 38 nm one finds a liquid crystal texture with a lateral length scale on the order of micrometers (Fig. 18). Apparently, the film thickness is not the natural length scale of the lateral pattern of nematic orientation.

The tethering of the polymers to the surface of the solid has some very pronounced effects on the textures of the layers. If thin films of a nonattached polymer (e.g., obtained by spin-coating) are annealed below the isotropization temperature the domain size grows with time and depends strongly upon annealing conditions. LC brushes show a completely different behavior. The observed textures do not vary even if brushes are annealed close to the isotropization temperature for several hours. This constancy of domain sizes and locations is not due to a slow kinetics, but is caused by the surface-anchoring of the brush at the substrate.

The LC-brushes exhibit a very interesting "surface memory effect." After heating the film to temperatures well above the N—I transition (T = 120°C) for a period from a few seconds to more than half an hour, the same identical patterns are found again upon cooling back to the LC state (Figs. 18a,c). Close inspection of the micrographs shows that the locations and sizes of domains are not only similar to the ones before isotropization, but are identical, even though the LC phase has been completely destroyed during heating (Fig. 18b). Spin-cast LCP films in contrast to this do not memorize their domain pattern after heating to the isotropic state. The domain size in spin-cast LCP films depends strongly on the time length of the annealing procedure. The surface memory effect exhibited by the LC brushes is most likely caused by an irregular in-plane orientational

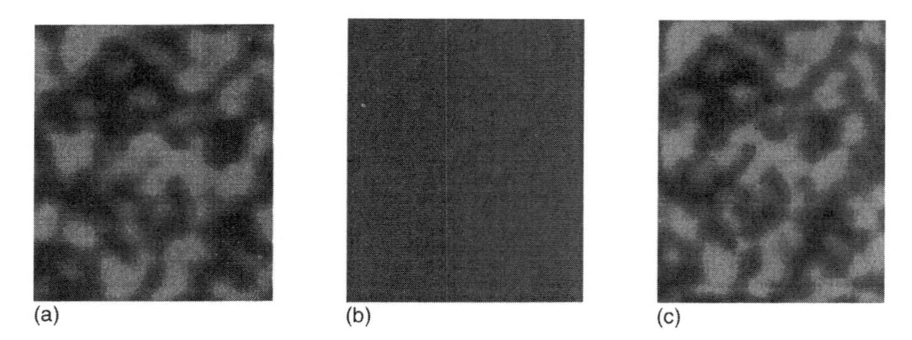

Figure 18 Surface memory effect of LCP brushes. The sample shown in (a) was heated to the isotropic phase (b) (T = 120°C, t = 30 min) and then returned to the LC state by cooling (c). The size of the area depicted by the micrographs is approximately 12 × 17 μm^2.

anchoring at the interface between substrate and brush. Selection of the in-plane easy axis is caused by rather minor symmetry breaking anisotropies in the substrate surface, the details of which are not completely understood so far. Such orienting factors could, for example, be microscratches or strongly physisorbed mesogenic units.

C. Introduction of Functional Groups Through Group Transformation

As pointed out above, a significant problem for the polymerization of monomers with functional units is that the thickness of the monolayer is directly proportional to the concentration of the monomer during polymerization. Additionally, it has to be considered that monomers with some functional units have a strong tendency for the occurrence of chain transfer reactions. Such processes also lower the molecular weight of the surface-attached brushes. If chain transfer becomes strong, which is the case of a sulfur group containing moieties in the monomer, only rather thin films can be obtained by this technique.

Such problems can be circumvented if the functional polymer monolayer is prepared by a two-step procedure. In the first step a simple and inexpensive precursor-polymer is generated by the ''grafting-from'' procedure, which is then transformed into the final structure by a subsequent polymer-analogous reaction. A schematic depiction of the concept is shown in Fig. 19. Thus, very thick monolayers can be easily grown, which are then transformed by a small amount of the functional compound into the desired structure. Also, the preparation of a thick monolayer from monomers that contain groups with strong transfer properties imposes no problem for such an approach as the transfer agents are not present during the polymerization reaction, but are introduced only at a later stage.

An additional advantage of such an approach is that if it is intended to change the surface chemistry of the system slightly it is no longer required to

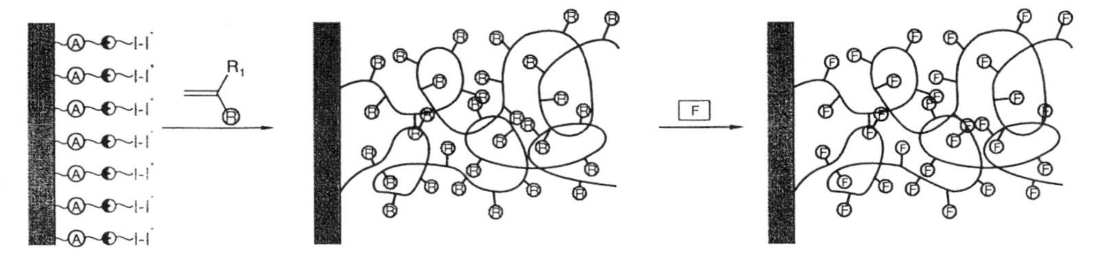

Figure 19 Schematic illustration of the concept for the preparation of functional polymer brushes via polymer-analogous transformation of reactive groups.

study the polymerization properties of each new monomer or monomer combination. Following such a two-step approach the polymerization kinetics of the system has to be studied only once. The obtained reactive precursor polymer then allows the introduction of a large number of functional units via simple chemical reactions.

In principle many different reactive monomers can be used for the preparation of the precursor polymers. Basic requirements for such systems are only that the corresponding monomer can be polymerized easily to a high molecular weight polymer and that the polymer-analogous transformation reaction can be carried out with high yield to a well-defined polymer structure. Some examples for such a strategy are shown in the following, one for the incorporation of thiol units into the polymer brush, the other for the incorporation of dye groups into the polymer. In both cases an N-hydroxy succinimide active ester (NHS ester) is chosen as the reactive precursor. This group is frequently used with great success for the reactions of amines with proteins or other biological molecules [51–54]. A reaction with alcohols to produce esters is also possible, although at a strongly reduced reaction rate. The system was applied by Ringsdorf and coworkers [55,56] in the early 1970s to polymers and copolymers with active ester groups in the sidechain for the preparation of polymers with functional sidechains.

1. Introduction of Sulfur Groups into Surface-Attached Polymer Brushes

Sulfur compounds can act as strong transfer agents during a free radical chain polymerization reaction [57,58]. In fact, such compounds are also used in technical processes to control the molecular weight of the polymers obtained in a radical chain polymerization process. A direct surface-initiated polymerization of monomers with sulfur groups in the sidechain allows only for the preparation of polymers with relatively low molecular weight and accordingly low film thickness. Polymer-analogous reaction of a surface-attached precursor polymer with the thiol groups (Fig. 20) avoids this problem and allows for the preparation of monolayers of high molecular weight polymer with high content of sulfur moieties. The copolymer monolayer shown in Fig. 20 was obtained by polymerizing a mixture of styrene and the active ester monomer. The incorporation of the sulfur groups into the surface-attached monolayers was confirmed by X-ray photoelectron spectroscopy (XPS) (not shown). Very likely the thiol groups were transformed into disulfide moieties, as thiols are very readily oxidized (even the presence of oxygen in air is sufficient). The generation of disulfide bridges will lead to a crosslinking of the polymer [59].

2. Incorporation of Dye Molecules

The described concept also allows the incorporation of more complex molecular building blocks. As an example, the same copolymer described above, based on

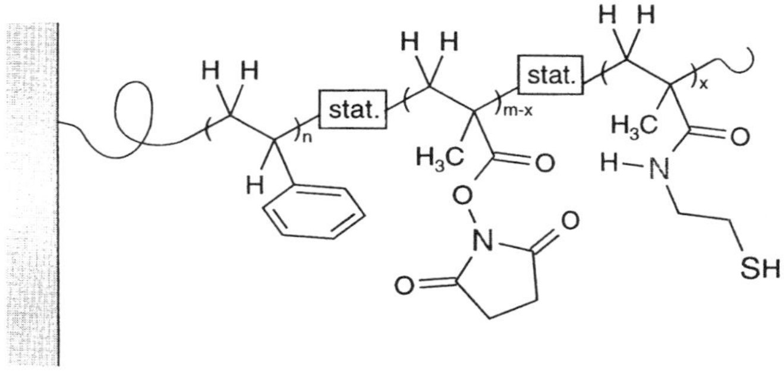

Figure 20 Structure of a polymer brush carrying thiol units that were attached to the polymer by reacting a poly(styrene-stat-methacrylic acid-N-hydroxysuccinimide ester) brush with cysteamine.

styrene and NHS active ester repeat units is reacted with the fluorescent dye 1-aminomethyl pyrene. The reaction is schematically depicted in Fig. 21. The monolayer thickness is approximately 42 nm. The active ester content of the copolymer is 30 mol%. A quartz slide is used as substrate. Figure 22 shows the UV/vis spectrum of the polymer monolayer after extraction. The absorption spectrum of the 1-aminomethyl pyrene is clearly visible. The absorption maximum is located at 340 nm, the maximum of the emission at 376 nm in good agreement with literature data for comparable compounds not attached to a solid surface [60,61]. The substrates show, after deposition of the monolayer, a blue fluorescence if the sample is irradiated with light having a wavelength of 360 nm.

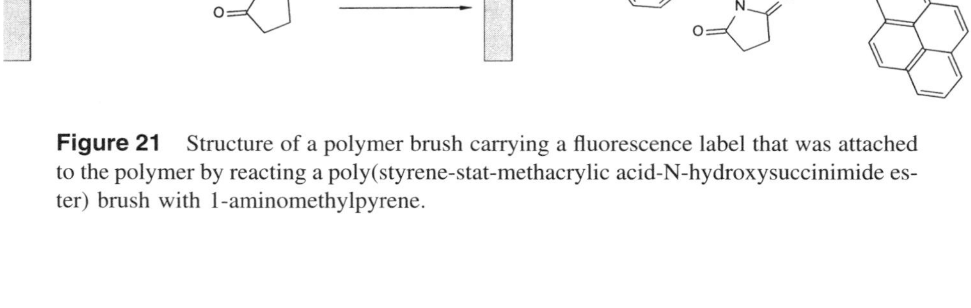

Figure 21 Structure of a polymer brush carrying a fluorescence label that was attached to the polymer by reacting a poly(styrene-stat-methacrylic acid-N-hydroxysuccinimide ester) brush with 1-aminomethylpyrene.

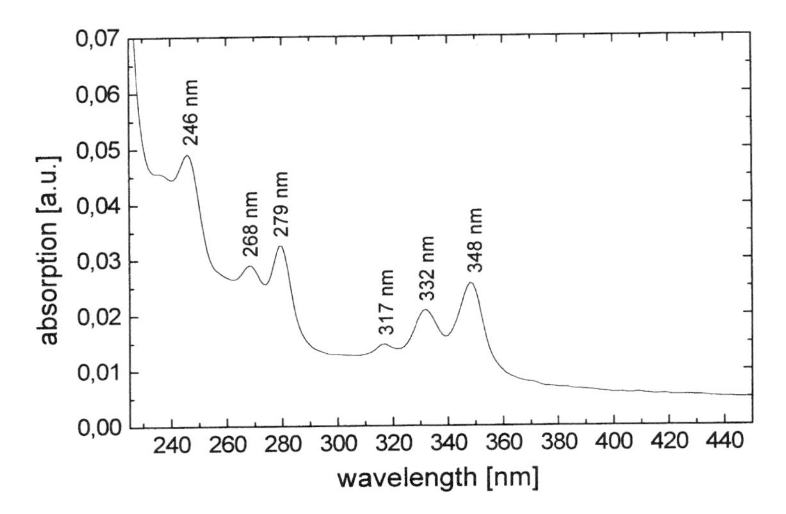

Figure 22 UV/vis spectrum of a pyrene labeled polymer monolayer prepared via subsequent attachment of 1-aminomethylpyrene to a polymer brush carrying N-hydroxysuccinimide groups; film thickness 42 nm on quartz glass.

V. TAILORING OF SURFACE PROPERTIES

Selection of an appropriate monomer and performance of a surface initiated polymerization allows one to link a large variety of different functional units to the surface of solid substrates. However, changing from one monomer to another also causes discrete changes in the physical properties of the monolayers and accordingly only allows for a relatively rough adjustment of the layer properties. Additionally, it has to be considered that changing the monomer from one system to the other does not only influence the property, which is intended to be optimized. In fact, many other physical properties of the coatings will be influenced by these changes of the monolayer chemistry as well. If one parameter (i.e., the polarity of the polymer) is changed, in most cases other parameters (glass transition temperature, swellability of the layer, mechanical properties, etc.) will also be altered.

One way to avoid such a problem is to tailor the properties of the monolayer by using two (or more) different monomers and to perform a (statistical) copolymerization reaction. Copolymerization allows the generation of an almost unlimited number of different polymers with very specific properties and a fine-tuning of the chemical composition of the layer simply by adjusting the comonomer ratio. Very precise tailoring of the coating properties can be achieved through a systematic variation of the monolayer composition.

A. Generation of Copolymer Monolayers

The composition of the surface-attached copolymer is determined by the composition of the monomer feed and the reactivity of the propagating species from the comonomers (homopropagation vs. crosspropagation). The chemical composition represented by the m_1/m_2 molar ratio of monomers 1 and 2 at low degrees of polymerization is described by the well-known copolymerization equation:

$$\frac{m_1}{m_2} = \frac{[M_1](r_1[M_1] + [M_2])}{[M_2]([M_1] + r_2[M_2])} \tag{5}$$

Here $[M_1]$ and $[M_2]$ denote the concentration of monomers M_1 and M_2 in the feed while r_1 and r_2 are the so-called monomer reactivity ratios.

The result of the quantitative evaluation of the growth of the copolymer brushes of styrene and methylmethacrylate is depicted in the copolymerization diagram in Fig. 23. For MMA = M_1 and styrene = M_2 the resulting copolymerization parameters were determined to be $r_1 = 0.29$ and $r_2 = 0.70$. The values are in good agreement with the parameters reported for free radical copolymerization of MMA and styrene in solution [57].

B. Fine-Tuning of Surface Layer Properties

As an example it has been shown that the refractive index of submicrometer thick coatings can be controlled by ''growing'' copolymer monolayer from surfaces.

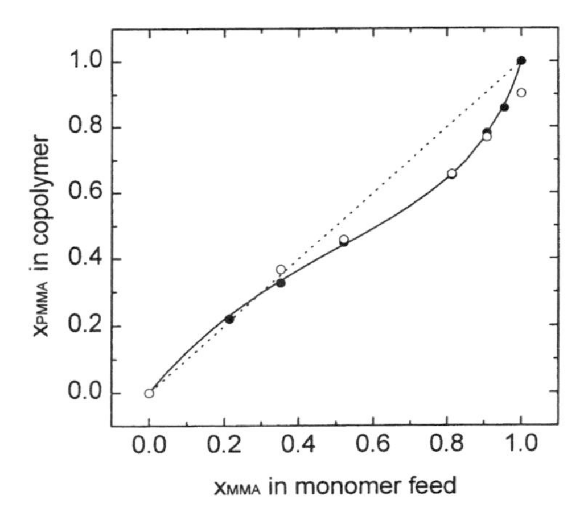

Figure 23 Copolymerization diagram for the copolymerization of MMA and styrene by ''grafting-from'' polymerization using immobilized azo initiators as obtained from FTIR spectroscopy (full circles) and XPS (open circles). The solid line gives the theoretical curve as calculated from the copolymerization equation, Eq. (5).

A typical one is that of the comonomer pair polystyrene/polymethyl methacrylate (PS/PMMA).

Recorded optical waveguide spectra and s- and p-polarized light (cf. Fig. 24) can be used to determine the layer thickness and the refractive index of these copolymer layers independently. The refractive indices of the layers are shown as a function of the monomer composition (Fig. 25). The refractive index n of the polystyrene homopolymer is n = 1.584; that of PMMA homopolymer is n = 1.489. Both values are in close agreement with those reported for the bulk polymer [57].

Included in Fig. 25 is a theoretical curve of the expected refractive indices calculated from the composition of the copolymers, only based on the assumption that the refractive index of the copolymer is simply the weighed average of the two values of the homopolymers. The very good agreement between the measured and predicted values indicate that it is possible to tailor the refractive index

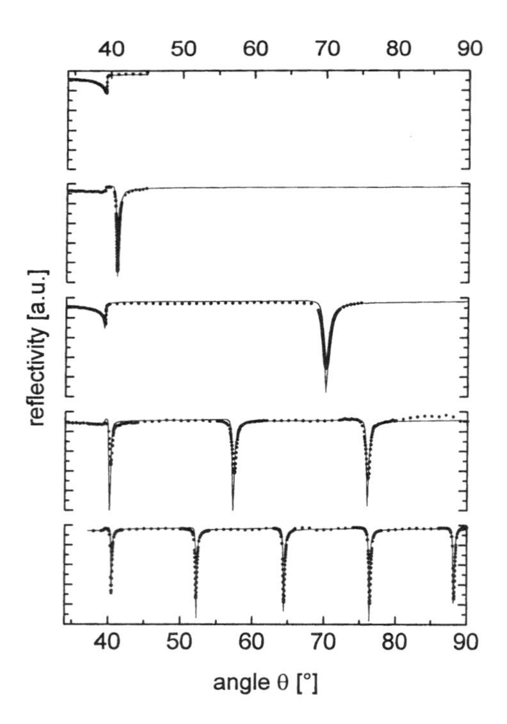

Figure 24 Optical wavequide spectra (p-polarized light) of copolymer monolayers prepared from styrene and MMA with varying PMMA content; top to bottom: x(PMMA) = 0.22, 0.45, 0.65, 0.86, and 1.00.

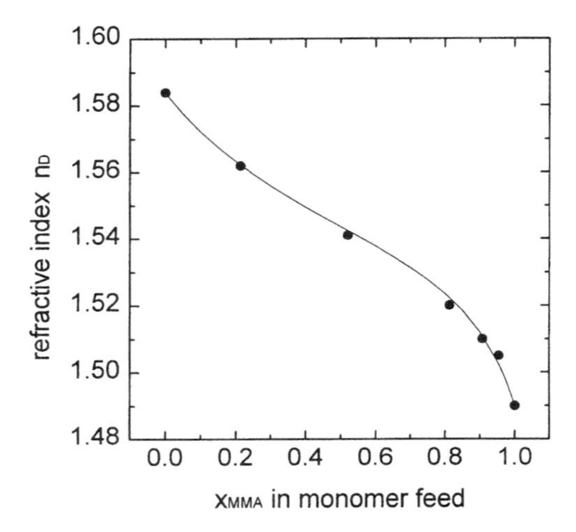

Figure 25 Refractive indices of copolymer monolayers (styrene/MMA) as a function of the PMMA content. The values were determined from the optical waveguide spectra shown in Fig. 24. The solid line was theoretically predicted from the copolymerization diagram (Fig. 23) and the bulk refractive indices of the two homopolymers.

of the polymer layers just by choosing the appropriate monomer composition in the feed. In turn, the determination of the refractive index of the copolymer layers allows the calculation of the copolymer composition if the refractive indices of the homopolymers are known (and are significantly different from each other).

VI. SWELLING OF POLYMER BRUSHES

The balance of the mutual interactions among the end-grafted chains, the substrate, and the solvent determines the degree to which the latter is incorporated into the brush. It is the resulting swelling of the polymer layer that determines its unique and exciting physical properties. Not only are the fundamental structural and dynamical features of a swollen brush of interest (e.g., the segment density profile or the rheological response of the interface), it is only through this swelling process that the incorporated functional groups become accessible for interactions or reactions with components (e.g., analytes from the solution phase).

The experimental techniques developed in order to characterize the swelling behavior of brushes, therefore, had to be compatible with the corresponding sample format, i.e., a thin polymer layer covalently attached to a solid support and in contact with a liquid phase. Neutron reflectometry (NR) turned out to be a particularly powerful method as it allows for the determination of scattering length density profiles normal to the surface from which the thickness of the layer, the segment density profile, the amount of solvent in the brush, etc. can be determined as a function of the quality of the solvent. We demonstrate this in the following for a polystyrene brush prepared by the ''grafting-from'' technique and brought into contact with mixtures of methanol and toluene of different molar fraction.

Figure 26 displays a series of NR scans (i.e., reflected intensity as a function of the momentum transfer q) from a PS-brush; a NR scan taken in air is displayed for comparison. The full curves are Fresnel simulations based on the scattering length density profiles given in Fig. 27. For all experiments, Si was modeled with a scattering length density of $\rho_{Si} = 2.07 \cdot 10^{-4}\ nm^{-2}$. Other than in the X-ray case where the electron density of SiO_2 is almost identical to that of Si and the oxide layer of the Si wafer can be ignored in the Fresnel fits, we introduced for the NR simulation a 2.5 nm thick native SiO_2 layer with a scattering length

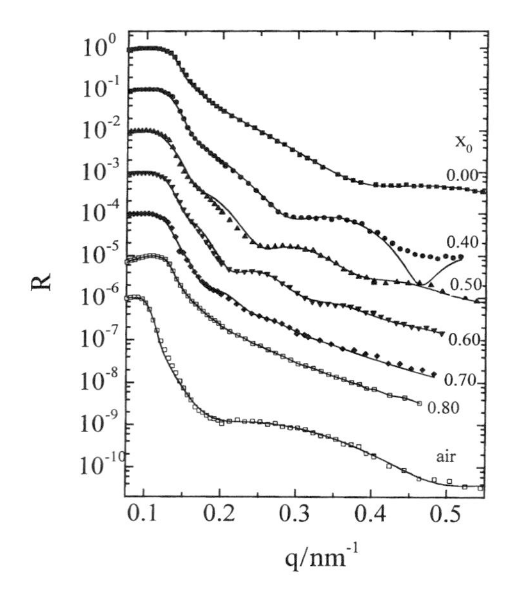

Figure 26 Neutron reflectivity curves R-versus-q for a polystyrene brush with a dry thickness of $d_0 = 19.5$ nm swollen in solvent mixtures of per-deuterated methanol and per-deuterated toluene of different volume fractions x_0 of toluene. Symbols are the data-points; the full curves are Fresnel fits based on the scattering length density profiles b/V-versus-z given in Fig. 27.

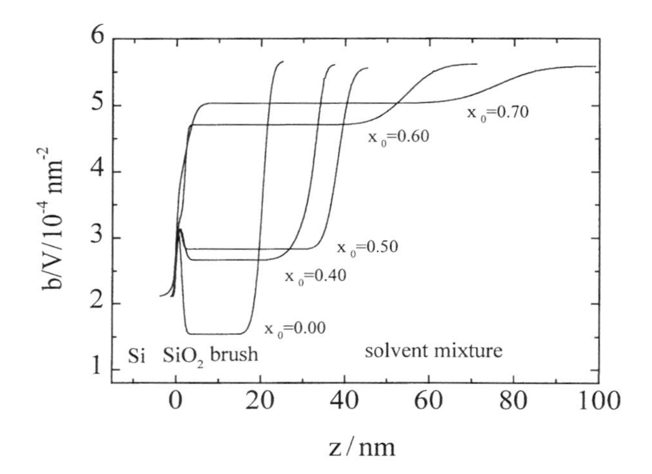

Figure 27 Scattering length density profiles b/V-versus-z as obtained from the neutron reflectivity curves given in Fig. 26. The sample is a polystyrene brush with a dry thickness $d_0 = 19.5$ nm x_0 denotes the volume content of toluene in the solvent mixture with methanol.

density of $\rho_{SiO2} = 3.4 \cdot 10^{-4}\ nm^{-2}$. The roughnesses of the Si/SiO_2 and SiO_2/PS interfaces were both assumed to be 0.6 nm.

Already in the original R(q)-data the swelling of the brush with increasing toluene content is obvious from the narrowing of the Kiessig fringes [62]. The thickness obtained for the collapsed layer in methanol agrees very well with the X-ray reference data as well as with a NR scan taken in air. Upon increasing the toluene content in the mixture the brush swells with the Kiessig fringes being more pronounced. However, only up to a certain limit: the NR data taken at 80 vol% toluene show no obvious Kiessig modulation, and hence the whole profile analysis has to be limited to the range $0 \leq x_0 \leq 0.70$. This means, in particular, that we cannot follow the swelling of the brush up to toluene contents in the mixed solvent that would correspond to the θ-mixture for free PS-chains which is reported as $x_0 = 0.80$.

It should be emphasized that only raw features of the scattering length density profiles can be deduced from the above reflectivity scans and further details would require measurement being extended to substantially higher q-values. However, various model simulations have shown that information on the average value of the scattering length density in the plateau region of the swollen brush as well as the obtained thicknesses do not depend significantly on such details and can be derived reliably from the observed (limited) range of the reflectivity curves. Perhaps more importantly, these calculations have also revealed that there is no reason to assume a brush profile more complex than the one displayed and based on a "box model" (a constant scattering length density profile across the brush of a certain thickness with a brush/solvent interface smeared by an error function).

The finding of a constant scattering length density profile is in excellent agreement with the theoretical expectations for dense brushes where the blob model predicts a flat chain segment density across the brush. The predicted depletion for noninteracting grafted polymers near the solid surface scales with the chain separation distance D which for our sample was in the range of 4 nm. This effect could not therefore be resolved by our reflectivity measurements. Experimentally, the smearing of the chain ends in contact with the solvent is certainly broadened compared to the theoretical predictions because of the unavoidable polydispersity due to this polymerization procedure. However, as shown by independent investigations the "grafting-from" scheme results in molecular weight distributions similar to those of the corresponding free radical polymerization in solution. The obtained polydispersity in the range $\overline{M}_W/\overline{M}_N = 1.5 - 2.5$ does not significantly interfere with the chain stretching phenomenon of dense brushes.

The obtained layer thickness increase with increasing toluene content x_0 in the solvent mixture is given in Fig. 28a. Shown is the swelling behavior of three different samples with different (dry) thicknesses. The maximum thickness increase that can be identified in the Kiessig fringes amounts to a factor of 3.7.

The increase in the scattering length density of the swollen brush resulting from the uptake of the deuterated solvent molecules is illustrated in Fig. 28b. Shown are the datapoints derived from the profiles given in Fig. 27. The solid curve corresponds to the expectation based on the thickness increase: ignoring possible changes of the apparent partial molar volumes of toluene and methanol in a mixture with grafted polystyrene it is possible to describe the scattering length density of the swollen film as the sum of two contributions, i.e., polystyrene and solvent, according to their respective volume fraction. The experimentally determined scattering length densities of the swollen brushes, b/V_{Brush}, show qualitatively the behavior expected from the thickness increase. However, this analogy only holds because the scattering length densities of perdeuterated toluene and methanol are almost identical. That means that $b/V_{Solvent}$, can be approximately calculated, irrespective of any preferential incorporation of one of the components of the solvent mixture. On the other hand, the use of solvent mixtures that contain both components in varying amounts of protonated or deuterated form allows for the determination of their preferential incorporation. By this strategy it is possible to determine the toluene content x_B of the mixed solvent incorporated into the brush and hence any deviation from the corresponding mole fraction x_0 outside.

The result for all investigated mixtures is given in Fig. 29. One can see

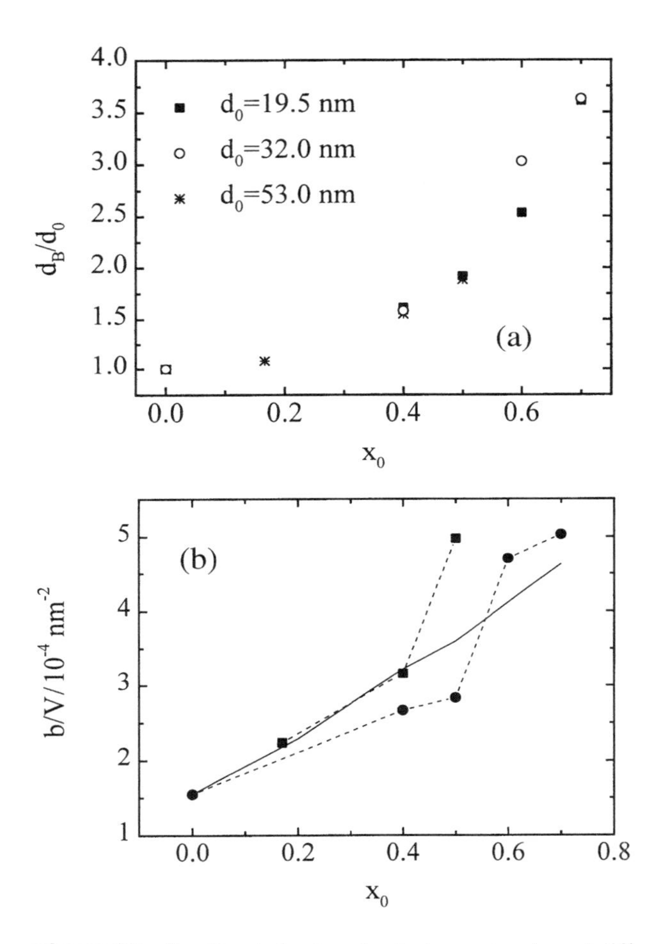

Figure 28 Swelling behavior of polystyrene brushes of different initial (dry) thickness d_0 as obtained from the scattering length density profiles (one example is given in Fig. 27): (a) gives the thickness values of the brush d_B scaled to their dry thickness d_0 as a function of the volume fraction x_0 of toluene in the solvent mixture; (b) shows the corresponding plateau values of the scattering length density of the swollen brush; symbols are as in (a). The full line gives the expected increase in the scattering length density based on the thickness increase (for details, see text).

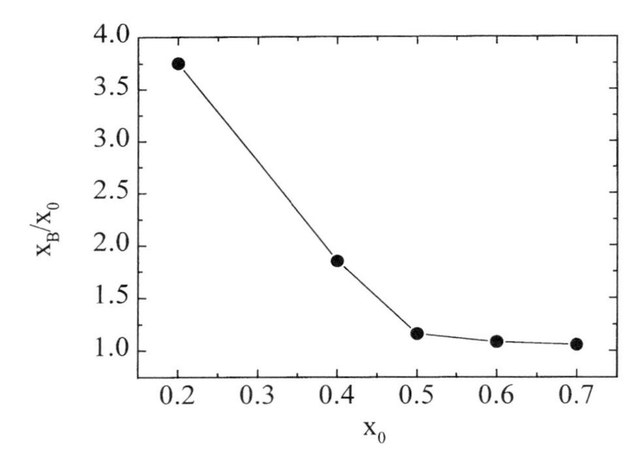

Figure 29 Preferential incorporation of toluene into the swollen brushes: given are the volume fractions x_B of toluene in the brushes, scaled to the fraction in the bulk mixture x_0 as a function of x_0.

indeed that for the mixtures with a low toluene content, $x_0 < 0.50$, a strong preference for the incorporation of toluene, i.e., the good solvent, into the brush is found. These data confirm quantitatively earlier qualitative observations of a preferential solvation of grafted PS by toluene. We point out, however, that this analysis is based on the assumption that the brush segment density profile parallels the solvent (composition) profile. This does not always need to be the case.

The second example concerns the optical determination of the swelling behavior of a polyelectrolyte brush in contact with air of different relative humidity. In this experiment the MePVP layers are exposed to water vapor. The humidity in the environment causes the monolayers to swell and the layer thickness increases strongly. To quantify this effect, optical waveguide experiments were carried out in a controlled humidity environment. An example is shown in Fig. 30, where waveguide spectra of such a layer are shown as a function of the relative humidity inside the measument cell at a constant temperature. It can be seen that with increasing humidity the layer thickness increases strongly, while the refractive index of the layer decreases due to incorporation of water into the monolayer. In Fig. 31 the film thickness is shown as a function of the moisture content of the environment. The changes in film thickness due to water uptake are strong enough to be detected by the unaided eye as the interference colors of the films change strongly during swelling and drying. In separate studies it was shown that the swelling process is completely reversible. Even changes in the relative humidity of less than 1% at room temperature can be detected using such monolayers.

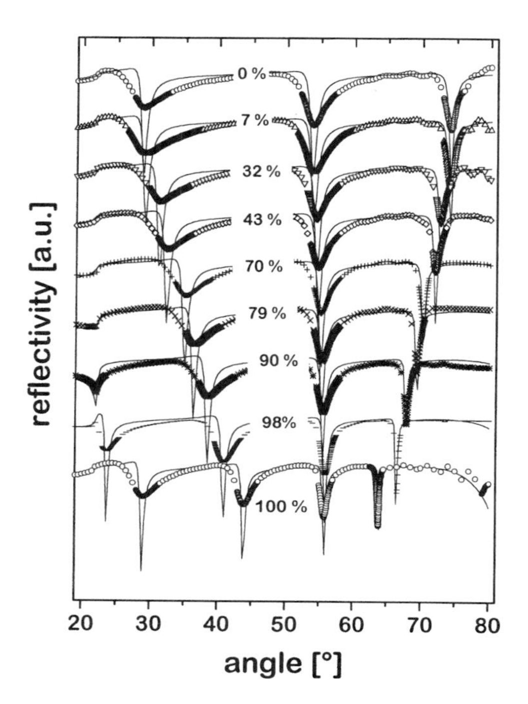

Figure 30 Waveguide spectra (p-pol) of a 750 nm thick (dry thickness) MePVP brush covalently attached to the surface of a LaSFN9/Au/SiO_x substrate. The spectra were measured at different constant relative humidities as shown in the at figure at T = 25°C. The solid lines are the calculated reflection curves obtained from Fresnel equations.

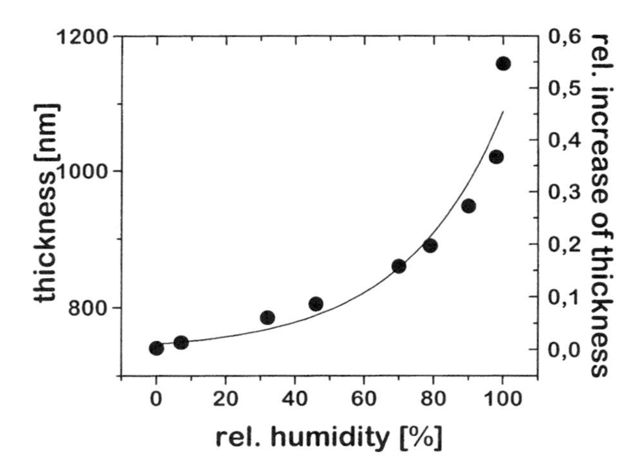

Figure 31 The thickness and the relative increase of thickness, related to the dry thickness of a 750 nm thick MePVP brush covalently attached to the surface of a LaSFN9/Au/SiO_x substrate as a function of the relative humidity of the environment. The solid line represents a guide for the eye.

VII. PHOTOPATTERNING OF POLYMER BRUSHES

In the examples discussed so far the radical polymerization of brushes was started by a thermal activation of the azo-initiators. However, the well-known possibility of starting the polymerization by photoactivation opens up a way to generate patterned surface coatings. Moreover, by step-and-repeat procedures one can prepare multielement patterns in which each domain represents a different functionality.

Three different strategies for patterning polymer brushes are summarized schematically in Fig. 32 [26]. The first procedure, equivalent to a dry-etch approach known from some photolithographic structuring protocols of photoresists in microelectronic device fabrication, is based on the ablation of a preformed polymer layer by irradiation with deep UV light. If this illumination which leads to bond breaks and hence to a volatilization of the oligomeric fragments is applied through a mask, the exposed areas gradually decrease in thickness with the protected areas remaining virtually unchanged (Fig. 32a).

A fundamental difference of laterally structured thin layers of grafted (collapsed) polymer chains compared to the normal spin-coated and hence only physisorbed photoresist films is their remarkable stability: even after extended exposure to good solvents the ''burned-in'' pattern remains on the substrate with excellent fidelity while the spin-cast film can be washed away completely. However, the high doses of UV photons needed to etch away the polymeric material makes this a rather uneconomic process.

Much more efficient is the procedure shown in Fig. 32b. Here, the irradiation of the sample is performed following the assembly of the initiator layer but before its activation and hence before the polymer layer is grown. This way, the initiator molecules in the exposed areas are passivated and/or destroyed by photodissociation and photo-oxidation (depending on the wavelength employed). The subsequent growth of a polymer layer can only be triggered in the protected areas where the initiators remained active. An example of a brush pattern generated by this procedure is given in Fig. 33. The pictures were taken with a surface plasmon microscope which is particularly sensitive for the imaging of lateral patterns of low contrast. The thickness dependence of the resonant excitation of surface plasmons at the metal–polymer film–air interface generates an image of the thin film when monitored with a camera in reflection. The picture shown in Fig. 33a was taken an angle of incidence ($\theta = 48.2$ deg) corresponding to the resonance of the bare initiator layer. The respective areas in the pattern appear dark while those with a polymer coating, and hence a slightly higher resonance angle, still reflect an appreciable fraction of the incident laser light. At a somewhat higher angle the contrast is reversed because now the coated areas ($d = 15.7$ nm polystyrene) are in resonance and hence appear dark, while the uncoated ones again gain intensity because they are tuned from their resonance condition

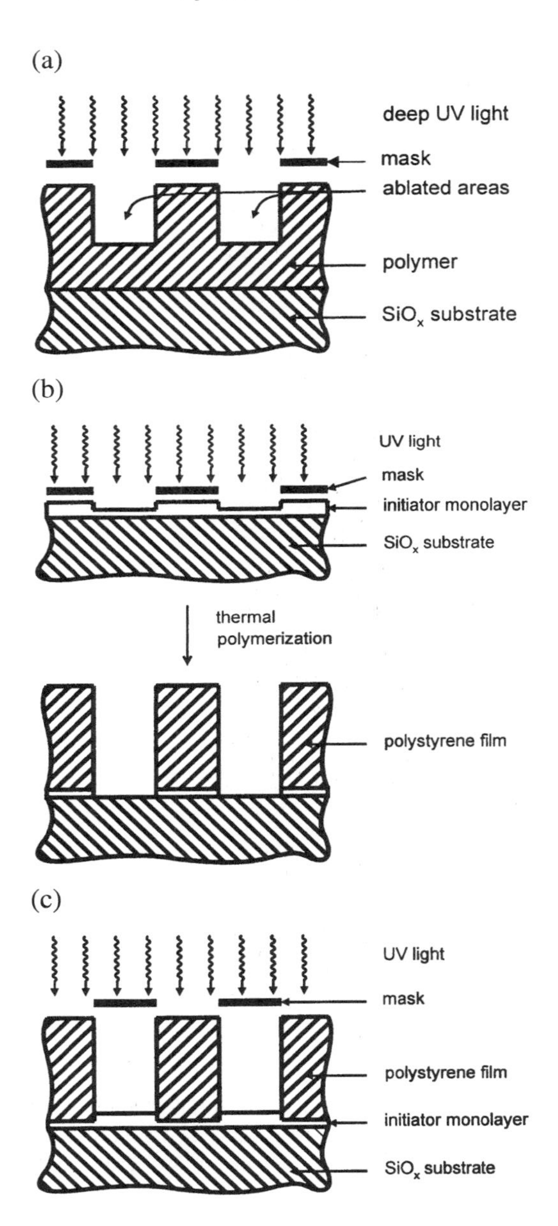

Figure 32 Schematic description of procedures used for patterning of covalently bonded polymer films: (a) polymer ablation; (b) ablation/passivation of immobilized initiator followed by thermal polymerization; (c) photopolymerization. (From Ref. 26.)

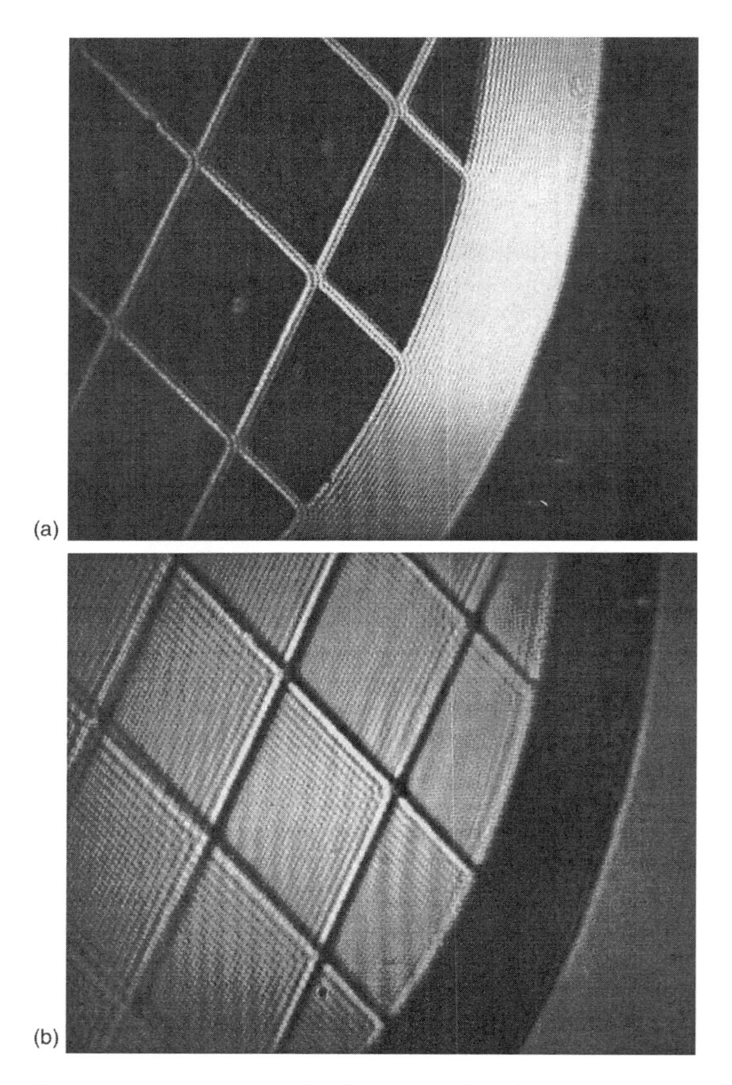

Figure 33 PSP micrographs of a structured PS film, prepared by partial ablation of the initiator monolayer followed by thermal polymerization at 60°C for 4 h (styrene/toluene 1:1 v/v): (a) micrograph taken at resonance angle of the irradiated areas (48.2°); (b) micrograph taken at resonance angle of the polymer modified areas (55.2°, 15.7 nm PS).

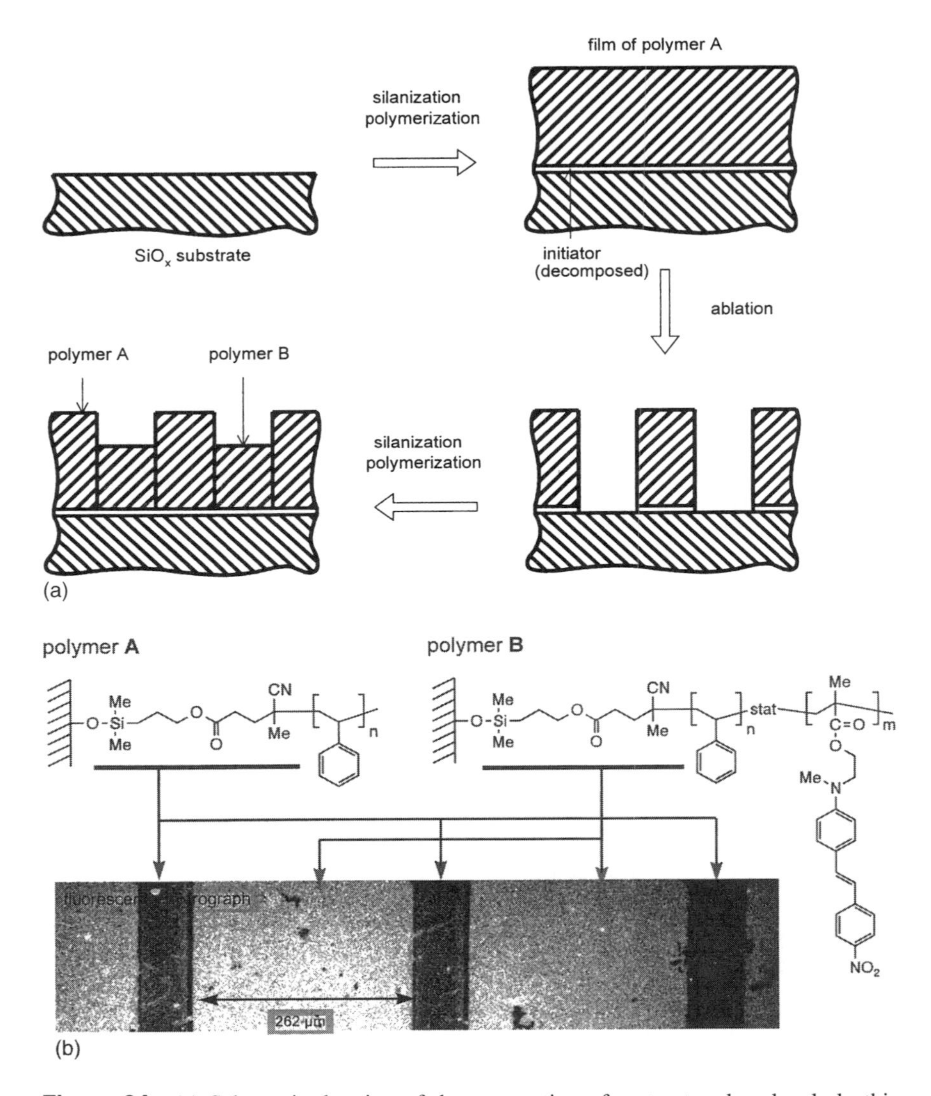

Figure 34 (a) Schematic drawing of the preparation of a structured molecularly thin polymer film with areas modified with covalently bonded poly(styrene) and areas with a monolayer of a statistical copolymer from styrene and a methacrylate carrying a fluorescence dye; (b) fluorescence micrograph of the sample described in (a); film thickness PS layer 22 nm film thickness of copolymer layer 2 nm. Details of the preparation are given in the text.

(Fig. 33b). As in the scheme of Fig. 32a, the polymer pattern obtained is again the negative image of the mask, this pattern is totally stable against solvent attack.

The most elegant way to generate a brush-pattern is shown in Fig. 32c. Here, the illumination with UV photons of a wavelength (λ = 342 nm) matching the initiator chromophore's absorption spectrum leads to the growth of a positive image of the mask pattern provided suitable monomers are available. Contrary to the two aforementioned processing steps which were done in air, the latter exposure is done in situ i.e., with the initiator monolayer being in contact with the monomer solution. This protocol is particularly attractive in the context of multifunctional patterns because the masked areas are still coated with intact initiator molecules that could be activated (by another mask or by heat) in the presence of another functional monomer solution.

One example is illustrated in Fig. 34. Here, a slightly modified version of the patterning scheme of Fig. 32a is employed. After the first pattern was grown by thermal activation of the protected initiator molecules in the masked areas, the formerly UV exposed areas were refunctionalized by a second treatment with initiator solution (cf. Fig. 34a). This way, a new layer of azo-initiator was formed, which was then thermally activated in the presence of another monomer solution. In fact, for this example, the second solution contained a monomer mixture of styrene and a fluorophore-modified methacrylate. This way, the ''holes'' in the polystyrene pattern were filled by a thin film of a copolymer, a fluorescently labeled polystyrene. The structural formula of this end-grafted copolymer is included in Fig. 34. The resulting pattern composed of two different types of polymer coatings could now be imaged with fluorescence microscopy [27]. The corresponding picture, shown in Fig. 34b, confirms that patterned polymer coatings can be grown with multifunctionalities, the diversity of which is nearly unlimited given the variety of suitable monomers. Practical limitations are only given by the number of possible step-and-repeat cycles.

VIII. BIOCOMPATIBILIZATION

The final example presented concerns the control of cell adhesion to solid supports through molecular design of the appropriate interfacial functionality. Two types of cell response to surface modifications are conceivable and actually needed for practical purposes in biomedical applications. One is the complete passivation of a substrate surface to prevent undesired cell adhesion. The opposite is the tailormade modification to stimulate cell adhesion and growth. Both methods for the manipulation of the biocompatibility of surfaces were recently tested using two different cell systems [63]. One example of the influence of terminally attached polymer molecules on cell adhesion, the interaction of neuroblastoma X glioma hybrid cells (NG108-15) with SiO_2 surfaces was studied. These cells show an unspecific adhesion behavior to many surfaces (especially glass and silicon wafers). Here the aim was to find a surface modification that lowered the cell-substrate interaction. This was achieved by attaching a monomolecular poly(styrene) film to the substrate surface. Even when only 5 to 10 nm of PS were generated over the surface, the adhesion of the cells to the substrate decreased very significantly. In order to obtain a patterned coating, a covalently bonded ultrathin poly(styrene) film was irradiated by deep UV (DUV) light through a lithographic mask so that some part of the polymer was removed by the UV light (and presumably the ozone generated). When NG 108-15 cells were brought into contact with the surfaces, the cells showed good adhesion to the parts of the surface where the polymer was removed and could not adhere on the nonirradiated parts where the polymer was still intact. Figure 35 shows NG108-15 cells grown for three days on a pattern produced in this way.

A second example concerns the study of the adhesion of neurons from the cerebellum of an embryonic rat (E 16) to glass and silicon surfaces. The development of these cells requires specific recognition groups from the substrate in vivo. Because these cells exhibit usually only very weak adhesion to glass or silicon substrate surfaces, poly(lysine) is generally used for neuronal culture. Although this is the most frequently applied surface coating for promotion of cell adhesion, it should be noted that the coating is not very well defined and it slowly desorbs

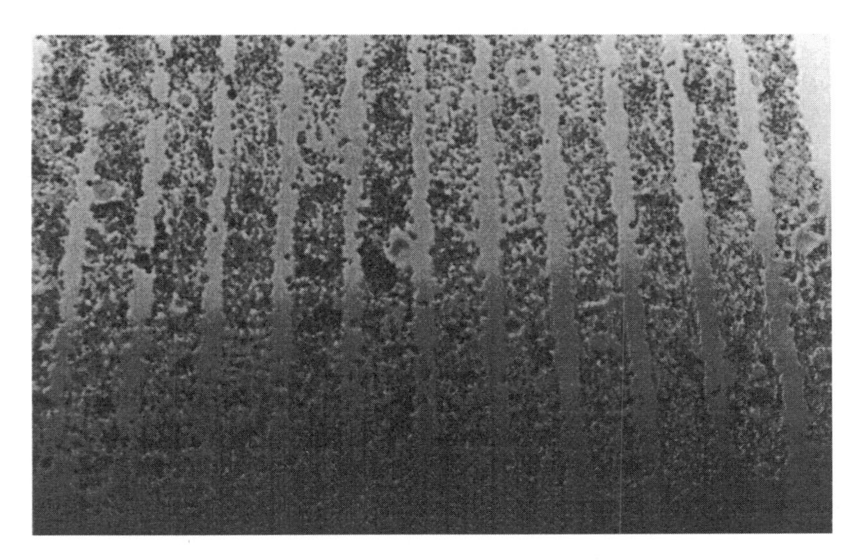

Figure 35 Patterned neuroblastoma x glioma cells (NG108-15) after 36 h in culture. The cells adhere well to the photocleaved regions of the substrate but cannot spread onto those regions where the polymer monolayer PS is intact. (Scale bar -20 μm.)

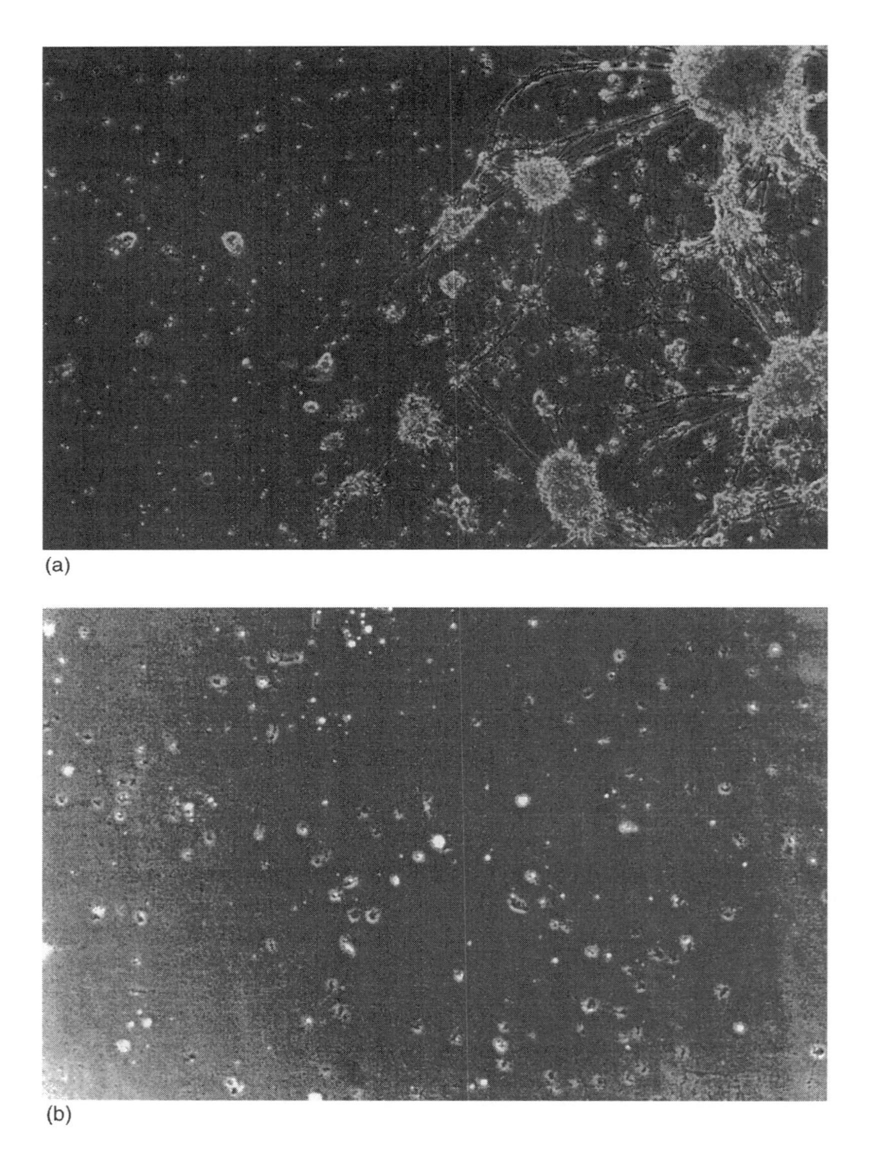

Figure 36 Experiments showing the development of neurons from the cerebellum of the rat on different substrates after 13 days in culture: (a) on an unmodified silicon oxide surface; (b) on a glass culture dish modified with a quaternary amine group in the side-chain.

from the surface, leading to insufficient long-term stability. A variety of styrene and methylmethacrylate polymers, some examples of which are depicted in Fig. 11, were generated on the substrate surfaces and the cell adhesion properties were studied. It is interesting to note that some surface modifications [e.g., poly(-methyl-acrylic acid)] showed good initial adhesion behavior and growth. The vitality of the neurons deteriorated with time, and after one week the cells were no longer vital. On the other hand, some other surface modifications, e.g., with poly(methylmethacrylate) led, surprisingly, to good adhesion and growth behavior. This proves that even small changes in the chemical structure of the attached film can lead to significant changes in cell surface interaction. In Fig. 36b one sees cells that have been brought into contact with a surface to which a polymer with a quarternary amine group in the sidechain has been attached. In contrast to an unmodified substrate (Fig. 36a) the cells show strong adhesion to the substrate and remain vital for prolonged periods of time. Systems with comparable adhesion properties to poly-L-lysine but with greatly improved long-term stability can thus be produced. The cells in the example of Fig. 36 have been cultured on the indicated surfaces for about two weeks in a chemically defined medium.

IX. CONCLUSIONS

The ''grafting-from'' approach allowed us to obtain functional endgrafted chains at densities that, for the first time, deserve the name ''brushes.'' We have shown that such polymer brushes can be prepared with a wide range of different functionalities. These brushes are well suited for fundamental studies of macromolecular systems in confined geometries because the covalent attachment of these long linear chains of high molecular weight (beyond 10^6 g/mole) to a solid support at high graft densities (chain–chain separation less than 3 nm) provides the experimental scenario for studies aimed at verifying the theoretical predictions of scaling concepts developed a long time ago.

From a more practical point of view these systems offer the possibility of fine-tuning the surface properties of solid components thus allowing for tailoring their performance in different applications. Here, the covalent attachment guarantees a long-term stability of the coating not reachable for merely physisorbed films.

ACKNOWLEDGMENTS

The authors wish to thank their coworkers Dr. O. Prucker, Dr. J. Habicht, Dr. M. Schimmel, Dr. M. Biesalski, Dr. N. Bunjes, F. Hofmann, I. Homann, Dr. H. Murata, Dr. S. Paul, B. Peng, Dr. D. Johannsmann, Dr. A. Offenhäusser, Dr. C.

Sprößler, Dr. M. Schmidt, and Dr. G. Tovar. Financial support by the Deutsche Forschungsgemeinschaft (Projects ''Polymerschichten,'' ''Polyelektrolytbürsten,'' ''Orientierungsschichten'') and the Max–Buchner Foundation are gratefully acknowledged.

REFERENCES

1. A Ulman. An Introduction to Ultrathin Organic Films, New York: Academic, 1991.
2. MC Petty. Langmuir–Blodgett Films: An Introduction. Cambridge: Cambridge University Press, 1996.
3. TR Lee, RI Carey, HA Biebuyck, GM Whitesides. Langmuir 10:741–745, 1994.
4. J Spinke, M Liley, FJ Schmitt, HJ Guder, L Angermaier, W Knoll. J Chem Phys 99:7012–7019, 1993.
5. S Arnold, ZQ Feng, T Kakiuchi, W Knoll, K Niki. J Electroanal Chem 438:91–97, 1997.
6. WJ Miller, VK Gupta, NL Abbott, H Johnson, MW Tsao, J Rabolt. Liq Cryst 23: 175–179, 1997.
7. FC Meldrum, J Flath, W Knoll. Langmuir 13:2033–2049, 1997.
8. GJ Fleer, MA Cohn-Stuart, JMHM Scheutjens, T Cosgrove, B Vincent. Polymers at Interfaces. London: Chapman & Hall, 1993.
9. KP Krenkler, R Laible, K Hamann. Angew Makromol Chem 53:101, 1978.
10. N Tsubokawa, M Hosoya, K Yanadori, Y Sone. J Macromol Sci Chem A27:445, 1990.
11. K Bridger, B Vincent. Eur Polym J 16:1017, 1980.
12. H Ben Ouada, H Hommel, AP Legrand, H Balard, E Papirer. J Colloid Interface Sci 122:441, 1988.
13. H Ishida, CH Chiang, JL Koenig. Polymer 23:251, 1982.
14. S Naviroj, JL Koenig, H Ishida. J Macromol Sci Phys B22, 291, 1983.
15. AM Zaper, JL Koenig. Polym Composites 6:156, 1985.
16. R Zajac, A Chakrabarti. Phys Rev E 52:6536, 1995.
17. A Kopf, J Baschnagel, J Wittmer, K Binder. Macromolecules 29:1433, 1996.
18. K Hamann, R Laible. Angew Makromol Chem 48:97, 1973.
19. G Boven, MLCM Oosterling, G Challa, AJ Schouten. Polymer 31:2377, 1990.
20. G Boven, R Folkersma, G Challa, AJ Schouten. Polym Commun 32:50, 1991.
21. E Carlier, A Goyot, A Revillon. React Polym 16:115, 1992.
22. O Prucker, J Rühe. Mat Res Soc Symp Proc 304:1675, 1993.
23. O Prucker, J Rühe. Macromolecules 31:592, 1998.
24. O Prucker, J Rühe. Macromolecules 31:602, 1998.
25. O Prucker, J Rühe. Langmuir 14:6893, 1998.
26. G Tovar, S Paul; W Knoll, O Prucker, J Rühe. Supramol Sci 2:89, 1995.
27. O Prucker, M Schimmel, G Tovar, W Knoll, J Rühe. Adv Mater 10:1073, 1998.
28. M Schimmel. PhD thesis, University of Mainz, 1998.
29. W Knoll. Ann Rev Phys Chem 49:569, 1998.
30. M Schimmel, J Rühe. 1999 (submitted for publication).

31. HGM van der Steeg, MA Cohen Stuart, A de Keizer, BH Bijsterbosch. Langmuir 8:2538, 1992.
32. MA Cohen Stuart, J Phys France 49:1001, 1988.
33. J Papenhuijzen, HA van der Schee, GJ Fleer. J Colloid Interface Sci 111:446, 1986.
34. MA Cohen Stuart, GJ Fleer, J Lyklema, W Norde, JMHM Scheutjens. Adv Colloid Interface Sci 34:477, 1991.
35. J Marra, HA van der Schee, GJ Fleer, J Lyklema. In: RH Ottewill, CH Rochester, AL Smith, eds. Adsorption From Solution. New York: Academic 1983.
36. R Ramachandran, P Somasundaran. J Colloid Interface Sci 120:184, 1987.
37. H Tanaka, L Ödberg, L Wagberg, T Lindström. J Colloid Interface Sci 134:229, 1990.
38. M Biesalski, J Rühe. 1999 (submitted for publication).
39. M Biesalski, J Rühe. Macromolecules 32:2309, 1999.
40. LS Luskin. In: R Yocum, E Nyquist, ed. Functional Monomers. New York: Marcel Dekker, 1974.
41. AG Boyes, UP Strauss. J Polym Sci 22:463, 1956.
42. RM Fuoss, P Strauss. J Polym Sci 3:246, 1948.
43. B Peng, D Johannsmann, J Rühe. Macromolecules 32:6759, 1999.
44. A Halperin, DRM Williams. Europhys Lett 21:575, 1993.
45. A Halperin, DRM Williams. J Physics Condensed Matt A297:6, 1994.
46. B Jéro//me. Rep Prog Phys 54:391, 1991.
47. H Yokoyama Mol Cryst Liq Cryst 165:265, 1988.
48. S Faetti. In: JC Khoo, F Simoni, eds. Physics of Liquid Crystalline Materials, ch. XII. Amsterdam: Gordon and Breach, 1991.
49. T Uchida, H Seki, H. In: B Bahadur, ed. Liquid Crystals and Uses, vol. 3, ch. 5. Singapore: World Scientific, 1990.
50. B Bahadur, ed. Liquid Crystals and Uses. Singapore: World Scientific, 1990.
51. W König, R Geiger. Chem Ber 103:2028, 1970.
52. GW Anderson, JE Zimmerman, FM Callahan. J Am Chem Soc 85:3039, 1963.
53. GW Anderson, JE Zimmerman, FM Callahan. J Am Chem Soc 86:1839, 1964.
54. KH Altmann, M Mutter. Chemie Unserer Zeit 27:274, 1993.
55. HG Batz, G Franzmann, H Ringsdorf. Angew Chem 84:1189, 1972.
56. HG Batz, G Franzmann, H Ringsdorf. Makromol Chem 172:27, 1973.
57. J Brandrup, EH Immergut, eds. Polymer Handbook. New York: Wiley, 1989.
58. KK Roy, D Pramanick, SR Palit. Makromol Chem 153:71, 1972.
59. S Uemura. In: BM Trost, I Fleming, eds. Comprehensive Organic Synthesis, vol. 7. Oxford: Pergamon, 1991.
60. RP Haughland. In: MTZ Spence, ed. Handbook of Fluorescence Probes and Research Chemicals. Eugene, OR: Molecular Probes, 1996.
61. UV-Atlas Organischer Verbindungen. Weinheim: VCH, 1966.
62. N Bunjes, S Paul, J Habicht, O Prucker, J Rühe, W Knoll. 1999 (submitted).
63. A Offenhäusser, J Rühe, W Knoll. J Vac Sci Technol A13:2606–2612.

14

Supramolecular Polymer Chemistry—Scope and Perspectives

Jean-Marie Lehn
Université Louis Pasteur, Strasbourg, and Collège de France, Paris, France

I. INTRODUCTION

Beyond molecular chemistry based on the covalent bond, supramolecular chemistry has developed as the chemistry of the entities generated via intermolecular noncovalent interactions [1–3]. The objects of supramolecular chemistry are thus defined by the nature of the molecular components and on the other hand by the type of interactions that hold them together (hydrogen bonding, electrostatic and donor–acceptor interactions, metal ion coordination, etc.). They may be divided into two broad, partially overlapping classes: *supermolecules*, well-defined oligomolecular species resulting from the specific intermolecular association of a few components, and *polymolecular assemblies*, formed by the spontaneous association of a large number of components into a large supramolecular architecture or a specific phase having more or less well-defined microscopic organization and macroscopic characteristics depending on its nature (films, layers, membranes, vesicles, micelles, mesophases, surfaces, solids, etc.).

The extension of the concepts of supramolecular chemistry [1–3] from supermolecules to polymolecular assemblies leads, in particular, to the implementation of molecular recognition as a means for controlling the evolution and the architecture of polymolecular species as they spontaneously build up from their components through self-organization [1–4]. Such recognition-directed self-assembly is of major interest in supramolecular design and engineering. In particular, its combination with the chemistry of macromolecules and of organized assemblies led to the emergence of the areas of supramolecular polymers and of supramolecular phases such as liquid crystals [1,3,5–11].

A very rich and active field of research thus developed involving the designed manipulation of molecular interactions and information through recognition processes to generate, in a spontaneous but controlled fashion, supramolecular polymers and phases by the self-assembly of complementary monomeric components, bearing two or more interaction/recognition groups. These systems belong to the realm of programmed supramolecular systems that generate organized entities following a defined plan based on molecular information and implemented through recognition events [1–3]. The main steps that may be distinguished in the process are:

1. Selective binding of complementary components via molecular recognition;
2. Growth through sequential binding of the components in the correct relative orientation; and
3. Termination requiring a builtin feature, a stop signal, that specifies the endpoint and signifies that the process has reached completion.

In addition, the reversibility of the connecting events (i.e., their kinetic lability) allows the exploration of the energy hypersurface of the system and confers to self-assembling systems the ability to undergo annealing and self-healing of defects. In contrast, covalently linked, nonlabile species cannot heal spontaneously and defects are permanent.

Since the previous chapters in this book provide a wide selection of relevant topics presented by some of the major actors in the domain, we emphasize here the conceptual and prospective aspects, beginning with a brief retrospective of our own work. It started with the implementation of the principles of supramolecular chemistry to generate supramolecular polymers and liquid crystals from molecular components interacting through specific hydrogen bonding patterns.

II. GENERATION OF HYDROGEN-BONDED SUPRAMOLECULAR POLYMERS AND LIQUID CRYSTALS

Intermolecular processes occurring in a material may markedly affect its properties. Thus, supramolecular polymerization might be expected to induce changes in phase organization, viscosity, optical features, etc. For instance, the interaction between molecular units that by themselves would not be mesogenic could lead to the formation of a supramolecular species presenting liquid-crystalline behavior. It might then be possible to take advantage of selective interactions so that the mesogenic supermolecule would form only from complementary components. This would amount to a macroscopic expression of molecular recognition, since recognition processes occurring at the molecular level would be displayed at the macroscopic level of the material by the induction of a mesomorphic phase.

A. Formation of Mesogenic Supermolecules by Association of Complementary Molecular Components

The most common type of molecular species that form thermotropic liquid crystals possesses an axial rigid core fitted with flexible chains at each end. One may then imagine splitting the central core into two complementary halves, whose association would generate the mesogenic supermolecule, as schematically represented in Fig. 1. This was realized with the derivatives P_1 and U_1 of the heterocyclic groups 2,6-diamino-pyridine P and uracil U presenting complementary arrays, DAD and ADA, respectively, of hydrogen bonding acceptor (A) and donor (D) sites. Whereas the pure compounds do not show liquid-crystalline behavior, 1:1 mixtures present a metastable mesophase of columnar hexagonal type as indicated by X-ray diffraction data. Its existence may be attributed to the formation of the mesomorphic supermolecule (**1**) via molecular recognition-directed association of the complementary components U_1 and P_1, followed by the self-organization into columns formed by stacks of disklike plates each containing two units of (**1**) (Fig. 2) [12].

Supramolecular discotic liquid crystals may be generated via the initial formation of disklike supermolecules. Thus, the tautomerism induced self-assembly of three units of the lactam–lactim form of disubstituted derivatives of phthalhydrazide yields of disklike trimeric supermolecule (Fig. 3). Thereafter, these disks self-organize into a thermotropic, columnar discotic mesophase [13].

Related processes are the formation of discotic mesogens from hydrogen-bonded phenanthridinone derivatives [14] and of columnar mesophases based on tetrameric cyclic arrangements (G-quartets) of guanine-related molecules [15].

All these cases represent overall examples of *hierarchical self-assembly*, where the initial assembly of molecular components into a disklike mesogenic supermolecule is a prerequisite for the subsequent formation of a discotic columnar architecture by stacking of the disks (Fig. 4; see also below).

Figure 1 Formation of a mesogenic supermolecule from two complementary components.

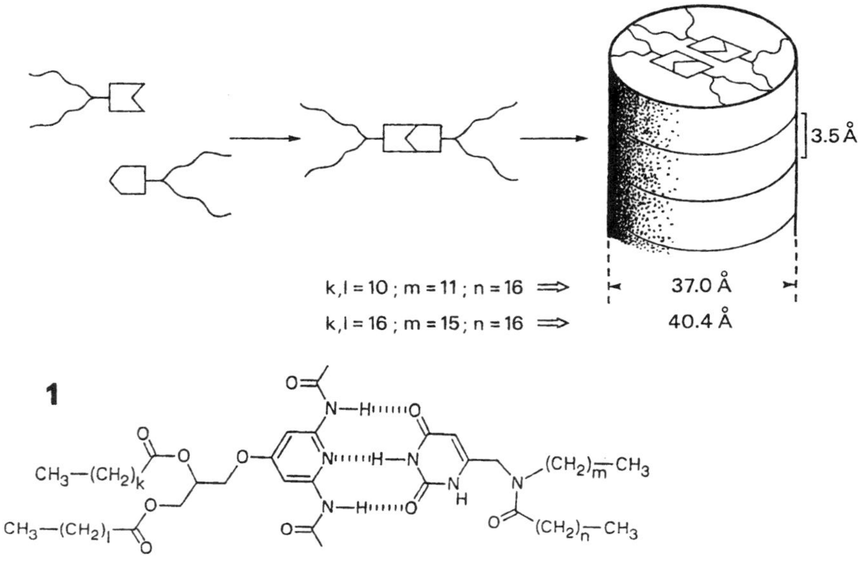

3.5 Å

k,l = 10 ; m = 11 ; n = 16 ⟹ 37.0 Å
k,l = 16 ; m = 15 ; n = 16 ⟹ 40.4 Å

1

$CH_3-(CH_2)_k$ O O O N—H O N H—N N H O N—H O O $(CH_2)_m-CH_3$ $(CH_2)_n-CH_3$ $CH_3-(CH_2)_l$ O O O

Figure 2 Columnar mesophases generated by the supermolecule **(1)** formed by the complementary components P_1 (left) and U_1 (right).

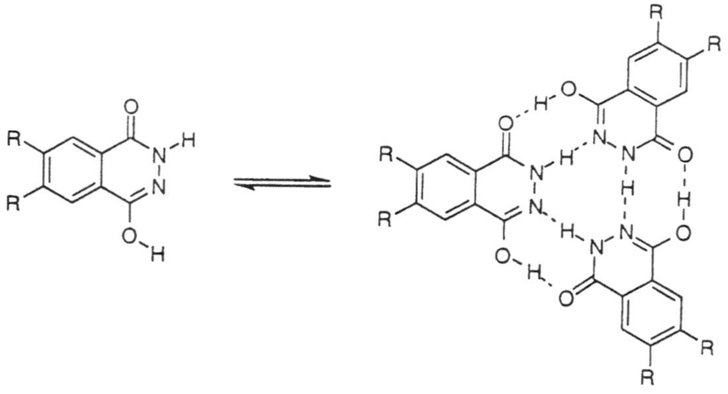

R R O N H N O H R R R R R R O H O N N H N H N O H O H O H N N H N O H O R R

Figure 3 Tautomerism induced self-assembly of a supramolecular cyclic trimer from the lactam–lactim form of phthalhydrazide derivatives.

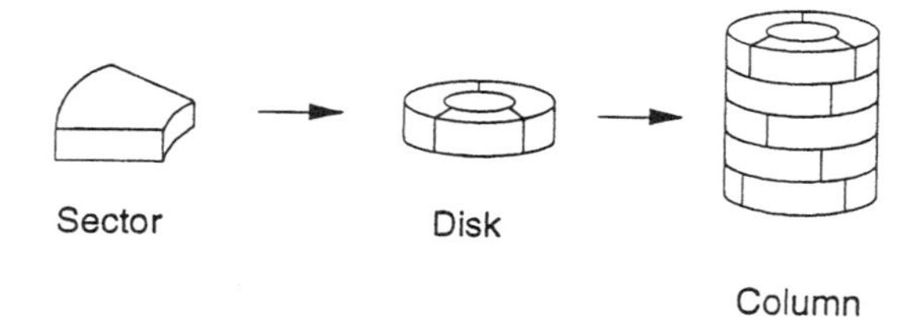

Figure 4 Hierarchical self-assembly. Self-assembly of sector components into a disk is a prerequisite for the subsequent self-organization of the disks into a discotic columnar architecture; the case illustrated is that of a trimeric mesogenic supermolecule; see Fig. 3.

B. Generation of Supramolecular Liquid-Crystalline Polymers

Mixing molecular monomers bearing two identical hydrogen bonding subunits should lead to the self-assembly of a linear "polymeric" supramolecular species via molecular recognition-directed association. Figure 5 schematically represents such a process. The resulting supramolecular polymeric material may be expected to present novel features resulting from its polyassociated nature, for instance, liquid-crystalline properties if suitable chains are introduced on the components.

Condensation of the complementary groups P and U with long chain derivatives of tartaric acid T (T = L, D, or meso M) gave substances TP_2 and TU_2 each bearing two identical units capable of undergoing supramolecular polymerization via triple hydrogen bonding [16]. Whereas the individual species LP_2, LU_2, DP_2, DU_2, MP_2, and MU_2 are solids, the mixtures (LP_2 + LU_2), (DP_2 + LU_2), and (MP_2 + MU_2) display thermotropic mesophases presenting an exceptionally wide domain of liquid crystallinity (from <25°C to 220–250°C) and a hexagonal columnar superstructure, with a total column diameter of about 37–38 Å. The materials have

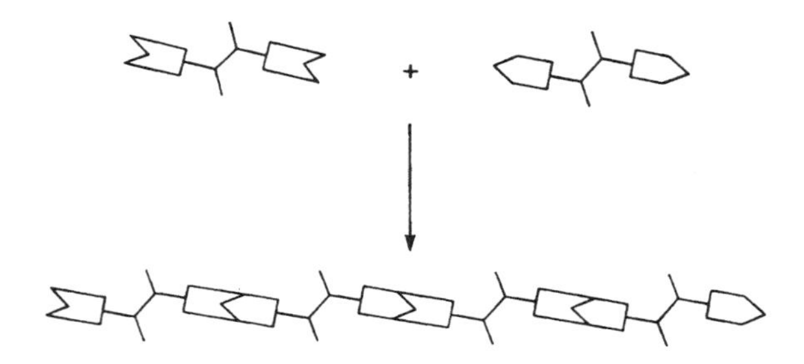

Figure 5 Formation of a polymeric supramolecular species by association of two complementary ditopic molecular components.

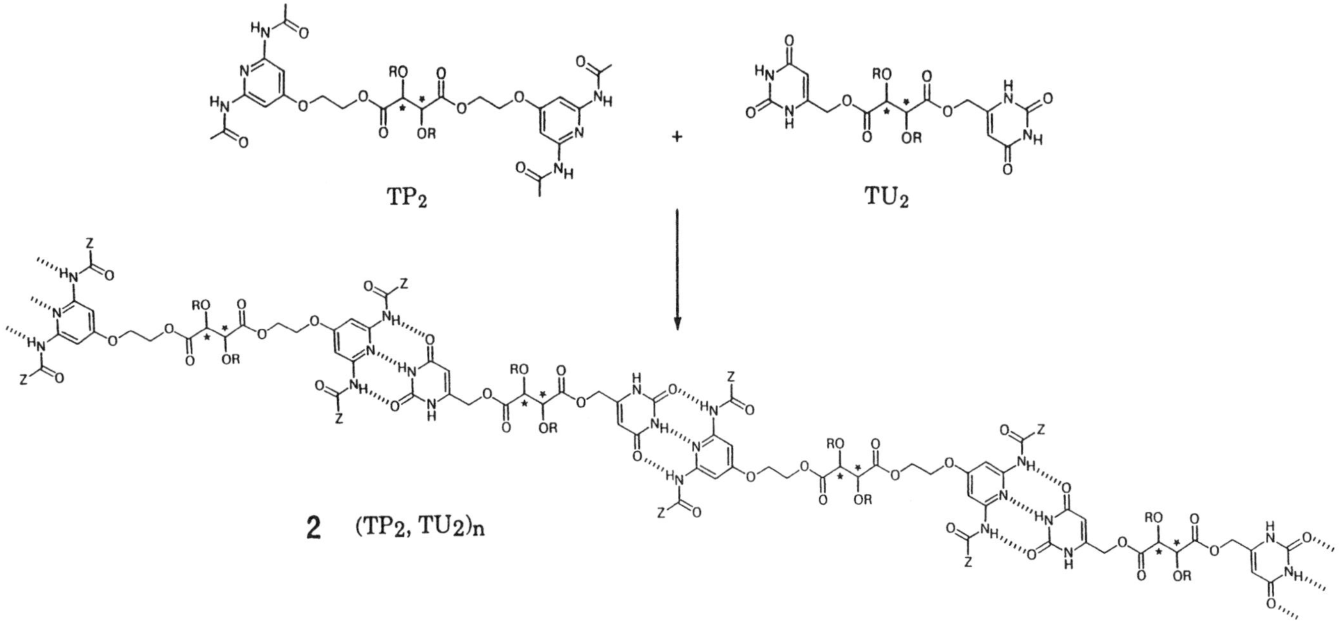

TP2 TU2

2 (TP2, TU2)n

Figure 6 Self-assembly of the polymolecular supramolecular species (TP_2, TU_2)$_n$ **(2)** from the complementary chiral components TP_2 and TU_2 via hydrogen bonding; T represents L−, D−, or meso(M)− tartaric acid; R = $C_{12}H_{25}$.

the aspect of a highly birefringent glue that forms fibers under spreading. The overall process may be described as the molecular recognition induced self-assembly of a supramolecular liquid-crystalline polymer $(TP_2, TU_2)_n$ **(2)** (Fig. 6).

The X-ray patterns for $(LP_2, LU_2)_n$ are consistent with columns formed by three polymeric strands having a triple helix superstructure, whereas those for the $(MP_2, MU_2)_n$ mixture fit a model built on three strands in a zigzag conformation (Fig. 7).

Electron microscopy studies revealed the successive states of self-assembly of **(2)** from nuclei to filaments and then to very long helical fibers of opposite chirality for the (L,L) **(2)** and (D,D) **(2)**, whereas the achiral (M,M) **(2)** material showed no helicity (Fig. 8) [17]. Thus, molecular chirality is transduced into supramolecular helicity which is expressed at the level of the material on nanometric and micrometric scales.

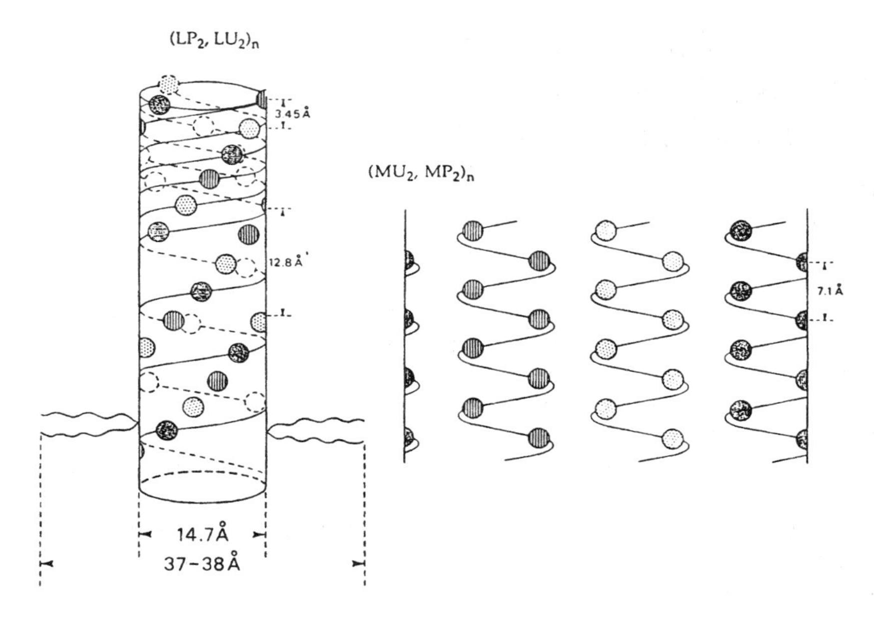

Figure 7 Schematic representation of the columnar superstructures suggested by the X-ray data for $(LP_2, LU_2)_n$, (left) and $(MP_2, MU_2)_n$ (right); each spot represents a PU base pair; spots of the same type belong to the same supramolecular strand; the dimensions are compatible with an arrangement of the PTP and UTU components along the strands indicated; the aliphatic chains stick out of the cylinder, more or less perpendicularly to its axis. For $(LP_2, LU_2)_n$, a single helical strand and the full triple helix are, respectively, represented on the bottom and at the top of the column. For $(MP_2, MU_2)_n$, the representation shown corresponds to the column cut parallel to its axis and flattened out.

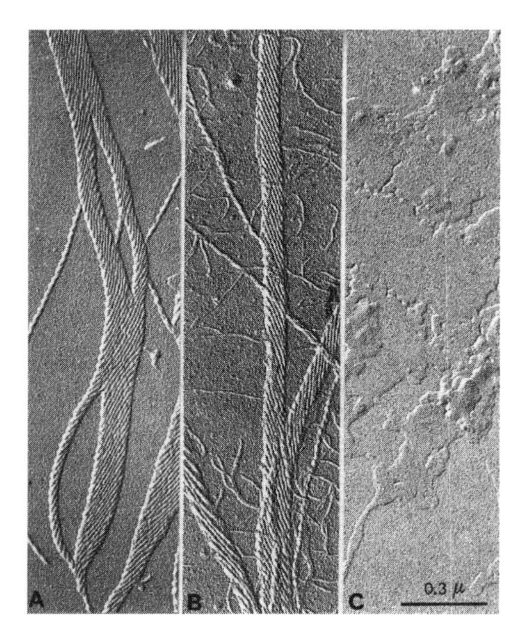

Figure 8 Helical textures observed by electron microscopy for the materials formed by the mixtures (A) LP_2 + LU_2, (B) DP_2 + DU_2, and (C) MP_2 + MU_2.

The racemic mixture of all four components LP_2, LU_2, DP_2, and DU_2 yielded long superhelices of opposite handedness that coexisted in the same sample, pointing to the occurrence of spontaneous resolution through chiral selection in molecular recognition-directed self-assembly of supramolecular liquid-crystalline polymers. Such chiral selection features of self-organized entities are of general significance in connection with the questions of spontaneous resolution and chirality amplification.

Supramolecular polymers have been obtained with other types of interaction patterns between monomers, from a single hydrogen bond between a carboxylic acid and a pyridine unit [9] to four hydrogen bonds between self-complementary heterocyclic groups [18]. Modifications of the liquid-crystalline properties have been induced in ternary mixtures by means of chiral additives [19].

C. Mesophases from Combination of Monotopic and Ditopic Complementary Components

In line with the processes described above, it may be possible to obtain liquid-crystalline materials from 2/1 mixtures of species containing, respectively, one and two recognition sites as represented schematically in Fig. 9.

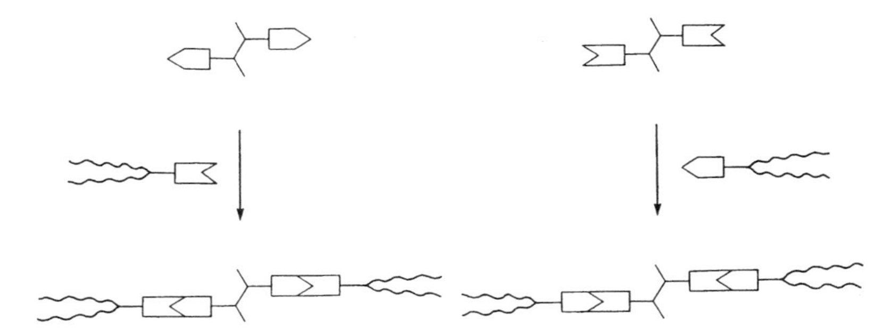

Figure 9 Formation of mesogenic species by 2/1 association of complementary monotopic and ditopic components.

Indeed, when a uracil component U_1 is combined with the complementary LP_2 unit in 2/1 ratio a mesophase is obtained; its occurrence may be attributed to the formation of mixed 2/1 supermolecules such as **(3)**, which forms a mesophase having a columnar structure of rectangular section and a very wide domain of liquid crystallinity [from $<20°C$ to 111–116°C, for $R = C_{12}H_{25}$, $m = C_{11}H_{22}$, $n = C_{16}H_{32}$ in **(3)**]. Similar observations were made for the 2/1 combination of a diaminopyridine unit P_1 with the complementary LU_2 component [20].

One may note that the U_1 and P_1 components (Fig. 2) represent chain termination groups inducing a decrease in chain length by depolymerization via end-capping on addition to the polymeric entities **(2)** (for such a process, see Ref. 21).

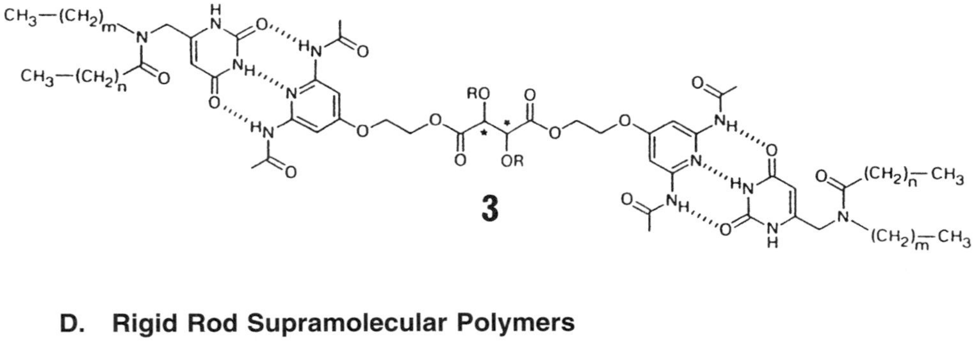

D. Rigid Rod Supramolecular Polymers

The introduction of rigid molecular units into macromolecular species has been extensively pursued in view of the novel physicochemical that the resulting rigid rods may present. Self-assembling rigid components may be designed by attaching recognition groups to a rigid core. Combination of two such complementary components may result in the formation of rigid rod supramolecular systems

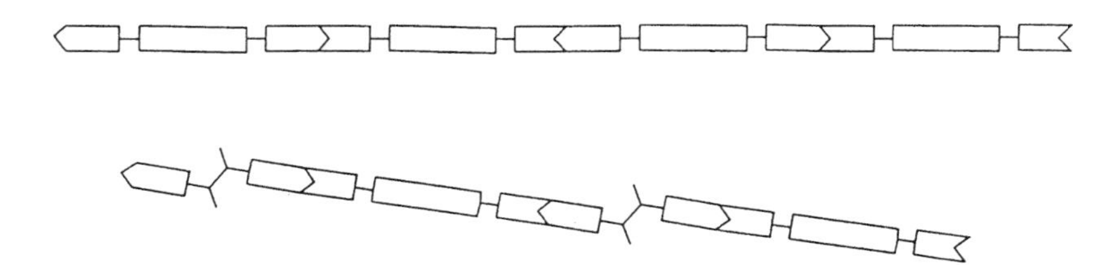

Figure 10 Schematic representation of: (top) a self-assembled rigid rod supramolecular system from two rigid complementary components; and (bottom) a self-assembled mixed system from a rigid unit and a complementary flexible one.

(Fig. 10). Mixed materials would be formed by combining a rigid unit with a complementary flexible one, such as the LP_2 or LU_2 species described above. The two complementary rigid components AP_2 and AU_2 each containing two identical recognition groups linked to an anthracenyl core self-assemble to yield the rigid rod supramolecular polymeric entity **(4)** which was found to present a lyotropic mesophase whereas AP_2 and AU_2 themselves are solids [22].

4 $(AP_2, AU_2)_n$

Hairy rigid rod polymers, in which flexible sidechains are attached to a rigid core, present attractive features [23]. A supramolecular version of such materials may be devised on the basis of components containing two recognition sites and capable of forming ''hairy'' ribbonlike structures. Thus, hydrogen bonding recognition between double-faced Janus-type recognition units, such as barbituric acid and triamino-pyrimidine derivatives, leads to ordered molecular solids through formation of polyassociated supramolecular strands (Fig. 11, left) [24a]. In the process molecular sorting out and left–right differentiation is occurring, so that it is possible to obtain extended structures bearing identical or different chains of various lengths on each side.

A similar type of species bearing identical chains on each side is formed from components containing autocomplementary recognition arrays of diamino-pyrimidone type (Fig. 11, right) [24b]. However, whereas the (barbituric acid,

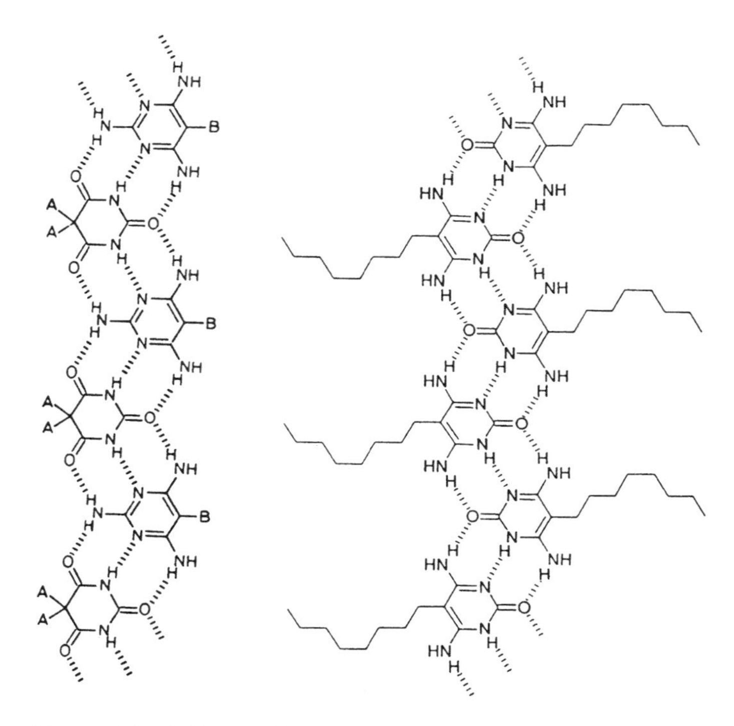

Figure 11 Self-assembled supramolecular ''hairy'' strands derived from components bearing sidechains and containing heterocomplementary (left) or self-complementary (right) recognition sites; A and B: aliphatic chains; for the corresponding crystal structures see Refs. 24a,b.

triamino-pyrimidine) combination may in principle yield either a linear strand or a cyclic entity [24b,c], this latter recognition group enforces the formation of a strand only. Furthermore, one may also point out that, in this case, the arrangement of interaction sites presents some relationship with the Oosawa model of chain growth [25].

One may note that the triple helical supramolecular species described above (Fig. 7, left) displays the features of a hairy cylinder. The self-assembled species **(4)** bearing long R chains represent supramolecular hairy rigid rods. Mesoscopic supramolecular assemblies of cylindrical rigid rod-type, of about 150 Å diameter and several tens of μm length, have been obtained from a tricyclic bis-imide Janus molecule and a long chain triaminotriazine derivative [26]. Numerous variations in the nature of the sidechains of these compounds may be envisaged, giving in principle access to a variety of materials.

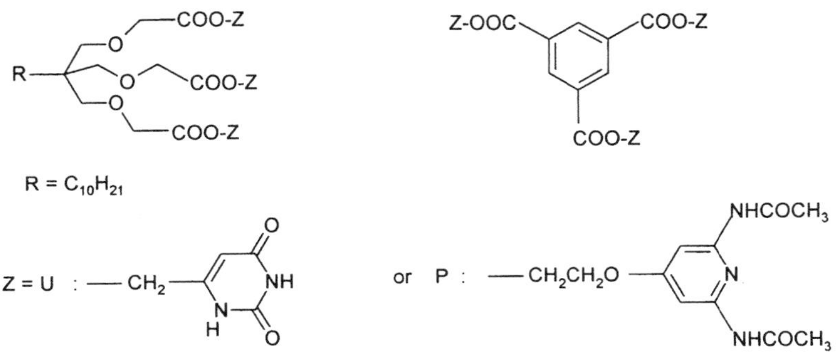

Figure 12 Ternary recognition components for the crosslinking of supramolecular polymeric species.

E. Crosslinking of Supramolecular Polymers

Extending further the procedures of polymer chemistry to supramolecular entities one may envisage devising supramolecular crosslinking agents. Thus, tripode-type components containing three equivalent recognition subunits may be expected to establish two-dimensional networks when mixed with linear polyassociated species such as those described above, yielding crosslinked supramolecular polymers (Fig. 12) [27]. Of course, one may also imagine corresponding tetravalent components bearing four interaction groups that would then yield formally three-dimensionally crosslinked entities. Such polytopic monomers may also generate supramolecular branched species in particular of dendrimeric type (see also below) [28].

F. "Ladder" and Two-Dimensional Supramolecular Polymers from Monomers Containing Janus-Type Recognition Groups

Janus-type recognition groups, such as barbituric or cyanuric acid and triaminopyrimidine or -triazine derivatives (see also Fig. 11), represent a special type of crosslinking unit by virtue of their ability to interact through their two hydrogen bonding faces. The incorporation of two such groups into ditopic molecular monomers provides entries towards the generation of "ladder" or double-ribbon polyassociations (when only a single ditopic component is used) or of two-dimensional supramolecular polymeric networks (from two complementary ditopic components). This is schematically illustrated in Fig. 13.

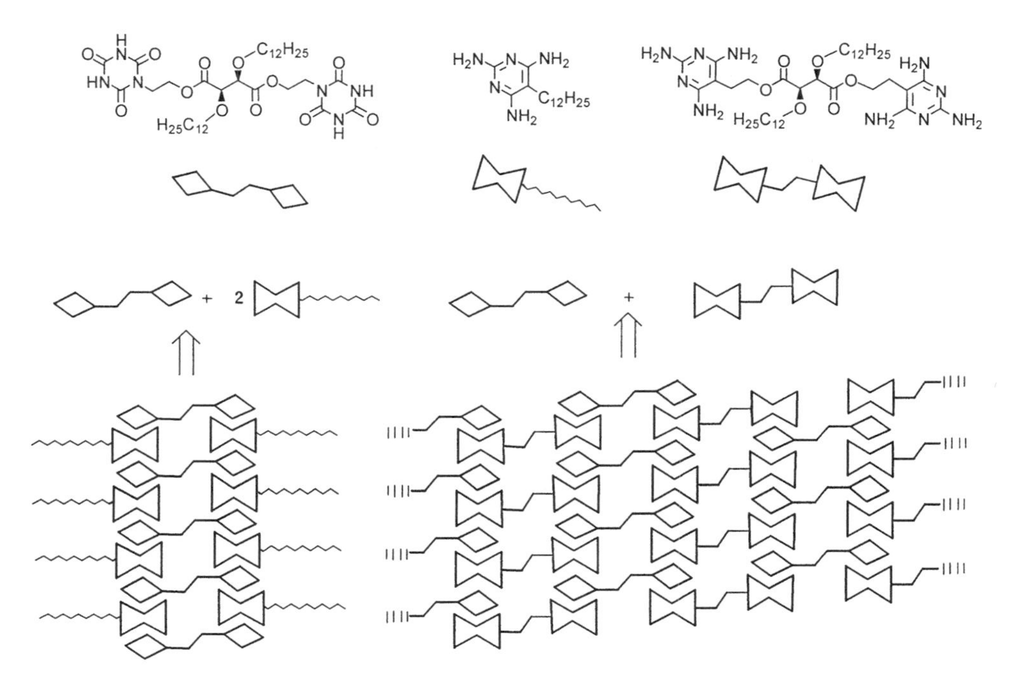

Figure 13 Schematic representation of the H-bond mediated self-assembly of monomer components bearing Janus-type complementary recognition groups (e.g., cyanuric acid and 2,4,6-triaminopyrimidine) forming supramolecular polymers: (left) of ''ladder'' or double-ribbon type from a double-Janus and a single Janus component; (right) of two-dimensional crosslinked nature from two double-Janus components.

Indeed, monomers in which two such groups are grafted onto tartaric acid units (e.g., replacing P and U units in TP_2 and TU_2; see Fig. 6) generate very high molecular weight aggregates that are characterized by various physical methods [29].

G. Supramolecular Coordination Polymers

When the monomeric components carry metal ion binding subunits, polyassociation occurs on addition of suitable ions, yielding supramolecular coordination polymers. This is the case for the L-tartaric acid derivative **(5)** bearing two methylated bipyridine groups. The binding of Cu (I) ions may inter alia generate chains where the components **(5)** are connected through [Cu (I) $(bipy)_2$] centers of tetrahedral coordination. Indeed, addition of $Cu(CH_3CN)_4\ PF_6$ to **(5)** resulted

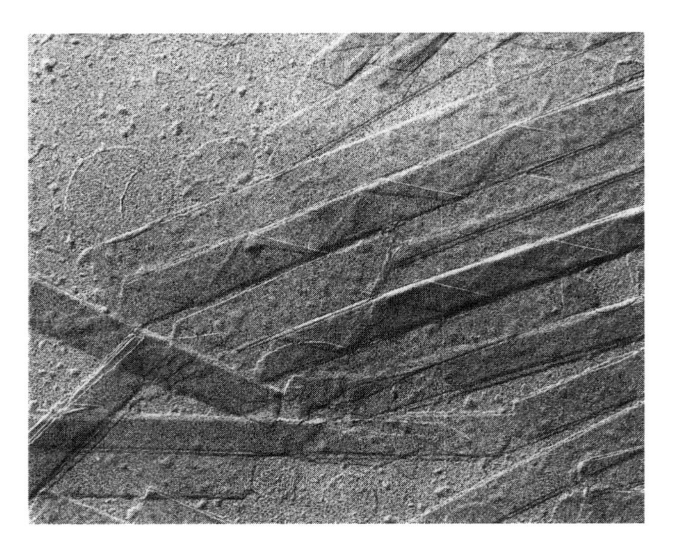

Figure 14 Inorganic nanotubes formed by the ditopic ligand **(5)** with $Cu(CH_3CH)_4PF_6$ observed by electron microscopy; the diameter of a tube is about 0.2 μ.

in the formation of organized phases, in particular, self-assembled inorganic nanotubes of very regular structure, resulting presumably from the helicoidal winding of a large tapelike entity, as revealed by electron microscopy (Fig. 14) [30a].

Soluble coordination polymers of variable degrees of polymerization have been obtained from bis-bipyridylketone and Cu (II) ions in different solvents; they have been characterized by electrospray mass spectrometry revealing molecular weights reaching >60,000 Daltons [30b].

5

III. BASIC FEATURES OF SUPRAMOLECULAR POLYMERS

The results presented above illustrate the rich domain that emerges from the combination of polymer chemistry with supramolecular chemistry. It involves the generation of polymeric superstructures by the designed use and manipulation of molecular interactions and information through molecular recognition processes. Figure 15 presents some of the different types of such ''informed'' supramolecular polymers that may be generated by recognition-directed self-assembly of complementary monomer species.

Broadening the scope, we may briefly consider a nonexhaustive panorama of various types and features of supramolecular polymers depending on their constitution, characterized by three main parameters: the nature of the core/framework of the monomers, the type of noncovalent interaction(s), and the eventual incorporation of functional subunits. The interactions may involve complementary arrays of hydrogen bonding sites, electrostatic forces, electronic donor–acceptor interactions, metal ion coordination, etc. The polyassociated structure itself may be of mainchain, sidechain, or branched, dendritic type, depending on the number and disposition of the interaction subunits. A central question is that of the size and the polydispersity of the polymeric supramolecular species formed. Of course their size is expected to increase with concentration and the polydispersity depends on the stability constants for successive associations.

The basic characteristics of *mainchain supramolecular polymers* are presented schematically in Fig. 16. Designating the monomer core residues by R_i, monomers bearing two identical interaction/recognition groups (homoditopic), may yield either *homopolymers*, when $R_i = R_j$, or regularly alternating *copolymers*, when $R_i \neq R_j$.

When several different core residues are used, a large number of polymeric objects may be generated. Thus, for a chain $(R_i\ R_j)_k$ of k pairs length, formed from m different monomers R_i and n different monomers R_j, the total number of different sequences is $(m \times n)^k$. Since chains of any length ($k = 1$ to p) can be formed, the total number of different objects that can be present, comprising all possible sequences of all lengths [i.e., the full virtual combinatorial diversity (see Section IV.C)] is $mn[mn^p - 1)/(mn - 1)]$. The fractions of the species as a function of chain length follow the size distribution curve of the system considered.

Initiation and chain growth occur on mixing of the complementary monomers. Chains of different compositions may be formed side by side, in principle without crossover, when several different pairs of complementary recognition groups are put to use (see top of Fig. 16). Heteroditopic monomers combining recognition groups from different pairs may act as chain-crossover components, allowing the combination of two different chains.

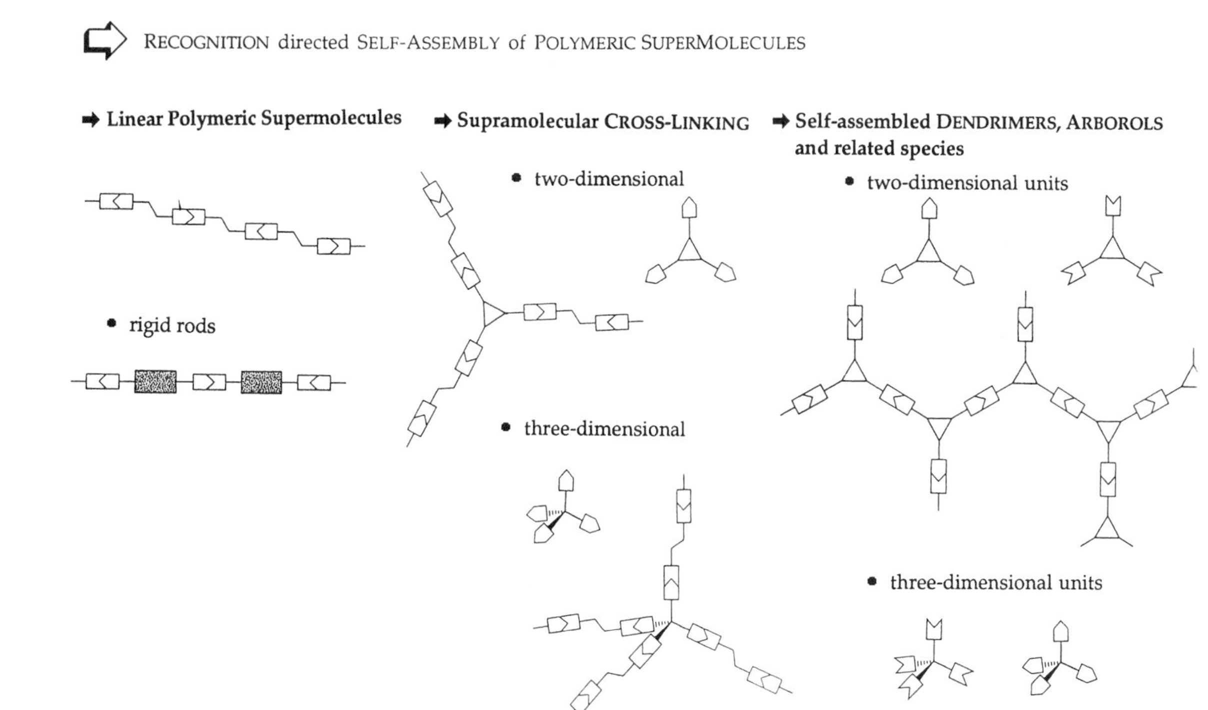

Figure 15 An aspect of the panorama of supramolecular polymer chemistry.

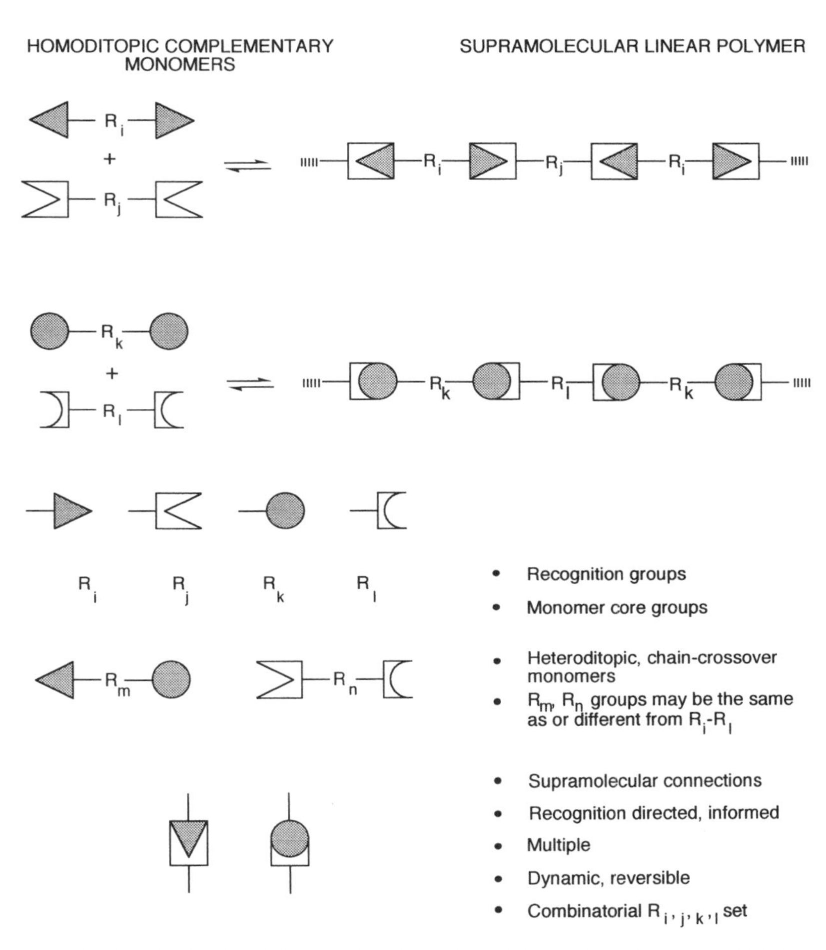

Figure 16 Schematic representation of the formation of linear mainchain supramolecular polymers from complementary homoditopic monomers, and of the constituting subunits.

Self-complementary monomers where the R units bear two complementary recognition groups yield homopolymers (only a single R_i) or random copolymers (two or more different core units R_i, R_j, R_k, . . .). Initiation and growth occur immediately after generation of the interaction groups.

In all cases, chain growth can be initiated by setting free one of the recognition groups by an external stimulus (e.g., light) from a derivative bearing a protecting group. Chain length may be altered/reduced by the addition of substances bearing a single recognition group, which acts as a chain termination component by end-capping [21].

Sidechain supramolecular polymers result from the binding of residues bearing recognition groups to complementary groups attached to the main chain of a covalent polymer (Fig. 17) (see, for instance, Ref. 31). Of course, such mainchain covalent polymers may be crosslinked in a supramolecular fashion by means of double-headed complementary additives establishing bridges between the side groups of two different chains.

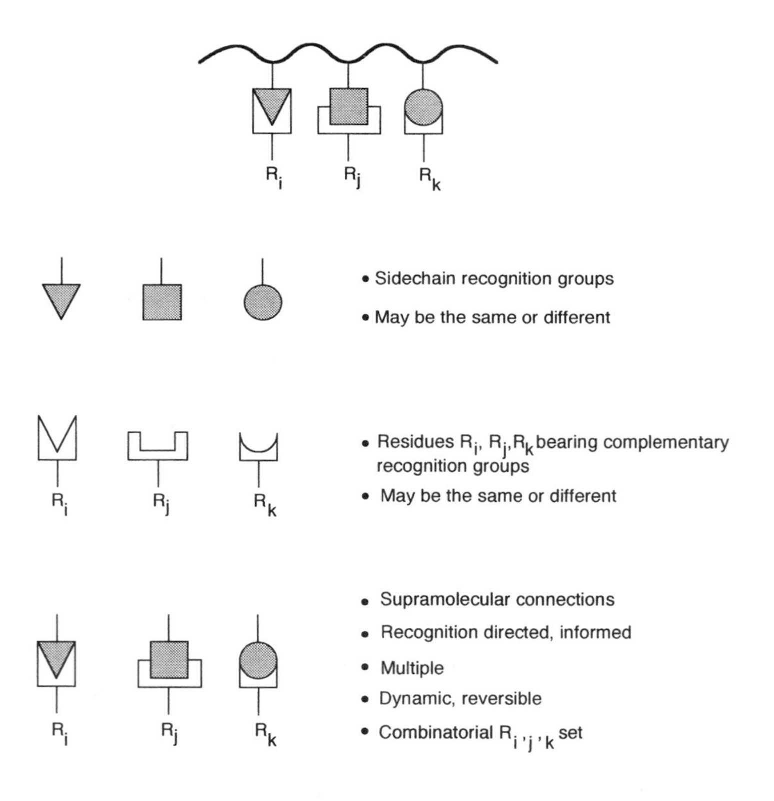

Figure 17 Schematic representation of the formation of sidechain supramolecular polymers from a covalent polymer bearing recognition groups binding complementary components, and of the constituting subunits.

Supramolecular coordination polymers represent a special class where the monomers are ditopic ligand molecules possessing two metal binding groups and where the connection is provided by metal ion coordination (Fig. 18). The metal ions play the role of association mediators, enabling one to select the ligand components and to direct the polyassociation according to the combination (metal ion/ligand binding site) in operation. One may take advantage of the vast set of metal binding units available and of their more or less selective coordination with specific ions. Initiation and growth occur only on addition of given metal

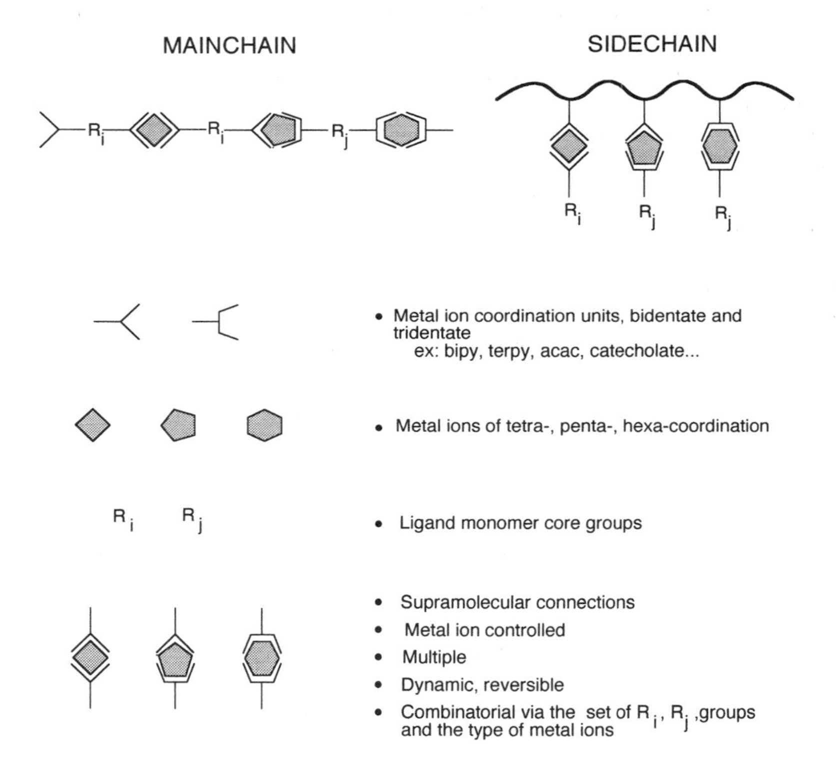

Figure 18 Schematic representation of the formation of supramolecular coordination polymers of mainchain or sidechain type from ligands containing bidentate and/or tridentate complexation subunits (such as bipyridine and terpyridine) binding metal ions of tetra-, penta-, or hexa-coordination, and of the constituting subunits.

ions and preferentially between specific ligands. Thus, in the presence of a mixture of ditopic ligands bearing different metal binding subunits, the nature of the ''monomers'' participating in the formation of mainchain coordination polymers may be determined/directed by the choice of the ion introduced. For instance, with bidentate (B) tridentate (T) metal binding sites, metal ions (M) of tetra-, penta-, and hexa-coordination are expected to yield (BMB), (BMT), and (TMT) connections, respectively (Fig. 18). Of course, sidechain coordination polymers can also be obtained, as well as discotic-type coordination assemblies, such as the columnar mesophases formed from trimeric gold complexes of pyrazole derivatives [32] in a fashion similar to the processes shown in Figs. 3 and 4. Components combining organic (e.g., hydrogen bonding, etc.) and metal ion binding sites may be expected to yield mixed organic/inorganic composite polymers. (For ligands combining hydrogen bonding and metal ion binding sites and their complexes see Ref. 33a; for a linear polymer of this type see Ref. 33b.)

Similar considerations hold for *supramolecular organometallic polymeric* entities [34].

Supramolecular crosslinking may be achieved by introducing molecular monomers bearing more than two organic or metal ion binding subunits. Such compounds provide links between chains in a way directed by the nature of the interacting groups. Figure 19 schematically represents crosslinking components of organic and inorganic types. An intriguing case is that of diamino-triazinone which possesses three different H-bonding faces DAA, ADD, and DAD, a sort of triple-Janus! Of course one may also envisage combining organic and inorganic interaction sites, which would allow the crosslinking of organic and inorganic supramolecular polymeric chains.

The use of suitable components containing multiple recognition groups should allow the designed generation of supramolecular species possessing a desired architecture. On the other hand, such components also greatly increase the dynamic combinatorial diversity of the system (see Section IV.C).

As in the case of covalent macromolecules, the supramolecular associations may also present internal (intrasupramolecular) interactions between sites located either in the mainchain or in sidechain appendages, thus leading to chain folding and structuration of the supramolecular entity.

Branched supramolecular polymers are obtained by means of crosslinking monomers. Of special significance is the fact that equimolar mixtures of complementary ternary or quaternary components lead in principle to the spontaneous generation of treelike species that represent recognition-directed, self-assembling supramolecular dendrimers of the usual dichotomic as well as of trichotomic types (see also Fig. 15). This holds both for organic and inorganic dendritic entities [28]. Mixed organic/inorganic dendrimers would be accessible from tri- and tetratopic monomers bearing both organic and metal ion binding sites.

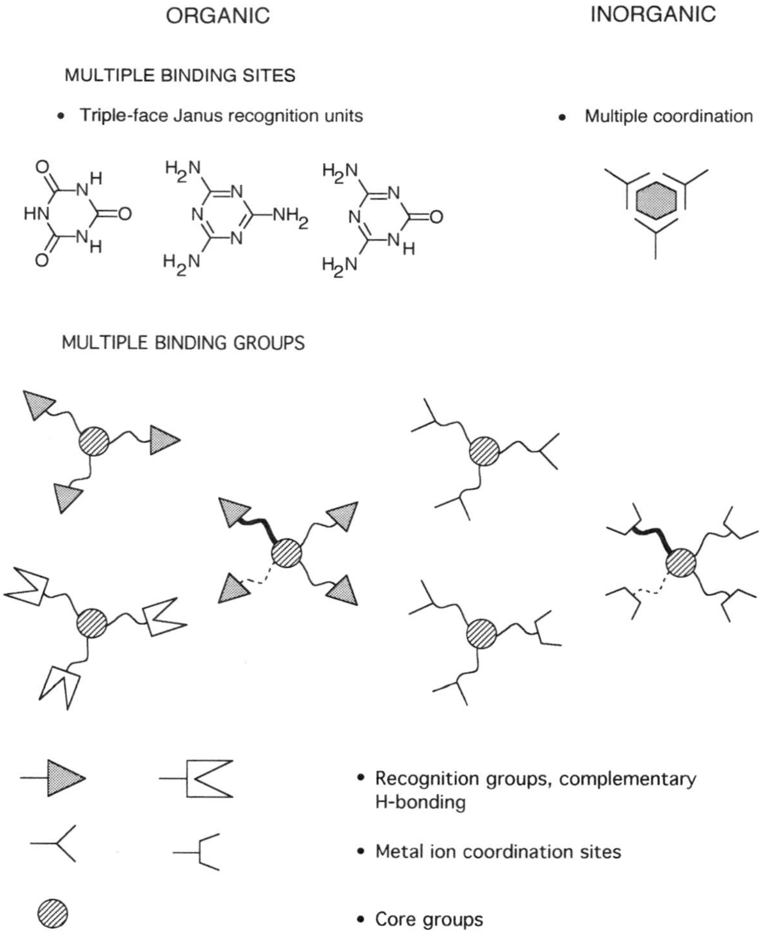

ORGANIC

INORGANIC

MULTIPLE BINDING SITES

- Triple-face Janus recognition units

- Multiple coordination

MULTIPLE BINDING GROUPS

- Recognition groups, complementary H-bonding
- Metal ion coordination sites
- Core groups

Figure 19 Schematic representation of supramolecular crosslinking agents of organic and inorganic types, and of the constituting subunits.

Hierarchical self-assembly takes place when several self-assembling events occur in sequence via a conditional process, a given step being a prerequisite for the subsequent one. This is, for instance, the case in the formation of discotic-type columnar liquid crystals from self-assembled supramolecular disks (see above; Fig. 4). One may distinguish three successive steps in the generation of the liquid-crystalline entities described in Section II.B: formation of a supramolecular strand (**2**); assembly of three strands into columnar superstructures (Fig. 7); and generation of fibers (Fig. 8) by association of several columnar entities.

A basic common feature of all these types of supramolecular polyassociations is that they are *reversible polymers* due to the lability of the noncovalent connections, and as a result they also possess the ability to generate dynamic diversity through scrambling of the monomer components (see below). Furthermore, since only correct recognition-directed complementary pairing is expected to occur, self-selection [35] between compatible units should take place and control of the self-assembly process is in principle possible.

In view of the lability of the associations, supramolecular polymers present features of ''living'' polymers capable of growing or shortening, or rearranging their interaction patterns, of exchanging components, and of undergoing annealing, healing, and adaptation processes.

Growth control and regulation of structure and composition may be achievable by means of external effectors (temperature, pH, metal ions, competing ligands, end-capping units, etc.). A relevant case is that of a molecular strand which undergoes a structural reorganization upon recognition-directed binding of a complementary effector to give a coiled disklike object which thereafter self-assembles into extended fibers [36].

These most interesting features raise important but difficult questions of characterization concerning composition, size, persistence length, shape, structure, etc. of the entities formed. To this end, an array of physicochemical methods is required and must be put to use, such as vapor phase osmometry, differential scanning calorimetry, electrospray mass spectrometry, NMR spectroscopy, gel permeation chromatography, light scattering, electron microscopy, near field microscopies, etc.). (For a relevant case, see Ref. 37.)

Supramolecular versions of the various species and procedures of molecular polymer chemistry may be imagined and implemented providing a wide field of future investigation that offers a wealth of novel entities and functionalities. Possible extensions concern, for instance, the introduction of various central cores, in particular those already known to yield molecular liquid crystals, the incorporation of photo-, electro-, or iono-active functional units, the potential use for detection devices as well as the extension to various other recognition components particularly those of biological nature (see, for instance, the case of the guanine-type quartets [15]).

IV. SUPRAMOLECULAR POLYMERS AS SUPRAMOLECULAR MATERIALS

A. Supramolecular Materials

The properties of a material depend both on the nature of the constituents and on the interactions between them. Supramolecular chemistry may thus be expected to have a strong impact on materials science via the explicit manipulation of the noncovalent forces that hold the constituents together. These interactions, and the recognition processes that they underlie, allow the design of materials and the control of their buildup from suitable units by self-assembly.

Through recognition-directed association, self-assembly, and self-organization processes, supramolecular chemistry opens new perspectives in materials science towards an area of supramolecular materials whose features depend on molecular information and which involve ''smart'' materials, network engineering, polymolecular patterning, etc. As shown above, liquid-crystalline polymers of supramolecular nature presenting various supramolecular textures may be obtained by the self-assembly of complementary subunits. This amounts to a macroscopic expression of molecular information via a phase change which, being a highly cooperative process, also corresponds to an amplification of molecular recognition and information from the microscopic to the macroscopic level.

Supramolecular engineering gives access to the molecular information-controlled generation of nanostructures* as well as of polymolecular architectures and patterns in molecular assemblies, layers, films, membranes, micelles, gels, colloids, mesophases, and solids as well as in large inorganic entities, polymetallic coordination architectures, and coordination polymers.

Molecular recognition processes may be used to induce and control processes between polymolecular assemblies such as organization of, and binding to, molecular layers and membranes [11,40], selective interaction of vesicles with molecular films [41], aggregation and fusion of vesicles bearing complementary recognition groups [42,43], etc.

The building up of supramolecular architectures and materials may involve several steps and proceed in particular via hierarchical self-assembly processes (see also above). Such sequential conditional processes enable the progressive buildup of more and more complex systems in a directed ordered fashion; on the other hand, they offer the intriguing possibility to intervene at each step so as either to suppress the following ones or to reorient the subsequent evolution of the system into another direction, towards another output entity.

Molecular recognition-directed processes also provide a powerful entry into supramolecular solid state chemistry and crystal engineering. The ability to control the way in which molecules associate may allow the designed generation of supramolecular architectures in the solid state. Modification of surfaces with recognition units leads to extended exo-receptors [1,3] that could display selective surface binding on the microscopic level, and to recognition-controlled adhesion on the macroscopic scale [41].†

* For the assembly of triblock copolymers into nanostructures see Ref. 38; also see, for instance, the recently described self-assembly of rodlike hydrogen-bonded nanostructures, Ref. 39.

† For the interaction between a molecule and a modified nanocrystalline solid bearing a complementary recognition group, see, for instance, Ref. 44.

B. Supramolecular Chemistry: Dynamic and Combinatorial

Supramolecular chemistry is intrinsically a dynamic chemistry in view of the lability of the interactions connecting the molecular components of a supramolecular entity. Moreover, and most significantly, the reversibility of the associations allows a continuous reorganization by both modification of the connections between the constituents and incorporation or extrusion of components by exchange with the surroundings, conferring therefore combinatorial features to the system.

Supramolecular chemistry has thus a direct relationship with the highly active area of combinatorial chemistry, however in a very specific fashion. Indeed, reversibility being a basic and crucial feature of supramolecular systems, the dynamic generation of supramolecular diversity from the reversible combination of noncovalently linked building blocks falls within the realm of the emerging area of dynamic combinatorial chemistry (DCC) which involves dynamic combinatorial libraries of either virtual (VCL) or real nature depending on the system and the conditions [45,46]. The concepts and perspectives of the DCC/VCL approach have been outlined recently [45].

Consequently, supramolecular materials are by nature dynamic materials, defined as materials whose constituents are linked through reversible connections (covalent or noncovalent) and undergo spontaneous and continuous assembly/disassembly processes in a given set of conditions [47]. Because of their intrinsic ability to exchange their constituents, they also have combinatorial character so that they may be considered as dynamic combinatorial materials (DCMs). Supramolecular materials thus are instructed, dynamic, and combinatorial; they may in principle select their constituents in response to external stimuli or environmental factors and therefore behave as adaptive materials [47].

C. Supramolecular Polymers as Dynamic Combinatorial Materials

It follows from the previous considerations that supramolecular polymer chemistry is both dynamic and combinatorial and that supramolecular polymers are therefore dynamic combinatorial materials based on dynamic libraries whose constituents have a combinatorial diversity determined by the number of different monomers (see Section III). Similar views apply to supramolecular liquid crystals.

The components effectively incorporated into the polyassociations depend in particular on the nature of the core groups and on the interactions with the environment, so that supramolecular polymers possess the possibility of adaptation by association/growth/dissociation sequences. The selection of components may occur on the basis of size commensurability [19], of compatibility in chemical properties, in charge, in rigidity/flexibility, etc. An example is given by the formation of homochiral helical fibers with chiral selection from a racemic mixture of monomeric tartaric acid derivatives: $LU_2 + LP_2 + DU_2 + DP_2 \rightarrow (LU_2 \equiv LP_2)_n + (DU_2 \equiv DP_2)_n$ [17] (see Section II.B).

Depending on the nature and variety of core/interaction/functional groups in mixtures of several different monomeric components, the dynamic features give access to higher levels of behavior such as healing, adaptability, and response to external stimulants (heat, light, additives, etc.).

V. CONCLUSION

Molecular information-based recognition events represent a means of performing programmed materials engineering and processing of biomimetic or abiotic type and may lead to self-assembling nanostructures, organized and functional species of nanometric dimensions that define a supramolecular nanochemistry, an area to which supramolecular polymer chemistry is particularly well suited and able to make important contributions.

Nanoscience and nanotechnology have become and will remain very active areas of investigation, in view of both their basic interest and their potential applications. Here again, supramolecular chemistry may have a deep impact. Indeed, the spontaneous but controlled generation of well-defined, functional supramolecular architectures of nanometric size through self-organization offers a very powerful alternative to nanofabrication and to nanomanipulation, providing a chemical approach to nanoscience and technology [1]. One may surmise that rather than having to stepwise construct bottom-up or to top-down prefabricate nanostructures, it will become possible to devise more and more powerful self-fabrication methodologies resorting to self-organization from instructed components. The results described above give an aspect of possible ways towards self-organized nanostructures. The dynamic and combinatorial features of such supramolecular architectures confers to them the potential to undergo healing and adaptation, processes of great value for the development of ''smart'' nanomaterials.

Widening the perspectives, one may consider that the science of supramolecular materials in general and supramolecular polymer chemistry in particular will strongly contribute to the emergence and development of *adaptive chemistry* [47] on the way towards complex matter.

REFERENCES

1. J-M Lehn. Supramolecular Chemistry—Concepts and Perspectives. Weinheim: VCH, 1995.

2. JL Atwood, JED Davies, DM MacNicol, F Vögtle, J-M Lehn, eds. Comprehensive Supramolecular Chemistry. Oxford: Pergamon, 1996.
3. J-M Lehn. Angew Chem Int Ed Engl 29:1304, 1990; see also Ref. 1, Chapter 9.
4. (a) D Philp, JF Stoddart. Angew Chem Int Ed Engl 35:1155, 1996; GM Whitesides, JP Mathias, CT Seto. Science 254:1312, 1991. (b) DS Lawrence, T Jiang, M Levett. Chem Rev 95:2229, 1995.
5. J-M Lehn. Makromol. Chem., Macromol. Symp. 69:1, 1993.
6. V Percec, H Jonsson, D Tomazos. In: CM Paleos, ed. Polymerization in Organized Media. Philadelphia: Gordon and Breach, 1992, p 1.
7. CT Imrie. TRIP 3:22, 1995.
8. CM Paleos, D Tsiourvas. Angew. Chem. Int. Ed. Engl. 34:1696, 1995.
9. T Kato, JMJ Fréchet. Macromol. Symp. 98:311, 1995; T Kato, M Fujumasa, JMJ Fréchet. Chem. Mater. 7:368, 1995.
10. A Ciferri. TRIP 5:142, 1997.
11. A Reichert, H Ringsdorf, P Schuhmacher, W Baumeister, T Scheybani. In: JL Atwood, JED Davies, DM MacNicol, F Vögtle, J-M Lehn, eds. Comprehensive Supramolecular Chemistry, vol. 9, Oxford: Pergamon, 1996, p 313.
12. M-J Brienne, J Gabard, J-M Lehn, I Stibor. J. Chem. Soc. Chem. Commun. 1868, 1989.
13. M Suarez, J-M Lehn, SC Zimmerman, A Skoulios, B Heinrich, J. Am. Chem. Soc. 37:9526, 1998.
14. R Kleppinger, CP Lillya, C Yang. Angew. Chem. Int. Ed. Engl. 34:1637, 1995.
15. G Gottarelli, GP Spada, A Garbesi. In: JL Atwood, JED Davies, DM MacNicol, F Vögtle, J-M Lehn, eds. Comprehensive Supramolecular Chemistry, vol. 9, Oxford: Pergamon, 1996, p 483.
16. C Fouquey, J-M Lehn, A-M Levelut. Adv. Mater. 2:254, 1990.
17. T Gulik-Krzywicki, C Fouquey, J-M Lehn. Proc. Nat. Acad. Sci. USA 90:163, 1993.
18. RP Sijbesma, FH Beijer, L Brunsveld, BJB Folmer, JHKK Hirschberg, RF Lange, JKL Lowe, EW Meijer. Science 278:1601, 1997.
19. C He, C-M Lee, AC Griffin, L Bouteiller, N Lacoudre, S Boileau, C Fouquey, J-M Lehn. Mol. Cryst. Liq. Cryst. In press.
20. M-J Brienne, C Fouquey, A-M Levelut, J-M Lehn. Unpublished work.
21. BJB Folmer, E Cavini, RP Sijbesma, EW Meijer, Chem. Commun. 1847, 1998.
22. M Kotera, J-M Lehn, J-P Vigneron. J. Chem. Soc. Chem. Commun. 197, 1994.
23. G Wegner. Thin Solid Films 216:105, 1992, and references therein.
24. (a) J-M Lehn, M Mascal, A DeCian, F Fischer. J. Chem. Soc. Chem. Commun. 479, 1990. (b) J-M Lehn, M Mascal, A DeCian, J Fischer. J. Chem. Soc. Perkin Trans. 461, 1992. (c) For the enforced formation of supramolecular macrocycles see: A Marsh, M Silvestri, J-M Lehn. Chem. Commun. 1527, 1996 and references therein.
25. See Fig. 26 in Chapter 1, this volume.
26. N Kimizuko, S Fujikawa, H Kuwahara, T Kunitake, A Marsh, J-M Lehn. J. Chem. Soc. Chem. Commun. 2103, 1995.
27. C Fouquey, J-M Lehn. Unpublished work.
28. (a) This volume, Chapters 4 and 9; (b) GR Newkome, CN Moorefield, F Vögtle. Dendritic Molecules. Weinheim: VCH, 1996. (c) For inorganic metallodendritic structures see for instance: C Gorman. Adv. Mater. 10:295, 1998; D Astruc. Top.

Cur. Chem. 160:47, 1991; S Campagna, G Denti, S Serroni, A Juris, M Venturi, V Ricevuto, V Balzani. Chem. Eur. J. 1:211, 1995, and references therein.

29. M Krische, A Petitjean, E Pitsinos, D Sarazin, C Picot. Work in progress.
30. (a) C Fouquey, T Gulik-Krzywicki, J-M Lehn. Annuaire Collège de France, 305, 1992–93. (b) J Rojo, J-M Lehn, H Nierengarten, E Leize, A Van Dorsselaer. To be published. (c) For recent examples of solid state coordination polymers, see for instance: L Carlucci, G Ciani, D M Proserpio. Chem. Commun. 449, 1999; T Ezuhara, K Endo, Y Aoyama. J. Am. Chem. Soc. 121:3279, 1999; A Mayr, J Guo, Inorg. Chem. 38:921, 1999.
31. C Bamford, K Al-Lame. J. Chem. Soc. Chem. Commun. 1580, 1993; HA Asanuma, T Hishiya, T Bau, S Gotoh, M Komiyama. J. Chem. Soc. Perkin Trans. 2:1915, 1998.
32. J Barberá, A Elduque, R. Giménez, LA Oro, JL Serrano. Angew. Chem. Int. Ed. Engl. 35:2832, 1996.
33. (a) AD Burrows, C-W Chan, MM Chowdry, JE McGrady, DMP Mingos. Chem. Soc. Rev. 25:329, 1996. (b) Z Qin, HA Jenkins, SJ Coles, KW Muir, RJ Puddephatt. Can. J. Chem. 77:155, 1999.
34. FT Edelmann, I Haiduc. Supramolecular Organometallic Chemistry. Weinheim: Wiley-VCH, 1999.
35. R Krämer, J-M Lehn, A Marquis. Proc. Natl. Acad. Sci. USA 90:5394, 1993.
36. V Berl, MJ Krische, I Huc, J-M Lehn. To be published.
37. EE Simanek, X Li, IS Choi, GM Whitesides. In: JL Atwood, JED Davies, DM MacNicol, F Vögtle, J-M Lehn, eds. Comprehensive Supramolecular Chemistry, vol. 9, Oxford: Pergamon, 1996, p 595.
38. SI Stupp, V LeBonheur, K Walker, LS Li, KE Huggins, M Keser, A Amstutz. Science 276:384, 1997.
39. HA Klok, KA Jolliffe, CL Schauer, LJ Prins, JP Spatz, M Möller, P Timmerman, DN Reinhoudt. J. Am. Chem. Soc. 121:7154, 1999.
40. (a) T. Kunitake. In: JL Atwood, JED Davies, DM MacNicol, F Vögtle, J-M Lehn, eds. Comprehensive Supramolecular Chemistry, vol. 9, Oxford: Pergamon, 1996, p 351. (b) K Ariga, T Kunitake. Acc. Chem. Res. 31:371, 1998.
41. (a) V Marchi-Artzner, F Artzner, O Karthaus, M Shimomura, K Ariga, T Kunitake, J-M Lehn. Langmuir 14:5164, 1998. (b) V Marchi-Artzner, J-M Lehn, T Kunitake. Langmuir 14:6470, 1998.
42. V Marchi-Artzner, L Jullien, T Gulik-Krzywicki, J-M Lehn. Chem. Commun. 117, 1997.
43. S Chiruvolu, S Walker, J Israelachvili, F-J Schmitt, D Leckband, JA Zasadinski. Science 264:1753, 1994.
44. L Cusack, SN Rao, D Fitzmaurice. Chem. Eur. J. 3:202, 1997.
45. J-M Lehn. Chem. Eur. J. 5:2455, 1999, and references therein.
46. (a) AV Eliseev. Current Opinion Drug Discov. Develop. 1:106, 1998. (b) AV Eliseev, J-M Lehn. Current Top. Microbiol. Immuno. 243:159, 1999.
47. J-M Lehn. In: R Ungaro, E Dalcanale (eds.) Supramolecular Science: Where It Is and Where It Is Going. Dordrecht: Kluwer, 1999, p 287.

15

Protein Polymerization and Polymer Dynamics Approach to Functional Systems

Fumio Oosawa
Aichi Institute of Technology, Toyota-Shi, Japan

I. INTRODUCTION

We presented, about 40 years ago, a theoretical framework for understanding the polymerization of protein molecules to helical polymers, based on the study on the G–F transformation of actin, a muscle protein [1, 2, 3]. (See Chapter 1.)

In pure water, all actin molecules are in the G-actin (globular actin monomer) state and at the physiological concentration of salts, almost all are in the F-actin (fibrous actin polymer) state. In the intermediate condition of salts, at low concentrations of actin no F-actin exists. At a certain critical concentration, F-actin begins to be formed. With increasing concentration of actin, the amount of F-actin increases, coexisting with G-actin, which is kept at the critical concentration. In this condition, each monomer undergoes a cycle between two states, G- and F-actin. The macroscopic G–F balance is maintained by microscopic cycling. The critical concentration decreases with increasing salt concentration. Thus, the G–F transformation has similar features to gas–liquid condensation or crystallization. Actually, the transformation consists of two processes, nucleation and growth. Usually, nucleation is rate limiting.

A theory of helical polymerization was proposed to explain such crystallization-like features. F-actin was assumed to be a helical polymer of G-actin, where each actin monomer (G-actin) is bound with four neighboring monomers through two kinds of bonds: one along the longitudinal strands and the other between two strands or along the genetic helix.

Soon after this proposal, electron micrographs showed that F-actin is a two-stranded helical polymer [4]. The bonding pattern predicted was finally confirmed by the structural analysis at atomic resolution. The 3-D structure of the actin molecule in a crystal of the complex with DNase I was determined by X-ray crystallography, and using this molecular structure, the structure of F-actin was built up to give the best fit to the X-ray diffraction data from an oriented gel of F-actin [5, 6]. Now, the amino acid residues taking parts in the two kinds of bonds in F-actin can be identified.

G-actin, when extracted from a muscle fiber into water, had bound ATP. During polymerization to F-actin, this ATP was hydrolyzed to ADP and inorganic phosphate [7]. The ADP was kept bound in F-Actin. In the process of depolymerization of F-actin to G-actin, rephosphorylation of ADP did not happen. After depolymerization, bound ADP was replaced with ATP in solution. Later it was found that G-actin having ADP instead of ATP also polymerizes to F-actin, although the rate of polymerization is much slower than G-actin having ATP [8]. Even G-actin without ATP or ADP can polymerize, if denaturation of this nucleotide-free G-actin is inhibited by a high concentration of sucrose [9]. The G–F transformation of actin was described by the scheme shown in Fig. 1.

Since then, the polymerization of various protein molecules was found to have similar features. There is a critical concentration for polymerization and the polymerization consists of nucleation and growth. Bacterial flagella are helically curved tubular polymers of flagellin molecules. Flagella are formed from purified flagellin molecules in solution, although spontaneous nucleation of flagella can hardly occur [10]. The polymerization of tubulin molecules to microtubules is

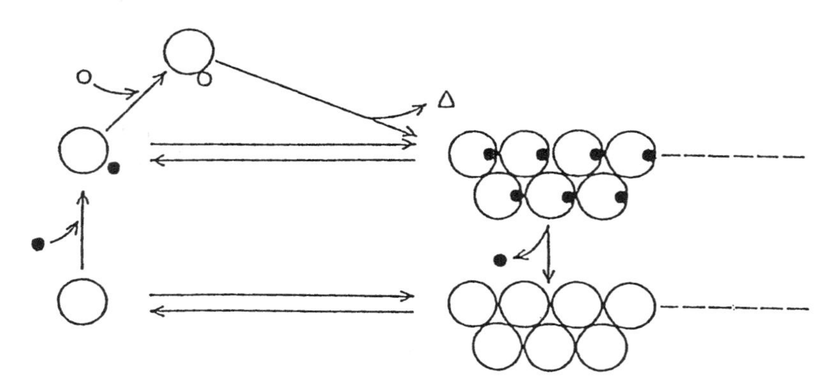

Figure 1 The G–F transformation of actin with ATP (open circle), ADP (closed circle), or no nucleotide; a classical scheme.

another example of tubular polymerization and in this case, the hydrolysis of GTP bound to monomers is associated with the polymerization process [11].

When we reached the idea of helical polymerization, we imagined immediately that the partial breaking or weakening of monomer–monomer bonds may produce a large conformational change of F-actin, keeping its filamentous continuity, as illustrated in Fig. 2 [12, 13]. Since then, we have been very interested in dynamic behaviors of helical polymers.

In this chapter, I briefly describe further development of the study on the polymerization of actin and other proteins and discuss regulation of the monomer–monomer bond and polymer conformation. Such regulation generates vari-

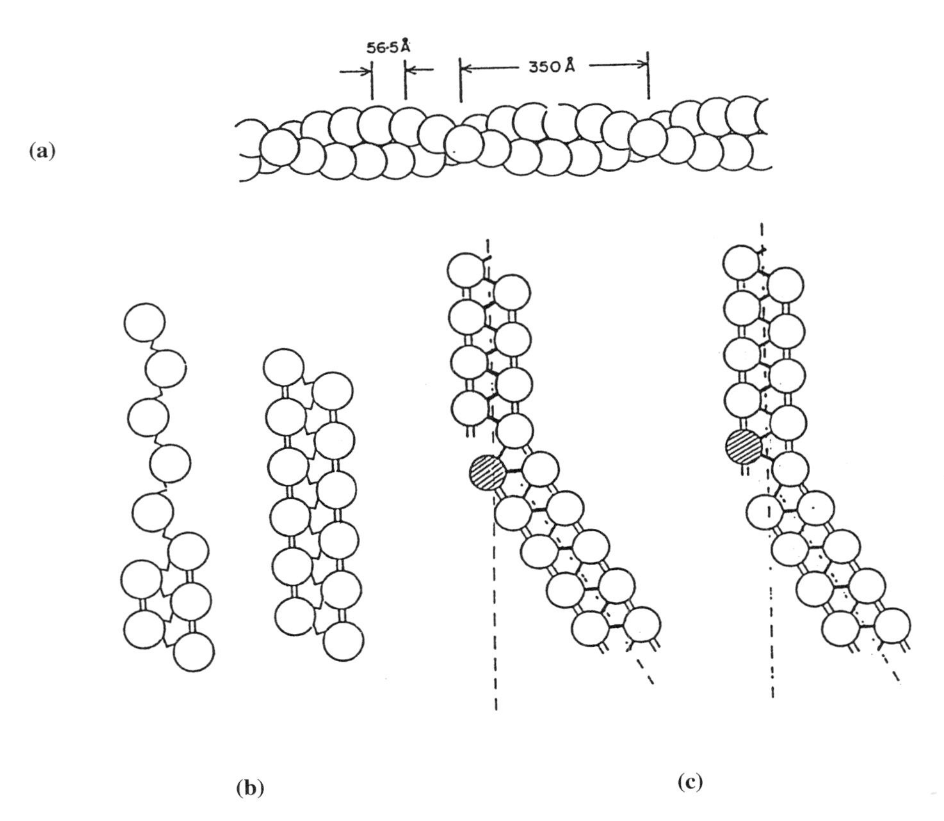

Figure 2 (a) A two-strand helical polymer structure of F-actin. (b) and (c) Illustration of hypothetical conformational changes of F-actin caused by weakening of the monomer–monomer bond possibly coupled with the ATP hydrolysis cycle.

ous dynamic functions of protein polymers. This chapter should help in understanding the design principle of protein polymers as a functional system.

II. MONOMER–MONOMER BOND FORMATION

In the case of polymerization of actin having bound ADP, an equilibrium state between G- and F-actins is established. Under this condition the concentration of G-actin coexisting with F-actin, the critical concentration, is related to the free energy of bonding of an actin monomer to the end of F-actin. It is estimated to be −6 to −13 kcal/mole [3]. (The bond free energy includes contributions of the two kinds of bonds.) The free energy depends on the concentration of salt ions, pH, and other environmental conditions. The polymerization is endothermic and the critical concentration decreases with rising temperature [14]. The addition of monomers to F-actin is driven by the entropy increase. Application of pressure promotes depolymerization [15]. The polymerization is associated with volume increase. The rearrangement of water molecules around monomers is probably involved in bonding. (Both too high temperature and too high pressure induce irreversible denaturation of actin molecules.)

F-actin is a semirigid filament. The flexural rigidity of F-actin was first estimated by quasielastic light scattering and later the thermal bending movement of F-actin was made directly visible by optical microscopy [16, 17]. The average amplitude of bending movement was about 40 nm in F-actin of the length of 1 μm. Extensibility was examined by mechanically stretching the F-actin, both ends of which were bound to thin glass needles [18, 19]. A force of a few hundred pN (piconewton) stretched F-actin of the length of 1 μm by a few Å. The comparison of flexibility and extensibility suggests that the elasticity of the monomer–monomer bond and the intramonomer structure has no special anisotropy.

The rigidity decreased with rising temperature [20]. This means that the short-range recovery force of deformed F-actin comes from the enthalpy decrease, in contrast with the entropy increase for the monomer–monomer bonding. The monomer–monomer interaction free energy is a complex function of the distance and angle between monomers. The bond formation must be a multistep process. Structural analysis indicates that the interaction occurs between certain areas of the monomer surface. Bond formation is not simply attributable to a single pair of specific sites of monomers.

The polymerization of flagellin to flagellum and that of tubulin to microtubule are also endothermic. The monomer–monomer bond free energy is of the same order as that in F-actin, although the dependency on the salt concentration is different in each.

F-actin has a structural polarity, as demonstrated by the manner of binding of myosin fragments. The growth rate is much larger at the end named B-end (barbed end) than at the other end, named P-end (pointed end) [21]. The rate of depolymerization was also different at two ends. However, in the case of polymerization that is not accompanied by ATP hydrolysis, the ratio of the rates of association and dissociation of actin monomers must be equal at two ends, giving the critical concentration of G-actin in equilibrium with F-actin.

The growth of bacterial flagella is unidirectional [22]. In a living cell, the growth occurs only at the distal end; the in vitro growth also occurs at the same end. The binding of a flagellin monomer to the end of a flagellum was found to be associated with a large conformational change of the monomer, from a partly unfolded conformation to a folded one [23]. Such a conformational change is catalyzed by interaction of the monomer with monomers at the distal end. This makes spontaneous nucleation difficult. Accordingly, formation of free flagella from flagellin molecules inside the cell is inhibited. Correlation of the directionality of growth with the conformational change required was examined using flagellin monomers in which a part of the unfolded chain was chemically removed. The growth was not exclusively unidirectional.

Also, in the case of actin the polymerization is probably associated with a (small) conformational change of each monomer, which may be related to directionality of the growth. The conformational difference between free monomers and monomers in the polymer is one of the regulatory factors of the manner of polymerization.

III. BOND REGULATION COUPLED WITH THE CHEMICAL REACTION

Nucleotides bound to G-actin change the polymerization rate and the monomer–polymer balance. They have special characteristics as regulators of the monomer–monomer bond strength.

In the state of G-actin, the bound nucleotide molecule, ATP or ADP, is quickly exchangeable with the free molecule in solution. An equilibrium is established between bound and free molecules. On the other hand, in F-actin the bound nucleotide molecule is not easily exchangeable. According to the 3-D structural analysis, G-actin appears to be composed of four domains. ATP or ADP is found in a deep cleft in the middle of the molecule. Probably, in G-actin the cleft has some flexibility to expose bound ATP or ADP to the solvent, whereas in F-actin such flexibility is limited. Therefore, whether each actin monomer in F-actin has ATP or ADP depends on the history of the monomer.

Previously ATP hydrolysis was thought to be directly coupled with the polymerization of G-ATP-actin. However, it was found later that ATP hydrolysis occurs after polymerization [24]. At the end of F-actin, a newly incorporated actin monomer keeps ATP for a while. The actin monomer in F-actin has an ability to hydrolyze ATP bound to it. Near the growing end of F-actin, there are three kinds of actin monomers: those having ATP, those having ADP and inorganic phosphate, and those having ADP alone. The fraction of three kinds of monomers near the ends changes depending on the rate of growth, as shown in Fig. 3 [25].

The hydrolysis of bound ATP weakens the monomer–monomer bond in F-actin. The bond free energy depends on whether the interacting monomers have ATP or ADP. The polymerization of G-actin without nucleotide can be performed in a sucrose solution. In this case, the critical concentration is very low; the monomer–monomer bond must be strong [26]. Various kinds of pairs of monomers having ATP, ADP and inorganic phosphate, ADP alone, or no nucleotide can be formed in F-actin. (Bound inorganic phosphate is exchangeable). The bond free energy and the bond elasticity are different in different pairs. Quantitative comparison has not yet been fully carried out.

In the case of polymerization of G-ATP-actin, the ratio of the rates of association and dissociation of monomers is different at two ends. The critical concentration of G-ATP-actin is defined by the condition that the sum of the growth rates at the two ends and the sum of the depolymerization rates are equal. At one of the two ends, the depolymerization may be faster than the growth and at the other end, the growth may be faster. Then a cycling of actin monomers takes place from one end of F-actin to the free monomer to the other end. This phenomenon was named ''tread-milling'' [27]. As a result, the polymer translates. Consumption of the free energy of ATP hydrolysis is required for such unidirectional translation of F-actin. The coupling of the polymerization with an irreversible chemical reaction produces a dynamic function of the polymer.

Tubulin monomers having GTP polymerize to microtubules having GDP. In this case also, the hydrolysis of GTP in microtubules occurs with a time lag after polymerization. The growing ends of microtubules have monomers keeping GTP. Then, those GTP are hydrolyzed. This hydrolysis makes the monomer–monomer bond weaker. At the end of the polymer, if monomers have GDP instead of GTP, they are more quickly dissociated from the end than those having

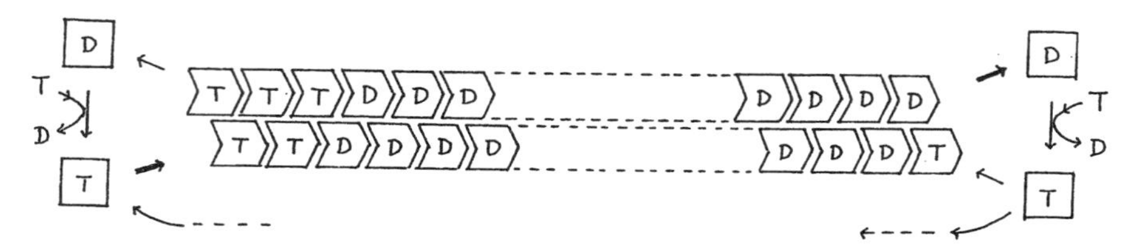

Figure 3 The monomers having ATP (T) or ADP (D) near two ends of F-actin during polymerization of G-actin having ATP; tread-milling may happen.

GTP. If all monomers at the ends have GDP, quick depolymerization occurs. When new monomers having GTP are added to the end, depolymerization is stopped and polymerization is made favorable. In an apparent balancing state of tubulin monomers having GTP and microtubules having mainly GDP, each polymer may repeat growth to long ones and depolymerization to short ones, as shown in Fig. 4 [28, 29]. This phenomenon was actually observed and named "dynamic instability." This is another dynamic function of the polymer coupling with an irreversible chemical reaction. In both cases of tread-milling and dynamic instability, the coupling is indirect.

Divalent cations bound to actin molecules also change the polymerization rate and the monomer–monomer bond strength. Usually, G-actin tightly binds a magnesium ion or a calcium ion nearby ATP or ADP. The polymerization of G-actin having a magnesium ion is much faster than that of G-actin having a calcium ion. These cations are quickly exchangeable in G-actin but nonexchangeable in F-actin. Therefore, F-actin having magnesium ions and that having calcium ions can be formed separately. The bond free energy and the bond elasticity must be different. Actually, a structural difference was reported between these two kinds of F-actin and their structural dynamics was discussed [30].

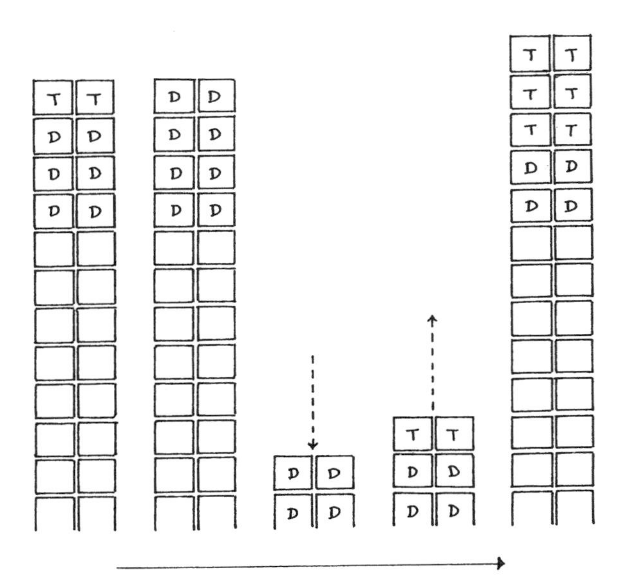

Figure 4 Dynamic instability of microtubules associated with the hydrolysis of bound GTP (T) to GDP (D) at the end; repetition of quick depolymerization and growth; a simplified model; microtubules are composed of more than 10 strands.

IV. MONOMER AND POLYMER CONFORMATIONS

As described in the previous section, the conformation of each actin monomer or each tubulin monomer in F-actin or microtubule is more or less different depending on the environmental condition and the species of bound nucleotides and bound cations. Does each monomer in the polymer assume the same conformation in the same environmental condition even when it has the same bound nucleotide and divalent cation? The monomer may take multiple conformations.

An interesting example was found in the case of bacterial flagella. Flagellin molecules form tubular polymers of a helical shape. The tube is composed of 11 longitudinal strands. In a straight tube, all monomers are in an equivalent position except those in the ends. However, in a helical tube, monomers in the innermost strand and those in the outermost strand are in different situations. They may take different conformations. To make the helical tube stable, it is likely that there are two free energy minima in the monomer conformation [3, 31]. Inner and outer strands in the helical tube are composed of monomers in two different conformations, respectively. Helices of different shapes, handedness, pitch, and amplitude are constructed by different combinations of these two kinds of strands, as shown in Fig. 5. All observed helical shapes of flagella can be explained by this idea. If all strands are composed of monomers in the same conformation, straight flagella are formed. In fact, two kinds of straight flagella exist. Flagellin monomers in these flagella have different bonding patterns [32, 33].

The free energies of the monomer in two conformations depend on the environmental condition. A polymorphic transition of flagella between different helices occurs with changes in the salt concentration, pH, or temperature [34]. The transition propagates throughout the whole length of each flagellum in a definite direction, as shown in Fig. 6. The transition of handedness can be caused also by a mechanical force [35]. Bacterial flagella give a typical case where polymorphism of polymers is generated from dimorphism of monomers.

Let us consider the case of actin. Actin molecules (from the same source) have the same amino acid sequence and take a specific 3-D structure in the crystal. Does each actin monomer in F-actin assume exactly the same conformation? Is it in the same state? The conformation of actin monomers in F-actin has been investigated by various methods. However, data obtained were only on an ensemble average of their conformation.

Recently, a new optical microscopic technique has been developed to investigate the conformation of a specified actin monomer in F-actin (36). An actin molecule was labeled with two kinds of small fluorescent molecules at different amino acid residues, and their distance was measured by the method of fluorescence energy transfer. The result has shown that the distance is not kept constant, but changes with time, as in Fig. 7. The change happens discontinuously and reversibly, indicating that actin monomer repeats transitions between two differ-

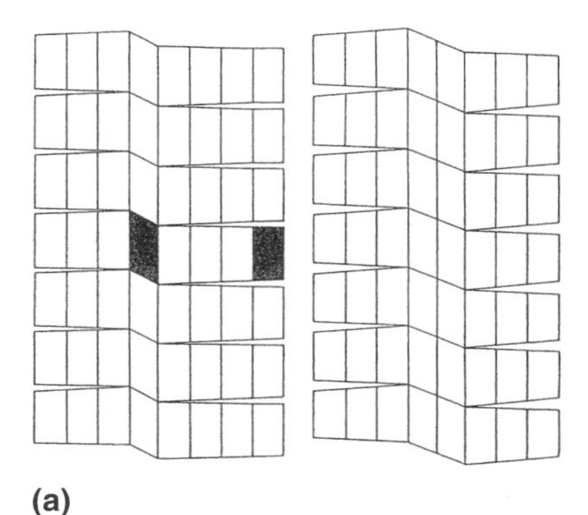

(a)

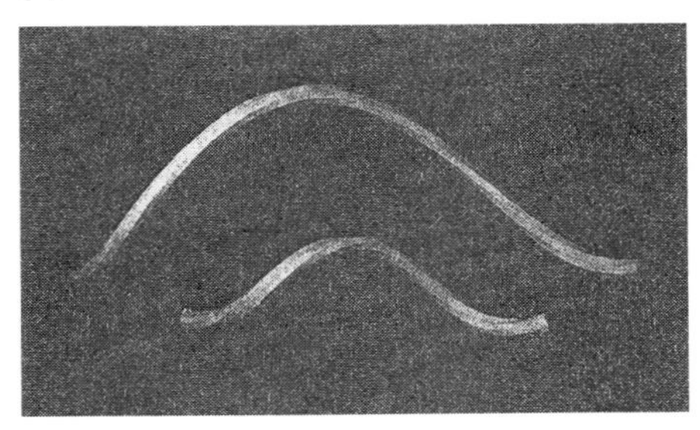

(b)

Figure 5 (a) A model showing the structure of helical tubes formed by combination of two kinds of strands, in which monomers are in two different conformations; different helical shapes are produced by changing the ratio of two kinds of strands. (b) Helical tubes made of the sheets of (a).

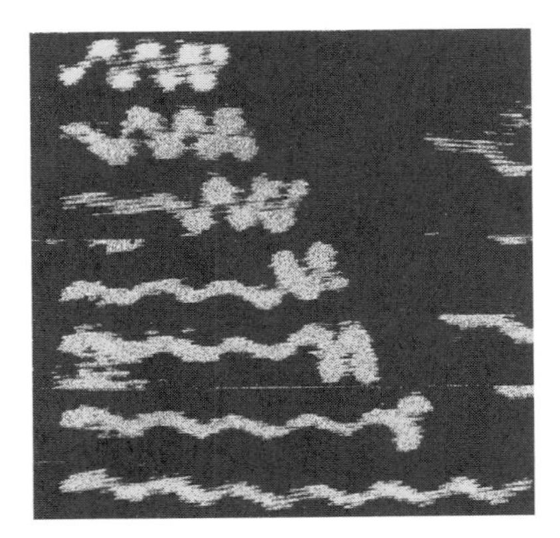

Figure 6 Unidirectional propagation of a polymorphic transition in a bacterial flagellum. (Observation by H. Hotani.)

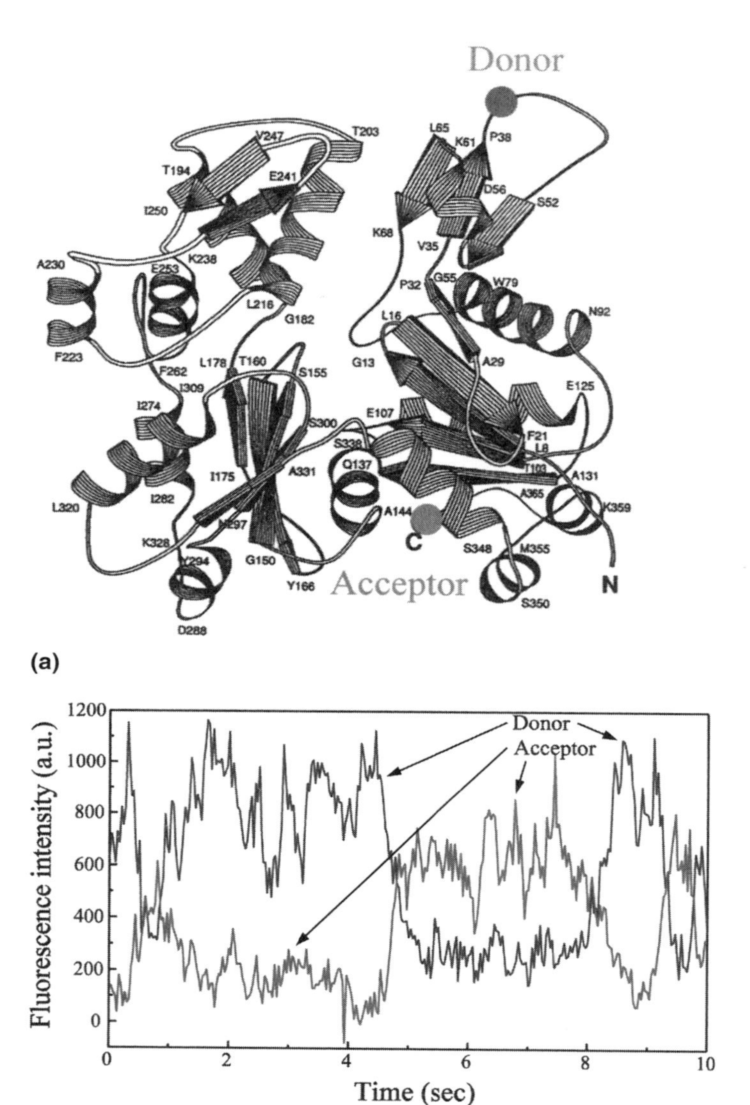

Figure 7 Transition of an actin monomer in F-actin between two conformations detected by fluorescence energy transfer: (a) donor and acceptor of fluorescence energy transfer in the 3-D structure of actin monomer; (b) record of fluorescence intensity from donor and acceptor; estimated distance is about 5 nm and 6 nm in two conformations, respectively. (Figures presented by Y. Ishii.)

ent conformations, compact and loose. The lifetime of each conformation is rather long. This experiment has given evidence that each monomer in F-actin does not always take the same conformation but possibly fluctuates among two or more conformations. It is very likely that the monomer conformation and the species of bound nucleotides or divalent cations have no one-to-one correspondence.

If monomers in F-actin make transitions between different conformations cooperatively, a remarkable change is expected to occur in the overall conformation of F-actin. Up to now, no such change has been observed. However, it was reported recently that the mobility of actin monomers in F-actin is remarkably changed by binding another protein molecule to its end [37]. The conformation of each monomer and the monomer–monomer bond seem to make a cooperative transition. The influence is spread over the whole F-actin.

Also, in the case of tubulin monomers in microtubules, the monomer conformation may not be uniquely determined by the species of bound nucleotides. Previous interpretation about dynamic instability seems to be too simple. A transition of conformation may propagate along the microtubule.

Block copolymers of bacterial flagellin molecules give another example. The helix of one block is influenced by that of the other block. Sometimes, flagellin molecules are forced to form a helix different from their own helix in the same environmental condition. Selection of the monomer conformation and the monomer–monomer bonding is cooperatively controlled in the polymer.

V. DYNAMICS OF POLYMERS INTERACTING WITH OTHER PROTEINS

Since single F-actin filaments were made directly visible under an optical microscope, various experiments were undertaken to observe dynamic behaviors of F-actin interacting with myosin and other actin-binding proteins.

When soluble myosin fragments and ATP were added to a solution of F-actin, F-actin showed large and fast bending and twisting movements, as shown in Fig. 8. Without ATP such movements did not occur [17]. The free energy of ATP hydrolysis was transferred to the bending and twisting movements. The effective temperature of these degrees of freedom was estimated to be three or four times higher than room temperature. The free energy was injected into the monomer–monomer bonds. On the other hand, when F-actin was put on myosin fragments fixed on a plate, the addition of ATP induced sliding movements of F-actin on myosin fragments, as illustrated in Fig. 9 [38].

Previously, concerning the mechanism of sliding movements between F-actin and myosin, we proposed the idea of loose coupling between the chemical reaction and the mechanical event. Namely, we assumed that the ATP hydrolysis and the sliding had no definite one-to-one correspondence [39, 40, 41]. Recent

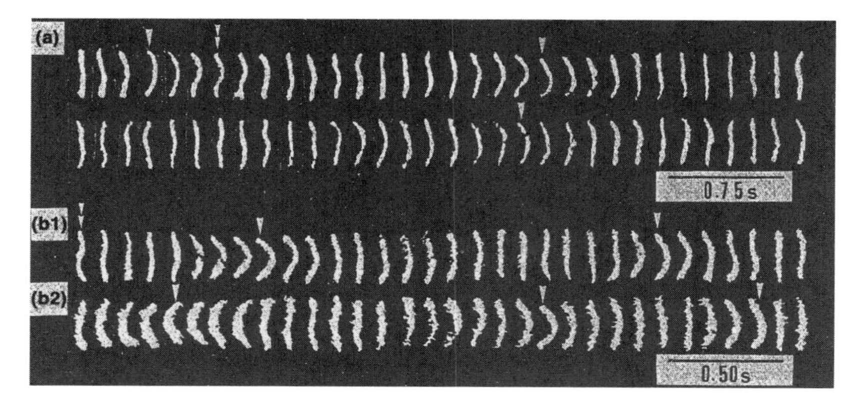

Figure 8 Bending movements of a thin filament (a complex of F-actin with tropomyosin and troponin) interacting with myosin fragments in the presence of ATP, where F-actin was labeled with fluorescent dye molecules to make it visible under a fluorescence microscope: (a) inactive state in the absence of calcium ions; (b) active state in the presence of calcium ions undergoing fast and large bending movements. The sequential micrographs were taken on 10 μm filaments at interval of 0.15 s (a) and 0.10 s (b1,b2). (From Ref. 17.)

experiments have given evidence of such loose coupling. Long-distance multistep sliding of myosin on fixed F-actin or of F-actin on fixed myosin was produced during or after hydrolysis of one ATP molecule on myosin [42]. The sliding speed and distance are widely variable depending on the external force applied against sliding. There is often a time lag between start of sliding and ending of the hydrolysis reaction [43]. Therefore, it is very likely that the free energy of ATP hydrolysis is transferred once and stored somewhere in myosin and/or F-actin and gradually released for sliding [44].

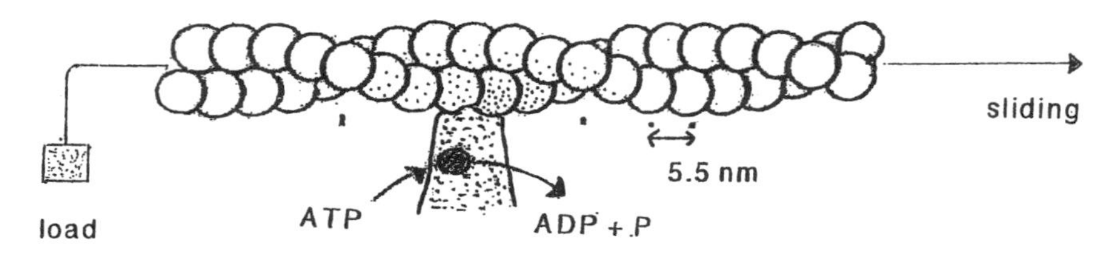

Figure 9 Illustration of sliding movements of F-actin on fixed myosin coupled with the ATP hydrolysis; the conformation of actin monomers during and after interaction with myosin may be changing.

What kind of conformational change is induced in F-actin by interaction with myosin hydrolyzing ATP? How is the influence of myosin spread and maintained in F-actin? The helical polymer structure of F-actin may be useful to realize storage and gradual release of high free energy. The experiment described in the previous section showed that the monomer conformation in F-actin makes transitional changes and the lifetime of each conformation is rather long. Is a high free energy state stably formed in F-actin?

Combination of kinesin or dynein molecules and microtubules also generates sliding movements upon addition of ATP. How is the polymer structure of the microtubule influenced by interaction with kinesin or dynein hydrolyzing ATP? Recent experiments have suggested a long-range influence of binding of the kinesin molecule hydrolyzing ATP on a microtubule [45].

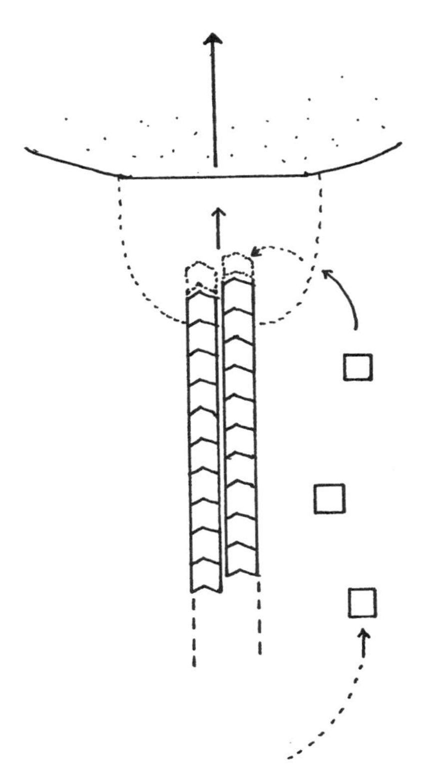

Figure 10 Illustration of translational movements produced by directional growth of F-actin; a simplified model, actually many kinds of proteins are involved to generate fast translation.

A large number of actin-binding proteins have been discovered from different cells. Their functions have been identified based on the scheme of helical polymerization. Some proteins bind to G-actin, inhibit the polymerization, and shift the G–F balance. Some bind to the end of F-actin and inhibit its growth. Some of them make nuclei for polymerization and break F-actin into fragments. Some proteins bind to the side of F-actin and change its interaction with other proteins. There are also proteins that make bundles or crosslinks of F-actin and control the 3-D structure of the F-actin network.

In cooperation with many kinds of actin-binding proteins, the G–F transformation of actin generates translational movement of a bacterial cell in a host cell, as in Fig. 10 [46]. Anchoring and nucleation proteins, depolymerizing proteins, and crosslinking proteins work to make possible a fast cycle of actin molecules from one end of F-actin to the other end, that is, tread-milling. This system has been artificially reconstructed in vitro [47].

The cytoplasmic streaming in a plant cell is due to interaction of F-actin bundles under the cell membrane with myosin in the streaming cytoplasm. The

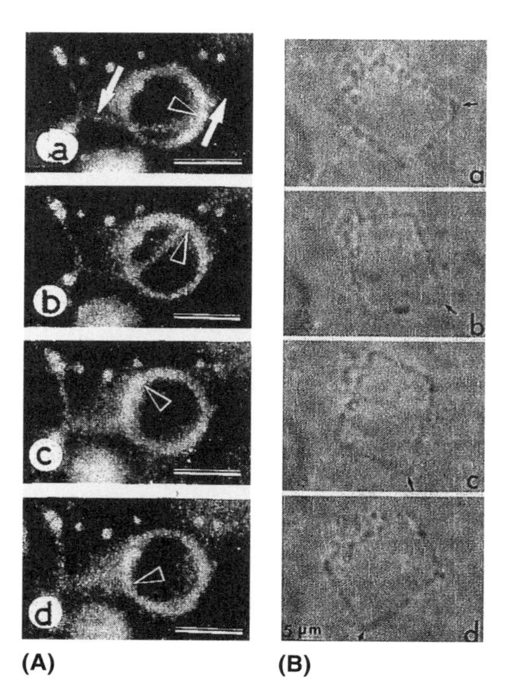

Figure 11 Active movements of F-actin bundles isolated from a plant cell: (A) rotational movements of a ring (presented by S. H. Fujime); (B) movements of corners of a polygon, where the polygon itself does not rotate (presented by K. Kuroda).

bundles, when squeezed out of the cell, often form circular rings. The ring shows active rotation interacting with myosin fixed on a plate [48]. Sometimes, polygons are formed and the positions of corners of the polygon move along the polygon, as shown in Fig. 11 [49].

Besides sliding movements on myosin, we can suppose various kinds of active movements of F-actin: translation, rotation, undulation, and so on. It is valuable to investigate if these movements occur in living cells.

Similar questions must be asked in the case of microtubules. For example, active undulation of a bundle of microtubules is observed in some kinds of cells. Dynamics of microtubules, particularly depolymerization at the ends, has been supposed to be a mechanism of chromosome movements for cell division [50].

VI. DESIGN OF THE MONOMER STRUCTURE AND REGULATION OF INTERACTION

Regulation of the monomer conformation and the monomer–monomer bonding is a key process for dynamic behavior of protein polymers. How are protein monomers designed? Recent research has revealed that most globular protein molecules are constructed from several structural units called modules, like a building of bricks [51, 52]. Each module has a compact 3-D structure which is formed by the folding of a short polypeptide chain, usually a sequence of 10–30 amino acid residues. The sequence is encoded in an exon in the gene DNA and the boundaries of a module correspond to the position of introns in DNA. It is likely that in the process of evolution, new protein molecules were produced by various combinations of modules, small ancient protein molecules.

A method to define modules in the 3-D structure of a protein molecule based on its distance map (a map giving distances between amino acid residues, alpha carbons) has been established. By application of this method to actin, its 3-D structure is expressed as an assembly of a number of modules. It appears to be composed of a central core and several modules attached to the core surface, as shown in Fig. 12 [53]. The core itself is composed of many modules. However, in this figure, only modules involved in the monomer–monomer interaction to form F-actin are specified.

The 3-D structure of the core of actin molecule is very similar to that of hsp (heat-shock-protein)-70, another protein molecule, suggesting that these two proteins came from the same ancestor. Addition of new modules to the core gives a new function to the molecule, the ability of helical polymerization. Added modules have to work together.

The design principle is as follows. Protein molecules are constructed by additive assembly of modules. Each module has its own function. By combination of modules, a new function is generated. Regulation of the function greatly depends on communication and cooperation between modules.

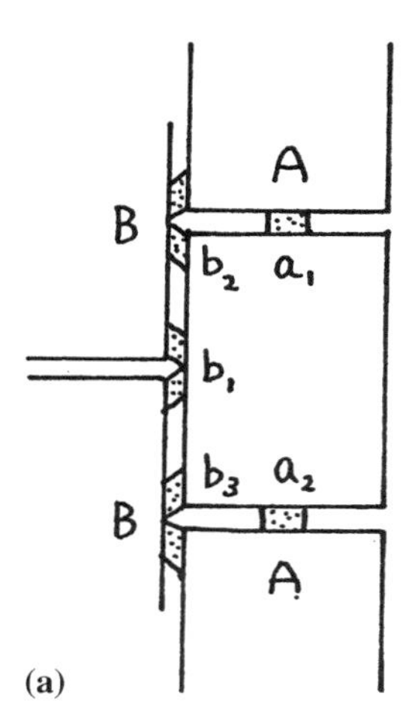

(a)

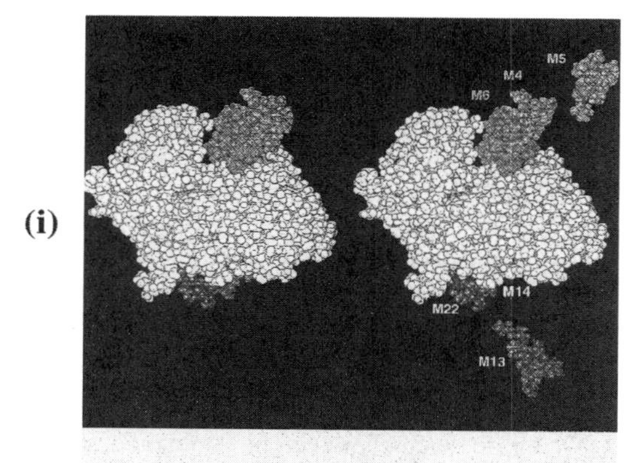

(i)

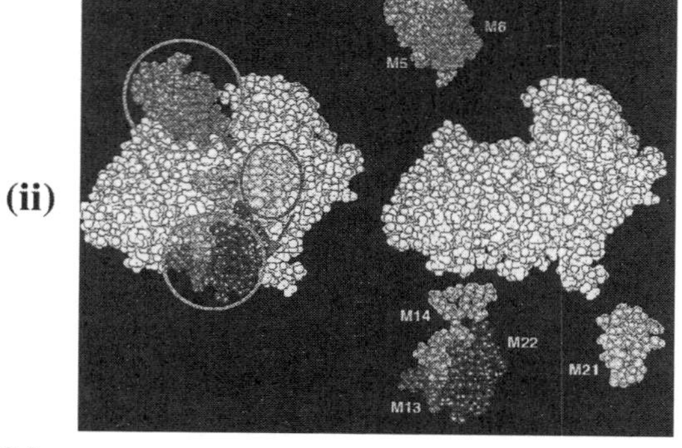

(ii)

(b)

In the case of actin, the monomer–monomer bond strength depends on ATP or ADP bound in the cleft. The conformation of modules in the central core has influences on the modules for the monomer–monomer bonding. The monomer–monomer bonding changes the conformation of modules in the core.

The expression of the 3-D structure of actin as an assembly of modules is useful for understanding the interaction of actin with many different kinds of actin-binding proteins. Probably, the whole surface of the actin molecule is utilized for such interaction. Overlapping or interference of interaction of actin with different proteins may be well expressed in terms of modules.

Including the monomer–monomer interaction in F-actin, the interaction of actin with many kinds of proteins has remarkable characteristics. In most cases, the mode of interaction is not unique but variable with some freedom. For example, the binding of myosin with F-actin has at least two modes, strong binding and weak binding; it probably has even more different modes. The mode of binding of tropomyosin to F-actin is also variable; it is changed by another protein, troponin, in the presence and absence of calcium ions. Such variability seems to be important for F-actin to exhibit its function. Its structural basis is not yet fully understood.

The structural analyses of tubulin monomers in microtubules or flagellin monomers in flagella at atomic resolution are now in progress. The design of these monomers will be made clear very soon.

VII. CONCLUDING REMARKS

As described in previous sections, protein polymers such as F-actin, microtubules, bacterial flagella, and the like show various dynamic behaviors. With development of new experimental techniques, I expect we will be able to observe further many different kinds of dynamic behaviors of protein polymers. Protein molecules and their polymers are a complex system having a large number of

---

Figure 12 (a) The pattern of monomer–monomer bonding in F-actin. Each monomer is bound with four neighboring monomers by two kinds of bonds, A and B; a1: M5; a2: M13; b1: M21; b2: M4, 5, 6; b3: M13, 14, 22. (b) The 3-D structure of actin monomer built up from a number of modules: bond A, the contact of neighboring monomers in each longitudinal strand occurs between M5 of one monomer and M13 of the other monomer; (ii) bond B, the contact of monomers of different longitudinal strands occurs between M12 of one monomer and M4, 5, and 6 of the other monomer and also M13, 14, and 22 of the third monomer. (Figures presented by M. Go.)

degrees of freedom. The energy transfer among different degrees of freedom and the energy exchange with the environment are involved in their dynamics.

To end this chapter, I emphasize that in addition to chemistry and structural analysis, we have to accumulate information about energetics and thermodynamics or statistical mechanics of protein molecules and their polymers. We have to know what kind of state protein molecules and their polymers assume to exhibit dynamic behaviors, not only in structural terms but also in thermodynamic and statistical mechanical terms. This is important for finding a clue to designing functional systems similar to protein polymers.

REFERENCES

1. F Oosawa, S Asakura, K Hotta, N Imai, T. Ooi. J Polym Sci 37: 323, 1959.
2. F Oosawa, M Kasai. J Mol Biol 4: 10, 1962.
3. F Oosawa, S Asakura. Thermodynamics of the Polymerization of Protein. Academic Press, N.Y., 1975.
4. J Hanson, J Lowy. J Mol Biol 6: 46, 1963.
5. W Kabsch, H. Mannherz, D Suck, P Pai, K Holmes. Nature 347: 37, 1990.
6. K Holmes, D Popp, W Gebhard, W Kabsch. Nature 347: 44, 1990.
7. F Straub, G Feuer. Biochim Biophys Acta 4: 455, 1950.
8. S Higashi, F Oosawa. J Mol Biol 12: 843, 1965.
9. M Kasai, E Nakano, F Oosawa: Biochim Biophys Acta 94: 494, 1965.
10. S Asakura, G Eguchi, T Iino. J Mol Biol 10: 42, 1964.
11. G Borisy, J Olmsted, R Klugman. Proc Natl Acad Sci 69: 2890, 1972.
12. F Oosawa, S Asakura, T Ooi. Prog Theor Phys Suppl 17: 14, 1960.
13. S Asakura, M Taniguchi, F Oosawa. J Mol Biol 7: 55, 1963.
14. S Asakura, M Kasai, F Oosawa. J Polym Sci 44: 35, 1960.
15. T Ikkai, T Ooi. Biochem 5: 1551, 1966.
16. S Fujime. J Phys Soc Jpn 29: 751, 1970.
17. T Yanagida, M Nakase, N Nishiyama, F Oosawa. Nature 307: 58, 1984.
18. A Kishino, T Yanagida. Nature 334: 74, 1988.
19. H Kojima, A Ishijima, T Yanagida. Proc Natl Acad Sci 91: 12962, 1994.
20. T Takebayashi, Y Morita, F Oosawa. Biochim Biophys Acta 492: 35, 1977.
21. H Kondo, S Ishiwata. J Biochem 79: 159, 1976.
22. S Asakura, G Eguchi, T Iino. J Mol Biol 35: 227, 1968.
23. Y Uratani, S Asakura, K Imahori. J Mol Biol 67: 85, 1972.
24. M Carlier, D Pantaloni, E Korn. J Biol Chem 259: 9983, 1984.
25. M Carlier. Adv Biophys 26: 51, 1990.
26. EM De La Cruz, T Pollard. Presented at the Actin Meeting, Maui, 1997.
27. A Wegner. J Mol Biol 108: 139, 1976.
28. T Mitchison, M Kirschner. Nature 312: 232, 1984.
29. T Horio, H Hotani. Nature 321: 605, 1986.

30. A Orlova, E Prochniewicz, E Egelman. J Mol Biol 245: 598, 1995.
31. S Asakura. Adv Biophys 1: 99, 1970.
32. K Namba, F Vonderviszt. Quart Rev Biophys 30: 1, 1997.
33. I Yamashita, K Hasegawa, H Suzuki, F Vonderviszt, Y Mimori-Kiyosue, K Namba. Nature Str Biol 5: 125, 1998.
34. R Kamiya, S Asakura. J Mol Biol 108: 513, 1976.
35. H Hotani. J Mol Biol 156: 791, 1982.
36. H Yokota, Y Ishii, T Wazawa, T Funatsu, T Yanagida. Biophys J 74: A46, 1998.
37. E Prochiniewicz, Q Zhang, P Janmey, D Thomas. J Mol Biol 260: 756, 1996.
38. S Kron, J Spudich. Proc Natl Acad Sci 83: 6272, 1986.
39. T Yanagida, T Arata, F Oosawa. Nature 316: 366, 1985.
40. F Oosawa, S Hayashi. Adv Biophys 22: 151, 1986.
41. R Vale, F Oosawa. Adv Biophys 26: 97, 1990.
42. K Kitamura, M Tokunaga, A Iwane, T Yanagida. Nature 397: 129, 1999.
43. A Ishijima, H Kojima, T Funatsu, M Tokunaga, H Higuchi, H Tanaka, T Yanagida. Cell 92: 161, 1998.
44. F Oosawa. Genes to Cells 5, 2000.
45. E Muto, T Miyamoto, T Funatsu, Y Harada, A Iwane, A Ishijima, T Yanagida. To be submitted.
46. L Tilney, D DeRosier, M Tilney. J Cell Biol 118: 71, 1992.
47. T Loisel, R Boujemaa, D Pantaloni, M Carlier. Nature 401: 613, 1999.
48. S Higashi-Fujime. J Cell Biol 87: 569, 1980.
49. K Kuroda. Int Rev Cytol 121: 267, 1990.
50. S Inoue, E Salmon. Mol Biol Cell 6: 1619, 1995.
51. M Go. Nature 291: 90, 1981.
52. K Takahashi, T Noguti, H Hojo, K Yamauchi, M Kinoshita, S Aimoto, T Ohkubo, M Go. Protein Eng 12: 673, 1999.
53. Y Niimura, M Go. To be submitted.

Index

Abiotic supramolecular chemistry, 362, 639
Actin, 4, 7, 26, 37, 38, 39, 43, 44, 643, 648–659
Adaptive materials, 638, 639
Adsorption kinetics, 516, 517, 534–536
Aging, 548
Aggregates
  cyclic/disklike, 72, 162, 170, 171, 617
  linear, 72, 74
  spherelike, 72
Aggregation number, 72, 74, 75, 77, 101
Alignment layers, 588–591
Amphiphilic molecules, 95, 96, 99
  bolaform, 95, 534–545
  dipolar, 95
Amphiphilic polymers, 93, 97, 98, 263–321
Amylose containing block copolymers, 317
Anchor groups, 438, 568, 571
Antracenyl residues, 157, 624
Archaea, 178–180, 183, 185, 189, 195, 209
Argon fluoride excimer laser, 203, 205
Aromatic rod-coil copolyamide, 299–301
Associating polymers, 138, 169
Atomic force microscopy (AFM), 298, 454, 498, 500, 517, 540, 546, 576
ATP hydrolysis, 44, 644, 647, 654, 655
Axial ratio, 28, 63
Azobenzene chromophore, 477, 482, 486

Bacteria, 177, 178, 181–185, 190, 192, 195, 209
Barbituric acid, 624, 625
Bilayers, 5, 9, 10, 22, 171, 172, 192, 193, 198, 202, 406, 474
Biocompatible surfaces, 177, 608–611
Biosensor, 198, 205, 209
Blends, 168, 169, 245
Bragg peak, 512–514, 542–545
Branched polymers, 626, 629, 634
Bridged dendrimers, 411
Brushes, 106, 139, 567

Calixarenes, 165
Catenanes, 13, 323
Cell adhesion, 608, 609, 611
Cell division, 657
Cell growth, 184
Charge density, 515, 528, 532, 534, 541, 544, 548

Charge reversal, 507, 515, 516, 527, 537, 547
Charge transfer, 15, 17
Chemisorption, 506–507, 536, 547, 567
Chirality, 620, 621, 622, 639
Chromophore, 521, 531, 533, 548
Clay, 532, 540
Closed surface crystal, 184, 185
Cohesive energy, 19
Colloid, 507, 512, 540, 544, 552
Columnar (hexagonal) phases, 3, 11, 36, 70, 78, 162, 617, 619
Comb copolymers, 49, 93
Combinatorial chemistry, 638
Comb-like nylons, 264–286
Comb-like poly(α-peptide)s, 264–286
Comb-like poly(β-peptide)s, 264–267
Compatibilizer, 219, 246
Computer simulations, 82, 599
Conductive polymers, 522
Conductivity, 489, 496, 498
Contact angle, 512, 513, 530, 542
Container molecules, 383
Cooperative effects, 14, 33, 40
Coordination polymers, 627, 633
Copolymers, 629, 631
  monolayers, 595–597
Core-shell architecture, 413
Corona
  charged, 107
  corona-core interface, 104
  polyelectrolyte, 107
  stretching, 108
  swelling, 107
Coulombic bonds, 4
Counterion, 513, 516, 523, 524, 527, 546, 547, 552
Critical atomic design parameters (CADPs), 360
Critical concentration, 28, 35, 37, 45, 63, 643
Critical micelle (aggregate) concentration, 22, 73, 101
Crosslinking (*see* Networks)
Crown ethers, 3, 6, 339
Crystal engineering, 481
Crystal growth, 195, 196
Crystallinity in rod-coil copolymers, 263–321
Cyanuric acid, 626
Cycling (actin monomers), 643, 648
Cyclodextrins, 337
Cyclophanes, 324
Cylindrical assemblies, 192
Cytoplasmic streaming, 656

de Gennes Dense Packing, 374
Deflection length, 29, 69
Degree of polymerization, 31
Dendrimer synthesis, 369
Dendrimers, 10, 27, 167, 365, 628, 630
Dendritic architectural copolymers, 388
Dendritic box, 386
Dendritic compositional copolymers, 388
Dendritic nanocomposites, 385
Dendritic nanodevices, 421
Dendritic rods/cylinders, 393
Dendritic state, 367, 420
Dendromesogens, 397
Depolymerization, 623
Deuteration, 512–514, 542
Diaminopyridine, 158, 617, 618
Diblock copolymers, 21, 23, 50, 93, 94, 97, 100, 110, 215, 225, 287–299
Diffusion, 517, 525, 546, 548, 549
Directional growth, 647, 651
Disclination, 185–189
Discotic assemblies, 6, 11, 162
Discotic molecules, 3, 5, 31, 36
Dislocation, 185–189
Dispersive forces, 17
Dissipative structures, 473, 492, 495, 498
Divergent/convergent synthesis, 369
DNA, 50, 487–492, 498, 507, 527, 539, 541
DNA-mimetics, 490
Drying, 511, 516, 519, 534–536, 542, 546
Dye, 507, 515, 533, 537, 543, 545, 547, 548, 552
Dynamic association-dissociation, 7, 37, 40, 45, 184, 616, 631, 632, 633, 636

Dynamic behavior, 645, 648, 653, 657, 659
Dynamic combinatorial chemistry, 638

Electrochemical reactions, 54, 485
Electron micrographs, 179, 180, 187, 188, 191, 196, 197, 200, 202, 206, 208, 221, 232, 235, 237, 241, 245, 251, 253, 295, 296, 305, 307, 311–313, 396, 459, 622, 628
Electrostatic interactions, 508, 510, 519, 520, 524–526
Ellipsometry, 512, 516, 545
Endoreceptor, 4, 12, 26, 383
Engineered assemblies, 6, 51–53
Engineered growth, 52
Excluded volume, 27, 33, 62, 75
Exoreceptor, 4, 26, 383

Fabrication of submicron structures, 205
Fibrin, 4, 7, 26, 45
Flagella/flagellin, 644, 646, 650, 651, 653, 659
Flexibility, 5, 31, 36, 64, 69, 70, 75
Flexural rigidity, 37, 645
FT-IR spectroscopy, 585, 586
Functional brushes, 567, 572, 591–593, 608–611
Functional materials, 487
Functional SAMs, 443–445, 453–463
Functional systems, 643–659
Functionalized surfaces, 205, 209, 565–611

G–F transformation, 43, 643, 646, 656
Glutamate dehydrogenase, 32
Glyco(dendrimers), 403
Grafts, 95, 140, 567–591
Growth mechanism, 31, 636
Growth rate, 646
GTP hydrolysis, 37, 648

Hairy polymers, 624, 625
Hard interaction, 28
H-bonded complexes (*See also* Supermolecules), 34, 35, 150–155
H-bonded mainchain polymers, 2, 3, 33, 51, 156–158, 170
H-bonded side chain polymers, 34, 160, 161
H-bonds, 16, 150, 151, 491, 508–511, 617
Helical growth/polymerization, 41, 643, 645
Helical rigidity, 263–299
Helical textures, 622
Helical tubes, 650, 651
Hemoglobin, 36, 81, 82
Hierarchical self assembly, 473, 495, 617, 619, 635, 637
Hydrophilic groups, 21, 95
Hydrophobic effect, 18
Hydrophobic tail, 21, 95
Hypercooperativity, 402
Hypervalency, 402

Immobilized bilayers, 474, 476, 481
Immunoassay, 207
Inorganic nanotubes, 628
Inorganic polymers, 635
Interdigitation, 25, 49
Interpenetration, 508, 513, 519, 540, 543
Intrasupramolecular, 634
Intussusceptive growth, 185, 186
Inverted transitions, 18, 20, 21
Ion exchange, 523–526, 528, 532
Ionic strength, 507, 513, 515–517, 519, 520, 524–527, 532, 534, 535
Isoelectric point, 510, 520, 539, 540, 542
Isotropic solutions, 3, 5, 33, 38, 45, 75, 163

Janus-type units, 624, 627, 634, 635

Ladder polymers, 626, 627
Lamellae, 95, 110
Langmuir–Blodgett (LB) films, 52, 73, 195, 483, 484, 505, 506, 536, 541, 544, 548, 551
Lattice growth, 185
Lattice orientation, 197
Layer properties, 594, 611

Layer thickness, 572, 574, 600, 603
Light emitting diode LED, 505, 550–552
Lipid film, 192, 194–196, 198, 199, 201
Lipid membrane, 198, 199, 201
Liquid crystalline block copolymers, 299–301, 314–317
Liquid crystalline polypeptides, 267–280
Liquid crystals
  molecular, 28, 30, 62
  supramolecular, 29–31, 34–37, 61, 76, 154, 619
Living polymerization, 223
Lyotropic phases, 3, 5, 11, 27, 62, 158, 624

Macrosegregation, 19
Mainchain polymers, 2, 156, 629, 630, 631
Mansfield–Tomalia–Rakesh equation, 414
Mechanical bonds, 13, 323
Membranes of block copolymers, 292
Mesogels, 121
Mesogenic side chains, 8, 13, 34
Mesoscopic structures, 89, 474, 492, 494, 625
Micelles
  crewcut, 106, 109
  cylindrical, 4, 9, 22, 101
  cylindrical core, 110
  diblock copolymers, 23, 104, 105, 217
  linear assembly, 8, 36, 37, 38
  rod-coil, copolymers, 111
  spherical, 22, 24, 101, 386
  starlike, 106, 109
  triblock copolymers, 118
Micromachining, 198
Microphase separation, 220, 226, 241, 244, 247
Microsegregation, 20, 125
Microtubules, 4, 7, 37, 38, 39, 40, 644, 649, 653, 657, 659
Mixed SAMs, 443–445, 446–451
Molar mass, 515, 516, 519, 525, 532, 534, 535
Molecular electronics, 177, 205, 209
Molecular imprinting, 26
Molecular orientation, 479
Molecular recognition, 2, 616
Monodendrons, 11, 399–404
Monolayer, 192–194, 195, 196, 198, 199, 201–203, 206
Morphology, 219, 291–299, 304–317
Multiblock copolymers, 126
Multistage open association, 31
Myosin, 653, 654

Nanobiological devices, 423
Nanoelectronics, 198
Nanooptics, 198
Nanoparticle, 207, 208, 507, 537, 539, 540, 544, 550, 552
Nanoscale amphiphiles, 388
Nanoscale tectons, 404
Nanotechnology, 177, 203, 639
Nematic phases, 3, 25, 35, 67, 69, 78, 80, 87, 155
Networks, 51, 158–160, 165, 168, 544, 546, 548, 626, 627, 630
Neutron reflectometry, 512–515, 542, 546, 597, 598
Noncentrosymmetry, 236, 249, 506–508, 547
Nonhelical rigidity, 299–317
Nonlinear optics (NLO), 506, 531, 545, 547, 548
Normal transition, 18
Nucleation and growth, 643

Order parameter, 28, 85
Organic/inorganic assemblies, 634
Orientation, 64, 229

Patterned ligands, 461–463
Patterned SAMs
  combination with chemical reactions, 451–452, 455, 458, 461
  microcontact printing, 451
Patterned thin films of polymers, 455–458
Patterning, 203, 205, 451–463, 532, 546, 550, 553, 637

Persistence length, 9, 11, 29, 35, 36, 37, 68, 78, 87, 88, 115, 636
pH effect, 507, 510, 515, 516, 520, 524, 525, 527, 528, 530, 534, 537, 539, 548, 549
Phase diagrams, 5, 62, 67, 78, 80, 228
Phase transition, 479, 482
Photoconductivity, 496
Photocurrent generation, 489
Photolithography, 571, 604–608
Photophysical properties, 301–303
Phthalydrazide derivatives, 617, 618
Physisorption, 567, 575
π–π interaction
  repulsive, 15
  stacking, 17
Planar assemblies, 26, 47
Planar growth, 46
Polarity (structural), 646
Polarizing microscopy, 269, 273, 276
Poly(*p*-benzamide), 30, 37, 38
Poly(4-vinylpyridine), 583
Poly(α-alkyl-β-L-aspartate)s, 274–280
Poly(α-peptide)s, 264–286
Poly(β-peptide)s, 264–267
Poly(γ-alkyl-α-L-glutamate)s, 267–274
Poly(L-lysine) complexes, 283–286
Poly(propylene imine) dendrimers, 406
Polycarbodiimide block copolymers, 299
Polycatenanes, 323
Polydispersity (*see* Size distribution)
Polyelectrolyte assemblies, 50–53, 280–286, 505–553
Polyelectrolyte brushes, 582
Polyelectrolyte surfactant assemblies, 50, 51, 135, 281, 475–487
Polyion complex, 482–487, 492, 495
Polyisocyanate block copolymers, 291–297
Polyisocyanide block copolymers, 297–299
Polypeptide block copolymers, 287–291
Polyrotaxanes, 12, 323
Polysaccharide, 532, 538
Polysoaps, 94, 97, 126, 128, 135
Proteins, 507, 510, 520, 527, 531, 532, 538, 539, 541, 542, 546
Protein structure, 181, 184
Pseudopolyrotaxanes, 3, 4, 336
Pseudorotaxanes, 336

Radiolabel, 525, 527, 549
Rate of polymerization/depolymerization, 644, 647
Resonance energy transfer, 516, 543
Reversibility, 2, 61, 159, 525–527, 548, 616
Rigid-flexible copolymers, 263–321
Rod-coil copolymers, 100, 263–321
Rotaxanes, 324
Roughness, 512, 513, 517, 527, 532, 534, 535, 542, 543, 546

Scrambling of components, 636
Second harmonic generation SHG, 526, 531, 537, 545, 547, 552
Segmented polymers, 28, 35
Self assembled monolayers (SAMs), 13, 47, 435–464, 565
  chemical transformations, 445–463
  comparison with polymers, 436–437
  computational studies, 440, 442–443
  definition, 436
  formation, mechanism of, 439–440
  structures, 440–443
  synthesis, 439
  systems, 438–439
Self assembly, 2–5, 26, 62, 177, 215, 263, 363, 435, 615
Self-healing, 508, 532, 538
Semifluid membrane, 195, 201, 203
Semirotaxanes, 346
Shape effects, 13, 24, 26, 636
Shape recognition, 25, 26
Sidechain polymers, 34, 160, 632
Signal processing, 177, 206, 209
Size distribution, 22, 32, 33, 36, 42, 44, 64, 74, 76, 77, 79, 80, 629

S-layers, 177–209
  assembly in suspension, 190, 193
  assembly on lipids, 194, 199
  assembly on liposomes, 202
  assembly on solids, 198
  protein structure, 181, 184
  ultrastructure, 177, 190
Sliding movements, 654, 655
Smart materials, 637, 639
Smectic phases, 24, 66, 67, 155
Soft interaction, 18, 28, 34
Solvophobic bonds, 13
Spectral switching, 480
Star copolymer, 242
Sterically induced stoichiometry (SIS), 374, 414
Stoichiometric complexes, 280–286
Strong segregation limit, 219, 227
Structural polarity, 646
Superlattice, 207, 208
Supermolecules, 4, 35, 615
Supramolecular categories, 360–362
Supramolecular chemistry, 1, 615
Supramolecular polymerization, 1, 2, 6, 8
Surface coverage, 580
Surface monolayer, 483, 487, 491
Surface plasmon resonance, 574
Surfaces, 47, 197, 571, 615, 637
Surfactant-polymer assemblies, 50, 280–286, 471–501
Swelling of brushes, 597–603

Tartaric acid, 619, 620
Tecto(dendrimers), 413–419
Templates, 10, 208, 323
Termination, 9, 44, 616, 623, 632
Thermotropic phases, 3, 11, 27, 34, 158, 619
Three-dimensional assemblies, 49, 50
Three-dimensional growth, 46
TMV, 4, 7
Tread-milling, 648
Triamino-pyrimidine, 624, 625, 626
Triblock copolymers, 116, 124, 216, 227, 231
Tubular growth, 41
Tubular polymers, 4, 650
Tubulin, 7, 644, 646
Two-dimensional polymers, 436–438

Uracil, 157, 158, 617, 618, 623
Ureidopyrimidone, 3, 33, 51, 163, 164
UV/vis spectroscopy, 512, 517, 518, 526, 531

van der Waals bonds, 13
Vesicles, 9, 22, 190, 202, 304–306, 615

Water content, 515, 520, 526, 549
Wave guide spectra, 579, 584, 585, 587, 596, 603
Weak segregation limit, 113, 219, 226
Wetting, 459–461,
Wormlike chain, 29, 34, 68

X-ray, 274, 495, 621, 644
X-ray photoelectron spectroscopy, 512, 513, 515, 517, 592
X-ray reflectometry, 512, 513, 517–519, 529, 540, 542–544, 546

ζ-potential, 512, 516, 527, 547